Lecture Notes in Computer Science 2285

Edited by G. Goos, J. Hartmanis, and J. van Leeuwen

Springer
Berlin
Heidelberg
New York
Barcelona
Hong Kong
London
Milan
Paris
Tokyo

Helmut Alt   Afonso Ferreira (Eds.)

STACS 2002

19th Annual Symposium
on Theoretical Aspects of Computer Science
Antibes - Juan les Pins, France, March 14-16, 2002
Proceedings

Springer

Series Editors

Gerhard Goos, Karlsruhe University, Germany
Juris Hartmanis, Cornell University, NY, USA
Jan van Leeuwen, Utrecht University, The Netherlands

Volume Editors

Helmut Alt
Free University of Berlin, Institute for Computer Science
Takustraße 9, 14195 Berlin, Germany
E-mail: alt@inf.fu-berlin.de

Afonso Ferreira
CNRS – I3S – INRIA Sophia Antipolis
2004 Route des Lucioles, 06902 Sophia Antipolis, France
E-mail: Afonso.Ferreira@sophia.inria.fr

Cataloging-in-Publication Data applied for

Die Deutsche Bibliothek - CIP-Einheitsaufnahme

STACS <19, 2002, Antibes; Juan-les-Pins>:
Proceedings / STACS 2002 / 19th Annual Symposium on Theoretical Aspects of
Computer Science, ANTIBES - Juan les Pins, France, March 14 - 16, 2002.
Helmut Alt ; Afonso Ferreira (ed.). - Berlin ; Heidelberg ; New York ;
Barcelona ; Hong Kong ; London ; Milan ; Paris ; Tokyo : Springer, 2002
    (Lecture notes in computer science ; Vol. 2285)
    ISBN 3-540-43283-3

CR Subject Classification (1998): F, E.1, I.3.5, G.2

ISSN 0302-9743
ISBN 3-540-43283-3 Springer-Verlag Berlin Heidelberg New York

This work is subject to copyright. All rights are reserved, whether the whole or part of the material is
concerned, specifically the rights of translation, reprinting, re-use of illustrations, recitation, broadcasting,
reproduction on microfilms or in any other way, and storage in data banks. Duplication of this publication
or parts thereof is permitted only under the provisions of the German Copyright Law of September 9, 1965,
in its current version, and permission for use must always be obtained from Springer-Verlag. Violations are
liable for prosecution under the German Copyright Law.

Springer-Verlag Berlin Heidelberg New York
a member of BertelsmannSpringer Science+Business Media GmbH

http://www.springer.de

© Springer-Verlag Berlin Heidelberg 2002
Printed in Germany

Typesetting: Camera-ready by author, data conversion by PTP-Berlin, Stefan Sossna
Printed on acid-free paper   SPIN: 10846319   06/3142   5 4 3 2 1 0

Preface

The Symposium on Theoretical Aspects of Computer Science (STACS) has become one of the most important annual meetings in Europe for the theoretical computer science community. It covers a wide range of topics in the area of foundations of computer science: algorithms and data structures, automata and formal languages, computational and structural complexity, logic, verification, and current challenges.

STACS 2002, the 19th in this series, was held in Antibes - Juan les Pins, on the French Riviera, March 14–16, 2002. Previous STACS symposia took place in Paris (1984), Saarbrücken (1985), Orsay (1986), Passau (1987), Bordeaux (1988), Paderborn (1989), Rouen (1990), Hamburg (1991), Cachan (1992), Würzburg (1993), Caen (1994), München (1995), Grenoble (1996), Lübeck (1997), Paris, (1998), Trier (1999), Lille (2000), and Dresden (2001). The proceedings of all these symposia have been published in the Lecture Notes in Computer Science series of Springer-Verlag.

STACS 2002 received 209 submissions from 30 countries, one of the highest numbers of submissions ever, for a European conference on theoretical computer science. They were dispatched to the Program Committee members, and underwent a thorough review process, where more than 900 reports were written. These reports then served as guidance to the Program Committee, which met on November 16–17, 2001, at the INRIA Sophia Antipolis (near Antibes), France, in order to select the conference program.

Fortunately, quantity yielded high scientific quality, and the task of selecting the best submissions was very difficult indeed. During the 2-day, very intense meeting of the Program Committee, less than 25% of the submissions were selected for oral presentation at the conference. The only drawback of this highly selective process was that many good, relevant papers with positive reviews could not be included in the final program. On the other hand, such a rigorous selection process was a guarantee that STACS 2002 was bound to be a scientifically rich and very fruitful forum for theoretical computer science.

We thank the three invited speakers at this symposium, Gilles Dowek (INRIA Rocquencourt), Michael O. Rabin (Harvard University), and Christian Scheideler (Johns Hopkins University) for accepting our invitation to share their insights on new developments in their research areas.

We would like to express our sincere gratitude to all the members of the local organizing committee who invested their time and energy to organize this conference. In particular, we thank J. Durand-Lose (Local Organizing Chair), D. Coudert (web site), S. Choplin (submissions server), and E. Deriche, B. Martin, H. Rivano, B. Madeline, D. Sergeant, and M.-H. Zeitoun.

Finally, we acknowledge the various sources of financial support for STACS 2002, namely the French CNRS, ERCIM, ESSI, the INRIA Sophia Antipolis, I3S, and the French Ministries of Research and Foreign Affairs.

December 2001

Helmut Alt
Afonso Ferreira

Program Committee

H. Alt (Berlin), Co-chair
H. Buhrman (Amsterdam)
B. Chlebus (Warsaw)
T. Erlebach (Zürich)
A. Ferreira (Sophia-Antipolis), Chair
H. Ganzinger (Saarbrücken)
D. Lugiez (Marseille)
Y. Métivier (Bordeaux)
C. Moore (Albuquerque/Santa Fe)
A. Muscholl (Paris)
G. Pucci (Padova)
G. Schnitger (Frankfurt)
T. Schwentick (Jena)
D. Trystram (Grenoble)
B. Vöcking (Saarbrücken)

Organizing Committee

S. Choplin
D. Coudert
E. Deriche
J. Durand-Lose, Chair
S. Julia
C. Julien
A. Ferreira, Conference Chair
B. Martin
H. Rivano
D. Sergeant
M.-H. Zeitoun

External Referees[1]

Hosam Abdo	Danielle Beauquier	Peter Brass
Luca Aceto	Bernd Becker	Stefan Brass
Dimitris Achlioptas	Richard Beigel	Jacques Briat
Micah Adler	Michael Benedikt	Stefan Burkhardt
Jochen Alber	Petra Berenbrink	Edson Caceres
Eric Allender	Anna Bernasconi	Cristian Calude
Jean-Paul Allouche	Karell Bertet	Iannis Caragiannis
Noga Alon	Jean Betrema	Olivier Carton
Ernst Althaus	Mauro Bianco	Julien Cassaigne
Andris Ambainis	Markus Blaeser	Didier Caucal
Klaus Ambos-Spies	Johannes Bloemer	Frederic Cazals
Christoph Ambühl	Hans Bodlaender	Julien Cervelle
Eric Angel	Bernard Boigelot	Samarjit Chakraborty
Dana Angluin	Mikolaj Bojanczyk	Witold Charatonik
Eugene Asarin	Paolo Boldi	Victor Chepoi
Jose Balcazar	Benedikt Bollig	Christian Choffrut
Evripidis Bampis	Ahmed Bouajjani	Marek Chrobak
David Barrington	Olivier Bournez	Andrea Clementi
Klaus Barthelmann	Patricia Bouyer	Richard Cleve
Hannah Bast	Julian Bradfield	Livio Colussi

---

[1] This list has been automatically compiled from the conference's database. We apologize for any omissions or inaccuracies.

Robert Cori
Ricardo Correa
Stavros Cosmadakis
Solange Coupet
Bruno Courcelle
Nadia Creignou
Pierluigi Crescenzi
Maxime Crochemore
Tristan Crolard
Janos Csirik
Felipe Cucker
Pierre-Louis Curien
Artur Czumaj
Georges Da Costa
Silvano Dal Zilio
Vincent Danos
Bhaskar DasGupta
Jürgen Dassow
Mart De Graaf
Ronald De Wolf
Gianna Del Corso
Joerg Derungs
Josee Desharnais
Luc Devroye
Pietro Di Gianantonio
Krzysztof Diks
Jürgen Dix
David Dobkin
Yevgeniy Dodis
Dirk Draheim
Maksymilian Dryja
Bertrand Ducourthial
Serge Dulucq
Jean-Guillaume Dumas
Arnaud Durand
Jerome Durand-Lose
Pierre-Francois Dutot
David Eppstein
Zoltan Esik
Javier Esparza
Kousha Etessami
Rolf Fagerberg
Carlo Fantozzi
Sandor Fekete
Stefan Felsner

Stephen Fenner
Henning Fernau
Paolo Ferragina
Alain Finkel
Paul Fischer
Michele Flammini
Paola Flocchini
Riccardo Focardi
Florence Forbes
Enrico Formenti
Lance Fortnow
Pierre Fraigniaud
Denis Francois
Martin Fuerer
Stefan Funke
Maurizio Gabbrielli
Didier Galmiche
Jérôme Galtier
Anna Gambin
Leszek Gasieniec
Thierry Gautier
Ricard Gavaldá
Remi Gilleron
Christian Glaßer
Andreas Goerdt
Karol Golab
Alfredo Goldman
Massimiliano Goldwurm
Erich Grädel
Martin Grohe
Roberto Grossi
André Große
Josef Gruska
Isabelle Guérin Lassous
Concettina Guerra
Fréderic Guinand
Jens Gustedt
Marc Gyssens
Peter Habermehl
Torben Hagerup
Alexander Hall
Eran Halperin
Joe Halpern
Laura Heinrich-Litan
Harald Hempel

Vollmer Heribert
Holger Hermanns
Uli Hertrampf
Mika Hirvensalo
John Hitchcock
Jaap-Henk Hoepman
Frank Hoffmann
Hendrik Jan Hoogeboom
Peter Hoyer
Juraj Hromkovich
Guillaume Huard
Neil Immerman
Sandra Irani
Giuseppe Italiano
Manfred Jaeger
Andreas Jakoby
Peter Jancar
David Janin
Claude Jard
Aubin Jarry
Emmanuel Jeannot
Philippe Jorrand
Sandrine Julia
Yan Jurski
Vanessa Kääb
Valentine Kabanets
Jarkko Kari
Juha Kärkkäinen
Marek Karpinski
Rainer Kemp
Claire Kenyon
Lefteris Kirousis
Daniel Kirsten
Hartmut Klauck
Bartek Klin
Adam Klivans
Christian Knauer
Pascal Koiran
Ulrich Kortenkamp
Darek Kowalski
Miroslaw Kowaluk
Marcus Kracht
Matthias Krause
Klaus Kriegel
Danny Krizanc

Piotr Krysta
Marcin Kubica
Antonin Kucera
Gregory Kucherov
Ravi Kumar
Orna Kupferman
Dietrich Kuske
Martin Kutrib
Miroslaw Kutylowski
Gregory Lafitte
Manuel Lameiras
Elmar Langetepe
Denis Lapoire
Francois Laroussinie
Slawomir Lasota
James Leifer
Giacomo Lenzi
Stefano Leonardi
Ming Li
Leonid Libkin
Marek Libura
Andrzej Lingas
Giuseppe Liotta
Markus Lohrey
Rita Loogen
Antoni Lozano
Fabrizio Luccio
Flaminia Luccio
Jack Lutz
Mukund Madhavan
Adam Malinowski
Vincenzo Manca
Heikki Mannila
Giovanni Manzini
Alberto Marchetti-
  Spaccamela
Luciano Margara
Bruno Martin
Elvira Mayordomo
Richard Mayr
Jacques Mazoyer
Lutz Meissner
Wolfgang Merkle
Marino Miculan
Michael Mitzenmacher

Angelo Montanari
Frank Morawietz
Malika More
Rémi Morin
Grégory Mounie
Jean-Frederic Myoupo
Markus Nebel
Frank Neven
Phong Nguyen
Gaia Nicosia
Peter Niebert
Rolf Niedermeier
Frank Niessner
Damian Niwinski
Mitsunori Ogihara
Friedrich Otto
Martin Otto
Leszek Pacholski
Linda Pagli
Aris Pagourtzis
Prakash Panangaden
Gilles Parmentier
Marco Pellegrini
Bernard Penz
Adriano Peron
Giuseppe Persiano
Holger Petersen
Antoine Petit
Andrea Pietracaprina
Nicholas Pippenger
Wojciech Plandowski
Andreas Podelski
Christopher Pollett
Christophe Prieur
Pavel Pudlak
Alexander Rabinovich
Rajeev Raman
Ramanujam
Edgar Ramos
Christophe Rapine
Andre Raspaud
Antoine Rauzy
Rahul Ray
Michel Raynal
Klaus Reinhardt

Giovanni Resta
Andrea Richa
Herve Rivano
Amadio Roberto
Mike Robson
Hein Roehrig
José Rolim
Yves Roos
Laurent Rosaz
Francesca Rossi
Gianluca Rossi
Peter Rossmanith
Guenter Rote
Michael Rusinowitch
Alex Russell
Marcin Rychlik
Zenon Sadowski
Nasser Saheb
Anand Sai
Jacques Sakarovitch
Alex Samorodnitsky
Peter Sanders
Martin Sauerhoff
Marcus Schaefer
Christian Scheideler
Christian Schindelhauer
Schmidt-Schauss
Georg Schnitger
Philippe Schnoebelen
Elmar Schoemer
Rainer Schuler
Götz Schwandtner
Nicole Schweikardt
Michael Seel
Steve Seiden
Helmut Seidl
Dietmar Seipel
Peter Selinger
Alan Selman
Geraud Senisergues
Jiri Sgall
Jeffrey Shallit
Boris Shapiro
Yaoyun Shi
Mihaela Sighireanu

Fred Sobik
Viorica Sofronie-
  Stokkermans
Christian Sohler
Siang Song
Eric Sopena
Ewald Speckenmeyer
Ludwig Staiger
Jukna Stasys
Stamatis Stefanakos
Angelika Steger
Benjamin Steinberg
Frank Stephan
Andrzej Szalas
Mario Szegedy
Géraud Sénizergues
Alain Terlutte
Denis Thérien
Jacques Theys
P.S. Thiagarajan
Thomas Thierauf
Rick Thomas
Sophie Tison
Ioan Todinca

Jacobo Toran
Leen Torenvliet
Csaba David Tóth
Stavros Tripakis
Dieter Van Melkebeek
Marc Van Kreveld
Peter Van emde Boas
Moshe Vardi
Nikolai Vereshchagin
Yann C. Verhoeven
Robert Veroff
Stephane Vialette
Laurent Viennot
Sebastiano Vigna
Laurent Vigneron
Gilles Villard
Eelco Visser
Paul Vitanyi
Joerg Vogel
Heribert Vollmer
Sergei Vorobyov
Imrich Vrto
Laurent Vuillon
Danica Vukadinovic

Klaus Wagner
Uwe Waldmann
Igor Walukiewicz
John Watrous
Klaus Weihrauch
Pascal Weil
Maik Weinard
Volkmar Welker
Carola Wenk
Matthias Westermann
Sue Whitesides
Peter Widmayer
Thomas Wilke
Gerhard Woeginger
Philipp Woelfel
Thomas Worsch
Sergio Yovine
Marc Zeitoun
Yunhong Zhou
Wieslaw Zielonka
Alexandre Zvonkine

Table of Contents

Current Challenges

Logic in Computer Science

Hyper-Encryption and Everlasting Security
Extended Abstract

Yan Zong Ding and Michael O. Rabin

[1] College of Computing, Georgia Institute of Technology, Atlanta, GA 30332
[2] DEAS, Harvard University, Cambridge MA 02138, USA

Abstract. We present substantial extensions of works [1], [2], and all previous works, on encryption in the bounded storage model introduced by Maurer in [25]. The major new result is that the shared secret key employed by the sender Alice and the receiver Bob can be re-used to send an exponential number of messages, against strong adaptive attacks. This essential step enhances the usability of the encryption method, and also allows strong authentication and non-malleability described below.

We give an encryption scheme that is provably secure against adaptive attacks by a computationally unbounded adversary in the bounded storage model. In the model, a sender Alice and a receiver Bob have access to a public random string α, and share a secret key s. Alice and Bob observe α on the fly, and by use of s extract bits from which they create a one-time pad X used to encrypt M as $C = X \oplus M$. The size of the secret key s is $|s| = k \log_2 |\alpha|$, where k is a security parameter. An Adversary $\mathcal{AD}$ can compute and store any function $A_1(\alpha) = \eta$, subject to the bound on storage $|\eta| \leq \gamma \cdot |\alpha|$, $\gamma < 1$, and captures C. Even if $\mathcal{AD}$ later gets the key s and is computationally unbounded, the encryption is provably secure. Assume that the key s is repeatedly used with successive strings $\alpha_1, \alpha_2, \ldots$ to produce encryptions $C_1, C_2, \ldots$ of messages $M_1, M_2, \ldots$. $\mathcal{AD}$ computes $\eta_1 = A_1(\alpha_1)$, obtains C_1, and gets to see the first message M_1. Using these he computes and stores $\eta_2 = A_1(\alpha_2, \eta_1, C_1, M_1)$, and so on. When he has stored η_l and captured C_l, he gets the key s (but not M_l). The main result is that the encryption C_l is provably secure against this adaptive attack, where l, the number of time the secret key s is re-used, is exponentially large in the security parameter k. On this we base non-interactive protocols for authentication and non-malleability. Again, the shared secret key used in these protocols can be securely re-used an exponential number of times against adaptive attacks. The method of proof is stronger than the one in [1], [2], and yields ergodic results of independent interest. We discuss in the Introduction the feasibility of the bounded storage model, and outline a solution. Furthermore, the existence of an encryption scheme with the provable strong security properties presented here, may prompt other implementations of the bounded storage model.

1 Introduction

A fundamental problem in cryptography is that of secure communication over an insecure channel, where a sender Alice wishes to communicate with a receiver

H. Alt and A. Ferreira (Eds.): STACS 2002, LNCS 2285, pp. 1–26, 2002.
© Springer-Verlag Berlin Heidelberg 2002

Bob, in the presence of a powerful Adversary $\mathcal{AD}$. The primary goal of encryption is to protect the privacy of the conversation between Alice and Bob against $\mathcal{AD}$. Modern cryptographic research has identified additional essentially important criteria for a secure encryption scheme. Namely that the encryption be non-malleable, be resistant to various chosen plaintext and ciphertext attacks, and if so desired, will allow the receiver to authenticate the received message and its sender. All these issues are settled in the present paper for the case that the Adversary $\mathcal{AD}$ is computationally unbounded.

If we wish to achieve unconditionally secure communication against an all-powerful adversary, we must use private-key cryptosystems.[1] Private-key encryption, in which the sender and the receiver share a common secret key for the encryption and decryption of messages, were proposed already in antiquity, dating back to as early as Julius Caesar, and are widely used today. A provably secure example of private-key encryption is the simple *one-time pad* scheme of Vernam [47]. In the one-time pad scheme, for the encryption and transmission of each message M, the sender Alice and the receiver Bob establish a shared random secret key X called a *one-time pad*, with $|X| = |M|$, where $|\cdot|$ denotes the length (i.e. number of bits) of a binary string. Alice encrypts M by computing $C = M \oplus X$, where $\oplus$ denotes bit-wise XOR. Bob decrypts C by computing $C \oplus X = M \oplus X \oplus X = M$.

The one-time pad scheme achieves information-theoretic secrecy, provided that for each transmission of a message, a *new* independent, uniformly random one-time pad, whose size equals that of the message, is established between Alice and Bob. In fact, a single careless re-use of the same one-time pad to encrypt a second message can result in the decryption of both ciphertexts. The non-reusability of the shared key renders this method impractical except for special situations where a large code book can be securely shipped from a sender to a receiver ahead of the transmissions. Furthermore, if an adversary captures and stores an encrypted message $C = M \oplus X$, where X is the one-time pad, and later on gets X (steals the code book), then he can decode: $M = X \oplus C$. The same holds for any existing private or public key encryption scheme.

In a seminal work, Shannon [43] proved that if the Adversary $\mathcal{AD}$ has complete access to the communication line and is unbounded in computational power, then information-theoretically secure private-key communication is only possible if the entropy of the space of secret keys is as large as that of the plaintext space. Essentially this implies that for achieving provable security, the shared one-time pad method, with its drawbacks, is optimal. Thus to overcome the limitations imposed by Shannon's result and have provable information theoretic security against a computationally unbounded Advesary, some other restriction has to be imposed.

---

[1] In a public-key cryptosystem, the encryption function using the public key, and the decryption function using the private key, are inverses of each other, and the public key is publicly known. Thus, a computationally unbounded adversary can, from the public key, compute the corresponding private key, using unlimited computing power.

Striving to produce a provably secure encryption scheme, which has the additional property that revealing the decryption key to an adversary after transmission of the ciphertext does not result in the decryption of the ciphertext, is important. Computational complexity based cryptosystems rely on unproven assumptions. If an adversary captures and stores ciphertexts, subsequent advances in algorithms or in computing power, say the introduction of quantum computers, may allow him to decrypt those ciphertexts.

A moment's thought shows that if the adversary can capture all the information visible to the sender and receiver, and later obtains the decryption key, then he will be able to decrypt the ciphertext just as the receiver does. Thus one is led to consider an adversary who is computationally unbounded but is limited in his storage capacity.

1.1 The Bounded Storage Model

We consider the bounded storage model, where the security guarantees are based on a *bound on the storage capacity* of the adversary. This model, introduced by Maurer in his influential work [25], assumes that there is a known bound B, possibly very large but fixed at any given time, on the Adversary $\mathcal{AD}$'s storage capacity. It is important to differentiate between this model and the *bounded space model* (see for example [10], [9], [23], [11], [14], [15], [3]). The bounded space model considers situations where the space available to the code breaker for *computation* is limited, e.g. log space or linear space. Thus, the space bound is in effect a limitation on the computational power of the Adversary. The bounded *storage* model, however, stipulates no limitation on the computational power of the Adversary $\mathcal{AD}$ who tries to subvert the security of the protocol. Imposing the bound B on the Adversary's storage allows us to construct efficient encryption/decryption schemes that are *provably information-theoretically secure*, and require of the sender Alice and the receiver Bob very modest computations and storage space.

In the basic form, the encryption schemes of [1], [2], [25] work as follows. Alice and Bob utilize a publicly accessible string α of n random bits, where for a fixed $\gamma < 1$, γn is larger than the bound B on the Adversary $\mathcal{AD}$'s storage capacity. In a possible implementation, the string α may be one in a stream of public random strings $\alpha_1, \alpha_2, \ldots$, each of length n, continually created in and beamed down from a satellite system, and available to all. Alice and Bob share a randomly chosen secret key s. For transmitting an encrypted message M, $|M| = m$, Alice and Bob listen or have listened to some $\alpha = \alpha_i$ (possibly a previously beamed α_i, see below), and by use of s create a common one-time pad $X(s, \alpha)$. In the schemes of [1], [2], [25], computing $X(s, \alpha)$ requires just fast XOR operations and very little memory space of Alice and Bob. Alice encrypts M as $C = M \oplus X(s, \alpha)$, and Bob decrypts C by $M = C \oplus X(s, \alpha)$. Every α is used only once. But the same α can be simultaneously used by multiple Sender/Receiver pairs without degrading security.

The Adversary $\mathcal{AD}$ listens to α, computes a *recording function* $A_1(\alpha) = \eta$, where $|\eta| = B \leq \gamma n$, $\gamma < 1$, and stores η. The Adversary also captures the ciphertext C. Later $\mathcal{AD}$ is given s, and applies a decoding algorithm $A_2(\eta, s, C)$ to gain information about M. There is no limitation on the space or work required for computing $A_1(\alpha)$ and $A_2(\eta, s, C)$. Theorem 1 of [1] and Theorem 2 of [2] say that the Aumann-Rabin and Aumann-Ding-Rabin schemes are *absolutely semantically secure*. Namely, for $B = 0.3n$ (later we will use $\gamma = 1/6$ without loss of generality),[2] for every two messages $M_0, M_1 \in \{0,1\}^m$, for an encryption $C_\delta = M_\delta \oplus X(s, \alpha)$, where $\delta \in \{0, 1\}$ is randomly chosen,

$$|\Pr[A_2(A_1(\alpha), s, C_\delta) = \delta] - 1/2| \le m \cdot 2^{-k/3}. \tag{1}$$

Here, k is a security parameter, e.g. $k = 200$. The security obtained in the above results is the absolute version (i.e. in a model allowing a computationally unbounded adversary) of *semantic security* pioneered by Goldwasser and Micali [18]. Note that the storage bound B is needed only at the time α is broadcast, and the recording function $\eta = A_1(\alpha)$ is computed. The decoding algorithm $A_2(\eta, s, C)$ can use unbounded computing power and storage space, and thus subsequent increase in storage does not help the Adversary $\mathcal{AD}$.

Feasibility of the Model. The present paper is mathematical and intends to establish the strong provable security properties of our encryption method, but we shall briefly discuss feasibility issues. The method requires an intense source of random bits. The second author and the Harvard VLSI group led by Woody Yang are implementing a physical device for this. One way of distributing the stream of random bits is to beam it from a satellite. Here the main limiting factor is the power per bit and the total power consumed by the satellite. The power/bit is roughly *inversely* proportional to the area of the receiving dish antenna, and proportional to the *square* of the orbit height. With a Geo-satellite (stationary orbit) and a ten meter dish, a rate of 50 Gigabytes per second can be achieved. With a system of low flying (700km) satellites, much higher rates are possible. The bit stream can be bounced between satellites by a focused low energy laser beam, and beamed down from several satellites to far apart locations. There is no need for time synchronization for the sender and receiver to enable them to select the same bits from the stream. Even with the cheapest DVD optical disk storage, a year's worth of a Terabits/sec stream will cost billions of dollars to store, exceeding by orders of magnitude the cost of the satellites, and that of producing the stream that can be simultaneously used by the whole world. Also, several systems of satellites can be used to proportionally increase the rate. The secret key s includes agreement between the sender and the receiver as to which bits to pick from each individual satellite. We are grateful to Lyman Hazelton for his advice on this analysis.

It should be noted that the sender and receiver need not read the bit stream and compute the common one-time pad $X(s, \alpha)$ at the time they encrypt and transmit a message. They can pre-process offline, by secret agreement and using

---

[2] The pedagogical choice of $\gamma = 0.3$ is for convenience only, and leads to the probability $2^{-k/3}$ in (1). Similar results are obtainable for any constant $\gamma < 1$.

long stretches of the stream, numerous one-time pads $X(s, \alpha_1), X(s, \alpha_2), \ldots$, and use each just once to encrypt a message when needed. To counter this mode, an adversary would have to continuously read the stream and store as many original or computed bits, as he has read from the stream.

1.2   Our Results and Methods

We now come to the main innovation of the present paper. Namely we allow an *adaptive* adversary who can conduct $l - 1$ rounds of Chosen Plaintext Attack (CPA) or Chosen Ciphertext Attack (CCA) on the secret key, where l is exponentially large in a security parameter k.

Consider a scenario where the same secret key s is repeatedly used with l successive public random strings $\alpha_1, \alpha_2, \ldots, \alpha_l$ to produce encryptions $C_1, C_2, \ldots C_l$ of messages $M_1, M_2, \ldots, M_l$, where for each i, $|M_i| = m$ and $|\alpha_i| = n$. The Adversary $\mathcal{AD}$ proceeds adaptively as follows: $\mathcal{AD}$ first observes α_1, computes any recording function $\eta_1 = A_1(\alpha_1)$, with $|\eta_1| = B \leq \gamma n$, $\gamma < 1$, and stores η_1. Then $\mathcal{AD}$ captures both C_1 and M_1, and thus obtains the one-time pad $X_1 = X(s, \alpha_1) = M_1 \oplus C_1$ for the encryption of M_1. The (plaintext, ciphertext) pair could be obtained by a Chosen Plaintext Attack (CPA) or a Chosen Ciphertext Attack (CCA). Using η_1 and X_1, $\mathcal{AD}$ computes and stores $\eta_2 = A_1(\alpha_2, \eta_1, X_1)$, and so on. In general, for $i = 2, \ldots, l-1$, in the i-th communication round, $\mathcal{AD}$ has available η_{i-1} and $X_1, \ldots, X_{i-1}$, where $X_j = X(s, \alpha_j)$ for each j, and computes any *adaptive* recording function

$$\eta_i = A_1(\alpha_i, \eta_{i-1}, X_1, \ldots, X_{i-1}), \qquad |\eta_i| = B \tag{2}$$

where η_i depends on not only α_i, but also η_{i-1}, and all the previous one-time pads $X_1, \ldots, X_{i-1}$. Then $\mathcal{AD}$ obtains (M_i, C_i) by a CPA or CCA, and thus obtains the i-th one-time pad $X_i = X(s, \alpha_i) = M_i \oplus C_i$. This adaptive attack can potentially be of help to $\mathcal{AD}$ because η_i and $X_1, \ldots, X_{i-1}$ are functions of the secret key s and the previous public random strings, and thus enable $\mathcal{AD}$ to increasingly collect more information on the secret key s in each round. In the l-th round, $\mathcal{AD}$ computes $\eta_l = A_1(\alpha_l, \eta_{l-1}, X_1, \ldots, X_{l-1})$, captures the current ciphertext C_l (but not the plaintext M_l), gets the secret key s (after computing η_l), and employs any decoding algorithm $A_2(\eta_l, s, C_l)$ to gain information on the corresponding plaintext M_l. Theorem 1 of Section 2 shows that our encryption method is secure under this adaptive attack, even if ml is exponentially large in the security parameter k. No previous result is of this nature.

The practical implication of Theorem 1 is that Alice and Bob can establish, once and for all, a shared secret key s, and re-use it an exponentially large number $l = 2^{O(k)}$ of times under the strong adaptive attack described above, without degrading the security of the encryption. Even if the secret key s is divulged at any time, all past secret messages remain forever secret. This is what is meant by *everlasting security*, a property not enjoyed by any previous encryption method. From now on, we shall refer to our encryption scheme as the Hyper-Encryption scheme. Combined with public-key cryptography, the scheme also gives rise to an efficient and secure pragmatic method for initializing a re-usable common secret key between two parties.

The strong secrecy of Hyper-Encryption against adaptive and computationally unbounded adversaries allows additional important applications. The encryption method enables repeated re-use of shared keys for message and user authentication, whereas all previous information-theoretically secure schemes for message authentication require a *new* shared random secret key per additional small number of authentications. The strong authentication in turn enables, for example, the construction of a provably secure Kerberos-like key distribution protocol [44], [21]. Furthermore, applying our authentication scheme to the ciphertext of a message, we have a simple solution for *non-malleable* encryption [13]. All these protocols have the same strong security properties against an adaptive computationally unbounded adversary, as the Hyper-Encryption scheme.

The Hyper-Encryption scheme, which elaborates on the Aumann-Ding-Rabin scheme [2], is a shared secret-key block-cipher scheme. In the rest of the paper, we shall consider encryptions of *message blocks* of size m (typically $m = 64$ bits), and use the words "message" and "block" interchangeably. Longer messages are encrypted block by block, using the same key.

For the encryption of an m-bit message block, the plain Hyper-Encryption scheme employs a random secret key of $mk \log_2 n$ bits long. However, using a *key amplification* method that will be described in the full paper, the sender and receiver can start with a shared random key of size only $k \log_2 n$ bits (independent of m), and generate in $\log_2 m$ rounds, once and for all, a re-usable shared key of size $mk \log_2 n$ bits that is *indistinguishable* from a truly random key of $mk \log_2 n$ bits long. If, say, $k = 200$, and $\log_2 n = 50$, then the size of an initial shared key is only 1,250 bytes. Using this amplified key instead of a truly random $mk \log_2 n$-bit key, we achieve essentially the same security properties, with an initial key of only $k \log_2 n$ bits long.

The method of proof for the security of Hyper-Encryption are related to, but differs in important respects from, and leads to stronger results than those in [1], [2], and yields ergodic results of independent interests.

1.3   Related Work

In a ground breaking work, Maurer [25] proposed the bounded storage model, and presented the first protocol for secure private-key communication in this model. However, in [25], the proof of the security is provided only for the case where a storage-bounded Adversary $\mathcal{AD}$ can only store B *original bits* of the public random string α. The analysis of [25] does not provide a proof for the general bounded storage model, where the Adversary can access all the bits of α, perform any computation on α, and store the result of the computation in the bounded storage available to him. The subsequent work of Cachin and Maurer [8] provided a scheme for which they proved security for the general bounded storage model. However, this scheme is considerably more complicated, employing sophisticated privacy amplification techniques [6], [31], yet gives a result that is not as strong and efficient as desired. To assure that the probability of revealing information of the message to $\mathcal{AD}$ is smaller than ϵ, the Cachin-Maurer scheme requires Alice and Bob to store $l \log n$ bits, and transmit l bits, where $n = |\alpha|$ and $l = 3/\epsilon^2$. Thus, if we modestly require $\epsilon = 10^{-5}$, we get that $l = 3 \times 10^{10}$. Say $n = 2^{45}$, so $\log n = 45$. Thus for this choice of ϵ, Alice and Bob each have to store 1.35×10^{12} bits, and transmit 3×10^{10} bits. Furthermore, the protocol calls for multiplications of elements in the field F_{2^l}, which are costly for l as large as 3×10^{10}.

Information-theoretic key agreement was studied by Maurer and colleagues (e.g. [6], [26], [27], [30], [28], [31], [32], [33]).

Cachin, Crépeau, and Marcil [7], and Ding [12] studied Oblivious Transfer in the bounded storage model.

The work of [1], [2] on encryption in the bounded storage model is referenced and discussed earlier in Section 1.1. As pointed out above, none of the previous work on encryption in this model deals with the important case of an adaptive adversary that was described in the Section 1.2.

The idea of a publicly available random string in cryptographic protocols was first introduced by Rabin [39].

Message authentication was introduced by Gilbert, MacWilliams and Sloan [17]. Message authentication based on universal hashing was introduced by Wegman and Carter [48]. Since then, there has been a vast literature on message authentication codes [45]. All previous information-theoretically secure schemes for message authentication require a *new* shared random secret key per additional small number of authentications.

Chosen Ciphertext Attack (CCA1) was introduced by M. Naor and Yung [38]. Adaptive Chosen Ciphertext Attack (CCA2) was introduced by Rackoff and Simon [41]. See [34] for references to CPA. Relations among security notions was studied by Bellare etal [4], and Katz and Yung [20].

Non-malleable encryption was introduced by D. Dolev, C. Dwork and M. Naor [13], and further studied by Bellare etal [4], Bellare and Sahai [5], Sahai [42], and Katz and Yung [20].

The bounded *space* model for Zero-Knowledge Proofs was studied by Condon and Ladner [10], [9], Kilian [23], De-Santis, Persiano, and Yung [11], Dwork and Stockmeyer [14], [15], and Aumann and Fiege [3]. Nisan and Zuckerman [35], [36] studied pseudorandomness in the bounded space model.

1.4   Organization of the Paper

Section 2 - The Scheme, the Attack, and the Main Result. This section presents the Hyper-Encryption scheme, formally defines the adaptive attack, and gives the main results, i.e. Theorems 1 and 2, on the everlasting security of the Hyper-Encryption scheme.

Section 3 - Proof of Everlasting Security: An Outline. This section outlines a proof of the everlasting security of Hyper-Encryption. The method of proof yields results of independent interests, in particular Main Lemma 1.

Section 4 - Authentication and Non-malleability. As an application of Hyper-Encryption, this section describes a message authentication scheme that is secure under the same strong adaptive attack, and extends the Hyper-Encryption scheme to be non-malleable. Based on these results, the section also describes an implementation of a provably secure Kerberos-like key distribution protocol.

2   The Scheme, the Attack, and the Main Result

2.1   Notations and Conventions

Throughout the paper, k is the security parameter, and m is the size (in number of bits) of a message block. The Greek letter α, often with a subscript, denotes a public random string, and $n = |\alpha|$ is the size of a public random string. For a bit string β, $\beta[j]$ denotes the j-th bit of β.

The Adversary $\mathcal{AD}$'s storage bound is denoted by B, and $B = \gamma n$ for some fixed $\gamma < 1$. For convenience and without loss of generality, in the paper we derive our results for $B = n/6$, i.e. the particular constant $\gamma = 1/6$. It shall be clear from the proofs that any fixed $\gamma < 1$ would lead to similar results, and only yield different constants in the formulae.

The following convenient definitions and notation will be used for the rest of the paper.

Definition 1. *Define $\mathcal{S} \overset{\mathrm{d}}{=} \{1, \ldots, n\}^k$. I.e. $\mathcal{S}$ is the set of all n^k k-tuples over $\{1, \ldots, n\}$.*

Definition 2. *Let $\alpha \in \{0,1\}^n$. For $s = (\sigma_1, \ldots, \sigma_k) \in \mathcal{S}$, define the bit $s(\alpha) \overset{\mathrm{d}}{=} \bigoplus_{j=1}^{k} \alpha[\sigma_j]$, where $\oplus$ denotes exclusive-or. For any integer t and a t-tuple $\boldsymbol{s} = (s_1, \ldots, s_t) \in \mathcal{S}^t$, define the t-bit string $\boldsymbol{s}(\alpha) \overset{\mathrm{d}}{=} (s_1(\alpha), \ldots, s_t(\alpha))$.*

Notation: For a finite set F, $x \overset{R}{\longleftarrow} F$ denotes choosing x uniformly at random from F.

2.2   The Scheme

We now describe the Hyper-Encryption scheme. Alice and Bob establish, once and for all, a shared secret key $\boldsymbol{s} = (s_1, \ldots, s_m) \in \mathcal{S}^m$, where $\mathcal{S}$ is defined in Definition 1, and m is the message block size. Thus the secret key $\boldsymbol{s}$ consists of mk integers[3] in $\{1, \ldots, n\}$, each interpreted as an index into a public random string of length n. The $mk \log_2 n$-bit secret key $\boldsymbol{s}$ can be either chosen randomly, or generated from an initial common random key that is $k \log_2 n$ bits long, by a *key amplification* method that was mentioned in the Introduction. In this section,

---

[3] In [2], it was required in the secret key $\boldsymbol{s} = (s_1, \ldots, s_m)$, for each $i \neq j$, s_i and s_j share no index in common. Here we relax this requirement.

we state and prove the results in the case of a randomly chosen key $s \xleftarrow{R} \mathcal{S}^m$. The results for an amplified key will be given and proved in the full paper.

The encryption scheme proceeds as follows. For each $i = 1, \ldots, m$, denote $s_i = (\sigma_{i1}, \ldots, \sigma_{ik})$, where s_i is the i-th component of the secret key s. For the encryption of a message $M \in \{0, 1\}^m$, a public random string $\alpha \in \{0, 1\}^n$ is broadcast. Alice and Bob listen to α, retain the mk bits $\alpha[\sigma_{ij}]$, i.e. the bits of α whose indices appear in s, and compute the m-bit string $s(\alpha)$ that is defined in Definition 2. That is, for each $i = 1, \ldots, m$, $s_i(\alpha) = \bigoplus_{j=1}^{k} \alpha[\sigma_{ij}]$, and $s(\alpha) = (s_1(\alpha), \ldots, s_m(\alpha))$. Using $s(\alpha)$ as a *one-time pad*, Alice encrypts M as $C = M \oplus s(\alpha)$, and sends C to Bob, who decrypts by computing $M = C \oplus s(\alpha)$. A description of the Hyper-Encryption scheme is also given in Figure 1.

Message Block $M \in \{0, 1\}^m$.

Secret key $s = (s_1, \ldots, s_m)$, $s_i = (\sigma_{i1}, \ldots, \sigma_{ik}) \in \mathcal{S} = \{1, \ldots, n\}^k$.

Public random string $\alpha \xleftarrow{R} \{0, 1\}^n$.

Both Alice and Bob listen while α is being broadcast.

1. Alice and Bob store $\alpha[\sigma_{ij}] \; \forall 1 \le i \le m, 1 \le j \le n$.

2. **for** $i = 1$ to m **do**
3.   Alice and Bob set $X_i = s_i(\alpha) = \bigoplus_{j=1}^{k} \alpha[\sigma_{ij}]$.

4. Alice and Bob set $X = s(\alpha) = (X_1, \ldots, X_m)$.

5. Alice encrypts $C = X \oplus M$, and sends C to Bob.

6. Bob decrypts $M = C \oplus X$.

Fig. 1. The Hyper-Encryption Scheme

2.3   The Adaptive Attack and Main Result

We now describe the adaptive attack. Consider a scenario where public random strings $\alpha_1, \alpha_2, \ldots$ are employed to create the ciphertexts $C_1, C_2, \ldots$ of messages $M_1, M_2, \ldots$ respectively. The Adversary $\mathcal{AD}$ listens to $\alpha_1, \alpha_2, \ldots$, observes the pairs $(M_1, C_1), (M_2, C_2), \ldots$ of corresponding messages and ciphertexts, and adapts his attacking strategy accordingly. Each pair (M_i, C_i) could be obtained by either a Chosen Plaintext Attack (CPA) or a Chosen Ciphertext Attack (CCA1&2). From (M_i, C_i), $\mathcal{AD}$ obtains the corresponding one-time pad $s(\alpha_i) = M_i \oplus C_i$. Thus for Hyper-Encryption, Chosen Ciphertext security and Chosen Plaintext security are equivalent.

Formally, the Adversary $\mathcal{AD}$ attacks in two phases. Phase I consists of $l - 1$ rounds, for some parameter l. In Round 1, $\alpha_1 \xleftarrow{R} \{0,1\}^n$ is broadcast. The Adversary $\mathcal{AD}$ chooses any recording function $A_1^{(1)} : \{0,1\}^n \longrightarrow \{0,1\}^B$, where $B = n/6$, computes $\eta_1 = A_1^{(1)}(\alpha_1)$, and stores η_1. Then by a CPA or a CCA, $\mathcal{AD}$ obtains $s(\alpha_1)$.

In Round $i > 1$ of Phase I, $\mathcal{AD}$ has available η_{i-1} and $s(\alpha_1), \ldots, s(\alpha_{i-1})$ from the previous rounds, and chooses any *adaptive* recording function $A_1^{(i)} : \{0,1\}^n \times \{0,1\}^B \times \{0,1\}^{(i-1)m} \longrightarrow \{0,1\}^B$. While $\alpha_i \xleftarrow{R} \{0,1\}^n$ is broadcast, $\mathcal{AD}$ computes and stores

$$\eta_i = A_1^{(i)}(\alpha_i, \eta_{i-1}, s(\alpha_1), \ldots, s(\alpha_{i-1})). \tag{3}$$

Then by a CPA or CCA, the Adversary $\mathcal{AD}$ obtains the i-th one-time pad $s(\alpha_i)$.

In Phase II, the l-th message M_l is encrypted as $C_l = M_l \oplus s(\alpha_l)$. While α_l is broadcast, using his information η_{l-1} and $s(\alpha_1), \ldots, s(\alpha_{l-1})$ collected in Phase I, the Adversary $\mathcal{AD}$ again computes any adaptive recording function $\eta_l = A_1^{(l)}(\alpha_l, \eta_{l-1}, s(\alpha_1), \ldots, s(\alpha_{l-1}))$, and stores η_l. Then $\mathcal{AD}$ is given the ciphertext C_l (but not the plaintext M_l), and also the secret key s. On η_l, s, and C_l, $\mathcal{AD}$ applies any decoding algorithm $A_2(\eta_l, s, C_l)$ in an attempt to gain information on the message M_l, using unlimited computing power and storage space.

Theorem 1. *For $m, k < 0.001n$, $\forall M_0, M_1 \in \{0,1\}^m$, for any recording functions $A_1^{(1)}, \ldots, A_1^{(l)}$, where for each i, $A_1^{(i)} : \{0,1\}^n \times \{0,1\}^{n/6} \times \{0,1\}^{(i-1)m} \longrightarrow \{0,1\}^{n/6}$, for any decoding algorithm A_2, for uniformly random $s \xleftarrow{R} \mathcal{S}^m$, and $\alpha_1, \ldots, \alpha_l \xleftarrow{R} \{0,1\}^n$,*

$$|\Pr[A_2(\eta_l, s, C_1) = 1] - \Pr[A_2(\eta_l, s, C_0) = 1]|$$
$$< m \cdot \left(3l \cdot 2^{-k/3} + l \cdot 2^{-0.01n}\right), \tag{4}$$

where $C_\delta = s(\alpha_l) \oplus M_\delta$, η_l is defined recursively in (3), $\mathcal{S} = \{1, \ldots, n\}^n$ is defined in Definition 1, and $s(\alpha)$ is defined in Definition 2.

Remark: Note that in the i-th round of $\mathcal{AD}$'s attack, the one-time pad $s(\alpha_i)$ is the maximum information the $\mathcal{AD}$ can obtain by a CPA or CCA on the Hyper-Encryption scheme. Thus, Theorem 1 implies the security of Hyper-Encryption against CPA and CCA.

Remark: In Theorem 1, the condition that $m, k < 0.001n$ holds well in practice, since the block size m (typically $m = 64$), and the security parameter k (e.g. $k = 200$), are much smaller than n (e.g. $n = 2^{50}$), which is larger than the storage capacity of $\mathcal{AD}$. From now on, we take for granted that m and k are negligibly small compared to n, and omit this condition in the statements of the subsequent theorems and lemmas.

Functionally, Theorem 1 provides the following strong security guarantee. The same key s is used to encrypt l messages, each m bits long. The Adversary

$\mathcal{AD}$ gets to see, for $i = 1, \ldots, l - 1$, the i-th message and ciphertext, and adaptively store $B = n/6$ bits based on all past information. After the l-th message is encrypted and sent, $\mathcal{AD}$ captures the ciphertext and gets the secret key s. Even under these favorable conditions, and using unlimited computing power, the information $\mathcal{AD}$ can get on the l-th message is exponentially negligible. Furthermore, the number l of different messages securely sent by use of the same key s can be exponentially large in k.

Equivalently, the security of the Hyper-Encryption scheme can be expressed in terms of the security of the one-time pad extracted by the secret key from the public random string. Namely, after the adaptive attack, given the secret key s and using all the information η_l stored, $\mathcal{AD}$ cannot distinguish between the l-th one-time pad $s(\alpha_l)$, and a truly random string. In other words, to the storage bounded adversary $\mathcal{AD}$, $s(\alpha_l)$ is *provably pseudorandom*.

Theorem 2. *For any recording functions $A_1^{(1)}, \ldots, A_1^{(l)}$, where for each i, $A_1^{(i)}$: $\{0,1\}^n \times \{0,1\}^{n/6} \times \{0,1\}^{(i-1)m} \longrightarrow \{0,1\}^{n/6}$, for any distinguishing algorithm $\mathcal{D}$, for $s \xleftarrow{R} \mathcal{S}^m$, $\alpha_1, \ldots, \alpha_l \xleftarrow{R} \{0,1\}^n$, and $R \xleftarrow{R} \{0,1\}^m$,*

$$|\Pr[\mathcal{D}(\eta_l, s, s(\alpha_l)) = 1] - \Pr[\mathcal{D}(\eta_l, s, R) = 1]|$$
$$< m \cdot \left(3l \cdot 2^{-k/3} + l \cdot 2^{-0.01n}\right), \tag{5}$$

where η_l is defined recursively in (3).

2.4 Adaptive Security Does Not Follow from Non-adaptive Security

It is important to note that Theorems 1 and 2 on the security of the Hyper-Encryption scheme under the adaptive attack described in Section 2.3, does not follow its security against a non-adaptive adversary considered in [1], [2]. The main reason is again that in each round of the adaptive attack, the recording function $\eta_i = A_1^{(i)}(\alpha_i, \eta_{i-1}, X_1, \ldots, X_{i-1})$ depends on $X_1 = s(\alpha_1), \ldots, X_{i-1} = s(\alpha_{i-1})$, and η_{i-1}, each a function of the secret key s. Let's take a closer look. The non-adaptive security guarantees that the adversary's algorithm A_2 cannot distinguish between (η_1, s, X_1) and (η_1, s, R_1), where R_1 is a uniformly random m-bit string. From this, what can be deduced is that after i rounds, $(\eta_i, s, X_1, \ldots, X_i)$ and $(\tilde{\eta}_i, s, R_1, \ldots, R_i)$ are indistinguishable, where η_i is defined above, but $\tilde{\eta}_i = A_1^{(i)}(\alpha_i, \tilde{\eta}_{i-1}, R_1, \ldots, R_{i-1})$ (with $\tilde{\eta}_1 = \eta_1$), each R_j being a uniformly random m-bit string. This however does NOT imply that $(\eta_i, s, X_1, \ldots, X_i)$ and $(\eta_i, s, R_1, \ldots, R_i)$ are indistinguishable, a result essential for the proof of Theorems 1 and 2. Notice the difference between η_i and $\tilde{\eta}_i$: the former depends on s, but the latter does not.

3 Proof of Everlasting Security: An Outline

In this section we outline a proof of Theorem 1. We first note that it suffices to prove the theorem in the case that the Adversary's recording algorithms

$A_1^{(1)}, \ldots, A_1^{(l)}$ are deterministic. This does not detract from the generality of our results for the following reason. By definition, a randomized algorithm is an algorithm that uses a random help-string r for computing its output. A randomized algorithm A with each fixed help-string r gives rise to a *deterministic* algorithm A_r. Therefore, that Theorem 1 holds for any deterministic recording algorithm implies that for any randomized recording algorithm A, *for each fixed* help-string r, A using r does not help $\mathcal{AD}$ in distinguishing the ciphertexts of any two given messages. Hence, on average, A using a randomly chosen r cannot succeed on the same input. The reader might notice that the help-string r could be arbitrarily long since the Adversary has unlimited computing power. In particular, it could be that $|r| > B$, thereby giving rise to a deterministic recording algorithm with $|A_r| = |A| + |r| > B$. But the bounded storage model imposes *no restriction* on the Adversary's *program size*. The only restriction is that the length of the output $|A_r(\alpha)| \leq B$ for each r. In the formal model, A is an unbounded non-uniform Turing Machine whose output tape is bounded by B bits.

Therefore, from now on, we consider deterministic recording algorithms. In the proof of Theorem 1, as the essential step, we first prove the theorem in the case of a single-bit message, i.e. the case that $m = 1$. The security of multi-bit messages follows from the single-bit security, and a *hybrid argument* [18].

3.1   Sketch of Ideas

To assist in reading the detailed proofs, we outline here the flow of ideas. We deal with the security of encryption for the one-bit case. The results for encryption of a block of m bits follow from the one-bit case by standard methods.

Alice and Bob employ a public random bit string $\alpha \in \{0,1\}^n$ and a randomly chosen secret key $s = (\sigma_1, \ldots, \sigma_k) \in \mathcal{S}$ to produce the one-time pad $s(\alpha)$. By definition, $s(\alpha) = \alpha[\sigma_1] \oplus \cdots \oplus \alpha[\sigma_k]$. It is clear that for the Adversary to decipher $C = s(\alpha) \oplus M$, where M is the one-bit message, is equivalent to computing $s(\alpha)$.

The Adversary $\mathcal{AD}$ computes and stores a string $\eta = A_1(\alpha)$ of length $|\eta| = n/6$. Thus A_1 is a function $A_1 : \{0,1\}^n \longrightarrow \{0,1\}^{n/6}$. A simple counting argument shows that for all but at most $2^{n-0.01n}$ of the strings $\alpha \in \{0,1\}^n$, $|A_1^{-1}(A_1(\alpha))| = |A_1^{-1}(\eta)| \geq 2^{dn}$, where $d = 0.822$. Thus with probability greater than $1 - 2^{-0.01n}$, after storing $\eta = A_1(\alpha)$, all that $\mathcal{AD}$ knows about α is that $\alpha \xleftarrow{R} A_1^{-1}(\eta) = H = \{\alpha_1, \ldots, \alpha_L\}$, where $L \geq 2^{dn}$. Since $\mathcal{AD}$ is computationally unbounded, we may even assume that he can enumerate all of H. Now, after $\mathcal{AD}$ computed and stored η, he is given the secret key s. What can he deduce about $s(\alpha)$?

Main Lemma 1 shows that if $L \geq 2^{dn}$, then for all but at most $K \cdot 2^{-k/3}$ of the keys s, where $K = n^k$ is the number of all possible keys,

$$|\Pr[s(\alpha) = 1] - \Pr[s(\alpha) = 0]| < 2^{-k/3+1}. \tag{6}$$

The above probabilities are for a *fixed* s, and random $\alpha \xleftarrow{R} H$. The implication of (6) is that based just on the knowledge of H and s, if (6) holds, the cases

$s(\alpha) = 1$ and $s(\alpha) = 0$ are essentially indistinguishable and each occurs with probability $1/2 \pm \delta$ where $\delta < 2^{-k/3}$.

Now, s is a randomly chosen key, and s and α are independently chosen. Collecting the above results it follows that no matter what $\mathcal{AD}$'s decoding algorithm $A_2(\eta, s)$ is,

$$\Pr\left[A_2(A_1(\alpha), s) = s(\alpha)\right] < \frac{1}{2} + 2^{-k/3} + 2^{-0.01n}. \qquad (7)$$

Namely, the probability that $L < 2^{dn}$ is smaller than $2^{-0.01n}$, and if $L \geq 2^{dn}$ then the probability for a random s that (6) does not hold is smaller than $2^{-k/3}$.

Main Lemma 1 is proved by a new method combining probabilistic arguments with linear algebra and using at one point the Cauchy-Schwartz inequality.

Main Lemm 1 and the above results are sufficiently strong to entail the extension to security of the Hyper-Encryption under repeated use of the key s and under an adaptive attack by the Adversary. This is achieved by a counting argument in the proof of Main Lemma 2, presented in Section 3.4.

3.2   Preliminaries

Definition 3. For a bit $b \in \{0, 1\}$, define $\overline{b} \stackrel{\mathrm{d}}{=} (-1)^b$. Thus $\overline{0} = 1$ and $\overline{1} = -1$. For a vector $v = (b_1, \ldots, b_t) \in \{0, 1\}^t$, define $\overline{v} \stackrel{\mathrm{d}}{=} (\overline{b_1}, \ldots, \overline{b_t})$.

Definition 4. Let $K = n^k$. Let $\mathcal{S} = \{s_1, \ldots, s_K\}$ be a fixed enumeration of all the K elements of $\mathcal{S} = \{1, \ldots, n\}^k$. For a string $\alpha \in \{0, 1\}^n$, define

$$v(\alpha) \stackrel{\mathrm{d}}{=} \left(\overline{s_1(\alpha)}, \ldots, \overline{s_K(\alpha)}\right). \qquad (8)$$

Definition 5. For a vector $v = (v_1, \ldots, v_t) \in \{1, -1\}^t$, the discrepancy of v is defined as $d(v) \stackrel{\mathrm{d}}{=} |\sum_{i=1}^t v_i|$.

Thus, for $\alpha \in \{0, 1\}^n$, $d(v(\alpha))$ measures the excess of 0's over 1's, or vice-versa, in the vector $(s_1(\alpha), \ldots, s_K(\alpha))$.

Lemma 1. [1], [2], [16] Let $\alpha \in \{0, 1\}^n$. Let p be the fraction of 0's in α, and $q = 1 - p$ the fraction of 1's in α. Then for $s \xleftarrow{R} \mathcal{S}$,

$$|\Pr\left[s(\alpha) = 1\right] - \Pr\left[s(\alpha) = 0\right]| = |p - q|^k. \qquad (9)$$

Corollary 1. For each $\alpha \in \{0, 1\}^n$ such that the fraction of 1's and the fraction of 0's in α are both no less than $1/4$,

$$d(v(\alpha)) \leq K \cdot 2^{-k}. \qquad (10)$$

Proof. Since

$$d(v(\alpha)) = K \cdot |\Pr[s(\alpha) = 1] - \Pr[s(\alpha) = 0]|.$$

$\square$

Note that $d(v(\alpha)) \le K \ \forall \alpha \in \{0,1\}^n$. The next lemma, derived from Corollary 1, says that for almost all $\alpha \in \{0,1\}^n$, the discrepancy $d(v(\alpha)) \stackrel{d}{=} |\sum_{i=1}^K \overline{s_i(\alpha)}|$ is exponentially small compared to K.

Lemma 2. *Let* $D = \{\alpha \in \{0,1\}^n : d(v(\alpha)) > K \cdot 2^{-k}\}$. *Then* $|D| < 2^{cn}$, *where* $c = 0.812$.

Proof. For a string α, denote by $z(\alpha)$ the number of zeros in α. By Corollary 1, if $\alpha \in D$, then $z(\alpha) < n/4$ or $z(\alpha) > 3n/4$. Therefore, by Stirling's approximation (see [19]), for $\alpha \xleftarrow{R} \{0,1\}^n$,

$$\Pr[z(\alpha) < n/4 \text{ or } z(\alpha) > 3n/4] < 2 \cdot 2^{-n[1-h(1/4)]} \quad < \quad 2^{-0.188n},$$

where $h(\cdot)$ is the binary entropy. The lemma thus follows. $\square$

Lemma 3. [1] *For every* $\alpha, \beta \in \{0,1\}^n$, *and* $s \in \mathcal{S}$,

$$s(\alpha \oplus \beta) = s(\alpha) \oplus s(\beta),$$

and thus

$$\overline{s(\alpha)} \cdot \overline{s(\beta)} = \overline{s(\alpha \oplus \beta)}.$$

Corollary 2. *For every* $\alpha, \beta \in \{0,1\}^n$,

$$|v(\alpha) \cdot v(\beta)| = d(v(\alpha \oplus \beta)),$$

where $\cdot$ *denotes the inner product of two vectors.*

Let $H = \{\alpha_1, \ldots, \alpha_L\} \subset \{0,1\}^n$ be a subset of L n-bit strings such that $L = |H| \ge 2^{dn}$, where $d = 0.822$. Recall that $\mathcal{S} = \{s_1, \ldots, s_K\}$ is a fixed enumeration of all elements of $\mathcal{S}$. Let V be the $K \times L$ matrix whose columns are $v(\alpha_1), \ldots, v(\alpha_L)$, i.e.

$$V \stackrel{d}{=} \begin{pmatrix} \overline{s_1(\alpha_1)} & \overline{s_1(\alpha_2)} & \cdots\cdots & \overline{s_1(\alpha_L)} \\ \overline{s_2(\alpha_1)} & \overline{s_2(\alpha_2)} & \cdots\cdots & \overline{s_2(\alpha_L)} \\ \vdots & \vdots & & \vdots \\ \vdots & \vdots & & \vdots \\ \overline{s_K(\alpha_1)} & \overline{s_K(\alpha_2)} & \cdots\cdots & \overline{s_K(\alpha_L)} \end{pmatrix}. \tag{11}$$

The matrix V in (11) is a generalization of the matrix employed in [1], [2]. Lemma 2 implies that in the matrix V of ± 1 values, almost all columns are well balanced.

Lemma 4. *Let $U = V^T \cdot V$, where V is defined in (11), and V^T is the transpose of V. For each $i = 1, \ldots, L$, the number of entries $U_{i,j}$ in the i-th row of the matrix U such that $|U_{i,j}| > K \cdot 2^{-k}$, is at most 2^{cn}, where $c = 0.812$.*

Proof. It is clear that

$$|U_{i,j}| = |v(\alpha_i) \cdot v(\alpha_j)| = d(v(\alpha_i \oplus \alpha_j)),$$

where the second equality follows from Corollary 2. Thus, if $|U_{i,j}| > K \cdot 2^{-k}$, then $\alpha_i \oplus \alpha_j \in D$, where D is defined in Lemma 2.

For each fixed α_i, the sequence $\{\alpha_i \oplus \alpha_1, \ldots, \alpha_i \oplus \alpha_L\}$ consists of L *distinct* strings in $\{0,1\}^n$. Thus, for each fixed i, by Corollary 2,

$$|\{j : \alpha_i \oplus \alpha_j \in D\}| < 2^{cn},$$

where $c = 0.812$. Therefore, for each fixed i,

$$|\{j : |U_{i,j}| > K \cdot 2^{-k}\}| < 2^{cn}.$$

$\square$

Corollary 3. *Let $U = [U_{i,j}] = V^T \cdot V$. Then*

$$\sum_{i=1}^{L} \sum_{j=1}^{L} |U_{i,j}| = \sum_{i=1}^{L} \left(\sum_{j:U_{i,j}>K/2^k} |U_{i,j}| + \sum_{j:U_{i,j}\leq K/2^k} |U_{i,j}| \right)$$
$$\leq L \cdot (2^{cn} \cdot K + L \cdot K \cdot 2^{-k}), \quad c = 0.812. \tag{12}$$

Remark: It is clear that in the matrix $U = V^T \cdot V$, every diagonal entry $U_{i,i} = K$. Lemma 4 implies that in every row and every column of the matrix U, for all but a $2^{-(1-c)n} = 2^{-0.188n}$ fraction of off-diagonal entries $U_{i,j}$, $|U_{i,j}|/K < 2^{-k}$. Thus, informally speaking, the matrix V is "approximately" column-orthogonal. Note that the matrix V is related to the Hadamard Matrix, which gives rise to a theory of Discrete Fourier Transform of Boolean functions, with important applications in Learning Theory of Boolean Functions [24], Derandomization [46], [37], and Communication Complexity [22].

3.3   The Main Lemma

We have seen (Lemma 2) that in the matrix V defined in (11), if $L \geq 2^{dn}$, where $d = 0.822$, then almost all columns are well balanced. As explained in the proof outline in Section 3.1, we need for the proof of security of Hyper-Encryption that in V almost all *rows* are well balanced.

Definition 6. *Let $H = \{\alpha_1, \ldots, \alpha_L\} \subset \{0,1\}^n$. Let $s \in \mathcal{S}$. Denote*

$$u_H(s) \stackrel{d}{=} \left(\overline{s(\alpha_1)}, \ldots, \overline{s(\alpha_L)} \right). \tag{13}$$

We say that s is good for H if $d(u_H(s)) < L \cdot 2^{-k/3+1}$. I.e., s is good H if

$$\left| \frac{|\{\alpha \in H : s(\alpha) = 0\}| - |\{\alpha \in H : s(\alpha) = 1\}|}{L} \right| < 2^{-k/3+1}. \qquad (14)$$

Thus, s is good for H if and only if for $\alpha \xleftarrow{R} H$, for each bit $\delta \in \{0, 1\}$,

$$\left| \Pr[s(\alpha) = \delta] - \frac{1}{2} \right| < 2^{-k/3}. \qquad (15)$$

We say that s is bad for H if s is not good for H.

Definition 7. Let $H = \{\alpha_1, \ldots, \alpha_L\} \subset \{0, 1\}^n$. Define

$$G_H \stackrel{d}{=} \{s \in \mathcal{S} : s \text{ is good for } H\}; \quad \text{and} \qquad (16)$$

$$B_H \stackrel{d}{=} \mathcal{S} - G_H = \{s \in \mathcal{S} : s \text{ is bad for } H\}. \qquad (17)$$

We now prove the following Main Lemma, which says that as long as $|H| \geq 2^{dn}$, almost all $s \in \mathcal{S}$ are good for H. This implies that in the matrix V defined in (11), almost all *rows* are well balanced between 1 and -1. This does not follow from the fact (Lemma 1) on the columns of V, but is rather a consequence of the approximate column-orthogonality of V.

Main Lemma 1 Let $H = \{\alpha_1, \ldots, \alpha_L\} \subset \{0, 1\}^n$. If $L \geq 2^{dn}$ where $d = 0.822$, and $k < 0.01n$, then

$$|B_H| < |\mathcal{S}| \cdot 2^{-k/3} = K \cdot 2^{-k/3}. \qquad (18)$$

In other words, $|G_H| > |\mathcal{S}| \cdot (1 - 2^{-k/3})$.

Proof. Let $B_H^+ = \left\{ s \in \mathcal{S} : \sum_{i=1}^L \overline{s(\alpha_i)} \geq L \cdot 2^{-k/3+1} \right\}$, and $B_H^- = B_H - B_H^+$. We bound $|B_H^+|$. The proof for $|B_H^-|$ is analogous.

Let $\delta^+ = (\delta_1, \ldots, \delta_K) \in \{0, 1\}^K$ be the characteristic vector of B_H^+, with respect to the fixed enumeration $\mathcal{S} = \{s_1, \ldots, s_K\}$ of $\mathcal{S} = \{1, \ldots, n\}^k$. That is, $\delta_i = 1$ if and only if $s_i \in B_H^+$. Let

$$\mathbf{1}_L \stackrel{d}{=} (1, \ldots, 1) \in \{0, 1\}^L \qquad (19)$$

be the L-dimensional vector in which every coordinate is 1. Thus, for each $s \in S$,

$$\sum_{i=1}^L \overline{s(\alpha_i)} = u_H(s) \cdot \mathbf{1}_L, \qquad (20)$$

where $u_H(s)$ is defined in (13).

Consider the matrix V defined in (11). By the definitions of B_H^+ and δ^+, and (20),

$$\delta^+ \cdot V \cdot \mathbf{1}_L \geq |B_H^+| \cdot L \cdot 2^{-k/3+1}. \tag{21}$$

On the other hand, by the Cauchy-Schwartz inequality,

$$\delta^+ \cdot V \cdot \mathbf{1}_L \leq ||\delta^+|| \cdot ||V \cdot \mathbf{1}_L||, \tag{22}$$

where $||(x_1, \ldots, x_K)|| = \sqrt{x_1^2 + \cdots + x_K^2}$. Since δ^+ is the characteristic vector of B_H^+,

$$||\delta^+|| = \sqrt{|B_H^+|}. \tag{23}$$

Next,

$$||V \cdot \mathbf{1}_L||^2 = \mathbf{1}_L^T \cdot V^T V \cdot \mathbf{1}_L = \mathbf{1}_L^T \cdot U \cdot \mathbf{1}_L = \sum_{i=1}^L \sum_{j=1}^L U_{i,j}$$

$$\leq L \cdot (2^{cn} \cdot K + L \cdot K \cdot 2^{-k}) \quad \text{by Corollary 3,} \tag{24}$$

where $c = 0.812$. Combining (21), (22), (23), and (24), together with the fact that $L \geq 2^{0.822n}$, we obtain

$$|B_H^+| \cdot L \cdot 2^{-k/3+1} \leq \sqrt{|B_H^+|} \cdot \sqrt{LK \cdot (2^{0.812n} + L \cdot 2^{-k})},$$

$$2\sqrt{|B_H^+|} \cdot 2^{-k/3} \leq \sqrt{K \cdot \left(\frac{2^{0.812n}}{L} + 2^{-k}\right)} < \sqrt{K \cdot (2^{-0.01n} + 2^{-k})}. \tag{25}$$

For $k < 0.01n$, $2^{-0.01n} < 2^{-k}$. Therefore from (25), we have

$$2\sqrt{|B_H^+|} \cdot 2^{-k/3} < \sqrt{2K \cdot 2^{-k}},$$

$$|B_H^+| < \frac{1}{2} \cdot K \cdot 2^{-k/3}. \tag{26}$$

Similarly, $|B_H^-| < \frac{1}{2} \cdot K \cdot 2^{-k/3}$. Hence, $|B_H| = |B_H^+| + |B_H^-| < K \cdot 2^{-k/3}$. $\quad\square$

Definition 8. Let $H \subset \{0,1\}^n$. We say that H is fat if $|H| \geq 2^{dn}$, where $d = 0.822$. Let $\boldsymbol{H} = (H_1, \ldots, H_l)$, where $H_i \subset \{0,1\}^n$ $\forall 1 \leq i \leq l$. We say that $\boldsymbol{H}$ is fat if H_i is fat $\forall 1 \leq i \leq l$.

Definition 9. Let $\boldsymbol{H} = (H_1, \ldots, H_l)$, where $H_i \subset \{0,1\}^n$ $\forall 1 \leq i \leq l$. Define

$$G_{\boldsymbol{H}} \overset{d}{=} \bigcap_{i=1}^l G_{H_i} = \{s \in \mathcal{S} : s \in G_{H_i} \ \forall 1 \leq i \leq l\}. \tag{27}$$

Corollary 4. Let $\boldsymbol{H} = (H_1, \ldots, H_l)$, $H_i \subset \{0,1\}^n$, be fat. Then

$$|G_{\boldsymbol{H}}| > |\mathcal{S}| \cdot (1 - l \cdot 2^{-k/3}). \tag{28}$$

Proof. By Main Lemma 1, and the fact that $G_{\boldsymbol{H}} = \mathcal{S} - \bigcup_{i=1}^l B_{H_i}$. $\quad\square$

3.4   Single-Bit Security

We prove in this section the security of the Hyper-Encryption scheme for single-bit message blocks, that is, Theorem 1 in the case that $m = 1$. For $m = 1$, the secret key shared between Alice and Bob is a random $s \xleftarrow{R} \mathcal{S}$, where $\mathcal{S}$ is defined in Definition 1. It is clear that to the Adversary $\mathcal{AD}$, decoding a ciphertext $C = M \oplus s(\alpha)$ is equivalent to determining the one-time pad $s(\alpha)$. Thus we consider a decoding algorithm $A_2(\eta_l, s)$ of the Adversary $\mathcal{AD}$, that in Phase II, given η_l defined recursively in (3), and the secret key s, tries to compute the one-time pad bit $s(\alpha_l)$.

For technical reasons, in the single-bit case we consider a slightly stronger Adversary who can store $B = 0.168n$ bits. (Note that $0.168 > 1/6$). First, we introduce the following simple yet important lemma.

Lemma 5. *Let* $0 < \gamma, \nu < 1$ *and* $\nu < 1 - \gamma$. *For any function* $f : \{0,1\}^n \longrightarrow \{0,1\}^{\gamma n}$,

$$\left| \left\{ \alpha \in \{0,1\}^n : |f^{-1}(f(\alpha))| \geq 2^{(1-\gamma-\nu)n} \right\} \right| > 2^n \cdot (1 - 2^{-\nu n}). \qquad (29)$$

In other words, for $\alpha \xleftarrow{R} \{0,1\}^n$,

$$\Pr\left[|f^{-1}(f(\alpha))| \geq 2^{(1-\gamma-\nu)n} \right] > 1 - 2^{-\nu n}. \qquad (30)$$

Proof. By counting, and giving an upper bound on the number of $\alpha \in \{0,1\}^n$ such that $|f^{-1}(f(\alpha))| \geq 2^{(1-\gamma-\nu)n}$.   □

We now prove the single-bit security. Recall that $A_1^{(1)}, \ldots, A_1^{(l)}$ are the recording functions of the Adversary $\mathcal{AD}$, where l is the number of rounds in the adaptive attack. For the single-bit case, each $A_1^{(i)}$ is a function $A_1^{(i)} : \{0,1\}^n \times \{0,1\}^B \times \{0,1\}^{i-1} \longrightarrow \{0,1\}^B$, where $B = 0.168n$.

Remark: Note that the single-bit security of Hyper-Encryption against a *non-adaptive* adversary, whose recording functions depend *only* on the public random strings α_i, follows directly from Main Lemma 1 and corollaries, and Lemma 5: For almost all $\alpha \in \{0,1\}^n$, the preimage $H = A_1^{-1}(A_1(\alpha))$ is *fat*, and as long as H is fat, almost all secret keys $s \in \mathcal{S}$ are *good* for H. The security against a non-adaptive adversary thus follows from the fact that α is a uniformly random string in H, the *independence* between s and H, Corollary 4, and the definition of good-ness (Definition 6 and (15)). However, this argument does *not* hold for an *adaptive* adversary, since each recording function $A_1^{(i)}(\alpha_i, \eta_{i-1}, s(\alpha_1), \ldots, s(\alpha_{i-1}))$ depends on η_{i-1} and $s(\alpha_1), \ldots, s(\alpha_{i-1})$, each depending on s, and thus the preimage H_i *depends on* s.

Consider any given recording functions $A_1^{(1)}, \ldots, A_1^{(l)}$. Recall that $s \in \mathcal{S}$ is the random secret key, and $\alpha_1, \ldots, \alpha_l \in \{0,1\}^n$ are the public random strings, where α_i is used in the i-th round. Denote $\boldsymbol{\alpha}_i = (\alpha_1, \ldots, \alpha_i)$, $\boldsymbol{\alpha} = (\alpha_1, \ldots, \alpha_l)$, $X_i = s(\alpha_i)$, $\boldsymbol{X}_i = (X_1, \ldots, X_i)$, and $\boldsymbol{X}(s, \boldsymbol{\alpha}) = (X_1, \ldots, X_{l-1})$.

Definition 10. *Define the random variable*

$$Y_1 = Y_1(\alpha_1) \stackrel{d}{=} \left\{\beta \in \{0,1\}^n : A_1^{(1)}(\beta) = \eta_1\right\}, \quad \text{where } \eta_1 = A_1^{(1)}(\alpha_1). \tag{31}$$

Thus $Y_1 = \left(A_1^{(1)}\right)^{-1}\left(A_1^{(1)}(\alpha_1)\right).$

For $2 \leq i \leq l$, *define the random variable*

$$Y_i = Y_i(s, \boldsymbol{\alpha}_i) \stackrel{d}{=} \left\{\beta \in \{0,1\}^n : A_1^{(i)}(\beta, \eta_{i-1}, \boldsymbol{X}_{i-1}) = \eta_i\right\} \tag{32}$$

where $\eta_i = A_1^{(i)}(\alpha_i, \eta_{i-1}, \boldsymbol{X}_{i-1}).$

Thus, Y_i *is the subset of all strings* $\beta \in \{0,1\}^n$ *that are mapped to* η_i. *Define the random variable*

$$\boldsymbol{Y} = \boldsymbol{Y}(s, \boldsymbol{\alpha}) \stackrel{d}{=} (Y_1, \ldots, Y_l). \tag{33}$$

For a vector $\boldsymbol{H} = (H_1, \ldots, H_l)$ of subsets, for simplicity in notation denote $|\boldsymbol{H}| \stackrel{d}{=} |H_1 \times \cdots \times H_l|$. Recall from Definition 8 that $\boldsymbol{H} = (H_1, \ldots, H_l)$ is called *fat* if $\forall\, 1 \leq i \leq l$, $|H_i| \geq 2^{dn}$, $d = 0.822$.

As a critical step of the proof of the single-bit security, we show that $\Pr[s \in G_{\boldsymbol{Y}}]$ is almost 1, where $\boldsymbol{Y}$ is defined above, and $G_{\boldsymbol{Y}}$ is defined in Definition 9.

Main Lemma 2 *For any recording functions* $A_1^{(1)}, \ldots, A_1^{(l)}$, *where for each* i, $A_1^{(i)} : \{0,1\}^n \times \{0,1\}^B \times \{0,1\}^{i-1} \longrightarrow \{0,1\}^B$, *and* $B = 0.168n$, *for* $s \stackrel{R}{\longleftarrow} \mathcal{S}$, *and* $\alpha_1, \ldots, \alpha_l \stackrel{R}{\longleftarrow} \{0,1\}^n$,

$$\Pr\left[s \in G_{\boldsymbol{Y}} \wedge (\boldsymbol{Y} \text{ is fat})\right]$$
$$> \left(1 - 2^{-k/3+1}\right)^{l-1} \cdot \left(1 - l \cdot 2^{-k/3}\right) \cdot \left(1 - l \cdot 2^{-0.01n}\right)$$
$$> 1 - (3l - 2) \cdot 2^{-k/3} - l \cdot 2^{-0.01}, \tag{34}$$

where fatness is defined in Definition 8.

Proof Sketch: The underlying sample space is $\mathcal{S} \times (\{0,1\}^n)^l$, i.e. each sample point is of the form $(s, \alpha_1, \ldots, \alpha_l) \in \mathcal{S} \times (\{0,1\}^n)^l$. Clearly, $\left|\mathcal{S} \times (\{0,1\}^n)^l\right| = K \cdot 2^{nl}$, where $K = n^k$. We count, using Main Lemma 1 and Lemma 5, the number of sample points $(s, \alpha_1, \ldots, \alpha_l)$ such that $s \in G_{\boldsymbol{Y}} \wedge \boldsymbol{Y}$ is fat, where the random variable $\boldsymbol{Y}$ is defined in Definition 10, and $G_{\boldsymbol{Y}}$ is defined in Definition 7 and show that this number is $\geq K \cdot 2^{nl} \cdot \left(1 - (3l - 2) \cdot 2^{-k/3} - l \cdot 2^{-0.01}\right)$, from which the lemma follows. $\qquad\square$

Corollary 5. *For any recording functions* $A_1^{(1)}, \ldots, A_1^{(l)}$ *as above, for* $s \stackrel{R}{\longleftarrow} \mathcal{S}$, *and*
$\boldsymbol{\alpha} \stackrel{R}{\longleftarrow} (\{0,1\}^n)^l$,

$$\Pr\left[s \in G_{\boldsymbol{Y}}\right] > 1 - (3l - 2) \cdot 2^{-k/3} - l \cdot 2^{-0.01n}. \tag{35}$$

Proof. $\Pr[s \in G_Y] \geq \Pr[s \in G_Y \wedge Y \text{ is fat}]$.   □

From Corollary 5, we obtain the single-bit security of Hyper-Encryption as follows. For each *fixed* $s \in G_Y$, s is *good* for $Y = (Y_1, \ldots, Y_l)$, and thus good in particular for Y_l. Therefore, for essentially half of the strings $\beta \in Y_l$, $s(\beta) = \delta$ for each $\delta \in \{0, 1\}$. Given s, $X(s, \alpha) = (s(\alpha_1), \ldots, s(\alpha_{l-1}))$, and $\eta_1, \ldots, \eta_l$, all $\mathcal{AD}$ knows about $(\alpha_1, \ldots, \alpha_l)$ is that $(\alpha_1, \ldots, \alpha_l)$ is *uniformly random* in $Y_1 \times \cdots \times Y_l$, and in particular, α_l is uniformly random in Y_l. By Definition 6 and (15), for each $\delta \in \{0, 1\}$, for each *fixed* $s \in G_Y$, for random $\alpha_l \xleftarrow{R} Y_l$, $|\Pr[s(\alpha_l) = \delta] - \frac{1}{2}| < 2^{-k/3}$, i.e. the distribution of $s(\alpha_l)$ is $2^{-k/3}$-*statistically indistinguishable* from the uniform distribution over $\{0, 1\}$. Hence as long as $s \in G_Y$, any algorithm A_2 that is given s and η_l as input, cannot compute $s(\alpha)$ with an advantage $2^{-k/3}$ or greater, in other words, for any algorithm A_2, for $\alpha_l \xleftarrow{R} Y_l$,

$$\left| \Pr[A_2(\eta_l, s) = s(\alpha_l) \mid s \in G_Y] - \frac{1}{2} \right| < 2^{-k/3}. \tag{36}$$

We thus conclude, from (35) and (36), that the adaptive adversary $\mathcal{AD}$ cannot compute the one-time pad bit $s(\alpha_l)$, and obtain the single-bit security:

Lemma 6. *For $B = 0.168n$, for any recording functions $A_1^{(1)}, \ldots, A_1^{(l)}$, where $A_1^{(i)} : \{0, 1\}^n \times \{0, 1\}^B \times \{0, 1\}^{i-1} \longrightarrow \{0, 1\}^B$ for each i, for any decoding algorithm A_2, for $s \xleftarrow{R} \mathcal{S}$, and $\alpha_1, \ldots, \alpha_l \xleftarrow{R} \{0, 1\}^n$,*

$$\left| \Pr[A_2(\eta_l, s) = s(\alpha_l)] - \frac{1}{2} \right| < 3l \cdot 2^{-k/3-1} + l \cdot 2^{-0.01n-1}. \tag{37}$$

where $\eta_1 = A_1^{(1)}(\alpha_1)$, and $\eta_i = A_1^{(i)}(\alpha_i, \eta_{i-1}, s(\alpha_{i-1}))$ for $i = 2, \ldots, l$.

Remark: From the proof of the single-bit security, it is clear that indeed the Hyper-Encryption scheme can tolerate an even stronger Adversary $\mathcal{AD}$. Namely, even if in Round i of Phase I, $\mathcal{AD}$ is given all previous $\boldsymbol{\eta}_{i-1} = (\eta_1, \ldots, \eta_{i-1})$, and computes

$$\eta_i = A_1^{(i)}(\alpha_i, \boldsymbol{\eta}_{i-1}, X_{i-1}), \tag{38}$$

and even if $\mathcal{AD}$ is given η_l, s, and $X(s, \alpha_l)$ in Phase II, he can still learn nothing about the one-time pad $s(\alpha_l)$ by any decoding algorithm $A_2(\boldsymbol{\eta}_l, s, X(s, \alpha_l))$.

3.5   Proof of Theorem 1

Theorem 1, in the general case $m \geq 1$, now follows from the single-bit security, and the standard *hybrid argument* of [18]. Details will be provided in the full paper.

4   Authentication and Non-malleability

Having addressed the issue of message secrecy in Section 2, we devote this chapter to message integrity. We first describe a message authentication scheme, by which the receiver Bob of a message M can verify that M, sent in the clear, is an unmodified message from the claimed sender Alice. Implementing message authentication by use of a shared key used only once, is well known. Using complexity assumption, it is also possible to implement message authentication where the shared key is repeatedly used. The *new essential feature* of our authentication scheme is that in the bounded storage model, we can *re-use* a shared secret key an exponentially (in k) number of times against strong adaptive attacks, and yet prove the information-theoretic security of the scheme.

Applying our message authentication scheme to a ciphertext produced by the Hyper-Encryption scheme, we obtain an enhanced encryption scheme that is both secure and non-malleable against the adaptive attack described in Section 2.

4.1   Message Authentication

We first describe our message authentication scheme, which we refer to as the Hyper-Authentication scheme. Let t be a parameter such that $t \geq m$ and $t \geq k/3$, where m is the message block size, and k is the security parameter. The sender Alice and and the receiver Bob agree, once and for all, on a secret key (s_a, s_b), where $s_a, s_b \in S^t$, $S = \{1, \ldots, n\}^k$. Again, the key (s_a, s_b) can be either truly random, or generated from a short random secret by our key amplification protocol. In this chapter, we consider the case that the key (s_a, s_b) is truly random.

For Alice to send and authenticate to Bob a message $M \in \{0,1\}^m$, Alice and Bob listen to a public random string $\alpha \in \{0,1\}^n$, and by use of s_a, s_b extract a common *check vector* (a, b) [40], where $a = s_a(\alpha)$ and $b = s_b(\alpha)$ (recall Definition 2). Then Alice computes $r = aM + b$ in the finite field F_{2^t},[4] and sends (M, r) to Bob, who verifies that $r = aM + b$ in F_{2^t}, and accepts M if and only if the equation holds. Here we assume that there is a fixed public irreducible polynomial of degree t over F_2 that is used for the arithmetics in F_{2^t}. A description of the Hyper-Authentication scheme is also given in Figure 2. Our scheme is an adaptation to the bounded storage model of the universal hashing based scheme introduced by Wegman and Carter [48]. See also [40]. We again emphasize that the crucial new feature of the Hyper-Authentication scheme is the provable security and re-usability of the secret key under strong adaptive attacks.

We now describe the adaptive attack against the Hyper-Authentication scheme. Public random strings $\alpha_1, \alpha_2, \ldots$ are used for the authentication of messages $M_1, M_2, \ldots$. The Adversary $\mathcal{AD}$ again attacks adaptively in two phases.

---

[4] Here $a, b \in \{0,1\}^t$ and $M \in \{0,1\}^m$, with $m \leq t$, are interpreted as elements of the field F_{2^t} in the natural way.

Message Block $M \in \{0, 1\}^m$.

Secret key $(\boldsymbol{s}_a, \boldsymbol{s}_b)$, $\boldsymbol{s}_a, \boldsymbol{s}_b \in \mathcal{S}^t$, $t \geq m$, $t \geq k/3$.

Public random string $\alpha \xleftarrow{R} \{0, 1\}^n$.

1. Alice and Bob compute $a = \boldsymbol{s}_a(\alpha)$ and $b = \boldsymbol{s}_b(\alpha)$.

2. Alice computes $r = aM + b$ in F_{2^t}, and sends (M, r) to Bob.

3. Bob verifies that $r = aM + b$ in F_{2^t}.

Fig. 2. The Hyper-Authentication Scheme

Phase I consists of $l - 1$ rounds. In Round 1, α_1 is broadcast, and $\mathcal{AD}$ computes a recording function $\eta_1 = A_1^{(1)}(\alpha_1)$, $|\eta_1| = B = n/6$. Then by a Chosen Message Attack (CMA), $\mathcal{AD}$ obtains (M_1, r_1), where $r_1 = \boldsymbol{s}_a(\alpha_1) \cdot M_1 + \boldsymbol{s}_b(\alpha_1)$. In Round $i = 2, \ldots, l - 1$, $\mathcal{AD}$ has available η_{i-1}, and $(M_1, r_1), \ldots, (M_{i-1}, r_{i-1})$. As α_i is broadcast, $\mathcal{AD}$ computes

$$\eta_i = A_1^{(i)}(\alpha_i, \eta_{i-1}, M_1, r_1, \ldots, M_{i-1}, r_{i-1}), \qquad |\eta_i| = n/6. \tag{39}$$

Then by a CMA, $\mathcal{AD}$ obtains (M_i, r_i), where $r_i = \boldsymbol{s}_a(\alpha_i) \cdot M_i + \boldsymbol{s}_b(\alpha_i)$.

In Phase II, $\mathcal{AD}$ computes η_l that is defined in (39). He is then given the secret key $(\boldsymbol{s}_a, \boldsymbol{s}_b)$, and also a pair (M_l, r_l) with $r_l = \boldsymbol{s}_a(\alpha_l) \cdot M_l + \boldsymbol{s}_b(\alpha_l)$. Now $\mathcal{AD}$ applies a forgery algorithm $\mathcal{F}$, and attempts to compute a pair

$$(\tilde{M}, \tilde{r}) = \mathcal{F}(\eta_l, \boldsymbol{s}_a, \boldsymbol{s}_b, M_l, r_l) \tag{40}$$

such that $\tilde{M} \neq M_l$ and $\tilde{r} = \boldsymbol{s}_a(\alpha_l) \cdot \tilde{M} + \boldsymbol{s}_b(\alpha_l)$. In other words, $\mathcal{AD}$ attempts to forge a valid authentication tag $\tilde{r}$ for some different message $\tilde{M} \neq M$ in l-th round.

We will in fact tolerate a stronger adversary. For each $j = 1, \ldots, l$, denote

$$(a_j, b_j) \stackrel{d}{=} (\boldsymbol{s}_a(\alpha_j), \boldsymbol{s}_b(\alpha_j)). \tag{41}$$

Note that in any attack against the Hyper-Authentication scheme, the maximum information the adversary can obtain from (M_j, r_j), where $r_j = a_j \cdot M_j + b_j$, is (a_j, b_j). We can show that even if for $j = 1, \ldots, l - 1$, $\mathcal{AD}$ in Round j of Phase I is given (a_j, b_j) instead of (M_j, r_j), and in Round i, instead of (39) computes

$$\eta_i = A_1^{(i)}(\alpha_i, \eta_{i-1}, a_1, b_1, \ldots, a_{i-1}, b_{i-1}), \qquad |\eta_i| = n/6, \tag{42}$$

$\mathcal{AD}$'s forgery cannot succeed.

Theorem 3. *For $B = n/6$, for any recording functions $A_1^{(1)}, \ldots, A_1^{(l)}$, $\forall M_l \in$*
$\{0,1\}^m$, for any forgery algorithm $\mathcal{F}$, for $s_a, s_b \xleftarrow{R} \mathcal{S}^{(t)}$ with $t \geq m$, for
$\alpha_1, \ldots, \alpha_l \xleftarrow{R} \{0,1\}^n$,

$$\Pr\left[\tilde{r} = a_l \cdot \tilde{M} + b_l \text{ in } F_{2^t}\right] < 2^{-t} + 6tl \cdot 2^{-k/3} + 2tl \cdot 2^{-0.01n}, \quad (43)$$

where $(\tilde{M}, \tilde{r})$ is defined in (40), and $\tilde{M} \neq M_l$.

Theorem 3 follows from Theorem 2. The proof will be provided in the full paper.

4.2   Secure and Non-malleable Hyper-Encryption

Informally speaking, an encryption scheme $E(\cdot)$ is non-malleable if from a given ciphertext $C = E(M)$ of a message M, the adversary cannot create a ciphertext $\tilde{C} = E(\tilde{M})$ for any $\tilde{M} \neq M$. It is clear that the one-time pad scheme is malleable: given a ciphertext $C = E(X, M) = M \oplus X$ of message M, encrypted with one-time pad X, by computing

$$C \oplus Z = (M \oplus X) \oplus Z = (M \oplus Z) \oplus X,$$

one gets a valid ciphertext of $M \oplus Z$, with the same one-time pad X. Since the plain Hyper-Encryption scheme of Section 2 uses one-time pads, it is malleable. In this section, applying the authentication method of Section 4.1 to the ciphertext of the Hyper-Encryption, we enhance the Hyper-Encryption scheme so as to achieve both security and non-malleability.

The enhanced Hyper-Encryption scheme proceeds as follows. Alice and Bob establish, once and for all, a secret key (s, s_a, s_b), where $s \in \mathcal{S}^m$, and $s_a, s_b \in \mathcal{S}^t$, with $t \geq m$ and $t \geq k/3$. For the encryption of a message $M \in \{0,1\}^m$, a public random string $\alpha \xleftarrow{R} \{0,1\}^n$ is broadcast. Alice and Bob compute $X = s(\alpha)$, $a = s_a(\alpha)$, and $b = s_b(\alpha)$. The ciphertext of M is (C, r), where $C = M \oplus X$, and $r = aC + b$ in F_{2^t}. Bob accepts C if and only if $r = aC + b$. Figure 3 contains a description of this enhanced scheme.

The everlasting non-malleability of the enhanced scheme follows from the argument in the proof of Theorem 3: In order for $\tilde{M} \neq M_l$, we must have $\tilde{C} = \tilde{M} \oplus X_l \neq M_l \oplus X_l = C_l$. Since $\tilde{C} \neq C$, it follows from the argument in the proof of Theorem 3 that from (C_l, r_l), the Adversary $\mathcal{AD}$ cannot successfully forge a valid pair $(\tilde{C}, \tilde{r})$ such that $\tilde{r} = a_l \cdot \tilde{C} + b_l$.

The everlasting security of the enhanced scheme follows from Theorem 1, and a reduction argument, the obvious details of which we omit.

4.3   Application: A Provably Secure Kerberos-Like System

Our strong encryption and authentication enable the construction of a probably secure Kerboros-like key distribution system [44], [21]. The system involves a

Message Block $M \in \{0,1\}^m$.

Secret key (s, s_a, s_b), $s \xleftarrow{R} S^m$, $s_a, s_b \xleftarrow{R} S^t$, $t \geq m, t \geq k/3$.

Public random string $\alpha \xleftarrow{R} \{0,1\}^n$.

1. Alice and Bob compute $X = s(\alpha)$, $a = s_a(\alpha)$, and $b = s_b(\alpha)$.

2. Alice computes $C = M \oplus X$, $r = aC + b$ in F_{2^t}, and sends (C, r) to Bob.

3. Bob decrypts $M = C \oplus X$, and verifies that $r = aC + b$ in F_{2^t}.

Fig. 3. Secure and Non-malleable Hyper-Encryption

trusted key distribution center $\mathcal{T}$, and uses the Hyper-Authentication and enhanced Hyper-Encryption schemes. Each user U shares, once and for all, a common secret key $S_{\mathcal{T},U}$ with the center $\mathcal{T}$ for both encryption and authentication. Users initially share no keys among themselves. In order for Alice to communicate privately with Bob, Alice first sends to the key distribution center $\mathcal{T}$ a message in the clear that is authenticated with the key $S_{\mathcal{T},A}$ by the Hyper-Authentication scheme, requesting a common secret key with Bob. Then the center $\mathcal{T}$ sends to Alice and Bob a random secret key $S_{A,B}$, encrypted and authenticated with keys $S_{\mathcal{T},A}$ and $S_{\mathcal{T},B}$ respectively, by the enhanced Hyper-Encryption schemes. Unlike the one-time session keys distributed by the traditional Kerberos system, the secret key $S_{A,B}$ can be used from now on between Alice and Bob without renewal. The secrecy, integrity and non-malleability of the communications between the users and the center $\mathcal{T}$, and among the users, and the re-usability of the keys, against strong adaptive attacks, are all guaranteed by Theorems 1, 2, and 3.

References

1. Y. Aumann and M. O. Rabin. Information Theoretically Secure Communication in the Limited Storage Space Model. In *Advances in Cryptology - Crypto '99*, pages 65-79, 1999.
2. Y. Aumann, Y. Z. Ding, and M. O. Rabin. Everlasting Security in the Bounded Storage Model. Accepted to *IEEE Transactions on Information Theory*, 2000.
3. Y. Aumann and U. Feige. One message proof systems with known space verifier. In *Advances in Cryptology - Crypto '93*, 1993.
4. M. Bellare, A. Desai, D. Pointcheval, and P. Rogaway. Relations Among Notions of Security for Public-Key Encryption Schemes. In *Advances in Cryptology - Crypto '98*, 1998.
5. M. Bellare and A. Sahai. Non-Malleable Encryption: Equivalence between Two Notions, and an Indistinguishability-Based Characterization. In *Advances in Cryptology - Crypto '99*, 1999.

6. C. H. Bennett, G. Brassard, C. Crepeau, and U. Maurer. Generalized privacy amplification. *IEEE Transactions on Information Theory*, 41(6), 1995.
7. C. Cachin, C. Crépeau, and J. Marcil. Oblivious transfer with a memory-bounded receiver. In *Proc. 39th IEEE Symposium on Foundations of Computer Science*, 1998.
8. C. Cachin and U. Maurer. Unconditional security against memory bounded adversaries. In *Advances in Cryptology - Crypto '97*, 1997.
9. A. Condon. Bounded Space Probabilistic Games. In *Proc. Annual Conference on Structure in Complexity Theory*, 1988.
10. A. Condon, and R. Ladner. Probabilistic Game Automata. In *Proc. Annual Conference on Structure in Complexity Theory*, 1986.
11. A. De-Santis, G. Persiano, and M. Yung. One-message statistical zero-knowledge proofs with space-bounded verifier. In *Proc. 19th ICALP*, 1992.
12. Y. Z. Ding. Oblivious Transfer in the Bounded Storage Model. In *Advances in Cryptology - Crypto '01*, 2001.
13. D. Dolev, C. Dwork, and M. Naor. Non-malleable Cryptography. *SIAM J. Comp.*, 30(2): 391-437, 2000.
14. C. Dwork and L. J. Stockmeyer. Finite State Verifiers I: The Power of Interaction. *JACM* 39(4): 800-828, 1992
15. C. Dwork and L. J. Stockmeyer. Finite State Verifiers II: Zero Knowledge. *JACM* 39(4): 829-858, 1992.
16. R. G. Gallager. *Low-Density Parity-Check Codes*. MIT Press, 1963.
17. E. Gilbert, F. MacWilliams, and N. Sloane. Codes which detect deception. *Bell Sys. Tech. J.*, 53(3): 405-424, 1974.
18. S. Goldwasser and S. Micali. Probabilistic Encryption. *JCSS* 28: 270-299, 1984.
19. R. L. Graham, D. E. Knuth, and O. Patashnik. *Concrete Mathematics*. Addison Wesley, 1989.
20. J. Katz and M. Yung. Complete characterization of security notations for probabilistic private-key encryption. In *Proc. 32nd ACM Symposium on Theory of Computing*, 2000.
21. J.T. Kohl, B.C. Neuman, and T. Tso. The Evolution of the Kerberos Authentication System. *Distributed Open Systems*, IEEE Computer Soceity Press, 1994, pp. 78-94.
22. E. Kushilevitz and N. Nisan. *Communication complexity*. Cambridge University Press, New York, 1997.
23. J. Kilian. Zero-knowledge with Log-Space Verifiers. In *Proc. 29th IEEE Symposium on the Foundations of Computer Science*, 1988.
24. N. Linial, Y. Mansour, and N. Nisan. Constant Depth Circuits, Fourier Transform, and Learnability. *JACM* 40(3): 607-620, July 1993.
25. U. Maurer. Conditionally-perfect secrecy and a provably-secure randomized cipher. *Journal of Cryptology*, 5: 53-66, 1992.
26. U. Maurer. Secret key agreement by public discussion from common information. *IEEE Transactions on Information Theory*, 39: 733-742, 1993.
27. U. Maurer. A unified and generalized treatment of authentication theory. In *STACS'96*, 1996.
28. U. Maurer. Information-theoretically secure secret-key agreement by NOT authenticated public discussion. In *Advances in Cryptology - EUROCRYPT '97*, 1997.
29. U. Maurer. Information-Theoretic Cryptography. In *Advances in Cryptology - CRYPTO '99*, 1999.
30. U. Maurer and S. Wolf. Toward characterizing when information-theoretic secret key agreement is possible. In *Advances in Cryptology - ASIACRYPT'96*, 1996.

31. U. Maurer and S. Wolf. Privacy amplification secure against active adversaries. In *Advances in Cryptology - Crypto '97*, 1997.
32. U. Maurer and S. Wolf. Unconditional secure key agreement and the intrinsic conditional information. *IEEE Transaction on Information Theory*, 45(2): 499-514, 1999.
33. U. Maurer and S. Wolf. Information-Theoretic Key Agreement: From Weak to Strong Secrecy for Free. In *Advances in Cryptology - EUROCRYPT '00*, 2000.
34. A. J. Menezes, P. C. van Oorschot, and S. A. Vanstone. *Handbook of Applied Cryptography*. CRC Press, 1996.
35. N. Nisan. Pseudorandom generators for space-bounded computation. In *Proc. 22rd ACM Symposium on Theory of Computing*, 1990.
36. N. Nisan and D. Zuckerman. Randomness is linear in space. *JCSS* 52(1): 43-52, 1996.
37. J. Naor and M. Naor. Small-bias Probabilistic Spaces: Efficient Constructions and Applications. In *Proc. 22rd ACM Symposium on Theory of Computing*, 1990.
38. M. Naor and M. Yung. Public-key cryptosystems provably secure against chosen ciphertext attacks. In *Proc. 22nd ACM Symposium on Theory of Computing*, 1990.
39. M. O. Rabin. Transaction Protection by Beacons. *JCSS* 27(2): 256-267, 1983.
40. T. Rabin and M. Ben-Or. Verifiable Secret Sharing and Multiparty Protocols with Honest Majority. In *Proc. 21st ACM Symposium on Theory of Computing*, 1989.
41. C. Rackoff and D. Simon. Non-interactive zero-knowledge proof of knowledge and chosen ciphertext attacks. In *Advances in Cryptology - CRYPTO'91*, 1991.
42. A. Sahai. Non-Malleable Non-Interactive Zero Knowledge and Adaptive Chosen-Ciphertext Security. In *Proc. 40th IEEE Symposium on Foundations of Computer Science*, 1999.
43. C. E. Shannon. Communication theory of secrecy systems. *Bell Sys. Tech. J.*, 28: 656-715, 1949.
44. J.G. Steiner, B.C. Neuman, and J.I. Schiller. Kerberos: An Authentication Service for Open Network Systems. In *USENIX Conference Proceedings*, Feb 1988, pp. 191-202.
45. D. Stinson and R. Wei. Bibliography on Authentication Codes, 1998. http://www.cacr.math.uwaterloo.ca/dstinson/acbib.htm
46. U. V. Vazirani. *Randomness, Adversaries and Computation* EECS, UC Berkeley, 1986.
47. G. S. Vernam. Cipher printing telegraph systems for secret wire and radio telegraphic communications, 1926. *Journal of the American Institute for Electrical Engineers* 22: 109-115, 1926.
48. N. M. Wegman and J. L. Carter. New hash functions and their use in authentication and set equality. *JCSS* 22(3): 265-279, 1981.

Models and Techniques for Communication in Dynamic Networks

Christian Scheideler

Department of Computer Science
Johns Hopkins University
3400 N. Charles Street
Baltimore, MD 21218, USA
scheideler@cs.jhu.edu

Abstract. In this paper we will present various models and techniques for communication in dynamic networks. Dynamic networks are networks of dynamically changing bandwidth or topology. Situations in which dynamic networks occur are, for example: faulty networks (links go up and down), the Internet (the bandwidth of connections may vary), and wireless networks (mobile units move around). We investigate the problem of how to ensure connectivity, how to route, and how to perform admission control in these networks. Some of these problems have already been partly solved, but many problems are still wide open. The aim of this paper is to give an overview of recent results in this area, to identify some of the most interesting open problems and to suggest models and techniques that allow us to study them.

1 Introduction

In this paper we present various models and techniques for communication in dynamic networks. Generally speaking, dynamic networks are networks of dynamically changing network characteristics such as bandwidth or topology. There are many scenarios in which dynamic networks occur. For example, a fixed interconnection network might experience (adversarial or random) faults, or we may have a wireless network formed by users that move around. Also, the Internet can be seen as a dynamic network, since some virtual connections or communication links might experience large differences in their availability over time, offering a low bandwidth at one time and a high bandwidth at another time.

In the future, dynamic networks will play an increasingly important role, since the development and deployment of powerful and flexible communication systems that are accessible by large parts of the population is seen as a key component for the transition to an information society. While this has been dominated in the past 10 years by the establishment of wired connections to the Internet, this is now shifting more and more to wireless connections, since wireless connections offer much higher mobility and flexibility. Many companies and universities have already deployed wireless access points that allow employees and/or students to access their local area network or the Internet via

H. Alt and A. Ferreira (Eds.): STACS 2002, LNCS 2285, pp. 27–49, 2002.
© Springer-Verlag Berlin Heidelberg 2002

mobile devices such as laptops and palmtops. So far, wireless communication has mostly been restricted to small areas, mostly because currently used strategies have big problems with establishing and maintaining routes in large-scale, mobile networks. In fact, the development of protocols for communication in mobile, wireless networks that are *provably* efficient and robust even under severe circumstances (for example, quickly moving devices) is still in its infancy.

There are several reasons why only a few theoretical results are known for dynamic networks so far. First of all, it is quite challenging to analyze the behavior of communication strategies in dynamic networks. Furthermore, it is quite difficult to formulate models for dynamic networks that are on the one hand abstract enough to allow a mathematical analysis and on the other hand close enough to reality so that they are of practical value. In this paper, we will present some of the models that have recently been used and also suggest some new models to stimulate fundamental research in this field. We will also describe some important, recent results in the area of dynamic networks. The results demonstrate that despite the (seemingly) enormous mathematical challenges, one can prove very strong results about communication in dynamic networks even in fairly general models.

Before we state any models and techniques, we first introduce some basic notation and describe the basic framework underlying all of these models. Afterwards, we will describe various models and approaches for faulty networks, the Internet, wireless networks, and adversarial networks. We will not try to give an exhaustive overview of results in each of these areas, since this would certainly require to write a whole book. Instead, we will rather concentrate on some, to our mind important results in these fields.

1.1   The Basic Model

In this paper we will only consider communication networks whose topology can be modelled as a directed graph $G = (V, E)$ with a set of nodes V representing the processing units or routers and a set of edges $E \subseteq V \times V$ representing the communication links. In some of our models we may assume that each edge $e \in E$ has a cost associated with it, which will usually represent its *capacity* (i.e. the number of packets it can forward in a step). By default, the capacity of an edge is equal to 1.

In a *static* network the communication links (and their cost values) do not change, i.e. the topology can be described by a fixed directed graph. In this paper we will concentrate on *dynamic* networks, i.e. networks in which the connections (or their properties) between the processing units are continuously changing.

In practice, network communication is performed in many different layers:

- *physical layer*: defines the physical transport of data from one node to the other
- *link layer*: defines the mechanism by which data is transmitted from one node to another
- *network layer*: defines functions for route selection and congestion control
- *transport layer*: defines functions for the end-to-end transport of data

If a shared medium is used such as in wireless communication, the link layer is also called *medium access control (MAC) layer*. We will concentrate on the upper three layers. Issues of particular importance will be connectivity, routing, and admission control. By "routing" we understand the process of moving information across a network from a source to a destination. Routing involves two basic activities:

– the determination of routing paths, and
– the transport of information groups (typically called *packets*) along the paths.

The first item is called *path (or route) selection*, and the second item is usually called *switching* (or *scheduling*). (Depending on the research community or research field, both path selection algorithms and switching algorithms have sometimes been called routing algorithms. We want to avoid these ambiguities here.) Different switching models have been defined in the past such as (virtual) circuit, wormhole, or packet switching. Here, we will only concentrate on packet switching. In packet switching models it is usually assumed that all packets are atomic (i.e. unsplittable) objects of uniform size. Time proceeds in synchronous time steps. In each time step, each link of capacity c can forward up to c packets. If too many packets contend to use the same link at the same time, then the switching algorithm has to decide which of them to prefer. Once a packet reaches its destination it is discarded.

We will concentrate on routing protocols in which the nodes locally decide which packets to move forward in each step, i.e., a decision only depends on the routing information carried by the packets and on the local history of the execution. These algorithms are called *local control* or *online* algorithms.

In several of the examples below we will study the performance of online routing protocols under an adversarial injection model. For static networks, the following model has widely been used, which is an extension by Andrews et al. [4] of a model introduced by Borodin et al. [12]:

Suppose that we have an adversary that is allowed to demand network bandwidth up to a prescribed injection rate $\lambda < 1$. For any $w \in \mathbb{N}$, an adversary is called (w, λ)-*bounded* if for all edges e and all time intervals I of length w, it injects no more than $\lambda \cdot w$ packets during I that contain edge e in their routing path. A protocol is called *stable* for a given injection rate λ and network G if for any (w, λ)-bounded adversary the number of packets stored in injection or edge buffers in G does not grow unboundedly with time. Moreover, a protocol is called *universally stable* if it is stable for any $\lambda < 1$ and any network G.

2   Communication in Faulty Networks

We start with the problem of efficient communication in faulty networks. Different fault models have been studied in the literature: faults may be permanent or temporary, nodes and/or edges may break down, and faults may happen at random or may be caused by an adversary or attacker. In the following, the former faults are called *random faults*, and the latter faults are called *worst-case faults*

Central questions in the area of faulty networks are:

- How frequently does a faulty network have to be repaired so that (either with high probability of with certainty) the size of its largest connected component is always a constant fraction of the original size?
- Or better, how frequently does a faulty network have to be repaired so that (either with high probability of with certainty) its largest connected component still offers a connectivity and throughput that is similar to the original network?
- How can an uninterrupted service be offered in faulty networks?
- Are there online routing algorithms that achieve a throughput in faulty networks that is close to a best possible throughput?

These questions will be addressed in the following subsections. We will show that for several of these questions, partial answers have already been obtained. Moreover, we will point out what parts of the questions are still wide open.

2.1   Large Connected Components in Faulty Networks

Suppose that the information is distributed among the nodes of a network in such a way that as long as at least a constant fraction of the nodes stays connected, agreement can be achieved resp. no information is lost. (This can, for example, easily be reached with the help of the information dispersal approach by Rabin [37].) Then we only have to start repairing the network once the largest connected component in it is below a certain constant fraction. The question is, how long do we have to wait for that? In order to address this question, we start with considering random faults and afterwards consider worst-case faults.

Given a graph G and a probability value of p, let $G^{(p)}$ be the random graph obtained from G by keeping each edge of G independently at random with probability p (i.e. p is the *survival probability*). Given a graph G, let $\gamma(G) \in [0,1]$ be the fraction of nodes of G contained in a largest connected component.

Let $\mathcal{G} = \{G_n \mid n \in \mathbb{N}\}$ be any family of graphs with parameter n. Let p^* be the *critical probability* for the existence of a linear-size connected component in G, i.e. for every constant $\epsilon > 0$:

1. for every $p > (1 + \epsilon)p^*$ there exists a constant $c > 0$ with

$$\lim_{n \to \infty} \Pr[\gamma(G_n^{(p)}) > c] = 1$$

2. for all constants $c > 0$ and for all $p < (1 - \epsilon)p^*$:

$$\lim_{n \to \infty} \Pr[\gamma(G_n^{(p)}) > c] = 0$$

Of course, it is not obvious whether critical probabilities exist. However, the results by Erdős and Rényi [15] and its subsequent improvements (e.g. [11,34])

imply that for the complete graph on n nodes, $p^* = 1/(n-1)$, and that for a random graph with $d \cdot n/2$ edges, $p^* = 1/d$. For the 2-dimensional $n \times n$-mesh, Kesten showed that $p^* = 1/2$ [28]. Ajtai, Komlós and Szemerédi proved that for the hypercube of dimension n, $p^* = 1/n$ [3]. For the n-dimensional butterfly network, Karlin, Nelson and Tamaki showed that $0.337 < p^* < 0.436$ [27].

If we define the *robustness* of a graph of average degree d as $r = p^* \cdot d$, then it certainly holds: the smaller r, the more robust is the graph (and therefore the longer repairs on it can be postponed.) Using the results for p^* above, we obtain for the

- complete graph: $r = 1$
- n-dimensional hypercube: $r = 1$
- 2-dimensional mesh: $r = 2$
- n-dimensional butterfly: $r \geq 1.348$
- random graph with $d \cdot n/2$ edges: $r = 1$

So the best graph of regular structure that is the easiest to build is the hypercube. An interesting open question is whether there is also a constant degree graph of regular structure with $r = 1$ (or at least an r that is arbitrarily close to 1).

Also worst-case fault models have been investigated. Leighton and Maggs showed that there is a so-called *indirect* constant-degree network connecting n inputs with n outputs via $\log n$ levels of n nodes each, which they named multibutterfly, that has the following property: no matter how an adversary chooses f nodes to fail, there will be a connected component left in it with at least $n - O(f)$ inputs and at least $n - O(f)$ outputs [31]. (In fact, one can even still route packets between the inputs and outputs in this component in almost the same amount of time steps as in the ideal case.) Later, Upfal showed that there is also a *direct* constant-degree network on n nodes, a so-called expander, that fulfills: no matter how an adversary chooses f nodes to fail, there will be a connected component left in it with at least $n - O(f)$ nodes [40]. Both results are optimal up to constants. However, it is not known what the best constants are that can be reached, and whether similar results can also be reached by networks of regular structure (i.e. that are easy to build). For some classes of graphs such as the class of complete graphs, the exact constant is easy to determine: a complete graph fulfills that no matter how an adversary chooses f nodes to fail, there will be a connected component left with $n - f$ nodes.

2.2   Simulation of Fault-Free Networks by Faulty Networks

Next we look at the problem that we do not only demand that the remaining component is large, but that it can still be used for efficient communication (such as in the case of the multibutterfly above).

Consider the situation that there can be up to f worst-case node faults in the system at any time. One way to check whether the largest remaining component still allows efficient communication is to check whether it is possible to embed into the largest connected component of a faulty network a fault-free network of the same size and kind. An *embedding* of a graph G into a graph H maps the nodes of G to non-faulty nodes of H and the edges of G to non-faulty paths in H. An embedding is called *static* if the mapping of the nodes and edges is fixed. Both static and dynamic embeddings have been used. A good embedding is one with minimum load, congestion, and dilation, where the *load* of an embedding is the maximum number of nodes of G that are mapped to any single node of H, the *congestion* of an embedding is the maximum number of paths that pass through any edge e of H, and the *dilation* of an embedding is the length of the longest path. The load, congestion, and dilation of the embedding determine the time required to emulate each step of G on H. In fact, Leighton, Maggs, and Rao have shown [32] that if there is an embedding of G into H with load ℓ, congestion c, and dilation d, then H can emulate any communication step (and also computation step) on G with slowdown $O(\ell + c + d)$.

When demanding a constant slowdown, only a few results are known so far. In the case of worst-case faults, it was shown by Leighton, Maggs and Sitaraman (using of dynamic embedding strategies) that an n-input butterfly with $n^{1-\epsilon}$ worst-case faults (for any constant ϵ) can still emulate a fault-free butterfly of the same size with only constant slowdown [33]. Furthermore, Cole, Maggs and Sitaraman showed that an $n \times n$ mesh can sustain up to $n^{1-\epsilon}$ worst-case faults and still emulate a fault-free mesh of the same size with (amortized) constant slowdown [14]. It seems that the n-node hypercube can even achieve a constant slowdown for $n^{1-\epsilon}$ worst-case faults, but so far only partial answers have been obtained [33].

Also random faults have been studied. For example, Håstad, Leighton and Newman [24] showed that if each edge of the hypercube fails independently with any constant probability $p < 1$, then the functioning parts of the hypercube can be reconfigured to simulate the original hypercube with constant slowdown. Combining this with what we know from Section 2.1, the hypercube seems to be an extremely good candidate for a fault-tolerant network.

For a list of further references concerning embeddings of fault-free into faulty networks see the paper by Leighton, Maggs and Sitaraman [33].

The embedding results immediately imply that there are values f so that as long as there are at most f faults at any time, then for certain networks its largest component will basically have the same communication properties as the original network at any time. The problem is, how to exploit this knowledge for routing in networks of dynamically changing faults, since it may not be known where future faults occur, especially if they are under the control of an adversary. An interesting approach might be to combine the results mentioned here with the models and approaches in Section 2.4.

2.3 Fault-Tolerant Path Selection in Networks

Next we consider the problem of adapting paths selected for data streams (such as multimedia streams) to faults in networks. We first start with suitable models for online path selection in fault-free networks.

In the *edge-disjoint paths (EDP) problem* we are given an undirected graph $G = (V, E)$ and a set T of pairs of nodes (s_i, t_i) (also called *requests*), and the problem is to find a subset $T' \subseteq T$ of maximum size so that each request in T' has an edge-disjoint path.

A generalization of the EDP is the *unsplittable flow problem* (UFP): each edge $e \in E$ is given a capacity of c_e and each request (s_i, t_i) has a demand of d_i. The task is to choose a subset $T' \subseteq T$ such that each request (s_i, t_i) in T' can send d_i flow along a single path, all capacity constraints are kept, and the sum of the demands of the requests in T' is maximized. (More general variants of the UFP also assign a profit to each request and consider the problem of maximizing the sum of the profits of the accepted requests.)

The EDP and UFP are classical optimization problems in the routing area, and there is a substantial body of literature on these two problems (see, for example, [6,29,10,30]). The EDP and UFP are useful models to study, for example, the performance of admission control algorithms for multimedia streams in fault-free networks. However, the EDP and the UFP are only of limited use for practical purposes, since they do not allow a rapid adaptation to edge faults or heavy load conditions. In fact, if an edge fault occurs, then the requests using that edge may have to be cancelled, since only a single flow path was taken for each request and alternative paths might not be available at the time of the edge fault. A straightforward solution to this problem would be to reserve multiple paths for each accepted request that can flexibly be used to ensure rapid adaptability. However, the paths should be chosen so that not too much bandwidth is wasted under normal conditions. Motivated by these thoughts, we suggest the following two optimization problems: the k edge-disjoint paths problem, and the k-splittable flow problem.

In the *k edge-disjoint paths problem* (k-EDP), we are given an undirected graph $G = (V, E)$ and a set of terminal pairs (or requests) T, and the problem is to find a subset $T' \subseteq T$ of maximum size so that each pair in T' is connected by k disjoint paths and also all paths selected for T' are disjoint.

In the *k-splittable flow problem* (k-SFP), we are given an undirected network $G = (V, E)$ with edge capacities and a set T of terminal pairs (s_i, t_i) with demands d_i. The problem is to find a subset $T' \subseteq T$ of maximum size so that each pair $(s_i, t_i) \in T'$ is connected by k disjoint paths of flow value d_i/k and the paths selected for T' satisfy the capacity constraints.

It is not difficult to imagine how the k-SFP can be used to achieve fault tolerance while not wasting too much bandwidth: suppose it is ensured that in the given network G there will be no more than f edge faults at any time. Given a request with demand d, we try to reserve $k + f$ disjoint flow paths for it, $k \geq 1$, with a flow value of d/k for each path (i.e. we raise the total demand for the k-SFP from d to $d \cdot (k + f)/k$). Then it obviously holds that as long as at most f edges break down, it will always be possible to ship a demand of d along the remaining paths. Furthermore, under fault-free conditions, only a fraction of $((d/k)(k + f) - d)/d = f/k$ of the reserved bandwidth is wasted, which can be made sufficiently small by setting k sufficiently large (which may, of course, be limited by the network resources).

Another way of achieving fault tolerance is to select for each accepted request of demand d a path of flow value d and a collection of disjoint, alternative paths of flow value d that are allowed to share their bandwidth with alternative paths of other requests. This allows a better utilization of the links than the two models above but is harder to administer. (If more than one edge is allowed to brake down, *arbitrary* overlaps of alternative paths can lead to the disconnection of a request.) Models that allow bandwidth sharing were considered in [17].

One can study the problems above under two different scenarios: all requests are given in advance, or requests arrive one after the other. In the latter case it is usually assumed that the algorithm must make a decision before seeing future requests. An algorithm is called *c-competitive* if it can accept at least a fraction of $1/c$ of the maximum number of acceptable requests.

Recently, we were able to show that there are online algorithms for the k-EDP that are $O(k^3 \cdot \Delta \cdot \alpha^{-1} \log n)$ competitive, where Δ is the maximum degree, α is the edge expansion, and n is the number of nodes of the given network [9]. To the best of our knowledge, this is the first non-trivial result for the k-EDP. Nothing non-trivial is known for the k-SFP so far.

So many interesting questions in this field are still wide open.

2.4 Routing in Faulty Networks

There is a large body of literature about routing around faults in specific networks (such as butterflies [13] and multibutterflies [31]. However, not many *universally applicable* approaches are known for routing packets around faults in general networks. We will look at two general approaches that are able to handle certain kinds of faults in arbitrary networks.

The information dispersal approach. Suppose that each edge will be faulty in a time step with probability p, and that this cannot be detected by the system. If p is sufficiently small (we will specify later what this means), then one approach that can be used so that with high probability no message gets lost is to use information dispersal [37]: Suppose that each message consists of k packets. Then it can be encoded in $k + f$ packets so that as long as at least k of the $k + f$ packets reach their destination, the original message can be recovered.

Suppose further that we have the following adversarial injection model: an adversary controls the injection of messages into the network, but guarantees that for each injected message a path can be selected for each of its $k - f$ packets so that for every time interval I of size at least w and every edge e, less than w paths are chosen for messages injected in I that contain e. In this case, we fulfill the conditions of the adversarial injection model introduced in Section 1.1. It is known that if the paths are revealed to the system, then simple switching disciplines such as longest-in-system or furthest-to-go will ensure that all packets will reach their destination in a finite amount of time steps [4]. This result, however, only holds for fault-free networks. Since we allow edges to be faulty in a time step with probability p, some of the packets may get lost. Suppose now that the longest path taken by a packet has a length of L. Then the probability that any particular packet will get lost is at most $1 - (1 - p)^L \le L \cdot p$. Hence, if $k, f = \Theta(\log n)$ and $p \le \frac{1}{c \cdot L}$ for a sufficiently large constant c, it can easily be shown via Chernoff bounds that with high probability at least k of the $k + f$ packets of a message will reach their destination. In general, p must be in the order of $O(1/L)$ so that this property can be guaranteed for a constant amount of communication overhead (i.e. $f/k = O(1)$). If edge faults can be detected, then we can allow much higher failure rates, as demonstrated in the next subsection.

Routing via balancing. For the case that edge faults are detectable, another general approach for routing in faulty networks is to use load balancing. We assume that now simply packets are injected. As above, the injections are controlled by an adversary. Almost all of the papers using that model assume that the adversary reveals the paths to the online algorithm. In this case, the only problem that is left is to decide which packet to prefer if several packets content to use the same link at the same time step. Recently, also the case was studied that the adversary does not reveal the paths to the algorithm [2,18,7], i.e. the protocol has to determine the paths itself. If edge faults are not known in advance, then this is the only reasonable model.

In the case of edge faults, Aiello et al. [2] suggested the following modification in the definition of the adversary. For any edge e and any time interval I, let $u_{e,I}$ be the amount of time steps e is working in I, and let $i_{e,I}$ be the number of packets injected by an adversary during I that contain e in their path. An adversary is called a (w, ϵ)-adversary if for every edge e and every time interval I of size at least w, $u_{e,I} - i_{e,I} \ge \lceil \epsilon \cdot w \rceil$.

Aiello et al. [2] showed that the following protocol called BASIC will remain stable under this model even if the adversary does not reveal the paths. The protocol maintains several buffers in each node v. For each edge $e = (v, u) \in E$ adjacent to v and for every destination $d \in V$, the protocol maintains a buffer for packets bound for d. Thus, there is a buffer of packets for each triple (v, u, d), which is called $Q_{v,u,d}$. Denote the set of packets in $Q_{v,u,d}$ at time t by $Q_{v,u,d}^t$ and the size of the set $Q_{v,u,d}^t$ as $q_{v,u,d}^t$. Each $q_{d,u,d}^t$ is always assumed to be 0 (as packets that arrive at their destination are immediately removed). Initially, all $q_{v,u,d}^t$ are also assumed to be 0. The following protocol is performed by each node v at each time step t (it is assumed here that edge faults are detectable):

1. For each working $e = (v, u) \in E$, let $d \in V$ be chosen such that $q_{v,u,d}^t - q_{u,v,d}^t$ is maximal over all $d \in V$ (break ties arbitrarily). If $q_{v,u,d}^t - q_{u,v,d}^t > 0$ then send one packet over the edge e from $Q_{v,u,d}$ to $Q_{u,v,d}$.
2. Accept all packets injected by the adversary to the node v.
3. Remove any packets that arrive at their destination.
4. For every d, distribute the packets with destination d among the buffers $Q_{v,u,d}$ as evenly as possible.

This protocol has the following performance.

Theorem 1 ([2]). *For any (w, ϵ)-adversary with $\epsilon > 0$, the number of packets stored at any given time in any of the buffers of* BASIC *is at most* $O(m^{3/2}n^{3/2}w/\epsilon')$, *where* $\epsilon' = \min[\epsilon, 1/w]$.

The model in [2] appears to be too restrictive, since for every interval I of size at least w, an edge e has to be up more times *in* I than the number of packets with path through e injected *in* I, which can certainly be relaxed. A more general model that avoids the use of intervals is the following:

Suppose that the adversary just has to guarantee that for every injected packet there is a schedule for delivering it over non-faulty edges to its destination. A *schedule* $S = ((e_0, t_0), (e_1, t_1), \ldots, (e_\ell, t_\ell))$ consists of a sequence of movements by which the injected packet P can be sent from its source node to its destination node. For this the adversary must ensure that

- P is injected at the starting point of e_1 at time step t_0 (this is represented by the imaginary *injection edge* e_0),
- the edges $e_1, \ldots, e_\ell$ form a connected path, with the endpoint of e_ℓ being the destination of P,
- the time steps have the ordering $t_0 < t_1 < \ldots < t_\ell$, and
- edge e_i is active at time t_i for all $1 \leq i \leq \ell$.

Another necessary requirement is that two schedules are not allowed to *intersect*, that is, they are not allowed to have a pair (e, t) in common. This ensures that a schedule represents a valid routing strategy for a packet.

Note that this model allows $(w, 0)$-adversaries. Hence, the result by Aiello et al. does not hold any longer in this model.

In order to measure the performance of protocols in this model, we need some parameters. A schedule $S = ((e_0, t_0), (e_1, t_1), \ldots, (e_\ell, t_\ell))$ is called *active* at time t for every t with $t_0 \leq t \leq t_\ell$. Let A be the maximum number of schedules that are allowed to be active at any time step.

For the special case that all packets must have the same destination, Awerbuch et al. [7] proposed the following algorithm.

Suppose that every node has a single buffer. For any node v and any time step t let $h_{v,t}$ denote the amount of packets in node v at the beginning of time step t. For the destination d, we set $h_{d,t} = -T$ for some parameter T determined later. Initially, all buffers are empty. In each step t, each node v performs the following so-called *T-balancing protocol*:

1. For every edge $e = (v, w)$, check whether $h_{v,t} - h_{w,t} > T$. If so, send a packet from v to w (otherwise do nothing).
2. Receive incoming packets and newly injected packets, and absorb those that reached the destination.

Let Δ be the maximum degree of a node and n be the number of nodes in the system. Then the following result can be shown.

Theorem 2 ([7]). *For any $A, \Delta \in \mathbb{N}$ and any $T \geq 2(\Delta - 1)$, the T-balancing protocol ensures that there are at most $\max[A, A + T - 1] \cdot \binom{n-1}{2} + A$ packets in* *the system at any time. Furthermore, the maximum number of packets in a node at any time is at most* $\max[A, A + T - 1](n - 2) + A$.

It is still an interesting open question whether a similar result can also be shown for multiple destinations.

3   Communication in the Internet

Routing algorithms for the internet must fulfill several properties:

- *Optimality*: Optimality refers to the ability of the routing algorithm to select the "best" route. The best route depends on the metrics and metric weights used to make the calculation. For example, one routing algorithm might use the number of hops and the delay, but might weigh delay more heavily in the calculation. Naturally, routing protocols must strictly define their metric calculation procedure.
- *Simplicity and low overhead*: The routing algorithm must offer its functionality efficiently, with a minimum of software and utilization overhead. Efficiency is particularly important when the software implementing the routing algorithm must run on a computer with limited resources.
- *Robustness and stability*: The routing algorithm should perform correctly in the face of unusual or unforeseen circumstances such as hardware failures, high load conditions, and incorrect implementations. Because routers are located at network junction points, they can cause considerable problems when they fail to be robust.
- *Flexibility and rapid convergence*: Routing algorithms should quickly and accurately adapt to a variety of network circumstances. This is important, since algorithms that converge too slowly can cause routing loops or network outages.

Internet routing protocols can be separated into two different classes: link state algorithms and distance vector algorithms. Link state algorithms (also known as *shortest path first* algorithms) flood routing information to all nodes in the network. However, each router sends only that portion of the routing table that describes the state of its own links. The most prominent link state algorithm used in the Internet is OSPF (*Open Shortest Path First*). Distance vector algorithms (also known as *Bellman-Ford* algorithms) call for each router to send all or some portion of its routing table, but only to its neighbors. The most prominent distance vector algorithm is RIP (*Routing Information Protocol*). In essence, link state algorithms send small updates everywhere, while distance vector algorithms send larger updates only to neighboring routers.

Because they converge more quickly, link state algorithms are somewhat less prone to routing loops than distance vector algorithms. On the other hand, link state algorithms require more CPU power and memory than distance vector algorithms. Link state algorithms can therefore be more expensive to implement and support. Despite their differences, both algorithm types perform in an acceptable way in most circumstances (otherwise, they would not have been chosen for use in the Internet!).

As already indicated above, both link state algorithms and distance vector algorithms use routing tables. Routing tables contain information about the status of links that is used by the switching software to select the best route. But what kind of information should they contain? How do routing algorithms determine that one route is preferable to others?

Routing algorithms for the Internet have used many different metrics to determine the best route. Sophisticated routing algorithms can base route selection on multiple metrics, combining them in a single (hybrid) metric. All of the following metrics have been used: path length, reliability, delay, bandwidth, load, and communication cost. The path length is the most common routing metric. The easiest way to define the path length is simply by the number of routers and switches a packet has to pass. However, some routing protocols also allow to assign an arbitrary cost to each network link (which can be any one of the above parameters, such as delay, available bandwidth, or the price of using a link). In this case, the path length is the sum of the costs associated with each link traversed.

It is not difficult to prove that in the ideal situation (all information is up-to-date), both RIP and OSPF will select the path of smallest available cost, i.e. for each individual request an optimal path will be chosen. However, minimizing individual costs does *not* imply minimizing the global cost.

In order to study the global impact of cost functions, Andrews et al. [5] recently considered the following model.

Again, we assume that we have an adversary that chooses the injection time, source, destination, and route for each packet. A sequence of injections is called (w, λ)-admissible for a window size w and injection rate $\lambda < 1$ if in any time interval I of size at least w the total number of packets injected into the network whose paths pass through any link e is at most $|I| \cdot \lambda$. In this case, the corresponding paths are also called (w, λ)-*admissible*. We say that a set of packet paths is *weakly (w, λ)-admissible* if we can partition time into windows of length w such that for each window *in the partition* and each link e, the number of paths that pass through e and correspond to packets injected during the window is at most $w\lambda$.

Suppose now that the adversary does not reveal the paths to the system, but that instead always a best possible route is selected according to some cost function. Is it then possible to ensure that if the injected packets have paths that are (w, λ)-admissible, then the selection process ensures that the selected paths are (W, Λ)-admissible, possibly for a different window size W and a different $\Lambda < 1$?

Based on the ϵ-approximation algorithm by Garg and Könemann [19] for concurrent multicommodity flow problems, which in turn builds upon the work of Plotkin, Shmoys, and Tardos [36], Andrews et al. [5] use the following algorithm:

Suppose that control information is communicated instantly. Whenever a source node chooses a route for a packet, this information is immediately transmitted to all the links on the route. Whenever the congestion on a link changes, this fact is instantaneously transmitted to all the source nodes. (Andrews et al. also show how to relax these assumptions, but for simplicity we will stick to them here.)

Every injected packet is routed along the path whose total congestion is the smallest under the current congestion function $c(\cdot)$, i.e. we route along shortest paths with respect to $c(\cdot)$. Initially, the congestion along every link is set to δ where δ is defined in (2). Each time a route is selected for an injected packet, the congestion $c(e)$ of every link e along the chosen route is updated to $c(e)(1+\mu/w)$ where μ is defined in (1). We reset the congestion of every link to its initial value of δ at the beginning of each *phase*. A phase terminates in t windows of w steps, where t is an integer defined in (3).

The values of μ, δ and t are defined as follows. Let m be the number of links in the network. For any $R \in (r, 1)$ of our choice, let

$$\mu = 1 - \left(\frac{\lambda}{\Lambda}\right)^{1/3} \tag{1}$$

$$\delta = \left(\frac{1 - \lambda\mu}{m}\right)^{1/r\mu} \tag{2}$$

$$t = \lfloor \frac{1 - \lambda\mu}{\lambda\mu} \ln \frac{1 - \lambda\mu}{m\delta} \rfloor + 1 \tag{3}$$

Andrews et al. prove the following theorem.

Theorem 3 ([5]). *For all packets injected during one phase, at most $tw\Lambda$ of their routes chosen by the above procedure go through the same link for some $\Lambda < 1$. In other words, these routes are weakly (tw, Λ)-admissible.*

Furthermore, they show the following lemma that implies that the selected paths are not only weakly (tw, Λ)-admissible, but also (w', λ')-admissible for some $w' \geq w$ and $\lambda' \in [\Lambda, 1)$.

Lemma 1 ([5]). *If a set of paths is weakly (w, λ)-admissible then it is also (w', λ')-admissible for some $w' \geq w$ and $\lambda' \in [\lambda, 1)$.*

Hence, using a universally stable switching protocol on top of their route selection scheme (such as longest-in-system or furthest-to-go [4]) would result in a routing protocol that is universally stable.

As we saw, the protocol by Andrews et al. [5] allows to admit all packets if, as a precondition, all of them are admissible. However, in the Internet one may have the situation that much more packets are injected than are admissible. Hence, an interesting question is what to do if not all packets are admissible, and therefore some packets may have to be dropped. In this case, one would like to drop the packets in a way so as to optimize a certain cost function. The simplest cost function would certainly be to maximize the number of admissible packets. Here, the following strategy would work (if the λ_e below is known in advance): If one would use the same scheme as above and then would assign to each packet P a survival probability that is equal to the minimum $\min[1, tw\Lambda/\lambda_e]$ over all edges e along its path (where λ_e is the number of paths crossing e in the phase in which P is injected), then the result should be that each surviving packet is admissible with high probability and the number of surviving packets is close to the maximum number of packets any scheme could deliver. The question, however, is whether a knowledge of λ_e is necessary for reaching this goal. Certain results seem to indicate that this is the case [6]. Other cost functions could be to maximize the profit when selecting the admissible packets. Certainly, more work has to be done to come up with a protocol that would potentially be able to extend (or replace) existing strategies in the Internet.

4   Communication in Wireless Networks

Communication in wireless networks is currently a rapidly evolving, highly active research field, though dominated by experimental studies. A particularly hot field has been in recent years mobile ad-hoc networks, which are wireless networks that do not rely on any fixed infrastructure. Applications of ad hoc networks have mostly been limited to the military communication environment. However, recently also the commercial sector is increasingly interested in this technology. Evidence of this is the establishment of a MANET working group in the Internet Engineering Task Force (IETF). Central questions in the area of wireless ad hoc networks are:

- How to ensure connectivity?
- How to select and maintain routes between the mobile units?
- How to send packets along the selected routes?

The first project in which a collection of computers formed a wireless network was the ALOHA project, which was conducted around 1970 at the University of Hawaii. In this project all communication was performed via a single master. Hence, it was not necessary to search for suitable routes but only to know the addresses of the computers in the network. Of course, in order to ensure that the network is connected, all computers had to be within the range of the master. If a larger distance has to be bridged with a limited transmission range, then several intermediate transmissions (often called *hops*) are necessary to send a packet from a source to a destination. In this case, finding a suitable route becomes a serious problem. One of the first that suggested strategies that solve the route selection problem in this case were Gitman et al. [20].

Today, route selection protocols for wireless networks are usually divided into proactive and reactive protocols. In a *proactive* protocol every node tries to keep routing information for all nodes in the network up to date. This is usually achieved by updating the information automatically in regular time intervals. If a message has to be sent to a certain destination, then the stored information allows to immediately determine a suitable route to the destination (if such a route exists). The latency for the transmission of a message can therefore be kept small. The disadvantage of this method is that the maintenance of the routing information requires a lot of extra communication and that space even has to be allocated for those nodes that are rarely used as a destination. All traditional link-state and distance-vector routing protocols are proactive. Examples of proactive protocols that have been suggested for wireless networks are OLSR, DSDV, and WRP.

Also reactive protocols store routing information. However, this information is only updated when needed. This has the effect that parts of the information may be outdated. Thus, it might happen that in order to send a message to a destination, first a suitable route has to found in the network. This may cause a high latency for the transmission of the message. However, the advantage of this route-on-demand strategy is that it usually causes a much lower communication overhead compared to proactive protocols, which allows to use reactive protocols in much larger wireless networks. Examples of reactive protocols are DSR, AODV, and TORA.

Also algorithms are known that are a combination of both concepts. The usual approach is to keep the routing information for a certain area around a node up to date and to collect information about distant nodes only in the case that a message has to be sent there. Examples are ZRP and LANMAR.

Detailed descriptions of many of the protocols mentioned above can be found at

`http://www.ietf.org/html.charters/manet-charter.html`.

All of the protocols above have only been studied via simulations or experiments. Theoretical work in the area of wireless networks has mostly concentrated on the packet radio network (PRN) model. In this model, a wireless network is usually seen as an undirected graph $G = (V, E)$, where two nodes i and j are connected if and only if i is within the transmission radius of j and vice versa. One time step is defined as the time it takes to transmit a packet from one node to another. Since it is usually assumed that all nodes can only transmit at one frequency, a *transmission conflict* occurs at node i if two of its neighboring nodes transmit a packet at the same time. Communication problems have been studied by many authors for both the case that a conflict can be detected and the case that it cannot be detected.

The advantage of the PRN model is that it is reasonably simple to allow theoretical investigations. However, its drawback is that interference problems are modelled in a highly simplified way. More realistic wireless network models have been considered, for example, in [1,21,22,39]. Another drawback of the PRN model is that in its original form it only allows to study wireless communication with a fixed transmission range. Both in cellular and ad-hoc wireless networks, being able to use a variable transmission power has a noticeable advantage over a fixed transmission power. This has been demonstrated experimentally both for cellular networks [25] and ad-hoc networks [35]. One primary motivation for using a variable transmission power is that energy is a scarce resource in mobile units, and therefore the possibility of being able to adjust the signal strength to the given situation can significantly help to save energy. Another important aspect is that contention among the mobile units can be significantly reduced. Furthermore, it allows the units to have more time to adjust to a changing constellation and therefore to have a much smoother transition from one interconnection scenario to another.

In order to model wireless networks of units with variable transmission power, one can use a complete graph with a cost function $c : V \times V \to \mathbb{R}_+$ that determines the signal strength (resp. energy consumption) necessary to send a packet from one node to another. Changing the distance between these two nodes or moving an obstacle between these will then change the cost function accordingly. Furthermore, in order to model interference more accurately, one can use a parameter $\alpha > 1$ with the property that if a node v transmits a message with signal strength s, then all nodes requiring a signal strength of at most $\alpha \cdot s$ to receive a message from v are blocked. Such a model has, for example, been used in [1].

4.1   Connectivity

Allowing a variable transmission power in the wireless model above adds much more flexibility but also creates new problems. For example, the connectivity problem is trivial in the fixed transmission power case (nodes either can or cannot reach each other), whereas it is highly non-trivial to select the right transmission power to ensure connectivity for the case of variable transmission powers (unless the nodes always communicate with maximum power, which would be too wasteful).

Recently, an elegant, rigorous approach was found to solve this problem under certain circumstances. Consider some fixed dimension D and some norm $|\cdot|$ on $\mathbb{R}^D$. Consider a set of points $V = \{v_1, \ldots, v_n\}$ in $\mathbb{R}^D$ and let $G = (V, E)$ be an undirected graph connecting these points. For every path $p = (v_{i_1}, v_{i_2}, \ldots, v_{i_\ell})$ in G we define its *stretch factor* as

$$f(p) = \frac{\sum_{j=1}^{\ell-1} |v_{i_j} - v_{i_{j+1}}|}{|v_{i_1} - v_{i_\ell}|} .$$

G is called an f-*spanner* for V if every pair $s, t \in V$ has a path in G of stretch factor at most f. Yao [41] and subsequent papers (e.g., [38,16]) showed that spanners for a given set of points V can be obtained by a generalization of *proximity graphs* [26]: partition $\mathbb{R}^D$ into a collection $\mathcal{C}$ of $k \in \mathbb{N}$ convex cones $C_0, \ldots, C_{k-1}$. Then, from every point $p \in V$ and every $j \in \{1, \ldots, k\}$, draw directed edges to the closest point of V lying in the translated cone $p + C_j$. The resulting graph is called a *partitioned neighborhood graph* (PNG). For the special case that all cones have an angular diameter of at most Θ, the graph is also called a Θ-*graph* [38].

Suppose that $|\cdot|$ represents the Euclidean distance. In this case, Ruppert and Seidel [38] showed that if every $C \in \mathcal{C}$ has an angular diameter of at most $\Theta < \pi/3$, then the resulting PNG is an f-spanner with

$$f = \frac{1}{1 - 2\sin(\Theta/2)} \cdot$$

This implies that in the 2-dimensional case 7 cones suffice for the PNG to be a 7.57-spanner. Combining the bound for f above with a result by Hardin, Sloane and Smith [23], one can also show that in the 3-dimensional case 20 cones suffice for the PNG to be an 88.1-spanner. Further improvements, partly using different norms and so-called weak spanner properties, can be found in [16].

Note that a spanner is always strongly connected. Hence, we can always ensure connectivity among nodes with variable transmission ranges by simply choosing the transmission range in a way so that it can transmit messages to all nearest neighbors in all cones (if possible), where the cones are selected so that the resulting PNG is guaranteed to be a spanner. This approach even works if there is a limit on the maximum possible transmission range, unless that limit does not suffice to achieve connectivity by any means.

A limitation of the PNG approach seems to be that it does not seem to work well for higher dimensions (too many cones are needed) and that it may only work in (certain) metric spaces. Furthermore, there are problems realizing the PNG approach in practice, since it might be necessary to use devices such as GPS in order to know the location of a node and therefore being able to decide in which cone it is. It is an interesting open problem whether there are alternative approaches. For example, in situations in which units randomly move around, it might suffice just to connect to some constant number of nearest neighbors to ensure connectivity with high probability.

4.2   Path Updates in Wireless Networks

Suppose that we want to use a proactive scheme for refreshing paths between different nodes in a wireless network. The usual approach is that for every pair of nodes its route is updated in regular time intervals. But how large should these intervals be? And is it really necessary to have a route for *every* pair of nodes to ensure a good connectivity?

In order to address these questions, suppose we keep information in a redundant form in the network so that as long as a constant fraction of the nodes is connected, no information is lost. Then we are in the situation already studied in Section 2.1.

So a reasonable strategy could be that we simply embed a graph G that is as robust as possible in the wireless network H. Communication would only be performed along the paths simulating edges in G. In this case, H is connected if and only if G is still connected with regard to those edges that still have working paths in H. Each time a node walks away from its position, a certain number of paths in H representing edges in G may get disconnected. If this exceeds some threshold (for example, a constant fraction of the nodes in G, and therefore a constant fraction of the nodes in H, is not connected any more), then one has to refresh the embedding.

If movements of nodes happen at random and we had the situation that each node is only crossed by one path representing an edge in G, then we would be back to the model in Section 2.1. So the question is, how to transfer the results obtained there to wireless networks if this is not the case? In particular, how should the paths be selected in the wireless network so that one can prove lower bounds on the time intervals needed for refreshing routes in order to ensure connectivity of a constant fraction of the nodes (with high probability)?

4.3   Routing in Wireless Networks

There are two important issues when routing packets in wireless networks: the energy necessary to perform the routing should be kept as low as possible while keeping the throughput as high as possible (i.e. allowing many packets to reach their destination).

The issue of achieving a high throughput has already been addressed in [21, 22]. However, to the best of our knowledge, no results are known so far about achieving a high throughput with a minimum amount of energy. Consider, for example, the following model.

Recall the adversarial routing model considered for faulty networks in Section 2.4: suppose that the adversary has to guarantee that for every injected packet there is a schedule (i.e. a sequence of hops over the time) for delivering it to its destination. Would it then be possible to design an online routing protocol that can achieve (nearly) the same amount of deliveries with an energy consumption that is close to the optimum? Concerning the optimum, two alternatives would be interesting: we want the total amount of consumed energy to be as low as possible, or we want the maximum amount of energy spent in a node to be as low as possible.

5   Communication in Adversarial Networks

All previous scenarios have in common that we have dynamic networks with specific characteristics. In this section, we choose a high-level approach that is not tied to any particular application. Suppose that we have a dynamic network that can form *any* directed network at *any* point of time, and the way it changes is under adversarial control. In order to ensure a high throughput efficiency such kind of dynamic networks, several challenging tasks have to be solved:

  – *Routing*: What is the next edge to be traversed by a packet?
  – *Switching*: What is the next packet to be transmitted on an edge?
    In particular, which destination should be preferred?
  – *Admission control*: What is the packet to be dropped?

This seems to be impossible to decide without knowing the future, especially if both the topology and the packet injections are under adversarial control: choosing the wrong edge may lead a packet further away from its destination, and preferring packets with a destination that will be isolated in the future or dropping the wrong packets in case of overfull buffers may tremendously decrease the number of successfully delivered packets. However, we will show that the seemingly impossible task is possible, and this not only for unicasting but even for multicasting.

First, we specify our model. Each edge can transport one packet and can be used independently from the other edges. Each node may have a certain number of buffers to store different types of packets (for example, packets with different destinations). To ensure that our model is realistic, we assume that the size of every buffer (i.e. the number of packets it can hold) is finite. Furthermore, each node can only have a finite number of incoming and outgoing edges at a time. As a parameter for the former case we will use the *maximum buffer height H*, and as a parameter for the latter case we will use the *maximum in/out degree Δ*.

The adversary does not only control the topology of the network but also the injection of packets. We distinguish between two adversarial injection models: one for unicast injections and one for multicast injections.

The adversarial unicast model. Each unicast packet has a fixed destination, determined at the time of its injection. The adversary can inject an arbitrary number of packets and can activate an arbitrary number of edges in each time step as long as the number of incoming or outgoing edges at a node does not exceed Δ. In this case, only some of the injected packets may be able to reach their destination, even when using a best possible strategy. For each of the successful packets a schedule of the form $S = ((e_0, t_0), (e_1, t_1), \ldots, (e_\ell, t_\ell))$ can be specified. Of course, schedules are not allowed to intersect. When speaking about schedules in the following, we always mean a delivery strategy chosen by a best possible routing algorithm.

The adversarial multicast model. In the multicast model, a packet is allowed to have several destinations. In order to allow a higher efficiency than simply sending out one packet for each destination, we adopt the standard multicast model, which assumes that a multicast packet only takes one time step to cross an edge and that a multicast packet can split itself while in transit. Requiring an efficient use of this property make the multicasting problem considerably harder than the unicast problem. Especially when using dynamic networks, it seems to be formidable problem to decide when and where to split a packet.

As in the unicast model, the adversary is basically allowed to inject an unlimited number of multicast packets. However, only some of these packets may be able to reach their destinations, even when using a best possible strategy. For each multicast packet P that can reach k of its destinations, a schedule can be identified. Any such schedule can be reduced to a directed tree with k leaves representing the destinations. Hence, we can view all schedules to be of the form $S = ((e_0, t_0), (e_1, t_1), \ldots, (e_\ell, t_\ell))$, where

- P is injected at the endpoint of e_0 at time step t_0,
- the edges $e_1, \ldots, e_\ell$ form an directed tree with the source as the root and the k destinations as leaves,
- for every directed path $((e_{i_1}, t_{i_1}), \ldots, (e_{i_k}, t_{i_k}))$, the time steps have the ordering $t_{i_1} < t_{i_2} \ldots < t_{i_k}$, and
- edge e_i is active at time t_i for all $1 \le i \le \ell$.

As in the unicast model, the schedules must fulfill the property that they do not overlap.

Analytical approach. Each time a multicast (or unicast) packet reaches one of its destinations, we count it as one *delivery*. The number of deliveries that is achieved by an algorithm is called its *throughput*. Since the adversary is allowed to inject an unbounded number of packets, we will allow routing algorithms to drop packets so that a high throughput can be achieved with a buffer size that is as small as possible.

In order to compare the performance of a best possible strategy with our online strategies, one can use competitive analysis. Given any sequence of edge activations and packet injections σ, let $\text{OPT}_B(\sigma)$ be the maximum possible throughput (i.e. the maximum number of deliveries) when using a buffer size of B, and let $A_{B'}(\sigma)$ be the throughput achieved by some given online algorithm A with buffer size B'. "Maximum possible" means here maximum possible given the same type of resources as used by the online algorithm. Otherwise, we would not have a fair comparison. The only difference between the resources of the online algorithm and a best possible algorithm we allow is the buffer size. (However, we place *no* restrictions on how the packets are selected and moved by an optimal strategy.) We call A (c, s)-*competitive* if for all σ and all B,

$$A_{s \cdot B}(\sigma) \ge c \cdot \text{OPT}_B(\sigma) - r$$

for some value $r \ge 0$ that is independent of $\text{OPT}_B(\sigma)$. Note that $c \in [0, 1]$. Usually, the competitive ratio is defined so that $c \ge 1$. However, in our case we choose the inverse definition, since it is more intuitive: c represents the fraction of the optimal throughput that can be achieved by A.

Using these models, we recently proposed different variants of the T-balancing algorithm presented in Section 2.4 for unicasting and multicasting [8]. In the unicast situation, we assume that every node has a buffer for every destination. Let $q_{v,d}^t$ denote the size of the buffer for destination d in node v at step t. In every time step t the unicast T-balancing algorithm performs the following operations in every node v:

1. For every working edge $e = (v, w)$, determine the destination d with maximum $q_{v,d}^t - q_{w,d}^t$ and check whether $q_{v,d}^t - q_{w,d}^t > T$. If so, send a packet with destination d from v to w (otherwise do nothing).

2. Receive incoming packets and absorb all packets that reached the destination.
3. Receive all newly injected packets. If a packet cannot be stored in a node because its height is already H, then delete the new packet.

A weighted form of the protocol can also be formulated for multicasting. We obtained the following main results [8]:

- $Unicasting$: For every $T \geq B + 2(\Delta - 1)$, the unicast variant of the T-balancing algorithm is $(1 - \epsilon, 1 + (1 + (T + \Delta)/B)L/\epsilon)$-competitive, where L is the average path length used by successful packets in an optimal solution. The result is sharp up to a constant factor in the space overhead.
- $Multicasting$: For every $T \geq B + D(3\Delta - 1)$, the multicast variant of the T-balancing algorithm is $(1 - \epsilon, 1 + (1 + (T + D\Delta)/B)D \cdot L/\epsilon)$-competitive, where L is defined as above and D is the maximum number of destinations a packet can have.

Many open questions remain. Although the space overhead is already reasonably low (essentially, $O(L/\epsilon)$ for unicasting), the question is whether it can still be reduced. For example, could knowledge about the location of a destination or structural properties of the network (for instance, it has to form a planar graph) help to get better bounds? Or are there approaches completely different from the balancing approach used above that can achieve a lower space overhead? In the worst case, the multicasting result stated above may need an exponentially large number of buffers per node, since for every possible set of destinations a buffer has to be reserved in each node. Can this be reduced to a polynomial number without restricting a best possible strategy? We suppose that this is not possible in general. But under certain circumstances (such as certain restrictions on the network topology and the movement of packets), this should be possible.

6   Conclusion

We presented existing and suggested new models and techniques for the study of dynamic networks. Although in some of the areas covered in this paper already a significant progress has been made, many problems are still wide open, and a long way still has to be gone to ensure that the methods developed for dynamic networks are not only (close to) optimal in theory but also work well in practice.

References

1. M. Adler and C. Scheideler. Efficient communication strategies for ad hoc wireless networks. *Theory Comput. Systems*, 33:337–391, 2000.
2. W. Aiello, E. Kushilevitz, R. Ostrovsky, and A. Rosén. Adaptive packet routing for bursty adversarial traffic. In *Proc. of the 30th ACM Symp. on Theory of Computing (STOC)*, pages 359–368, 1998.
3. M. Ajtai, J. Komlós, and E. Szemerédi. Largest random component of a k-cube. *Combinatorica*, 2(1):1–7, 1982.

4. M. Andrews, B. Awerbuch, A. Fernández, J. Kleinberg, T. Leighton, and Z. Liu. Universal stability results for greedy contention-resolution protocols. In *Proc. of the 37th IEEE Symp. on Foundations of Computer Science (FOCS)*, pages 380–389, 1996.

5. M. Andrews, A. Fernández, A. Goel, and L. Zhang. Source routing and scheduling in packet networks. In *Proc. of the 42nd IEEE Symp. on Foundations of Computer Science (FOCS)*, pages 168–177, 2001.

6. B. Awerbuch, Y. Azar, and S. Plotkin. Throughput-competitive on-line routing. In *Proc. of the 34th IEEE Symposium on Foundations of Computer Science*, pages 32–40, 1993.

7. B. Awerbuch, P. Berenbrink, A. Brinkmann, and C. Scheideler. Simple routing strategies for adversarial systems. In *Proc. of the 42nd IEEE Symp. on Foundations of Computer Science (FOCS)*, pages 158–167, 2001.

8. B. Awerbuch, A. Brinkmann, and C. Scheideler. Unicasting and multicasting in adversarial systems. Unpublished manuscript, 2001.

9. A. Bagchi, A. Chaudhary, P. Kolman, and C. Scheideler. Algorithms for the maximum disjoint k-paths problem. Unpublished manuscript, November 2001.

10. A. Baveja and A. Srinivasan. Approximation algorithms for disjoint paths and related routing and packing problems. *Mathematics of Operations Research*, 25, 2000.

11. B. Bollobas. The evolution of random graphs. *Transactions of the AMS*, 286:257–274, 1984.

12. A. Borodin, J. Kleinberg, P. Raghavan, M. Sudan, and D. P. Williamson. Adversarial queueing theory. In *Proc. of the 28th ACM Symp. on Theory of Computing (STOC)*, pages 376–385, 1996.

13. R. Cole, B. Maggs, and R. Sitaraman. Routing on butterfly networks with random faults. In *Proc. of the 36th IEEE Symp. on Foundations of Computer Science (FOCS)*, pages 558–570, 1995.

14. R. Cole, B. Maggs, and R. Sitaraman. Reconfiguring arrays with faults part I: worst-case faults. *SIAM Journal on Computing*, 26(6):1581–1611, 1997.

15. P. Erdős and A. Rényi. On the evolution of random graphs. *Publ. Math. Inst. Hungar. Acad. Sci.*, 5:17–61, 1960.

16. M. Fischer, T. Lukovski, and M. Ziegler. Partitioned neighborhood spanners of minimal outdegree. In *Proc. of the 11th Canadian Conference on Computational Geometry*, 1999.

17. L. Fleischer, A. Meyerson, I. Saniee, and A. Srinivasan. Fast and efficient bandwidth reservation for robust routing. In *DIMACS Mini-Workshop on Quality of Service Issues in the Internet*, February 2001.

18. D. Gamarnik. Stability of adaptive and non-adaptive packet routing policies in adversarial queueing networks. In *Proc. of the 31st ACM Symp. on Theory of Computing (STOC)*, pages 206–214, 1999.

19. N. Garg and J. Könemann. Faster and simpler algorithms for multicommodity flow and other fractional packing problems. In *Proc. of the 39th IEEE Symp. on Foundations of Computer Science (FOCS)*, pages 300–309, 1998.

20. I. Gitman, R. Van Slyke, and H. Frank. Routing in packet-switching broadcast radio networks. *IEEE Transactions on Communication*, 24:926–930, 1976.

21. P. Gupta and P. Kumar. The capacity of wireless networks. *IEEE Transactions on Information Theory*, IT-46(2):388–404, 2000.

22. P. Gupta and P. Kumar. Internets in the sky: The capacity of three dimensional wireless networks. *Communications in Information and Systems*, 1(1):33–50, 2001.

23. R. Hardin, N. Sloane, and W. Smith. *A library of putatively optimal coverings of the sphere with n equal caps.* http://www.research.att.com/~njas/coverings/.
24. J. Håstad, T. Leighton, and M. Newman. Reconfiguring a hypercubein the presence of faults. In *Proc. of the 19th ACM Symp. on Theory of Computing (STOC)*, pages 274–284, 1987.
25. J. Jacobsmeyer. Congestion relief on power-controlled CDMA networks. *IEEE Journal on Selected Areas in Communications*, 14(9):1758–1761, 1996.
26. J. Jaromczyk and G. Toussaint. Relative neighborhood graphs and their relatives. *Proceedings of the IEEE*, 80:1502–1517, 1992.
27. A. Karlin, G. Nelson, and H. Tamaki. On the fault tolerance of the butterfly. In *Proc. of the 26th ACM Symp. on Theory of Computing (STOC)*, pages 125–133, 1994.
28. H. Kesten. The critical probability of bond percolation on the square lattice equals 1/2. *Communication in Mathematical Physics*, 74:41–59, 1980.
29. J. Kleinberg. *Approximation Algorithms for Disjoint Paths Problems.* PhD thesis, Department of Electrical Engineering and Computer Science, Massachusetts Institute of Technology, 1996.
30. P. Kolman and C. Scheideler. Improved bounds for the unsplittable flow problem. In *Proc. of the 13th ACM Symp. on Discrete Algorithms (SODA)*, 2002.
31. F. Leighton and B. Maggs. Fast algorithms for routing around faults in multibutterflies and randomly-wired splitter networks. *IEEE Transactions on Computers*, 41(5):578–587, 1992.
32. F. Leighton, B. Maggs, and S. Rao. Packet routing and job-shop scheduling in O(congestion + dilation) steps. *Combinatorica*, 14(2):167–186, 1994.
33. F. Leighton, B. Maggs, and R. Sitaraman. On the fault tolerance of some popular bounded-degree networks. *SIAM Journal on Computing*, 27(5):1303–1333, 1998.
34. T. Luczak, B. Pittel, and J. Wierman. The structure of a random graph at the point of the phase transition. *Transactions of the AMS*, 341:721–748, 1994.
35. J. Monks. *Transmission Power Control for Enhancing The Performance of Wireless Packet Data Networks.* PhD thesis, Department of Electrical and Computer Engineering, University of Illinois, Urbana, IL, March 2001.
36. S. Plotkin, D. Shmoys, and E. Tardos. Fast approximation algorithms for fractional packing and covering problems. In *Proc. of the 32nd IEEE Symp. on Foundations of Computer Science (FOCS)*, pages 495–504, 1991.
37. M. Rabin. Efficient dispersal of information for security, load balancing, and fault tolerance. *Journal of the ACM*, 36(2):335–348, 1989.
38. J. Ruppert and R. Seidel. Approximating the d-dimensional complete Euclidean graph. In *Proc. of the 3rd Canadian Conference on Computational Geometry*, pages 207–210, 1991.
39. S. Ulukus and R. Yates. Stochastic power control for cellular radio systems. *IEEE Transactions on Communication*, 46:784–798, 1998.
40. E. Upfal. Tolerating a linear number of faults in networks of bounded degree. *Information and Computation*, 115(2):312–320, 1994.
41. A. Yao. On constructing minimum spanning trees in k-dimensional spaces and related problems. *SIAM Journal on Computing*, 11(4):721–736, 1982.

What Is a Theory?

Gilles Dowek

INRIA-Rocquencourt, BP 105, 78153 Le Chesnay Cedex, France.
Gilles.Dowek@inria.fr http://logical.inria.fr/~dowek

Abstract. *Deduction modulo* is a way to express a theory using computation rules instead of axioms. We present in this paper an extension of deduction modulo, called *Polarized deduction modulo*, where some rules can only be used at positive occurrences, while others can only be used at negative ones. We show that all theories in propositional calculus can be expressed in this framework and that cuts can always be eliminated with such theories.

Mathematical proofs are almost never built in pure logic, but besides the deduction rules and the logical axioms that express the meaning of the connectors and quantifiers, they use something else - *a theory* - that expresses the meaning of the other symbols of the language. Examples of theories are equational theories, arithmetic, type theory, set theory, ...

The usual definition of a theory, as a set of axioms, is sufficient when one is interested in the provability relation, but, as well-known, it is not when one is interested in the structure of proofs and in the theorem proving process. For instance, we can define a theory with the axioms $a = b$ and $b = c$ (where a, b and c are individual symbols) and prove the proposition $a = c$. However, we may also define this theory by the computation rules $a \longrightarrow b$ and $c \longrightarrow b$ and then a proposition $t = u$ is provable if t and u have the same normal form using these computation rules. The advantages of this presentation are numerous.

- We know that all the symbols occurring in a proof of $t = u$ must occur in t or in u or one of their reducts. For instance, the symbol d need not be used in a proof of $a = c$. We get this way analyticity results.
- In automated theorem proving, we can use this kind of results to reduce the search space. In fact, in this case, we just need to reduce deterministically the terms and check the identity of their normal forms. We get this way decisions algorithms.
- Since the normal form of the proposition $a = d$ is $b = d$ and b and d are distinct, the proposition $a = d$ is not provable in this theory. We get this way independence results and, in particular, consistency results.
- In an interactive theorem prover, we can reduce the proposition to be proved, before we display it to the user. This way, the user is relieved from doing trivial computations.

To define a theory with computation rules, not any set of rules is convenient. For instance, if instead of taking the rules $a \longrightarrow b$, $c \longrightarrow b$ we take the rules

H. Alt and A. Ferreira (Eds.): STACS 2002, LNCS 2285, pp. 50–64, 2002.
© Springer-Verlag Berlin Heidelberg 2002

$b \longrightarrow a$, $b \longrightarrow c$, we lose the property that a proposition $t = u$ is provable if t and u have a common reduct. To be convenient, a rewrite system must be confluent. Confluence, and sometimes also termination, are necessary to have analyticity results, completeness of proof search methods, independence results, ...

When we have rules rewriting propositions directly, for instance

$$x \times y = 0 \longrightarrow x = 0 \vee y = 0$$

confluence is not sufficient anymore to have these results, but cut elimination is also required [7,4]. Confluence and cut elimination are related. For instance, with the non confluent system $b \longrightarrow a$, $b \longrightarrow c$, we can prove the proposition $a = c$ introducing a cut on the proposition $b = b$, but, because the rewrite system is not confluent, this cut cannot be eliminated. Confluence can thus be seen as a special case of cut elimination when only terms are rewritten [6], but in the general case, confluence is not a sufficient condition for cut elimination.

Computation rules are not the only alternative to axioms. Another one is to add non logical deduction rules to predicate logic either taking an introduction and elimination rule for the abstraction symbol in various formulations of set theory [15,2,10,1,3,9] or interpreting logic programs or definitions as deduction rules [11,16,17,13] or in a more general setting [14]. Non logical deduction rules and computation rules have some similarities, but we believe that computation rules have some advantages. For instance, non logical deduction rules may blur the notion of cut in natural deduction and extra proof reduction rules have to be added (see, for instance, [5]). Also with some non logical deduction rules, the contradiction $\perp$ may have a cut free proof and thus consistency is not always a consequence of cut elimination. In contrast, the notion of cut, the proof reduction rules and the properties of cut free proofs remain the usual ones with computation rules.

When a theory is given by a set of axioms, we sometimes want to find an alternative way to present it with computation rules, in such a way that cut elimination holds. From cut elimination, we can deduce analyticity results, consistency and various independence results, completeness of proof search methods and in some cases decision algorithms. Many theories have been presented in such a way, including various equational theories, several presentations of simple type theory (with combinators or lambda-calculus, with or without the axiom of infinity, ...), the theory of equality, arithmetic, ... However, a systematic way of transforming a set of axioms into a set of rewrite rules is still to be found. A step in this direction is Knuth-Bendix method [12] and its extensions, that permit to transform some equational theories into rewrite systems with the cut elimination property (i.e. with the confluence property). Another step in this direction is the result of S. Negri and J. Von Plato [14] that gives a way to transform some sets of axioms, in particular all quantifier free theories, into a set of non logical deduction rules in sequent calculus, preserving cut elimination. In this paper, we propose a way to transform any consistent quantifier free theory into a set of computation rules with the cut elimination property.

Our first attempt was to use Deduction modulo [7,8] or Asymmetric deduction Modulo [6] as a general framework where computation and deduction can be mixed. In Deduction modulo, the introduction rule of conjunction

$$\frac{A \quad B}{A \wedge B}$$

is transformed into a rule

$$\frac{A \quad B}{C} \text{ if } C \equiv A \wedge B$$

where $\equiv$ is the congruence generated by the computation rules, and the other deduction rules are transformed in a similar way. In Asymmetric deduction modulo, this rule is rephrased

$$\frac{A \quad B}{C} \text{ if } C \longrightarrow A \wedge B$$

where the congruence is replaced by the rewriting relation.

However, although we have no formal proof of it, it seems that the theory formed with the single axiom $P \Rightarrow Q$ (where P and Q are proposition symbols) cannot be expressed neither in Deduction modulo nor in Asymmetric deduction modulo (while the theory $P \Leftrightarrow Q$ can, as well as more complex theories such as arithmetic or type theory). Here we shall continue weakening deduction modulo and introduce *Polarized Deduction Modulo* where when rewriting C to $A \wedge B$ we shall distinguish negative and positive occurrences of C. This way we will be able to transform the axiom $P \Rightarrow Q$ into the negative rule $P \longrightarrow Q$ where P can be rewritten into Q at negative occurrences only, or into the positive rule $Q \longrightarrow P$ where Q can be rewritten into P at positive occurrences only.

1   Polarized Deduction Modulo

Definition 1 (Polarized rewrite system). *A rewrite rule is a pair $P \longrightarrow A$ where P is an atomic proposition and A an arbitrary proposition. A polarized rewrite system $\langle \mathcal{R}_-, \mathcal{R}_+ \rangle$ is a pair of sets of rewrite rules. The rules of $\mathcal{R}_-$ are called negative and those of $\mathcal{R}_+$ are called positive.*

Definition 2 (Rewriting). *Given a polarized rewrite system, we define the one step rewriting relations $\longrightarrow^1_-$ and $\longrightarrow^1_+$*

- *if $P \longrightarrow A$ is a negative rule then $P \longrightarrow^1_- A$,*
- *if $(A \longrightarrow^1_+ A'$ and $B = B')$ or $(A = A'$ and $B \longrightarrow^1_- B')$, then $A \Rightarrow B \longrightarrow^1_- A' \Rightarrow B'$,*
- *if $(A \longrightarrow^1_- A'$ and $B = B')$ or $(A = A'$ and $B \longrightarrow^1_- B')$, then $A \wedge B \longrightarrow^1_- A' \wedge B'$ and $A \vee B \longrightarrow^1_- A' \vee B'$,*

- *if $P \longrightarrow A$ is a positive rule then $P \longrightarrow_+^1 A$,*
- *if $(A \longrightarrow_-^1 A'$ and $B = B')$ or $(A = A'$ and $B \longrightarrow_+^1 B')$,*
  then $A \Rightarrow B \longrightarrow_+^1 A' \Rightarrow B'$,
- *if $(A \longrightarrow_+^1 A'$ and $B = B')$ or $(A = A'$ and $B \longrightarrow_+^1 B')$,*
  then $A \wedge B \longrightarrow_+^1 A' \wedge B'$ and $A \vee B \longrightarrow_+^1 A' \vee B'.$

Then the rewriting relations $\longrightarrow_-$ and $\longrightarrow_+$ are defined as the transitive closures of the relations $\longrightarrow_-^1$ and $\longrightarrow_+^1$.

The deduction rules of *Polarized natural deduction modulo* are those of figure 1. Those of *Polarized sequent calculus modulo* are those of figure 2.

$$\frac{}{\Gamma \vdash_{\mathcal{R}} B} \text{ axiom if } A \in \Gamma \text{ and } A \longrightarrow_- C \longleftarrow_+ B$$

$$\frac{\Gamma, A \vdash_{\mathcal{R}} B}{\Gamma \vdash_{\mathcal{R}} C} \Rightarrow\text{-intro if } C \longrightarrow_+ (A \Rightarrow B)$$

$$\frac{\Gamma \vdash_{\mathcal{R}} C \quad \Gamma \vdash_{\mathcal{R}} A}{\Gamma \vdash_{\mathcal{R}} B} \Rightarrow\text{-elim if } C \longrightarrow_- (A \Rightarrow B)$$

$$\frac{\Gamma \vdash_{\mathcal{R}} A \quad \Gamma \vdash_{\mathcal{R}} B}{\Gamma \vdash_{\mathcal{R}} C} \wedge\text{-intro if } C \longrightarrow_+ (A \wedge B)$$

$$\frac{\Gamma \vdash_{\mathcal{R}} C}{\Gamma \vdash_{\mathcal{R}} A} \wedge\text{-elim if } C \longrightarrow_- (A \wedge B)$$

$$\frac{\Gamma \vdash_{\mathcal{R}} C}{\Gamma \vdash_{\mathcal{R}} B} \wedge\text{-elim if } C \longrightarrow_- (A \wedge B)$$

$$\frac{\Gamma \vdash_{\mathcal{R}} A}{\Gamma \vdash_{\mathcal{R}} C} \vee\text{-intro if } C \longrightarrow_+ (A \vee B)$$

$$\frac{\Gamma \vdash_{\mathcal{R}} B}{\Gamma \vdash_{\mathcal{R}} C} \vee\text{-intro if } C \longrightarrow_+ (A \vee B)$$

$$\frac{\Gamma \vdash_{\mathcal{R}} D \quad \Gamma, A \vdash_{\mathcal{R}} C \quad \Gamma, B \vdash_{\mathcal{R}} C}{\Gamma \vdash_{\mathcal{R}} C} \vee\text{-elim if } D \longrightarrow_- (A \vee B)$$

$$\frac{\Gamma \vdash_{\mathcal{R}} B}{\Gamma \vdash_{\mathcal{R}} A} \perp\text{-elim if } B \longrightarrow_- \perp$$

Fig. 1. Polarized natural deduction modulo

As usual, the rules of natural deduction are those of intuitionistic logic, and the rules for classical logic are obtained by adding the excluded middle. The rules of sequent calculus are those of classical logic and those of intuitionistic logic are obtained by restricting the right hand part of sequents to have one proposition at most.

For simplicity, we have given only the rules of propositional logic, but the case of quantifiers is not more complicated. Notice also that there there is no rule for negation: the proposition $\neg A$ is an abbreviation for $A \Rightarrow \perp$.

In general, rewriting has two properties. First, it is oriented and for instance the term $2 + 2$ rewrites to 4, but the term 4 does not rewrite to $2 + 2$. Then, rewriting preserves provability. For instance, the proposition $even(2+2)$ rewrites to $even(4)$ that is equivalent. Thus we can always transform the proposition $even(2 + 2)$ to $even(4)$ and we never need to backtrack on this operation.

$$\frac{}{A \vdash_{\mathcal{R}} B} \text{ axiom if } A \longrightarrow_{-} C \longleftarrow_{+} B$$

$$\frac{\Gamma, A \vdash_{\mathcal{R}} \Delta \quad \Gamma \vdash_{\mathcal{R}} B, \Delta}{\Gamma \vdash_{\mathcal{R}} \Delta} \text{ cut if } A \longleftarrow_{-} C \longrightarrow_{+} B$$

$$\frac{\Gamma, B_1, B_2 \vdash_{\mathcal{R}} \Delta}{\Gamma, A \vdash_{\mathcal{R}} \Delta} \text{ contr-left if } A \longrightarrow_{-} B_1, A \longrightarrow_{-} B_2$$

$$\frac{\Gamma \vdash_{\mathcal{R}} B_1, B_2, \Delta}{\Gamma \vdash_{\mathcal{R}} A, \Delta} \text{ contr-right if } A \longrightarrow_{+} B_1, A \longrightarrow_{+} B_2$$

$$\frac{\Gamma \vdash_{\mathcal{R}} \Delta}{\Gamma, A \vdash_{\mathcal{R}} \Delta} \text{ weak-left}$$

$$\frac{\Gamma \vdash_{\mathcal{R}} \Delta}{\Gamma \vdash_{\mathcal{R}} A, \Delta} \text{ weak-right}$$

$$\frac{\Gamma \vdash_{\mathcal{R}} A, \Delta \quad \Gamma, B \vdash_{\mathcal{R}} \Delta}{\Gamma, C \vdash_{\mathcal{R}} \Delta} \Rightarrow\text{-left if } C \longrightarrow_{-} (A \Rightarrow B)$$

$$\frac{\Gamma, A \vdash_{\mathcal{R}} B, \Delta}{\Gamma \vdash_{\mathcal{R}} C, \Delta} \Rightarrow\text{-right if } C \longrightarrow_{+} (A \Rightarrow B)$$

$$\frac{\Gamma, A, B \vdash_{\mathcal{R}} \Delta}{\Gamma, C \vdash_{\mathcal{R}} \Delta} \wedge\text{-left if } C \longrightarrow_{-} (A \wedge B)$$

$$\frac{\Gamma \vdash_{\mathcal{R}} A, \Delta \quad \Gamma \vdash_{\mathcal{R}} B, \Delta}{\Gamma \vdash_{\mathcal{R}} C, \Delta} \wedge\text{-right if } C \longrightarrow_{+} (A \wedge B)$$

$$\frac{\Gamma, A \vdash_{\mathcal{R}} \Delta \quad \Gamma, B \vdash_{\mathcal{R}} \Delta}{\Gamma, C \vdash_{\mathcal{R}} \Delta} \vee\text{-left if } C \longrightarrow_{-} (A \vee B)$$

$$\frac{\Gamma \vdash_{\mathcal{R}} A, B, \Delta}{\Gamma \vdash_{\mathcal{R}} C, \Delta} \vee\text{-right if } C \longrightarrow_{+} (A \vee B)$$

$$\frac{}{\Gamma, A \vdash_{\mathcal{R}} \Delta} \perp\text{-left if } A \longrightarrow_{-} \perp$$

Fig. 2. Polarized sequent calculus modulo

When rewriting is polarized, the first property is kept, but not the second. For instance, if we have the negative rule $P \longrightarrow Q$, the sequent $P \vdash_{\mathcal{R}} P$ can be proved with the axiom rule, but its normal form $Q \vdash_{\mathcal{R}} P$ cannot (because the positive occurrence of P cannot be rewritten). Thus, proof search in polarized deduction modulo may require backtracking on rewriting.

We shall use a functional notation for proofs as terms.

Definition 3. *(Proof-terms) Proof-terms are defined inductively as follows.*

$$
\begin{aligned}
\pi ::=\quad & \alpha \\
& | \ \lambda\alpha \ \pi \ | \ (\pi_1 \ \pi_2) \\
& | \ \langle \pi_1, \pi_2 \rangle \ | \ fst(\pi) \ | \ snd(\pi) \\
& | \ i(\pi) \ | \ j(\pi) \ | \ \delta(\pi_1, \alpha\pi_2, \beta\pi_3) \\
& | \ \delta_{\perp}(\pi)
\end{aligned}
$$

Each proof-term construction corresponds to a natural deduction rule: terms of the form α express proofs built with the axiom rule, terms of the form $\lambda\alpha \ \pi$ and $(\pi_1 \ \pi_2)$ express proofs built respectively with the introduction and elimination rules of the implication, terms of the form $\langle \pi_1, \pi_2 \rangle$ and $fst(\pi), snd(\pi)$ express proofs built with the introduction and elimination rules of the conjunction, terms of the form $i(\pi), j(\pi)$ and $\delta(\pi_1, \alpha\pi_2, \beta\pi_3)$ express proofs built with the introduction and elimination rules of the disjunction, terms of the form $\delta_\perp(\pi)$ express proofs built with the elimination rule of the contradiction.

2   Proof Reduction

2.1   Reduction Rules

As in pure logic, a cut in polarized natural deduction modulo is a sequence formed by an introduction rule followed by an elimination rule. For instance, the proof

$$\cfrac{\cfrac{\cfrac{\pi_1}{\Gamma \vdash_{\mathcal{R}} A} \quad \cfrac{\pi_2}{\Gamma \vdash_{\mathcal{R}} B}}{\Gamma \vdash_{\mathcal{R}} C} \wedge\text{-intro } C \longrightarrow_- A \wedge B}{\Gamma \vdash_{\mathcal{R}} A'} \wedge\text{-elim } C \longrightarrow_+ A' \wedge B'$$

is a cut. Eliminating this cut consists in replacing this proof by the simpler proof π_1. Expressed on proof-terms, this rule is rephrased

$$fst(\langle \pi_1, \pi_2 \rangle) \triangleright \pi_1$$

Similar rules can be designed for the other forms of cut, leading to the proof rewrite system of figure 3.

$$(\lambda\alpha \ \pi_1 \ \pi_2) \triangleright [\pi_2/\alpha]\pi_1$$
$$fst(\langle \pi_1, \pi_2 \rangle) \triangleright \pi_1$$
$$snd(\langle \pi_1, \pi_2 \rangle) \triangleright \pi_2$$
$$\delta(i(\pi_1), \alpha\pi_2, \beta\pi_3) \triangleright [\pi_1/\alpha]\pi_2$$
$$\delta(j(\pi_1), \alpha\pi_2, \beta\pi_3) \triangleright [\pi_1/\beta]\pi_3$$

Fig. 3. Proof reduction rules

2.2   Subject Reduction

In the example above, the proof π_1 is a proof of A. For the reduction rule to be correct, we have to make sure that it is also a proof of A'. This is not the case in general, but this is the case if the relations relations $\longrightarrow_-$ and $\longrightarrow_+$ commute, i.e. if whenever $A \longleftarrow_- B \longrightarrow_+ C$ then there is a proposition D such that $A \longrightarrow_+ D \longleftarrow_- C$. Notice that when the relations $\longrightarrow_-$ and $\longrightarrow_+$ are identical, this property is just confluence.

Proposition 1. *If* $\longrightarrow_-$ *and* $\longrightarrow_+$ *commute,* π *is a proof-term of* $\Gamma \vdash_{\mathcal{R}} A$ *and* $A \longrightarrow_- A'$ *then* π *is also a proof-term of* $\Gamma \vdash_{\mathcal{R}} A'$.

Proof. We prove, by induction on the structure of π that, more generally, if π is a proof-term of $\Gamma \vdash_\mathcal{R} A$, $\Gamma \longrightarrow_+ \Gamma'$ and $A \longrightarrow_- A'$, then π is also a proof-term of $\Gamma' \vdash_\mathcal{R} A'$.

- (axiom) If π is a variable α, we have B in Γ and $B' \longleftarrow_+ B \longrightarrow_- C \longleftarrow_+$ $A \longrightarrow_- A'$ thus we have a proposition D such that $B' \longrightarrow_- D \longleftarrow_+ A'$ and π is a proof of $\Gamma' \vdash_\mathcal{R} A'$.
- ($\Rightarrow$-intro) If $\pi = \lambda\alpha\ \pi_1$, then π_1 is a proof of $\Gamma, B \vdash_\mathcal{R} C$ and $A' \longleftarrow_-$ $A \longrightarrow_+ B \Rightarrow C$. Thus there is a proposition $B' \Rightarrow C'$ such that $A' \longrightarrow_+$ $B' \Rightarrow C' \longleftarrow_- B \Rightarrow C$. Thus $B \longrightarrow_+ B'$ and $C \longrightarrow_- C'$ and, by induction hypothesis, π_1 is a proof of $\Gamma', B' \vdash_\mathcal{R} C'$. Thus π is a proof of $\Gamma' \vdash_\mathcal{R} A'$.
- ($\Rightarrow$-elim) If $\pi = (\pi_1\ \pi_2)$, then π_1 is a proof of $\Gamma \vdash_\mathcal{R} C$ and $C \longrightarrow_- B \Rightarrow A$. Thus $C \longrightarrow_- B \Rightarrow A'$ and, by induction hypothesis, π_1 is a proof of $\Gamma' \vdash_\mathcal{R}$ $B \Rightarrow A'$ and π_2 is a proof of $\Gamma' \vdash_\mathcal{R} B$. Thus π is a proof of $\Gamma' \vdash_\mathcal{R} A'$.
- ($\wedge$-intro) If $\pi = \langle\pi_1, \pi_2\rangle$, then π_1 is a proof of $\Gamma \vdash_\mathcal{R} B$ and π_2 is a proof of $\Gamma \vdash_\mathcal{R} C$ and $A' \longleftarrow_- A \longrightarrow_+ B \wedge C$. Thus there is a proposition $B' \wedge C'$ such that $A' \longrightarrow_+ B' \wedge C' \longleftarrow_- B \wedge C$. Thus $B \longrightarrow_- B'$ and $C \longrightarrow_- C'$ and, by induction hypothesis, π_1 is a proof of $\Gamma' \vdash_\mathcal{R} B'$ and π_2 of $\Gamma' \vdash_\mathcal{R} C'$. Thus π is a proof of $\Gamma' \vdash_\mathcal{R} A'$.
- ($\wedge$-elim) If $\pi = fst(\pi_1)$, then π_1 is a proof of $\Gamma \vdash_\mathcal{R} C$ and $C \longrightarrow_- A \wedge B$. Thus $C \longrightarrow_- A' \wedge B$ and, by induction hypothesis, π_1 is a proof of $\Gamma' \vdash_\mathcal{R} A' \wedge B$. Thus π is a proof of $\Gamma' \vdash_\mathcal{R} A'$. The same holds if $\pi = snd(\pi_1)$.
- ($\vee$-intro) If $\pi = i(\pi_1)$, then π_1 is a proof of $\Gamma \vdash_\mathcal{R} B$ and $A' \longleftarrow_- A \longrightarrow_+ B \vee$ C. Thus there is a proposition $B' \vee C'$ such that $A' \longrightarrow_+ B' \vee C' \longleftarrow_- B \vee C$ and, by induction hypothesis, π_1 is a proof of $\Gamma' \vdash_\mathcal{R} B'$. Thus π is a proof of $\Gamma' \vdash_\mathcal{R} A'$. The same holds if $\pi = j(\pi_1)$.
- ($\vee$-elim) If $\pi = \delta(\pi_1, \alpha\ \pi_2, \beta\pi_3)$ then π_1 is a proof of $\Gamma \vdash_\mathcal{R} D$, $D \longrightarrow_- B \vee C$, π_2 is a proof of $\Gamma, B \vdash_\mathcal{R} A$ and π_3 a proof of $\Gamma, C \vdash_\mathcal{R} A$. By induction hypothesis, π_1 is a proof of $\Gamma' \vdash_\mathcal{R} B \vee C$, π_2 is a proof of $\Gamma', B \vdash_\mathcal{R} A'$ and π_3 a proof of $\Gamma', C \vdash_\mathcal{R} A'$. Thus π is a proof of $\Gamma' \vdash_\mathcal{R} A'$.
- ($\perp$-elim) If $\pi = \delta_\perp(\pi_1)$ then π_1 is a proof of $\Gamma \vdash_\mathcal{R} B$ and $B \longrightarrow_- \perp$. By induction hypothesis, π_1 is a proof of $\Gamma' \vdash_\mathcal{R} \perp$. Thus π is a proof of $\Gamma' \vdash_\mathcal{R} A'$.

Proposition 2. *If $\longrightarrow_-$ and $\longrightarrow_+$ commute, π is a proof-term of $\Gamma \vdash_\mathcal{R} A$ and $A \longleftarrow_+ A'$ then π is also a proof-term of $\Gamma \vdash_\mathcal{R} A'$.*

Proof. By induction on the structure of π.

- (axiom) If π is a variable α, we have B in Γ and $B \longrightarrow_- C \longleftarrow_+ A \longleftarrow_+ A'$ thus we have $B \longrightarrow_- C \longleftarrow_+ A'$ and π is a proof of $\Gamma \vdash_\mathcal{R} A'$.
- ($\Rightarrow$-intro) If $\pi = \lambda\alpha\ \pi_1$, then π_1 is a proof of $\Gamma, B \vdash_\mathcal{R} C$ and $A' \longrightarrow_+ A \longrightarrow_+$ $B \Rightarrow C$. Hence $A' \longrightarrow_+ B \Rightarrow C$. Thus π is a proof of $\Gamma \vdash_\mathcal{R} A'$.
- ($\Rightarrow$-elim) If $\pi = (\pi_1\ \pi_2)$, then π_1 is a proof of $\Gamma \vdash_\mathcal{R} C$ and $C \longrightarrow_- B \Rightarrow A$. Thus, by proposition 1, π_1 is a proof of $\Gamma \vdash_\mathcal{R} B \Rightarrow A$. We have $B \Rightarrow A' \longrightarrow_+$ $B \Rightarrow A$. Hence by induction hypothesis π_1 is a proof of $\Gamma \vdash_\mathcal{R} B \Rightarrow A'$. Thus π is a proof of $\Gamma \vdash_\mathcal{R} A'$.

- ($\land$-intro) If $\pi = \langle \pi_1, \pi_2 \rangle$, then π_1 is a proof of $\Gamma \vdash_{\mathcal{R}} B$ and π_2 is a proof of $\Gamma \vdash_{\mathcal{R}} C$ and $A' \longrightarrow_+ A \longrightarrow_+ B \land C$. Hence $A' \longrightarrow_+ B \land C$. Thus π is a proof of $\Gamma \vdash_{\mathcal{R}} A'$.
- ($\land$-elim) If $\pi = fst(\pi_1)$, then π_1 is a proof of $\Gamma \vdash_{\mathcal{R}} C$ and $C \longrightarrow_- A \land B$. Thus, by proposition 1, π_1 is a proof of $\Gamma \vdash_{\mathcal{R}} A \land B$. We have $A' \land B \longrightarrow_+ A \land B$. Hence by induction hypothesis π_1 is a proof of $\Gamma \vdash_{\mathcal{R}} A' \land B$. Thus π is a proof of $\Gamma \vdash_{\mathcal{R}} A'$. The same holds if $\pi = snd(\pi_1)$.
- ($\lor$-intro) If $\pi = i(\pi_1)$, then π_1 is a proof of $\Gamma \vdash_{\mathcal{R}} B$ and $A' \longrightarrow_+ A \longrightarrow_+ B \lor C$. Hence $A' \longrightarrow_+ B \lor C$. Thus π is a proof of $\Gamma \vdash_{\mathcal{R}} A'$. The same holds if $\pi = j(\pi_1)$.
- ($\lor$-elim) If $\pi = \delta(\pi_1, \alpha\, \pi_2, \beta\pi_3)$ then π_1 is a proof of $\Gamma \vdash_{\mathcal{R}} D$, $D \longrightarrow_- B \lor C$, π_2 is a proof of $\Gamma, B \vdash_{\mathcal{R}} A$ and π_3 a proof of $\Gamma, C \vdash_{\mathcal{R}} A$. By induction hypothesis, π_2 is a proof of $\Gamma, B \vdash_{\mathcal{R}} A'$ and π_3 of $\Gamma, C \vdash_{\mathcal{R}} A'$. Thus π is a proof of $\Gamma \vdash_{\mathcal{R}} A'$.
- ($\bot$-elim) If $\pi = \delta_\bot(\pi_1)$ then π_1 is a proof of $\Gamma \vdash_{\mathcal{R}} B$, $B \longrightarrow_- \bot$. By proposition 1, π_1 is a proof of $\Gamma \vdash_{\mathcal{R}} \bot$. Thus π is a proof of $\Gamma \vdash_{\mathcal{R}} A'$.

Proposition 3. *(Subject reduction) If $\longrightarrow_-$ and $\longrightarrow_+$ commute, π is a proof of $\Gamma \vdash_{\mathcal{R}} A$ and $\pi \triangleright \pi'$ then π' is a proof of $\Gamma \vdash_{\mathcal{R}} A$.*

Proof. By induction over the length of the reduction. For the one step case, we consider the different cases according to the form of the redex.

- If $\pi = (\lambda\alpha\pi_1\ \pi_2)$, then π_1 is a proof of $\Gamma, B' \vdash_{\mathcal{R}} A'$. The term $\lambda\alpha\pi_1$ is a proof of $\Gamma \vdash_{\mathcal{R}} C$ with $C \longrightarrow_+ B' \Rightarrow A'$. The term π_2 is a proof of $\Gamma \vdash_{\mathcal{R}} B$ and $(\lambda\alpha\pi_1\ \pi_2)$ is a proof of A with with $C \longrightarrow_- B \Rightarrow A$. By commutation, we have $B' \Rightarrow A' \longrightarrow_- B'' \Rightarrow A'' \longleftarrow_+ B \Rightarrow A$. Thus $B' \longrightarrow_+ B'' \longleftarrow_- B$ and $A' \longrightarrow_- A'' \longleftarrow_+ A$. By propositions 1 and 2, π_1 is a proof of $\Gamma, B' \vdash_{\mathcal{R}} A$ and π_2 is a proof of $\Gamma \vdash_{\mathcal{R}} B'$. Thus π' is a proof of $\Gamma \vdash_{\mathcal{R}} A'$.
- If $\pi = fst(\langle \pi_1, \pi_2 \rangle)$, then π_1 is a proof of $\Gamma \vdash_{\mathcal{R}} A'$, and π_2 a proof of $\Gamma \vdash_{\mathcal{R}} B'$. The term $\langle \pi_1, \pi_2 \rangle$ is a proof of $\Gamma \vdash_{\mathcal{R}} C$ with $C \longrightarrow_+ A' \land B'$ and $fst(\langle \pi_1, \pi_2 \rangle)$ is a proof of $\Gamma \vdash_{\mathcal{R}} A$ with $C \longrightarrow_- A \land B$. By commutation, there is a proposition $A'' \land B''$ such that $A' \land B' \longrightarrow^- A'' \land B'' \longleftarrow^+ A \land B$. Hence $A' \longrightarrow^- A'' \longleftarrow^+ A$ and by propositions 1 and 2, π' is a proof of A. The same holds if $\pi = snd(\langle \pi_1, \pi_2 \rangle)$.
- If $\pi = \delta(i(\pi_1), \pi_2, \pi_3)$, then π_1 is a proof of $\Gamma \vdash_{\mathcal{R}} B$, the term $i(\pi_1)$ is a proof of $\Gamma \vdash_{\mathcal{R}} D$ with $D \longrightarrow_+ B \lor C$, the term π_2 is a proof of $\Gamma, B' \vdash_{\mathcal{R}} A$ and π_3 a proof of $\Gamma, C' \vdash_{\mathcal{R}} A$ with $D \longrightarrow_- B' \lor C'$. By commutation, there is a proposition $B'' \lor C''$ such that $B \lor C \longrightarrow^- B'' \lor C'' \longleftarrow^+ B' \lor C'$. Hence $B \longrightarrow^- B'' \longleftarrow^+ B'$ and by propositions 1 and 2, π_1 is a proof of $\Gamma \vdash_{\mathcal{R}} B'$ Thus π' is a proof of A. The same holds if $\pi = \delta(j(\pi_1), \pi_2, \pi_3)$.

2.3 Termination

This section is an adaptation to polarized deduction modulo of the cut elimination proof, *à la* Tait and Girard, of [8].

Definition 4. *(Neutral proof)*
A proof is said to be neutral *if its last rule is an axiom or an elimination, but not an introduction.*

Definition 5. *(Reducibility candidate)*
A set R *of proofs is a* reducibility candidate *if*

- *if $\pi \in R$, then π is strongly normalizable,*
- *if $\pi \in R$ and $\pi \rhd \pi'$ then $\pi' \in R$,*
- *if π is neutral and if for every π' such that $\pi \rhd^1 \pi'$, $\pi' \in R$ then $\pi \in R$.*

Let $\mathcal{C}$ be the set of all reducibility candidates.

Definition 6. *(Pre-model) Consider a language $\mathcal{L}$, a pre-model for $\mathcal{L}$ is a function associating a reducibility candidate $\hat{P}$ to each atomic proposition P.*

Definition 7. *Let A be a proposition. We define the set $|A|$ of proofs by induction over the structure of A.*

- *If P is atomic then $|P| = \hat{P}$.*
- *A proof π is element of $|A \Rightarrow B|$ if it is strongly normalizable and when π reduces to a proof of the form $\lambda \alpha \pi_1$ then for every π' in $|A|$, $[\pi'/\alpha]\pi_1$ is an element of $|B|$.*
- *A proof π is an element of $|A \wedge B|$ if it is strongly normalizable and when π reduces to a proof of the form $\langle \pi_1, \pi_2 \rangle$ then π_1 and π_2 are elements of $|A|$ and $|B|$.*
- *A proof π is an element of $|A \vee B|$ if it is strongly normalizable and when π reduces to a proof of the form $i(\pi_1)$ (resp. $j(\pi_2)$) then π_1 (resp. π_2) is an element of $|A|$ (resp. $|B|$).*
- *A proof π is an element of $|\bot|$ if it is strongly normalizable.*

Proposition 4. *For every proposition A, $|A|$ is a reducibility candidate.*

Proof. See [8].
In deduction modulo, a pre-model is a pre-model of a rewrite system $\mathcal{R}$, if for each rule $P \longrightarrow A$, we have $|P| = |A|$. In the polarized case, we take the following weaker condition.

Definition 8. *A pre-model is a pre-model of a polarized rewrite system $\mathcal{R}$ if*

- *for each negative rule $P \longrightarrow A$, we have $|P| \subseteq |A|$,*
- *for each positive rule $P \longrightarrow A$, we have $|A| \subseteq |P|$.*

Proposition 5. *Let $\mathcal{R}$ be a polarized rewrite system, in a pre-model of $\mathcal{R}$ we have*

- *if $A \longrightarrow_- B$ then $|A| \subseteq |B|$,*
- *if $A \longrightarrow_+ B$ then $|B| \subseteq |A|$.*

Proof. By induction over the structure of A.

Theorem 1. *Let $\mathcal{R}$ be a polarized rewrite system such that $\longrightarrow_-$ and $\longrightarrow_+$ commute and that has a pre-model. Let A be a proposition, π be a proof of A modulo $\mathcal{R}$ and σ a substitution mapping proof variables of propositions B to elements of $|B|$. Then $\sigma\pi$ is an element of $|A|$.*

Proof. By induction over the structure of π. For sake of brevity, we detail only the cases of the axiom rule and the rules of implication.

- (axiom) If π is a variable α, then $\sigma\pi = \sigma\alpha$. If α is bound by σ then $\sigma\pi$ is an element of $|C|$ and $|C| \subseteq |B| \subseteq |A|$. Hence $\sigma\pi \in |A|$.
  If α is not bound by σ then $\sigma\pi = \alpha$ and thus it is in $|A|$.
- ($\Rightarrow$-intro) The proof π has the form $\lambda\alpha\rho$ where α is a proof variable of some proposition B and ρ a proof of some proposition C. We have $\sigma\pi = \lambda\alpha\sigma\rho$, consider a reduction sequence issued from this proof. This sequence can only reduce the proof $\sigma\rho$. By induction hypothesis, the proof $\sigma\rho$ is an element of $|C|$, thus the reduction sequence is finite.
  Furthermore, every reduct of $\sigma\pi$ is of the form $\lambda\alpha\rho'$ where ρ' is a reduct of $\sigma\rho$. Let then τ be any proof of $|B|$, the proof $[\tau/\alpha]\rho'$ can be obtained by reduction from $([\tau/\alpha]\circ\sigma)\rho$. By induction hypothesis, the proof $([\tau/\alpha]\circ\sigma)\rho$ is an element of $|C|$. Hence, as $|C|$ is a reducibility candidate, the proof $[\tau/\alpha]\rho'$ is an element of $|C|$.
  Hence, the proof $\sigma\pi$ is an element of $|B \Rightarrow C|$. As $A \longrightarrow_+ B \Rightarrow C$, we have $|B \Rightarrow C| \subseteq |A|$, hence $\sigma\pi \in |A|$.
- ($\Rightarrow$-elim) The proof π has the form $(\rho_1\ \rho_2)$ and ρ_1 is a proof of some proposition C such that $C \longrightarrow_- B \Rightarrow A$ and ρ_2 a proof of the proposition B. We have $\sigma\pi = (\sigma\rho_1\ \sigma\rho_2)$. By induction hypothesis $\sigma\rho_1$ and $\sigma\rho_2$ are in the sets $|C|$ and $|B|$. As $C \longrightarrow_- B \Rightarrow C$ we have $|C| \subseteq |B \Rightarrow A|$ and thus $\sigma\rho_1 \in |B \Rightarrow A|$. Hence these proofs are strongly normalizable. Let n be the maximum length of a reduction sequence issued from $\sigma\rho_1$ and n' the maximum length of a reduction sequence issued from $\sigma\rho_2$. We prove by induction on $n + n'$ that $(\sigma\rho_1\ \sigma\rho_2)$ is in the set $|A|$. As this proof is neutral we only need to prove that every of its one step reducts is in $|A|$. If the reduction takes place in $\sigma\rho_1$ or in $\sigma\rho_2$ then we apply the induction hypothesis. Otherwise $\sigma\rho_1$ has the form $\lambda\alpha\ \rho'$ and the reduct is $[\sigma\rho_2/\alpha]\rho'$. By the definition of $|B \Rightarrow A|$ this proof is in $|A|$.
  Hence, the proof $\sigma\pi$ is an element of $|A|$.

Corollary 1. *Every proof of A is in $|A|$ and hence strongly normalizable*

Using the same technique as in [8] we can extend this cut elimination result to intuitionistic polarized sequent calculus modulo, provided we extend the proof reduction rules with the ultra-reduction rules

$$\delta(\pi_1, \alpha\pi_2, \beta\pi_3) \triangleright \pi_2$$

$$\delta(\pi_1, \alpha\pi_2, \beta\pi_3) \triangleright \pi_3$$

We can also extend the result to classical sequent calculus, defining a classical pre-model of a rule $P \longrightarrow A$ as a pre-model of the rule $P \longrightarrow A''$ where A'' is the light double negation of A defined by

- $A'' = A$ if A is atomic,
- $(A \Rightarrow B)'' = A' \Rightarrow B'$,
- $(A \wedge B)'' = A' \wedge B'$,
- $(A \vee B)'' = A' \vee B'$,
- $\perp'' = \perp$,

with

- $A' = \neg\neg A$ if A is atomic,
- $(A \Rightarrow B)' = \neg\neg(A' \Rightarrow B')$,
- $(A \wedge B)' = \neg\neg(A' \wedge B')$,
- $(A \vee B)' = \neg\neg(A' \vee B')$,
- $\perp' = \neg\neg\perp$.

3   The Equivalence Lemma

We are now ready to relate theories defined by axioms and by polarized rewrite rules.

Proposition 6 (Equivalence). *Let $\mathcal{R}$ be a polarized rewrite system such that $\longrightarrow_-$ and $\longrightarrow_+$ commute. Let $\mathcal{T}$ be the set of axioms formed with, for each negative rule $P \longrightarrow A$ of $\mathcal{R}$ the axiom $P \Rightarrow A$ and for each positive rule $P \longrightarrow A$ of $\mathcal{R}$ the axiom $A \Rightarrow P$. Then*

$$\Gamma \vdash_{\mathcal{R}} A$$

if and only if

$$\mathcal{T}, \Gamma \vdash A$$

Proof. Notice, first, that for every proposition $A \Rightarrow B$ of $\mathcal{T}$, the sequent $A \vdash_{\mathcal{R}} B$ is provable with the axiom rule, using either the rule $A \longrightarrow_- B$ or the rule $B \longrightarrow_+ A$ and thus the sequent $\vdash_{\mathcal{R}} A \Rightarrow B$ is provable. Using propositions 1 and 2 for every proposition D such that $A \Rightarrow B \longrightarrow_- C \longleftarrow_+ D$, the sequent $\vdash_{\mathcal{R}} D$ is provable. Thus, by induction over the structure of a proof of $\mathcal{T}, \Gamma \vdash A$, we build a proof of $\Gamma \vdash_{\mathcal{R}} A$ replacing by a proof all invocations to the axioms of $\mathcal{T}$.

Conversely, we first prove, by induction over the structure of A that if $A \longrightarrow_- B$ then $\mathcal{T} \vdash A \Rightarrow B$ and that if $A \longrightarrow_+ B$ then $\mathcal{T} \vdash B \Rightarrow A$. Thus, by induction over the structure of a proof of $\Gamma \vdash_{\mathcal{R}} A$, we build a proof of $\mathcal{T}, \Gamma \vdash A$. As an example, we give the case of the $\wedge$-intro rule. The proof has the form

$$\frac{\dfrac{\pi_1}{\Gamma \vdash_{\mathcal{R}} B} \quad \dfrac{\pi_2}{\Gamma \vdash_{\mathcal{R}} C}}{\Gamma \vdash_{\mathcal{R}} A} \wedge\text{-intro where } A \longrightarrow_+ B \wedge C$$

By the induction hypothesis we have proofs π_1' and π_2' of $\mathcal{T}, \Gamma \vdash B$ and $\mathcal{T}, \Gamma \vdash C$. We first build the proof:

$$\dfrac{\dfrac{\pi_1'}{\mathcal{T}, \Gamma \vdash B} \quad \dfrac{\pi_2'}{\mathcal{T}, \Gamma \vdash C}}{\mathcal{T}, \Gamma \vdash B \wedge C} \wedge\text{-intro}$$

Then, we have $A \longrightarrow_+ (B \wedge C)$, thus $\mathcal{T} \vdash (B \wedge C) \Rightarrow A$ and we can build a proof of the proposition A.

4   Expressing Axioms as Rewrite Rules

Now, we consider a theory given by a set of quantifier free axioms and we want to present it with a polarized rewrite system such that $\longrightarrow_-$ and $\longrightarrow_+$ commute and such that polarized deduction modulo this rewrite system has the cut elimination property.

In polarized deduction modulo, there is no cut free proof of $\bot$. Thus cut elimination implies consistency and consistency is a necessary condition for a set of axioms to be transformed into a polarized rewrite system. As we shall see, this condition is sufficient. We shall prove that any consistent theory can be presented with a polarized rewrite system such that the left hand sides of the negative rules and positive rules are disjoint (i.e. no atomic proposition can be rewritten both by a negative and a positive rule). We shall also see that with such a rewrite system the relations $\longrightarrow_-$ and $\longrightarrow_+$ always commute and that cut elimination always holds.

Proposition 7. *Consider a rewrite system such that the left hand sides of the negative rules and positive rules are disjoint, then the relation $\longrightarrow_-$ and $\longrightarrow_+$ commute and cut elimination holds.*

Proof. Commutation is a simple consequence of the absence of critical pairs. To prove cut elimination, we build a pre-model. Left hand sides of negative rules by the smallest reducibility candidate (the intersection of all reducibility candidates) and all the other atomic propositions by the largest (the set of all strongly normalizable proof-terms). The conditions of definition 8 hold obviously, thus cut elimination holds for intuitionistic natural deduction. This results extends to intuitionistic sequent calculus and classical sequent calculus.

Theorem 2. *If a quantifier free theory is consistent, then it can be presented as a rewrite system such that the relation $\longrightarrow_-$ and $\longrightarrow_+$ commute and cut elimination holds.*

Proof. Let Γ be a consistent set of quantifier free axioms. We prove that Γ can be presented as a rewrite system such that the left hand sides of the negative rules and positive rules are disjoint.

Let ν be a model of Γ. Following [14], we consider the conjunctive-disjunctive (clausal form) Γ' of Γ.

We pick a clause of Γ'. In this clause there is either a literal of the form P such that $\nu(P) = 1$ or a literal of the form $\neg P$ such that $\nu(P) = 0$.

In the first case, we pick all the clauses of Γ' where P occurs positively

$$P \vee A_1, ..., P \vee A_n$$

we replace these clauses by the proposition

$$(\neg A_1 \vee ... \vee \neg A_n) \Rightarrow P$$

and then by the positive rule

$$P \longrightarrow_+ \neg A_1 \vee ... \vee \neg A_n$$

In the second, we pick all the clauses of Γ' where P occurs negatively

$$\neg P \vee A_1, ..., \neg P \vee A_n$$

we replace these clauses by the proposition

$$P \Rightarrow (A_1 \wedge ... \wedge A_n)$$

and then by the negative rule

$$P \longrightarrow_- A_1 \wedge ... \wedge A_n$$

We repeat this process with all clauses of Γ'. We obtain this way a polarized rewrite system $\mathcal{R}$. All the rules have a left hand side whose interpretation in ν is 1 if the rule is positive and 0 if the rule is negative. Hence the left hand sides of the negative rules and positive rules are disjoint.

Example 1. In deduction modulo, it is well-known that consistency does not imply cut elimination. For instance, the theory defined by the rewriting rule (Crabbé's rule)

$$A \longrightarrow (B \wedge \neg A)$$

is consistent but does not have the cut elimination property [2,8].

However this rule is not the only way to present this theory in deduction modulo and the algorithm above can be used to find another presentation. The proposition $A \Leftrightarrow (B \wedge \neg A)$ has a model $\nu(A) = \nu(B) = 0$ and from its clausal form

$$(\neg A) \wedge (\neg A \vee B) \wedge (A \vee \neg B)$$

we get the rules

$$A \longrightarrow_- \bot$$

$$B \longrightarrow_- A$$

Notice that in this case, the theory can be also presented with the simpler, non polarized rewrite system $A \longrightarrow \bot$, $B \longrightarrow A$. It remains to be investigated which theories can be expressed with a non polarized system and which ones cannot.

Conclusion

With any proof search method (such as resolution or tableaux), a lot of work is duplicated when we search for a proof of a proposition A and then of a proposition B in the same axiomatic theory Γ. This suggests that before searching for proofs in some theory, we should "prepare" the theory and express it in such a way that this duplication is avoided. This preparation of a theory can be compared to the compilation of a program: a program is compiled once and not each time it is executed.

An extreme case is when the theory Γ is contradictory, proof search in the prepared theory should then be trivial. When the theory is consistent, the search for a proof in the prepared theory should restrict to analytic proofs, i.e. the search for a proof of a proposition A should involve only the sub-formulas of A (including instances and reducts) and thus the search of a proof of the contradiction $\perp$ should fail immediately.

Transforming an axiomatic theory into a rewrite system such that deduction modulo this rewrite system has the cut elimination property is an example of such a preparation. Knuth-Bendix method permits to do this for some equational theories. We have proposed here a similar preparation method for quantifier free theories. Of course, the general case still needs to be investigated.

Since preparation seems to involve a consistency check, and consistency is undecidable in general, such a preparation method can only be partial.

At last, it is well-known that when a theory has the cut elimination property, then it is consistent, but that the converse does not hold: there are consistent theories that do not have the cut elimination property, for instance set theory or the theory $A \longrightarrow B \wedge \neg A$ [2,8]. However, we have seen that various presentations of the same theory may or may not have the cut elimination property. Thus, if we take the definition that an axiomatic theory has the cut elimination property if *one* of its presentation in deduction modulo has the cut elimination property, then the theory $A \longrightarrow B \wedge \neg A$ has the cut elimination property and the problem is open for set theory (while [1] seems to suggest that the result might be positive). Then, it is not so obvious that there are consistent theories that do not have the cut elimination property, and in particular we have seen that for quantifier free theories, consistency and cut elimination coincide. The generality of this result also remains to be investigated.

Acknowledgements. I want to thank Thérèse Hardin, Hélène Kirchner, Claude Kirchner and Benjamin Werner for many discussions on deduction modulo, cut elimination and Knuth-Bendix method.

References

1. S.C. Bailin. A normalization theorem for set theory. *The Journal of Symbolic Logic*, 53, 3, 1988, pp. 673-695.
2. M. Crabbé. Non-normalisation de la théorie de Zermelo. Manuscript, 1974.

3. M. Crabbé. Stratification and cut-elimination. *The Journal of Symbolic Logic*, 56, 1991, pp. 213-226.
4. G. Dowek. Axioms vs. rewrite rules: from completeness to cut elimination. H. Kirchner and Ch. Ringeissen (Eds.), *Frontiers of Combining Systems*, Lecture Notes in Artificial Intelligence 1794, Springer-Verlag, 2000, pp. 62-72.
5. G. Dowek. About folding-unfolding cuts and cuts modulo. *Journal of Logic and Computation* 11, 3, 2001, pp. 419-429.
6. G. Dowek. Confluence as a cut elimination property. *Workshop on Logic, Language, Information and Computation*, 2001.
7. G. Dowek, Th. Hardin, and C. Kirchner. Theorem proving modulo. *Journal of Automated Reasoning* (to appear). *Rapport de Recherche INRIA* 3400, 1998.
8. G. Dowek and B. Werner. Proof normalization modulo. *Types for proofs and programs*, T. Altenkirch, W. Naraschewski, and B. Rues (Eds.), *Lecture Notes in Computer Science* 1657, Springer-Verlag, 1999, pp. 62-77. *Rapport de Recherche* 3542, INRIA, 1998.
9. J. Ekman. *Normal proofs in set theory*. Doctoral thesis, Chalmers University of Technology and University of Göteborg, 1994.
10. L. Hallnäs. *On normalization of proofs in set theory*. Doctoral thesis, University of Stockholm, 1983.
11. L. Hallnäs and P. Schroeder-Heister. A proof-theoretic approach to logic programming. I. Clauses as rules. *Journal of Logic and Computation* 1, 2, 1990, pp. 261-283. II. Programs as definitions. *Journal of Logic and Computation* 1, 5, 1991, pp. 635-660.
12. D.E. Knuth and P.B. Bendix. Simple word problems in universal algebras. J. Leech (Ed.), *Computational Problems in Abstract Algebra*, Pergamon Press, 1970, pp. 263-297.
13. R. McDowell and D. Miller. Cut-Elimination for a logic with definitions and induction. *Theoretical Computer Science* 232, 2000, pp. 91-119.
14. S. Negri and J. Von Plato. Cut elimination in the presence of axioms. *The Bulletin of Symbolic Logic*, 4, 4, 1998, pp. 418-435.
15. D. Prawitz. *Natural deduction, a proof-theoretical study*. Almqvist & Wiksell, 1965.
16. P. Schroeder-Heister. Cut elimination in logics with definitional reflection. D. Pearce and H. Wansing (Eds.), *Nonclassical Logics and Information Processing*, Lecture Notes in Computer Science 619, Springer-Verlag, 1992, pp. 146-171.
17. P. Schroeder-Heister. Rules of definitional reflection. *Logic in Computer Science*, 1993, pp. 222-232.

A Space Lower Bound for Routing in Trees

Pierre Fraigniaud[1] and Cyril Gavoille[2]

[1] CNRS, Laboratoire de Recherche en Informatique, Université Paris-Sud, 91405 Orsay cedex, France. *pierre@lri.fr*

[2] Laboratoire Bordelais de Recherche en Informatique, Université Bordeaux I, 33405 Talence cedex, France. *gavoille@labri.fr*

Abstract. The design of compact routing schemes in trees form the kernel of sophisticated strategies for compact routing in arbitrary graphs. This paper focuses on the space complexity for routing messages along shortest paths in trees. It was recently shown that the family of n-node trees supports routing schemes using addresses and routing tables of size $O(\log^2 n/\log\log n)$ bits per node, if the output port numbers of each node are chosen by an adversary. This paper shows that this result is tight, that is the sum of the sizes of the address and of the local routing table is at least $\Omega(\log^2 n/\log\log n)$ bits for some node of some tree.

Keywords. Algorithms and data structures, computational and structural complexity, compact routing.

1 Introduction

It is well known, a *routing scheme* consists of two parts:

1. a pre-processing algorithm in charge of, giving a network G, setting data structures (e.g., routing tables, node addresses, port labeling, etc.) on which the routing decisions in G will be based;
2. a routing protocol, i.e., a distributed algorithm, running at each node of G, which mainly determines, for every message entering the node, the output port on which the message has to be forwarded. Note that the routing protocol may also be in charge of other tasks such as updating the headers, etc.

For instance, a classical routing scheme is the so-called *routing table* which stores a look-up table at each node. This table returns, for each possible destination address of a message, the output port on which the message has to be forwarded. To be more precise, upon reception of a message M, a node x considers the *header* attached to M. The header contains the *address* of the destination, say node y. This address is denoted by address(y). Node x then extracts from its look-up table the *output port number* p corresponding to the entry address(y). Finally M is forwarded on the incident link of x labeled by p. We note that the header is not modified here, and the entire routing of M is determined by the

H. Alt and A. Ferreira (Eds.): STACS 2002, LNCS 2285, pp. 65–75, 2002.
© Springer-Verlag Berlin Heidelberg 2002

information address(y) originally stored in the header of M by the source node. The pre-processing algorithm is in charge of constructing the look-up tables. Giving different addresses between 1 and n to the nodes of an n-node network, and labeling the incident links of each node x by different integers between 1 and $\deg(x)$, yields look-up tables of size $O(n \log n)$. The question is: can we do better? I.e., can we design routing schemes requiring less space at each node of the network? The answer is "yes", and the main tool to achieve that improvement is a combination of clustering and tree covers. In particular, we refer to the hierarchical routing schemes presented in [1,2,13], and to the more recent schemes presented in [3,4,14,15]). Although the stretch factors (i.e., the maximum, over all pairs of nodes, of the ratio between the length of the route over the length of a shortest path between two nodes) resulting from these schemes might be not optimal, the improvement in term of memory space is often significant. All these schemes use as subroutine a shortest path routing schemes in some (partial) spanning trees of the network. This motivates the design of shortest path routing schemes for trees.

There are several complexity issues related to the design of routing schemes. Here we list three of them. (1) The size of the addresses and of the headers. For instance, routing tables use addresses and headers of $\lceil \log n \rceil$ bits[1], but other schemes may use longer addresses – e.g., the scheme in [13] uses addresses on $O(\log^2 n)$ bits, though headers on $O(\log n)$ bits only. (2) The size of the local data structure stored at each node. This size plus the size of the addresses is informally called *memory requirement*. And (3) The length of the routes, i.e., the stretch factor. In [14] several trade-offs between the memory requirement and the stretch factor have been given, improving the previous results of the seminal paper [13]. Namely, they show that any network support a routing scheme with stretch $4k - 5$ using local data structures of size[2] $\tilde{O}(kn^{1/k})$, and using addresses and headers of size $O(k \log^2 n / \log \log n)$, for every integral parameter $k \geqslant 2$. Using a *handshaking* technique (an agreement of $O(\log n)$ bits between the source and the destination) they even reduced the stretch to $2k - 1$. Results and well established Erdös conjectures [5] from structural graph theory allow to believe that the space-time tradeoff is optimal. However, it is let open whether one could decrease the size of the addresses and headers.

As shown in [6,15], these complexity issues strongly depend on whether or not we have the ability to set the output port numbers during the pre-processing phase. The *fixed-port* model specifies that the port numbers are fixed (say by an adversary), as opposed to the *designer-port* model which specifies that the port numbers can be fixed by the designer of the scheme. The routing schemes in [13,15] are based on sub-routing schemes on trees in the fixed-port model. Indeed, every node uses several routing protocols and hence its incident links (i.e., ports) must be labeled once for all the protocols. In [6,14], it is shown that, under the fixed-port model, there is a routing scheme using addresses and local data structures of size $O(\log^2 n / \log \log n)$ bits.

---

[1] All logarithms are in base two.

[2] $\tilde{O}(f(n))$ stands for $O(f(n)(\log n)^{O(1)})$.

Our Result

In this paper, we prove a tight $\Omega(\log^2 n / \log\log n)$ lower bound for the memory requirement of shortest path routing in trees under the fixed-port model. An consequence of this result is that the routing schemes in [6,14] are optimal as far as memory requirement is concerned. Another consequence is that the sizes of the headers and addresses used in [13,15] cannot be improved based on routing on spanning trees. Therefore, other techniques must be derived in order to decrease these sizes, say to $O(\log n)$, if possible. Our lower bound is obtained via sophisticated counting arguments presented in the forthcoming sections.

Previous Works Related to Routing in Trees

Several previous works addressed the routing problem in trees (results on compact routing in arbitrary networks have been recently surveyed in [8]). Under the fixed-port model, in [6] it is shown that every routing scheme that selects addresses in the range $[1, n]$, has an $n - o(n)$ bit local routing table for some n-node tree, whereas a scheme using $n + o(n)$ bits and $[1, n]$ range addresses exists [6]. Better performances are achieved under the designer-port model. In [7] a scheme using $O(\sqrt{n})$ bit local routing tables and $[1, n]$ range addresses is constructed, and this is a tight bound according to [4]. On the other hand, slightly augmenting the address size is worth since, for instance, under the fixed-port model, [3] describes a routing scheme that uses $3 \log n$ bits for the addresses and $O(\sqrt{n} \log n)$ bits for each local data structure. This is worth also for the designer-port model, since n-node trees have a routing scheme with only $c \log n$ bits of memory requirement [6], for a small constant $c > 1$. It is even proved in [15] that c can be reduced to $c = 1 + O(1/\log\log n)$. As said before, the significance of the fixed-port model has been shown in [13,14], and a routing scheme for trees using addresses and local data structures of size $O(\log^2 n / \log\log n)$ bits has been independently presented in [6] and [14].

2 Forwarding Labeling Scheme

We are given arbitrary trees in which the edges incident to every node x are labeled by distinct integers in $\{1, \ldots, \deg(x)\}$. Note that an edge (x, y) receives two labels, one at x and one at y, and these two labels may be distinct. We often refer to these labels by the use of *port*, port i of node x referring to the incident edge of x labeled i (at x). In this section, unless specified otherwise, all trees and graphs are assumed to be port-labeled as described above.

Consider an arbitrary family $\mathcal{F}$ of graphs. In the fixed-port model, every routing scheme on $\mathcal{F}$ assigns to every graph $G \in \mathcal{F}$, and to every node x of G, two entities: (1) an address for x, denoted by address(x); and (2) a data structure used for routing at x, denoted by data(x). Moreover the scheme defines a routing protocol, i.e., an algorithm *send* (which usually depends on the family $\mathcal{F}$) that is in charge of routing. For every pair of distinct nodes x and y of G, send(data(x), address(y)) returns the label of an edge incident to x on a shortest path[3] from x to y.

Hereafter, we denote by length(address(x)) and by length(data(x)) the size (in bits) of address(x) and of data(x) respectively.

Theorem 1. *In the fixed-port model, for every shortest path routing scheme on n-node trees there exists a tree and a node x of that tree, for which*

$$\text{length}(\text{address}(x)) + \text{length}(\text{data}(x)) = \Omega(\log^2 n / \log \log n).$$

The remaining of the paper is dedicated to proving Theorem 1. The proof is based on the following definition from the informative labeling theory (cf. [10, 11,12] and [9] for an overview).

Definition 1. *Given a family $\mathcal{F}$ of graphs, a forward labeling on $\mathcal{F}$ is a pair (L, R) where R is called the routing function, and L is called the labeling function. For every $G \in \mathcal{F}$ and every node x of G, L assigns a non-empty binary string $L(x)$ to x, and, for any two distinct nodes x, y of G, $R(L(x), L(y))$ returns the label of an edge incident to x on a shortest path from x to y.*

If there exists a routing scheme with length(address(x)) + length(data(x)) $\leq$ α, then, we can derive a forward labeling with labels of length $\alpha + O(\log \alpha)$ as follows. Each label $L(x)$ is obtained by the concatenation of address(x) and data(x), preceded by a sequence $s(x)$ of $O(\log \alpha)$ bits describing the length of address(x). One may use for $s(x)$ a binary encoding of length(address(x)) and prepend each bit with 0 and terminate $s(x)$ with 11. The routing function R can be computed from the routing protocol *send* by extracting the fields data(x) and address(y) from the labels $L(x)$ and $L(y)$, by the use of $s(x)$ and $s(y)$. Hence a lower bound on the size of the labels of any forward labeling yields a lower bound on length(address(x)) + length(data(x)) of any routing scheme, up to an additive factor bounded by the logarithm of this bound.

To compute a lower bound on the size of the labels in a forward labeling on trees, we proceed as follow. For any family $\mathcal{T}$ of trees, and for any forward labeling (L, R) on $\mathcal{T}$, we construct the set $\mathcal{L}$ consisting of all the labels assigned by L on all the trees of $\mathcal{T}$, i.e., $\mathcal{L} = \bigcup_{T \in \mathcal{T}} \{L(x) \mid x \in T\}$. There exists some label $\ell \in \mathcal{L}$ with length(ℓ) $\geq \log(|\mathcal{L}| + 2) - 1$. Indeed, there are at most $\sum_{i=1}^{k} 2^i = 2^{k+1} - 2$ non-empty binary strings of length at most k, so that every set $\mathcal{L}$ of non-empty binary strings requires some strings of length at least $\log(|\mathcal{L}| + 2) - 1$. Hence our aim is to derive a lower bound on $|\mathcal{L}|$, independent of (L, R). For that purpose, let h, d, t be three positive integers (depending on n, and to be specified later) such that (1) $t < d$, (2) $d \cdot t^h = O(n)$, and (3) $t = \Theta(h) = \Theta(\log n / \log \log n)$. Let T_0 be the rooted tree obtained from a complete t-ary tree $\overline{T}_0$ of depth h in which we add $d - t - 1$ leaves at every internal node, and $d - t$ leaves at

---

In the case of trees the shortest path is unique. However we speak about "shortest path routing in trees" because it is possible to define compact routing scheme on trees that are not along the shortest route, e.g. see [2].

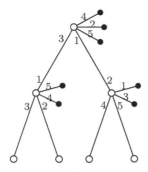

Fig. 1. A port-labeled tree $T \in \mathcal{T}$ for $h = 2$, $d = 5$, and $t = 2$.

the root. We consider the family $\mathcal{T}$, depending on h, d, and t, composed of all the possible port-labeled trees obtained from T_0 (see Fig. 1). For every tree $T \in \mathcal{T}$, we denote by $\overline{T}$ the underlying labeled t-ary tree of depth h. $\overline{T}$ has $|\overline{T}| = \sum_{i=0}^{h} t^i = (t^{h+1} - 1)/(t - 1)$ nodes, and T has $|\overline{T}| + (|\overline{T}| - t^h)(d - t - 1) + 1$ nodes, that is $|T| = (t^h - 1)(d - 1)/(t - 1) + 2$.

To prove Theorem 1 we need the following definition that is a specific variant of forward labeling, valid for trees in $\mathcal{T}$ only:

Definition 2. *A* relaxed forwarding labeling *on $\mathcal{T}$ is a forward labeling where the condition on the routing function in Definition 1 is required to apply only for pairs (x, y) where the current node x and the destination node y belong to $\overline{T}_0$, and x is an ancestor of y.*

Let (L, R) be a relaxed forward labeling on $\mathcal{T}$, and let $\mathcal{L} = \bigcup_{T \in \mathcal{T}} \{L(x) \mid x \in T\}$. We assume that (L, R) is chosen such that $|\mathcal{L}|$ is minimal. By definition, every forward labeling is a relaxed forward labeling, therefore it suffices to obtain a lower bound on the size of $\mathcal{L}$. This is the main key point of the proof given in the next section.

3   Proof of Theorem 1

The whole section consists, given the family $\mathcal{T}$ described in the previous section, and a relaxed forward labeling (L, R) on $\mathcal{T}$, to derive an $\Omega(\log^2 n / \log \log n)$ lower bound for $\log |\mathcal{L}|$.

For every non-leaf node x of the (non labeled) tree $\overline{T}_0$, let $x_1, \ldots, x_t$ be its t children. For every non-leaf node x of a port-labeled tree $\overline{T}$, let ϕ_x denotes the ordered sequence of port-labels such that $\phi_x[i]$ is the label of the edge leading to x_i. Let $\mathcal{S}(d)$ be the set of all possible t-sequences of distinct integers in $1, \ldots, d$, i.e.,

$$\mathcal{S}(d) = \{(s_1, \ldots, s_t) \mid 1 \leqslant s_i \neq s_j \leqslant d \text{ for all } i \neq j\}.$$

We have $|\mathcal{S}(d)| = d!/(d - t)!$, and, for every non-leaf $x \in \overline{T}$, $\phi_x \in \mathcal{S}(d)$. For every $s \in \mathcal{S}(d)$, we define $\mathcal{T}(s)$ as the set of port-labeled trees T of $\mathcal{T}$ such that $\phi_r = s$, where r is the root of T. Observe that $\mathcal{T} = \bigcup_{s \in \mathcal{S}(d)} \mathcal{T}(s)$. Now, for every port-labeled tree $\overline{T}$, and every $i \in \{1, \ldots, t\}$, we denote by $\overline{T}^{(i)}$ the labeled subtree of $\overline{T}$ rooted at the ith child of the root of $\overline{T}_0$. Let us also define, for every $s \in \mathcal{S}(d)$,

$$\mathcal{W}(s) = \bigcup_{T \in \mathcal{T}(s)} \left\{ \Big(L(r), L(x_1), \ldots, L(x_t) \Big) \text{ where } r \text{ is the root of } T \text{ and } x_i \in \overline{T}^{(i)} \text{ for all } i \right\}.$$

Clearly,

$$\bigcup_{s \in \mathcal{S}(d)} \mathcal{W}(s) \subset \mathcal{L} \times \ldots \times \mathcal{L}, \tag{1}$$

where $\mathcal{L}$ appears $t + 1$ times. On the other hand, we have:

Lemma 1. *For all $s \ne s' \in \mathcal{S}(d)$, $\mathcal{W}(s) \cap \mathcal{W}(s') = \varnothing$.*

Proof. Assume that there exists $(\ell_0, \ell_1, \ldots, \ell_t) \in \mathcal{W}(s) \cap \mathcal{W}(s')$. Therefore, there exist two port-labeled trees $T \in \mathcal{T}(s)$ and $T' \in \mathcal{T}(s')$, of roots r and r' respectively, such that: (1) $\ell_0 = L(r) = L(r')$; and (2) for every $i \in \{1, \ldots, t\}$, $\ell_i = L(x_i) = L(x_i')$ for some x_i's in $\overline{T}^{(i)}$ and some x_i''s in $\overline{T}'^{(i)}$. Since $s \ne s'$, let i be such that $s[i] \ne s'[i]$. The routing function R is such that $R(L(r), L(x_i)) = \phi_r[i] = s[i]$ because $T \in \mathcal{T}(s)$. Thus, $R(\ell_0, \ell_i) = s[i]$. On the other hand, considering now T', we obtain that $R(L(r'), L(x_i')) = \varphi_{r'}[i] = s'[i]$, that is $R(\ell_0, \ell_i) = s'[i]$, a contradiction. ∎

A direct consequence of Lemma 1 is that $|\bigcup_s \mathcal{W}(s)| = \sum_s |\mathcal{W}(s)|$. Hence, from Equation 1,

$$\sum_{s \in \mathcal{S}(d)} |\mathcal{W}(s)| \leqslant |\mathcal{L}|^{t+1}. \tag{2}$$

We have:

Lemma 2. *For all $d \geqslant t$ where t divides d, $|\mathcal{S}(d)| \geqslant (d/t)^t$ and $\left| \mathcal{S}\left((d/t)^t \right) \right| \leqslant |\mathcal{S}(d)|^t$.*

Proof. We have $|\mathcal{S}(d)| = t!\binom{d}{t}$, and $\binom{d}{t} \geqslant (d/t)^t$ for all $d \geqslant t$. So, $|\mathcal{S}(d)| \geqslant (d/t)^t$ and $|\mathcal{S}(d)|^t \geqslant (d/t)^{t^2}$. On the other hand, $|\mathcal{S}(d)| \leqslant d^t$, thus $|\mathcal{S}((d/t)^t)| \leqslant (d/t)^{t^2}$. It follows that $\left| \mathcal{S}\left((d/t)^t \right) \right| \leqslant |\mathcal{S}(d)|^t$. ∎

The next lemma gives an inductive lower bound on $|\mathcal{L}|$ as a function of the height h and the degree d of the trees. Let $f(h, d) = |\mathcal{L}|$. We have $f(0, d) = 1$ for all d.

Lemma 3. *For all $d \geqslant t$ where t divides d, and for every $h \geqslant 1$,*

$$f(h, d)^{t+1} \geqslant |\mathcal{S}(d)| \cdot f\left(h - 1, (d/t)^t \right).$$

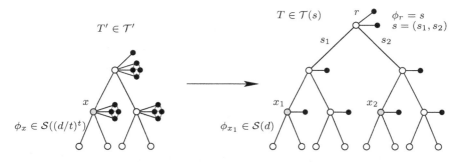

Fig. 2. Transformation of $T' \in \mathcal{T}'$ into a tree $T \in \mathcal{T}(s)$.

Proof. From Equation 2, $|\mathcal{L}|^{t+1} = f(h,d)^{t+1} \geq \sum_{s \in \mathcal{S}(d)} |\mathcal{W}(s)|$. We will show that $|\mathcal{W}(s)| \geq f\left(h-1, (d/t)^t\right)$ for every $s \in \mathcal{S}(d)$. For that purpose, we construct a relaxed forward labeling (L', R') on the family $\mathcal{T}'$ of parameter $h' = h - 1$ and $d' = (d/t)^t$ that uses labels in $\mathcal{W}(s)$ only. Let $s \in \mathcal{S}(d)$, and let $\sigma : \mathcal{S}\left((d/t)^t\right) \to \mathcal{S}(d)^t$ be a one-to-one function to be specified later (such a function exists from Lemma 2). Thanks to σ, we associate a port-labeled tree $T \in \mathcal{T}(s)$ to every port-labeled tree $T' \in \mathcal{T}'$, as follows (cf. Fig. 2). To every node x of $\overline{T}'$ is associated a sequence $(x_1, \ldots, x_t)$ of nodes of $\overline{T}$ where x_i is the node corresponding to x in the subtree $\overline{T}^{(i)}$. We set the port-labels of the x_i's such that $\phi_{x_i} = \sigma(\phi_x)[i]$. The port-labels of the root r of T is set such that $\phi_r = s$. Finally, we fix arbitrarily all remaining port-labels of T.

Let us consider a tree $T' \in \mathcal{T}'$, and a node x of $\overline{T}'$. By the previous transformation, let T be the associated tree in $\mathcal{T}(s)$, with root r. Let $(x_1, \ldots, x_t)$ be the associated sequence of nodes in the subtrees $\overline{T}^{(1)}, \ldots, \overline{T}^{(t)}$. We set

$$L'(x) = (L(r), L(x_1), \ldots, L(x_t)).$$

We can easily check that $L'(x) \in \mathcal{W}(s)$, thus the labeling function L' uses at most $|\mathcal{W}(s)|$ labels for the family $\mathcal{T}'$. To construct R' we need to specify the one-to-one mapping σ. Let

$$\mu_1 : \{1, \ldots, d/t\}^t \to \left\{1, \ldots, (d/t)^t\right\}.$$

and

$$\mu_2 : \{1, \ldots, t\} \times \{1, \ldots, d/t\} \to \{1, \ldots, d\}$$

be two arbitrary bijections. For instance $\mu_1(a_1, \ldots, a_t) = 1 + \sum_{i=1}^{t}(a_i - 1)(d/t)^{i-1}$ and $\mu_2(a, b) = (a-1)d/t + b$, but any bijections would fit as well, and the exact values of μ_1 and μ_2 do not really matter. For every pair $(i, j) \in \{1, \ldots, t\}^2$, let

$$p(i, j) = \mu_1^{-1}(\phi_x[j])[i],$$

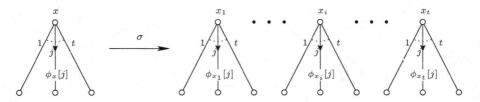

Fig. 3. The port-label transformation.

i.e., $p(i,j)$ is the ith element of the sequence $\mu_1^{-1}(\phi_x[j]) \in \{1,\ldots,d/t\}^t$ where $\phi_x[j]$ is the label of the port of x leading to its jth child. The function σ is then defined by

$$\sigma(\phi_x)[i] = (\phi_{x_i}[1],\ldots,\phi_{x_i}[t]) = (\mu_2(1,p(i,1)),\mu_2(2,p(i,2)),\ldots,\mu_2(t,p(i,t))). \tag{3}$$

In other words, by μ_1, the port-label $\phi_x[j] \in \left\{1,\ldots,(d/t)^t\right\}$ is decomposed into a sequence of t integers in $\{1,\ldots,d/t\}$, namely $p(1,j),\ldots,p(t,j)$. By μ_2, the pair $(j,p(i,j))$ is assigned to $\phi_{x_i}[j]$ (see also Fig. 3).

Observation 1. For any j, $\phi_x[j]$ can be obtained from the sequences of port-labels $\phi_{x_1}[j],\ldots,\phi_{x_t}[j]$ by the formula:

$$\phi_x[j] = \mu_1\Big(\mu_2^{-1}(\phi_{x_1}[j])[2],\ldots,\mu_2^{-1}(\phi_{x_t}[j])[2]\Big). \tag{4}$$

As for any sequence, the "[2]" stands here to designate the second element of the pair $\mu_2^{-1}(\phi_{x_i}[j]) \in \{1,\ldots,t\} \times \{1,\ldots,d/t\}$.

Observation 2. For any i, $\phi_{x_i}[j] \neq \phi_{x_i}[j']$ for all $j \neq j'$. Indeed, from Equation 3, $\phi_{x_i}[j] = \phi_{x_i}[j']$ implies $\mu_2(j,p(i,j)) = \mu_2(j',p(i,j'))$, that is $j = j'$.

The second observation shows that, for every i, $\phi_{x_i} \in \mathcal{S}(d)$. Therefore $\sigma(\phi_x) \in \mathcal{S}(d)^t$ as required. The first observation allows us now to define R'. Given $T' \in \mathcal{T}'$, consider two labels $L'(x) = (\ell_0,\ell_1,\ldots,\ell_t)$ and $L'(y) = (m_0,m_1,\ldots,m_t)$ where x is an ancestor of y, and y belongs to $\overline{T}'$. We define R' such that

$$R'(L'(x),L'(y)) = \mu_1\left(\mu_2^{-1}(R(\ell_1,m_1))[2],\ldots,\mu_2^{-1}(R(\ell_t,m_t))[2]\right),$$

that is Equation 4 in which each term $R(\ell_i,m_i)$ replaces $\phi_{x_i}[j]$. Let T be the tree associated to T', and let $x_1,\ldots,x_t$ and $y_1,\ldots,y_t$ be the nodes of T associated to nodes x and y of T', respectively. Since x is an ancestor of y in $\overline{T}'$, x_i is an ancestor of y_i in $\overline{T}$. Moreover, if the path from x to y goes through, say, the jth child of x in $\overline{T}'$, then, for every i, the path from x_i to y_i goes through the jth child of x_i in $\overline{T}$. In other words, in the tree $\overline{T}$, $R(L(x_i),L(y_i))$ returns $\phi_{x_i}[j]$ for all i. Since, for every i, ℓ_i and m_i are respectively the labels $L(x_i)$ and $L(y_i)$, it follows that

$$R'(L'(x),L'(y)) = \mu_1\left(\mu_2^{-1}(\phi_{x_1}[j])[2],\ldots,\mu_2^{-1}(\phi_{x_t}[j])[2]\right).$$

Therefore $R'(L'(x), L'(y)) = \phi_x[j]$, as desired. This completes the construction of the relaxed forward labeling in $\mathcal{T}'$, and hence the proof of the lemma.   ∎

Let $\alpha = 1/(t+1)$ and $\gamma(d) = (d/t)^t$. For every $i \geq 0$, let $\gamma^{(i)}$ be the ith iterate of the function γ, that is $\gamma^{(0)} = \mathrm{Id}$, and $\gamma^{(i)} = \gamma \circ \gamma^{(i-1)}$ for every $i \geq 0$.

Lemma 4. *For all $d > t \geq 2$ with d power of t, $|\mathcal{L}| \geq \prod_{i=1}^{h} \gamma^{(i)}(d)^{\alpha^i}$.*

Proof. From Lemma 3, $|\mathcal{L}| = f(h, d) \geq (|\mathcal{S}(d)| \cdot f(h-1, \gamma(d)))^\alpha$. By Lemma 2, $|\mathcal{S}(d)| \geq \gamma(d)$, thus $f(h, d) \geq (\gamma(d) \cdot f(h-1, \gamma(d)))^\alpha$. Let $d = t^k$ with $k \geq 2$. For every $t \geq 2$, $\gamma(d) = t^{(k-1)t} > t$ and t divides $\gamma(d)$. Similarly, for all $i > 0$, we have $\gamma^{(i)}(d) > t$ and t divides $\gamma^{(i)}(d)$. So, one can reapply Lemma 3, and we obtain,

$$
\begin{aligned}
f(h, d) &\geq (\gamma(d) \cdot f(h-1, \gamma(d)))^\alpha \geq \gamma(d)^\alpha \cdot \left(\gamma^{(2)}(d) \cdot f(h-2, \gamma^{(2)}(d))\right)^{\alpha^2} \\
&\geq \gamma(d)^\alpha \cdot \gamma^{(2)}(d)^{\alpha^2} \cdots \gamma^{(h)}(d)^{\alpha^h} \cdot f(0, \gamma^{(h)}(d))^{\alpha^h}
\end{aligned}
$$

∎

Lemma 5. *For all $d > t \geq 2$ with d power of t, $\log|\mathcal{L}| \geq (1 - e^{-h/t}) \cdot t \cdot \log(d/t^2)$.*

Proof. We have $d \geq t^2$ and $t \geq 2$. Thus for every $i > 0$ we obtain

$$
\gamma^{(i)}(d) = \frac{d^{t^i}}{t^{t+t^2+\ldots+t^i}} = \frac{d^{t^i}}{t^{t(t^i-1)/(t-1)}} \geq \frac{d^{t^i}}{t^{2t^i}} = \left(\frac{d}{t^2}\right)^{t^i} .
$$

From Lemma 4, we get

$$
\log|\mathcal{L}| \geq \sum_{i=1}^{h} \alpha^i \log\left(\gamma^{(i)}(d)\right) \geq \sum_{i=1}^{h} (\alpha t)^i \log\left(\frac{d}{t^2}\right) .
$$

Now, $\sum_{i=1}^{h}(\alpha t)^i = t\left(1 - \left(\frac{t}{t+1}\right)^h\right)$. Since $\left(\frac{t}{t+1}\right)^h \to e^{-h/t}$ as $t \to +\infty$, the result follows.   ∎

To complete the proof of Theorem 1, it just remains to set suitable values for d, h, and t. This is the goal of this last lemma:

Lemma 6. *For n large enough, for $d = t^h$ and $t = h = \lfloor (\log\sqrt{n})/\log\log\sqrt{n} \rfloor$, every tree in $\mathcal{T}$ has at most n nodes, and $\log|\mathcal{L}| > (\log n)^2/(7\log\log n)$.*

Proof. Observe that for every $x \geq 2$, $(x/\log x)^{x/\log x} \leq 2^x$. So, for $x = \log\sqrt{n}$, we have

$$
t^h \leq \left(\frac{x}{\log x}\right)^{x/\log x} \leq 2^x = \sqrt{n} .
$$

Thus the number of nodes of any tree of $\mathcal{T}$ is

$$(d-1)\left(\frac{t^h-1}{t-1}\right)+2 \ \leqslant\ d\cdot t^h\ \leqslant\ n.$$

From Lemma 5,

$$\log|\mathcal{L}| \ \geqslant\ c\cdot t\cdot\log\left(d/t^2\right)\ \geqslant\ c\cdot t\cdot\log d - O(\log(t^t))\ \geqslant\ c\cdot t^2\cdot\log t - O(\log n),$$

where $c = 1 - e^{-1} > 4/7$. Therefore,

$$\log|\mathcal{L}| \ \geqslant\ c\cdot\left(\frac{\log\sqrt{n}}{\log\log\sqrt{n}}\right)^2\cdot\log\left(\frac{\log\sqrt{n}}{\log\log\sqrt{n}}\right) - O(\log n)$$

$$\geqslant\ \frac{c}{4}\cdot\frac{(\log n)^2}{\log\log n} - O(\log n)\ >\ \frac{(\log n)^2}{7\log\log n}\ .$$

This completes the proof of Theorem 1.   ∎

4   Conclusion

In this paper, we have shown that, in the fixed-port model, shortest path routing in trees has a memory requirement of at least $\Omega(\log^2 n/\log\log n)$, and this bound is tight. Beside the significance of the result in regard with previous upper bounds for routing in trees and in arbitrary networks, this result must also be compared to the best upper bound on the memory requirement for shortest path routing in graphs of bounded treewidth. Up to our knowledge, the best result is due to [12] where it is shown that there exists a shortest path routing scheme for graphs on treewidth k using addresses and local data structures of size $O(k\log^2 n)$. For tree (treewidth 1), this bound is not tight, by a factor $\log\log$. We hence leave as an open problem the question of whether one can design (under the fixed-port model) routing schemes with memory requirement $O(k\log^2 n/\log\log n)$ for graphs of treewidth k. For instance, can we route on shortest paths in outer-planar graphs (treewidth 2) using addresses and local data structures of $O(\log^2 n/\log\log n)$ bits only?

References

1. Baruch Awerbuch and David Peleg. Sparse partitions. In 31th *Annual IEEE Symposium on Foundations of Computer Science (FOCS)*, pages 503–513. IEEE Computer Society Press, October 1990.
2. Baruch Awerbuch and David Peleg. Routing with polynomial communication-space trade-off. *SIAM Journal on Discrete Mathematics*, 5(2):151–162, May 1992.
3. Lenore J. Cowen. Compact routing with minimum stretch. *Journal of Algorithms*, 38:170–183, 2001.

4. Tamar Eilam, Cyril Gavoille, and David Peleg. Compact routing schemes with low stretch factor. In 17th *Annual ACM Symposium on Principles of Distributed Computing (PODC)*, pages 11–20. ACM PRESS, August 1998.
5. Paul Erdös. Extremal problems in graph theory. In *Publ. House Cszechoslovak Acad. Sci., Prague*, pages 29–36, 1964.
6. Pierre Fraigniaud and Cyril Gavoille. Routing in trees. In 28th *International Colloquium on Automata, Languages and Programming (ICALP)*, volume 2076 of Lecture Notes in Computer Science, pages 757–772. Springer, July 2001.
7. Cyril Gavoille. A survey on interval routing. *Theoretical Computer Science*, 245(2):217–253, 2000.
8. Cyril Gavoille. Routing in distributed networks: Overview and open problems. *ACM SIGACT News - Distributed Computing Column*, 32(1):36–52, March 2001.
9. Cyril Gavoille and David Peleg. Compact and localized distributed data structures. Research Report RR-1261-01, LaBRI, University of Bordeaux, 351, cours de la Libération, 33405 Talence Cedex, France, August 2001. To appear in Journal of Distributed Computing.
10. Cyril Gavoille, David Peleg, Stéphane Pérennes, and Ran Raz. Distance labeling in graphs. In 12th *Symposium on Discrete Algorithms (SODA)*, pages 210–219. ACM-SIAM, January 2001.
11. David Peleg. Informative labeling schemes for graphs. In 25th *International Symposium on Mathematical Foundations of Computer Science (MFCS)*, volume 1893 of Lecture Notes in Computer Science, pages 579–588. Springer, August 2000.
12. David Peleg. Proximity-preserving labeling schemes. *Journal of Graph Theory*, 33:167–176, 2000.
13. David Peleg and Eli Upfal. A trade-off between space and efficiency for routing tables. *Journal of the ACM*, 36(3):510–530, July 1989.
14. Mikkel Thorup and Uri Zwick. Approximate distance oracles. In 33rd *Annual ACM Symposium on Theory of Computing (STOC)*, pages 183–192, Hersonissos, Crete, Greece, July 2001.
15. Mikkel Thorup and Uri Zwick. Compact routing schemes. In 13th *Annual ACM Symposium on Parallel Algorithms and Architectures (SPAA)*, pages 1–10, Hersonissos, Crete, Greece, July 2001. ACM PRESS.

Labeling Schemes for Dynamic Tree Networks

(Extended Abstract)

Amos Korman, David Peleg*, and Yoav Rodeh

Department of Computer Science and Applied Mathematics,
The Weizmann Institute of Science, Rehovot 76100, Israel
{pandit,peleg,yrodeh}@wisdom.weizmann.ac.il

Abstract. The paper concerns distributed distance labeling schemes on dynamic trees. The paper first presents a labeling scheme for distances in the dynamic tree model, with amortized message complexity $O(\log^2 n)$ per operation, where n is the size of the tree at the time the operation takes place. The protocol maintains $O(\log^2 n)$ bit labels, where n is the current tree size. This label size is known to be optimal even in the static scenario. A more general labeling scheme is then introduced for the dynamic tree model, based on extending an existing *static* tree labeling scheme to the dynamic setting. The approach fits a number of natural tree functions, such as distance, separation level and flow. The main resulting scheme incurs an overhead of a $O(\log n)$ multiplicative factor in both the label size and amortized message complexity in the case of *dynamically growing* trees (with no vertex deletions). In the fully-dynamic model the scheme incurs also an increased *additive* overhead in amortized communication, of $O(\log^2 n)$ messages per operation.

1 Introduction

Network representations play an extensive role in the areas of distributed computing and communication networks. Their goal is to cheaply store useful information about the network and make it readily and conveniently accessible. This is particularly significant when the network is large and geographically dispersed, and information about its structure must be accessed from various local points in it.

The current paper deals with a network representation method based on assigning *informative labels* to the vertices of the network. Whereas most traditional network representations rely on maintaining a *global* description of the network, the informative localized labeling schemes studied here are based on associating a label with each vertex, allowing us to infer information about any two vertices *directly* from their labels, without using *any* additional information sources. Hence the entire representation is based on the set of labels alone. Naturally, our focus is on labeling schemes using relatively *short* labels (say, of length poly-logarithmic in n). Labeling schemes of this type have been developed for a

---

* Supported in part by a grant from the Israel Science Foundation.

H. Alt and A. Ferreira (Eds.): STACS 2002, LNCS 2285, pp. 76–87, 2002.
© Springer-Verlag Berlin Heidelberg 2002

variety of information types, including vertex adjacency [9], distance [14,13,8,5, 6,11], tree routing [4,17], flow and connectivity [12], tree ancestry [1,10,3], and various other tree functions, such as center, least common ancestor, separation level, and Steiner weight of a given subset of vertices [15]. See [7] for a survey.

These studies provide a reasonable understanding of informative localized labeling schemes and their basic combinatorial properties for *static* (fixed topology) networks. However, they are somewhat limited when it comes to handling applications such as distributed systems and communication networks. First, in these application areas the typical setting is that of a dynamic network, whose topology undergoes repeated changes. Therefore, for a representation scheme to be practically useful, it should be capable of reflecting online the current up-to-date picture in a dynamic setting. Moreover, the schemes described in the above cited papers are *centralized*, in the sense that they are based on a sequential algorithm which given a description of the entire graph outputs the entire set of vertex labels. Hence while the resulting labels reflect local knowledge and can be *used* locally, their *generation* process is still centralized and global.

Consequently, our interest in the current paper is in the development of localized informative labeling schemes in the *dynamic distributed* setting, using distributed online protocols. The current paper makes a first step in that direction, by studying informative labeling schemes for a *dynamic tree network*. The model is that of a network with a tree topology, where the leaves of the tree can be removed from the network and new leaves can be added at any time.

Our main results are as follows. Throughout the paper, denote by n the current number of vertices in the dynamic tree, at any given time. We first present a labeling scheme for distances in the dynamic tree model, with amortized message complexity $O(\log^2 n)$ per operation. The protocol maintains $O(\log^2 n)$ bit labels. This label size is optimal even in the static scenario [8].

It is worth noting an interesting point concerning the amortization method. Most of the literature on dynamic trees usually assumes a static graph spanned by the dynamic tree, and n is taken to be the size of the underlying graph. In contrast, our approach allows for dispensing with the underlying graph and basing the analysis on the size of the tree alone, at the expense of slightly more involved definitions for the amortized complexity measures.

We then introduce a rather general labeling scheme for the dynamic tree model, based on extending an existing *static* tree labeling scheme to the dynamic setting. The approach fits a number of natural tree functions, such as distance [14], separation level [15] and flow [12]. Intuitively, the resulting scheme incurs an overhead of an $O(\log n)$ multiplicative factor in both the label size and amortized message complexity in the case of *dynamically growing* trees (where no vertex deletions are allowed). In the special case where an upper bound on n is known in advance, this method can yield a different tradeoff, with an $O(\log^2 n / \log \log n)$ multiplicative overhead factor on the label size but an overhead factor of only $O(\log n / \log \log n)$ on the amortized message complexity. In the general model where vertex deletions are allowed as well, the scheme incurs also an increased *additive* overhead in amortized communication, of $O(\log^2 n)$ messages per operation.

2   Preliminaries

Our communication network model is restricted to tree topologies. The network is assumed to dynamically change via vertex additions and deletions. It is assumed that the *root* of the tree is never deleted. The following events may occur:
1. A new vertex u is added as a child of an existing vertex v. Subsequently, v is informed of this event, and assigns a unique *child-number* to u, in the sense that no currently existing child of v has the same child-number.
2. A tree leaf is deleted. In this case, the leaf's parent is informed of this event.

To implement child-numbers, each vertex keeps a counter that is increased by one for every new child. If vertex deletions are allowed, this counter may become quite large, although the size of the tree is small. This problem will be handled in detail later.

We use the following notation. For a vertex u, denote by $\omega(u)$ the *weight* of u, representing the number of vertices in u's subtree. For a nonroot vertex u, denote by $p(u)$ its parent in the tree. The *ancestry* relation in the tree is defined as the transitive closure of the parenthood relation. In particular, a vertex is its own ancestor.

A *labeling scheme* $\pi = \langle \mathcal{M}_\pi, \mathcal{D}_\pi \rangle$ for a binary function F on pairs of vertices of a tree is composed of the following components:
1. A *marker* algorithm $\mathcal{M}_\pi$ that given a tree, assigns labels to its vertices.
2. A polynomial time *decoder* algorithm $\mathcal{D}_\pi$ that given the labels $l(u)$ and $l(v)$ of two vertices u and v, outputs $F(u,v)$.

In this paper we are interested in distributed networks where each vertex in the tree is a processor. This does not affect the definition of the decoder algorithm of the labeling scheme, but the marker algorithm changes into a distributed marker protocol.

Let us first consider static networks, where no changes in the topology of the network are allowed. For these networks we define *static* labeling schemes, where the marker protocol $\mathcal{M}$ is initiated at the root of a tree network and assigns static labels to all the vertices once and for all. We use the following complexity measures to evaluate a static labeling scheme $\pi = \langle \mathcal{M}, \mathcal{D} \rangle$.
1. *Label Size*, $\mathcal{LS}(\mathcal{M}, n)$: the maximum size of a label assigned by $\mathcal{M}$ to a vertex on any n-vertex tree.
2. *Message Complexity*, $\mathcal{MC}(\mathcal{M}, n)$: the maximum number of messages sent by $\mathcal{M}$ during the labeling process on any n-vertex tree.
3. *Bit Complexity*, $\mathcal{BC}(\mathcal{M}, n)$: the maximum number of bits sent by $\mathcal{M}$ during the labeling process on any n-vertex tree.

Example 1. A possible static labeling scheme STATDFS for the ancestry relation is to simply perform a DFS starting at the root, assigning each vertex the interval $[a, b]$ where a is its DFS number and b is the largest DFS number given to any of its descendents. As described in [8], using the corresponding decoder, this is a correct labeling scheme for the ancestry relation. Clearly, $\mathcal{MC}(\textsc{StatDFS}, n) = O(n)$, $\mathcal{BC}(\textsc{StatDFS}, n) = O(n \log n)$ and $\mathcal{LS}(\textsc{StatDFS}, n) = O(\log n)$.

A labeling scheme for routing is presented in [4]. The scheme of [4] is designed as a sequential algorithm, but examining the details reveals that this algorithm can be easily transformed into a distributed protocol, and so we get a static labeling scheme for routing with label size and communication complexity similar to those of the $\textsc{StatDFS}$ static labeling scheme.

The second setting considered is that of dynamically growing tree networks, where only vertex additions are allowed. For these networks we define *semi-dynamic* labeling schemes which involve a marker protocol $\mathcal{M}$ which is activated after every change in the network topology. The protocol $\mathcal{M}$ maintains the labels of all vertices in the tree so that the corresponding decoder algorithm will work correctly. We assume that the topological changes occur sequentially. Moreover, we assume that the changes are sufficiently spaced, so that the protocol has enough time to complete its operation in response to a given topological change before the occurrence of the next change. In the full paper we will quantify these assumptions and discuss the performance of our schemes in more general settings.

For a semi-dynamic labeling scheme $\pi = \langle \mathcal{M}, \mathcal{D} \rangle$, the definition of $\mathcal{LS}(\mathcal{M}, n)$ remains as for static schemes, and we are interested in the following measures.
1. $\mathcal{MC}(\mathcal{M}, n_0, n_+)$: the maximum number of messages sent by $\mathcal{M}$, where n_0 is the initial tree size (when $\mathcal{M}$ is invoked) and n_+ is the number of additions made to the tree.
2. $\mathcal{BC}(\mathcal{M}, n_0, n_+)$: the maximum number of bits sent by $\mathcal{M}$, for n_0, n_+ as above.

Finally, we consider dynamic networks as defined above, where both vertex additions and deletions are allowed. For more explicit time references, we use the notation $\bar{n} = (n_1, n_2, \ldots, n_t)$, where n_i is the size of the tree after the ith topological change. For simplicity, we assume $n_1 = 1$ unless stated otherwise. We also denote $n_f = n_0 + n_+$. The definition of $\mathcal{LS}(\mathcal{M}, n)$ remains as before, and our complexity measures are modified as follows.
1. $\mathcal{MC}(\mathcal{M}, \bar{n})$: the maximum number of messages sent by $\mathcal{M}$.
2. $\mathcal{BC}(\mathcal{M}, \bar{n})$: the maximum number of bits sent by $\mathcal{M}$.

3 A Dynamic Labeling Scheme for Distance

This section presents an efficient dynamic distance labeling scheme called DL. To this aim, we first present (in Subsect. 3.1) a static distance labeling scheme, called $\textsc{StatDL}$, we then describe (in Subsect. 3.3) a semi-dynamic distance labeling scheme based on $\textsc{StatDL}$, called $\textsc{SemDL}$, and finally we show (in Subsect. 3.4) how to use $\textsc{SemDL}$ for constructing the dynamic labeling scheme DL.

3.1 The $\textsc{StatDL}$ Static Labeling Scheme

A basic ingredient in both Protocols $\textsc{StatDL}$ and $\textsc{SemDL}$, is that each vertex u holds a pointer $\mu(u)$ to one of its children, henceforth referred to as the *marked child* of u. Denote $D(u) = \{v \mid v$ is a child of u, $v \neq \mu(u)\}$. We refer to $v \in D(u)$ as a *common child* of u. Mark all edges $(u, \mu(u))$ in the tree. At any given time, the pointers are required to satisfy the following *balance property*: for every vertex u in the tree, the path from the root to u contains at most $O(\log n)$ unmarked edges. This property holds in both STATDL and SEMDL.

The label structure: A simple representation of the path from the root to some vertex u is based on consecutively listing all the child-numbers along the path. Towards compacting this representation, we now do the following: for every marked edge along the path, replace the corresponding child-number in the list by a $\star$. The resulting list is called the *full label* of u, and is denoted by $L^{\text{full}}(u)$. Given two such labels $L^{\text{full}}(u)$ and $L^{\text{full}}(v)$, describing the path from the root to two vertices u and v, we can calculate the distance between u and v by finding the first place where the lists differ.

As the full label $L^{\text{full}}(u)$ is of length $depth(u)$, we now compact it by replacing any maximal sublist of d consecutive stars with the condensed notation $\star^d$. The resulting list is called the *compact label* of u, denoted $L^{\text{comp}}(u)$. By the balance property of the pointers in the tree, the compact label of any vertex in an n-vertex tree is of length at most $O(\log n)$.

Since we do not allow deletions in our network, every child-number used is in the range $\{1 \ldots, n\}$, and therefore can be encoded using at most $O(\log n)$ bits. Clearly, elements of the form $\star^d$ in the list can also be encoded using $O(\log n)$ bits, since $d \leq n$. Therefore, since the list consists of at most $O(\log n)$ elements, the total label size is $O(\log^2 n)$. Note that from the compact label we can easily extract its full form. Therefore we will always assume the vertices store their compact labels, but may freely use their full label.

The decoder algorithm: Given two full labels, $L^{\text{full}}(u)$ and $L^{\text{full}}(v)$, of vertices u and v, the decoder algorithm calculates the distance between the two vertices u and v as follows.

1. Find the largest prefix z that is common to both $L^{\text{full}}(u)$ and $L^{\text{full}}(v)$.
2. Return $|L^{\text{full}}(v)| + |L^{\text{full}}(u)| - 2 \cdot |z|$.

To see that the decoder operates correctly, note that the prefix z is the full label of u's and v's least common ancestor w, i.e., $z = L^{\text{full}}(w)$. The sought distance satisfies $d(u, v) = d(u, w) + d(w, v)$. But $d(u, w)$ is simply the length of $L^{\text{full}}(u)$ minus that of $L^{\text{full}}(w)$. The same method is used to calculate $d(w, v)$, and the claim follows.

The marker protocol: The marker protocol of the static scheme STATDL is initiated at the root of a tree of size n, where the vertices may or may not already have child-numbers. The protocol operates as follows.

1. The root broadcasts a message requesting each vertex to assign its children new child-numbers. Each vertex orders its children arbitrarily, and assigns them the child-numbers $1, 2, \ldots$, and so on.
2. The root broadcasts a message requesting all vertices to (recursively) calculate their weight. This is done by a simple convergecast protocol (cf. [16]), whereby each vertex collects the weights of all of its children, sums their

weights and adds one, thereby obtaining a weight equal to the number of vertices in its subtree.

3. Each vertex u sets its pointer $\mu(u)$ to point at its heaviest child v.
4. The distance labels are now assigned recursively starting at the root r. $L^{\text{full}}(r)$ is set to the empty list Λ. Once a vertex u computes its own compact label $L^{\text{comp}}(u)$, it sends it to each of its children. For a child v of u with child-number s, the full label of v is defined as u's full label plus s appended at the end, or $\star$ if u is pointing at v.

Clearly, the labels assigned by this protocol are of the form defined in Subsection 3.1. Also, it is easy to verify that the new child-numbers are of size at most $\log n$ bits. Since every common child v of a vertex u has $\omega(v) \leq \omega(u)/2$, the balance property is also clearly satisfied. Therefore we have the following. (Throughout, most proofs are omitted from this extended abstract.)

Proposition 1. *(1)* $\mathcal{LS}(\text{STATDL}, n) = O(\log^2 n)$, *(2)* $\mathcal{MC}(\text{STATDL}, n) = O(n \log n)$, *(3)* $\mathcal{BC}(\text{STATDL}, n) = O(n \log^2 n)$.

3.2 Dynamic Distributed Tools

Protocol SEMDL attempts to mimic Protocol STATDL and assign the same labels. Since the vertices cannot maintain their exact weight in a dynamic network without incurring a considerable overhead in communication, we will only keep an estimate of the weight, using Protocol WW. As a result, a vertex does not exactly know which of its children is the heaviest. Therefore, we require only that the common children be small. Note that it is necessary to minimize the number of times a vertex pointer changes, since each such change results in extra communication. Pointer changes are handled by the "heavy child" Protocol HC.

Protocol WW: Using the method of [2], one can easily derive a *weight estimation* protocol, hereafter named RWW, for the dynamically growing tree model, with the following property.

Property RWW: For constant $\delta > 0$, Protocol RWW enables the root to maintain an approximation $\tilde{n}$ to n, such that at any time, $n \leq \tilde{n} \leq \delta n$.

Moreover, we observe that by letting each vertex monitor the protocol messages that pass through it, one can achieve a slightly modified protocol, named WW, for which the following stronger property holds.

Property WW: For constant $\delta > 0$, Protocol WW enables every vertex v to maintain a weight approximation $\tilde{\omega}(v)$ s.t. at any time, $\omega(v) \leq \tilde{\omega}(v) \leq \delta \cdot \omega(v)$.

This results from a slight change in the third item of Lemma 4.2 of [2], namely, replacing U by $\omega(v)$. We note that this generalization does not work in the case that vertex deletions are allowed. Similar to the proof in [2], we have

Lemma 1. *(1)* $\mathcal{MC}(\text{WW}, n_0, n_+) = O(n_f \log^2 n_f)$,   *(2)* $\mathcal{BC}(\text{WW}, n_0, n_+) = O(n_f \log^2 n_f \log \log n_f)$.

Protocol HC: Protocol HC uses Protocol WW with $\delta = 5/4$. We further assume that each vertex knows not only its own estimated weight $\tilde{\omega}(v)$ but also all of its children's estimated weights. This requirement will at most multiply the communication complexity of Protocol WW by 2. For each vertex u, define the *weight ratios* $\varrho(u) = \omega(u)/\omega(p(u))$ and $\tilde{\varrho}(u) = \tilde{\omega}(u)/\tilde{\omega}(p(u))$. We have

$$\varrho(u)/\delta \ \leq \ \tilde{\varrho}(u) \ \leq \ \delta \cdot \varrho(u). \tag{$*$}$$

Each vertex can calculate $\tilde{\varrho}(u)$ for all of its children. Protocol HC is initiated on a tree where vertices already have pointers, and operates as follows. Whenever a vertex v observes that one of its common children u satisfies $\tilde{\varrho}(u) > 3/4$, it changes its pointer $\mu(v)$ to u. This event is called a *shuffle* γ of the vertex v, and its *weight* is defined as $\omega(\gamma) = \omega(v)$.

Our choice of δ and Eq. $(*)$ imply that if a vertex u has $\tilde{\varrho}(u) > 3/4$ then $\varrho(u) > 1/2 + c_1$ for some $c_1 > 0$, and hence at most one child u of a vertex v can have $\tilde{\varrho}(u) > 3/4$. Also, for every vertex v, and every common child u of v, $\varrho(u) < 1 - c_2$ for some $c_2 > 0$. This yields the following.

Claim. [**Balance Property**] For every vertex v in an n-vertex tree, the path from the root to v has at most $O(\log_{1-c_2} n) = O(\log n)$ unmarked edges.

It follows that the balance property of the pointers is satisfied at all times. As mentioned, the communication complexity of Protocol HC is at most twice that of Protocol WW.

3.3   Protocol SEMDL

Protocol SEMDL deals with the semi-dynamic model allowing only vertex additions. The first step of Protocol SEMDL is to run Protocol STATDL. It then initiates Protocol HC and monitors its behavior.

1. If a new vertex v is added as a child of a vertex u with child-number s, then v's label is defined as u's full label plus s appended at the end, or $\star$ if u is pointing at v.
2. If a shuffle γ of a vertex v occurs in Protocol HC, and v shifts its pointer from u^{old} to u^{new}, whose child numbers are s^{old} and s^{new} respectively, then v initiates the following changes:
   a) It instructs all vertices in the subtree of u^{new} to change the entry in position $depth(v)$ of their full label from s^{new} to $\star$.
   b) It instructs all vertices in the subtree of u^{old} to change the entry in position $depth(v)$ of their full label from $\star$ to s^{old}.
   As these changes are applied to the full label, each vertex needs to recompact its label after performing the change.

Clearly, Protocol SEMDL maintains the labels as defined in Subsection 3.1, and therefore since Protocol HC maintains the balance property, we have that $\mathcal{LS}(\text{SEMDL}, n) = O(\log^2 n)$.

Complexity: The operation of step 1 of the protocol involves sending one message of size $O(\log^2 n)$, and the operation of step 2 requires $O(\omega(v)) = O(\omega(\gamma))$

messages of size $O(\log n)$. Since the current tree size satisfies $n < n_f$ at all times, the communication complexity of Protocol SEMDL is bounded above by
$$\mathcal{MC}(\text{SEMDL}, n_0, n_+) = O(n_f \log^2 n_f) + O(\sum_{\gamma \in \Gamma} \omega(\gamma)),$$
$$\mathcal{BC}(\text{SEMDL}, n_0, n_+) = O(n_f \log^2 n_f \log \log n_f) + O(\sum_{\gamma \in \Gamma} \omega(\gamma) \log n_f).$$

Lemma 2. *If Protocol HC is initiated on a tree of size n_0 with pointers satisfying the balance property, then $\sum_{\gamma \in \Gamma} \omega(\gamma) = O(n_f \log n_f)$.*

Proposition 2. *(1) $\mathcal{LS}(\text{SEMDL}, n) = O(\log^2 n)$, (2) $\mathcal{MC}(\text{SEMDL}, n_0, n_+) = O(n_f \log^2 n_f)$, (3) $\mathcal{BC}(\text{SEMDL}, n_0, n_+) = O(n_f \log^2 n_f \log \log n_f)$.*

3.4 The DL **Dynamic Distance Labeling Scheme**

We now present our algorithm for the fully dynamic case (allowing vertex deletions). The general idea is to run Protocol SEMDL while ignoring deletions, i.e., treating deleted vertices as if they were never deleted. However, at some point or another in the execution, deletions must be accounted for, since if the tree size falls significantly, then the old labels (and child numbers) which were small with respect to the original tree size, might become too large. We therefore run, in parallel to Protocol SEMDL, a protocol for estimating the number of topological changes in the tree. Every $\Theta(n)$ topological changes, we restart Protocol SEMDL again. Clearly, the labels will be as defined in Subsection 3.1, and the only remaining question concerns the resulting communication complexity.

Denote by τ the number of topological changes made to the tree during the execution. Fix $\delta = 5/4$. We use the "change watch" Protocol CW, which is an instance of the protocol of [2] in which the root maintains an estimate $\tilde{\tau}$ of τ satisfying $\tau \le \tilde{\tau} \le \delta \cdot \tau$. As stated in [2],
1. $\mathcal{MC}(\text{CW}, \bar{n}) = O((n_1 + t) \log^2(n_1 + t))$,
2. $\mathcal{BC}(\text{CW}, \bar{n}) = O((n_1 + t) \log^2(n_1 + t) \log \log(n_1 + t))$.
Here n_1 is not necessarily 1, since the protocol is initiated on trees of larger size.

Protocol DL.

1. The root calculates and records the initial number of vertices in the tree, n_0.
2. Protocols SEMDL and CW are started. If a vertex v is deleted during this run, then its parent u simulates its behavior as if v and all its descendents were never deleted. In particular:
   a) For Protocols WW and HC the simulation requires no extra messages.
   b) For Protocol SEMDL, u will simply not pass to v any label changing messages resulting from a shuffle.
   c) Protocol CW is in fact designed to account for vertex deletions.
3. When the topological changes estimate becomes $\tilde{\tau} \ge n_0/4$, return to Step 1.

Theorem 1. *(1) $\mathcal{LS}(\text{DL}, n) = O(\log^2 n)$, (2) $\mathcal{MC}(\text{DL}, \bar{n}) = O(\sum_i \log^2 n_i)$, (3) $\mathcal{BC}(\text{DL}, \bar{n}) = O(\sum_i \log^2 n_i \log \log n_i)$.*

4   A General Dynamic Labeling Scheme

In this section we present a dynamic labeling scheme for a wide variety of binary functions on trees. This scheme takes an existing *static* labeling scheme for the same function and uses it as a subroutine. Specifically, we proceed as follows. Assuming we are given a static labeling scheme $\pi = \langle \mathcal{M}_\pi, \mathcal{D}_\pi \rangle$ for the function F, we first describe a semi-dynamic labeling scheme SEMGL for F, and later show how it can be transformed into a dynamic labeling scheme GL. We assume that the message complexity of the static labeling scheme π is polynomial[1] in the size of the tree. We then extend this scheme into a fully-dynamic scheme.

Example 2. If we take the STATDFS static labeling scheme for the ancestry relation, presented in Example 1, if n is not known (Part 1), we can get a semi-dynamic labeling scheme SEMDFS with $\mathcal{LS}(\text{SEMDFS}, n) = \log^2 n$, and $\mathcal{MC}(\text{SEMDFS}, 1, n_+) = n_+ \log n_+$. We can also get a dynamic labeling scheme DFS (using Part 3), with $\mathcal{LS}(\text{DFS}, n) = \log^2 n$, and $\mathcal{MC}(\text{DFS}, \bar{n}) = O(\sum_i \log n_i) + O(\sum_i \log^2 n_i) = O(\sum_i \log^2 n_i)$.

As mentioned in Example 1, there is static labeling scheme for routing with the same label size and communication complexity as that of STATDFS, and so we can create a semi-dynamic and dynamic routing labeling schemes with same label size and communication complexity as the SEMDFS and DFS schemes respectively.

The family of functions to which our method is applicable can be characterized as all the functions satisfying the following two conditions:

(C1) For two vertices u and v in the tree, $F(u, v)$ depends only on the path between them (including the child-numbers of the edges in the path).

(C2) For three vertices u, w and v in the tree, such that w is on the path between u and v, $F(u, v)$ can be calculated in polynomial time from $F(u, w)$ and $F(w, v)$.

In particular, the ancestry relation and the labeling for routing are functions our method can be applied to. We further assume, for simplicity of presentation, that F is symmetric,i.e., $F(u, v) = F(v, u)$. A slight change to the suggested protocols handles the more general case, without affecting the asymptotic complexity.

We assume that the labels given by the static labeling scheme are unique; i.e., given a tree, no two vertices get the same static label. Given any static labeling scheme, this additional requirement can be ensured at an extra cost of at most n to $\mathcal{MC}(n)$ and $\log n$ to $\mathcal{LS}(n)$, and therefore will not affect the asymptotic complexities of our scheme.

---

[1]   We actually require that the message complexity is bounded above by some function f which satisfies $f(\alpha \cdot n) = \Theta(f(n))$ for any constant $\alpha > 0$. We also require $f(a+b) \geq f(a) + f(b)$. These two requirements are satisfied by most natural relevant functions, such as $c \cdot n^\alpha \log^\beta n$, where $c > 0$, $\alpha \geq 1$ and $\beta > 0$. For simplicity, we assume $\mathcal{MC}(\cdot, n)$ itself satisfies these requirements.

4.1   Protocol SEMGL

In this description, we assume the initial tree size is 1, and consists of just the root. To handle initial tree sizes different than 1, we simply simulate Protocol SEMGL, starting from the root of the initial tree, and as if vertices of the initial tree were added one after the other. Therefore $\mathcal{MC}(\text{SEMGL}, n_0, n_+) = \mathcal{MC}(\text{SEMGL}, 1, n_0 + n_+ - 1)$. For simplicity of notation, we denote $\mathcal{MC}(\text{SEMGL}, n) = \mathcal{MC}(\text{SEMGL}, 1, n - 1)$ throughout this section.

The label structure: The label of each vertex can be represented as a 2-dimensional unbounded matrix, where each element in the matrix is either empty or a triplet $\langle l, R, s \rangle$, where l, R, s will be defined later on. In each row i, the nonempty elements are aligned to the left, followed by empty entries from the right. A row is *empty* if all its entries are. We say that a row is *full* if the number of elements in it is at least $d - 1$, where d is a positive integer fixed in advance. For a label L we use the following notation.
1. NF(L) is the first *nonfull* row of L.
2. NE(L) is the first *nonempty* row of L. (NE(L) = ∞ for an empty matrix L).
3. Least(L) is the last nonempty element on row NE(L) (undefined for empty L).

The shuffle operation: Our protocol repeatedly engages the static labeling scheme, and uses the labels it outputs to construct our dynamic label. In doing so, the protocol occasionally applies to the already labeled portion of the tree a *shuffle* operation. A *level i shuffle* of a subtree rooted at w does the following:
1. The vertex w invokes the static labeling scheme on its subtree, to give a new static label $l(v)$ to each vertex v in the subtree.
2. Mark by L the label of $z = p(w)$. Give each vertex v in the subtree the label L modified by inserting an element $\langle l, R, s \rangle$ instead of the first empty element of row i in L, where $l = l(v)$, $R = F(v, z)$, and s is w's child-number.
We define the *root* and *p-root* of this shuffle operation to be w and z, respectively.

Adding a vertex: The label of the root is the empty matrix. Suppose a vertex v is a added as a child of vertex u. Then the labeling algorithm acts as follows.

1. Let $i = \text{NF}(L(u))$.
2. Search up in the tree from u until reaching the first vertex z (including u) such that $\text{NE}(L(z)) \geq i$. Let w be z's child on the path to v. (In particular, if $z = u$ then $w = v$.)
3. Perform a level i-shuffle rooted at w. We say u is the *initiator* of this shuffle.

The decoder algorithm: For a label L, denote by $L{\uparrow}$ the label obtained from L by removing the element in position Least(L), leaving it empty. For two labels L_1 and L_2, we say that $L_1 \subseteq L_2$ if $L_2 \underbrace{{\uparrow}{\uparrow} \ldots {\uparrow}}_{i} = L_1$ for some $i \geq 0$.

The decoder procedure receives as input the labels $L(u)$ and $L(v)$ of two vertices u and v. It proceeds in one of two ways, either calculating $F(u, v)$ directly using the decoder $\mathcal{D}_\pi$ of the static scheme π or finding a label L of an intermediate vertex y such that y is on the path between u and v, and recursively calculating

$F(u, y)$ and $F(y, v)$, and then calculating $F(u, v)$ using Condition (C2) on F. The procedure operates as follows.

1. If $L(u)\!\uparrow = L(v)$ then return the R field of $\mathrm{Least}(L(v))$ as $F(u, v)$. Analogously if $L(v)\!\uparrow = L(u)$.
2. Else, if $L(v) \subseteq L(u)$, then let $L = L(u)\!\uparrow$, and proceed recursively as explained above. Analogously if $L(u) \subseteq L(v)$.
3. Else, if $L(u)\!\uparrow \not\subseteq L(v)$, then let $L = L(u)\!\uparrow$, and proceed recursively. Analogously if $L(v)\!\uparrow \not\subseteq L(u)$.
4. Else $L(u)\!\uparrow \subseteq L(v)$ and $L(v)\!\uparrow \subseteq L(u)$, and therefore $L(v)\!\uparrow = L(u)\!\frown$. Then,
   a) If the placing of element $\mathrm{Least}(L(u))$ is different than that of $\mathrm{Least}(L(v))$, then set $L = L(v)\!\uparrow$ and proceed recursively.
   b) If the placing is the same, and the s field of $\mathrm{Least}(L(u))$ and $\mathrm{Least}(L(v))$ is different, then set $L = L(v)\!\uparrow$ and proceed recursively.
   c) If the placing is the same and the s field is also the same, then denote by l_v and l_u the l fields of $\mathrm{Least}(L(v))$ and $\mathrm{Least}(L(u))$ respectively, and return $F(u, v) = \mathcal{D}_\pi(l_u, l_v)$.

We omit from this extended abstract the correctness proof of the decoder and the complexity analysis of the scheme. In the full paper we show the following. First, setting $d = 2$,

Theorem 2. *(1)* $\mathcal{LS}(\mathrm{SEMGL}, n) = O(\log n \cdot \mathcal{LS}(\pi, n))$, *(2)* $\mathcal{MC}(\mathrm{SEMGL}, n_0, n_+) = O(\log n_f \cdot \mathcal{MC}(\pi, n_f))$.

If n_f is known in advance, then setting $d = \log n_f$ we achieve a different tradeoff between the communication complexity and label size.

Theorem 3. *If n_f is known in advance, then (1)* $\mathcal{LS}(\mathrm{SEMGL}, n) = O(\frac{\log n_f \log n}{\log \log n_f} \mathcal{LS}(\pi, n))$, *(2)* $\mathcal{MC}(\mathrm{SEMGL}, n_0, n_+) = O(\frac{\log n}{\log \log n_f} \mathcal{MC}(\pi, n_f))$.

4.2   Protocol GL

Protocol GL uses Protocol SEMGL in much the same way as Protocol DL uses Protocol SEMDL. Again, the idea is to ignore deletions completely, while counting topological changes, until the number of topological changes becomes $\Theta(n)$, and then start Protocol SEMGL from scratch. In this case, the size of the tree during each sub-execution of Protocol SEMGL, is $\Theta(n_0)$, where n_0 is the size of the tree at the beginning of the sub-execution. We can therefore either set $d = 2$ for all sub-executions, or set $d = \log n_0$ with the appropriate n_0 for each sub-execution.

Protocol GL operates as follows.

1. The root calculates and records the number of vertices in the tree n_0.
2. Protocols SEMGL and CW are started, where in Protocol SEMGL, d is set to either 2 or $\log n_0$. The fact that a vertex v gets deleted does not change anything in the execution of Protocol SEMGL, including the sub-executions of Protocol π.

3. When the estimated number of topological changes becomes at least $n_0/4$, return to Step 1.

Theorem 4. *Let* GL^1 *be* GL *setting* $d = 2$ *and* GL^2 *be* GL *setting* $d = \log n_0$. *Then (1)* $\mathcal{LS}(\mathrm{GL}^1, n) = O(\log n \cdot \mathcal{LS}(\pi, n))$,   *(2)* $\mathcal{MC}(\mathrm{GL}^1, \bar{n}) = O(\sum_i \log^2 n_i)$ $+ O(\sum_i \log n_i \mathcal{MC}(\pi, n_i)/n_i)$,   *(3)* $\mathcal{LS}(\mathrm{GL}^2, n) = O(\frac{\log^2 n}{\log \log n} \mathcal{LS}(\pi, n))$, *(4)* $\mathcal{MC}(\mathrm{GL}^2, \bar{n}) = O(\sum_i \frac{\log n_i}{\log \log n_i} \mathcal{MC}(\pi, n_i)/n_i) + O(\sum_i \log^2 n_i)$.

References

1. S. Abiteboul, H. Kaplan and T. Milo. Compact labeling schemes for ancestor queries. In *Proc. 12th ACM-SIAM Symp. on Discrete Algorithms*, Jan. 2001.
2. Y. Afek, B. Awerbuch, S.A. Plotkin and M. Saks. Local management of a global resource in a communication. *J. of the ACM*, pages 1–19, 1989.
3. S. Alstrup, C. Gavoille, H. Kaplan and T. Rauhe. Identifying nearest common ancestors in a distributed environment. IT-C Technical Report 2001-6, The IT University, Copenhagen, Denmark, Aug. 2001.
4. P. Fraigniaud and C. Gavoille. Routing in trees. In *Proc. 28th Int. Colloq. on Automata, Languages & Prog.*, LNCS 2076, pages 757–772, July 2001.
5. C. Gavoille, M. Katz, N.A. Katz, C. Paul and D. Peleg. Approximate Distance Labeling Schemes. In *9th European Symp. on Algorithms*, Aug. 2001, Aarhus, Denmark, SV-LNCS 2161, 476–488.
6. C. Gavoille and C. Paul. Split decomposition and distance labelling: an optimal scheme for distance hereditary graphs. In *Proc. European Conf. on Combinatorics, Graph Theory and Applications*, Sept. 2001.
7. C. Gavoille and D. Peleg. Compact and Localized Distributed Data Structures. Research Report RR-1261-01, LaBRI, Univ. of Bordeaux, France, Aug. 2001.
8. C. Gavoille, D. Peleg, S. Pérennes and R. Raz. Distance labeling in graphs. In *Proc. 12th ACM-SIAM Symp. on Discrete Algorithms*, pages 210–219, Jan. 2001.
9. S. Kannan, M. Naor, and S. Rudich. Implicit representation of graphs. In *Proc. 20th ACM Symp. on Theory of Computing*, pages 334–343, May 1988.
10. H. Kaplan and T. Milo. Parent and ancestor queries using a compact index. In *Proc. 20th ACM Symp. on Principles of Database Systems*, May 2001.
11. H. Kaplan and T. Milo. Short and simple labels for small distances and other functions. In *Workshop on Algorithms and Data Structures*, Aug. 2001.
12. M. Katz, N.A. Katz, A. Korman and D. Peleg. Labeling schemes for flow and connectivity. In *Proc. 19th ACM-SIAM Symp. on Discrete Algorithms*, Jan. 2002.
13. M. Katz, N.A. Katz, and D. Peleg. Distance labeling schemes for well-separated graph classes. In *Proc. 17th Symp. on Theoretical Aspects of Computer Science*, pages 516–528, February 2000.
14. D. Peleg. Proximity-preserving labeling schemes and their applications. In *Proc. 25th Int. Workshop on Graph-Theoretic Concepts in Computer Science*, pages 30–41, June 1999.
15. D. Peleg. Informative labeling schemes for graphs. In *Proc. 25th Symp. on Mathematical Foundations of Computer Science*, volume LNCS-1893, pages 579–588. Springer-Verlag, Aug. 2000.
16. D. Peleg. *Distributed Computing: A Locality-Sensitive Approach*. SIAM, 2000.
17. M. Thorup and U. Zwick. Compact routing schemes. In *Proc. 13th ACM Symp. on Parallel Algorithms and Architecture*, pages 1–10, Hersonissos, Crete, Greece, July 2001.

Tight Bounds for the Performance of Longest-in-System on DAGs

Micah Adler[1] and Adi Rosén[2]

[1] Department of Computer Science, University of Massachusetts, Amherst, MA
01003, USA.
micah@cs.umass.edu [***]
[2] Department of Computer Science, Technion, Haifa 32000, Israel.
adiro@cs.technion.ac.il [†]

Abstract. A growing amount of work has been invested in recent years
in analyzing packet-switching networks under worst-case scenarios rather
than under probabilistic assumption. Most of this work makes use of
the model of "adversarial queuing theory" proposed by Borodin et al.
[6], under which an adversary is allowed to inject into the network any
sequence of packets as long as – roughly speaking – it does not overload
the network.
We show that the protocol Longest-In-System, when applied to directed
acyclic graphs, uses buffers of only linear size (in the length of the longest
path in the network). Furthermore, we show that any packet incurs only
linear delay as well. These results separate LIS from other common universally stable protocols for which there exist exponential lower bounds
that are obtained on DAGs. Our upper bounds are complemented by
linear lower bounds on buffer sizes and packet delays.

1 Introduction

The behavior of packet-switching networks, in which packets are injected into the
network in a continuous manner, has been the subject of considerable amount of
work in recent years. See e.g., [8,7,14,15,19,16,9,6,3,1,11,13,2,4]. In such networks
packets are transmitted between adjacent switches over links, in discrete time
steps, a prescribed number of packets over each link in any time step. New
packets are injected into the network dynamically at any time. Each packet is
injected into a source node and has to reach a destination node over a prescribed
path. The packets travel to their destinations in a "store-and-forward" manner,
being stored in buffers at the tail of outgoing links in intermediate switches.

As the capacity of the links in the network is limited, two important questions
arise in this setting: what are the delays incurred by the packets, and what are

---

[***] Part of this research was done while with the Dept. of Computer Science, University
of Toronto.
[†] Research supported in part by a grant from the Fund for the Promotion of Research
at the Technion. Part of this research was done while with the Dept. of Computer
Science, University of Toronto.

H. Alt and A. Ferreira (Eds.): STACS 2002, LNCS 2285, pp. 88–99, 2002.
© Springer-Verlag Berlin Heidelberg 2002

the sizes of the buffers used. A crucial question is the question of *stability*, i.e., whether the maximum buffer size grows with time. (In other words, is there a finite upper bound, which is independent of time, that bounds the size of the buffers in the network). The answers to these questions depend on the topology of the network, the injection pattern of the packets, and the contention-resolution protocol (used when more packets than the capacity of an edge attempt to cross that edge at a given time).

Considerable amount of research has been published on these questions under certain probabilistic assumptions on the injection of packets [8,7,14,15,17,19,16]. In particular the assumptions are usually that the packets are generated at the different nodes by independent Poisson processes, and are destined to uniformly distributed destinations. More recently, a growing amount of research has been invested in an attempt to answer the same questions without resorting to probabilistic assumptions. Rather, the questions of stability, queue size, and packets delay are studied under *worst-case* scenarios [6,3,1,13,11,12,2,4], in an effort to prove stability and small queue size even when the packets are injected by an adversary and not by an oblivious random process. To formulate this adversary we use the model of "adversarial queuing theory" introduced by Borodin et al. [6]. Informally, in this model an adversary can inject any sequence of packets, under the condition that the paths that the packets have to follow do not accumulate on any edge at a rate higher than the capacity of this edge.[1] For a formal definition of the adversary see Section 1.1.

The question of stability under this model received rather detailed answers. A number of natural protocols are known to be universally stable (i.e., they are stable on any network topology), while others are known not to be always stable [6,3]. The set of universally stable networks (i.e., networks that are stable with any greedy protocol) is well characterized [13,12]. While stability ensures that there are some upper bounds, independent of time, on the sizes of the buffers, it does not guarantee small bounds on them. In contrast to the question on stability, the question of good upper bounds on queue sizes (and delays of packets) for specific and universally stable protocols remains rather open. The only known upper bounds on queue sizes on general networks are exponential [6, 3]. The only known sub-exponential bounds on queue sizes hold only for the line and the cycle. Moreover, several natural and universally stable protocols have exponential lower bounds [3]. Thus, determining whether there exists a protocol with polynomial queue sizes (and packet delays) is an important open question in this area [3]. The protocol Longest-In-System (henceforth LIS), in which a packet has the highest priority to cross an edge if its time of injection is the earliest, has been suggested as a good candidate: no exponential lower bound is known for it, and there is a *centralized* polynomial algorithm that works in the spirit of LIS [3]. Determining the specific behavior of LIS was thus also raised as an important open question [6,3].

---

[1] Note that it is necessary to limit the power of the adversary. Otherwise an adversary can inject into the network a sequence of packets for which buffers inherently grow to infinity.

In this paper we show that the protocol LIS, when applied to Directed Acyclic Graphs (henceforth DAGs) is polynomial. Previously, for DAGs, it was only known that any greedy protocol is stable, but only with exponential upper bounds. We show that on DAGs, LIS has linear (in the longest path in the network) buffer sizes and linear packet delays. This is the first sub-exponential deterministic upper bound on buffer sizes obtained in the framework of adversarial queuing theory, other than on the line and the cycle. [2] Moreover, this result separates LIS from other natural stable protocols (such as Shortest-in-System), as these have exponential lower bounds that are obtained on DAGs [3].

In addition to our upper bounds, we show that these bounds are best (or almost best) possible, up to constant factors. We give a matching linear lower bound on the maximum delay incurred by the packets. We also give an almost matching linear lower bound on the maximum buffer size used by LIS on DAGs. Our results thus establish the behavior of LIS on DAGs as being linear.

1.1   The Model

We model a communication network as a graph $G = (V, E)$, $|V| = n$, $|E| = m$. Each node $v \in V$ represents a communication switch, and each edge $e \in E$ represents a link between two switches. The switches store and forward packets; the packets are stored in the switches in *buffers*. Every switch has a buffer for each outgoing link, and stores in this buffer the packets to be sent on the corresponding link. We model here the network as a *directed graph*. Thus each edge $e \in E$ is a directed edge, capable of delivering packets in the direction it which it is oriented. The network is *synchronous*. Each edge can transmit a single packet in each time step (in the direction it is oriented).

New packets are injected into the network at any time step. Each packet p is injected into an arbitrary source node s_p, and has to reach an arbitrary destination node d_p. To reach d_p from s_p, the packet has to follow a prescribed path $\pi_p = (s_p = v_0, v_1, \ldots, v_k = d_p)$.

Each time step, in each node, is divided into two sub-steps: in the first sub-step packets are sent from the buffers in which they are stored (on the edge that corresponds to each buffer). At most one packet is sent per edge. In the second sub-step each nodes receives the packets sent to it over its incoming edges, and the packets injected into it from the outside. Those packets that reach their destination are absorbed. The other packets are placed in the buffers that correspond to the next edge on their path.

The injection of packets into the network is controlled by an *adversary*. We use the model of Adversarial Queuing Theory introduced by Borodin et al. [6].

Definition 1. *We say that the adversary injecting packets is an $A(w, \varepsilon)$ adversary, for some $\varepsilon \geq 0$ and some integer $w \geq 1$, if the following holds: for any time $t \in \mathcal{N}$, let $\mathcal{I}^t$ be the set of packets injected during the w time steps from t*

---

[2] Scheideler and Vöcking [18] consider a related model to the present one, and give polynomial bounds on layered graphs. See below.

to $t + w - 1$, *inclusive. Let Π^t be the set of paths that the packets in $\mathcal{I}^t$ have to follow: $\Pi^t = \{\pi_p : p \in \mathcal{I}^t\}$. Then, every edge $e \in E$ is used by the paths of Π^t at most $\lfloor (1 - \varepsilon)w \rfloor$ times.*

A *protocol* is a set of n algorithms, one per each switch and run locally in it, that control the sending of the packets from the buffers on the corresponding outgoing links. In particular each algorithm uses only information available in its specific switch and decides if a packet is to be sent over an outgoing link and which packet it will be. A *greedy protocol* is a protocol that always sends a packet over an outgoing edge e from node v, if there is at least one packet in v that has e on its path.

1.2   Our Results

Our results concentrate on the performance of the Longest-In-System (LIS) protocol, run on networks which are DAGs. We show tight (or almost tight), up to constant factors, linear bounds on both the delay incurred by the packets, and the size of the buffers used by the protocol.

In particular, we show that the size of the buffers used by LIS on DAGs, when the packets are injected by an $A(w, \varepsilon)$ adversary, is bounded from above by $O(\ell w(1 - \varepsilon))$, where ℓ is the length of the longest path in the network. More precisely, let the *level* of a node in a DAG be the length of the longest path that leads to it. We show that the buffers in a node v of level i never exceed $O(iw(1 - \varepsilon))$. As mentioned above this to the best of our knowledge the first polynomial upper bound on buffer sizes in the present model, other than on the line and the cycle. Moreover, this upper bound separates LIS from other universally stable protocols such as Shortest-in-System, for which there are known exponential lower bounds on DAGs [3]. Our upper bound is complemented by a lower bound of $\Omega(iw(1 - \varepsilon)^2)$ on the size of the buffers of LIS in a level-i node of a DAG. We observe that for a fixed adversary this lower bound matches our upper bound (up to constant factors).

We also provide a linear upper bound of $O(iw(1 - \varepsilon))$ on the delay incurred by any packet destined to a node of level i, and give matching lower bounds. We thus show that the maximum delay incurred by a packet destined to a node v of level i is $\Theta(iw(1 - \varepsilon))$.

1.3   Related Work

Two papers in the literature on adversarial analysis of packet-switching networks are in particular close to the present work.

Andrews and Zhang [4] give a lower bound for LIS in the framework of adversarial queuing theory. They give a lower bound on the maximum delay of packets which is exponential in d, the length of the longest path followed by any packet. This lower bound is obtained on a tree of depth exponential in d. Thus, our upper bounds also show that there is an exponential gap between the two measures (longest directed path in the network, and longest path followed by any packet).

Scheideler and Vöcking [18] consider layered graphs and a model related to the present model. In their model each node has buffers on *incoming* edges, and in addition a separate buffer for packets injected into the specific node. They give a protocol for layered graphs, which is in the spirit of LIS, but uses also additional control packets. This protocol achieves for layered graphs linear delay for the packets. Their protocol uses buffers of constant size (on any of the incoming edges), while they do not give upper bounds on the size of the buffers used for injected packets (although such bounds can be inferred from the upper bounds on the delay of the packets).

2   Upper Bounds for LIS on DAGs

We first give linear upper bounds on the maximum delay incurred by the packets, and the maximum size of the buffers in the nodes.

For the purpose of the proofs we order the nodes in the graph by their topological order as induced by the DAG, assigning each node a *level*, starting with level 0, and ending at level ℓ. A node of level i is a node such that the longest path leading to it is of length i.

The main theorem of this section is the following.

Theorem 1. *When LIS is used on a DAG, and packets are injected by a $A(w, \varepsilon)$ adversary, every packet is delivered to its destination within $\ell w(1-\varepsilon)$ time steps, where ℓ is the length of the longest path in the network.*

This theorem immediately follows from the slightly stronger lemma below.

Lemma 1. *Let p be a packet injected at time t, and let v be a node on its path, which is not the destination node. If node v is of level i, then p clears node v by time step $t + \lfloor (i + 1)w(1 - \varepsilon) \rfloor$.*

Proof. We prove the lemma by double induction, on i (the level of the node), and on t (the time of the injection of the packet). Let $e = (v, u)$ be the edge on which p has to leave v.

For $i = 0$, any packet that has a node v of level 0 on its path, must be injected into node v, since no edge leads into this node. This includes the packet p under consideration, and any other packet. Thus, if p is injected into v at time t, it can be delayed in v only by other packets injected into v at times $t' \leq t$.

We prove the lemma for a node v of level 0 by induction on t. For $t = 1$ packet p can be delayed only by packets injected at time 1 and use e on their path. By the definition of the adversary there are at most $\lfloor w(1-\varepsilon) \rfloor$ such packets (including p itself). It follows that p will leave v by time step $1 + \lfloor w(1-\varepsilon) \rfloor$. For $t > 1$, consider any packet injected into v at time $t' \leq t - w$. By the induction hypothesis any such packet clears node v by time step $t' + \lfloor w(1 - \varepsilon) \rfloor \leq t - w + \lfloor w(1 - \varepsilon) \rfloor \leq t$. Thus, packet p can be delayed in v only by packets injected in the time interval

$[t - w + 1, t]$. There are at most $\lfloor w(1 - \varepsilon) \rfloor$ such packets (including p itself). It follows that p will leave v by time step $t + \lfloor w(1 - \varepsilon) \rfloor$.

For $i > 0$ we will use the induction hypothesis that for $k < i$, every packet injected at time $t' \geq 1$, and has a node of level k on its path, clears this node by time step $t' + \lfloor (k + 1)w(1 - \varepsilon) \rfloor$.

We now prove the claim for $i > 0$, and any $t \geq 1$, by induction on t. If a packet p is injected at time $t = 1$, then it can be delayed at v only by packets injected at time 1 (and that have e on their path). We distinguish between two cases. The first one is when p is injected into v and the other one is when p arrives to v over an edge. In the first case, p is clearly at v at the end of time step 1. Since there are at most $\lfloor w(1 - \varepsilon) \rfloor$ packets that use e and are injected at time 1, packet p will clear v by time step $1 + \lfloor w(1 - \varepsilon) \rfloor$. If p has to arrive to v over an edge, then it has to arrive over some edge $e' = (w, v)$, for a node w of level $k < i$. By the induction hypothesis (on i) packet p clears node w by time $1 + \lfloor (k+1)w(1-\varepsilon) \rfloor$. This means that it is in node v by the end of this time step. Since it can be delayed in v only by packets injected at time 1, it follows that packet p will clear v by time step $1 + \lfloor (k+1)w(1-\varepsilon) \rfloor + \lfloor w(1-\varepsilon) \rfloor \leq 1 + \lfloor (k+2)w(1-\varepsilon) \rfloor \leq 1 + \lfloor (i+1)w(1-\varepsilon) \rfloor$.

For packets injected at time $t > 1$, we again distinguish between the case where the packet is injected into node v, and the case where it has to arrive into v over an edge. If the packet is injected into node v, then clearly it is in v by the end of time step t. Otherwise it has to arrive into v over an edge $e' = (w, v)$, for a node w of level $k < i$. By the induction hypothesis (on i) packet p clears node w by time step $t + \lfloor (k + 1)w(1 - \varepsilon) \rfloor$, i.e., it arrives into v by the end of this time step. In any case we know that packet p arrives in v by the end of time step $T = t + \lfloor iw(1 - \varepsilon) \rfloor$.

Now consider any packet injected at time step $t' \leq t - w$, and use edge e. By the induction hypothesis on t, we have that any such packet clears node v by time step $t' + \lfloor (i+1)w(1-\varepsilon) \rfloor \leq t - w + \lfloor (i+1)w(1-\varepsilon) \rfloor \leq t + \lfloor iw(1-\varepsilon) \rfloor = T$. Therefore, if packet p does not clear node v by time T, it must be at v at this time, and can further be delayed in v only by packets injected in time interval $[t - w + 1, t]$ (and use e). By the definition of the adversary there are at most $w(1 - \varepsilon)$ such packets (including p). It follows that p clears v by time $T + \lfloor w(1 - \varepsilon) \rfloor = t + \lfloor iw(1 - \varepsilon) \rfloor + \lfloor w(1 - \varepsilon) \rfloor \leq t + \lfloor (i + 1)w(1 - \varepsilon) \rfloor$. ☐

Proof of Theorem 1. Any packet injected at t with destination v of level i, has to arrive to v over an edge $e' = (w, v)$ for a node w of level $k < i$. Therefore, by Lemma 1 such packet clears node w (and arrives at v) by time step $t + \lfloor (k + 1)w(1 - \varepsilon) \rfloor \leq t + \lfloor iw(1 - \varepsilon) \rfloor \leq t + \lfloor \ell w(1 - \varepsilon) \rfloor$. ☐

Lemma 1 also provides upper bounds on the sizes of the buffers used by LIS on DAGs. We have the following corollary.

Corollary 1. *Consider LIS when run on a DAG, where the packets are injected by a $A(w, \varepsilon)$ adversary. Then the buffer at the tail of any edge $e = (v, u)$, for node v of level i, stores at any time at most $(i + 1)w(1 - \varepsilon)$ packets.*

Proof. Assume by way of contradiction that there is a time when the buffer at the tail of an edge $e = (v, u)$, for a node v of level i stores more than $(i+1)w(1-\varepsilon)$

packets. Then there is at least one packet that will clear node v (of level i) more than $(i+1)w(1-\varepsilon)$ time steps after its injection. A contradiction.   □

3   Lower Bounds

We now give lower bounds on the maximum delay incurred by the packets when LIS is run on DAGs, and the maximum buffer size used by the nodes. We begin with a lower bound on the delay of the packets.

Theorem 2. *For any $i, w,$ and ε, there is a DAG and an $A(w, \varepsilon)$ adversary, such that some packet, destined to a node v of level i, is delayed $\Omega(iw(1-\varepsilon))$ time steps.*

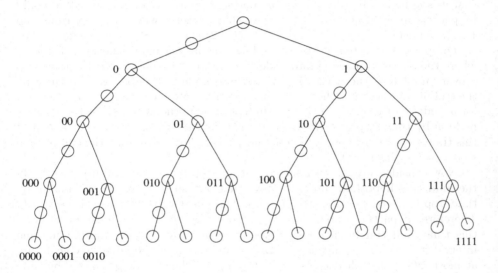

Fig. 1. The tree network used for the lower bound.

Proof. Consider the following network: start with a complete binary tree of height k, with all edges directed towards the root. We assign a binary string $s(v)$ to every node v of this network, where the mth node in the left to right ordering at distance j from the root receives the j-bit binary representation of the number $m - 1$. Thus, every internal node receives the longest common prefix of its children, and the root is assigned the empty string. We also modify the network as follows: for every internal node v of the tree, remove the edge pointing to v from its left child v_l, (v_l, v), and add a node v'_l and edges (v_l, v'_l) and (v'_l, v). The resulting network N is a subnetwork of the complete binary tree of height $2k$.

For any node v, let $Z(v)$ be the number of 0's in the string $s(v)$. Note that the distance in N of node v from the root of the tree is $|s(v)| + Z(v)$. Also, for any leaf node v, let $D(v)$ be the node on the path from v to the root such that $s(D(v))$ is obtained by removing least significant bits from $s(v)$ until either the empty string is obtained, or one bit after the first 0 is removed. Note that in the original binary tree, $D(v)$ is the first node for which v is a descendant of, or equal to, the left child of a child of $D(v)$, or the root, if no such node exists.

We can now describe the adversary. The adversary injects $\frac{w(1-\epsilon)}{2}$ packets to each leaf node v at time $k - Z(v)$. For simplicity, we assume here that $\frac{w(1-\epsilon)}{2}$ is an integer, but this assumption is easy to remove.[3]

Each such packet has a destination of $D(v)$. We now see that this is in fact a valid adversary. Note that for a leaf v to send a packet that, for nodes u_1 and u_2 of the original tree, traverses the edge (u_1, u_2), or the two edges (u_1, u_1') and (u_1', u_2), the $k - |s(u_1)| - 1$ least significant bits of $s(v)$ must all be 1's, and the $|s(u_1)|$ most significant bits of $s(v)$ must all match the corresponding bits of $s(u_1)$. Thus, at most two leaves send packets that traverse any edge, and therefore any edge is used by at most $w(1 - \varepsilon)$ packets.

Claim. For any node v from the original tree such that $|s(v)| \leq k - 2$, all packets that pass through v on their way to another node arrive between time $(k - |s(v)| - 1)\frac{w(1-\varepsilon)}{2} + 2k - Z(v) - |s(v)| + 1$ and time $(k - |s(v)|)\frac{w(1-\varepsilon)}{2} + 2k - Z(v) - |s(v)|$. During this time interval, exactly one packet arrives at v from each child of v during each time step.

Proof. We prove this by induction on $k - |s(v)|$, i.e., on the height of the node v. For the base case, let v be any node such that $|s(v)| = k - 2$, let v_r be the right child of v, and let v_ℓ be the node other than v that is adjacent to the left child of v (i.e., the left child of v in the original tree). Note that $s(v_\ell) = s(v)0$. Let v_ℓ'' be the right child of v_ℓ, and let v_ℓ' be the node other than v_ℓ that is adjacent to the left child of v_ℓ. We see that $\frac{w(1-\varepsilon)}{2}$ packets are injected by the adversary to v_ℓ' at time $k - Z(v_\ell')$. These packets arrive, one per time step, to v_ℓ starting at time $k - Z(v_\ell') + 2$. Since $s(v_\ell') = s(v_\ell)0$, $Z(v_\ell') = Z(v_\ell) + 1$, and so $k - Z(v_\ell') + 2 = k - Z(v_\ell) + 1$. The packets that v_ℓ receives from v_ℓ'' are injected by the adversary to v_ℓ'' at time $k - Z(v_\ell'') = k - Z(v_\ell)$, and thus the packets from v_ℓ'' arrive to v_ℓ at exactly the same time steps as the packets from v_ℓ'. However, since the packets from v_ℓ' have been in the system longer, when being sent on to v, they will all have priority over the packets from v_ℓ''. All of the packets from v_ℓ' have v as a final destination. Thus, the first of the packets that v receives from v_ℓ that will be forwarded on arrives at v at time $[k - Z(v_\ell) + 1] + \frac{w(1-\varepsilon)}{2} + 2$, and the remainder arrive one per time step for the next $\frac{w(1-\varepsilon)}{2} - 1$ time steps. Since $Z(v_\ell) = Z(v) + 1$ and $|s(v)| = k - 2$, this means that exactly one of those packets arrives at every time step between $(k - |s(v)| - 1)\frac{w(1-\varepsilon)}{2} + 2k - Z(v) - |s(v)| + 1$,

---

[3] Without the assumption, the adversary injects $\lfloor \frac{w(1-\varepsilon)}{2} \rfloor$ packets to each leaf and the lower bound becomes $\Omega(i\lfloor \frac{w(1-\varepsilon)}{2} \rfloor)$.

and time $(k - |s(v)|)\frac{w(1-\varepsilon)}{2} + 2k - Z(v) - |s(v)|$. A similar argument shows the analogous fact for the packets that arrive to v from v_r and are forward onward by v.

For the inductive step, we assume the claim for all v' such that $k - |s(v')| \le j$. Choose any v such that $k - |s(v)| = j + 1$, let v_r be the right child of v, and let v_ℓ be the node other than v that is adjacent to the left child of v. By induction, the packets that v_ℓ forwards to v arrive at v_ℓ between time $(j-1)\frac{w(1-\varepsilon)}{2} + 2k - Z(v_\ell) - |s(v_\ell)| + 1$ and time $j\frac{w(1-\varepsilon)}{2} + 2k - Z(v_\ell) - |s(v_\ell)|$. Of these packets, the packets that v_ℓ receives from its left child originated at a node v'_ℓ such that $s(v'_\ell) = s(v_\ell)01^{j-1}$, and the packets that v_ℓ receives from its right child originated at a node v''_ℓ such that $s(v'_\ell) = s(v_\ell)1^j$. Since $Z(v'_\ell) = Z(v''_\ell) + 1$, all the packets that v_ℓ receives from its left child have been in the system longer than the packets that v_ℓ receives from its right child.

Thus, all the packets that v_ℓ receives from its left child and passes on to v (which must also have v as their final destination) have priority over any packet that v_ℓ receives from its right child and passes on to v. Since v_ℓ receives one packet to forward per time step from each of its children, the effect of this is that all of the packets destined for v are forwarded before any packet that has a further destination. There are $\frac{w(1-\varepsilon)}{2}$ packets that v_ℓ forwards to v that have v as a final destination, and thus the packets that v receives from v_ℓ that v will forward arrive between times $j\frac{w(1-\varepsilon)}{2} + 2k - Z(v_\ell) - |s(v_\ell)| + 3$, and time $(j+1)\frac{w(1-\varepsilon)}{2} + 2k - Z(v_\ell) - |s(v_\ell)| + 2$. Since $|s(v_\ell)| = |s(v)| + 1$ and $Z(v_\ell) = Z(v) + 1$, these steps are between $(k - |s(v)| - 1)\frac{w(1-\varepsilon)}{2} + 2k - Z(v) - |s(v)| + 1$ and time $(k - |s(v)|)\frac{w(1-\varepsilon)}{2} + 2k - Z(v) - |s(v)|$. It is easy to see that exactly one of those packets arrives at v at each time step. A similar argument shows that at each time step where v receives a packet from v_ℓ that v will forward onward, v also receives a packet from v_r that it will forward. $\qquad\square$

The theorem now follows from the fact that the longest path that leads to any node v is $i = 2(k - |s(v)|)$. By claim 3, packets destined for v reach v_r, the right child of v, no earlier than time $(k - |s(v_r)| - 1)\frac{w(1-\varepsilon)}{2} + 2k - Z(v_r) - |s(v_r)| + 1$. The last of these packets will reach v at time $(k - |s(v_r)|)\frac{w(1-\varepsilon)}{2} + 2k - Z(v_r) - |s(v_r)| + 2$. These packets are inserted at a leaf v' such that $s(v') = s(v_r)01^{k-|s(v_r)|-1}$, and thus are inserted at time step $k - Z(v') = k - Z(v_r) - 1$. Thus, the total time spent in the system by these packets is $(k - |s(v_r)|)\frac{w(1-\varepsilon)}{2} + k - |s(v_r)| + 3$. Substituting $|s(v)| + 1$ for $|s(v_r)|$, this is $(k - |s(v)| - 1)(\frac{w(1-\varepsilon)}{2} + 1) + 3 = \Omega(iw(1 - \epsilon))$. $\quad\square$

For the adversary described in the above proof, the largest queue required is $\frac{w(1-\varepsilon)}{2}$. However, we can also demonstrate the following:

Theorem 3. *For any w, ε, and $i = \Omega(\frac{1}{1-\epsilon})$, there is a DAG and an $A(w, \varepsilon)$ adversary, such that LIS requires queues of size $\Omega(iw(1 - \varepsilon)^2)$, in nodes of level i.*

Proof. Call the network N constructed in the previous proof an *adversary tree* of depth k. Let s be the largest power of two such that $2s\lceil\frac{1}{1-\epsilon}\rceil + \log s - 1 \le i$. For the lower bound on queue size, we shall use s adversary trees, one for each height that is a multiple of $\lceil\frac{1}{1-\varepsilon}\rceil$ between 1 and $m = \lceil\frac{1}{1-\varepsilon}\rceil s$. Each of the roots of these adversary trees also serves as a leaf of a complete binary tree of height $\log s$. Call this binary tree the *root tree*. As before, the adversary will inject $\frac{w(1-\varepsilon)}{2}$ packets to each leaf of every adversary tree.[4] Each such packet is destined for the same node in its respective adversary tree as before, unless that packet is injected at a leaf v for which $s(v) = 1^*$, in which case the packet is now destined for the root of the root tree (instead of the root of the adversary tree). All the packets that are injected to a given leaf are injected at the same time step, where packets that are injected to a leaf v, in an adversary tree with height j and with $Z(v) = h$ in its adversary tree, are injected at time step $m - h + (m - j)(\frac{w(1-\varepsilon)}{2} + 2)$.

We first demonstrate that this is a valid adversary. We saw in the proof of Theorem 2 that this adversary does not inject more than $w(1 - \varepsilon)$ packets that cross any edge of any adversary tree.

As to edges of the root tree observe that for every adversary tree, only packets from one leaf travel into the root tree. The adversary injects into each leaf $\frac{w(1-\varepsilon)}{2}$ packets. Moreover, the injection times of packets into two leaves belonging to two different adversary trees are more than $\lceil\frac{1}{1-\varepsilon}\rceil\frac{w(1-\varepsilon)}{2} \geq w/2$ time steps apart.

We see from Claim 3 that all of the packets that arrive at the leaves of the root tree, and are destined for the root of the root tree, do so at a rate of one per leaf, where all leaves start receiving these packets at time $T = (m-1)\frac{w(1-\varepsilon)}{2}+2m+1$. It is easy to see that any node at distance d from the leaves of the root tree will then forward one packet per time step starting at time $T + d + 1$, and continue doing so until time $T+d+2^d$. Let c be either child of the root. The node c receives $\lfloor\frac{w(1-\varepsilon)s}{4}\rfloor$ packets destined for the root in an interval of length $\lfloor\frac{w(1-\varepsilon)s}{8}\rfloor$. Since c can only forward one packet per time step, there is a time step where it has a queue of size $\lfloor\frac{w(1-\varepsilon)r}{8}\rfloor$. The level of node c is $\ell = 2s\lceil\frac{1}{1-\varepsilon}\rceil + \log s - 1$. Since $s = \Omega(\ell(1 - \varepsilon))$, in terms of ℓ, the lower bound of $\lfloor\frac{w(1-\varepsilon)s}{8}\rfloor$ on the buffer size of c is $\Theta(\ell w(1 - \varepsilon)^2)$. Using $i = \Omega(\frac{1}{1-\varepsilon})$ and the definition of s, we have that $i/3 \leq \ell \leq i$. If $\ell < i$, we can add dummy nodes until the level of c is i, and thus, we obtain a node with level exactly i with queue size of $\Omega(iw(1 - \varepsilon)^2)$.   □

4   Conclusions

We considered in this paper the protocol Longest-In-System when run on DAGs. We obtained linear upper and lower bounds on the queue sizes and maximum delay of packets, when the packets are injected according to the adversarial queuing model. We thus prove the first polynomial bounds (other than on the line and cycle) in the context of this model, and establish the behavior of LIS on DAGs to be linear. Furthermore, these results separate LIS from other universally stable protocols for which there are exponential lower bounds on DAGs. Comparing our results to the results in [4], also demonstrates that there is an

---

[4] We assume again that $\frac{w(1-\varepsilon)}{2}$ is an integer; otherwise we use the value $\lfloor\frac{w(1-\varepsilon)}{2}\rfloor$ instead.

exponential gap between measuring the performance of protocols in terms of the longest path in the network and in terms of the longest path followed by any packet.

As mentioned above, we obtain that LIS is linear on DAGs. It remains however open if LIS is polynomial on general networks, and more generally if there exists a universally stable polynomial protocol.

Acknowledgments. We thank Matthew Andrews for discussions on the results in [4]. We also thank Eyal Kushilevitz for useful discussions.

References

1. W. Aiello, E. Kushilevitz, R. Ostrovsky, A. Rosén. Adaptive packet Routing for Bursty Adversarial Traffic. *Journal of Computer and System Sciences*, Special issue for STOC 98, Vol. 60, No. 3, pp. 482–509, 2000.
2. M. Andrews, Instability of FIFO in Session Oriented Networks. In *Proc. of the 11th SODA*, 2000.
3. M. Andrews, B. Awerbuch, A. Fernández, J. Kleinberg, T. Leighton, and Z. Liu, "Universal Stability Results for Greedy Contention-Resolution Protocols", JACM 48(1), pp. 39–69, 2001.
4. M. Andrews, and L. Zhang, The Effect of Temporary Sessions on Network Performance. In *Proc. of the 11th SODA*, 2000.
5. M. Andrews, A. Fernández, M. Harchol-Balter, and T. Leighton, L. Zhang, "General Dynamic Routing with Per-Packet Delay Guarantees of O(distance + 1/session rate)," In *Proc. of 38th FOCS*, pp. 294–302, 1997.
6. A. Borodin, J. Kleinberg, P. Raghavan, M. Sudan, and D. Wiliamson, "Adversarial Queuing Theory", JACM 48(1), pp. 13–38, 2001.
7. A.Z. Broder, A.M. Frieze, and E. Upfal, "A General Approach to Dynamic Packet Routing with Bounded Buffers", In *Proc. of 37th FOCS*, pp. 390-399, 1996.
8. A.Z. Broder, and E. Upfal, "Dynamic Deflection Routing on Arrays," In *Proc. of 28th STOC*, pp. 348–355, 1996.
9. R. Cruz, "A Calculus for Network Delay, Part I: Network Elements in Isolation," *IEEE Transactions on Information Theory*, pp. 114–131, 1991.
10. R. Cruz, "A Calculus for Network Delay, Part II: Network Analysis," *IEEE Transactions on Information Theory*, pp. 132–141, 1991.
11. D. Gamarnik, Stability of Adversarial Queues via Fluid Model. In Proc. of the 39th FOCS, pp. 60–70, 1998.
12. D. Gamarnik. Stability of Adaptive and Non-Adaptive Packet Routing Policies in Adversarial Queuing Networks. In *Proc. of the 31st STOC*, pp. 206–214, 1999.
13. A. Goel, Stability of Networks and Protocols in the Adversarial Queuing Model for Packet Routing. Stanford University Technical Notes STAN-CS-TN-97-59, 1997.
14. M. Harchol-Balter and P. Black, "Queuing Analysis of Oblivious Packet-Routing Algorithms," In *Proc. of 5th SODA*, pp. 583–592, 1994.
15. M. Harchol-Balter and D. Wolfe "Bounding Delays in Packet Routing Networks," *Proc. of 27th STOC*, pp. 248-257, 1995.
16. M. Mitzenmacher, "Bounds on the Greedy Routing Algorithm for Array Networks", J. Comput. System Sci. 53 (1996), No. 3, pp. 317–327.

17. C. Scheideler and B. Vöcking, "Universal Continuous Routing Strategies," In *Proc. of 8th SPAA*, 1996.
18. C. Scheideler and B. Vöcking, "From Static to Dynamic Routing: Efficient Transformations of Store-and-Forward Protocols", In *Proc. of the 31st ACM Symposium on the Theory of Computing*, pp. 215–224, 1999.
19. G. Stamoulis and J. Tsitsiklis, "The Efficiency of Greedy Routing in Hypercubes and Butterflies," *IEEE Transactions on Communications*, 42 (11), pp. 3051–208, 1994.

Approximate Strong Separation with Application in Fractional Graph Coloring and Preemptive Scheduling[*]

Klaus Jansen

Institut für Informatik und praktische Mathematik, Christian-Albrechts-Universität zu Kiel, 24 098 Kiel, Germany, kj@informatik.uni-kiel.de

Abstract. In this paper we show that approximation algorithms for the weighted independent set and s-dimensional knapsack problem with ratio a can be turned into approximation algorithms with the same ratio for fractional weighted graph coloring and preemptive resource constrained scheduling. In order to obtain these results, we generalize known results by Grötschel, Lovasz and Schrijver on separation, non-emptiness test, optimization and violation in the direction of approximability.

1 Introduction

The fractional graph coloring is defined as follows (see also [10,15,24,26,27,28, 29]). Let $G = (V, E)$ be a graph with a positive weight w_v for each vertex $v \in V$. Let $\mathcal{I}$ be the set of all independent sets of G. The weighted fractional coloring problem consists of assigning a non-negative real value x_I to each independent set I of G such that each vertex $v \in V$ is completely covered by independent sets containing v (i.e. the sum of their values is at least w_v) and the total value $\sum_I x_I$ is minimized. Grötschel et al. [15] proved that the weighted fractional coloring problem is NP-hard for general graphs, but can be solved in polynomial time for perfect graphs. They have proved the following interesting result: For any graph class $\mathcal{G}$, if the problem of computing $\alpha(G, w)$ (the weight of the largest weighted independent set in G) for graphs $G \in \mathcal{G}$ is NP-hard, then the problem of determining the weighted fractional chromatic number $\chi_f(G, w)$ is also NP-hard. This gives a negative result of the weighted fractional coloring problem even for planar cubic graphs and unit disk graphs. Furthermore, if the weighted independent set problem for graphs in $\mathcal{G}$ is polynomial-time solvable, then the weighted fractional coloring problem for $\mathcal{G}$ can also be solved in polynomial time.

The first inapproximability result for the unweighted version of the problem (i.e. where $w_v = 1$ for each vertex $v \in V$) was obtained by Lund and Yannakakis [24] who proved that there exists a $\delta > 0$ such that there is no

---

[*] Research of the author was supported in part by the EU Thematic Network AP-POL, Approximation and Online Algorithms, IST-1999-14084 and by the EU Research Training Network ARACNE, Approximation and Randomized Algorithms in Communication Networks, HPRN-CT-1999-00112.

H. Alt and A. Ferreira (Eds.): STACS 2002, LNCS 2285, pp. 100–111, 2002.
© Springer-Verlag Berlin Heidelberg 2002

polynomial-time approximation algorithm for the problem with approximation ratio n^δ, unless P=NP. Feige and Kilian [10] showed that the fractional chromatic number $\chi_f(G)$ cannot be approximated within $\Omega(|V|^{1-\epsilon})$ for any $\epsilon > 0$, unless ZPP=NP. Recently, the authors [19] proved that fractional coloring is NP-hard even for graphs with $\chi_f(G) = 3$ and constant degree 4. Similarly, as it was shown by Gerke and McDiarmid [14], the problem remains NP-hard even for triangle-free graphs. Regarding the approximability of the fractional chromatic number, Matsui [26] gave a polynomial-time 2-approximation algorithm for unit disk graphs.

The general preemptive resource constrained scheduling problem denoted by $P|res...,pmtn|C_{max}$ is defined as follows: There are given n tasks $\mathcal{T} = \{T_1,...,T_n\}$, m identical machines and s resources such that each task $T_j \in \mathcal{T}$, is processed by one machine requiring p_j units of time and r_{ij} units of resource i, $i = 1,...,s$, from which only c_i units are available at each time. The objective is to compute a preemptive schedule of the tasks minimizing the maximum completion time C_{max}. The three dots in the notation indicates that there are no restrictions on the number of resources s, the largest possible capacity o and resource requirement r values, respectively. If any of these is limited, the corresponding fixed limit replaces the corresponding dot in the notation (e.g. if $s \leq 1$, then $P|res1..,pmtn|C_{max}$ is used, or if $r_{ij} \leq r$, then $P|res ..r, pmtn|C_{max}$).

Resource constrained scheduling is one of the classical scheduling problems. Garey and Graham [12] proposed approximation algorithms for the non-preemptive variant $P|res...|C_{max}$ with approximation ratio $s + 1$ (if the number of machines is unbounded, $m \geq n$) and $\min(\frac{m+1}{2}, s + 2 - \frac{2s+1}{m})$ (if $m \geq 2$). Further results are known for some special cases: Garey et al. [13] proved that if $m \geq n$ and each task T_j has unit-execution time, i.e. $p_j = 1$, the problem (denoted by $P|res..., p_j = 1|C_{max}$) can be solved by First Fit and First Fit Decreasing heuristics providing asymptotic approximation ratio $s + \frac{7}{10}$ and a ratio between $s + ((s - 1)/s(s + 1))$ and $s + \frac{1}{3}$, respectively. De la Vega and Lueker [7] gave a linear-time algorithm with asymptotic approximation ratio $s + \epsilon$ for each fixed $\epsilon > 0$. Further results and improvements for the non-preemptive variant are given in [6,30,31]. For the preemptive variant substantially less results are known: Blazewicz et al. [3] proved that for m is fixed, the problem $Pm|pmtn, res...|C_{max}$ (with identical machines) and even the variant $Rm|pmtn, res...|C_{max}$ (with unrelated machines) can be solved in polynomial time. Krause et al. [23] studied $P|pmtn, res1..|C_{max}$, i.e. the case where there is only one resource ($s = 1$), and proved that both First Fit and First Fit Decreasing heuristics can guarantee $3 - 3/n$ asymptotic approximation ratio.

In this paper we show new approximability results for fractional coloring and preemptive resource constrained scheduling based on methods in [16]. Grötschel, Lovasz and Schrijver [16] proved that an oracle polynomial time algorithm for the strong separation problem can be used to solve the strong non-emptiness problem, and that the strong separation, violation and optimization problems are equivalent for well described polyhedra. In this paper we generalize the results in the context of approximability for a fractional packing polytope (a general bounded linear program with many implicit given inequalities). A fractional packing polytope $K \subset \mathbb{R}^N$ is given by the intersection of two polyhedra:

- a polyhedron A that is the solution set of M inequalities $a_i^T x \leq b_i$, where $b_i > 0$ for $1 \leq i \leq M$,
- a polyhedron B that is solution set of further inequalities.

We assume that the separation problem is NP-hard for A and solvable in polynomial time for B. Similar to [16] we suppose that an upper bound φ on the facet complexity for K is known, but that the inequalities of A and B are not given explicitly. For our applications the separation oracle for A works only approximatively. For an approximation ratio $a \geq 1$ we denote by K^a the packing polytope where b_i is replaced by $a\,b_i$. The ratio is either a constant or depends on the instance (for example $\log N$ or $N^{1/2}$). Then the approximate separation problem for K is: given a vector y either assert that $y \in K^a$ or find a hyperplane that separates y from K. In the paper we study this approximate separation problem and further basic algorithmic problems like weak approximate separation, strong approximate non-emptiness test, strong approximate optimization, and strong approximate violation. We show how oracle polynomial time algorithms for one problem can be used for other ones. The proofs are based on ideas in [16].

After that we use these techniques to show new approximability results for the two optimization problems above. We show that approximation algorithms for the maximum weighted independent set problem or the $(s + 1)$-dimensional knapsack problem with ratio a can be turned into approximation algorithms with ratio a for the fractional weighted chromatic number problem or preemptive resource constrained scheduling problem, respectively. The key idea here is to solve first approximatively the dual optimization problem with many implicit given constraints and then to turn the dual solution into a primal solution. The last step is based on ideas by Karp and Karmarkar for bin packing [21]. The results imply a polynomial approximation scheme for weighted fractional coloring of disk graphs and a fully polynomial time approximation scheme for the resource constrained scheduling problem $P|pmtn, res1..|C_{max}$ (improving the results by Matsui [26] and Krause et al. [23]).

2   Strong Approximate Separation

We say that the polyhedron K has facet complexity at most φ if there is a system of inequalities with rational coefficients that has solution set K and such that the encoding length of each inequality is at most φ. In case $K = \mathbb{R}^N$ we require $\varphi \geq N + 1$. A polyhedron has vertex complexity v if there is a finite sets V of rational vectors such that $K = conv(V)$ and such that each of the vectors in V has encoding length at most v. In case $K = \emptyset$ we require $v \geq N$. A well described polyhedron is a triple $(K; N, \varphi)$ where $K \subset \mathbb{R}^N$ is a polyhedron with facet complexity at most φ. The encoding length of a well described polyhedron

K is $\varphi + n$. Clearly, if a polyhedron $K \subset \mathbb{R}^N$ has facet complexity at most φ and vertex complexity at most v, then $\varphi \geq N + 1$ and $v \geq N$.

In our paper we study a fractional packing polytope K (bounded polyhedron) of the following form:

$$a_i^T x \leq b_i \quad i = 1, \ldots, M,$$
$$x \in B.$$

Let $S(K, \delta)$ be the set $\{x \in \mathbb{R}^N \mid \|x - y\| \leq \delta \text{ for some } y \in K\}$. Interestingly, if K has facet complexity at most φ, then K has vertex complexity at most $4N^2\varphi$ and all vertices of K are contained in the ball $S(0, 2^{4N^2\varphi})$. In our case K is bounded and $K \subset S(0, 2^{4N^2\varphi})$.

Let K be a fractional packing polytope and let a be an approximation ratio ≥ 1. The **Strong Approximate Separation Problem (S-APP-SEP)** is defined by:

- given a vector $y \in \mathbb{R}^N$; either
  (i) assert that $y \in K^a$, or
  (ii) find a hyperplane that separates y from K (i.e. find a vector $c \in \mathbb{R}^N$ such that $c^T x < c^T y$ for each $x \in K$).

Now we can define also a weak version of APPSEP as follows. Let $S(K, -\delta) = \{x \in K \mid S(x, \delta) \subset K\}$. The **Weak Approximate Separation Problem (W-APP-SEP)** is:

- given a vector $y \in \mathbb{R}^N$ and a rational number $\delta > 0$; either
  (i) assert that $y \in S(K^a, \delta)$, or
  (ii) find a vector $c \in \mathbb{Q}^N$ with $\|c\|_\infty = 1$ such that $c^T x \leq c^T y + \delta$ for every $x \in S(K, -\delta)$ (i.e. find an almost separating hyperplane).

For $a = 1$ the two basic problems are known as the strong and weak separation problems. Given a vector $y \in B$ we consider the problem to compute the following value:

$$OPT(A, y) := \max\{a_i^T y / b_i \mid 1 \leq i \leq M\}.$$

In general the corresponding optimization problem is NP-hard. Consider for example the fractional clique polytope with inequalities $\sum_{v \in I} x_v \leq 1$ for independent sets I in a graph $G = (V, E)$ and set $B = \{(x_v) \mid x_v \geq 0\}$. In this case the vector y corresponds to vertex weights and $OPT(A, y)$ is the weight of a maximum weighted independent set in G. This problem is NP-complete for graphs. The strong optimization problem over the fractional clique polytope is the dual problem of the fractional weighted chromatic number problem.

Suppose now that there is an a - approximation algorithm R for the general problem that finds a row i with value

$$R(A, y) = a_i^T y / b_i \geq \frac{1}{a} OPT(A, y)$$

in time polynomial in N and φ. The final goal is to show that for any vector d there is also an a - approximation algorithm for the strong optimization problem:

$$\max d^T x$$
$$\text{s.t.} \quad a_i^T x \leq b_i \ i = 1, \ldots, M,$$
$$x \in B.$$

As first step we show that the strong approximate separation problem with parameter a is solvable in polynomial time.

Given a vector $y \in \mathbb{R}^N$ the approximate separation problem for K can be solved as follows: First we test whether $y \in B$. By assumption, the strong separation problem can be solved in polynomial time for B (either we get a hyperplane that separates y from B or $y \in B$). If the algorithm finds a hyperplane then we are done (the same hyperplane separates y also from K). If $y \in B$ then we apply algorithm R to A, y and compute a row i with $a_i^T y / b_i = R(A, y) \geq \frac{1}{a} OPT(A, y)$. If $R(A, y) > 1$ then R gives us a hyperplane that separates y from K. Otherwise (for $R(A, y) \leq 1$) we can show that $y \in K^a$. To prove this consider the following two cases:

Case 1: $OPT(A, y) > a$. In this case R would find a row i with weight $R(A, y) \geq \frac{1}{a} OPT(A, y) > 1$. This gives a contradiction.

Case 2: $OPT(A, y) \leq a$. In this case we know that $a_i^T y / b_i \leq a$ for $1 \leq i \leq M$ or in other words that $y \in K^a$.

This gives the first result:

Theorem 1. *If there is an approximation algorithm R that finds a row a_i with $a_i^T y / b_i \geq \frac{1}{a} \max\{a_j^T y / b_j | 1 \leq j \leq M\}$ for each $y \in B$ in time polynomial in N, then there is a polynomial time algorithm to solve the strong approximate separation problem for K and y with parameter a.*

The fractional chromatic number problem can be formulated as the following linear program LP_1:

$$\min \sum_{I \in \mathcal{I}} x_I$$
$$\text{s.t.} \sum_{I:v \in I} x_I \geq w_v \quad \forall v \in V, \tag{1}$$
$$x_I \geq 0 \qquad \forall I \in \mathcal{I}.$$

The dual problem DLP_1 has the following form:

$$\max \sum_{v \in V} w_v y_v$$
$$\text{s.t.} \sum_{v:v \in I} y_v \leq 1 \quad \forall I \in \mathcal{I}, \tag{2}$$
$$y_v \geq 0 \qquad \forall v \in V.$$

The corresponding polytope is called the fractional clique polytope. In this case the polyhedron A is the solution set of the inequalities $\sum_{v:v \in I} y_v \leq 1$ and the polyhedron B is equal to $\mathbb{R}_+^N$.

Corollary 1. *Let $\mathcal{G}$ be a graph class. If there is an a - approximation algorithm to solve the maximum weighted independent set problem for a graph $G \in \mathcal{G}$ with positive weights y in polynomial time, then there is a polynomial time algorithm to solve the strong approximate separation problem (with parameter a) for the fractional clique polytope corresponding to G.*

For our resource constrained scheduling problem, a *configuration* is a compatible (or feasible) subset of tasks that can be scheduled simultaneously. Let F be the set of all configurations, and for every $f \in F$, let x_f denote the length (in time) of configuration f in the schedule. Clearly, $f \in F$ iff $\sum_{j \in f} r_{ij} \leq c_i$, for $i = 1, \ldots, s$ and $|f| \leq m$.

By using these variables, the problem of finding a preemptive schedule of the tasks with smallest makespan value (subject to the resource constraints) can be formulated as the following linear program [18]:

$$
\begin{aligned}
\text{Min } & \sum_{f \in F} x_f \\
\text{s.t. } & \sum_{f \in F : j \in f} x_f \geq p_j, \quad j = 1, \ldots, n, \\
& x_f \geq 0, \quad \forall f \in F.
\end{aligned}
\tag{3}
$$

The dual problem for the scheduling problem has the form:

$$
\begin{aligned}
\text{Max } & \sum_{j=1}^{n} p_j y_j \\
\text{s.t. } & \sum_{j : j \in f} y_j \leq 1 \quad \forall f \in F, \\
& y_j \geq 0 \quad \forall j = 1, \ldots, n.
\end{aligned}
\tag{4}
$$

The optimization problem among all configurations is equivalent to a $(s+1)$ dimensional knapsack problem:

$$
\begin{aligned}
\text{Max } & \sum_{j=1}^{n} y_j x_j \\
\text{s.t. } & \sum_{j=1}^{h} r_{ij} x_j \leq c_i, \quad i = 1, \ldots, s, \\
& \sum_{j=1}^{h} x_j \leq m, \\
& x_j \in \{0, 1\}, \quad j = 1, \ldots, n.
\end{aligned}
\tag{5}
$$

Corollary 2. *If there is an a - approximation algorithm to solve the maximum $(s+1)$ - dimensional knapsack problem that runs in polynomial time, then there is a polynomial time algorithm to solve the strong approximate separation problem (with parameter a) for the dual of the resource constrained scheduling problem.*

3   Ellipsoid Algorithm Based on Approximate Oracle

A set $E \subset \mathbb{R}^N$ is an ellipsoid if there exist a vector $b \in \mathbb{R}^N$ and a positive definite matrix B such that

$$
E = E(B, b) := \{x \in \mathbb{R}^N | (x - b)^T B^{-1}(x - b) \leq 1\}.
$$

In oder to solve the optimization problem approximately we use the central cut ellipsoid algorithm based on an approximate oracle.

Given: rational number $\epsilon > 0$, a fractional packing polytope $(K; N, \varphi)$ with $K \subset S(\{0\}, R)$ given by an oracle W-APP-SEP that, for any $y \in \mathbb{Q}^N$ and any rational number $\delta > 0$ either decides that $y \in S(K^a, \delta)$ or finds a vector $c \in \mathbb{Q}^N$ with $||c||_\infty = 1$ such that $c^T x \leq c^T y + \delta$ for every $x \in K$.

Output: either

   (i) a vector $z \in S(K^a, \epsilon)$, or

   (ii) a positive definite matrix $B \in \mathbb{Q}^{N \times N}$ and a point $b \in \mathbb{Q}^N$ such that $K \subset E(B, b)$ and $vol(E(B, b)) \leq \epsilon$.

Theorem 2. *There exists an oracle-polynomial time algorithm that solves the above problem (based on a weak approximate separation oracle).*

The algorithm and analysis is based directly on the method in [16]. A complete description can be found in the full paper.

4   Approximate Non-emptiness Test, Optimization, and Violation

In this section we discuss approximate version of the non-emptiness test, optimization and violation problem. We show that the approximate versions of optimization, violation and separation are polynomial equivalent. The proofs are based on the non-approximate versions in [16] and are omitted due to space limitations.

The **approximate non-emptiness problem (APP-NEMPT)** is defined by

  – given a fractional packing polytope K, a number $a \geq 1$; either

   (i) find a vector $x \in K^a$, or

   (ii) assert that K is empty.

Theorem 3. *There exists an oracle-polynomial time algorithm that, for any fractional packing polytope $(K; N, \varphi)$ given by a strong approximate separation oracle with parameter $a \geq 1$, solves the approximate non-emptiness problem.*

The **strong approximate optimization problem (S-APP-OPT)** is defined by

  – given a fractional packing polytope K (that contains the zero vector), a number $a \geq 1$ and a vector $d \in \mathbb{R}^N$,

   • find a vector $z \in K$ such that $d^T z \geq \frac{1}{a} \cdot \max_{x \in K} d^T x$.

Theorem 4. *There exists an oracle-polynomial time algorithm that, for any fractional packing polytope $(K; N, \varphi)$ given by a strong approximate separation oracle with parameter $a \geq 1$ (that contains the zero vector), solves the strong approximate optimization problem.*

Instead of the general vector $d \in \mathbb{Q}^N$ we can also use a vector $\tilde{d}$ such that the encoding length $< \tilde{d} >$ is bounded by a polynomial in φ and N and that $\max\{d^T x | x \in K\} = \max\{\tilde{d}^T x | x \in K\}$ [16], Lemma (6.2.9) and Theorem (6.6.5). Then, we use the algorithm with objective function $\tilde{d}^T x$. This shows that the number of calls to the approximate separation oracle, and the number of elementary arithmetic operations executed by the algorithm are bounded by a polynomial in φ.

As consequence we obtain the main result below and two corollaries:

Theorem 5. *If there is an approximation algorithm R that finds a row a_i with $a_i^T y / b_i \geq \frac{1}{a} max\{a_j^T y / b_j | 1 \leq j \leq M\}$ for each $y \in B$ and if K contains the zero vector, then there is a polynomial time algorithm to solve the strong approximate optimization problem for K with ratio a.*

Corollary 3. *Let $\mathcal{G}$ be a graph class. If there is an a - approximation algorithm to solve the maximum weighted independent set problem for a graph $G \in \mathcal{G}$, then there is a polynomial time algorithm to solve the strong approximate optimization problem for the fractional clique polytope with ratio a.*

Corollary 4. *If there is an a - approximation algorithm to solve the $(s+1)$ dimensional knapsack problem, then there is a polynomial time algorithm to solve the strong approximate optimization problem for the dual of the preemptive resource constrained scheduling problem with ratio a.*

The **strong approximate violation problem (S-APP-VIO)** can be stated as follows:

- given a fractional packing polytope K, a number $a \geq 1$, a vector $c \in \mathbb{Q}^n$ and number $\gamma \in \mathbb{Q}$, either
  (i) assert that $c^T x \leq a \cdot \gamma$ for each $x \in K$, or
  (ii) find a $y \in K$ such that $c^T y > \gamma$.

Theorem 6. *The strong approximate violation problem can be solved in oracle - polynomial time for any fractional packing polytope K given by a strong approximate optimization oracle.*

Theorem 7. *The strong approximate separation problem can be solved in oracle - polynomial time for any fractional packing polytope $K \subset \mathbb{R}_+^N$ (that contains the zero vector) given by a strong approximate violation oracle.*

As consequence we obtain the following result:

Theorem 8. *There is an a - approximation algorithm to solve the maximum weighted independent set problem for a graph G with weights $y \in \mathbb{Q}^N$ that runs in time polynomial in N, if and only if there is a polynomial time algorithm to solve the strong approximate optimization problem for the fractional clique polytope with ratio a.*

An analog of the result above by means of optimal algorithms was known before [15]: There is an optimal polynomial time algorithm to compute a maximum weighted independent set, if and only if there is an optimal algorithm to compute the fractional clique number in polynomial time.

5   The Dual Problem to the Fractional Packing Problem

The next step is to show that we can obtain also a solution of the dual problem with the same approximation ratio a. Let DLP_1 be our linear program

$$\text{Max } d^T y$$
$$\text{s.t. } a_i^T y \leq b_i \quad \forall 1 \leq i \leq M, \tag{6}$$
$$y \in B$$

where B is given by further (in)-equalities. Furthermore, let LP_1 be the corresponding primal linear program. Here we obtain the main result:

Theorem 9. *If there is an approximation algorithm R that finds a row a_i with $a_i^T y / b_i \geq \frac{1}{a} \max\{a_j^T y / b_j | 1 \leq j \leq M\}$ for each $y \in B$ in time polynomial in N, then there is a polynomial time algorithm to solve the strong approximate optimization problem for the linear program dual to K with ratio a.*

Proof. The approximation algorithm R gives us a strong approximate separation algorithm for the fractional packing polytope K. Using Theorem 5 the strong optimization problem for the fractional packing polytope K can be solved with approximation ratio a in polynomial time. In other words, we can compute a vector z in the corresponding packing polytope with $d^T z \geq OPT(DLP_1)/a$, $z \in B$ and $a_i^T z \leq b_i$ for each $1 \leq i \leq M$. The algorithm based on the approximate separation oracle uses only a polynomial number of separation hyperplanes. Furthermore, each hyperplane corresponds to one of these rows or inequalities in B. Let $\mathcal{M} = \{i_1, \ldots, i_m\}$ be set of rows of A and let B' be the polyhedron corresponding to the inequalities in B used by the algorithm above. Notice that the total number of inequalities is bounded by a polynomial in φ. Let DLP_2 be the linear program of the form:

$$\text{Max } d^T y$$
$$\text{s.t. } a_i^T y \leq b_i \quad \forall i \in \mathcal{M}, \tag{7}$$
$$y \in B'.$$

Furthermore, let LP_2 be the corresponding restricted dual linear program.

First, we have $d^T z \geq OPT(DLP_1)/a = OPT(LP_1)/a$ and $d^T z \leq OPT(LP_1) = OPT(DLP_1)$. Second we know that the computation of the algorithm to compute z is also a valid application of the algorithm for DLP_2. In other words, the approximate oracle gives same answers as before. This implies that $d^T z \geq OPT(DLP_2)/a$. By duality we have again $OPT(DLP_2) = OPT(LP_2)$.

This together implies that

$$\frac{OPT(LP_2)}{a} = \frac{OPT(DLP_2)}{a} \leq d^T z \leq OPT(LP_1) = OPT(DLP_1).$$

Now we can compute the optimum solution x^* of LP_2 using the standard algorithm by Khachiyan; LP_2 has only a polynomial number of variables and constraints. The objective value of x^* is equal to $OPT(LP_2) \leq a \cdot OPT(LP_1)$. This means that we have an approximation algorithm with ratio a.

The argument above was used first by Karp and Karmarkar for the bin packing problem [21]. Notice that φ is bounded by the size of the instance for the fractional weighted coloring and preemptive resource constrained scheduling problem (by considering each possible independent set or configuration). Therefore, we obtain the following results:

Theorem 10. *Let $\mathcal{G}$ be a graph class. If there is an a - approximation algorithm to solve the maximum weighted independent set problem for each graph $G \in \mathcal{G}$, then there is a polynomial time algorithm to solve the fractional weighted chromatic number problem for G with ratio a.*

Corollary 5. *Let $\mathcal{G}$ be a graph class. If there is a (F)PTAS to compute the weighted independent set for each graph $G \in \mathcal{G}$ and weights w, then we obtain a (F)PTAS for the fractional weighted coloring problem for graphs $G \in \mathcal{G}$.*

Using a recent result in [9] on the maximum weighted independent set problem for intersection graphs of disks in the plane, we obtain the following:

Corollary 6. *There is a PTAS for the computation of the fractional weighted chromatic number for intersection of disks in the plane.*

Since this graph class contains planar graphs and unit disk graphs, we get the next corollary:

Corollary 7. *There is a PTAS for the computation of the fractional weighted chromatic number for planar and unit disk graphs.*

This improves an approximation algorithm with ratio 2 by Matsui [26] for fractional coloring on unit disk graphs.

Theorem 11. *Let $\mathcal{I}$ be a set of instances of the resource constrained scheduling problem, and let $\mathcal{I}'$ be the set of corresponding instances of the $(s + 1)$ dimensional knapsack problem. If there is an a - approximation algorithm to solve the $(s + 1)$ dimensional knapsack problem for any instance $I' \in \mathcal{I}'$, then there is a polynomial time algorithm to solve the preemptive resource constrained scheduling problem with ratio a.*

Corollary 8. *For any fixed number s of resources, there is a PTAS for $P|res\,s..,$ $pmtn|C_{max}$. Furthermore, there is a FPTAS for $P|res1..,pmtn|C_{max}$.*

6 Conclusion

We mention that a different approach by Jansen and Porkolab [20] gives a related result with a faster running time. It transforms an approximation algorithm with ratio a for the maximum weighted independent set and s-dimensional knapsack problem into an approximation algorithm with ratio $a(1 + \epsilon)$ for the minimum fractional weighted coloring and preemptive resource constrained scheduling problem, respectively. The number of calls to the subroutine is only $O(n(\ln n + \epsilon^{-2} + \ln a\epsilon^{-3}))$ where n is the number of vertices or tasks.

We expect that further results in [16] can be extended in the direction of approximability and inapproximabilty. Furthermore the techniques in the paper can be used probably also for other applications like fractional path coloring in WDM networks [5]. An interesting open question is whether it is possible to derive an approximation algorithm for the maximum weighted independent set problem from an approximation algorithm for the fractional weighted chromatic number problem. A partial answer in this direction is given by Erlebach and Jansen [8].

Acknowledgement. We thank Thomas Erlebach and Lorant Porkolab for helpful comments and discussions.

References

1. J. Blazewicz, W. Cellary, R. Slowinski and J. Weglarz, Scheduling under resource constraints - deterministic models, *Annals of Operations Research* 7 (1986).
2. J. Blazewicz, K.H. Ecker, E. Pesch, G. Schmidt and J. Weglarz, Scheduling in Computer and Manufacturing Systems, Springer Verlag, Berlin (1996).
3. J. Blazewicz, J.K. Lenstra and A.H.G. Rinnooy Kan, Scheduling subject to resource constraints: Classification and Complexity, *Discrete Applied Mathematics* 5 (1983), 11-24.
4. A. Caprara, H. Kellerer, U. Pferschy and D. Pisinger, Approximation algorithms for knapsack problems with cardinaliy constraints, *European Journal of Operational Research* 123 (2000), 333-345.
5. I. Caragiannis, A. Ferreira, C. Kaklamanis, S. Perennes and H. Rivano, Fractional path coloring with applications to WDM networks, *28th International Colloquium on Automata, Languages and Programming* ICALP 2001, LNCS 2076, Springer Verlag, 732-743.
6. C. Chekuri and S. Khanna, On multi-dimensional packing problems, *Proceedings 10th ACM-SIAM Symposium on Discrete Algorithms* SODA 1999, 185-194.
7. W.F. de la Vega and C.S. Lueker, Bin packing can be solved within $1 + \epsilon$ in linear time, *Combinatorica* 1 (1981), 349-355.
8. T. Erlebach and K. Jansen, Conversion of coloring algorithms into maximum weight independent set algorithms, *Workshop on Approximation and Randomization Algorithms in Communication Networks* (2000), Carleton Scientific, 135-146.
9. T. Erlebach, K. Jansen and E. Seidel, Polynomial-time approximation schemes for geometric graphs, *Proceedings of the 12th ACM-SIAM Symposium on Discrete Algorithms* (2001), 671-679.
10. U. Feige and J. Kilian, Zero knowledge and the chromatic number, *Journal of Computer and System Sciences* 57 (1998), 187-199.
11. A.M. Frieze and M.R.B. Clarke, Approximation algorithms for the m-dimensional $0 - 1$ knapsack problem, *European Journal of Operational Research* 15 (1984), 100-109.

12. M.R. Garey and R.L. Graham, Bounds for multiprocessor scheduling with resource constraints, *SIAM Journal on Computing* 4 (1975), 187-200.
13. M.R. Garey, R.L. Graham, D.S. Johnson and A.C.-C. Yao, Resource constrained scheduling as generalized bin packing, *Journal Combinatorial Theory A* 21 (1976), 251-298.
14. S. Gerke and C. McDiarmid, Graph imperfection, unpublished manuscript, 2000.
15. M. Grötschel, L. Lovász and A. Schrijver, The ellipsoid method and its consequences in combinatorial optimization, *Combinatorica* 1 (1981), 169-197.
16. M. Grötschel, L. Lovász and A. Schrijver, Geometric Algorithms and Combinatorial Optimization, Springer Verlag, Berlin (1988).
17. K. Jansen and L. Porkolab, Linear-time approximation schemes for scheduling malleable parallel tasks, *Proceedings 10th ACM-SIAM Symposium on Discrete Algorithms* SODA 1999, 490-498, and *Algorithmica*, to appear.
18. K. Jansen and L. Porkolab, Computing Optimal Preemptive Schedules for Parallel Tasks: Linear Programming Approaches, *Proceedings 11th Annual International Symposium on Algorithms and Computation* ISAAC 2000, LNCS 1969, Springer Verlag, 398-409, and *Mathematical Programming*, to appear.
19. K. Jansen and L. Porkolab, Preemptive scheduling on dedicated processors: applications of fractional graph coloring *Proceedings 25th International Symposium on Mathematical Foundations of Computer Science* MFCS 2000, LNCS 1893, Springer Verlag, 446-455.
20. K. Jansen and L. Porkolab, On preemptive resource constrained scheduling: polynomial time approximation schemes, unpublished manuscript, 2001.
21. N. Karmarkar and R.M. Karp, An efficient approximation scheme for the one-dimensional bin-packing problem, *Proceedings of 23rd IEEE Symposium on Foundations of Computer Science* FOCS 1982, 206-213.
22. K. Kilakos and O. Marcotte, Fractional and integral colourings, *Mathematical Programming* 76 (1997), 333-347.
23. K.L. Krause, V.Y. Shen and H.D. Schwetman, Analysis of several task scheduling algorithms for a model of multiprogramming computer systems, *Journal of the ACM* 22 (1975), 522-550 and Errata, *Journal of the ACM* 24 (1977), 527.
24. C. Lund and M. Yannakakis, On the hardness of approximating minimization problems, *Journal of the ACM* 41 (1994), 960-981.
25. O. Oguz and M.J. Magazine, A polynomial time approximation algorithm for the multidimensional $0-1$ knapsack problem, *working paper, University of Waterloo*, 1980.
26. T. Matsui, Approximation algorithms for maximum independent set problems and fractional coloring problems on unit disk graphs, *Proceedings Symposium on Discrete and Compuational Geometry*, LNCS 1763 (2000), Springer Verlag, 194-200.
27. T. Niessen and J. Kind, The round-up property of the fractional chromatic number for proper circular arc graphs, *Journal of Graph Theory* 33 (2000), 256-267.
28. E.R. Schreinerman and D.H. Ullman, Fractional graph theory: A rational approach to the theory of graphs, Wiley Interscience Series in Discrete Mathematics (1997).
29. P.D. Seymour, Colouring series-parallel graphs, *Combinatorica* 10 (1990), 379-392.
30. A. Srivastav and P. Stangier, Algorithmic Chernoff-Hoeffding inequalities in integer programming, *Random Structures and Algorithms* 8(1) (1996), 27-58.
31. A. Srivastav and P. Stangier, Tight approximations for resource constrained scheduling and bin packing, *Discrete Applied Mathematics* 79 (1997), 223-245.

Balanced Coloring: Equally Easy for All Numbers of Colors?

Benjamin Doerr

Mathematisches Seminar II, Christian-Albrechts-Universität zu Kiel,
Ludewig-Meyn-Str. 4, D–24098 Kiel, Germany,
bed@numerik.uni-kiel.de,
http://www.numerik.uni-kiel.de/~bed/

Abstract. We investigate the problem to color the vertex set of a hypergraph $\mathcal{H} = (X, \mathcal{E})$ with a fixed number of colors in a balanced manner, i.e., in such a way that all hyperedges contain roughly the same number of vertices in each color (discrepancy problem). We show the following result:

Suppose that we are able to compute for each induced subhypergraph a coloring in c_1 colors having discrepancy at most D. Then there are colorings in arbitrary numbers c_2 of colors having discrepancy at most $\frac{11}{10} c_1^2 D$. A c_2–coloring having discrepancy at most $\frac{11}{10} c_1^2 D + 3c_1^{-k}|X|$ can be computed from $(c_1 - 1)(c_2 - 1)k$ colorings in c_1 colors having discrepancy at most D with respect to a suitable subhypergraph of $\mathcal{H}$.

A central step in the proof is to show that a fairly general rounding problem (linear discrepancy problem in c_2 colors) can be solved by computing low-discrepancy c_1–colorings.

1   Introduction and Results

This paper deals with the problem of balanced hypergraph colorings (or equivalently, balanced partitions). A coloring in c colors is called balanced, if all hyperedges contain roughly the same number of vertices in each color. More precise, we define the discrepancy of a coloring to be the maximum deviation (taken over all hyperedges E and all colors d) of vertices in color d contained in E compared to the fair value $\frac{1}{c}|E|$.

Whereas the discrepancy problem was mostly studied in the context of 2 colors (see e. g. Beck and Sós [BS95], Matoušek [Mat99] or Chazelle [Cha00]), there has recently been work on the general problem (e. g. [DS99,BCC+00]). In this paper, we are interested in the relation between the discrepancy problem in different numbers of colors. Since [Doe01a] gave a class of hypergraphs having very different discrepancies in different numbers of colors, one might be pessimistic. On the other hand, there are several classes of hypergraphs having similar discrepancies in all numbers of colors, cf. [DS01]. This paper tries to solve this dichotomy.

A first result of this type already appeared in the paper [DS99]. There a recursive algorithm was presented that computes c–colorings from low-discrepancy

H. Alt and A. Ferreira (Eds.): STACS 2002, LNCS 2285, pp. 112–120, 2002.
© Springer-Verlag Berlin Heidelberg 2002

2–colorings. In this paper we extend this result to arbitrary numbers of colors. Unfortunately, it is not possible to use similar methods. Roughly speaking, it is relatively easy to compute c_2–colorings from c_1–colorings for $c_2 > c_1$ by a recursive partitioning scheme. Imbalances inflicted in earlier rounds of the recursion are split up in the following ones in a balanced manner. Therefore the final discrepancy can be estimated by something similar to a geometric series.

For the case $c_2 < c_1$ things are different. Of course, the case that c_2 divides c_1 is relatively trivial, but the remaining situations need more effort. The problem becomes visible already if we try to compute 2–colorings from 3–colorings. A natural approach would be to find a low-discrepancy 3–coloring and then to recursively recolor the vertices in color 3 according to further 3-colorings having low discrepancy on these vertices. If we organize this in a suitable way, at most $O(1)$ vertices in color 3 are left after $O(\log n)$ iterations (assuming n to be the number of vertices), which may be colored arbitrarily.

The draw-back of this approach is that the imbalances of the 3–colorings might accumulate. Thus for the final 2–coloring we cannot obtain a better discrepancy guarantee than $O(\log |X|)$ times the maximum discrepancy of the 3–colorings (cf. Theorem 2 for a precise version of this statement).

Our objective in this paper though is to show that the size of the hypergraph does not matter: Suppose that one can color a hypergraph and all its subgraphs with a fixed number of colors such that the corresponding discrepancies are at most D. Then this hypergraph can be colored with any number of colors such that the discrepancy is at most a constant factor larger than D. More precisely, we show:

Theorem 1. *Let $\mathcal{H} = (X, \mathcal{E})$ be a hypergraph. Let $c_1, c_2 \in \mathbb{N}_{\geq 2}$ be arbitrary numbers of colors. Suppose that for each induced subhypergraph of $\mathcal{H}$ there is a c_1–coloring of the vertex set having discrepancy at most D. Then there exists a c_2–coloring of $\mathcal{H}$ having discrepancy at most $\frac{11}{10} c_1^2 D$.*

A c_2–coloring of $\mathcal{H}$ having discrepancy at most $\frac{11}{10} c_1^2 D + 3 c_1^{-k} |X|$ can be computed from $(c_1 - 1)(c_2 - 1)k$ colorings in c_1 colors having discrepancy at most D with respect to a suitable induced subhypergraph of $\mathcal{H}$.

The key idea to prove Theorem 1 is to show an even more general result. Roughly speaking, in the setting of Theorem 1 it is also possible to round any floating coloring to an ordinary one with small discrepancy. This will be made precise in Section 3, where the necessary definitions introduced in Section 2 are available.

In a sense, Theorem 1 is best possible: Fix two numbers c_1, c_2 of colors. Since Theorem 1 works for arbitrary numbers of colors, we may apply it also with the roles of c_1 and c_2 interchanged. This shows that for all hypergraphs $\mathcal{H}$ the maximum discrepancy among the induced subhypergraphs is the same in c_1 and c_2 colors (apart from constant factors depending on c_1, c_2 only).

The reason why we have these strong results (compared to the negative example of [Doe01a]) is that we use the stronger assumption that all induced subhypergraphs admit a low-discrepancy coloring in c_1 colors. This excludes the class of hypergraphs exhibited in [Doe01a]. On the other hand, most results known in discrepancy theory are hereditary (refer to the maximum discrepancy among all subhypergraphs) since they relate the discrepancy to another hereditary property like maximum degree (Beck and Fiala [BF81], Srinivasan [Sri97]), VC-dimension and shatter functions (Matoušek, Welzl and Wernisch [MWW84], Matoušek [Mat95]) or total unimodularity (Ghouila-Houri [GH62], Doerr [Doe01b]).

Our main result admits several corollaries. We present two in Section 4, one concerning hypergraph having more vertices than hyperedges, the other one showing a connection between multi-color discrepancies and the problem of integral approximate solutions of linear equations.

2   Notation and Preliminaries

2.1   Multi-color Discrepancies

Let $\mathcal{H} = (X, \mathcal{E})$ be a finite hypergraph, that is, X is a finite set and $\mathcal{E} \subseteq 2^X$. Throughout this section let $c \in \mathbb{N}_{\geq 2}$ denote the number of classes we want to partition the vertices of $\mathcal{H}$ into. It is natural to represent the partition by a coloring. The partition classes then are formed by the sets of equally colored vertices. A c–coloring of $\mathcal{H}$ is a mapping $\chi : X \to M$, where M is any set of cardinality c. For convenience, normally one has $M = [c] := \{1, \dots, c\}$. Sometimes a different set M will be advantageous.

The basic idea of measuring the deviation of a given partition from the ideal one motivates these definitions: The *discrepancy of a hyperedge $E \in \mathcal{E}$ in color $d \in M$ with respect to χ* is

$$\operatorname{disc}_{\chi,d}(E) := \left| |\chi^{-1}(d) \cap E| - \frac{|E|}{c} \right|,$$

the *discrepancy of $\mathcal{H}$ with respect to χ* is

$$\operatorname{disc}(\mathcal{H}, \chi) := \max_{d \in M, E \in \mathcal{E}} \operatorname{disc}_{\chi,d}(E)$$

and the *discrepancy of $\mathcal{H}$ in c colors* is

$$\operatorname{disc}(\mathcal{H}, c) := \min_{\chi : X \to [c]} \operatorname{disc}(\mathcal{H}, \chi).$$

For a subset $X_0 \subseteq X$ of vertices denote by $\mathcal{H}_{|X_0} = (X_0, \mathcal{E}_{|X_0})$ the hypergraph induced by X_0, i. e., $\mathcal{E}_{|X_0} := \{E \cap X_0 | E \in \mathcal{E}\}$. The *hereditary discrepancy* in c colors is defined by

$$\operatorname{herdisc}(\mathcal{H}, c) := \max_{X_0 \subseteq X} \operatorname{disc}(\mathcal{H}_{|X_0}, c).$$

As in 2 colors, the notion of multi-color discrepancy has a natural extension to matrices. Let $A \in \mathbb{R}^{m \times n}$ be any real matrix and $\chi : [n] \to M$, where M is again an arbitrary set of cardinality c. Then the *discrepancy of A with respect to χ* is defined by

$$\mathrm{disc}(A, \chi) := \max_{d \in M, i \in [m]} \left| \sum_{j \in \chi^{-1}(d)} a_{ij} - \frac{1}{c} \sum_{j \in [n]} a_{ij} \right|.$$

The *discrepancy of A in c colors* is

$$\mathrm{disc}(A, c) := \min_{\chi:[n] \to [c]} \mathrm{disc}(A, \chi).$$

Immediately we see that $\mathrm{disc}(A, c) = \mathrm{disc}(\mathcal{H}, c)$ if A is the incidence matrix of $\mathcal{H}$. The extension to matrices is justified by two reasons. Firstly, it is an interesting optimization problem on its own right to partition the column vectors $a^{(j)}, j \in [n]$ of A into balanced classes, i. e., in a way that for all partition classes $\chi^{-1}(d)$ the sums $\sum_{j \in \chi^{-1}(d)} a^{(j)}$ are roughly equal. Secondly, even for some hypergraph problems the matrix notion is more convenient, e. g., to prove the Beck–Fiala theorem.

For the problems we are concerned with in this article it makes no difference whether we restrict ourselves to the special case of hypergraphs or the generalization to matrices. Hence from now on we will deal with the matrix case only. The notion of hereditary discrepancy translates to matrices in the obvious way: Write $A_{|J}$ to denote the submatrix of A containing the columns with index in J only. Then

$$\mathrm{herdisc}(A, c) := \max_{J \subseteq [n]} \mathrm{disc}(A_{|J}, c).$$

If one allows a logarithmic dependence on the size of the hypergraph, an elementary proof shows some relation between the discrepancy problem in different numbers of colors:

Theorem 2. *Let $A \in \mathbb{R}^{m \times n}$ and $c_2 < c_1$. If c_2 divides c_1, then any c_1–coloring $\chi_1 : [n] \to [c_1]$ yields a c_2–coloring χ_2 for A such that*

$$\mathrm{disc}(A, \chi_2) \leq \tfrac{c_1}{c_2} \mathrm{disc}(A, \chi_1).$$

Otherwise, a c_2–coloring χ such that

$$\mathrm{disc}(A, \chi_2) \leq \left\lfloor \tfrac{c_1}{c_2} \right\rfloor \log_{1/(1 - \frac{c_2}{c_1} \lfloor \frac{c_1}{c_2} \rfloor)}(n) D$$

$$\leq \left\lfloor \tfrac{c_1}{c_2} \right\rfloor \log_2(n) D$$

can be computed from $\log_2 n$ colorings in c_1 colors having discrepancy at most D with respect to suitable submatrices of A.

For small hypergraphs, this result is even superior to our main result. In general, of course, it leads into the wrong direction (suggesting that the size of the hypergraph might have an influence on how different hypergraphs behave in the discrepancy problem in different numbers of colors).

2.2 Linear Discrepancies

A rather general concept is the one of *linear discrepancy*. Here every vertex has an individual weight which describes in which ratio the vertex in average should contribute to each color class of each hyperedge it belongs to. A less obscure way of viewing this problem is to recognize it as rounding problem: For a given floating coloring assigning each vertex not a single color, but a weighted mixture of colors, we are looking for a 'pure' coloring (assigning each vertex a single color) such that each hyperedge in total receives every color roughly in the same amount by both colorings. This rounding aspect will be a central theme of the main proof.

At this point it will be convenient to choose a different set of colors. Denote by $E^c = \{e^{(1)}, \dots, e^{(c)}\}$ the standard basis of $\mathbb{R}^c$. Denote by $\overline{E}^c$ the convex hull of E^c, which is nothing more than the set of all $p \in [0,1]^c$ such that $\|p\|_1 = 1$. We call these vectors *c–color weights*. In the hypergraph case our objective hence is to 'round' a floating coloring $p : X \to \overline{E}^c$ to a coloring $q : X \to E^c$ in such a way that the imbalances $\left\| \sum_{x \in E} p(x) - \sum_{x \in E} q(x) \right\|_\infty$ are small for all $E \in \mathcal{E}$.

For matrices we define: A mapping $p : [n] \to \overline{E}^c$ is called a *floating coloring*. Denote by $\overline{C}^c$ the set of all floating colorings and by $C^c := \{p \,|\, p : [n] \to E^c\}$ the set of all pure colorings. For $p, q \in \overline{C}^c$ put

$$d_A(p,q) := \max_{i \in [m]} \left\| \sum_{j=1}^{n} a_{ij}(p(j) - q(j)) \right\|_\infty .$$

It is clear that d_A is pseudo-metric on $\overline{C}^c$, in particular it satisfies the triangle inequality. The *linear discrepancy of A with respect to $p \in \overline{C}^c$* now is

$$\mathrm{lindisc}(A, p) := d_A(p, C^c) = \min_{q \in C^c} d_A(p, q).$$

Let $\overline{p} : [n] \to \overline{E}^c; j \mapsto \frac{1}{c}\mathbf{1}_c$. Then $d_A(\overline{p}, q) \le D$ just means that q is a c–coloring such that $\mathrm{disc}(A, q) \le D$. Thus the linear discrepancy problem is a direct generalization of the discrepancy problem.

2.3 Types

Let $c_1, c_2 \in \mathbb{N}_{\ge 2}$. A vector $t \in \{0, \dots, c_1 - 1\}^{c_2}$ shall be called (c_1, c_2)–*type* in this paper if $\|t\|_1 = \sum_{i \in [c_2]} t_i = c_1$. Denote by T_{c_1, c_2} the set of all (c_1, c_2)–types. Put $n_{c_1, c_2} := |T_{c_1, c_2}|$ and $s_{c_1, c_2} := \sum_{t \in T_{c_1, c_2}} t$. The following three lemmata (proofs omitted) give some properties of these types.

Lemma 1. *The number of (c_1, c_2)–types is*

$$n_{c_1, c_2} = \binom{c_1 + c_2 - 1}{c_1} - c_2.$$

Lemma 2. *The sum of all (c_1, c_2)–types is $s_{c_1,c_2} = \frac{c_1}{c_2} n_{c_1,c_2} 1_{c_2}$.*

Lemma 3. *Let $v \in \{0, \dots, c_1-1\}^{c_2}$ such that c_1 divides $\|v\|_1$. Then v is the sum of (c_1, c_2)–types each thereof occurring just once, i.e., there are $\varepsilon_t \in \{0, 1\}, t \in T_{c_1,c_2}$ such that $v = \sum_{t \in T_{c_1,c_2}} \varepsilon_t t$. These $\varepsilon_t, t \in T_{c_1,c_2}$, can be found efficiently by a Greedy-Algorithm.*

3 Linear and Hereditary Discrepancy in Arbitrary Numbers of Colors

In this section we show how linear discrepancies in c_2 colors can be bounded in terms of the c_1–color hereditary discrepancy. Recall from Section 2 that the linear discrepancy problem in particular solves the ordinary discrepancy problem. For a matrix A we write $\|A\|_\infty := \max_{i \in [m]} \sum_{j \in [n]} |a_{ij}|$ to denote the operator norm induced by the maximum norm.

Theorem 3. *Let A be any real matrix, $c_1, c_2 \in \mathbb{N}_{\geq 2}$ and $p \in \overline{\mathcal{C}^{c_2}}$. Then there is a $q \in \mathcal{C}^{c_2}$ such that*

$$d_A(p, q) \leq \frac{c_1^2}{(c_1-1)c_2} n_{c_1,c_2} \operatorname{herdisc}(A, c_1).$$

A $q \in \mathcal{C}^{c_2}$ satisfying

$$d_A(p, q) \leq \frac{c_1^2}{(c_1-1)c_2} n_{c_1,c_2} D + c_1^{-k} \|A\|_\infty$$

can be computed from kn_{c_1,c_2} colorings in c_1 colors having discrepancy at most D with respect to some submatrix of A.

Note that for $c_1 = c_2 = 2$, Theorem 3 is just the result $\operatorname{lindisc}(A, 2) \leq 2 \operatorname{herdisc}(A, 2)$ of Beck and Spencer [BS84] and Lovász, Spencer and Vesztergombi [LSV86].[1] We recall the fact that for every real number $x \in [0, 1]$ that has a finite c–ary expansion $x = \sum_{k=0}^{l} b_k c^{-k}$ for some $b_k \in \{0, \dots, c-1\}, b_l \neq 0$, this expansion is unique (among all finite expansions). Denote by $l_c(x)$ the length of this c–ary expansion of x. Put $M_{c,l} := \{x \in [0, 1] \mid l_c(x) \leq l\}$ and $\mathcal{C}^{c_2}_{c_1,l} := \{p \mid p : [n] \to \overline{E}^{c_2} \cap M^{c_2}_{c_1,l}\}$ for all $l \in \mathbb{N}_0$.

The following lemma is the heart of our proof. It analyzes how well (with respect to $d_A(\cdot, \cdot)$) a floating coloring having a c_1–ary expansion of length l can be rounded to one of length $l-1$.

Lemma 4. *Let $p \in \mathcal{C}^{c_2}_{c_1,l}$ for some $l \in \mathbb{N}$. Then a $q \in \mathcal{C}^{c_2}_{c_1,l-1}$ such that*

$$d_A(p, q) \leq \frac{1}{c_2} c_1^{-l+2} n_{c_1,c_2} D$$

can be computed from n_{c_1,c_2} colorings in c_1 colors having discrepancy at most D with respect to some submatrix of A.

---

[1] By $\operatorname{lindisc}(A, c)$ we denote the maximum value $\operatorname{lindisc}(A, p)$ among all floating c–colorings $p \in \overline{\mathcal{C}^c}$. For two colors, this is equivalent to $\operatorname{lindisc}(A, 2) := \max_{p \in [0,1]^n} \min_{q \in \{0,1\}^n} \|A(p - q)\|_\infty$.

Proof. This algorithm solves the problem. Set $q^{(0)} = p$. Let $t^{(1)}, \ldots, t^{(n_{c_1,c_2})}$ be an enumeration of T_{c_1,c_2} such that $\|t^{(r)}\|_\infty \geq \|t^{(r+1)}\|_\infty$ for all $r \in [n_{c_1,c_2} - 1]$. For all $r = 1, \ldots, n_{c_1,c_2}$ do:

Iteration r: For every $j \in [n]$ let $q^{(r-1)}(j) = \sum_{k=0}^{l} b_k^{(r-1)}(j) c_1^{-k}$ for some $b_k^{(r-1)}(j) \in \{0, \ldots, c_1 - 1\}^{c_2}$ denote the c_1–ary expansion of the vector $q^{(r-1)}(j)$. Set $J^{(r)} := \{j \in [n] | b_l^{(r-1)}(j) \geq t^{(r)}\}$. Choose a coloring $\chi^{(r)} : J^{(r)} \to [c_1]$ such that $\mathrm{disc}(A_{|J^{(r)}}, \chi^{(r)}) \leq D$. Choose a function $f^{(r)} : [c_1] \to [c_2]$ such that $|(f^{(r)})^{-1}(d)| = t_d^{(r)}$ for all $d \in [c_2]$. For all $j \in [n], d \in [c_2]$ put

$$q^{(r)}(j)_d := \begin{cases} q^{(r-1)}(j)_d & \text{if } j \notin J^{(r)} \\ q^{(r-1)}(j)_d + (c_1 - t_d^{(r)}) c_1^{-l} & \text{if } j \in J^{(r)}, d = f^{(r)}(\chi^{(r)}(j)) \\ q^{(r-1)}(j)_d - t_d^{(r)} c_1^{-l} & \text{if } j \in J^{(r)}, d \neq f^{(r)}(\chi^{(r)}(j)). \end{cases}$$

Finally set $q := q^{(n_{c_1,c_2})}$.

We defer the correctness proof to the full version of this paper. □

The proof of Theorem 3 (also omitted) mainly consist of a repeated application of Lemma 4. As linear discrepancies are a generalization of the discrepancy problem, this already shows $\mathrm{disc}(A, c_2) = O_{c_1,c_2}(\mathrm{herdisc}(A, c_1))$. Unfortunately, the implicit constants are exponential in the numbers of colors.

4   Proof of the Main Results

In this section, we replace the exponential dependency on the number of colors by a quadratic dependency on c_1 only.

Theorem 4. *Let A be an m by n matrix. Let $c_1, c_2 \in \mathbb{N}_{\geq 2}$ be arbitrary numbers of colors. Suppose that for each submatrix of A there is a c_1–coloring having discrepancy at most D. Then there exists a c_2–coloring for A having discrepancy at most $\frac{11}{10} c_1^2 D$.*

A c_2–coloring for A having discrepancy at most $\frac{11}{10} c_1^2 D + 3c_1^{-k} \|A\|_\infty$ can be computed from $(c_1 - 1)(c_2 - 1)k$ colorings in c_1 colors having discrepancy at most D with respect to a suitable submatrix of A.

This improvement is made possible by a detour through 2 colors. Already by applying Theorem 3 first on the numbers of colors c_1 and 2 and then a second time on 2 and c_2, we can lower the constant to $O(c_1^2 c_2)$. We do slightly better (completely removing the dependence on c_2) by invoking the result of Srivastav and the author [DS99]. Using a recursive approach, they show that c–color discrepancies can be bounded in terms of 2–color discrepancies: For any hypergraph $\mathcal{H}$ and any number c of colors, a c–coloring χ satisfying

$$\mathrm{disc}(\mathcal{H}, \chi) \leq 2.0005\, D$$

can be computed from at most $(c-1)$ 2–colorings for induced subhypergraphs of $\mathcal{H}$ having discrepancy at most D with respect to a suitable weight. From the proof it is clear that an analogous result holds as well for discrepancies of matrices. We will use this fact without further proof.

Proof (of Theorem 4, sketched). By Theorem 3, we can compute good 2–colorings with respect to weights, since the weighted discrepancy problem is just the linear discrepancy one restricted to constant floating colorings (in the language of hypergraphs: All vertices have the same weight). Note that Theorem 3 works for induced submatrices as well. Therefore, combining Theorem 3 with the result cited above, we have Theorem 4 and its hypergraph version Theorem 1.   □

We end this section with two corollaries. If A has more columns than rows, a reduction by linear algebra due to Spencer [Spe87] can be applied: In two colors, the linear discrepancy is at most the maximum linear discrepancy among all submatrices containing at most m columns of A. This yields:

Corollary 1. *For any $m \times n$ matrix A and any number of colors $c \in \mathbb{N}_{\geq 2}$, we have*

$$\mathrm{disc}(A, c) \leq \tfrac{11}{10} c^2 \max_{\substack{J \subseteq [n] \\ |J| \leq m}} \mathrm{disc}(A_{|J}, c).$$

A c–coloring χ for A having discrepancy $\mathrm{disc}(A, \chi) \leq \tfrac{11}{10} c^2 D + 3c^{-k} \|A\|_\infty$ can be computed from $(c-1)^2 k$ colorings in c colors having discrepancy at most D for a suitable submatrix of A having at most m columns.

Proof (sketched). We use Theorem 3 to step down to 2 colors, apply Spencer's reduction and return to c colors again (using [DS99]).   □

The linear discrepancy in two colors also describes how well a solution of a linear system can be rounded to an approximate integer one (this is a folklore result easily being deduced from the definition). Combined with Theorem 3 we derive:

Corollary 2. *Let $A \in \mathbb{R}^{m \times n}$ and $b \in \mathbb{R}^m$ such that the linear system $Ax = b$ has a solution x. Then there is a $z \in \mathbb{Z}^n$ such that $\|x - z\|_\infty \leq 1$ and $\|Az - b\|_\infty \leq \tfrac{11}{10} c^2 \, \mathrm{herdisc}(A, c)$ for all $c \in \mathbb{N}_{\geq 2}$.*

5   Concluding Remarks

In this paper we showed that the hereditary discrepancy in nearly independent of the numbers of colors. This strongly contrasts the ordinary discrepancy. Our result suggests that the hereditary discrepancy is a very general measure of how well a hypergraph behaves in partitioning problems.

To prove the main result $\mathrm{herdisc}(A, c_2) = \Theta_{c_1}(\mathrm{herdisc}(A, c_1))$ we needed a detour through linear discrepancies. It seems to be an interesting question whether this is necessary, or if a bound of type $\text{herdisc}(A, c_2) = \Theta_{c_1, c_2}(\text{herdisc}(A, c_1))$ can be proven more directly. The best result avoiding the detour we have contains a logarithmic dependence of the number of columns, cf. Theorem 2.

A second open problem is the precise influence of the numbers of colors. Our bound contains a factor of $O(c_1^2)$, whereas we only know examples justifying a factor of $\Omega(c_1)$.

References

[BCC+00]   T. Biedl, E. Cenek, T. M. Chan, E. D. Demaine, M. L. Demaine, R. Flei-scher, and Ming-Wei Wang. Balanced k-colorings. In *Proceedings of the 25th International Symposium on Mathematical Foundations of Computer Science (MFCS 2000)*, volume 1893 of *Lecture Notes in Computer Science*, pages 202–211, Berlin–Heidelberg, 2000. Springer Verlag.

[BF81]   J. Beck and T. Fiala. "Integer making" theorems. *Discrete Applied Mathematics*, 3:1–8, 1981.

[BS84]   J. Beck and J. Spencer. Integral approximation sequences. *Math. Programming*, 30:88–98, 1984.

[BS95]   J. Beck and V. T. Sós. Discrepancy theory. In R. Graham, M. Grötschel, and L. Lovász, editors, *Handbook of Combinatorics*. 1995.

[Cha00]   B. Chazelle. *The Discrepancy Method*. Princeton University, 2000.

[Doe01a]   B. Doerr. Discrepancy in different numbers of colors. *Discrete Mathematics*, 2001. To appear.

[Doe01b]   B. Doerr. Lattice approximation and linear discrepancy of totally unimodular matrices. In *Proceedings of the 12th Annual ACM-SIAM Symposium on Discrete Algorithms*, pages 119–125, 2001. To appear in *Combinatorica*.

[DS99]   B. Doerr and A. Srivastav. Approximation of multi-color discrepancy. In D. Hochbaum, K. Jansen, J. D. P. Rolim, and A. Sinclair, editors, *Randomization, Approximation and Combinatorial Optimization (Proceedings of APPROX-RANDOM 1999)*, volume 1671 of *Lecture Notes in Computer Science*, pages 39–50, Berlin–Heidelberg, 1999. Springer Verlag.

[DS01]   B. Doerr and A. Srivastav. Recursive randomized coloring beats fair dice random colorings. In A. Ferreira and H. Reichel, editors, *Proceedings of the 18th Annual Symposium on Theoretical Aspects of Computer Science (STACS) 2001*, volume 2010 of *Lecture Notes in Computer Science*, pages 183–194, Berlin–Heidelberg, 2001. Springer Verlag.

[GH62]   A. Ghouila-Houri. Caractérisation des matrices totalement unimodulaires. *C. R. Acad. Sci. Paris*, 254:1192–1194, 1962.

[LSV86]   L. Lovász, J. Spencer, and K. Vesztergombi. Discrepancies of set-systems and matrices. *Europ. J. Combin.*, 7:151–160, 1986.

[Mat95]   J. Matoušek. Tight upper bound for the discrepancy of half-spaces. *Discr. & Comput. Geom.*, 13:593–601, 1995.

[Mat99]   J. Matoušek. *Geometric Discrepancy*. Springer-Verlag, Berlin, 1999.

[MWW84]   J. Matoušek, E. Welzl, and L. Wernisch. Discrepancy and approximations for bounded VC–dimension. *Combinatorica*, 13:455–466, 1984.

[Spe87]   J. Spencer. *Ten Lectures on the Probabilistic Method*. SIAM, 1987.

[Sri97]   A. Srinivasan. Improving the discrepancy bound for sparse matrices: better approximations for sparse lattice approximation problems. In *Proceedings of the Eighth Annual ACM-SIAM Symposium on Discrete Algorithms (New Orleans, LA, 1997)*, pages 692–701, New York, 1997. ACM.

The Complexity of Graph Isomorphism for Colored Graphs with Color Classes of Size 2 and 3

Johannes Köbler[1] and Jacobo Torán[2]

[1] Institut für Informatik, Humboldt-Universität zu Berlin, D-10099 Berlin, Germany,
koebler@informatik.hu-berlin.de
[2] Abt. Theoretische Informatik, Universität Ulm, D-89069 Ulm, Germany,
toran@informatik.uni-ulm.de

Abstract. We prove that the graph isomorphism problem restricted to colored graphs with color multiplicities 2 and 3 is complete for symmetric logarithmic space SL under many-one reductions. This result improves the existing upper bounds for the problem.
We also show that the graph automorphism problem for colored graphs with color classes of size 2 is equivalent to deciding whether an undirected graph has more than a single connected component and we prove that for color classes of size 3 the graph automorphism problem is contained in SL.

1 Introduction

The graph isomorphism problem GI consists in deciding whether there is a bijection between the nodes of two given graphs, G and H, preserving the edge relation. GI is one of the most intensively studied problems in Theoretical Computer Science. Besides its many applications, the main source of interest in GI has been the evidence that this problem is probably neither in P nor NP-complete. Other sources of interest include the sophistication of the tools developed to attack the problem (for example [3,14]), and the connections between GI and structural complexity (see [10]).

Understandably, many GI restrictions have been considered. For example, P upper bounds are known in the cases of planar graphs [7] or graphs of bounded valence [14]. In some cases, like trees [12,4] or graphs with colored vertices and bounded color classes [15], even NC algorithms for isomorphism exist (NC^k is defined to be the class of problems solvable by a uniform family of bounded indegree Boolean circuits with $O(\log^k n)$ depth and polynomial size, and NC = $\bigcup_{k \geq 0} NC^k$).

However, until recently none of these GI restrictions were known to be complete for a natural complexity class, and it seemed that these problems lack the structure needed for a hardness result. Several results in the last years have changed this situation. In [9] it was shown that the isomorphism problem for trees is hard for NC^1 and for L (logarithmic space) depending on the encoding

H. Alt and A. Ferreira (Eds.): STACS 2002, LNCS 2285, pp. 121–132, 2002.
© Springer-Verlag Berlin Heidelberg 2002

of the input. Moreover in [17] it was proved that isomorphism for general graphs is hard for NL (nondeterministic logarithmic space) and for every logarithmic space class based on counting. Regarding completeness, two versions of tree isomorphism are respectively complete for NC^1 and L [12,4,9] and until now this was the only GI restriction to a special graph family known to be complete for a complexity class.

We continue in this paper with this line of research concentrating on the family of graphs with colored nodes and with a small number of nodes with the same color. We will denote by b-GI the isomorphism problem for colored graphs with color multiplicities bounded by b. Isomorphisms between colored graphs are color preserving and this reduces the number of possible isomorphisms. But observe that even for the case $b = 2$ the number of color-preserving mappings between n-vertex graphs could be as large as $2^{\frac{n}{2}}$. This problem has a fairly long tradition in the area of algorithms. Babai gave in [2] a random polynomial-time algorithm for testing graph isomorphism in the case that the color multiplicities are bounded by a constant b. The algorithm was improved to be deterministic in [6] and, by using sophisticated algebraic tools, Luks proved in [15] that b-GI lies in fact in NC.

We focus in this paper in the family of colored graphs with color multiplicities bounded by the constants 2 and 3 proving that 2-GI and 3-GI are many-one complete for symmetric logarithmic space SL under logarithmic space reductions. The complexity class SL introduced in [11] has different characterizations, but the easiest way to define it is the following: SL is the class of problems that are logarithmic space reducible to the reachability problem for undirected graphs, UGAP. Our results improve on the one hand the upper bound for the problem given by Luks from NC to SL (SL is included in NL and in NC^2). On the other hand the results provide new natural examples for complete problems in SL showing that reachability questions can be expressed in a natural way as isomorphisms in a class of graphs.

Closely related to graph isomorphism is the graph automorphism problem GA that consists in deciding whether a given graph has a non-trivial automorphism, or in other words, whether there is a permutation of the nodes, different from the identity, preserving the adjacency relation. The relationship between GA and GI is not completely clear. It is known that GA is many-one reducible to GI [13] but a reduction in the other direction is not known and GA seems to be an easier problem. We show that this situation also occurs when we restrict the problem to colored graphs with color classes of size 2 and 3. We prove that 2-GA and 3-GA belong also to SL and moreover 2-GA is equivalent to the problem $\overline{UCC}$ of deciding whether a given graph has more than one connected component. $\overline{UCC}$ belongs to SL but it seems to be easier than UGAP and it is not known to be complete for SL [1].

For simplicity we have proven our hardness and equivalence results for logarithmic space many-one reducibility, but in fact they hold for stronger reducibilities, like for example DLOGTIME uniform NC^1 many-one reducibility. All the graphs considered in this paper are undirected graphs.

2   Upper Bound for 2-GA

We show in this section that 2-GA is reducible to the problem $\overline{UCC}$ of deciding whether a given graph has more than a single connected component. For a graph G we denote by $\text{Aut}(G)$ the set of automorphisms in G. The next lemma shows that we can restrict ourselves to graphs with at most two edges between any two colors. The proof of this result is straightforward.

Lemma 1. *Let $G = (V, E)$ be a graph with colored vertices, and let C_i and C_j two color classes in G. Then $\text{Aut}(G) = \text{Aut}(G')$, where $G' = (V, E')$ is a copy of G but with the dual set of edges between vertices of C_i and C_j (for every pair (u, v) with $u \in C_i$ and $v \in C_j$, it holds $(u, v) \in E \Leftrightarrow (u, v) \notin E'$).*

Theorem 1. *The Graph Automorphism problem restricted to graphs with color classes of size at most 2 can be reduced to $\overline{UCC}$.*

Proof. Let G be a graph with color classes of size at most 2. By the above lemma we can assume that for every pair of colors, there are at most two edges connecting the nodes of these colors. Also, w.l.o.g. we can assume that there is no edge between the nodes of the same color in G (these edges do not change the automorphism group). The possible connections that have to be considered between the nodes of two different colors are shown in Figure 1.

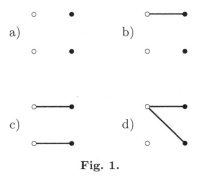

Fig. 1.

We reduce the question of whether G has a nontrivial automorphism to a reachability question in an undirected graph $f(G) = (V', E')$. V' contains one special node plus one node for each color,

$$V' = \{v_{id}\} \cup \{v^i \mid i \text{ is a color in } G\}.$$

The idea of the reduction is to place the edges in such a way that for every color i, node v_{id} is reachable from node v^i in $f(G)$ if and only if every automorphism in $\text{Aut}(G)$ fixes the nodes of color i. For every pair (i, j) of colors, the edges between two color classes induce edges in E' in the following way:

Type of connection between i and j Edges in E'

a)   None
b) $(v^i, v_{id}), (v^j, v_{id})$
c) (v^i, v^j)
d) (v^i, v_{id}).

Also, if there is only one node of color i, the edge (v^i, v_{id}) is included in E'.

Lemma 2. *For each color i, one can reach node v_{id} from v^i in $f(G)$ if and only if every automorphism in $\mathrm{Aut}(G)$ fixes the nodes of color i.*

Proof. The proof from left to right is by induction on the number of edges in a path between v^i and v_{id}. If the path has length one, then the connections between the nodes of color i and some other color class in G must be like in cases b) or d). In any of these cases the nodes of color i must be fixed under any automorphism in G. If the path of minimal length has $k + 1$ edges then there must be a color j and a path with k edges from v^j to v_{id} in $f(G)$. Also there must be an edge (v^i, v^j) in $f(G)$ which means that the connections between color classes i and j in G must be like in case c). By induction hypothesis every automorphism in $\mathrm{Aut}(G)$ fixes the nodes of color j, and the connection c) forces the nodes of color i to be fixed also.

For the other direction, suppose that for some color i node v_{id} is not reachable from v^i in $f(G)$. We will show that in this case there is some non-trivial automorphism in G. W.l.o.g. let $\{1, \ldots, k\}$ be the set of colors j such that v^j is connected to v^i in $f(G)$. Since v_{id} is not reachable from v^i, in case there are edges between two classes with colors in $\{1, \ldots, k\}$ in G, these edges form a connection of type c), and the only possible edges between classes of colors l and m, with $l \notin \{1, \ldots, k\}$ and $m \in \{1, \ldots, k\}$ are connections of type d). We claim that the color fixing permutation φ interchanging the two nodes of the classes of colors $j \in \{1, \ldots, k\}$ and fixing the rest of the nodes is a non-trivial automorphism in G. Let u, v be two nodes in V. If neither u nor v have colors in the set $\{1, \ldots, k\}$ then φ acts like the identity on (u, v). If both u and v have colors l, m in the set $\{1, \ldots, k\}$ then either there is no connection or the connections between the color classes l and m must be of type c). It follows that $(u, v) \in E$ if and only if $(u', v') \in E$, where u' and v' are the other nodes with colors l and m respectively. By the definition of φ, $(u', v') = (\varphi(u), \varphi(v))$. Finally, if u has color $l \notin \{1, \ldots, k\}$ and v has color $m \in \{1, \ldots, k\}$ and there is a connection between nodes of color l and m then it must be of type d) (a connection of type c) would force color l to be in $\{1, \ldots, k\}$). $(\varphi(u), \varphi(v)) = (u, v')$. As in the previous two cases we have $(u, v) \in E$ if and only if $(\varphi(u), \varphi(v)) \in E$. $\square$ (Lemma 2)

Observe that Lemma 2 implies that the graph G has a nontrivial automorphism if and only if $f(G)$ has more than one connected component. $\square$

3 Upper Bounds for 3-GI and 3-GA

We deal in this section with graphs having up to 3 different nodes of each color. For a pair (G, H) of graphs, we will denote by $\mathrm{Iso}(G, H)$ the set of isomorphisms between both graphs. We denote by B_3 the set of 6 bijections defined over a domain set of 3 elements. An isomorphism between graphs with color classes of size 3 can be decomposed in a product of bijections from B_3.

Theorem 2. *The Graph Isomorphism problem restricted to graphs with color classes of size at most 3 is in the class* SL.

Proof. Let G and H be the input graphs. By Lemma 1 we can assume that for any pair (i, j) of colors there are at most 4 edges in each one of the input graphs having as endpoints nodes of colors i and j. If there are more than 4 edges we consider the dual connections. The three nodes of a color i in G are labeled i_1, i_2 and i_3, and in H by i'_1, i'_2 and i'_3.

We will reduce graphs G and H to a single undirected graph $f(G, H)$ and translate the isomorphism question to a reachability question in $f(G, H)$. For each color i we will consider 6 nodes in $f(G)$ corresponding to the possible bijections in B_3. Additionally there is one extra node w that will be used to indicate that some isomorphisms between subgraphs of G and H are not possible. The set of nodes in $f(G, H)$ is

$$\{v_\varphi^i \mid \varphi \in B_3 \text{ and } i \text{ is a color in } G\} \cup \{w\}.$$

The set of edges in $f(G, H)$ is constructed according to the following 2 rules. Observe that both rules can be applied in logarithmic space since the color classes have constant size 3 and there can only be at most 6 potential isomorphisms between the nodes of a given color.

Let C_i (resp. C'_i) denote the set of nodes of color i in G (resp. in H). For each color i the edges with both endpoints in C_i or C'_i might imply that some of the bijections from C_i to C'_i cannot be extended to an isomorphism between G and H. The restrictions in the set of possible bijections can also be induced by the connections with a different color class. In these cases we include some edges in the graph $f(G, H)$ as in Rule 1, indicating that a bijection between the nodes of a color cannot be extended to an isomorphism:

Rule 1: For every pair of colors i, j and every bijection $\varphi \in B_3$, we include in $f(G, H)$ the edge (v_φ^i, w) if the edges between C_i and C_j in G and the edges between C'_i and C'_j in H imply that no isomorphism in $\mathrm{Iso}(G, H)$ can map the nodes of C_i into C'_i like φ, that is, if for every $\psi \in B_3$, $\varphi \times \psi \notin \mathrm{Iso}(C_i \cup C_j, C'_i \cup C'_j)$, where $(C_i \cup C_j)$ is the graph made by the set of nodes C_i and C_j and the edges between them. For the trivial case in which there is only one color i in G, the edge (v_φ^i, w) is included in $f(G, H)$ when $\varphi \notin \mathrm{Iso}(C_i, C'_i)$.

We also include in $f(G, H)$ some edges between nodes corresponding to different color classes indicating that a partial isomorphism between the nodes of some color forces a partial isomorphism between the nodes of another color.

Rule 2: For every pair of colors i, j, and $\varphi \in B_3$. If for a pair of nodes $a, b \in \{i_1, i_2, i_3\}$ and a pair of nodes $a', b' \in \{i'_1, i'_2, i'_3\}$ and two bijections $\eta, \pi \in B_3$ the edges between the sets of nodes $\{i_a, i_b\}$ and $\{j_{\eta(a)}, j_{\eta(b)}\}$ are exactly the two edges $\{(i_a, j_{\eta(a)}), (i_b, j_{\eta(b)})\}$ and the edges between the sets of nodes $\{i'_a, i'_b\}$ and $\{j'_{\pi(a)}, j'_{\pi(b)}\}$ are exactly $\{(i'_a, j'_{\pi(a)}), (i'_b, j'_{\pi(b)})\}$ and $\varphi \times \psi \in \mathrm{Iso}(C_i \cup C_j, C'_i \cup C'_j)$ (for $\psi = \pi \varphi \eta^{-1}$) then we include in E' the edge (v^i_φ, v^j_ψ) (see Figure 2).

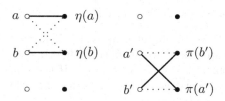

Fig. 2. A situation in which Rule 2 is applied. The dotted lines indicate that these edges do not exist.

We show with the following lemmas that a reachability question in $f(G, H)$ can be used to decide whether G and H are isomorphic.

Lemma 3. *For a pair of colors i, j and $\varphi, \psi \in B_3$, if $(v^i_\varphi, w) \notin f(G, H)$ and $(v^j_\psi, w) \notin f(G, H)$ and $(v^i_\varphi, v^j_\pi) \notin f(G, H)$ for any $\pi \in B_3$, then $\varphi \times \psi$ is an isomorphism between the subgraphs $C_i \cup C_j$ and $C'_i \cup C'_j$.*

Proof. Let $G = (V, E)$ and $H = (W, E)$ be the input graphs and let us denote by a, b and c the nodes of color i in G, by d, e and f the color j nodes. We denote by the same symbols with ' the nodes of these colors in H. W.l.o.g we can suppose that for $l \in \{a, b, c\} \varphi(l) = l'$ and for $m \in \{d, e, f\}$ $\psi(m) = m'$. Let us suppose by contradiction that $\varphi \times \psi \notin \mathrm{Iso}(C_i \cup C_j, C'_i \cup C'_j)$. It follows that for a pair of nodes a, d,

$$(a, d) \in E \Leftrightarrow (\varphi(a), \psi(d)) \notin F.$$

We consider that $(a, d) \in E$, the other case is analogous. Since v^i_φ and v^j_ψ are not connected to w, by Rule 1 follows that there are two bijections η and $\pi \in B_3$ such that $\varphi \times \eta$ and $\pi \times \psi \in \mathrm{Iso}(C_i \cup C_j, C'_i \cup C'_j)$. Because of these facts we have that $\eta(d) \neq d'$ and $\pi(a) \neq (a')$. Again w.l.o.g. we can suppose $\eta(d) = e'$ and $\pi(a) = b'$. It follows $(\varphi(a), \eta(d)) = (a', e') \in F$ and $(\pi(a), \psi(d)) = (b', d') \in F$. The rest of the proof consists in considering the different possibilities for the bijections η and π showing that in each case we reach a contradiction.

Case 1: Suppose $\eta(d) = e', \eta(e) = d'$ and $\eta(f) = f'$. $(a', d') \notin F$ implies $(\varphi^{-1}(a'), \eta^{-1}(d')) = (a, e) \notin E$ and therefore $(\pi(a), \psi(e)) = (b', e') \notin F$ and also $(\varphi^{-1}(b'), \eta^{-1}(e')) = (b, d) \notin E$. If $\pi(b) = a'$ then from $(a', e') \in F$ follows $(\pi^{-1}(a'), \psi^{-1}(e')) = (b, e) \in E$. But then we have that the edges between the nodes a, b, d, e in G and between their counterparts in H are exactly as is Rule 2, and there should be an edge in $f(G, H)$ from v_φ^i to some node of color j contradicting the hypothesis. The other possible situation in when $\pi(b) = c'$ and $\pi(c) = a'$ but then from $(b, d) \notin F$ follows $(\varphi^{-1}\pi(b), \eta^{-1}\psi(d)) = (c, e) \notin E$, and from $(a', e') \in F$ follows $(\pi^{-1}(a'), \psi^{-1}(e')) = (c, e) \in E$ which is a contradiction.

Case 2: Suppose $\eta(d) = e', \eta(e) = f'$ and $\eta(f) = d'$. By the same arguments as in Case 1, we have $(b, f) \in E$, $(a, f) \notin E$ $(b', f') \notin F$ and $(b', e') \notin E$. If $\pi(b) = a'$ and $\pi(c) = c'$ then from $(b, e) \notin E$ follows $(\pi(b), \psi(e)) = (a', e') \notin F$ which is a contradiction. Finally, if $\pi(b) = c'$ and $\pi(c) = a'$ then it follows $(a, d), (b, f), (c, e) \in E$, $(a, f), (b, e), (c, d) \notin E$, $(a', e'), (b', d'), (c', f') \in F$ and $(a', d'), (b', f'), (c', e') \notin F$. Consider the pair of nodes a and e. If $(a, e) \in E$ then $(a', f'), (b', e') \in F$ contradicting the fact that there are at most 4 edges in H connecting the i and j nodes. On the other hand, if $(a, e) \notin E$ then the set of edges between the i and j color nodes are exactly $(a, d), (b, f), (c, e)$ in E and $(a', e'), (b', d'), (c', f')$ in F. By Rule 2 there should be some edge in $f(G, H)$ from v_φ^i to some node of color j, contradicting the hypothesis. $\square$ (Lemma 3)

Lemma 4. *For each pair of colors i, j, and $\varphi, \psi \in B_3$, if there is a path from v_φ^i in $f(G, H)$ to v_ψ^j not having w as an intermediate node then every isomorphism in $\mathrm{Iso}(G, H)$ that maps the nodes of color i like φ, is forced to map the nodes of color j like ψ.*

Proof. We use induction on the length of a minimal path from v_φ^i to v_ψ^j in $f(G, H)$. If this path has length 1 then the colors i and j are different and the edge (v_φ^i, v_ψ^j) in $f(G, H)$ has been placed by Rule 2. This implies that for some $a, b \in \{1, 2, 3\}$ and some $c, d \in \{1, 2, 3\}$ the edges $(i_a, j_{\eta(a)}), (i_b, j_{\eta(b)})$ (for some $\eta \in B_3$) are the only edges between the sets of nodes $\{i_a, i_b\}$ and $\{j_{\eta(a)}, j_{\eta(b)}\}$ in G and the edges $(i_a, j_{\pi(a)}), (i_b, j_{\pi(b)})$ (for some $\pi \in B_3$) are the only edges between the sets of nodes $\{i'_c, i'_d\}$ and $\{j_{\pi(c)}, j_{\pi(d)}\}$ in H. Moreover, because of Rule 2, $\psi = \pi\varphi\eta^{-1}$. If an isomorphism in $\mathrm{Iso}(G, H)$ maps the i nodes to the i' nodes like φ, then for $l \in \{\eta(a), \eta(b)\}$ j_l must be mapped to a node j' in H connected to $i'_{\varphi\eta^{-1}(l)}$, and this node is $j_{\pi\varphi\eta^{-1}(l)} = j_{\psi(l)}$. It follows also for $c \notin \{a, b\}$, j_c is mapped to $j_{\pi\varphi\eta^{-1}(c)}$.

For the induction step, if the number of edges in the path from v_φ^i and v_ψ^j in $f(G, H)$ is $k+1$, let m be the first color after i in the path. There has to be some bijection $\phi \in B_3$ and an edge (v_φ^i, v_ϕ^m) between the i nodes and the j nodes in $f(G, H)$ and this edge must be introduced by Rule 2. By induction hypothesis, every isomorphism in $\mathrm{Iso}(G, H)$ mapping the i nodes like φ must map the m nodes like ϕ, and every isomorphism in $\mathrm{Iso}(G, H)$ mapping the m nodes like ϕ must map the j nodes like ψ. Both conditions together imply the result. $\square$ (Lemma 4)

Observe that Lemma 4 implies that for any color i, and $\varphi, \psi \in B_3$, if $\varphi \neq \psi$, and v_φ^i is reachable from v_ψ^i in $f(G, H)$ then there is no isomorphism in $\mathrm{Iso}(G, H)$ mapping the nodes of color i like φ. Another consequence of the lemma (together with the definition of Rule 1, is that for any color i and $\varphi \in B_3$, if node w can be reached from v_φ^i in $f(G, H)$ then there is no isomorphism in $\mathrm{Iso}(G, H)$ mapping the nodes of color i like φ.

Lemma 5. *Suppose that there are k different colors in G and H. If there is a set of nodes in $f(G, H)$, $v_{\varphi_1}^1, \ldots, v_{\varphi_k}^k$ one of each color, and such that no other node in $f(G, H)$ is reachable from this set then $\varphi_1 \times \ldots \times \varphi_k$ is an isomorphism between G and H.*

Proof. The proof is by induction on the number of color classes k in G and H. If there is only one color the result is trivial. For the case of two colors i and j, consider that from the set of nodes v_φ^i and v_ψ^j one cannot reach any other node in $f(G, H)$. This implies that the only possible edge in $f(G, H)$ with an endpoint in this set is the one connecting both nodes. If this edge does not exist the result follows by Lemma 3. On the other hand, if (v_φ^i, v_ψ^j) is an edge in $f(G, H)$ then the edge was placed by Rule 2 and $\varphi \times \psi \in \mathrm{Iso}(G, H)$.

For the induction step, consider that there are k colors in G and H and there is a set of nodes one of each color $v_{\varphi_1}^1, \ldots, v_{\varphi_k}^k$ in $f(G, H)$. Consider the graphs G' and H' obtained by deleting the nodes of color k in G and H and all the edges having an endpoint of this color. Since eliminating one color can only reduce the set of local restrictions for isomorphisms, there is no new edge in $f(G', H')$ that was not already present in $f(G, H)$ and therefore from $v_{\varphi_1}^1, \ldots, v_{\varphi_{k-1}}^k$ no other node is reachable in $f(G', H')$. By induction hypothesis $\varphi_1 \times \ldots \times \varphi_{k-1} \in \mathrm{Iso}(G', H')$. We claim that this isomorphism between G' and H' can be extended to an isomorphism in $\mathrm{Iso}(G, H)$ by mapping the nodes in C_k to C_k' as in φ_k.

To see that this is an isomorphism we will show that for every $j < k$ it holds $\varphi_k \times \varphi_j \in \mathrm{Iso}(C_k \cup C_j, C_k' \cup C_j')$. If the edge $(v_{\varphi_k}^k, v_{\varphi_j}^j)$ belongs to $f(G, H)$ then the edge was placed by Rule 2 and $\varphi_k \times \varphi_j \in \mathrm{Iso}(C_k \cup C_j, C_k' \cup C_j')$. On the other hand if this edge does not exist, there is no other edge in $f(G, H)$ between k and j nodes having $v_{\varphi_k}^k$ or $v_{\varphi_j}^j$ as endpoint (from the set $v_{\varphi_1}^1, \ldots, v_{\varphi_k}^k$ no other node can be reached). The result follows then by Lemma 3.

$\square$ (Lemma 5)

It follows by Lemmas 4 and 5 that there is an isomorphism from G to H if and only if there is a set of nodes in the graph $f(G, H)$, one of each color, such that from this set no other node in $f(G, H)$ can be reached. In order to transform this into a question in SL we need the following lemma:

Lemma 6. *Let A and B be two connected components in $f(G, H)$ satisfying the following conditions:*

i)   *The nodes in A have different colors and the nodes in B have different colors.*
ii) *Node w does not belong to $A \cup B$ and.*
iii) *The intersection of colors in A and B is not empty.*

Then the set of colors present in A is the same as the set of colors present in B.

Proof. We show that for any pair of colors i and j, and for two bijections φ and $\xi \in B_3$, if the node w in $f(G, H)$ cannot be reached neither from v_φ^i nor from

v_ξ^i and if v_φ^i has a neighbor of color j in $f(G, H)$, then so does v_ξ^i. The result follows from this fact.

Suppose that (v_φ^i, v_ψ^j) is an edge in $f(G, H)$. This edge was set by Rule 2 and therefore there is a pair of nodes $a, b \in \{i_1, i_2, i_3\}$ and a pair of nodes $a', b' \in \{i'_1, i'_2, i'_3\}$ and two bijections $\eta, \pi \in B_3$ satisfying the conditions of Rule 2 and such that $\psi = \pi\varphi\eta^{-1}$. Since v_ξ^i is not connected to w, there must be a bijection $\gamma \in B_3$ satisfying $\xi \times \gamma \in \mathrm{Iso}(C_i \cup C_j, C'_i \cup C'_j)$. η and π describe the connections between i and j nodes in G and H and therefore we have $\gamma = \pi\xi\eta^{-1}$. We are again in the conditions of Rule 2, and the edge (v_ξ^i, v_γ^j) belongs to $f(G, H)$.
$\square$ (Lemma 6)

We know that there is an isomorphism from G to H if and only if there is a set of nodes in $f(G, H)$, one of each color and such that from this set no other node in $f(G, H)$ can be reached. By Lemma 6 in order to test this it suffices to check for each color i that there is one node v_φ^i in $f(G, H)$ from which neither w nor two nodes of the same color can be reached (this question can be solved within the class SL). In order to see this observe that if G and H are isomorphic, such a set must exist. By Lemma 6 if such a set exists then there is a set of color disjoint connected components in $f(G, H)$ containing all colors. $\square$

We show now that the graph automorphism problem for colored graphs with color classes of size 3 lies also in the class SL. Although a direct proof similar to the one in Theorem 2 is possible, it is easier to give a reduction from 3-GA to UGAP based on the fact that 3-GI $\in$ SL.

Theorem 3. *The Graph Automorphism problem restricted to graphs with color classes of size at most 3 is in the class* SL.

Proof. We will show that 3-GA is logarithmic space many-one reducible to UGAP. This implies that 3-GA lies in SL. Let $G = (V, E)$ be a graph with its nodes partitioned into color classes of size at most three. We denote by $G_{[i]}$ a copy of G but with node i marked with a new special color. There is a non-trivial automorphism in G if and only if for a pair of distinct nodes of the same color i and j in V, there is an automorphism mapping i to j, if and only if for such a pair of nodes $G_{[i]}$ is isomorphic to $G_{[j]}$. Since the color classes in $G_{[i]}$ have size 3 at most, this means that 3-GA is reducible to a set of disjunctive queries to 3-GI. By Theorem 2, 3-GI lies in SL and can be reduced to UGAP. The list of queries to 3-GI can then be reduced to a list of reachability queries in undirected graphs. The disjunctive list of queries to UGAP can be reduced to a single one by connecting the graphs in parallel. This provides a many-one reduction from 3-GA to UGAP. $\square$

4 Lower Bounds

We prove now the hardness results for 2-GI and 2-GA.

Theorem 4. 2-GI *is hard for* SL *under logarithmic space many-one reductions.*

Proof. We show that the graph accessibility problem for undirected graphs, UGAP, is reducible to the complement of 2-GI. The result follows since UGAP is logarithmic space complete for SL, and this class is closed under complementation [16].

Let $G = (V, E)$ be an undirected graph with two designated nodes $s, t \in V$. Consider the new graph $G' = G_1 \cup G_2$ where G_1 and G_2 are two copies of G, and for a node $v \in V$ let us call v_1 and v_2 the copies of v in G_1 and G_2 respectively. Furthermore, we color each pair of nodes v_1, v_2 with color i_v, and color t_1 with a special color 1 and t_2 with another color 2. We claim that there is no path from s to t in G if and only if there is automorphism φ in G' mapping s_1 to s_2. Clearly if there is no path between s and t in G, these two nodes belong to different connected components. The desired automorphism can be obtained by mapping the nodes of the connected component of s_1 in G_1 to the corresponding nodes in G_2 and mapping the rest of the nodes in G' (and in particular t_1) to themselves.

Conversely, if there is a path between s and t in G, the mentioned automorphism φ does not exist since the nodes s_2 ($\varphi(s_1)$) and t_1 ($\varphi(t_1)$) should be in the same connected component, but there are no edges between G_1 and G_2 in G'.

The question of whether there is an automorphism in G' with the mentioned properties, can in turn be reduced to 2-GI. Let H_1 be a copy of G' with node s_1 having a special color 3, and H_2 be another copy with s_2 having color 3. There is an automorphism mapping s_1 to s_2 in G' if and only if H_1 and H_2 are isomorphic. The size of the color classes in each of the graphs H_1 and H_2 is at most 2. □

Corollary 1 (of the proof). 2-GA *is hard for* $\overline{\text{UCC}}$ *under logarithmic space many-one reductions.*

Proof. The reduction for this result is the same as in the previous theorem. Observe that in G' the color classes are of size at most 2 and there is a nontrivial automorphism in G' if and only if G has more than a single connected component. □

5   Conclusions

The following results follow directly from the upper and lower bounds given in Sections 2, 3 and 4.

- 2-GA is equivalent to $\overline{\text{UCC}}$ under logarithmic space many-one reductions.
- 2-GI and 3-GI are complete for SL under logarithmic space many-one reductions.
- 3-GA belongs to SL.

A natural way to continue this research is to study the situation for other bounds $b \geq 4$ on the size of the color classes. From the results in [17] it can easily be derived that for $b \geq 2$, b^2-GI is hard for the modular class $\mathrm{Mod}_b\mathrm{L}$, and $2b^2$-GA is also hard for $\mathrm{Mod}_b\mathrm{L}$. From this follows that 4-GI is hard for $\oplus \mathrm{L}$ and therefore a proof of the fact that 4-GI belongs to SL would imply that $\oplus \mathrm{L}$ is included in SL which is something we do not expect. It remains an interesting open problem whether better upper bounds than the ones given in [15] can be obtained for b-GI and b-GA for other special values of b ($b \geq 4$).

We observe that the blow-up in the complexity of the problem when going from color classes of size 3 to size 4 also happens in the related area of graph identification using first-order formulas with counting. Immerman and Lander show in [8] that 3 variables suffice to identify all colored graphs of color size 3, while $\Omega(n)$ variables are needed to identify all graphs of color size 4 using first order formulas with counting, as proved in [5].

Acknowledgments. We would like to thank V. Arvind, Pascal Tesson and Denis Thérien for interesting discussions on the topics of the paper. The second author would also like to thank Klaus-Jörn Lange for his invitation to Tübingen, where part of this research was done.

References

1. Álvarez, C., Greenlaw, R.: A compendium of problems complete for symmetric logarithmic space. Journal of Computational Complexity **9** (2000) 73–95
2. Babai, L.: Monte carlo algorithms for graph isomorphism testing. Technical Report 79-10, Dép. Math. et Stat., Univ. de Montréal (1979)
3. Babai, L.: Moderately exponential bounds for graph isomorphism. In: Proc. International Symposium on Fundamentals of Computing Theory 81. Volume 117 of Lecture Notes in Computer Science., Springer-Verlag, Berlin Heidelberg New York (1981) 34–50
4. Buss, S.: Alogtime algorithms for tree isomorphism, comparison, and canonization. In: Computational Logic and Proof Theory, 5th Kurt Gödel Colloquium'97. Volume 1289 of Lecture Notes in Computer Science., Springer-Verlag, Berlin Heidelberg New York (1997) 18–33
5. Cai, J., Fürer, M., Immerman, N.: An optimal lower bound for the number of variables for graph identification. Combinatorica **12** (1992) 389–410
6. Furst, M., Hopcroft, J., Luks, E.: Polynomial time algorithms for permutation groups. In: Proc. 21st IEEE Symposium on the Foundations of Computer Science, IEEE Computer Society Press (1980) 36–41
7. Hopcroft, J.E., Tarjan, R.E.: A V^2 algorithm for determining isomorphism of planar graphs. Information Processing Letters **1** (1971) 32–34
8. Immerman, N., Lander, E.: Describing graphs: a first order approach to graph canonization. In Selman, A.L., ed.: Complexity Theory Retrospective. Springer-Verlag, Berlin Heidelberg New York (1990) 59–81
9. Jenner, B., McKenzie, P., Torán, J.: A note on the hardness of tree isomorphism. In: Proc. 13th Annual IEEE Conference on Computational Complexity, IEEE Computer Society Press (1998) 101–106

10. Köbler, J., Schöning, U., Torán, J.: The Graph Isomorphism Problem: Its Structural Complexity. Birkhäuser, Boston (1993)
11. Lewis, H., Papadimitriou, C.: Symmetric space-bounded computation. Theoretical Computer Science **19** (1982) 161–187
12. Lindell, S.: A logspace algorithm for tree canonization. In: Proc. 24th ACM Symposium on Theory of Computing, ACM Press (1992) 400–404
13. Lozano, A., Torán, J.: On the non-uniform complexity of the graph isomorphism problem. In Ambos-Spies, K., Homer, S., Schöning, U., eds.: Complexity Theory, Current Research. Cambridge University Press (1993) 245–273 (also in Proceedings of the 7th Structure in Complexity Theory Conference, 118–129, 1992).
14. Luks, E.: Isomorphism of bounded valence can be tested in polynomial time. Journal of Computer and System Sciences **25** (1982) 42–65
15. Luks, E.: Parallel algorithms for permutation groups and graph isomorphism. In: Proc. 27th IEEE Symposium on the Foundations of Computer Science, IEEE Computer Society Press (1986) 292–302
16. Nisan, N., Ta-Shma, A.: Symmetric logspace is closed under complement. In: Proc. 27th ACM Symposium on Theory of Computing, ACM Press (1995) 140–146
17. Torán, J.: On the hardness of graph isomorphism. In: Proc. 41st IEEE Symposium on the Foundations of Computer Science, IEEE Computer Society Press (2000) 180–186

On the Complexity of Generating Maximal Frequent and Minimal Infrequent Sets[*]

E. Boros[1], V. Gurvich[1], L. Khachiyan[2], and K. Makino[3]

[1] RUTCOR, Rutgers University, 640 Bartholomew Road, Piscataway New Jersey 08854-8003; {boros,gurvich}@rutcor.rutgers.edu.
[2] Department of Computer Science, Rutgers University, 110 Frelinghuysen Road, Piscataway, New Jersey, 08854-8019; leonid@cs.rutgers.edu
[3] Division of Systems Science, Graduate School of Engineering Science, Osaka University, Toyonaka, Osaka, 560-8531, Japan; makino@sys.es.osaka-u.ac.jp

Abstract. Let A be an $m \times n$ binary matrix, $t \in \{1, \ldots, m\}$ be a threshold, and $\varepsilon > 0$ be a positive parameter. We show that given a family of $O(n^\varepsilon)$ maximal t-frequent column sets for A, it is NP-complete to decide whether A has any further maximal t-frequent sets, or not, even when the number of such additional maximal t-frequent column sets may be exponentially large. In contrast, all minimal t-infrequent sets of columns of A can be enumerated in incremental quasi-polynomial time. The proof of the latter result follows from the inequality $\alpha \le (m - t + 1)\beta$, where α and β are respectively the numbers of all maximal t-frequent and all minimal t-infrequent sets of columns of the matrix A. We also discuss the complexity of generating all closed t-frequent column sets for a given binary matrix.

Keywords. Data mining, frequent sets, infrequent sets, independent sets, hitting sets, transversals, dualization.

1 Introduction

Let us consider an $m \times n$ binary matrix $A : \mathcal{R} \times \mathcal{C} \rightarrow \{0, 1\}$, and an integral threshold value $t \in \{1, \ldots, m\}$. To each subset $C \subseteq \mathcal{C}$ of the columns, let us associate the subset $R(C) \subseteq \mathcal{R}$ of all those rows $r \in \mathcal{R}$, for which $A(r, c) = 1$ in every column $c \in C$. The cardinality $|R(C)|$ is called the *support* of the set C. Let us call a subset $C \subseteq \mathcal{C}$ of the columns *frequent* (or more precisely, *t-frequent*) if its support is at least the given integral threshold t, i.e. if $|R(C)| \ge t$, and let us denote by $\mathcal{F}_t$ the family of all t-frequent subsets of the columns of the given binary matrix A. Let us further call a subset $C \subseteq \mathcal{C}$ *infrequent* (or *t-infrequent*)

---
[*] This research is supported in part by the National Science Foundation (Grant IIS-0118635), the Office of Naval Research (Grant N00014-92-J-1375), and Grants-in-Aid for Scientific Research of the Ministry of Education, Culture, Sports, Science and Technology of Japan. Visits of the second author to Rutgers University were also supported by DIMACS, the National Science Foundation's Center for Discrete Mathematics and Theoretical Computer Science.

H. Alt and A. Ferreira (Eds.): STACS 2002, LNCS 2285, pp. 133–141, 2002.
© Springer-Verlag Berlin Heidelberg 2002

if its support does not exceed the given threshold t, i.e. if $|R(C)| < t$. Clearly, subsets of frequent sets are also frequent, and supersets of infrequent sets are also infrequent. Let us denote by $\mathcal{M}_t \subseteq \mathcal{F}_t$ the family of all maximal t-frequent sets (i.e. those which are t-frequent, but no superset of them is t-frequent), and by $\mathcal{I}_t$ the family of all minimal t-infrequent sets (i.e. those which are infrequent but all proper subsets of them are t-frequent.)

The generation of frequent sets of a given binary matrix A is an important task of knowledge discovery and data mining, e.g. it is used for mining association rules [1,2,16,21,22], correlations [8], sequential patterns [3], episodes [23], emerging patterns [10], and appears in many other applications. Most practical procedures to generate $\mathcal{F}_t$ are based on the anti-monotone *Apriori* heuristic (see [2]) and build frequent sets in a bottom-up way, running in time proportional to the number of frequent sets. It was also demonstrated recently in [9] that these methods are inadequate in practice when there are (many) frequent sets of large size (see also [4,13,19]), due the fact that $|\mathcal{F}_t|$ can be exponentially larger than $|\mathcal{M}_t|$.

These results show that it is perhaps more important to find the *boundary* of the frequent sets, i.e. the families of maximal frequent and minimal infrequent sets $\mathcal{M}_t \cup \mathcal{I}_t$ (proposed e.g. in [26]), and use those as condensed representation of the data set, as suggested in [21]. Furthermore, no algorithm using membership queries "$X \in \mathcal{F}_t$?" can generate all (maximal) frequent sets in fewer than $|\mathcal{M}_t \cup \mathcal{I}_t|$ steps (see e.g. [16]). There were several other examples presented in [21] to show the usefulness of maximal frequent sets and minimal infrequent sets, e.g. providing error bounds for the confidence of an arbitrary Boolean rule, in terms of minimal infrequent sets.

In this short paper we prove the following inequality.

Theorem 1. *If $\mathcal{I}_t \neq \emptyset$ then*

$$|\mathcal{M}_t| \leq (m - t + 1)|\mathcal{I}_t|. \tag{1}$$

Note that the requirement that $\mathcal{I}_t$ must be non-empty is necessary because for $|\mathcal{I}_t| = 0$ we would have $\mathcal{M}_t = \{\mathcal{C}\}$ and hence $|\mathcal{M}_t| = 1$, in contradiction with (1). The condition $\mathcal{I}_t \neq \emptyset$ thus excludes the degenerate case, when the entire column set of A is t-frequent.

Before proceeding further, let us mention some algorithmic implications of (1). It follows from the results of [5,16,17] that the incremental complexity of generating $\mathcal{M}_t \cup \mathcal{I}_t$ is equivalent with that of the transversal hypergraph problem (for definitions and related results see e.g. [11]). The latter problem is known to be solvable in incremental quasi-polynomial time [14], implying thus the same for the joint generation of maximal frequent and minimal infrequent sets. Specifically, it follows from [14] that for each $k \leq |\mathcal{M}_t \cup \mathcal{I}_t|$, we can generate k sets in $\mathcal{M}_t \cup \mathcal{I}_t$ in $poly(n,m) + k^{o(\log k)}$ time. The above inequality (1) clearly implies that if we can generate $\mathcal{M}_t \cup \mathcal{I}_t$ in time $T(|\mathcal{M}_t \cup \mathcal{I}_t|)$, then the entire set $\mathcal{I}_t$ can be generated in time $T((m - t + 2)|\mathcal{I}_t|)$. We thus conclude that the family of minimal infrequent sets $\mathcal{I}_t$ can be generated in output quasi-polynomial time, i.e. in time bounded by a quasi-polynomial in $|\mathcal{I}_t|$. This can be further improved to show that the incremental complexity of generating $\mathcal{I}_t$ is also equivalent with that of the transversal hypergraph problem (see [6,7] for more detail). Hence

Corollary 1. *For each* $k \leq |\mathcal{I}_t|$, *we can compute* k *minimal* t-*infrequent sets of* A *in* $poly(n, m) + K^{o(\log K)}$ *time, where* $K = \max\{k, m\}$.

Let us note next that the matrix A can also be interpreted as the adjacency matrix of a bipartite graph $G = (\mathcal{R} \cup \mathcal{C}, E)$, i.e., in which $(r, c) \in E$ iff $A(r, c) = 1$. Then, maximal frequent sets of A correspond to maximal complete bipartite subgraphs $K_{R,C} \lhd G$, where $R \subseteq \mathcal{R}$, and $C \subseteq \mathcal{C}$. It is known (see e.g., [18]) that determining the number of maximal complete bipartite subgraphs of a bipartite graph is a very difficult #P-complete problem, and hence by the above equivalence, determining $| \cup_{t \geq 1} \mathcal{M}_t|$ is also #P-complete. (In [12] an $O(l^3 2^{2l}(m + n))$ algorithm was presented to generate all maximal complete bipartite subgraphs of a bipartite graph on $m + n$ vertices, or equivalently, to generate all maximal frequent sets of A, where l denotes the maximum of $|C||R(C)|/(|C|+|R(C)|-1)$, with the maximum taken over all maximal frequent sets C.) Strengthening these (negative) results and a statement of [20], we can show the following:

Theorem 2. *Given an* $m \times n$ *matrix* A, *a threshold* t, *and a subfamily* $\mathcal{S} \subseteq \mathcal{M}_t$, *it is NP-hard to decide if* $\mathcal{S} \neq \mathcal{M}_t$, *even if* $|\mathcal{S}| = O(n^\varepsilon)$ *and* $|\mathcal{M}_t|$ *is exponentially large in* n *whenever* $\mathcal{S} \neq \mathcal{M}_t$, *where* $\varepsilon > 0$ *can be arbitrarily small.*

Yet, it is easy to show that determining whether or not $\mathcal{S} \neq \mathcal{M}_t$ for polylog-arithmically large $|\mathcal{S}|$ can be done in polynomial time.

Finally, let us remark that the inequality (1) is best possible, as the following examples show. Let A be an $m \times (m - t + 1)$ matrix, in which every entry is 1, except the diagonal entries in the first $m - t + 1$ rows, which are 0. Then any $m - t$ element subset of the columns is a maximal t-frequent set, while the set $\mathcal{C}$ of all columns is the only minimal t-infrequent set. Thus we have equality in (1) for such matrices.

It is also worth mentioning that (1) stays accurate, up to a factor of $\log m$, even if $m \gg n$ and $|\mathcal{I}_t|$ is arbitrarily large. To see this, let us consider a binary matrix A with $m = 2^k$ rows and $n = 2k$ columns ($k \geq 1$, integer), such that each row contains exactly one 0 and one 1 in each pair of the adjacent columns $\{1, 2\}$, $\{3, 4\}, \ldots, \{2k - 1, 2k\}$, and in all 2^k possible ways in the $m = 2^k$ rows. It is not difficult to see that for $t = 1$ there are 2^k maximal 1-frequent sets (every row of the matrix is the characteristic vector of a maximal 1-frequent set), and that there are only k minimal 1-infrequent sets, namely $\{2i - 1, 2i\}$ for $i = 1, ..., k$. Thus, for such examples we have $|\mathcal{M}_t| = (m/\log m)|\mathcal{I}_t|$. The same examples also show that $|\mathcal{M}_t|$ cannot be bounded by a polynomial function in $|\mathcal{I}_t|$ and n, the number of columns of A. Needless to say that in general, $|\mathcal{I}_t|$ cannot be bounded by a polynomial in $|\mathcal{M}_t|$, n and m.

2   Closed Frequent Sets

Following [27], let us call a subset $C \subseteq \mathcal{C}$ of the columns *closed* if $R(C') \subsetneqq R(C)$ for all $C' \supsetneqq C$, or in other words, if $c \in C$ exactly when $A(r, c) = 1$ for all $r \in R(C)$ (see also [24,25]). Let us further denote by $\mathcal{D}_t$ the family of all closed t-frequent column sets. Clearly, we have

$$\mathcal{M}_t \subseteq \mathcal{D}_t \subseteq \mathcal{F}_t$$

for all $t = 1, ..., m$.

For the converse direction, it is also easy to see by the definitions that every closed t-frequent set is also a maximal t'-frequent set for some $t' \geq t$, implying the following claim.

Proposition 1. $\mathcal{D}_t = \cup_{t' \geq t} \mathcal{M}_{t'}$.   □

Let us note next that for $C \in \mathcal{D}_t \setminus \mathcal{D}_{t+1}$ we either have a subset $C' \subset C$, $C' \neq \emptyset$ for which $C' \in \mathcal{D}_{t+1}$, or $A(r, c) = 0$ for all $c \in C$ and $r \notin R(C)$. Since the number of subsets of the latter type is limited by n and it is easy to identify those in $O(mn)$ time, all sets in $\mathcal{D}_t \setminus \mathcal{D}_{t+1}$ can be obtained in $O(nm + n|\mathcal{D}_{t+1}|)$ time by trying to increment all sets of $\mathcal{D}_{t+1}$ in all possible ways. Denoting by τ the maximum number of 1s in a column of A, we can claim that $\mathcal{D}_t = \emptyset$ for all $t > \tau$, and that $\mathcal{D}_\tau$ can easily be generated in $O(nm)$ time. Putting all these together, we can conclude that, in contrast to Theorem 2, closed frequent sets can be generated efficiently.

Proposition 2. *The family $\mathcal{D}_t$ can be generated in incremental polynomial time for any $t \in \{1, ..., m\}$.*   □

Let us finally remark that in many examples we can have $|\mathcal{D}_t|$ exponentially larger than $|\mathcal{M}_t|$ and $|\mathcal{F}_t|$ exponentially larger than $|\mathcal{D}_t|$, simultaneously. To see such an infinite family of examples, let us choose k, $l > k$ and t as arbitrary positive integers, set $m = kt$, $n = kl$, and define the matrix A as follows. Let $U_i = \{(i-1)l + j \mid j = 1, ..., l\}$ for $i = 1, ..., k$, let $\mathcal{C} = \cup_{i=1}^{k} U_i$, and let $a_i \in \{0, 1\}^n$ $(1 \leq i \leq k)$ be a binary vector in which $a_{ij} = 0$ if $j \in U_i$, and $a_{ij} = 1$ otherwise. Finally, let $A \in \{0, 1\}^{m \times n}$ be the matrix formed by t copies, as rows, of each of the vectors a_i, $i = 1, ..., k$.

It is now easy to see that the maximal t-frequent sets in this matrix are exactly the column subsets C of the form $C = \mathcal{C} \setminus U_i$ for some $1 \leq i \leq k$. Thus, $|\mathcal{M}_t| = k$. Furthermore, the column subsets of the form $C = \mathcal{C} \setminus \cup_{i \in S} U_i$ for some nonempty subset $S \subseteq \{1, ..., k\}$ are exactly the closed t-frequent sets of A, therefore we have $|\mathcal{D}_t| = 2^k - 1$. Finally, any subset $C \subseteq \mathcal{C}$ of the columns, disjoint from at least one of the sets $U_1, ..., U_k$, is a t-frequent set, implying that $|\mathcal{F}_t| > 2^{(l-1)k} > |\mathcal{D}_t|^k$.

3   Proofs of Theorems 1 and 2

For the proof of Theorem 1 we shall need the following combinatorial lemma.

Lemma 1. *Given a base set V of size $|V| = m$ and a threshold $t \in \{1, \dots, m\}$, let $\mathcal{S} = \{S_1, \dots, S_\alpha\}$ and $\mathcal{T} = \{T_1, \dots, T_\beta\}$ be two families of subsets of V such that*

(i) $|S| \geq t$ for all $S \in \mathcal{S}$, while $|T| < t$ for all $T \in \mathcal{T}$, and
(ii) for each of the $\alpha(\alpha - 1)/2$ pairs $S', S'' \in \mathcal{S}$ there exists a $T \in \mathcal{T}$, such that $S' \cap S'' \subseteq T$.

Then $\alpha \leq (m - t + 1)\beta$, whenever $\alpha \geq 2$.

Let us remark first that if $\alpha = 1$ then the family $\mathcal{T}$ might be empty, which would violate the inequality $\alpha \leq (m-t+1)\beta$. Let us also mention that by (ii) $\beta \geq 1$ must hold whenever $\alpha \geq 2$. In addition, conditions (i) and (ii) together imply that $\mathcal{S}$ is a Sperner family, i.e. $S_i \not\subseteq S_j$ whenever $i \neq j$ (since otherwise $S_i = S_i \cap S_j \subseteq T_k$ would follow by (ii) for some $T_k \in \mathcal{T}$, contradicting condition (i).) Without loss of generality we can assume that $\mathcal{T}$ is also Sperner, for otherwise we can replace $\mathcal{T}$ by the family of all maximal sets of $\mathcal{T}$.

Proof of Lemma 1. We shall prove the Lemma by induction on t. If $t = 1$ then $\mathcal{T} = \{\emptyset\}$ by condition (i). In view of (ii), this implies that the sets of $\mathcal{S}$ are pairwise disjoint, and hence $\alpha \leq m = (m - t + 1)\beta$.

In a general step, let us define subfamilies $\mathcal{S}_v = \{S \setminus \{v\} \mid S \in \mathcal{S}, \; v \in S\}$ and $\mathcal{T}_v = \{T \setminus \{v\} \mid S \in \mathcal{T}, \; v \in T\}$ for each $v \in V$. Let us further introduce the notations $\alpha_v = |\mathcal{S}_v|$, and $\beta_v = |\mathcal{T}_v|$.

For vertices $v \in V$ for which $\alpha_v \geq 2$ (and thus $\beta_v \geq 1$) the families $\mathcal{S}_v$ and $\mathcal{T}_v$ satisfy all the assumptions of the Lemma with $m' = m - 1$ and $t' = t - 1$, and hence

$$\alpha_v \leq (m' - t' + 1)\beta_v = (m - t + 1)\beta_v \tag{2}$$

follows by the inductive hypothesis. Let us then consider the partition $V = V_1 \cup V_2$, where $V_1 = \{v \in V \mid \alpha_v \leq 1\}$, and $V_2 = \{v \in V \mid \alpha_v \geq 2\}$. Summing up the inequalities (2) for all $v \in V_2$, we obtain

$$\sum_{v \in V_2} \alpha_v \leq (m - t + 1) \sum_{v \in V_2} \beta_v. \tag{3}$$

On the left hand side, using condition (i) and the definition of α_v we obtain

$$\alpha t - |V_1| \leq \sum_{S \in \mathcal{S}} |S| - |V_1| \leq \sum_{S \in \mathcal{S}} (|S| - |S \cap V_1|) = \sum_{S \in \mathcal{S}} |S \cap V_2| = \sum_{v \in V_2} \alpha_v, \tag{4}$$

where the first inequality follows by $|S| \geq t$ for $S \in \mathcal{S}$, while the second one is implied by $|V_1| \geq \sum_{S \in \mathcal{S}} |S \cap V_1|$, which follows from the definition of V_1.

On the right hand side of (2) we can write

$$\sum_{v \in V_2} \beta_v = \sum_{T \in \mathcal{T}} |T \cap V_2| \le \sum_{T \in \mathcal{T}} |T| \le \beta(t-1), \tag{5}$$

where the first equality follows by the definition of β_v and $\mathcal{T}_v$, and the last inequality follows by the conditions $|T| < t$ for $T \in \mathcal{T}$.

Putting together (3),(4) and (5) we obtain $t\alpha - |V_1| \le (m - t + 1)(t - 1)\beta$, or equivalently that

$$\alpha \le \frac{|V_1|}{t} + \frac{t-1}{t}(m - t + 1)\beta. \tag{6}$$

If $|V_1| \le m - t + 1$, then

$$\frac{|V_1|}{t} + \frac{t-1}{t}(m - t + 1)\beta \le (m - t + 1)\beta,$$

and hence $\alpha \le (m - t + 1)\beta$ by (6). On the other hand, if $|V_1| > m - t + 1$, then for each set $S \in \mathcal{S}$ we have $|S \cap V_1| \ge |S| - |V_2| \ge t - |V_2| > 1$. Now by the definition of the set V_1 we obtain $\alpha \le |V_1|/(t - |V_2|) = (m - |V_2|)/(t - |V_2|) \le m - t + 1 \le (m - t + 1)\beta$. $\qquad \square$

Proof of Theorem 1. Assume without loss of generality that $|\mathcal{M}_t| \ge 2$, for otherwise (1) readily follows from the assumption of the theorem that $|\mathcal{I}_t| \ge 1$. Let us recall that to any subset $C \subseteq \mathcal{C}$ of the columns we have associated the subset $R(C)$ of those rows $r \in \mathcal{R}$ for which $A(r, c) = 1$ for every column $c \in C$. Thus, by definition we have $R(C) = \bigcap_{y \in C} R(\{y\})$, implying

$$R(C' \cup C'') = R(C') \cap R(C'') \quad \text{for all} \quad C', C'' \subseteq \mathcal{C}. \tag{7}$$

In its turn, (7) implies that the mapping $C \mapsto R(C)$ is anti-monotone, i.e. $R(C') \supseteq R(C'')$ whenever $C' \subseteq C''$. Furthermore, $|R(F)| \ge t$ for every maximal t-frequent set $F \in \mathcal{M}_t$, while $|R(U)| < t$ for every minimal t-infrequent set $U \in \mathcal{I}_t$. It is also easy to see that the restriction of the above mapping on $\mathcal{M}_t$ is injective, i.e. $R(F') \ne R(F'')$ for any two distinct maximal t-frequents sets of columns $F', F'' \in \mathcal{M}_t$. If $F', F'' \in \mathcal{M}_t$ then their union $F' \cup F''$ is not t-frequent, and hence there exists a minimal t-infrequent set $U \in \mathcal{I}_t$, for which $R(F') \cap R(F'') = R(F' \cup F'') \subseteq R(U)$. Thus, the families $\mathcal{S} = \{R(F) \mid F \in \mathcal{M}_t\}$ and $\mathcal{T} = \{R(U) \mid U \in \mathcal{I}_t\}$ satisfy the conditions of Lemma 1 with $V = \mathcal{R}$, which implies the inequality $|\mathcal{S}| \le (m - t + 1)|\mathcal{T}|$. Since the mapping $C \mapsto R(C)$ is a one-to one correspondence between $\mathcal{M}_t$ and $\mathcal{S}$, we have $|\mathcal{S}| = |\mathcal{M}_t|$. Now (1) follows from the trivial inequality $|\mathcal{T}| \le |\mathcal{I}_t|$. $\qquad \square$

Proof of Theorem 2. We reduce our problem from the following well-known NP-complete problem: Given a graph $G = (V, E)$ and an integer threshold t, determine if G contains an independent vertex set of size at least t. Let us first substitute every vertex $v \in V$ of G by two new vertices v' and v'' connected by an edge, i.e., consider the graph $G' = (V', E')$, where $V' = \{v', v'' \mid v \in V\}$ and $E' = \{\{(v', v'') \mid v \in V\} \cup \{(v', u'), (v', u''), (v'', v'), (v'', u'') \mid (u, v) \in E\}$. Clearly, $G' = (V', E')$ has an independent set of size t if and only if G has one, moreover, if G' has one, then it has at least 2^t.

Let us now associate a matrix A to G' as follows. Let $\mathcal{C} = V'$ be the set of columns of the matrix A. To every edge $(v, w) \in E'$ we assign $t - 2$ identical rows in A containing 0 in the columns v and w, and 1 in all other columns. Furthermore, to every vertex $v \in V'$ we assign one row containing 0 in the column v and 1 in all other columns. Thus, A has $m = (t-2)|E'|+|V'| = t|V|+4(t-2)|E|$ rows, and $n = |V'| = 2|V|$ columns.

Clearly, for every edge $e = (v, w) \in E'$ the set $C_e = \mathcal{C} \setminus \{v, w\}$ is a maximal t-frequent set of A. Let $\mathcal{S} = \{C_e \mid e \in E'\}$. We claim that $\mathcal{S} \neq \mathcal{M}_t$ for this matrix, if and only if there exists an independent set I of size $|I| \geq t$ in the graph G'.

To see this claim, let assume first that $I \subseteq V'$ is an independent set of G', $|I| \geq t$. Then $R(V' \setminus I)$ contains all rows corresponding to vertices $v \in I$, and hence $|R(V' \setminus I)| \geq |I| \geq t$. Since I does not contain an edge of the graph, the set $C = V' \setminus I$ is not a subset of the member of $\mathcal{S}$, and thus it is contained by a maximal t-frequent set of the matrix A, which does not belong to $\mathcal{S}$.

For the other direction, let us assume now that $C \subseteq \mathcal{C} = V'$ is a maximal t-frequent set of A, not contained by members of $\mathcal{S}$. This latter implies that $I = V' \setminus C$ does not contain an edge of G', i.e. that I is an independent set of G'. This also implies that $R(C)$ cannot contain any of the rows corresponding to an edge of G', and hence $|R(C)| = |V' \setminus C| = |I|$. Thus, $|I| \geq t$ follows by our assumption that C is a t-frequent set, i.e. I is an independent set of size at least t.

Let us recall finally that the maximum independent set problem remains NP-hard, even if the input is restricted to cubic planar graphs (see e.g. [15]), i.e. we can assume $|E| = O(|V|)$. Therefore, we have $|\mathcal{S}| = |E'| = |V| + 4|E| = O(|V|)$, and either we have $\mathcal{S} = \mathcal{M}_t$, or $|\mathcal{M}_t| \geq |\mathcal{S}| + 2^t$. Since we can assume without loss of generality that $t = \Theta(|V|)$, we obtain the the statement of the theorem for $|S| = O(n)$, that is for $\varepsilon = 1$. For smaller values of ε it suffices to add $n^{1/\varepsilon}$ isolated vertices to G'. $\square$

References

1. R. Agrawal, T. Imielinski and A. Swami. Mining associations between sets of items in massive databases. In: *Proceedings of the 1993 ACM-SIGMOD International Conference on Management of Data*, pp. 207-216.

2. R. Agrawal, H. Mannila, R. Srikant, H. Toivonen and A. I. Verkamo, Fast discovery of association rules, In U. M. Fayyad, G. Piatetsky-Shapiro, P. Smyth and R. Uthurusamy eds., *Advances in Knowledge Discovery and Data Mining*, 307-328, AAAI Press, Menlo Park, California, 1996.

3. R. Agrawal and R. Srikant. Mining sequential patterns. In: *Proceedings of the 11th International Conference on Data Engineering, 1995*, pp.3-14.

4. R.J. Bayardo, Efficiently mining long patterns from databases. In: *Proceedings of the 1998 ACM-SIGMOD International Conference on Management of Data*, pp. 85-93.
5. J. C. Bioch and T. Ibaraki, Complexity of identification and dualization of positive Boolean functions, *Information and Computation* 123 (1995) 50-63.
6. E. Boros, V. Gurvich, L. Khachiyan and K.Makino, Generating partial and multiple transversals of a hypergraph. In: *Proceedings of the 27th International Colloquium on Automata, Languages and Programming (ICALP)*, (U. Montanari, J.D.P. Rolim and E. Welzl, eds.) Lecture Notes in Computer Science **1853** pp. 588-599, (Springer Verlag, Berlin, Heidelberg, New York, 2000).
7. E. Boros, V. Gurvich, L. Khachiyan and K.Makino, Generating Weighted Transversals of a Hypergraph, DIMACS Technical Report 00-17, Rutgers University, 2000. (http://dimacs.rutgers.edu/TechnicalReports/2000.html)
8. S. Brin, R. Motwani, and C. Silverstein. Beyond market basket: Generaliing association rules to correlations. In: *Proceedings of the 1997 ACM-SIGMOD Conference on Management of Data*, pp. 265-276.
9. S. Brin, R. Motwani, J. Ullman, and S. Tsur. Dynamic itemset counting and implication rules for market basket data. In: *Proceedings of the 1997 ACM-SIGMOD Conference on Management of Data*, pp. 255-264.
10. G. Dong and J. Li. Efficient mining of emerging patterns. In: *Proceeding of the 1999 ACM SIGKDD International Conference on Knowledge Discovery and Data Mining*, pp. 43-52.
11. T. Eiter and G. Gottlob, Identifying the minimal transversals of a hypergraph and related problems, *SIAM Journal on Computing*, 24 (1995) 1278-1304.
12. D. Eppstein, Arboricity and bipartite subgraph listing algorithms, *Information Processing Letters* **51** (1994), pp. 207-211.
13. J. Han, J. Pei, and Y. Yin, Mining frequent patterns without candidate generation, In: *Proceedings of the 2000 ACM-SIGMOD Conference on Management of Data*, pp. 1-12.
14. M. L. Fredman and L. Khachiyan, On the complexity of dualization of monotone disjunctive normal forms. *J. Algorithms*, 21 (1996) 618-628.
15. M. R. Garey and D. S. Johnson, *Computers and Intractability*, Freeman, New York, 1979.
16. D. Gunopulos, R. Khardon, H. Mannila, and H. Toivonen, Data mining, hypergraph transversals and machine learning. In: *Proceedings of the 16th ACM-SIGACT-SIGMOD-SIGART Symposium on Principles of Database Systems*, (1997) pp. 12-15.
17. V. Gurvich and L. Khachiyan, On generating the irredundant conjunctive and disjunctive normal forms of monotone Boolean functions, *Discrete Applied Mathematics*, 1996-97, issue 1-3, (1999) 363-373.
18. S. O. Kuznetsov, Interpretation on graphs and complexity characteristics of a search for specific patterns, *Nauchn. Tekh. Inf., Ser. 2 (Automatic Document. Math. Linguist.)* **23**(1), (1989) pp. 23-37.
19. D. Lin and Z.M. Kedem. Pincer-search: a new algorithm for discovering the maximum frequent set. In: *Proceedings of the Sixth European Conference on Extending Database Technology*, to appear.
20. K. Makino and T. Ibaraki, Inner-core and outer-core functions of partially defined Boolean functions, *Discrete Applied Mathematics*, 1996-97, issue 1-3 (1999), 307-326.

21. H. Mannila and H. Toivonen, Multiple uses of frequent sets and condensed representations. In: *Proceedings of the 2nd International Conference on Knowledge Discovery and Data Mining*, (1996) pp. 189-194.

22. H. Mannila and H. Toivonen, Levelwise search and borders of theories in knowledge discovery. Series of Publications C C-1997-8, University of Helsinki, Department of Computer Science (1997).

23. H. Mannila, H. Toivonen, and A. I. Verkamo. Discovery of frequent episodes in event sequences. *Data Mining and Knowledge Discovery*, 1 (1997), 259-289.

24. N. Pasquier, Y. Bastide, R. Taouil, and L. Lakhal, Discovering frequent closed itemsets for association rules. *Proc. of the 7th ICDT Conference*, Jerusalem, Israel, January 10-12, 1999; *Lecture Notes in Computer Science*, **1540**, pp. 398-416, Springer Verlag, 1999.

25. N. Pasquier, Y. Bastide, R. Taouil, and L. Lakhal, Closed Set Based Discovery of Small Covers for Association Rules, *Proc. 15emes Journees Bases de Donnees Avancees, BDA*, pp. 361-381, 1999.

26. R. H. Sloan, K. Takata, G. Turan, On frequent sets of Boolean matrices, *Annals of Mathematics and Artificial Intelligence* 24 (1998) 1-4.

27. M.J. Zaki and M. Ogihara, Theoretical foundations of association rules, *3rd SIGMOD Workshop on Research Issues in Data Mining and Knowledge Discovery*, June 1998.

On Dualization in Products of Forests

Khaled M. Elbassioni[*]

Department of Computer Science, Rutgers University, 110 Frelinghuysen Road,
Piscataway NJ 08854-8003; (elbassio@paul.rutgers.edu).

Abstract. Let $\mathcal{P} = \mathcal{P}_1 \times \ldots \times \mathcal{P}_n$ be the product of n partially ordered sets, each with an acyclic precedence graph in which either the in-degree or the out-degree of each element is bounded. Given a subset $\mathcal{A} \subseteq \mathcal{P}$, it is shown that the set of maximal independent elements of $\mathcal{A}$ in $\mathcal{P}$ can be incrementally generated in quasi-polynomial time. We discuss some applications in data mining related to this dualization problem.

Keywords. Data mining, dualization, forest, incremental algorithms, poset.

1 Introduction

Given a finite set V and a hypergraph $\mathcal{H} \subseteq 2^V$, the *hypergraph dualization* problem calls for enumerating all maximal independent sets of $\mathcal{H}$, i.e., all maximal subsets of V that do not contain any hyperedge of $\mathcal{H}$. This problem has important applications in combinatorics [19], artificial intelligence [12], reliability theory [11], database theory [4,9], integer programming [7,9], and learning theory [5].

In this paper, we consider a natural generalization of this dualization problem which replaces edges of a hypergraph by a finite set of vectors over products of partially ordered sets (posets). Specifically, let $\mathcal{P} \overset{\text{def}}{=} \mathcal{P}_1 \times \ldots \times \mathcal{P}_n$ be the product of n posets. Let us use $\preceq$ to denote the precedence relation in $\mathcal{P}$ and also in $\mathcal{P}_1, \ldots, \mathcal{P}_n$, i.e., if $p = (p_1, \ldots, p_n) \in \mathcal{P}$ and $q = (q_1, \ldots, q_n) \in \mathcal{P}$, then $p \preceq q$ in $\mathcal{P}$ if and only if $p_1 \preceq q_1$ in $\mathcal{P}_1$, , $\ldots$, $p_n \preceq q_n$ in $\mathcal{P}_n$. For $\mathcal{A} \subseteq \mathcal{P}$, denote by $\mathcal{A}^+ = \{x \in \mathcal{P} \mid x \succeq a, \text{ for some } a \in \mathcal{A}\}$ and $2\,\mathcal{A}^- = \{x \in \mathcal{P} \mid x \preceq a, \text{ for some } a \in \mathcal{A}\}$, the ideal and filter generated by $\mathcal{A}$. Any element in $\mathcal{P} \setminus \mathcal{A}^+$ is called *independent of* $\mathcal{A}$. Let $\mathcal{I}(\mathcal{A})$ be the set of all maximal independent elements for $\mathcal{A}$ (also referred to as the *dual* of $\mathcal{A}$):

$$\mathcal{I}(\mathcal{A}) \overset{\text{def}}{=} \{p \in \mathcal{P} \mid p \notin \mathcal{A}^+ \text{ and } (q \in \mathcal{P}, q \succeq p, q \neq p \Rightarrow q \in \mathcal{A}^+)\}.$$

Then for any $\mathcal{A} \subseteq \mathcal{P}$, we have the following decomposition of $\mathcal{P}$:

$$\mathcal{A}^+ \cap \mathcal{I}(\mathcal{A})^- = \emptyset, \quad \mathcal{A}^+ \cup \mathcal{I}(\mathcal{A})^- = \mathcal{P}. \tag{1}$$

---

[*] Partially supported by DIMACS, the National Science Foundation's Center for Discrete Mathematics and Theoretical Computer Science.

H. Alt and A. Ferreira (Eds.): STACS 2002, LNCS 2285, pp. 142–153, 2002.
© Springer-Verlag Berlin Heidelberg 2002

Given $\mathcal{A} \subseteq \mathcal{P}$, we consider the problem of incrementally generating all elements of $\mathcal{I}(\mathcal{A})$:

$DUAL(\mathcal{P}, \mathcal{A}, \mathcal{B})$: *Given a subset $\mathcal{A} \subseteq \mathcal{P}$ in a poset $\mathcal{P}$ and a collection of maximal independent elements $\mathcal{B} \subseteq \mathcal{I}(\mathcal{A})$, either find a new maximal independent element $p \in \mathcal{I}(\mathcal{A}) \setminus \mathcal{B}$, or prove that $\mathcal{A}$ and $\mathcal{B}$ form a dual pair: $\mathcal{B} = \mathcal{I}(\mathcal{A})$.*

Clearly, the entire set $\mathcal{I}(\mathcal{A})$ can be generated by initializing $\mathcal{B} = \emptyset$ and iteratively solving the above problem $|\mathcal{I}(\mathcal{A})| + 1$ times. If $\mathcal{P}$ is the Boolean cube, i.e., $\mathcal{P}_i = \{0, 1\}$ for all $i = 1, \ldots, n$, we obtain the hypergraph dualization problem whose complexity is still an important open question. The best known algorithm runs in quasi-polynomial time $poly(n, m) + m^{o(\log m)}$, where $m = |\mathcal{A}| + |\mathcal{B}|$, see [13]. For products of general partially ordered sets, it is not known whether the problem is NP-hard. In this note, it will be shown that the problem is unlikely to be NP-hard for products of posets whose precedence graphs are forests of (directed) trees with bounded degrees. Specifically, for $x \in \mathcal{P}_i$, denote by $x^{\perp}$ the set of immediate predecessors of x, i.e., $x^{\perp} = \{y \in \mathcal{P}_i \mid y \prec x, \quad \nexists z \in \mathcal{P}_i : y \prec z \prec x\}$, and let in-deg$(\mathcal{P}_i) = \max\{|x^{\perp}| : x \in \mathcal{P}_i\}$. Similarly, denote by $x^{\top}$ the set of immediate successors of x, and let out-deg$(\mathcal{P}_i) = \max\{|x^{\top}| : x \in \mathcal{P}_i\}$. Throughout, let $m \overset{\text{def}}{=} |\mathcal{A}| + |\mathcal{B}|$, $[n] \overset{\text{def}}{=} \{1, \ldots, n\}$, $d \overset{\text{def}}{=} \max_{i \in [n]} \min\{\text{in-deg}(\mathcal{P}_i), \text{out-deg}(\mathcal{P}_i)\}$, and $\mu \overset{\text{def}}{=} \max\{|\mathcal{P}_i| : i \in [n]\}$. Our main result is the following:

Theorem 1. *Problem $DUAL(\mathcal{P}, \mathcal{A}, \mathcal{B})$ can be solved in $poly(n, m, \mu) + m^{d.o(\log m)}$ time, if $\mathcal{P}$ is a product of forests.*

In the next section, we discuss an application of Theorem 1 related to the generation of subsets of a poset that satisfy a certain monotone property. The proof of the theorem will be given in Sections 3, 4, 5, and 6.

2   Generating Maximal Frequent and Minimal Infrequent Elements in a Database

Let $\mathcal{P} = \mathcal{P}_1 \times \ldots \times \mathcal{P}_n$ be the product of n posets. Consider a database $\mathcal{D} \subseteq \mathcal{P}$ of transactions, each of which is an n-dimensional vector of attributes over $\mathcal{P}$. For an element $p \in \mathcal{P}$, let us denote by $S(p) = S_{\mathcal{D}}(p) \overset{\text{def}}{=} \{q \in \mathcal{D} \mid q \succeq p\}$, the set of transactions in $\mathcal{D}$ that *support* $p \in \mathcal{P}$. Note that, by this definition, the function $|S(.)| : \mathcal{P} \mapsto \mathbb{Z}_+$ is an anti-monotone function, i.e., $|S(p)| \leq |S(q)|$, whenever $p \succeq q$. Given $\mathcal{D} \subseteq \mathcal{P}$ and an integer threshold t, let us say that an element $p \in \mathcal{P}$ is t-frequent if it is supported by at least t transactions in the database, i.e., if $|S_{\mathcal{D}}(p)| \geq t$. Conversely, $p \in \mathcal{P}$ is said to be t-infrequent if $|S_{\mathcal{D}}(p)| < t$. Since the function $|S_{\mathcal{D}}(.)|$ is anti-monotone, we may restrict our attention only to *maximal frequent* and *minimal infrequent* elements. Denote by $\mathcal{F}_{\mathcal{D},t}$ the set of all minimal t-infrequent elements of $\mathcal{P}$ with respect to the database $\mathcal{D}$. Then $\mathcal{I}(\mathcal{F}_{\mathcal{D},t})$ is the set of all maximal t-frequent elements. Consider the following problem of incrementally generating all elements of $\mathcal{F}_{\mathcal{D},t}$:

$GEN(\mathcal{P}, \mathcal{F}_{\mathcal{D},t}, \mathcal{X})$: *Given a subset* $\mathcal{X} \subseteq \mathcal{F}_{\mathcal{D},t} \subseteq \mathcal{P}$, *either prove that* $\mathcal{X} = \mathcal{F}_{\mathcal{D},t}$ *or find a new element in* $\mathcal{F}_{\mathcal{D},t} \setminus \mathcal{X}$.

The problem $GEN(\mathcal{P}, \mathcal{I}(\mathcal{F}_{\mathcal{D},t}), \mathcal{Y})$ can be analogously defined. Although it might be hard in general to generate each of $\mathcal{F}_{\mathcal{D},t}$ and $\mathcal{I}(\mathcal{F}_{\mathcal{D},t})$ separately (see [15,17]), the following joint generation problem may be easier:

$GEN(\mathcal{P}, \mathcal{F}_{\mathcal{D},t}, \mathcal{I}(\mathcal{F}_{\mathcal{D},t}), \mathcal{X}, \mathcal{Y})$: *Given two explicitly listed collections* $\mathcal{X} \subseteq \mathcal{F}_{\mathcal{D},t}$ *and* $\mathcal{Y} \subseteq \mathcal{I}(\mathcal{F}_{\mathcal{D},t})$ *in a poset* $\mathcal{P}$, *either find a new element in* $(\mathcal{F}_{\mathcal{D},t} \setminus \mathcal{X}) \cup (\mathcal{I}(\mathcal{F}_{\mathcal{D},t}) \setminus \mathcal{Y})$, *or prove that these collections are complete:* $(\mathcal{X}, \mathcal{Y}) = (\mathcal{F}_{\mathcal{D},t}, \mathcal{I}(\mathcal{F}_{\mathcal{D},t}))$.

In fact, under the assumption that the precedence graph of each poset $\mathcal{P}_i$ is a rooted tree, Theorem 1, combined with the results of [6,8,15], implies that the two problems GEN$(\mathcal{P}, \mathcal{F}_{\mathcal{D},t}, \mathcal{I}(\mathcal{F}_{\mathcal{D},t}), \mathcal{X}, \mathcal{Y})$ and GEN$(\mathcal{P}, \mathcal{F}_{\mathcal{D},t}, \mathcal{X})$ are unlikely to be NP-hard:

Theorem 2. *If the precedence graph of each poset* $\mathcal{P}_i$ *is a forest in which the in-degree (or, alternatively, the out-degree) of each element is bounded by a constant, then problem* $GEN(\mathcal{P}, \mathcal{F}_{\mathcal{D},t}, \mathcal{I}(\mathcal{F}_{\mathcal{D},t}), \mathcal{X}, \mathcal{Y})$ *is solvable in incremental quasi-polynomial time. If, further, each poset* $\mathcal{P}_i$ *is a join semi-lattice, (that is, the precedence graph of* $\mathcal{P}_i$ *is a rooted tree for which* $d = 1$*), then problem* $GEN(\mathcal{P}, \mathcal{F}_{\mathcal{D},t}, \mathcal{X})$ *is also solvable in incremental quasi-polynomial time.*

The separate and joint generation of maximal frequent and minimal infrequent elements of a poset are important tasks in knowledge discovery and data mining. If each poset $\mathcal{P}_i = \{0, 1\}$, then these problems reduce to generating maximal frequent and minimal infrequent sets, which is used for mining association rules [1,2,14], correlations [10], sequential patterns [3], episodes [18], and many other applications. If the database $\mathcal{D}$ contains categorical (e.g., make of car), or quantitative (e.g., age, income) attributes, and the corresponding posets $\mathcal{P}_i$ are total orders, then the above generation problems can be used to mine the so called *quantitative* association rules [21]. More generally, each attribute a_i in the database can assume values belonging to some partially ordered set $\mathcal{P}_i$. For example, [20] describes applications where items in the database belong to sets of *taxonomies* (or *is-a hierarchies*), and proposes several algorithms for mining association rules among these hierarchical data (see also [16]). Note that, in this last example, each poset $\mathcal{P}_i$ has a tree precedence graph (in which all out-degrees are 1), and therefore, Theorem 2 applies.

3   Preliminaries

Let $\mathcal{P} = \mathcal{P}_1 \times \ldots \times \mathcal{P}_n$, where the precedence graph of each poset $\mathcal{P}_i$ is a forest. Given two subsets $\mathcal{A} \subseteq \mathcal{P}$, and $\mathcal{B} \subseteq \mathcal{I}(\mathcal{A})$, we say that $\mathcal{B}$ is *dual to* $\mathcal{A}$ if $\mathcal{B} = \mathcal{I}(\mathcal{A})$, i.e., if $\mathcal{B}$ contains all the maximal elements of $\mathcal{P} \setminus \mathcal{A}^+$. Let us remark that, by (1), the latter condition is equivalent to $\mathcal{A}^+ \cup \mathcal{B}^- = \mathcal{P}$.

Given any $\mathcal{Q} \subseteq \mathcal{P}$, let us denote by

$$\mathcal{A}(\mathcal{Q}) = \{a \in \mathcal{A} \mid a^+ \cap \mathcal{Q} \neq \emptyset\}, \qquad \mathcal{B}(\mathcal{Q}) = \{b \in \mathcal{B} \mid b^- \cap \mathcal{Q} \neq \emptyset\},$$

the subsets of $\mathcal{A}, \mathcal{B}$ whose ideal and filter respectively intersect $\mathcal{Q}$. A simple but an important observation, which will be used frequently in the algorithm below, is that

$$\mathcal{Q} \subseteq \mathcal{A}^+ \cup \mathcal{B}^- \iff \mathcal{Q} \subseteq \mathcal{A}(\mathcal{Q})^+ \cup \mathcal{B}(\mathcal{Q})^-. \qquad (2)$$

Note that, for $a \in \mathcal{A}$ and $\mathcal{Q} = \mathcal{Q}_1 \times \ldots \times \mathcal{Q}_n$, $a^+ \cap \mathcal{Q} \neq \emptyset$ if and only if $a_i^+ \cap \mathcal{Q}_i \neq \emptyset$, for all $i \in [n]$. Thus, the sets $\mathcal{A}(\mathcal{Q})$ and $\mathcal{B}(\mathcal{Q})$ can be found in $O(nm\mu)$ time.

To solve problem $DUAL(\mathcal{P}, \mathcal{A}, \mathcal{B})$, we shall use the same general approach used in [13] to solve the hypergraph dualization problem, by decomposing it into a number of smaller subproblems which are solved recursively. In each such subproblem, we start with a subposet $\mathcal{Q} = \mathcal{Q}_1 \times \ldots \times \mathcal{Q}_n \subseteq \mathcal{P}$ (initially $\mathcal{Q} = \mathcal{P}$), and two subsets $\mathcal{A}(\mathcal{Q}) \subseteq \mathcal{A}$ and $\mathcal{B}(\mathcal{Q}) \subseteq \mathcal{B}$, and we want to check whether $\mathcal{A}(\mathcal{Q})$ and $\mathcal{B}(\mathcal{Q})$ are dual in $\mathcal{Q}$. As mentioned before, the latter condition is equivalent to checking whether $\mathcal{Q} \subseteq \mathcal{A}(\mathcal{Q})^+ \cup \mathcal{B}(\mathcal{Q})^-$. To estimate the reduction in problem size from one level of the recursion to the next, we measure the change in the "volume" of the problem defined as $v = v(\mathcal{A}, \mathcal{B}) \stackrel{\text{def}}{=} |\mathcal{A}||\mathcal{B}|$. Since $\mathcal{B} \subseteq \mathcal{I}(\mathcal{A})$ is assumed, the following condition holds, by (1) for the original problem and all subsequent subproblems:

$$a \not\leq b, \quad \text{for all } a \in \mathcal{A}, b \in \mathcal{B}. \qquad (3)$$

Let $C(\mathcal{A}, \mathcal{B}) = C(v(\mathcal{A}, \mathcal{B}))$ denote the number of subproblems that have to be solved in order to solve the original problem. We assume that $C(\mathcal{A}, \mathcal{B}) \leq R(v(\mathcal{A}, \mathcal{B}))$ where $R(v)$ is a super-additive function of v (i.e., $R(v) + R(v') \leq R(v + v')$ for all $v, v' \geq 0$).

We start with three propositions: Proposition 1 is useful for decomposing dualization on products of posets with disconnected precedence graphs into a number subproblems in which every poset has a connected precedence graph. Proposition 2 provides the base case for recursion. Proposition 3 states that a problem, closely related to the dualization problem, is NP-hard.

Proposition 1. *Let* $\mathcal{P} = \mathcal{P}_1 \times \ldots \times \mathcal{P}_n$ *and* $\mathcal{A}, \mathcal{B} \subseteq \mathcal{P}$. *Suppose that each poset* $\mathcal{P}_i$, $i \in [n]$, *can be partitioned into independent posets* $(\mathcal{P}_i)_1, \ldots, (\mathcal{P}_i)_{k_i}$ *(where two subposets* $\mathcal{Q}$ *and* $\mathcal{R}$ *are called independent if* $q \not\leq r$ *and* $q \not\geq r$ *for all* $q \in \mathcal{Q}, r \in \mathcal{R}$*). For every* $y = (y_1, \ldots, y_n) \in [k_1] \times \ldots \times [k_n]$, *if we let* $\mathcal{P}^y = (\mathcal{P}_1)_{y_1} \times \ldots \times (\mathcal{P}_n)_{y_n}$, $\mathcal{A}_y = \mathcal{A}(\mathcal{P}^y)$, *and* $\mathcal{B}_y = \mathcal{B}(\mathcal{P}^y)$, *then*

$$\mathcal{P} \subseteq \mathcal{A}^+ \cup \mathcal{B}^- \iff \forall y \in [k_1] \times \ldots \times [k_n] : \mathcal{P}^y \subseteq \mathcal{A}_y^+ \cup \mathcal{B}_y^-. \qquad (4)$$

Moreover, if for each $y \in [k_1] \times \ldots \times [k_n]$, $C(\mathcal{A}_y, \mathcal{B}_y) \leq R(v(\mathcal{A}_y, \mathcal{B}_y))$, *then* $C(\mathcal{A}, \mathcal{B}) \leq R(v(\mathcal{A}, \mathcal{B}))$.

Proposition 2. *Suppose that* $\min\{|\mathcal{A}|, |\mathcal{B}|\} \leq const$, $\mathcal{A}, \mathcal{B} \subseteq \mathcal{P}$, *then problem* $DUAL(\mathcal{P}, \mathcal{A}, \mathcal{B})$ *is solvable in* $poly(n, m, \mu)$ *time.*

Note that it is necessary to maintain the condition $\mathcal{A}, \mathcal{B} \subseteq \mathcal{P}$ in Proposition 2. Without this condition, the problem becomes NP-hard even for $\mathcal{B} = \emptyset$:

Proposition 3. *Given a subposet $\mathcal{Q}$ of a poset $\mathcal{P}$ and a subset $\mathcal{A} \subseteq \mathcal{P}$, it is coNP-complete to decide if $\mathcal{Q} \subseteq \mathcal{A}^+$.*

Clearly, $\mathcal{A}, \mathcal{B} \subseteq \mathcal{P}$ holds initially, but might not hold after decomposing $\mathcal{P}$. To solve this problem, we shall maintain the property that each poset $\mathcal{P}_i$ has a connected precedence graph. This will allow us to project the elements of $\mathcal{A}$ and $\mathcal{B}$ on the poset $\mathcal{P}$ without increasing their number. More precisely, if there is an $a \in \mathcal{A}$, $k \in [n]$ such that $a_k^+ \cap \mathcal{P}_k \neq \emptyset$, but $a_k \notin \mathcal{P}_k$, we replace a_k by $\min(a_k^+ \cap \mathcal{P}_k) \stackrel{\text{def}}{=} \min\{x \in \mathcal{P}_k \mid x \succeq a_k\}$. Note that the existence of such a unique minimum is guaranteed by the fact that $\mathcal{P}_k$ has a connected precedence graph. Similarly, if there is a $b \in \mathcal{B}$, $k \in [n]$ such that $b_k^- \cap \mathcal{P}_k \neq \emptyset$, but $b_k \notin \mathcal{P}_k$, we replace b_k by $\max(b_k^- \cap \mathcal{P}_k) \stackrel{\text{def}}{=} \max\{x \in \mathcal{P}_k \mid x \preceq b_k\}$. Note also that the duality condition (3) continues to hold after such replacements.

In the next section we develop several rules for decomposing a given dualization problem into smaller subproblems. The algorithm will select between these rules in such a way that the total volume is reduced from one iteration to the next.

4   Decomposition Rules

In general, the algorithm will decompose a given problem by selecting an $i \in [n]$ and partitioning $\mathcal{P}_i$ into two subposets $\mathcal{P}_i'$ and $\mathcal{P}_i''$, defining accordingly two poset products $\mathcal{P}'$ and $\mathcal{P}''$. Specifically, let $a^o \in \mathcal{A}$, $b^o \in \mathcal{B}$. By (3), there exists an $i \in [n]$, such that $a_i^o \not\preceq b_i^o$. Let us assume, without loss of generality, that $i = 1$ and set $\mathcal{P}_1' \leftarrow \mathcal{P}_1 \cap (a_1^o)^+$, $\mathcal{P}_1'' \leftarrow \mathcal{P}_1 \setminus \mathcal{P}_1'$ (we may alternatively set $\mathcal{P}_1'' \leftarrow \mathcal{P}_1 \cap (b_1^o)^-$, and $\mathcal{P}_1' \leftarrow \mathcal{P}_1 \setminus \mathcal{P}_1''$, see Step 3 of the algorithm below). For brevity, we shall denote by $\overline{\mathcal{P}}$ the product $\mathcal{P}_2 \times \ldots \times \mathcal{P}_n$, and accordingly by $\overline{q}$ the vector $(q_2, \ldots, q_n)$, for an element $q = (q_1, q_2, \ldots, q_n) \in \mathcal{P}$.

Defining $\mathcal{P}' = \mathcal{P}_1' \times \overline{\mathcal{P}}$ and $\mathcal{P}'' = \mathcal{P}_1'' \times \overline{\mathcal{P}}$ to be the two subposets induced by the above partitioning (see Fig. 1-a), and letting $\mathcal{A}' \stackrel{\text{def}}{=} \mathcal{A}(\mathcal{P}') = \{a \in \mathcal{A} \mid a_1 \succeq a_1^o\}$, $\mathcal{A}'' \stackrel{\text{def}}{=} \mathcal{A} \setminus \mathcal{A}'$, $\mathcal{B}' \stackrel{\text{def}}{=} \mathcal{B}(\mathcal{P}') = \{b \in \mathcal{B} \mid b_1 \succeq a_1^o\}$, $\mathcal{B}'' \stackrel{\text{def}}{=} \mathcal{B} \setminus \mathcal{B}'$, we conclude by (2) that $\mathcal{A}, \mathcal{B}$ are dual in $\mathcal{P}$ *if and only if*

$$\mathcal{A}, \mathcal{B}' \text{ are dual in } \mathcal{P}', \quad i.e., \quad \mathcal{P}' \subseteq \mathcal{A}^+ \cup (\mathcal{B}')^-, \quad and \tag{5}$$

$$\mathcal{A}'', \mathcal{B} \text{ are dual in } \mathcal{P}'', \quad i.e., \quad \mathcal{P}'' \subseteq (\mathcal{A}'')^+ \cup \mathcal{B}^-. \tag{6}$$

As described above, it is required to maintain the property that each poset $\mathcal{P}_i$ has a connected precedence graph. Clearly, if $\mathcal{P}_1$ has a connected graph, then so does $\mathcal{P}_1'$ by the above definitions (since $a_1^o \in \mathcal{P}_1'$). However, this might not be the case for $\mathcal{P}_1''$, and thus let us denote the connected components of its precedence graph by $(\mathcal{P}_1'')_1, (\mathcal{P}_1'')_2 \ldots, (\mathcal{P}_1'')_h$. Let $\mathcal{A}_j'' \stackrel{\text{def}}{=} \mathcal{A}'' ((\mathcal{P}_1'')_j \times \overline{\mathcal{P}}) = \{a \in \mathcal{A}'' \mid a_1^+ \cap (\mathcal{P}_1'')_j \neq \emptyset\}$ and $\mathcal{B}_j'' \stackrel{\text{def}}{=} \mathcal{B}'' ((\mathcal{P}_1'')_j \times \overline{\mathcal{P}}) = \{b \in \mathcal{B}'' \mid b_1^- \cap (\mathcal{P}_1'')_j \neq \emptyset\}$

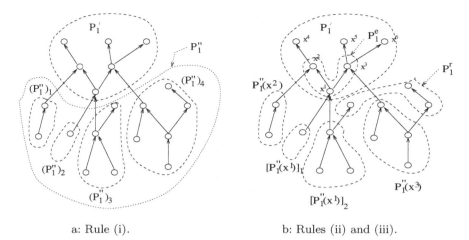

a: Rule (i).   b: Rules (ii) and (iii).

Fig. 1. Decomposing the poset $\mathcal{P}_1$.

for $j = 1, \ldots, h$. Then checking (6) becomes equivalent to checking whether

$$(\mathcal{P}_1'')_j \times \overline{\mathcal{P}} \subseteq (\mathcal{A}_j'')^+ \cup (\mathcal{B}' \cup \mathcal{B}_j'')^-, \qquad j = 1, \ldots, h. \qquad (7)$$

Thus we obtain our first decomposition rule:

Rule (i) Solve subproblem (5) together with h subproblems (7).

Clearly, subproblems (5) and (6) are not independent. Once we know that (5) is satisfied, we gain some information about the solution of subproblem (6). To use this dependence, suppose that subproblem (5) has no solution, i.e., there exists no $q \in \mathcal{P} \setminus [\mathcal{A}^+ \cup (\mathcal{B}')^-]$. Let us define $\mathcal{P}_1^e = \{x \in \mathcal{P}_1' \mid x^\perp \cap \mathcal{P}_1'' \neq \emptyset\}$ to be the set of elements in $\mathcal{P}_1'$ with immediate predecessors in $\mathcal{P}_1''$ (see Fig. 1-b). Let, for each $x \in \mathcal{P}_1^e$, the set $\mathcal{P}_1''(x)$ be the subtree of x^-, lying in $\mathcal{P}_1''$ and whose root in $\mathcal{P}_1'$ is x, i.e., $\mathcal{P}_1''(x) = (x^\perp \cap \mathcal{P}_1'')^-$. Observe that $\mathcal{P}_1''(x)$ and $\mathcal{P}_1''(y)$ are independent posets for $x \neq y$, $x, y \in \mathcal{P}_1^e$ since the precedence graph of $\mathcal{P}_1$ is a forest. Let us further define, for each $x \in \mathcal{P}_1^e$, the sets

$$\mathcal{A}''(x) = \mathcal{A}''(\mathcal{P}_1''(x) \times \overline{\mathcal{P}}), \widetilde{\mathcal{A}}(x) = \{a \in \mathcal{A} \mid a_1 \preceq x\},$$
$$\mathcal{B}'(x) = \mathcal{B}'(\mathcal{P}_1''(x) \times \overline{\mathcal{P}}), \quad \mathcal{B}''(x) = \mathcal{B}''(\mathcal{P}_1''(x) \times \overline{\mathcal{P}}).$$

The following simple lemma is implied by the above definitions.

Lemma 1. *For every $x \in \mathcal{P}_1^e$ and every $b \in \mathcal{B}'(x)$, we must have $b_1 \succeq x$.*

Continuing, let $\mathcal{P}_1^r = \mathcal{P}_1'' \setminus \left(\bigcup_{x \in \mathcal{P}_1^e} \mathcal{P}_1''(x)\right)$, let $(\mathcal{P}_1^r)_1, \ldots, (\mathcal{P}_1^r)_k$ be the connected components of $\mathcal{P}_1^r$, and let $\mathcal{A}_j^r = \mathcal{A}''((\mathcal{P}_1^r)_j \times \overline{\mathcal{P}})$, $\mathcal{B}_j^r = \mathcal{B}''((\mathcal{P}_1^r)^j \times \overline{\mathcal{P}})$, for $j = 1, \ldots, k$. We can now decompose subproblem (6) into

$$\mathcal{P}_1''(x) \times \overline{\mathcal{P}} \subseteq \mathcal{A}''(x)^+ \cup (\mathcal{B}''(x) \cup \mathcal{B}'(x))^-, \qquad x \in \mathcal{P}_1^e, \qquad (8)$$
$$(\mathcal{P}_1^r)_j \times \overline{\mathcal{P}} \subseteq (\mathcal{A}_j^r)^+ \cup (\mathcal{B}_j^r)^-, \qquad j = 1, \ldots, k. \qquad (9)$$

Given that (5) is satisfied, we claim that for each $x \in \mathcal{P}_1^e$, (8) is equivalent to

$$\forall a \in \widetilde{\mathcal{A}}(x) : \mathcal{P}_1''(x) \times (\overline{\mathcal{P}} \cap \overline{a}^+) \subseteq \mathcal{A}''(x)^+ \cup \mathcal{B}''(x)^-, \qquad (10)$$

where $\overline{\mathcal{P}} \cap \overline{a}^+ = (\mathcal{P}_2 \cap a_2^+) \times \ldots \times (\mathcal{P}_n \cap a_n^+)$. To see (10), we make use of the following:

Lemma 2. *Given $z \in \mathcal{P}_1$, let $\mathcal{P}_1' = \mathcal{P}_1 \cap z^+$, $\mathcal{P}_1'' \subseteq \mathcal{P}_1 \cap z^- \setminus \{z\}$ be two disjoint subsets of $\mathcal{P}_1$. Define*

$$\mathcal{A}'' = \{a \in \mathcal{A} \mid a_1^+ \cap \mathcal{P}_1'' \neq \emptyset\}, \qquad \mathcal{A}' = \{a \in \mathcal{A} \setminus \mathcal{A}'' \mid a_1^+ \cap \mathcal{P}_1' \neq \emptyset\},$$
$$\mathcal{B}' = \{b \in \mathcal{B} \mid b_1^- \cap \mathcal{P}_1' \neq \emptyset\}, \qquad \mathcal{B}'' = \{b \in \mathcal{B} \setminus \mathcal{B}' \mid b_1^- \cap \mathcal{P}_1'' \neq \emptyset\}.$$

Suppose further that we know that $\mathcal{P}_1' \times \overline{\mathcal{P}} \subseteq (\mathcal{A}' \cup \mathcal{A}'')^+ \cup (\mathcal{B}')^-$, then

$$\mathcal{P}_1'' \times \overline{\mathcal{P}} \subseteq (\mathcal{A}'')^+ \cup (\mathcal{B}' \cup \mathcal{B}'')^- \iff \forall a \in \widetilde{\mathcal{A}} : \mathcal{P}_1'' \times (\overline{\mathcal{P}} \cap \overline{a}^+) \subseteq (\mathcal{A}'')^+ \cup (\mathcal{B}'')^-,$$

where $\widetilde{\mathcal{A}} = \{a \in \mathcal{A}' \cup \mathcal{A}'' \mid a_1 \preceq z\}$.

Proof. Suppose first that $\mathcal{P}_1'' \times \overline{\mathcal{P}} \subseteq (\mathcal{A}'')^+ \cup (\mathcal{B}' \cup \mathcal{B}'')^-$. Let $(q_1, \overline{q}) \in \mathcal{P}_1'' \times (\overline{\mathcal{P}} \cap \overline{a}^+)$ for some $a \in \widetilde{\mathcal{A}}$, then $(q_1, \overline{q}) \in (\mathcal{A}'')^+ \cup (\mathcal{B}' \cup \mathcal{B}'')^-$. If $(q_1, \overline{q}) \preceq (b_1, \overline{b}) \in \mathcal{B}'$, then by the definition of $\mathcal{B}'$, there is a $y \in \mathcal{P}_1'$ such that $y \preceq b_1$. But then, $a \in \widetilde{\mathcal{A}}$, $\overline{q} \in \overline{\mathcal{P}} \cap \overline{a}^+$ and $y \in \mathcal{P}_1'$ imply that $(a_1, \overline{a}) \preceq (z, \overline{q}) \preceq (y, \overline{q}) \preceq (b_1, \overline{b})$, which contradicts the assumed duality condition (3). This shows that $(q_1, \overline{q}) \in (\mathcal{A}'')^+ \cup (\mathcal{B}'')^-$.

For the other direction, let $(q_1, \overline{q}) \in (\mathcal{P}_1'' \times \overline{\mathcal{P}}) \setminus (\mathcal{B}')^-$. Since $x \preceq y$ for all $x \in \mathcal{P}_1''$, $y \in \mathcal{P}_1'$, we must have $(y, \overline{q}) \notin (\mathcal{B}')^-$ for all $y \in \mathcal{P}_1'$, for otherwise we get the contradiction $(q_1, \overline{q}) \preceq (y, \overline{q}) \preceq (b_1, \overline{b})$ for some $b \in \mathcal{B}'$. Now we use our assumption that $\mathcal{P}_1' \times \overline{\mathcal{P}} \subseteq (\mathcal{A}' \cup \mathcal{A}'')^+ \cup (\mathcal{B}')^-$ to conclude that $(y, \overline{q}) \in (\mathcal{A}' \cup \mathcal{A}'')^+$ for all $y \in \mathcal{P}_1'$. In particular, we have $(z, \overline{q}) \succeq (a_1, \overline{a})$ for some $(a_1, \overline{a}) \in \mathcal{A}' \cup \mathcal{A}''$. But this implies that $a \in \widetilde{\mathcal{A}}$ and hence that $(q_1, \overline{q}) \in \mathcal{P}_1'' \times (\overline{\mathcal{P}} \cap \overline{a}^+)$ for some $a \in \widetilde{\mathcal{A}}$. This gives $(q_1, \overline{q}) \in (\mathcal{A}'')^+ \cup (\mathcal{B}'')^-$. $\qquad\square$

By considering the dual poset of $\mathcal{P}$ (that is, the poset $\mathcal{P}^*$ with the same set of elements as $\mathcal{P}$, but such that $x \prec y$ in $\mathcal{P}^*$ whenever $x \succ y$ in $\mathcal{P}$), and exchanging the roles of $\mathcal{A}$ and $\mathcal{B}$, we get the following symmetric version of Lemma 2.

Lemma 3. *Let $\mathcal{P}_1'' = \mathcal{P}_1 \cap z^-$, $\mathcal{P}_1' \subseteq \mathcal{P}_1 \cap z^+ \setminus \{z\}$ be two disjoint subsets of $\mathcal{P}_1$ where $z \in \mathcal{P}_1$. Let $\mathcal{A}'', \mathcal{A}', \mathcal{B}'', \mathcal{B}'$ be defined as in Lemma 2, and let $\widetilde{\mathcal{B}} = \{b \in \mathcal{B}' \cup \mathcal{B}'' \mid b_1 \succeq z\}$. Suppose we know that $\mathcal{P}_1'' \times \overline{\mathcal{P}} \subseteq (\mathcal{A}'')^+ \cup (\mathcal{B}' \cup \mathcal{B}'')^-$, then*

$$\mathcal{P}_1' \times \overline{\mathcal{P}} \subseteq (\mathcal{A}' \cup \mathcal{A}'')^+ \cup (\mathcal{B}')^- \iff \forall b \in \widetilde{\mathcal{B}} : \mathcal{P}_1' \times (\overline{\mathcal{P}} \cap \overline{b}^-) \subseteq (\mathcal{A}')^+ \cup (\mathcal{B}')^-.$$

Clearly, the equivalence of (8) and (10), given (5), is immediate from Lemma 2 by taking $z \leftarrow x$ and $\mathcal{P}_1'' \leftarrow \mathcal{P}_1''(x)$. Then $\mathcal{P}_1' \leftarrow x^+ \cap \mathcal{P}_1$, $\mathcal{A}'' \leftarrow \mathcal{A}''(x)$, $\mathcal{A}' \leftarrow \mathcal{A}'(\mathcal{P}_1' \times \overline{\mathcal{P}})$, $\widetilde{\mathcal{A}} \leftarrow \widetilde{\mathcal{A}}(x)$, $\mathcal{B}'' \leftarrow \mathcal{B}''(x)$, and by Lemma 1, $\mathcal{B}' \leftarrow \mathcal{B}'(x)$.

Now for each $x \in \mathcal{P}_1^e$, denoting by $[\mathcal{P}_1''(x)]_1, \ldots, [\mathcal{P}_1''(x)]_{k(x)}$ the connected components of $\mathcal{P}_1''(x)$, each problem of the form (10) can be further decomposed into the $k(x)$ subproblems:

$$[\mathcal{P}_1''(x)]_j \times (\overline{\mathcal{P}} \cup \overline{a}^+) \subseteq \mathcal{A}_j''(x)^+ \cup \mathcal{B}_j''(x)^-, \quad j = 1, \ldots, k(x), \quad a \in \widetilde{\mathcal{A}}(x), \quad (11)$$

where $\mathcal{A}''_j(x) = \{a \in \mathcal{A}''(x) \mid a_1^+ \cap [\mathcal{P}''_1(x)]_j \neq \emptyset\}$, and $\mathcal{B}''_j(x) = \{b \in \mathcal{B}''(x) \mid b_1^- \cap [\mathcal{P}''_1(x)]_j \neq \emptyset\}$. Thus we arrive at the following decomposition rule:

Rule (ii) Solve subproblem (5). If it has a solution then we get an element $q \in \mathcal{P} \setminus (\mathcal{A}^+ \cup \mathcal{B}^-)$. Otherwise, we solve $|\widetilde{\mathcal{A}}(x)|$ subproblems (11), for each $x \in \mathcal{P}_1^e$ and $j = 1, \ldots, k(x)$, and finally subproblems (9) for $j = 1, \ldots, k$.

Suppose finally that subproblem (6), or its equivalent (7), has no solution (i.e. there is no $q \in \mathcal{P}'' \setminus [(\mathcal{A}'')^+ \cup \mathcal{B}^-]$). We proceed in this case as follows. For $x \in \mathcal{P}_1$, let $\widetilde{\mathcal{A}}(x) = \{a \in \mathcal{A} \mid a_1 \preceq x\}$, $\widetilde{\mathcal{B}}(x) = \{b \in \mathcal{B} \mid b_1 \succeq x\}$, and $\hat{\mathcal{A}}'(x) = \{a \in \mathcal{A}' \mid a_1 = x\}$. Let us use $x^1, \ldots, x^w$ to denote the elements of $\mathcal{P}'_1$ and assume, without loss of generality, that they are topologically sorted in this order, that is, $x^j \prec x^l$ implies $j < l$ (see Fig. 1-b). Now we can use the following rule to solve our problem:

a. Solve subproblems (7), then

b. (*decompose (5):*) for $j = 1$ to w, solve

$$\{x^j\} \times \overline{\mathcal{P}} \subseteq \left[\left(\bigcup_{y \in (x^j)^\perp} \widetilde{\mathcal{A}}(y)\right) \cup \hat{\mathcal{A}}'(x^j)\right]^+ \cup \widetilde{\mathcal{B}}(x^j)^-. \tag{12}$$

The following lemma will allow us to eliminate the contribution of the set $\mathcal{A}''$ in subproblems (12) at the expense of possibly introducing at most $|\mathcal{B}|^d$ additional subproblems.

Lemma 4. *Given* $x^j \in \mathcal{P}'_1$, *suppose we know that* $(y^- \cap \mathcal{P}_1) \times \overline{\mathcal{P}} \subseteq \widetilde{\mathcal{A}}(y)^+ \cup \mathcal{B}^-$ *for all* $y \in (x^j)^\perp$, *then (12) is equivalent to*

$$\{x^j\} \times \left[\overline{\mathcal{P}} \cap \left(\bigcap_{y \in (x^j)^\perp} \overline{b}(y)^-\right)\right] \subseteq \hat{\mathcal{A}}'(x^j)^+ \cup \widetilde{\mathcal{B}}(x^j)^-, \tag{13}$$

for all collections $\{b(y) \in \widetilde{\mathcal{B}}(y) \mid y \in (x^j)^\perp\}$.

Proof. We prove by induction on $|Y|$, where $Y \subseteq (x^j)^\perp$, that

$$\{x^j\} \times \overline{\mathcal{P}} \subseteq \left[\left(\bigcup_{y \in (x^j)^\perp} \widetilde{\mathcal{A}}(y)\right) \cup \hat{\mathcal{A}}'(x^j)\right]^+ \cup \widetilde{\mathcal{B}}(x^j)^- \qquad \Longleftrightarrow$$
$$\{x^j\} \times \left[\overline{\mathcal{P}} \cap \left(\bigcap_{y \in Y} \overline{b}(y)^-\right)\right] \subseteq \left[\left(\bigcup_{y \in (x^j)^\perp \setminus Y} \widetilde{\mathcal{A}}(y)\right) \cup \hat{\mathcal{A}}'(x^j)\right]^+ \cup \widetilde{\mathcal{B}}(x^j)^-, \tag{14}$$

for all collections $\{b(y) \in \widetilde{\mathcal{B}}(y) \mid y \in Y\}$. This trivially holds for $Y = \emptyset$ and will prove the lemma for $Y = (x^j)^\perp$. To show (14), assume that it holds for some $Y \subset (x^j)^\perp$ and let $x \in (x^j)^\perp \setminus Y$. Consider a subproblem of the form

$$\{x^j\} \times \left[\overline{\mathcal{P}} \cap \left(\bigcap_{y \in Y} \overline{b}(y)^-\right)\right] \subseteq \left[\widetilde{\mathcal{A}}(x) \cup \left(\bigcup_{y \in (x^j)^\perp \setminus (Y \cup \{x\})} \widetilde{\mathcal{A}}(y)\right) \cup \hat{\mathcal{A}}'(x^j)\right]^+ \cup \widetilde{\mathcal{B}}(x^j)^-,$$

for some collection $\{b(y) \in \widetilde{\mathcal{B}}(y) \mid y \in Y\}$. Now we apply Lemma 3 with $z \leftarrow x$, $\mathcal{P}_1'' \leftarrow x^- \cap \mathcal{P}_1$, $\mathcal{P}_1' \leftarrow \{x^j\}$, $\mathcal{A}'' \leftarrow \widetilde{\mathcal{A}}(x)$, $\mathcal{A}' \leftarrow \left(\bigcup_{y \in (x^j)^\perp \setminus (Y \cup \{x\})} \widetilde{\mathcal{A}}(y)\right) \cup \hat{\mathcal{A}}'(x^j)$, and $\widetilde{\mathcal{B}} \leftarrow \widetilde{\mathcal{B}}(x)$ to get the required result. $\qquad \square$

Informally, Lemma 4 says that, given $x^j \in \mathcal{P}_1'$, if the dualization subproblems for all sub-forests that lie below x^j have been already verified to have no solution, then we can replace the solution to subproblem (12) by solving at most $\prod_{y \in (x^j)^\perp} |\widetilde{\mathcal{B}}(y)|$ subproblems of the form (13). Observe that it is important to check subproblems (12) in the topological order $j = 1, \ldots, w$ to be able to use Lemma 4. Thus we get:

Rule (iii) Solve subproblems (7), and if they do not have a solution, then solve subproblems (13), for all collections $\{b(y) \in \widetilde{\mathcal{B}}(y) \mid y \in (x^j)^\perp\}$, for $j = 1, \ldots, w$.

Finally it remains to remark that all the decomposition rules described above result, indeed, in posets with connected precedence graphs.

5   The Algorithm

Given subsets $\mathcal{A}, \mathcal{B} \subseteq \mathcal{P} = \mathcal{P}_1 \times \ldots \times \mathcal{P}_n$ that satisfy the necessary duality condition (3), we proceed as follows:

Step 0. If the precedence graph of $\mathcal{P}_i$ is not connected, for some $i \in [n]$ (only can happen initially), use Proposition 1 to decompose the problem into a number of subproblems over posets with connected precedence graphs.

Step 1. For each $k \in [n]$:

　　1. (*eliminate:*) if $a_k^+ \cap \mathcal{P}_k = \emptyset$ for some $a \in \mathcal{A}$ ($b_k^- \cap \mathcal{P}_k = \emptyset$ for some $b \in \mathcal{B}$), then by (2), a (respectively, b) can be discarded from further consideration;

　　2. (*project:*) if $a_k \prec \min(a_k^+ \cap \mathcal{P}_k)$ for some $a \in \mathcal{A}$ ($b_k \succ \max(b_k^- \cap \mathcal{P}_k)$ for some $b \in \mathcal{B}$), then set $a_k \leftarrow \min(a_k^+ \cap \mathcal{P}_k)$ (respectively, $b_k \leftarrow \max(b_k^- \cap \mathcal{P}_k)$).

Thus we may assume for next steps that $\mathcal{A}, \mathcal{B} \subseteq \mathcal{P}$.

Step 2. If $\min\{|\mathcal{A}|, |\mathcal{B}|\} \leq 3$, then the dualization problem can be solved in polynomial time using Proposition 2.

Step 3. Let a^o, b^o be arbitrary elements of $\mathcal{A}, \mathcal{B}$ respectively. Find an $i \in [n]$ such that $a_i^o \not\preceq b_i^o$. Assume, without loss of generality, that $i = 1$. If in-deg($\mathcal{P}_1$) $\leq$ out-deg($\mathcal{P}_1$), we set $\mathcal{P}_1' \leftarrow \mathcal{P}_1 \cap (a_1^o)^+$, $\mathcal{P}_1'' \leftarrow \mathcal{P}_1 \setminus \mathcal{P}_1'$, otherwise, we set $\mathcal{P}_1'' \leftarrow \mathcal{P}_1 \cap (b_1^o)^-$, and $\mathcal{P}_1' \leftarrow \mathcal{P}_1 \setminus \mathcal{P}_1''$. In the latter case, we should use the symmetric versions of the decomposition rules (i)-(iii), listed above, which we obtain by exchanging the roles of $\mathcal{A}$ and $\mathcal{B}$ in these rules and replacing $\mathcal{P}$ by its dual poset $\mathcal{P}^*$. We assume therefore, without loss of generality, that the former case holds.

Step 4. Let $\mathcal{A}', \mathcal{A}'', \mathcal{B}', \mathcal{B}'', \ldots$ be as defined in the previous section, and define

$$\epsilon^{\mathcal{A}} = \frac{|\mathcal{A}'|}{|\mathcal{A}|}, \qquad \epsilon^{\mathcal{B}} = \frac{|\mathcal{B}''|}{|\mathcal{B}|}.$$

Observe that $\epsilon^{\mathcal{A}} > 0$ and $\epsilon^{\mathcal{B}} > 0$ since $a^o \in \mathcal{A}'$ and $b^o \in \mathcal{B}''$.

Step 5. Define

$$\epsilon(v) = 1/\chi(v), \quad \text{where } \chi(v)^{\chi(v)} = v^d, \ v = v(\mathcal{A}, \mathcal{B}).$$

If $\min\{\epsilon^{\mathcal{A}}, \epsilon^{\mathcal{B}}\} > \epsilon(v)$, we use decomposition rule (i) which amounts to solving recursively subproblem (5) of volume $v(\mathcal{A}, \mathcal{B}') = |\mathcal{A}||\mathcal{B}'|$, and subproblems (7) of volumes $v(\mathcal{A}''_j, \mathcal{B}' \cup \mathcal{B}''_j) \leq |\mathcal{A}''_j||\mathcal{B}|$, for $j = 1, \ldots, h$. This gives rise to the recurrence

$$C(v(\mathcal{A}, \mathcal{B})) \leq 1 + C(|\mathcal{A}||\mathcal{B}'|) + \sum_{j=1}^{h} C(|\mathcal{A}''_j||\mathcal{B}|). \tag{15}$$

Step 6. If $\epsilon^{\mathcal{B}} \leq \epsilon(v)$, we apply rule (ii) and get the recurrence

$$C(v) \leq 1 + C(|\mathcal{A}||\mathcal{B}'|) + \sum_{x \in \mathcal{P}_1^e} \sum_{j=1}^{k(x)} |\widetilde{\mathcal{A}}(x)| C(|\mathcal{A}''_j(x)||\mathcal{B}''_j(x)|) + \sum_{j=1}^{k} C(|\mathcal{A}^r_j||\mathcal{B}^r_j|). \tag{16}$$

Step 7. Finally, if $\epsilon^{\mathcal{A}} \leq \epsilon(v) < \epsilon^{\mathcal{B}}$, we use rule (iii) which gives

$$C(v(\mathcal{A}, \mathcal{B})) \leq 1 + \sum_{j=1}^{h} C(|\mathcal{A}''_j||\mathcal{B}|) + \sum_{j=1}^{w} \left(\prod_{y \in (x^j)^{\perp}} |\widetilde{\mathcal{B}}(y)| \right) C(|\hat{\mathcal{A}}'(x^j)||\widetilde{\mathcal{B}}(x^j)|). \tag{17}$$

6   Proof of Theorem 1

The proof essentially goes along the same lines as in [13]. We show by induction on $v = v(\mathcal{A}, \mathcal{B})$, that recurrences (15)-(17) imply that $C(v) \leq R(v) \overset{\text{def}}{=} v^{\chi(v)}$. Since, for $\min\{|\mathcal{A}|, |\mathcal{B}|\} \leq 3$, Step 2 of the algorithm implies that $C(v) = 1$, we may assume that $\min\{|\mathcal{A}|, |\mathcal{B}|\} \geq 4$, i.e., $v \geq 16$.

Let us consider first recurrence (15). Observe that $\mathcal{A}''_1, \ldots, \mathcal{A}''_h$ partition $\mathcal{A}''$ since $\mathcal{P}_1$ is a forest, and therefore, we get by the induction hypothesis and the super-additivity of $R(.)$

$$C(v) \leq 1 + R(|\mathcal{A}||\mathcal{B}'|) + \sum_{j=1}^{h} R(|\mathcal{A}''_j||\mathcal{B}|) \leq 1 + R(|\mathcal{A}||\mathcal{B}'|) + R(\sum_{j=1}^{h} |\mathcal{A}''_j||\mathcal{B}|)$$
$$= 1 + R((1 - \epsilon^{\mathcal{B}})v) + R((1 - \epsilon^{\mathcal{A}})v).$$

Now using the facts that $\epsilon^{\mathcal{A}} > \epsilon(v), \epsilon^{\mathcal{B}} > \epsilon(v)$, and $v \geq 16$, we obtain by the monotonicity of $\mathcal{X}(v)$

$$C(v) \leq 1 + 2[(1 - \epsilon(v))v]^{\chi((1-\epsilon(v))v)} \leq 1 + 2[(1 - \epsilon(v))v]^{\chi(v)}$$
$$= 1 + 2 \left(1 - \tfrac{1}{\chi(v)}\right)^{\chi(v)} v^{\chi(v)} \leq 1 + \tfrac{2}{e} v^{\chi(v)} \leq v^{\chi(v)} = R(v), \tag{18}$$

concluding the induction proof for this case.

Let us consider next (16) and observe that the sets $\mathcal{A}_j''(x)$, $j = 1, \ldots, k(x)$, $x \in \mathcal{P}_1^e$, are disjoint (again, since $\mathcal{P}_1$ is a forest), that $\mathcal{B}_j''(x) \subseteq \mathcal{B}''$ for all $x \in \mathcal{P}_1^e$ and $j = 1, \ldots, k(x)$, and that the sets $\mathcal{B}_j^r$, $j = 1, \ldots, k$ are disjoint and each is a subset of $\mathcal{B}''$. Consequently,

$$C(v) \le 1 + R(|\mathcal{A}||\mathcal{B}'|) + |\mathcal{A}|R(\sum_{x \in \mathcal{P}_1^e} \sum_{j=1}^{k(x)} |\mathcal{A}_j''(x)||\mathcal{B}_j''(x)|)$$
$$+ R(\sum_{j=1}^{k} |\mathcal{A}_j^r||\mathcal{B}_j^r|) \le 1 + R((1 - \epsilon^{\mathcal{B}})v) + (|\mathcal{A}| + 1)R(\epsilon^{\mathcal{B}}v).$$

Since $\min\{|\mathcal{A}|, |\mathcal{B}|\} \ge 4$ is assumed, we have $|\mathcal{A}| + 1 \le |\mathcal{A}||\mathcal{B}|/3 = v/3$ and thus

$$C(v) \le 1 + R((1 - \epsilon^{\mathcal{B}})v) + \frac{v}{3}R(\epsilon^{\mathcal{B}}v) \le R((1 - \epsilon^{\mathcal{B}})v) + \frac{v}{2}R(\epsilon^{\mathcal{B}}v) \qquad (19)$$
$$\le [(1 - \epsilon^{\mathcal{B}})v]^{\chi(v)} + \frac{v}{2}[\epsilon^{\mathcal{B}}v]^{\chi(v)} = \psi(\epsilon^{\mathcal{B}})R(v),$$

where $\psi(\epsilon) \stackrel{\text{def}}{=} (1 - \epsilon)^{\chi(v)} + \frac{v}{2}\epsilon^{\chi(v)}$. Since $\psi(\epsilon)$ is convex in $\epsilon \in [0, 1]$, $\psi(0) = 1$, $\epsilon^{\mathcal{B}} \le \epsilon(v)$, and

$$\psi(\epsilon(v)) = \left(1 - \frac{1}{\chi(v)}\right)^{\chi(v)} + \frac{v}{2}\left(\frac{1}{\chi(v)}\right)^{\chi(v)} \le \frac{1}{e} + \frac{1}{2v^{d-1}} < 1,$$

it follows that $\psi(\epsilon^{\mathcal{B}}) < 1$ and hence, $C(v) \le R(v)$.

Let us consider finally recurrence (17) and observe that $|(x^j)^{\perp}| \le d$ for every $x^j \in \mathcal{P}_1'$, by our assumption that in-deg$(\mathcal{P}_1) \le$ out-deg$(\mathcal{P}_1)$ (see Step 3 of the algorithm). Thus (17) gives by induction

$$C(v(\mathcal{A}, \mathcal{B})) \le 1 + \sum_{j=1}^{h} R(|\mathcal{A}_j''||\mathcal{B}|) + |\mathcal{B}|^d \sum_{j=1}^{w} R(|\hat{\mathcal{A}}'(x^j)||\widetilde{\mathcal{B}}(x^j)|). \qquad (20)$$

Note that this the only place in which the bound d on the degrees appears. Since the sets $\hat{\mathcal{A}}'(x^j)$, $j = 1, \ldots, w$, partition $\mathcal{A}'$, and $|\mathcal{B}|^d \le v(|\mathcal{A}|, |\mathcal{B}|)^d/3$ for $|\mathcal{A}| \ge 3$, we get $C(v) \le 1 + R((1 - \epsilon^{\mathcal{A}})v) + \frac{v^d}{3}R(\epsilon^{\mathcal{A}}v)$. This implies by symmetry to (19) that $C(v) \le R(v)$.

Note that $\chi(v) < \chi(m^2) < 2\chi(m) \sim 2d \log m / \log \log m$, and we get the bound stated in Theorem 1. $\square$

Acknowledgment. The author thanks Endre Boros, Vladimir Gurvich, and Leonid Khachiyan for many helpful discussions that led to this work.

References

1. R. Agrawal, T. Imielinski and A. Swami, Mining association rules between sets of items in massive databases. In *Proc. the 1993 ACM-SIGMOD Int. Conf. Management of Data*, pp. 207-216.
2. R. Agrawal and R. Srikant, Fast algorithms for mining association rules in large databases . In *Proc. 20th Int. Conf. Very Large Data Bases*, pp. 487-499, 1994.

3. R. Agrawal and R. Srikant, Mining sequential patterns. In *Proc. 11th Int. Conf. Data Engineering*, pp. 3-14, 1995.
4. R. Agrawal, H. Mannila, R. Srikant, H. Toivonen and A. I. Verkamo, Fast discovery of association rules. In *Advances in Knowledge Discovery and Data Mining*, pp. 307-328, AAAI Press, Menlo Park, California, 1996.
5. M. Anthony and N. Biggs, *Computational Learning Theory*, Cambridge University Press, 1992.
6. J. C. Bioch and T. Ibaraki, Complexity of identification and dualization of positive Boolean functions, *Information and Computation*, 123 (1995) pp. 50-63.
7. E. Boros, K. Elbassioni, V. Gurvich, L. Khachiyan and K.Makino, On generating all minimal integer solutions for a monotone system of linear inequalities. In *Automata, Languages and Programming, 28th Int. Colloquium, ICALP 2001*, LNCS 2076, pp. 92-103 (Springer Verlag, Berlin, Heidelberg, New York , July 2001).
8. E. Boros, V. Gurvich, L. Khachiyan and K. Makino, An inequality limiting the number of maximal frequent sets, DIMACS tech. report 2000-37, Rutgers Univ.
9. E. Boros, V. Gurvich, L. Khachiyan and K. Makino, Generating weighted transversals of a hypergraph, in *Proceedings of the 2nd Japanese-Hungarian Symposium on Discrete Mathematics and Its Applications*, pp. 13-22, Hungarian Academy of Sciences, Budapest, Hungary, April 20-23, 2001.
10. S. Brin, R. Motwani and C. Silverstein, Beyond market basket: Generalizing association rules to correlations. In *Proc. the 1997 ACM-SIGMOD Int. Conf. Management of Data*, pp. 265–276.
11. C. J. Colbourn, *The combinatorics of network reliability*, Oxford Univ.Press, 1987.
12. T. Eiter and G. Gottlob, Identifying the minimal transversals of a hypergraph and related problems, *SIAM Journal on Computing*, 24 (1995) pp. 1278-1304.
13. M. L. Fredman and L. Khachiyan, On the complexity of dualization of monotone disjunctive normal forms, *Journal of Algorithms*, 21 (1996) pp. 618-628.
14. D. Gunopulos, R. Khardon, H. Mannila, and H. Toivonen, Data mining, hypergraph transversals and machine learning. In *Proc. 16th ACM-SIGACT-SIGMOD-SIGART Symposium on Principles of Database Systems*, (1997) pp. 12-15.
15. V. Gurvich and L. Khachiyan, On generating the irredundant conjunctive and disjunctive normal forms of monotone Boolean functions, *Discrete Applied Mathematics*, 1996-97, issue 1-3, (1999) pp. 363-373.
16. J. Han and Y. Fu, Discovery of multiple-level association rules from large databases. In *Proc. 21st Int. Conf. Very Large Data Bases*, pp. 420-431, 1995.
17. K. Makino and T. Ibaraki, Interior and exterior functions of Boolean functions, *Discrete Applied Mathematics*, 69 (1996) pp. 209-231.
18. H. Mannila, and H. Toivonen, Discovery of frequent episodes in event sequences, *Data Mining and Knowledge Discovery*, 1 (1997) pp. 259-289.
19. R. C. Read, Every one a winner, or how to avoid isomorphism when cataloging combinatorial configurations, *Annals of Discrete Math.* 2 (1978) pp. 107-120.
20. R. Srikant and R. Agrawal, Mining generalized association rules. In *Proc. 21st Int. Conf. Very Large Data Bases*, pp. 407-419, 1995.
21. R. Srikant and R. Agrawal, Mining quantitative association rules in large relational tables. In *Proc. of the ACM-SIGMOD 1996 Int. Conf. Management of Data*, pp. 1-12, 1996.

An Asymptotic $\mathcal{O}(\ln \rho / \ln \ln \rho)$-Approximation Algorithm for the Scheduling Problem with Duplication on Large Communication Delay Graphs

R. Lepere[1] and C. Rapine[2]

[1] ID-IMAG, avenue Jean Kuntzman - ZIRST, 38 330 Montbonnot Saint Martin, France. `Renaud.Lepere@imag.fr`
[2] GILCO-ENSGI-INPG, 46 avenue Félix Viallet, 38 031 Grenoble Cedex 1, France. `Christophe.Rapine@gilco.inpg.fr`

Abstract. This article is concerned with the problem of scheduling a parallel application depicted by a precedence graph in presence of large communication delays. The target architecture is constituted of a bounded number m of identical processors linked together by an interconnection network. Communication delays represent the time of data transfer between two tasks of the application allocated to different processors. Our objective is to find an allocation of tasks to the processors and an execution order on each machine such that the overall completion time is minimized. We consider the special case of unit execution time for all computation tasks and a uniform communication delay ρ. We present a new approach based on the reduction of the problem to the successive schedulings of "small graphs", roughly speaking graphs which can be scheduled in time at most $\rho + 1$ on an unbounded number of processors. Allowing duplication, corresponding to the recomputation of some of the tasks, this technique allows us to derive an asymptotic $\mathcal{O}(\ln \rho / \ln \ln \rho)$-approximation algorithm for general precedence graph structure.

Keywords. Scheduling – duplication – communication times – performance guarantee – makespan – approximation algorithm.

1 Introduction

Networks of workstations and PC-clusters are becoming popular parallel architectures. One of the main difficulties in obtaining high performances on those parallel systems comes from the important cost of communications, which may be very slow operations relatively to processor computations. Communication latency and overhead occur when tasks assigned on different processors exchange data. Taking into account the high communication costs in the scheduling decision appears to be a challenge and a necessity to get the full efficiency of those systems.

H. Alt and A. Ferreira (Eds.): STACS 2002, LNCS 2285, pp. 154–165, 2002.
© Springer-Verlag Berlin Heidelberg 2002

We consider in this article a parallel application to execute on such a parallel system composed of m identical processors. The application is depicted by an acyclic precedence task graph where nodes represent computations and arcs represent data dependencies between the tasks. We consider the special case of unit execution time for all the tasks and a uniform (potentially large) integer communication delay ρ. The delay ρ represents the time of a data transfer between two tasks on the target architecture. For the execution, we assume the non-preemptive delay model. If there is a precedence constraint between tasks x and y, task y can not start its execution before the completion of task x, plus the delay ρ if the two tasks are not executed on the same processor. In this setting, task duplication appears as a way to reduce the penalty of high interprocessor communication delays. It corresponds to the execution of the same task on different processors to avoid a data transfer. In this article, we focus on the scheduling problem under this model allowing task duplication on a bounded number of processors; our objective is the minimization of the maximal completion time of the tasks, called *makespan*.

Scheduling with communication delays has been shown to be a NP-hard problem [17], even in the case of an unbounded number of processors [16], and several non-approximability results have been established in the past decade [8, 2]. List scheduling algorithms, first introduced by Graham [5], form an efficient and popular class of heuristics for scheduling problems [12]. Unfortunately they may perform quite poorly in presence of communication delays: no better performance guarantee than $\mathcal{O}(\rho)$ are known [9,10] for the problem. Another popular approach is *clustering* algorithms [4], corresponding to the scheduling problem on an unbounded number of processors. For large communication delays ($\rho \geq 1$) the article of Papadimitriou and Yannakakis [14] was the pioneer work in the domain, proposing a 2-approximation for the scheduling problem with duplication on an unbounded number of processors. The model they used is the one adopted in this article. Several authors [13,1,3] have improved and extended this result. However, those algorithms are not widely used in practice since the work generated by the duplication is too important: the number of tasks executed cannot be bounded relatively to the total number of tasks of the application.

Hence, if duplication appears to be a useful technique to get approximation algorithms for the clustering problem, very few results exist for the scheduling problem on a bounded number m of processors: Hanen and Munier [6] proposed a $(2 - 1/m)$-approximation for unit execution and communication time applications, showing that the use of duplication allow to obtain the same performance guarantee as Graham's List Scheduling algorithms without communication delays. For large communication delays, Munier [11] gave the first constant approximation ratio, for the special case of tree precedence graphs.

We focus in this article on the case of large communication delays when potentially $\rho \gg 1$. We present a technique based on the reduction of the problem to the scheduling of "small" graphs. Intuitively this notion of small graphs corresponds to graphs which can be scheduled in less than $\rho + 1$ time units on an unbounded number of processors using duplication. We show that if one can schedule small graphs efficiently, then an asymptotic approximation algorithm can be derived for arbitrary graphs. An algorithm has an asymptotic performance λ is its performance ratio tends towards λ when the optimum objective value becomes sufficiently large. Combining this result with a quite simple covering algorithm for the scheduling problem of small graphs leads to an algorithm with an asymptotic performance guarantee in $\mathcal{O}(\ln \rho / \ln \ln \rho)$ for arbitrary graphs. We also show that constant performance guarantee can be obtained for a class of graphs containing in-forests and interval-order graphs.

2 A Lower Bound of the Earliest Execution Time

Recall that we are interested in scheduling an application depicted by a precedence graph G on a bounded number m of processors. If there exists a precedence constraint $u \to x$ in G, we call as usual task u a predecessor of task x, and x a successor of u. More generally, if there exists a directed path in G from u to x, u is called an ancestor of task x. We denote by $A(x)$ the set of ancestors of a task x. When duplication is allowed, a task x may be executed several times; we call each of its duplicates a *copy* of x. We introduce in this section a lower bound of the earliest execution time of a task on an unbounded number of processors. We call this quantity the height of a task.

Definition 1. *The height $h_G(x)$ of a task x is defined inductively by:*

- *If x is a root of the graph, we set $h_G(x) = 0$.*
- *Otherwise let $x_1, \ldots, x_{|A(x)|}$ be the ancestors of x indexed by non-increasing height. Then we set*

$$h_G(x) = \max\{ h_G(x_i) + i \mid i = 1, \ldots, \min\{|A(x)|, \rho + 1\}\}$$

This bound is a generalization of the one used by Papadimitriou & Yannakakis, and also a straightforward generalization of the critical path. When there is no ambiguity on the graph considered, we denote simply $h_G(x)$ by $h(x)$. The following lemma proves that the height of a task is a lower bound of its earliest execution time:

Lemma 1. *In any schedule, task x can not start its execution before time $h(x)$.*

Proof. The proof is based on induction on the structure of the graph. Consider any schedule for the graph, and let $t(x)$ be the starting time of x in this schedule. If x is a root, the result clearly holds. Assume now that the lemma is true for all the ancestors $A(x) = \{x_1, \ldots, x_{|A(x)|}\}$ of task x indexed by non-increasing order of their heights. For short we denote by $\tau = \min\{|A(x)|, \rho+1\}$. If x does not start its execution before time $h(x_k) + \rho + 1$ for any of its ancestors for $k = 1, \ldots, \tau$, it involves directly that $t(x)$ is greater than $\max_{k=1,\tau}\{h(x_k)\} + \rho + 1 \geq h(x)$. Otherwise we can consider the ancestor x_i with the largest index in $\{1, \ldots, \tau\}$ such that x starts its execution strictly before time $h(x_i) + \rho + 1$. It implies that a copy of x_i is executed on the same processor than x in the schedule. Since for each ancestor x_j with $j \leq i$ we have $h(x_j) \geq h(x_i)$, a copy must also be scheduled on the processor where x is executed. We are in the situation where copies of ancestors $x_1, \ldots, x_i$ are executed on the same processor than x, none of them being able to start before time $h(x_i)$. The result follows.

We define the height $h(G)$ of a graph G as $\max_x \{h_G(x)\} + 1$. Due to lemma 1, the height of a graph is a lower bound of the makespan of any feasible schedule for G. In order to reduce the problem of scheduling graphs to the scheduling of graphs of height smaller than $\rho + 1$, we establish that the suppression of all the tasks of height lower than ρ makes the height of a graph diminish by at least $\rho + 1$. In the following of this article, we say that a graph is a *small graph* iff its height is at most $\rho + 1$. Lemma 2 is the key point of the algorithm presented in section 3.

Lemma 2. *For a graph G, we denote by G_ρ the graph induced by tasks of height smaller than ρ, $\{x \in G \mid h_G(x) \leq \rho\}$, and by $G \setminus G_\rho$ the graph induced by the remaining tasks, $\{x \in G \mid h_G(x) > \rho\}$. Then, if $G \setminus G_\rho$ is not empty, we have:*

$$h(G \setminus G_\rho) \leq h(G) - (\rho + 1)$$

Proof. For short we denote by G' the graph $G \setminus G_\rho$, and respectively by $h(x) = h_G(x)$ and $h'(x) = h_{G'}(x)$ the height of task x in graphs G and G'. We are going to prove a slightly stronger result, namely that $h'(x) \leq h(x) - (\rho + 1)$ for any task x of G'. The proof is based on induction on the structure of the graph G'. If task x is a root in G', the result clearly holds, since $h'(x) = 0$, and by definition $x \notin G_\rho$ implies $h(x) \geq \rho + 1$.

Consider now a task x in G' which is not a root. Let $A'(x) = \{x_1', \cdots, x_{|A'(x)|}'\}$ be the ancestors of task x in G' indexed by non-increasing height in this graph. In the same way let $A(x) = \{x_1, \cdots, x_{|A(x)|}\}$ be the ancestors of task x in G sorted by non-increasing height in the graph G. We assume according to our induction hypothesis that for any task y in $A'(x)$, $h'(y) \leq h(y) - (\rho + 1)$. We can notice here that the set $A'(x)$ is equal to the set $A(x) \setminus \{x \mid h(x) \leq \rho\}$, which corresponds to the set $\{x_i \mid i = 1, \ldots, |A'(x)|\}$. However the $(x_i')_i$ and the $(x_i)_i$ are not necessarily sorted in the same order. More precisely, there exists a permutation σ on the set $\{1, \cdots, |A'(x)|\}$ such that $x_i' = x_{\sigma(i)}$. In order to prove that $h(x) - h'(x) \geq \rho + 1$ we establish first that for all i in $\{1, \cdots, |A'(x)|\}$, $h(x_i) - h'(x_i') \geq \rho + 1$. For a given index i in $\{1, \cdots, |A'(x)|\}$, we distinguish two cases:

- If $i \leq \sigma(i)$, then, due to our indexing, $h(x_i) \geq h(x_{\sigma(i)}) = h(x_i')$. Since the induction hypothesis imposes that $h(x_i') - h'(x_i') \geq \rho + 1$, we can conclude that $h(x_i) - h'(x_i') \geq \rho + 1$.
- If $i > \sigma(i)$, then by the pigeonhole principle there exists an index $j < i$ such that $\sigma(j) \geq i$. So we have $h'(x_j') \geq h'(x_i')$ and $h(x_j') = h(x_{\sigma(j)}) \leq h(x_i)$. Using the induction hypothesis, we have also $h(x_j') - h'(x_j') \geq \rho + 1$. Finally we can conclude that $h(x_i) - h'(x_i') \geq \rho + 1$.

Thus, for all i in $\{1, \ldots, |A'(x)|\}$, the inequality $h(x_i) - h'(x_i') \geq \rho + 1$ is verified. By definition of $h(x)$, we obtain $h(x) - (h'(x_i') + i) \geq \rho + 1$ for all i in $\{1, \cdots, \min(|A'(x)|, \rho + 1)\}$. By definition of $h'(x)$, we can conclude that $h(x) - h'(x) \geq \rho + 1$.

3   Scheduling Graphs Using Small Graphs

Lemma 2 of the previous section suggests to schedule a graph G by scheduling iteratively the subgraphs G_ρ. It means that our algorithm first schedules the graph G_ρ, and then treats the remaining graph $G' = G \setminus G_\rho$ as a new graph, scheduling G'_ρ, and so on. The main point is to ensure that the scheduling of G' is independent of the one of G_ρ. This is done by inserting a communication phase of ρ units of time after the completion of G_ρ. Here lemma 2 ensures that the quality of the schedule is not deteriorated too much by the insertion of this communication phase, since G can not be scheduled in time less than $\rho + 1 + h(G')$.

We assume in this section that we are given a scheduling algorithm $\mathcal{A}$ for the "small graphs", i.e. graphs of height smaller than $\rho + 1$. The way to obtain such an algorithm is described in detail in section 4. A sketch of our algorithm to schedule a graph of arbitrary height using algorithm $\mathcal{A}$ as a black-box for small graphs is depicted in figure 1. Lemma 3 establishes the performance of such a phase by phase algorithm relatively to the efficiency of algorithm $\mathcal{A}$. By notation abuse, $|G|$ refers for a graph G to its number of tasks, which is also its sequential execution time in our model.

INPUT: A task graph G, and a scheduling algorithm $\mathcal{A}$ for graphs of height smaller than $\rho + 1$.

   ▷ Set $H := G$
   ▷ While $H \neq \emptyset$

      ▷ Use algorithm $\mathcal{A}$ to schedule H_ρ.
      ▷ Insert a communication phase of ρ time units.
      ▷ Set $H := H \setminus H_\rho$.

   ▷ End While

OUTPUT: A schedule of graph G.

Fig. 1. The phase by phase algorithm using a sub-scheduling algorithm $\mathcal{A}$ for small graphs.

Lemma 3. *Assuming that algorithm $\mathcal{A}$ can schedule any graph H of height smaller than $\rho + 1$ in time less than $\alpha \frac{|H|}{m} + \beta(\rho + 1)$, then the phase by phase algorithm can schedule any graph G in time less than $\alpha \frac{|G|}{m} + (\beta + 1)h(G) + \beta(\rho + 1)$.*

Proof. Consider a graph G and let k be the number of phases of the algorithm, i.e. the number of calls to algorithm $\mathcal{A}$. If $H_1, \ldots, H_k$ refer to the successive subgraphs of G on which algorithm $\mathcal{A}$ is applied, then the makespan of the schedule computed by our algorithm is lower than $\alpha(|H_1| + \ldots + |H_k|)/m + k\beta(\rho + 1) + (k - 1)\rho$. The last term corresponds to the delay ρ inserted between the successive schedules. Clearly the H_i's form a partition of G, which implies that $|H_1| + \ldots + |H_k| = |G|$. In addition, according to lemma 2, the height of the graph decreases at least of $\rho + 1$ at each step. This allows to bound the number k of phases. More precisely we have $k \leq \lceil \frac{h(G)}{\rho + 1} \rceil \leq \frac{h(G)}{\rho + 1} + 1$, which establishes the bound of lemma 3 for the makespan of the algorithm.

Notice that lemma 3 holds even if duplication is not allowed; but designing an algorithm $\mathcal{A}$ for small graphs without using duplication certainly leads to very poor values of α and β, since, intuitively, $h(G)$ appears to be a very poor lower bound of the optimal makespan of a graph G in this case.

As a corollary, this algorithm clearly leads to an asymptotic guarantee of $\alpha + \beta + 1$. Notice that if the height of the graph is reasonably large, let's say $h(G) \geq \rho + 1$, the asymptotic guarantee of the algorithm can be easily converted into a performance guarantee of $\alpha + 2\beta + 1$. However no reasonable performance guarantee can be obtained in the case of very small graph, i.e. $h(G) \ll \rho$, which is one drawback of our approach. One way to deal with this problem would be to require more from algorithm $\mathcal{A}$, namely to deliver a schedule whose makespan is bound in $|H|/m$ and $h(H)$, instead of $\rho + 1$. In this setting, it is not hard to see that the phase by phase algorithm directly provides a performance guarantee of $\alpha + \beta + 1$. In the next section we try to justify our approach, and why requiring a guarantee in $h(H)$ appears to be a much more difficult challenge. We then present an algorithm $\mathcal{A}$ for small graphs, based on a greedy covering technique.

4   Scheduling Small Graphs

We consider in this section the problem of scheduling a graph H of height smaller than $\rho + 1$. One natural arising question may be roughly formulated as : is this problem really easier than scheduling graph of arbitrary height? Of course lemma 3 seems to justify the reduction of our scheduling problems to the one of small graphs. But it may also appear that the optimal makespan of a small graph H can not be bounded by a reasonable function in $|H|/m$ and $\rho + 1$, which would condemn our search for an efficient algorithm $\mathcal{A}$. However 2 facts has motivated us in this approach. First, and this is in the same time a strong point and a weakness, we do not require $\mathcal{A}$ to be an approximation algorithm: as we have noticed, when $h(G) \ll \rho$, the performance ratio of $\mathcal{A}$ can be arbitrarily large. In a sense we have relaxed our problem. In particular it will allow us to insert communication phases of ρ time units in our algorithm.

Secondly, this approach matches Papadimitriou & Yannakakis' algorithm [14] on an unbounded number of processors. Indeed a graph H of height $h(H) \leq \rho + 1$

has the following nice property: H can be scheduled (optimally) in time $h(H)$ on an unbounded number of processors simply by alloting each one of its leaves, together with all its ancestors, to a different processor. This is a direct consequence of the definition of the height $h(x)$ (recall that $h(H) = \max_x\{h(x)\} + 1$) of a node: if $h(x) \leq \rho$, we have simply $h(x) = |A(x)|$, which corresponds to the sequential execution time of all the ancestors of x. Hence the completion time of x is $h(x) + 1$. Using this algorithm in the phase by phase scheme, we obtain as Papadimitriou & Yannakakis a 2-approximation for the clustering problem. This remark made us believe that our reduction to the small graph scheduling problem is not a too much brute force approach.

We present now the principle of our algorithm $\mathcal{A}$ to schedule a small graph H on m processors, ensuring a makespan bounded by $\alpha|H|/m + \beta(\rho + 1)$. As suggested by the clustering case, we only need to focus on the scheduling of the sets $A_1, \ldots, A_l$, where l is the number of leaves of H and each A_i is constituted by a leaf together with all its ancestors. Each set A_i contains at most $h(H) \leq \rho + 1$ tasks and can be scheduled independently of the others using duplication. However simply using a list scheduling algorithm on the A_i's may lead to a very poor solution, since clearly two different sets A_i and A_j may have almost all their tasks in common, leading to a prohibitive number of duplicates if not executed on the same processor. One natural approach consists in packing the sets sharing a lot of tasks together to limit the use of duplication. This is certainly a promising approach to explore to get constant coefficients α and β.

The approach presented in the following is in a sense totally different. Surprisingly, we do not try at all to pack together the sets A_i's. They will be scheduled just as independent tasks of duration $|A_i|$ using duplication. Once more we use a phase by phase approach, inserting an idle period of ρ time units between 2 phases in order that communications can take place. Each phase itself consists in selecting a set S of tasks defined as a subset of the A_i's and such that the work (i.e. number of distinct tasks) represents at least a fraction $1/\gamma$ of the total number of tasks, where $\gamma > 1$ is a parameter of the algorithm. The technique used to achieve this selection is closed to a greedy covering algorithm [7]. Such a greedy covering algorithm iteratively selects a set A_i with the maximal number of uncovered elements. We can built a convenient set S using this technique, with the halting criteria that current candidate A to be included into S must verify $|A \setminus S| \geq |A|/\gamma$. In fact we use an even simpler strategy: we simply scan in an arbitrary order all the sets A, adding them to current selection S only if $|A \setminus S| \geq |A|/\gamma$. All sets A included in S are then scheduled as independent jobs using any list scheduling algorithm, and a delay of ρ units of time is added. The key point of the algorithm consists in "updating" the remaining sets A_i at the end of the step, deleting all their tasks that belong to S. Indeed, communication can take place between these tasks and any of their successors scheduled in further steps, which makes a future duplication useless. The algorithm is depicted in figure 2. In the algorithm, γ is simply a tuning parameter to adapt the compromise between the bounds $|H|/m$ and $\rho + 1$. One can simply set $\gamma = 2$ to have a more convenient look at the code.

INPUT: a small graph H, m processors, and a parameter $\gamma > 1$.
- ▷ Set $\mathcal{V}$:= $\{A_1, \dots, A_l\}$ the sets of the leaves of H together with their ancestors
- ▷ While $\mathcal{V} \neq \{\emptyset\}$
  - ▷ Set $S := \emptyset$;
  - ▷ For all $A \in \mathcal{V}$
    - ▷ If $|A \setminus S| \geq |A|/\gamma$ Then
      - ▷ Set $S := S \cup A$; $\mathcal{V} := \mathcal{V} \setminus \{A\}$; End If
  - ▷ End For
  - ▷ Schedule the sets $A_i \subseteq S$ using a list scheduling algorithm
  - ▷ Insert a communication phase of ρ time units.
  - ▷ For all $A \in \mathcal{V}$ do $A := A \setminus S$.
- ▷ End While

OUTPUT: A schedule of graph H on m processors

Fig. 2. The algorithm to schedule small graphs

4.1   Performance Analysis of the Algorithm

The algorithm is built as a succession of steps, each one consisting of the scheduling of a set S of tasks followed by a idle period of ρ units of time. Consider an execution of the algorithm resulting in K such steps, indexed from 0 to $K - 1$. At beginning of step k, we have a collection of sets A_i yet to be scheduled. We denote by h_k the largest cardinality of these sets. At the end of step k a set S_k of tasks has been selected for execution, and the remaining sets A_i are updated for next step, which makes their cardinality decrease. The performance analysis of the algorithm is based on lemmas 4 and 5 which respectively bound the duration of a step and the value h_k.

Lemma 4. *The time duration of step k is bounded by $\gamma |S_k|/m + h_k + \rho$*

Proof. The proof is done using the classical Graham's analysis of list scheduling algorithm. Let t be the starting time of step k. In our algorithm, set S_k is built as the union of different sets A_i, say $A_1^k, \dots, A_p^k$. Using duplication, we schedule these sets just as independent jobs of respective time duration $|A_1^k|, \dots, |A_p^k|$ according to an arbitrary list scheduling strategy. Notice that any of these time durations is by definition smaller than h_k. Let A be the finishing job of the schedule, and let t_A be its starting time. By construction the duration of the step is equal to $(t_A - t) + |A| + \rho \leq (t_A - t) + h_k + \rho$.

Due to the greedy list scheduling strategy, all the processors are continuously busy in the time interval $[t, t_A[$. It implies that $t_A - t \leq \sum_j |A_j^k|/m$. But clearly among these tasks we have counted quite many duplicates, which are nothing but redundant work. The real work corresponds to the number of different tasks executed in the step, which is exactly $|S_k|$. Our selection strategy ensures the very useful property that $|S_k| = |\cup_{j=1}^p A_j^k| \geq 1/\gamma \sum_{j=1}^p |A_j^k|$, which concludes the proof.

Lemma 4 allows us to give an upper bound of the makespan ω_A of our schedule. Noticing that the idle period of ρ units of time is clearly useless at the last step, we obtain:

$$\omega_A \leq \sum_{k=0}^{K-1} (\gamma |S_k|/m + h_k) + (K-1)\rho$$

The simple lemma 5 will allow us to bound the number of steps of the algorithm together with the value of h_k:

Lemma 5. *At the beginning of step k of the algorithm, any remaining set A to schedule contains at most $h(H)/\gamma^k$ tasks.*

Proof. Consider any step k, except the last one, and A one of the sets of tasks not selected to belong to S at this step. Let $h_k(A)$ be its cardinality at the beginning of step k. At the beginning of step $k + 1$, the cardinality of A will be equal to $h_{k+1}(A) = |A \setminus S|$ due to the update operation, i.e. A will contain only the subset of its tasks not scheduled in S. The fact that A is not selected to belong to S implies that $|A \setminus S| < |A|/\gamma$. Hence $h_{k+1}(A) < h_k(A)/\gamma$. Since at step 0 any set A contains at most $h(H)$ tasks, a direct induction on k gives the result.

We now have all the keys to give the performance of our algorithm to schedule small graphs.

Lemma 6. *For any value $\gamma > 1$, the algorithm can schedule a graph H of height at most $\rho + 1$ in time at most*

$$\gamma |H|/m + (\log_\gamma(\rho + 1) + \frac{\gamma}{\gamma - 1})(\rho + 1)$$

Proof. The sets $(S_k)_{k=0, K-1}$ form by construction of partition of the tasks of H. Hence we have $\sum_{k=0}^{K-1} |S_k| = |H|$. In addition lemma 5 implies that each h_k is bounded by $h(H)/\gamma^k \leq (\rho + 1)/\gamma^k$. It results from lemma 4 that the makespan ω_A of our scheduling is bounded by:

$$\gamma |H|/m + (\rho + 1) \sum_{k=0}^{K-1} 1/\gamma^k + (K-1)\rho$$

To conclude we bound the number of steps of the algorithm using lemma 5. Assume that at a step k we are in the situation where $h_k \leq \gamma$. In this case all the remaining sets A_i's are selected, since necessarily $|A_i \setminus S| \geq 1 \geq |A_i|/\gamma$. It involves that this step happens to be the last step of the algorithm. If $K = 1$, i.e. the algorithm has only one step, the result clearly holds. Otherwise consider the preceding step $K - 2$. Since this is not the last step we must have $h_{K-2} > \gamma$. In addition lemma 5 provides the inequality $(\rho + 1)/\gamma^{K-2} \geq h_{K-2}$. It implies that $\rho + 1 > \gamma^{K-1}$. The result follows.

We conclude this section giving the final result of the paper, combining lemmas 3 and 6.

Theorem 1. *For any value $\gamma > 1$, we can schedule a graph G of arbitrary height in time less than*

$$\gamma \frac{|G|}{m} \;+\; [\log_\gamma(\rho + 1) + \frac{2\gamma - 1}{\gamma - 1}]h(G) \;+\; [\log_\gamma(\rho + 1) + \frac{\gamma}{\gamma - 1}](\rho + 1)$$

Parameter γ permits to modify the compromise between the workload term $|G|/m$ and the term $h(G)$ which plays the part of the critical path value in a Graham-like analysis. Natural value $\gamma = 2$ leads to an asymptotic guarantee of $5 + \log(\rho + 1)$ for our algorithm, and a performance guarantee of $7 + 2\log(\rho + 1)$ for any graph of height at least $\rho + 1$. One can play with this parameter to obtain an asymptotic guarantee slightly better than $\mathcal{O}(\ln\rho)$ taking for instance $\gamma = \sqrt{\ln\rho}$. We obtain then an asymptotic guarantee in $\mathcal{O}(\ln\rho/\ln\ln\rho)$.

5   Special APX Graph Structures

We conclude by giving some examples of graph structures for which our approach leads to simple approximation algorithms. As noticed in section 3, we only need to concentrate on the problem of graphs of height smaller than $\rho + 1$, designing an algorithm to schedule any small graph H in time at most $\alpha|H|/m + \beta h(H)$, for some constants α and β.

For in-forests, a simple algorithm consists in scheduling each in-tree as an independent job using any list scheduling algorithm. Notice that, clearly, no duplication is needed. The reader can convinced himself of the following result:

Lemma 7. *Any in-forest F can be scheduled, without duplication, in time at most $|F|/m + 2h(F)$.*

We focus here on a generalization of in-forests. We define the class $\mathscr{F}$ as the set of graphs G such that any two nodes either have no ancestor in common, or all the ancestors of one of the nodes are also ancestors of the other:

$$\mathscr{F} = \{G = (V, E) \mid \forall x, y \in V, A(x) \cap A(y) = \emptyset \vee A(x) \subseteq A(y) \vee A(y) \subseteq A(x)\}$$

In-forests clearly belongs to the class $\mathscr{F}$. This class is however a bit wider, in particular $\mathscr{F}$ contains interval-order graphs. Papadimitriou & Yannakakis [15] showed that scheduling interval-order graphs with unit execution time and without communication on m processors is polynomial. We present now a simple algorithm to schedule a small graph of $\widehat{\mathcal{H}}$.

Consider H such a small graph, and let L be the set of its leaves (its nodes with no successor). We build upon L a graph $H_L = (L, A)$ where $l \to_A l'$ iff $A(l) \neq \emptyset$ and $A(l) \subseteq A(l')$, where $A()$ refers to the set of ancestors of a task in the graph H. Due to the definition of $\widehat{\mathcal{H}}$, graph H_L is an in-forest. Consider a tree T of H_L, and let l be its (unique) leaf. By construction the ancestors in H of any node of T belong to $A(l)$. In addition, if l' is the leaf of another tree T', then $A(l') \cap A(l) = \emptyset$. We associate to each tree T_i of H_L the set B_i defined as the union of the nodes of T_i together with the ancestors in H of its leaf. The B_i's then form a partition of the nodes of H, and can be scheduled just as independent jobs, without using duplication. However some of these sets may be too large to get a guarantee relatively to $|H|/m$. To tackle with this problem we decompose each set B, associated to a tree T, into sets $B^1, B^2, \ldots$, each one containing at most $2h(H)$ tasks. Set B^1 is initialized with $A(l)$, the ancestors in H of the leaf of T, and completed by nodes of T till it contains $2h(H)$ tasks or all nodes are yet selected. This procedure is repeated to build $B^2, \ldots$ while all the nodes of T have not been selected. The important fact in this decomposition is that $\sum_j |B^j| \leq 2|B|$. To schedule H we then simply schedule the B_i^j's as independent jobs, applying any list scheduling algorithm. Since the size of a job is bounded by $2h(H)$, and the total workload is at most $2|H|$, we obtain, using classical Graham's analysis, a makespan bounded by $2|H|/m + 2h(H)$. As a consequence we have the following lemma:

Lemma 8. *Any graph G of $\widehat{\mathcal{H}}$ can be scheduled in time at most $2|G|/m + 3h(G)$.*

6   Concluding Remarks

In this paper we have presented an asymptotic $\mathcal{O}(\ln \rho / \ln \ln \rho)$-approximation in the special case of unitary processing time and a uniform communication delay ρ. This result improves the state of the art in the approximability of scheduling using duplication, and may be considered as satisfying in practice, when communications are typically at most 1000 times slower than computations, except in meta-computing via internet. However the theoretical question to know if this scheduling problem is APX or not remains open.

References

1. I. Ahmad and Y.-K. Kwok. On exploiting task duplication in parallel program scheduling. *IEEE Transactions on Parallel and Distributed Systems*, 9(9):872–892, 1998.
2. E. Bampis, A. Giannakos, and J-C. Konig. On the complexity of scheduling with large communication delays. *European Journal of Operations Research*, 94:252–260, 1996.

3. D. Darbha and D.P. Agrawal. Optimal scheduling algorithm for distributed-memory machines. *IEEE Transactions on Parallel and Distributed Systems*, 9(1):87–95, 1998.
4. A. Gerasoulis and T. Yang. On the granularity and clustering of directed acyclic graphs. *IEEE Transaction on Parallel and Distributed Systems*, 4:186–201, 1993.
5. R.L. Graham. Bounds on multiprocessing timing anomalies. *SIAM Journal on Applied Mathematics*, 17(2):416–429, March 1969.
6. C. Hanen and A. Munier. Using duplication for scheduling unitary tasks on m processors with communication delays. *Theoretical Computer Science*, 178:119–127, 1997.
7. D. S. Hochbaum and A. Pathria. Analysis of the greedy approach in covering problems. *Naval Research Quaterly*, 45:615–627, 1998.
8. J. Hoogeveen, J.-K. Lenstra, and B. Veltman. Three, four, five, six, or the complexity of scheduling with communication delays. *Operations Research Letters*, 16:129–137, 1994.
9. C-Y. Lee, J-J. Hwang, Y-C. Chow, and F.D. Anger. Scheduling precedence graphs in systems with interprocessor communication times. *SIAM Journal on Computing*, 18(2):244–257, April 1989.
10. Z. Liu. Worst-case analysis of scheduling heuristics of parallel systems. *Parallel Computing*, 24(5-6):863–891, 1998.
11. A. Munier. Approximation algorithms for scheduling trees with general communication delays. *To appear in Parallel Computing*, –. Postscript version available at http://www-poleia.lip6.fr/ munier/arbres.ps.gz.
12. A. Munier, M. Queyranne, and A. Schulz. Approximation bounds for a general class of precedence constrained parallel machine scheduling problems. In *Proceedings of Integer Programming and Combinatorial Optimization (IPCO)*, volume 1412 of *Lecture Notes in Computer Science*, pages 367–382, 1998.
13. M.A. Palis, J-C. Liou, and D.S.L. Wei. Task Clustering and Scheduling for Distributed Memory Parallel Architectures. *IEEE Transactions on Parallel and Distributed Systems*, 7(1):46–55, 1996.
14. C. Papadimitriou and M. Yannakakis. Towards an architecture-independent analysis of parallel algorithms. *SIAM Journal on Computing*, 19(2):322–328, 1990.
15. C.H. Papadimitriou and M. Yannakakis. Scheduling interval-ordered tasks. *SIAM Journal on Computing*, 8(3):405–409, August 1979.
16. C. Picouleau. Two new $\mathcal{NP}$-complete scheduling problems with communication delays and unlimited number of processors. *Discrete Applied Mathematics*, 60:331–342, 1995.
17. V.J. Rayward-Smith. UET scheduling with unit interprocessor communication delays. *Discrete Applied Mathematics*, 18:55–71, 1987.

Scheduling at Twilight the Easy Way[*]

Hannah Bast

Max-Planck-Institut für Informatik
66123 Saarbrücken, Germany
hannah@mpi-sb.mpg.de

Abstract. We investigate particularly simple algorithms for optimizing the trade-off between load imbalance and assignment overheads in dynamic multiprocessor scheduling scenarios, when the information that is available about the processing time of a task before it is completed is vague. We describe a simple and elegant generic algorithm that, in a very general model, always comes surprisingly close to the theoretical optimum, and the performance of which we can analyze exactly with respect to constant factors. In contrast, we prove that algorithms that assign tasks in equal-sized portions perform far from optimal in general. In fact, we give evidence that the performance of our generic scheme cannot be improved by any constant factor without sacrificing the simplicity of the algorithm. We also give lower bounds on the performance of the various decreasing-size heuristics that have typically been used so far in concrete applications.

1 Introduction

Many scheduling problems fit into the following abstract scenario: n tasks of similar complexity are to be scheduled in arbitrary order on p identical processors such that the completion time of the last task, the so-called makespan of the schedule, becomes as small as possible. Assignments may be at runtime, in *chunks* of several tasks at a time, and for each such assignment an overhead is incurred, at least part of which is independent of the size of the chunk, that is, the number of tasks in it. The larger these chunks are chosen, the fewer assignments are necessary, and the lower the total assignment overhead will be. On the other hand, only vague information on the processing times of the tasks is available in advance, and the larger a chunk, the harder it is to predict its processing time. To avoid that some processors finish long before others, chunks should therefore be as small as possible, especially towards the end of the scheduling process. It is not hard to see (and is explained in Section 2) that the optimal makespan is achieved by minimizing the *total wasted time* of the schedule, which is the sum of all the overheads plus the idle times of those processors waiting for the last processor to finish.

Previous work concerning this basic scheduling problem can be roughly classified as follows: on the one hand, a large number of heuristics have been designed and investigated [7][12][4][9][13][5], on the other hand, many extensions of the basic scenario and also real systems have been developed, all of which rely on a good heuristic for the

[*] Partially supported by the Future and Emerging Technologies programme of the EU under contract number IST-1999-14186 (ALCOM-FT).

H. Alt and A. Ferreira (Eds.): STACS 2002, LNCS 2285, pp. 166–178, 2002.
© Springer-Verlag Berlin Heidelberg 2002

basic problem [8][3][10][6][11]. It is only relatively recently, however, that a rigorous theoretical study was presented [1], which, in a sense, does not leave much to be desired: in a general model, capturing all kinds of vague-information scenarios, matching upper and lower bounds are presented, together with a generic strategy achieving these bounds. This result is stated and explained in Section 2.

This paper puts emphasis on two more practical aspects, which played only a secondary role in [1]: the simplicity of the scheduling algorithm, and the constant factors in the performance bounds. Concerning simplicity, it is a matter of fact that most existing heuristics are quite elaborate, while, as far as we know, only the two simplest ones are actually in use: the *fixed-size* (FIX) heuristics, where the chunk size is kept fixed throughout the scheduling process, and what we call *geometric* (GEO) heuristics, where chunk sizes decrease at an exponential rate. It is exactly these two heuristics that we investigate in this paper.

The remainder of this introduction gives an account of our results. In short summary, our results suggest that, for any scenario within our framework, a suitably chosen geometric heuristic yields a hardly beatable combination of simplicity and efficiency, while, in comparison, the efficieny of any fixed-size heuristic is always off by an order of magnitude. Another, maybe important, aspect of this paper is that it provides a fast and easy access to the main result of the extensive study undertaken in [1].

We give our analysis in the general model of [1], and from this derive bounds for four specific settings: (1) the *independent-tasks* setting, where the processing times of the tasks are independent, identically distributed random variables; (2) the *bounded-ranges* setting, where the processing times of all tasks are inside of a given range $[T_{\min}, T_{\max}]$; (3) the *exponential-tails* setting, where for the processing time T of a single task, $T \geq T_{\min}$ and $\Pr(T \geq t) \sim e^{-t/\lambda}$, for some $T_{\min}, \lambda > 0$ (common for all tasks), but the processing times of different task may arbitrarily depend on each other; (4) the *heavy-tails* setting, where for the processing time T_w of a chunk of size w, $T \geq w \cdot T_{\min}$ and $\Pr(T_w \geq t) \sim (t/w)^{-\kappa}$, for some $T_{\min}, \kappa > 0$, and, again, dependencies may be arbitrary. The general model, and how it covers each of these settings, will be explained in Section 2.

In Section 3, we investigate the generic algorithm $\mathrm{GEO}(C, w_{\min})$, which, when W tasks are left, assigns a chunk of $\lceil W/(Cp) + w_{\min} \rceil$ tasks to the next available processor. As is easy to see, instances of this scheme have the nice property that, without the rounding, the chunk sizes exactly form a geometric series (as our analysis will show, rounding is not a serious issue in the context of this scheduling problem). We prove that for every setting of the parameters of the general model of [1], there is an instance of GEO, whose wasted time is within a logarithmic factor of the theoretical optimum; see Table 1. Moreover, our analysis is exact with respect to constant factors, which, from a practical perspective, somewhat compensates for the asymptotic deviation from the theoretical optimum. It should be clear at this point, that our actual objective function is the makespan, which, as explained in the next section, is one pth of the sum of the total processing time and the total wasted time. The magnitude of the wasted time should therefore always be seen in relation to the total processing time, which, by our choice of time scale, is always about n.

Table 1. Asymptotic wasted times that are achieved by the optimal algorithm, by an optimal geometric algorithm, and by an optimal fixed-size algorithm, respectively, for a variety of settings, where n is the number of tasks ($\kappa = 2$ for the heavy-tails setting here)

	Independent	Bounded	Exponential	Heavy-tails
OPT	$\log \log n$	$\log n$	$\log^2 n$	$\sqrt{n}$
GEO	$\log n$	$\log n$	$\log^2 n$	$\sqrt{n} \log n$
FIX	$\sqrt{n}$	$\sqrt{n}$	$\sqrt{n \log n}$	$n^{2/3}$

As a by-product of our analysis of GEO, we also obtain a tight bound for the performance of the popular FAC2 scheme of [4], which essentially equals $GEO(2, 0)$ except that it recomputes the chunk size only every pth chunk (keeping the last computed size for the intermediate chunks); this principle is called *factoring* in [4]. Our bound on the wasted time of FAC2 is a factor of $1/\ln 2 \approx 1.44$ off what can be achieved by a GEO heuristic, which suggests that the factoring principle is not meaningful regarding efficiency.

In Section 4, we investigate $FIX(w)$, which, whenever a processor is available, assigns a chunk of w tasks to it (possibly less for the very last chunk). We prove that even for an optimal choice of w, the deviation of the wasted time from the optimum is always by an order of magnitude larger for FIX than for a suitable GEO scheme; in particular, $FIX(w)$ is *never* optimal; again see Table 1. This adds to the known disadvantages of the fixed-size heuristic: that it is hard to compute the optimal chunk size for a given distribution of processing times, and that no natural default setting with a generally reasonable performance exists.

In Section 5, we give lower bounds for the GSS [12], FAC [4], and TAPER [9] heuristics, all of which are variants of GEO that have actually been used in concrete applications. We show that if the processing times of the tasks are independent, identically distributed random variables (which were indeed the assumptions underlying the design of these heuristics), for each of these heuristics the expected difference of the makespan from the optimum is on the order of $\sqrt{n/p}$. Asymptotically, this is as bad as the performance of a static scheme which simply assigns a single chunk of size n/p (suitably rounded) to each processor.

2   Preliminaries

We first make our scheduling scenario precise. Given are n tasks, ordered in a queue, to be processed on p identical processors, initially idle. Whenever a processor is idle, it may remove an arbitrary number of tasks, called a *chunk*, from the head of the queue. The processor is then working for a period of time, which consists of an *overhead* time, at least part of which is independent of the number of tasks in the chunk, plus the *processing time* of the chunk, which is just the sum of the processing times of the contained tasks. The processing time of a chunk is not known in advance, and at any time all that is known about a scheduled chunk is whether it is completed yet or not. Once a chunk has been

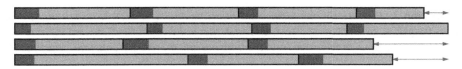

Fig. 1. A schedule on four processors and with fourteen chunks. Overheads are drawn in dark grey, processing times in light grey. The sum of the lengths of the arrows is the imbalance

assigned to a processor it has to be run to completion on that processor (nonpreemptive scheduling).

A *scheduling algorithm* is an algorithm that determines how many tasks an idle processor removes from the queue at which time. We will say that a chunk is *scheduled* (synonymously: *assigned, allocated*) by an algorithm, and we will refer to the number of tasks in a chunk as the *size* of that chunk. According to the above description, for determining a chunk size an algorithm may employ knowledge of the processing times of already completed chunks. However, all algorithms considered in this paper ignore this information, so that for them the partitioning of the tasks into chunks is independent of the task processing times.

An assignment of a collection of chunks to a set of processors will be called a *schedule*. We will often have to sum over all chunks $\mathcal{C}$ in a schedule $\mathcal{S}$, for which we will use the notation $\sum_{\mathcal{C} \in \mathcal{S}}$. The following definition names some characteristic properties of a schedule, which Figure 1 illustrates by an example.

Definition 1. *For a schedule $\mathcal{S}$ on p processors, let l_k be the number of chunks assigned to the kth processor, let h_k be the sum of their overheads, and let T_k be the sum of their processing times. Then define*

$$\text{chunks}(\mathcal{S}) = \sum_{k=1}^{p} l_k,$$

$$\text{imbalance}(\mathcal{S}) = \sum_{k=1}^{p} \left(\max\{h_1 + T_1, \ldots, h_p + T_p\} - (h_k + T_k) \right),$$

$$\text{waste}(\mathcal{S}) = \left(\sum_{k=1}^{p} h_k + \text{imbalance}(\mathcal{S}) \right) / p.$$

The last quantity is called the average wasted time *of $\mathcal{S}$; the (total) wasted time is just p times this quantity. In the context of a schedule $\mathcal{S}$, the* average overhead per chunk *is the quantity $\sum_{k=1}^{p} h_k / \text{chunks}(\mathcal{S})$.*

From the definition above, or more easily from Figure 1, it is straightforward to deduce that p times the makespan of a schedule is just its total wasted time plus the total processing time of all tasks. A scheduling algorithm has no influence on the latter, so in order to achieve a schedule with near-optimal makespan, it must take care to incur as little wasted time (total or average) as possible.

Following [1], we model the unpredictability of processing times by two parameters: the *variance estimator*, which represents an algorithm's a priori estimate of the processing times of chunks, and the *deviation*, which measures the deviation of the actual processing times from these estimates (and is not known in advance).

Definition 2. *For strictly increasing and bijective* $\alpha, \beta : \mathbb{R}^+ \to \mathbb{R}^+$ *such that for all* $w > 0, w/2 \le \alpha(w) \le w \le \beta(w)$, *and such that either both* $\beta(w)/w$ *and* $w/\alpha(w)$ *are nonincreasing in* w, *or* $\beta(w)/w$ *is nondecreasing and* $w/\alpha(w)$ *is constant,*[1] *the function*

$$[\alpha, \beta] : w \mapsto [\alpha(w), \beta(w)]$$

is called a variance estimator. *With respect to a variance estimator* $[\alpha, \beta]$, *we define, for a chunk* $\mathcal{C}$ *of size* w *with processing time* T *that is part of a schedule on* p *processors,*

$$\text{early}_\alpha(\mathcal{C}) = \max\{ 0, \alpha(w) - T \},$$
$$\text{late}_\beta(\mathcal{C}) = \max\{ 0, T - \beta(w) \},$$
$$\text{dev}_{\alpha,\beta}(\mathcal{C}) = \text{early}_\alpha(\mathcal{C}) + (p - 1) \cdot \text{late}_\beta(\mathcal{C}),$$

called the earliness, lateness, *and* deviation, *respectively, of* $\mathcal{C}$ *with respect to* $[\alpha, \beta]$. *In the context of a variance estimator* $[\alpha, \beta]$ *and a schedule* $\mathcal{S}$, *the* average deviation per chunk *is the quantity* $\sum_{\mathcal{C} \in \mathcal{S}} \text{dev}_{\alpha,\beta}(\mathcal{C})/\text{chunks}(\mathcal{S})$.

For the fine points of this definition, we refer the reader to [1]. In particular, it is clarified there why the monotonicity conditions are natural for a variance estimator, that the condition $\alpha(w) \le w \le \beta(w)$ is just a convention regarding the time scale, and that the 2 may be replaced by an arbitrary larger constant.

In the terminology thus introduced, we now state the Main Theorem of [1]. As the intuition behind the bound is not quickly explained, we refer the reader to Section 2.3 of [1], dedicated to exactly this task. The first row of Table 1 gives the asymptotic behaviour in n for a selection of variance estimators. The $*$ operator used below is defined as $f^*(x) = \min\{i : f^{(i)}(x) \le 0\}$, i.e., it "counts" how often a function has to be applied to reduce its argument to zero or below (always a finite number of times in the following).

Theorem 1 ([1]). *For every variance estimator* $[\alpha, \beta]$, *there is an algorithm that for all* $n, p \in \mathbb{N}$, *given* n *tasks (with arbitrary, unknown processing times) and* p *processors, produces a schedule* $\mathcal{S}$ *with*

$$\text{waste}(\mathcal{S}) = O\Big((h + \varepsilon) \cdot \gamma^*(n/p) \Big),$$

where h *and* ε *are the average overhead and the average deviation per chunk, and* $\gamma = \text{id} - \max\{ h + \varepsilon, \beta^{-1} \}$. *This bound is optimal in the sense that no algorithm can improve on it by more than a constant factor (for a precise statement of the lower bound, see [1]).*

To obtain from this theorem bounds for a specific setting, we have to provide a bound on the average deviation with respect to a suitably chosen variance estimator. As the bound of Theorem 1 is more sensitive to an increase in ε than to a widening of $[\alpha, \beta]$, this variance estimator should be chosen such that ε will be relatively small, that is, intuitively, such that it is unlikely that the processing time of a chunk of size

---

[1] The conditions on α and β in [1] are slightly weaker, but we do not need this generality here.

w lies outside of $[\alpha(w), \beta(w)]$. The following table gives a variance estimator and the corresponding expected average deviation and the order of growth of γ^* for each of the four settings described in the introduction. The first row has already been established in [1], the second row is trivial, and the third and fourth row are proven in the full paper [2].

Independent	$[\alpha, \beta] : w \mapsto [w - \Theta(\sqrt{w \ln w}), w + \Theta(\sqrt{w \ln w})]$	$\varepsilon = O(1)$	$\gamma^* \approx \log\log$
Bounded	$[\alpha, \beta] : w \mapsto [T_{\min} \cdot w, T_{\max} \cdot w]$	$\varepsilon = 0$	$\gamma^* \approx \log$
Exponential	$[\alpha, \beta] : w \mapsto [T_{\min} \cdot w, w \cdot \ln(pw^2)]$	$\varepsilon = O(1)$	$\gamma^* \approx \log^2$
Heavy	$[\alpha, \beta] : w \mapsto [T_{\min} \cdot w, pw^2]$	$\varepsilon = O(1)$	$\gamma^* \approx \sqrt{\ }$

When dealing with the independent-tasks setting, we will make frequent use of the following lemma, which is a generalization of a result in [1]; a proof can be found in the full paper [2].

Lemma 1. *Let Z be the sum of m independent random variables with finite variances $\sigma_1^2, \ldots, \sigma_m^2$ and finite absolute third central moments $\varrho_1^3, \ldots, \varrho_m^3$, and assume that Z has mean $\mu_Z \leq 0$ and variance $\sigma_Z^2 > 0$. Then for some constant $\vartheta > 0$, with $t = -\mu_Z/\sigma_Z$ and $\eta = \vartheta \cdot e^{t^2/2} \cdot \sum_{i=1}^m \varrho_i^3 \big/ \big(\sum_{i=1}^m \sigma_i^2 \big)^{3/2}$,*

$$2/3 - \eta \leq \frac{\mathbf{E} \max\{0, Z\}}{\sigma_Z \cdot \dfrac{1}{\sqrt{2\pi}} \dfrac{1}{1+t^2} e^{-t^2/2}} \leq 1 + \eta.$$

If $\mu_Z = 0$, the bound holds with $1 - \eta$ instead of $2/3 - \eta$. If Z is normal, it holds with $\eta = 0$.

3   Geometric Schemes

The scheme $\mathrm{GEO}(C, w_{\min})$ assigns $\lceil W/(Cp) + w_{\min} \rceil$ tasks to an idle processor, when W tasks are unassigned (in the end, when fewer tasks are left than specified by this number, simply all the remaining tasks are assigned). Without the rounding, the chunk sizes would exactly form a geometric sequence[2]; this follows from

$$(W - (W/(Cp) + w_{\min}))/(Cp) + w_{\min} = (W/(Cp) + w_{\min}) \cdot (1 - 1/(Cp)).$$

Theorem 2 below names the optimal instance of GEO for an arbitrary given variance estimator, and states a bound on its average wasted time that is specified exactly up to some lower order additive term. As explained in the remark, the bound is always within a logarithmic (and thus, as explained in the introduction, small) factor of the optimum.

For the (standard) independent-tasks setting, we can further strengthen our result as indicated in Theorem 3. As a by-product, we obtain a result for a variant of GEO where

---

[2] Note that rounding the elements of this geometric series does not give the same series, as if the original formula (with rounding) were applied for each assignment. For our analyses, however, the difference turns out to be insignificant, so that an implementation should simply choose the for it most convenient rounding (one could even round downwards).

the chunk size is recomputed only every pth assignment (and kept for the intermediate chunks); the *factoring* principle introduced by Flynn *et al.* [4]. Note that the constant factor for $C = 2$ (which corresponds to the FAC2 scheme advocated by [4]), is $1/\ln 2 \approx 1.44$, to be compared with 1 for the optimal GEO scheme; see also the remark following Theorem 3.

According to the theorems, an optimal choice for C for each of the independent-tasks, bounded ranges, exponential-tails, and heavy-tails setting is as indicated in the following table; a good choice for $w_{\min}$ for each of these settings would be h. However, the bound of Theorem 2 is quite robust against deviations from these optimal parameter settings.

independent	bounded	exponential	heavy
arbitrary $C > 1$	$C = T_{\max}/T_{\min}$	$C = 2\ln n/T_{\min}$	$C = 2\sqrt{n}/T_{\min}$

Theorem 2. *For a given variance estimator* $[\alpha, \beta]$, *and for* $n, p \in \mathbb{N}$, *let* $w_{\min} \geq 1$ *be arbitrary, and let* $C = \beta(w_{\min})/\alpha(w_{\min})$ *when* $\beta(w)/w$ *is nonincreasing, and* $C = 2 \cdot \alpha^{-1}(n/p)/\beta^{-1}(n/p)$ *otherwise. Then GEO(C,$w_{\min}$), given* n *tasks (with arbitrary, unknown processing times) and* p *processors, produces a schedule* $\mathcal{S}$ *with*

$$\text{waste}(\mathcal{S}) = (h + \varepsilon) \cdot C \cdot \big(\ln(n/p) + O(1)\big),$$

where h *and* ε *are the average overhead and the average deviation per chunk.*

Remark 1. For $w_{\min} = h + \varepsilon = O(1)$, it holds that $C = O(\gamma^*(n/p))$, where $\gamma = \text{id} - \max\{ h + \varepsilon, \beta^{-1} \}$ as in Theorem 1. When $C = O(1)$ this is trivial, otherwise this follows from the simple observation that for an arbitrary increasing function $f : \mathbb{R} \to \mathbb{R}^+$, $(\text{id} - f)^*(y) \geq y/f(y)$.

Theorem 3. *For* $n, p \in \mathbb{N}$, $w_{\min} \geq 1$, *and* $C > 1$, *let* $\mathcal{S}$ *be the schedule produced by GEO($C, w_{\min}$) given* n *tasks and* p *processors, when the task processing times are independent, identically distributed random variables with mean 1 and finite third moment, and let* h *be the average overhead per chunk. Then,*

$$\mathbf{E}\,\text{waste}(\mathcal{S}) = C \cdot h \cdot \ln(n/p) + O(\,p^{2/3}/(C - 1)\,).$$

Employing the factoring principle, i.e., keeping the chunk size for p *successive chunks, and taking* $C = 2$, *it holds that*

$$\mathbf{E}\,\text{waste}(\mathcal{S}) = \frac{1}{\ln 2} \cdot h \cdot \ln(n/p) + O(\,p^{5/3}\,).$$

Remark 2. The first bound is minimized by taking $C = 1 + \Theta(p^{1/3}/\sqrt{\ln(n/p)})$; then $\mathbf{E}\,\text{waste}(\mathcal{S}) = (1 + o(1)) \cdot h \cdot \ln(n/p)$. For $C = 1$, no logarithmic bound holds, but instead (by Lemma 4 from Section 5) the lower bound $\mathbf{E}\,\text{waste}(\mathcal{S}) \geq 1/3 \cdot \sigma \cdot \sqrt{n/p}$. Given the tightness of our analysis, this provides strong evidence (though does not prove) that no simple algorithm can achieve an average wasted time below $h \cdot \ln(n/p)$.

When factoring is applied, $C = 2$ is the smallest value for which our analysis yields a logarithmic bound. This goes a long way towards explaining why $C = 2$ turned out to give the best results in the experimental study of [4].

The following two lemmas are the keys to the proofs of Theorems 2 and 3. Throughout this section, in the context of a schedule $\mathcal{S}$, let w_j denote the size of the jth chunk, and W_j the number of unassigned tasks (W for work) before that chunk is assigned.

Lemma 2. *The number of chunks assigned by GEO($C, w_{\min}$), given n tasks and p processors, is bounded by $\lceil Cp \cdot \ln(1 + n/(Cpw_{\min})) \rceil$. If factoring is employed, it is bounded by $p \cdot \lceil \log_{1/(1-1/C)}(1 + n/(Cpw_{\min})) \rceil$.*

Proof (sketch). By the equation given at the beginning of this section, the quantity $W/(Cp) + w_{\min}$, where W is the amount of unassigned work, decreases by a factor of at least $c = 1/(1 - 1/(Cp))$ after each assignment (except the last, which reduces W to zero). This yields chunks$(\mathcal{S}) \leq \lceil \log_c((n/(Cp) + w_{\min})/w_{\min}) \rceil$, and since $\ln(1/(1 - 1/x)) \geq 1/x$, for all $x > 0$, the first bound follows. The bound for GEO with factoring follows by an analogous argument on the quantity $W/C + p \cdot w_{\min}$.

Lemma 3. *For an arbitrary schedule $\mathcal{S}$, if it holds that for all j, $w_j \leq W_j/(ABp) + K$, where $A = \max_j w_j/\alpha(w_j)$, $B = \max_j \beta(w_j)/w_j$, and $K \geq 0$ is some constant, then*

$$\text{imbalance}(\mathcal{S})/p \leq h_{\max} + BK + \sum_{\mathcal{C} \in \mathcal{S}} \text{early}_\alpha(\mathcal{C})/p + \max_{\mathcal{C} \in \mathcal{S}} \text{late}_\beta(\mathcal{C}),$$

where $h_{\max}$ is the maximal overhead.

Proof (sketch). We prove by induction that for $j = 0, \ldots, l$, the imbalance of the schedule $\mathcal{S}_j$ consisting of the first j chunks is bounded by

$$ph_{\max} + pBK + W_{j+1}/A + \sum_{\mathcal{C} \in \mathcal{S}_j} \text{early}_\alpha(\mathcal{C}) + (p - 1) \cdot \max_{i=1,\ldots,l} \text{late}_\beta(\mathcal{C}).$$

In the induction step $j - 1 \to j$, two cases have to be distinguished: if the jth chunk $\mathcal{C}_j$ is the latest to finish among the first j chunks, the imbalance of $\mathcal{S}_j$ can be bounded by $(p - 1) \cdot (h_{\max} + \beta(w_j) + \text{late}_\beta(\mathcal{C}_j))$, and it suffices to check that $(p - 1) \cdot \beta(w_j) \leq pBw_j - w_j/A \leq W_j/A + pBK - w_j/A = pBK + W_{j+1}/A$; in the opposite case, it suffices to observe that the imbalance decreases by at least the processing time of $\mathcal{C}_j$, which is at least $w_j/A - \text{early}_\alpha(\mathcal{C}_j)$. For more details, see the full paper [2].

Proof (of Theorem 2). Let A and B as in Lemma 3, and check that, by the choice of C in the theorem, and using that the maximal chunk size is $\lceil n/p/C + w_{\min} \rceil$, always $C \geq A \cdot B$ (without loss of generality assuming that $n/p \geq \beta(2w_{\min})$). Since further, by the definition of GEO, $w_j \leq W_j/(Cp) + w_{\min} + 1$, Lemma 3 gives us

$$\text{imbalance}(\mathcal{S})/p \leq h_{\max} + B(w_{\min} + 1) + \sum_{\mathcal{C} \in \mathcal{S}} \text{early}_\alpha(\mathcal{C})/p + \max_{\mathcal{C} \in \mathcal{S}} \text{late}_\beta(\mathcal{C}),$$

Bounding the sum of the last two terms by $\sum_{\mathcal{C} \in \mathcal{S}} \text{dev}_{\alpha,\beta}(\mathcal{C})/p = \varepsilon \cdot \text{chunks}(\mathcal{S})/p$, and adding $h \cdot \text{chunks}(\mathcal{S})/p$, we obtain

$$\text{waste}(\mathcal{S}) \leq (h + \varepsilon) \cdot \text{chunks}(\mathcal{S})/p + C(w_{\min} + 1) + h_{\max},$$

which together with the bound from Lemma 2 proves our theorem.

Proof (of Theorem 3). In this proof it is up to us to choose a variance estimator, and we take $\alpha(w) = w/A$ and $\beta(w) = B \cdot w$. Then for arbitrary $A, B > 1$ with $A \cdot B \leq C$, by an application of Lemma 3 like that in the previous proof,

$$\text{imbalance}(\mathcal{S})/p \leq h_{\max} + B(w_{\min} + 1) + \sum\nolimits_{\mathcal{C} \in \mathcal{S}} \text{early}_\alpha(\mathcal{C})/p + \max_{\mathcal{C} \in \mathcal{S}} \text{late}_\beta(\mathcal{C}).$$

In order to prove the first part of Theorem 3 we will first bound each of the last two terms and then determine A, B such that their sum becomes as small as possible.

Let $\mathcal{C}_1, \ldots, \mathcal{C}_l$ denote the sequence of chunks assigned by GEO$(C, w_{\min})$, let w_i and T_i denote the size and processing time of $\mathcal{C}_i$, respectively, and let σ denote the standard deviation of a single task's processing time. Then, for $a = 1 - 1/A$,

$$\text{early}_\alpha(\mathcal{C}_i) = \max\{0, w_i/A - T_i\} = \max\{0, w_i - aw_i - T_i\},$$

and verify that $w_i - aw_i - T_i$ is a random variable with mean $-aw_i$ and standard deviation $\sigma\sqrt{w_i}$. Concerning this random variable, the parameters t and η of Lemma 1 are just $t_i = a/\sigma \cdot \sqrt{w_i}$ and $\eta_i = \vartheta \cdot e^{t_i^2/2} \cdot \varrho^3/\sigma^3 \big/ \sqrt{w_i}$, where ϱ^3 is the third absolute central moment of a single task's processing time and $\vartheta > 0$ is the constant according to Lemma 1. Applying that lemma we thus obtain that, for $K = 1 + \vartheta \cdot \varrho^3/\sigma^3 \big/ \sqrt{w_{\min}}$,

$$\mathbf{E}\,\text{early}_\alpha(\mathcal{C}_i) \leq (1 + \eta_i) \cdot \sigma\sqrt{w_i} \cdot \frac{1}{\sqrt{2\pi}} \frac{1}{1 + t_i^2} e^{-t_i^2/2} \leq K \cdot \sigma\sqrt{w_i} \cdot \frac{1}{\sqrt{2\pi}} \frac{1}{1 + t_i^2},$$

and hence, writing $\sqrt{w_i}$ as $\sigma/a \cdot t_i$, and summing over all chunks,

$$\sum\nolimits_{\mathcal{C} \in \mathcal{S}} \mathbf{E}\,\text{early}_\alpha(\mathcal{C}) \leq K \cdot \sigma^2/a \cdot \frac{1}{\sqrt{2\pi}} \sum_{i=1}^{l} \frac{1}{t_i + 1/t_i}.$$

In order to bound the sum, let us make the simplifying assumption that the chunk sizes were computed without rounding (the difference to the exact analyis is negligible in the context of this proof, certainly at most a factor of two). In that case $w_{i+1} \leq w_i \cdot (1 - 1/(Cp))$, and thus $t_{i+1} \leq t_i \cdot \sqrt{1 - 1/(Cp)} \leq t_i \cdot (1 - 1/(2Cp))$, and with $I_1 = \{i : t_i < 1\}$ and $I_2 = \{i : t_i \geq 1\}$, we obtain that

$$\sum_{i=1}^{l} \frac{1}{t_i + 1/t_i} \leq \sum_{i \in I_1} t_i + \sum_{i \in I_2} t_i^{-1} \leq 2 \cdot \sum_{j=0}^{\infty} (1 - 1/(2Cp))^{-j} = 4Cp.$$

This gives us the following bound on the expected total earliness

$$\mathbf{E}\left[\sum\nolimits_{\mathcal{C} \in \mathcal{S}} \text{early}_\alpha(\mathcal{C})\right] \leq 2K \cdot \sigma^2/a \cdot Cp = 2K \cdot \sigma^2 \cdot \frac{Cp}{1 - 1/A}.$$

By an analogous calculation, making use of an additional trick (to be found in the full paper [2]), we can further prove

$$\mathbf{E}\left[p \cdot \max_{\mathcal{C} \in \mathcal{S}} \text{late}_\beta(\mathcal{C})\right] \leq 4K \cdot \sigma^2 \cdot \frac{(Cp)^{2/3}}{B - 1}.$$

A simple optimization now shows that under the constraints $A, B > 1$ and $A \cdot B \leq C$, the sum of the last two bounds is minimal for $A = \frac{(Cp)^{1/3}+C}{(Cp)^{1/3}+1}$ and $B = C \cdot \frac{(Cp)^{1/3}+1}{(Cp)^{1/3}+C}$, and plugging these into the bounds above we obtain that

$$\mathbf{E}\left[\sum_{\mathcal{C}\in\mathcal{S}} \text{early}_\alpha(\mathcal{C})/p + \max_{\mathcal{C}\in\mathcal{S}}\text{late}_\beta(\mathcal{C})\right] \leq 4K \cdot \frac{\sigma^2}{C-1} \cdot \left(C + (Cp)^{1/3}\right)^2.$$

Plugging this into the bound obtained from Lemma 3 at the beginning of this proof, this together with the bound from Lemma 2 proves the first part of Theorem 3.

The second part assumes that the factoring principle is employed, in which case the formula for the jth chunk size becomes

$$w_j = \lceil (W_j + i \cdot w_j)/(Cp) + w_{\min}\rceil,$$

where $i \in \{0,\ldots,p-1\}$ is the number of chunks assigned since the last chunk size was computed (0 if a new chunk size is computed for $\mathcal{C}_j$). We therefore have $(Cp-p+1)\cdot w_j \leq W_j + Cp(w_{\min} + 1)$, and hence

$$w_j \leq W_j/(Cp-p+1)+\frac{Cp}{Cp-p+1}\cdot(w_{\min}+1) = W_j/((C-1+1/p)\cdot p)+O(w_{\min}).$$

We can now apply Lemma 3 and bound total earliness and maximal lateness in the same way as above (now with $C - 1 + 1/p$ playing the role of C), to show that for $C - 1 + 1/p > 1$,

$$\mathbf{E}\,\text{imbalance}(\mathcal{S})/p = h_{\max} + O\left(C \cdot w_{\min} + \frac{\sigma^2}{C-2+1/p} \cdot \left(C + (Cp)^{1/3}\right)^2\right).$$

Together with the bound from Lemma 2, and taking $C = 2$ (which is the smallest value, for which $C - 1 + 1/p > 1$ holds in general), this proves the second part of Theorem 3.

4   Fixed-Size Schemes

The following theorem states an upper and a matching lower bound on the average wasted time of an arbitrary fixed-size scheme with respect to an arbitrary variance estimator $[\alpha, \beta]$. For every variance estimator, this bound is by an order of magnitude larger than what can be achieved by GEO; see Table 1 and the remark below.

Let $\text{FIX}(w)$ denote the strategy that, whenever a processor is available, assigns a chunk of w tasks to it (possibly less for the very last chunk). To simplify our presentation, we will assume throughout this section that each chunk has the same overhead, that the number p of processors divides the number n of tasks and that the chunk size w divides n/p. Without these restrictions, the bounds for FIX would rather be even worse than better.

Theorem 4. *Assume that the overhead for each chunk is h. For $n, p \in \mathbb{N}$ and $w \in \mathbb{N}$ with $n/(pw) \in \mathbb{N}$, let $\mathcal{S}$ be the schedule produced by FIX(w) given n tasks and p processors. Then, for an arbitrary given variance estimator $[\alpha, \beta]$,*

$$\text{waste}(\mathcal{S}) \leq (h + \varepsilon) \cdot n/(pw) + \min\{\, h + \beta(w), n/(pw) \cdot \big(\beta(w) - \alpha(w)\big) \,\},$$

where ε is the average deviation per chunk.

On the other hand, for every $\varepsilon \geq 0$, and for all $n, p \in \mathbb{N}$ with $n/(pw) \in \mathbb{N}$, there exist $T_1, \ldots, T_n \geq 0$ such that, given n tasks with processing times $T_1, \ldots, T_n$ and p processors, FIX(w) produces a schedule with average deviation ε and

$$\text{waste}(\mathcal{S}) \geq (h + \varepsilon) \cdot n/(pw) + \min\{\, h + \beta(w), n/(pw) \cdot \big(\beta(w) - \alpha(w)\big) \,\} \,/\, 4.$$

Remark 3. The order of growth of these bounds with respect to n/p is simply that of $\min_{w \in \mathbb{N}} (n/(pw) + \beta(w))$, for example

$[w - \sqrt{w}, w + \sqrt{w}]$	$[w/2, 2w]$	$[w, w \log w]$	$[w, w^2]$
$\sqrt{n/p}$	$\sqrt{n/p}$	$\sqrt{n/p \cdot \log(n/p)}$	$(n/p)^{2/3}$

Proof (sketch). For the proof of the upper bound define $\mathcal{E} = \sum_{\mathcal{C} \in \mathcal{S}} \text{dev}_{\alpha, \beta}(\mathcal{C})$. Since the time a processor has to wait for the last processor to finish is at most the overhead plus the processing time of the last chunk to finish, we have, denoting that chunk by $\mathcal{C}$,

$$\text{imbalance}(\mathcal{S}) \leq (p - 1) \cdot \big(h + \beta(w) + \text{late}_\beta(\mathcal{C})\big) \leq p \cdot \big(h + \beta(w)\big) + \mathcal{E}.$$

It is also not hard to show that p allocations can increase the imbalance by at most $(p - 1) \cdot (\beta(w) - \alpha(w))$ plus the deviations of the corresponding chunks (see [2] for details), which, via a simple induction, implies that

$$\text{imbalance}(\mathcal{S}) \leq n/w \cdot \big(\beta(w) - \alpha(w)\big) + \mathcal{E}.$$

From these two bounds the upper bound claimed in the theorem follows easily.

For the lower bound, given some $\varepsilon \geq 0$, we construct the following sequence of chunk processing times. Let $m = n/(pw)$ and $m' = \min\{m, \lfloor (h+\beta(w))/(\beta(w)-\alpha(w)) \rfloor\} \leq m$. The processing times of the first $m - m'$ chunks of each processor are all fixed to w. For the remaining chunks, we distinguish between the $p - 1$ first (say) processors and the pth processor. For each of the $p - 1$ first processors, the processing time of each chunk after the $(m - m')$th is fixed to $\alpha(w)$. For the pth processor the processing time of each chunk after the $(m - m')$th is fixed to $\beta(w)$, except for the last chunk, whose processing time is fixed to $\beta(w) + \varepsilon \cdot n/w \,/\, (p - 1)$. The corresponding schedule clearly has an average deviation of ε, and it is not hard to see that its wasted time can be bounded as claimed in the theorem. For more details, we again have to refer to the full paper [2].

5 Analysis of GSS, FAC, and TAPER

In this section, we will prove that in the independent-tasks setting, Guided Self-Scheduling (GSS) by Polychronopoulos and Kuck [12], the original Factoring scheme (FAC) of Flynn, Hummel, and Schonberg [4], and the Tapering scheme (TAPER) by Lucco [9], all incur a wasted time that is on the same asymptotic order as for the naive static scheme, which simply assigns a single chunk of size n/p (suitably rounded) to each processor.

The constant factors, as implied by Lemma 4 below, appear rather low, but there is a nonnegligible probability that the wasted time will exceed its mean by a significant factor. Therefore each of these strategies will indeed perform poorly every once in a while (which is also what we observed in experiments).

Theorem 5. *For $n, p \in \mathbb{N}$, let $\mathcal{S}$ be the schedule produced by GSS, FAC, or TAPER given n tasks and p processors, when the processing times of the tasks are independent, identically distributed random variables with mean 1, variance $\sigma^2 > 0$ and finite third moment. Then*

$$\mathbf{E}\,\mathrm{waste}(\mathcal{S}) = \Omega\left(\sigma \cdot \sqrt{n/p}\,\right).$$

To prove the theorem, it suffices to consider the size w_1 of the first chunk assigned by GSS, FAC, or TAPER. For GSS, we have $w_1 \geq n/p$, for FAC we have $w_1 + \sigma\sqrt{p/2}\sqrt{w_1} \geq n/p$, and for TAPER we have $w_1 + 1.3\sigma\sqrt{w_1} \geq n/p$. The theorem is therefore implied by the following general lemma, which, intuitively, says that one should be very careful so as not to choose the initial chunks too large.

Lemma 4. *If $w_1 + K \cdot \sigma\sqrt{w_1} \geq n/p$, for a fixed $K \geq 0$, it holds that for p and n/p sufficiently large (with respect to $K \cdot \sigma$)*

$$\mathbf{E}\,\mathrm{waste}(\mathcal{S}) \geq \frac{1}{4}\frac{1}{1+K^2}\,e^{-K^2/2} \cdot \sigma \cdot \sqrt{n/p}.$$

For $K = 0$, the $1/4$ may be replaced by $1/3$.

Proof (sketch). Let $\mathcal{C}_1$ denote any one of the first p chunks with a size w_1 satisfying $w_1 + K \cdot \sigma\sqrt{w_1} \geq n/p$. Denote by $T_{\mathcal{C}_1}$ its processing time, and by $T_{\mathcal{S}\backslash\mathcal{C}_1}$ the total processing time of the remaining chunks. It is not hard to see that $p \cdot \mathrm{waste}(\mathcal{S}) \geq (p-1) \cdot T_{\mathcal{C}_1} - T_{\mathcal{S}\backslash\mathcal{C}_1}$, and thus, with $Z = (p-1) \cdot T_{\mathcal{C}_1} - T_{\mathcal{S}\backslash\mathcal{C}_1}$, $\mathbf{E}\,\mathrm{waste}(\mathcal{S}) \geq \mathbf{E}\max\{0, Z\} / p$. A straightforward application of Lemma 1 now gives a lower bound on $\mathbf{E}\max\{0, Z\}$, from which the bound claimed in the lemma can be derived by a series of simple estimations. For the details, again see [2]. ∎

Acknowledgments. Thanks to Berthold Vöcking for prodding me to write this up. I'm also grateful to one of the referees for his detailed and knowledgeable comments on the original submission. I only wish his motivation had been to improve the final version and not to reject the paper.

References

1. BAST, H., On Scheduling Parallel Tasks at Twilight, *Theory of Computing Systems* **33**:5 (2000), pp. 489–563.
2. BAST, H., Scheduling at Twilight the Easy Way, full paper (2002), http://www.mpi-sb.mpg.de/~hannah/papers.
3. EAGER, D. L., AND SUBRAMANIAM, S., Affinity scheduling of unbalanced workloads, *in* Proceedings Supercomputing (SC 1994), pp. 214–226.

4. FLYNN, L. E., HUMMEL, S. F., AND SCHONBERG, E., Factoring: A method for scheduling parallel loops, *Communications of the ACM* **35**:8 (1992), pp. 90–101.
5. HAGERUP, T., Allocating Independent Tasks to Parallel Processors: An Experimental Study, *Journal of Parallel and Distributed Computing* **47** (1997), pp. 185–197.
6. HUMMEL, S. F., SCHMIDT, J., UMA, R. N., AND WEIN, J., Load-sharing in heterogeneous systems via weighted factoring, *in* Proceedings 8th Annual ACM Symposium on Parallel Algorithms and Architectures (SPAA 1996), pp. 318–328.
7. KRUSKAL, C. P., AND WEISS, A., Allocating independent subtasks on parallel processors, *IEEE Transactions on Software Engeneering* **11** (1985), pp. 1001–1016.
8. LIU, J., SALETORE, V. A., AND LEWIS, T. G., Safe self scheduling—a parallel loop scheduling scheme for shared-memory multiprocessors, *International Journal of Parallel Programming* **22**:6 (1994), pp. 589–616.
9. LUCCO, S., A dynamic scheduling method for irregular parallel programs, *in* Proceedings Conference on Programming Language Design and Implementation (PLDI 1992), pp. 200–211.
10. MARKATOS, E. P., AND LEBLANC, T. J., Using processor affinity in loop scheduling on shared-memory multiprocessors, *IEEE Transactions on Parallel and Distributed Systems* **5**:4 (1994), pp. 379–400.
11. ORLANDO, S., AND PEREGO, R., Scheduling data-parallel computations on heterogeneous and time-shared environments, *in* Proceedings European Conference on Parallel Computing (EURO-PAR 1998), Springer Lecture Notes in Computer Science **1470**, pp. 356–365.
12. POLYCHRONOPOULOS, C. D., AND KUCK, D. J., Guided self-scheduling: A practical scheduling scheme for parallel supercomputers, *IEEE Transactions on Computers* **36** (1987), pp. 1425–1439.
13. TZEN, T. H., AND NI, L. M., Trapezoid self-scheduling—a practical scheduling scheme for parallel compilers, *IEEE Transactions on Parallel and Distributed Systems* **4**:1 (1993), pp. 87–98.

Complexity of Multi-dimensional Loop Alignment

Alain Darte and Guillaume Huard

LIP, ENS-Lyon, F-69364 LYON Cedex 07, France
`Alain.Darte,Guillaume.Huard@ens-lyon.fr`

Abstract. Loop alignment is a classical program transformation that can enable the fusion of parallel loops, thereby increasing locality and reducing the number of synchronizations. Although the problem is quite old in the one-dimensional case (i.e., no nested loops), it came back recently – with a multi-dimensional form – when trying to refine parallelization algorithms based on multi-dimensional schedules. The main result of this paper is that, unlike the problem in 1D, finding a multi-dimensional shift of statements that makes an innermost loop parallel is strongly NP-complete. Nevertheless, we identify some polynomially-solvable cases that can occur in practice and we show that the general problem can be stated as a system of integer linear constraints.

Keywords. Program Transformation, Loop Optimization, Automatic Parallelization, Retiming, Complexity.

1 Introduction

The goal of this paper is to address, from a theoretical point of view, how loop *shifting*[1] interacts with loop parallelism. Code transformations for revealing loop parallelism have been extensively studied in the last twenty years, leading to several algorithms (see the books [25,24,6]). A loop is *parallel* if all iterations are independent, i.e., if all dependences between statements are *loop-independent* (in the body of the loop, for the same iteration). On the other hand, *loop-carried* dependences (i.e., between different iterations) prevent parallelization. Most programs, after being cleaned up, already contain explicitly parallel loops, or can be parallelized by simple loop distributions (as in Allen, Callahan, and Kennedy's algorithm [1]). Programs with more tightly-coupled computations are less common but they may need more complex transformations to be parallelized. Loop skewing (via Lamport's hyperplane method [13]) can be needed, more unimodular transformations (i.e., combination of loop reversal, loop interchange, and loop skewing) as in Wolf and Lam's algorithm [23] can be useful, or even

---

[1] Shifting is sometimes called alignment. But we prefer to make a distinction between *shifting*, a technique that consists in shifting some statements with respect to the loop iterations, and *alignment*, which is one particular goal achieved by this technique.

H. Alt and A. Ferreira (Eds.): STACS 2002, LNCS 2285, pp. 179–191, 2002.
© Springer-Verlag Berlin Heidelberg 2002

general affine transformations as in Feautrier's algorithm [8], or Lim and Lam's algorithm [16]. The more complex the algorithm, the less frequent it is needed in real codes: in other words, complex transformations are "niche transformations". But why not always use the most general algorithm if it subsumes the simplest one? There are several reasons:

Complexity. The complexity of the algorithm increases when more transformations are allowed. When writing a compiler one may wonder if it is worth paying the worst-case complexity if almost no program can benefit from the transformation. This is true for (current) general purpose compilers. Nevertheless, "niche transformations" are important, first as a theoretical point of view to better understand the limit of techniques, second if the compilation time can be increased and if an accurate analysis is desirable (e.g., in future compilers for embedded systems). The general (NP-complete) case of loop shifting we study here is also a "niche transformation", while frequent cases can be solved in polynomial-time.

Control of the solution. The more general the form of the transformations an algorithm considers, the less control it has on the simplicity of the solution. Even if a simple transformation was possible, the algorithm may miss it, except if it explicitly looks for it before running the general procedure. To give an extreme (but frequent) example, all algorithms based on the notion of schedule (which enforces dependences to be "carried" at the outermost possible level) automatically distribute an already-parallel loop with several statements into several parallel loops, possibly loosing locality and/or increasing synchronizations, while a basic but fast compiler just leaves the loop as it is. More generally, codes with data reuse in the loop body, such as parallel loops with internal (i.e., loop independent) dependences, cannot be obtained, surprisingly.

Code generation. The more general the selected transformation, the more complex the code generation. If the resulting code is too complicated, some further optimizations can now be impossible. For example, in the context of HPF compilation, applying loop skewing in the source code can limit communication optimizations if the resulting communication pattern is too complicated to analyze. Also new loop indices involve in general extra costly operations such as multiplications and divisions, floor and ceiling functions, minima and maxima, and this complexity is not due to the code generation algorithms [3,11,21] but to the transformation itself. Furthermore, these operations are not only time-consuming. If the loop transformation is used for circuit synthesis, they are also costly in terms of hardware. Avoiding complex operations saves gates [22].

Finer optimizations. When looking directly for general transformations, it is hard to target an accurate objective function. In other words, general algorithms do not really subsume simpler ones because the objective functions for simpler algorithms can be more sophisticated. For example, if the objective is just maximal parallelism, the parallelization can produce a transformation bad for locality (e.g., for OpenMP), or not adequate for a given data mapping (e.g., for HPF).

Only two loop transformations (used for parallelism detection) can be considered as "simple" in the sense that they keep the structure of the iteration domain: loop distribution and loop shifting. Loop distribution, which replicates the loop structure into several loops without changing the iteration domain, is the basic tool for Allen, Callahan, and Kennedy's algorithm [1]. This transformation is sufficient for parallelism detection in many practical cases, and its simplicity makes possible its study with other objectives: for example, partial loop distribution [4] can be used to derive parallel loops including loop-independent dependences, fusion techniques can derive codes with better locality [17], loop distribution can be extended for complex flow programs [12], etc. Our goal here is to make a similar study when shifting statements, the main question being: **Can we decide when loop shifting is sufficient for finding a parallel loop?** More generally, given the linear part(s) of an affine transformation (i.e., a particular axis choice for iteration domains), **can we decide if it can be completed by an adequate loop shifting for revealing parallelism?**

The idea to use shifting alone, for loop parallelism, was studied by Okuda [18] who noticed that by allowing parallel loops with loop-independent dependences, it was possible to obtain codes with fewer synchronizations than with a classical shifted-linear schedule. Instead of using such a schedule, he proposed to shift first and then to transform the code with a linear schedule (considering at this point only loop-carried dependences). Nevertheless, Okuda left open a question related to our problem: how to shift so that, after linear scheduling, the innermost loop is parallel and the outermost loop has as few iterations as possible? A similar technique, called loop alignment, was previously developed by Peir [19]: the goal is to shift so that all loops are parallel, which is a particular case of our problem, in one dimension. The idea to decompose the construction of an affine transformation in two steps (a step for the linear part, a step for the shifting part) was developed in [7]. But, again, they were not able to completely characterize the cases when a loop shifting exists that completes the unimodular part of the transformation. Finally in [14], our main problem is addressed and a polynomial-time algorithm is given, nevertheless, in the following sections, we end up with a different conclusion, and prove that the problem is strongly NP-complete.

The rest of the paper is organized as follows. We illustrate loop shifting in Section 2 and recall its basic properties in Section 3, mainly how to shift statements along the loops we want to parallelize (*internal shifting*): the problem is polynomially solvable. In Section 4, we study how shifting along surrounding loops (*external shifting*) can make internal shifting possible when it was not. We show that the problem is strongly NP-complete and that it can be formulated as a polynomial number of integer linear constraints. We also identify some polynomially-solvable cases that can occur in practice. We conclude in Section 5.

2   Loop Shifting: Examples

Loop shifting means to move statements across iterations of the loop while preserving dependences. The goal of this section is to illustrate loop shifting through classical examples and with various objectives.

Consider the code of Figure 1, a classical example from Peir and Cytron [20], whose dependence graph is given in Figure 2. To find parallelism is this code,

Peir and Cytron use a multi-dimensional *loop alignment* as a preliminary step, grouping by a shift into a single arc all the dependences formed by a recurrence circuit in the dependence graph. Here loop alignment leads to the code of Figure 3 where $S4$ is shifted by $(1, -4)$ (i.e., 1 iteration for the i-loop, -4 iterations for the j-loop), $S3$ by $(2, -1)$, $S2$ by $(1, 5)$, and statements are reordered in the loop body, following the new loop-independent dependences. The innermost loop is now parallel (see the dependences in Figure 4). Here, we could also use a simpler shifting technique. Indeed, we can just align the dependences that are not carried by the outermost loop to make them loop independent. No need to shift in both dimensions, a shift in the innermost dimension is sufficient, see Figures 5 and 6. Note that, for this code, a standard unimodular approach would have to apply a loop skewing to get one parallel loop surrounded by one sequential loop, but with a completely different result: the resulting code would have no prologue nor epilogue, but loop bounds and access functions to arrays would be much more complicated, and some locality would be lost. See the effect in Figures 7 and 8.

The problem of shifting statements that belong to different loop body so as to group several loops into as few sequential and parallel loops as possible (in other words, combining loop shifting with loop fusion while keeping parallelism) was addressed in [2]. For fusing parallel loops, a shift must be found so that all dependences between statements in the same loop body are loop independent, i.e., with dependence distance equal to 0 (otherwise, the loop would be either incorrect if a distance is negative, or sequential if it is positive). Determining if a shift exists so that a set of parallel loops can be fused in at most K parallel loops is NP-complete [2], if K is arbitrary. Thus, shifting and fusing parallel loops into as few parallel loops as possible is hard. But this is not true for $K = 1$, as in the previous example. Determining if a shift exists that results in a single parallel loop can be quickly checked as we will recall in Section 3. One just has to check that, in the graph formed by the dependences to be considered, all (undirected) cycles have a zero weight. We call this transformation *internal shifting*.

But, when internal shifting is not sufficient for parallelizing a loop, another question arises: **Can we take advantage of a shift along a surrounding loop so that the innermost loop becomes parallelizable by shifting?** This can indeed happen because shifting along a surrounding loop can change the dependences to be considered for parallelizing the innermost loop. If we are able to carry the "right" dependences with the surrounding loop, the innermost loop may become parallelizable. This transformation is what we call *external shifting* and it is the main problem addressed in this paper. The code of Figure 9 illustrates the interest of external shifting for innermost loop parallelization. The computation of `b[i,j]` depends on two values of the array `a` that are computed by S1 in the same iteration of the surrounding loop but in different iterations of the innermost loops (if they were fused). This situation forbids the parallel fusion of the two loops because the dependence graph to consider for the fusion of the two loops has two arcs from S1 to S2 with weights 0 and 1, thus an undirected cycle of nonzero weight. Now, with a simple shift of the surrounding loop, these two dependences become carried by the surrounding loop. First, we get two

```
DO i=-1,13
  DO j=-4,25
    S1: a[i,j]=b[i,j-6]+d[i-1,j+3]
    S2: b[i+1,j-1]=c[i+2,j+5]
    S3: c[i+3,j-1]=a[i,j-2]
    S4: d[i,j-1]=a[i,j-1]
  ENDDO
ENDDO
```

Fig. 1. Code of Peir and Cytron.

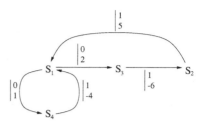

Fig. 2. Dependence Graph.

```
prologue (with a DO j...)
DO i=1,13
  prologue
  DOALL j=1,21
    S4: d[i-1,j+3]=a[i-1,j+3]
    S3: c[i+1,j]=a[i-2,j-1]
    S2: b[i,j-6]=c[i+1,j]
    S1: a[i,j]=b[i,j-6]+d[i-1,j+3]
  ENDDO
  epilogue
ENDDO
epilogue (with a DO j...)
```

Fig. 3. Peir and Cytron's Solution.

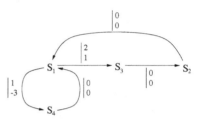

Fig. 4. Dependence Graph.

```
DO i=-1,13
  prologue
  DOALL j=-3,24
    S1: a[i,j-1]=b[i,j-7]+d[i-1,j+2]
    S2: b[i+1,j-1]=c[i+2,j+5]
    S3: c[i+3,j]=a[i,j-1]
    S4: d[i,j-1]=a[i,j-1]
  ENDDO
  epilogue
ENDDO
```

Fig. 5. Solution with Innermost Shifting.

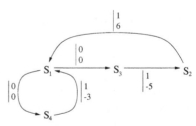

Fig. 6. Dependence Graph.

```
DO j=-11,45
  DOALL i=max(-1,ceil((j-25)/7)),
           min(13,floor((j+4)/7))
    S1: a[i,j-7i]=b[i,j-7i-6]+d[i-1,j-7i+3]
    S2: b[i+1,j-7i-1]=c[i+2,j-7i+5]
    S3: c[i+3,j-7i-1]=a[i,j-7i-2]
    S4: d[i,j-7i-1]=a[i,j-7i-1]
  ENDDO
ENDDO
```

Fig. 7. Solution with Loop Skewing.

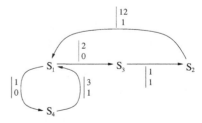

Fig. 8. Dependence Graph.

```
                                                                DOALL j=1,M
                                      DOALL j=1,M                 a[1,j]=b[0,j-1]
                                       a[1,j]=b[0,j-1]            ENDDO
                                      ENDDO                      DO i=2,N
                                      DO i=2,N                    S1': a[i,1]=b[i-1,0]
                                       DOALL j=1,M                DOALL j=2,M
                                        S2: b[i-1,j]=a[i-1,j]      S2: b[i-1,j-1]=a[i-1,j-1]
                                               +a[i-1,j-1]              +a[i-1,j-2]
 DO i=1,N                              ENDDO                      S1: a[i,j]=b[i-1,j-1]
  DOALL j=1,M                          DOALL j=1,M                ENDDO
   S1: a[i,j]=b[i-1,j-1]                S1: a[i,j]=b[i-1,j-1]      S2': b[i-1,M]=a[i-1,M]
  ENDDO                                ENDDO                             +a[i-1,M-1]
  DOALL j=1,M                         ENDDO                      ENDDO
   S2: b[i,j]=a[i,j]+a[i,j-1]          DOALL j=1,M                DOALL j=1,M
  ENDDO                                 b[N,j]=a[N,j]+a[N,j-1]     b[N,j]=a[N,j]+a[N,j-1]
 ENDDO                                 ENDDO                      ENDDO
```

Fig. 9. Code to be Fused.   **Fig. 10.** After Retiming.   **Fig. 11.** After Fusion.

nested loops (Figure 10) that are still sequentialized. But now internal shifting enables the parallel fusion. Indeed, the only dependence not carried by the outer loop is now from S2 to S1, with weight 1. The dependence graph has no cycle of nonzero weight since it has no cycle at all. In the final code (Figure 11), both statements are in the same loop: there is no synchronization anymore between the two parallel loops, which are now fused. (See also the graphs in Figure 12.)

3   Loop Shifting: Basic Properties

3.1   Definitions and Notations

For the sake of simplicity, we assume a sequence of loops (non necessarily perfectly nested) with unit steps, where each statement is surrounded by exactly n loops. We model this sequence of loops by a dependence graph of dimension n, which is a directed graph $G = (V, E, w)$ where the set of vertices V is the set of statements, the set of arcs E is the set of dependences between statements and the mapping $w : E \to \mathbb{Z}^n$ assigns to each arc an integral vector of dimension n: the i-th component $w(e)_i$ of the vector $w(e)$ is the dependence distance relative to loops at depth i. In other words, we assume uniform dependences as in all previous examples (uniform if all loops were fused, regardless of the validity of this fusion). We refer to [5] for how handling direction vectors for shifting.

A path from a vertex to itself, with arcs followed backwards or forwards, is a *cycle*. If all arcs are forwards, the cycle is a *circuit*. A dependence graph is *legal* if all its circuits have a lexico-positive weight, where the lexicographic order is the order of the dictionary. If G is the dependence graph of a correct code, then G is legal. $\tilde{G}$ is the graph obtained from G by keeping only the arcs e that are not carried by the outermost dimension, i.e., such that the first component of $w(e)$ is zero, and the weight of an arc e in $\tilde{G}$ as the vector formed by the $(n-1)$

last components of $w(e)$. Finally, G_i is the graph deduced from G by restricting the weights to their i-th component: G_i is a graph of dimension 1.

3.2   Shifting and Shortest Path Properties

In VLSI, shifting is a well-known technique, called *retiming* (see [15]), for solving circuit optimization problems, and we will use indifferently the words *shifting* or *retiming* as a function assigning an integer value $r(u)$ (scalar or vector) to each vertex u of the dependence graph. The principle of a scalar retiming is to move an operation u from iteration i to iteration $i + r(u)$. Applying a retiming $r : V \to \mathbb{Z}$ to a dependence graph G of dimension 1 changes the weights of its arcs: for an arc $e = (u, v)$, the dependence is now from iteration $i + r(u)$ to iteration $i + w(e) + r(v)$, therefore a dependence distance equal to $w(e) + r(v) - r(u)$. For a graph of dimension n, a (multi-dimensional) retiming is the composition of n scalar retimings, one for each graph G_i. We denote by $G_r = (V, E, w_r)$ the graph obtained from G after a retiming r: for an arc $e = (u, v)$, $w_r(e) = w(e) + r(v) - r(u)$. Following [15], we say that a retiming r is *legal* for G if all arc weights in G_r are nonnegative (in dimension 1), and lexico-nonnegative for a multi-dimensional retiming.

The following proposition is a direct extension of a classical and straightforward result in one dimension (see details in [5]).

Proposition 1. *Given a graph $G = (V, E, w)$, there exists a legal retiming r for G if and only if G has no circuit of lexico-negative weight.*

This legal retiming can be found with the Bellman-Ford algorithm in $O(|V||E|)$ steps, where each step involves the lexicographic order of dimension n, or with n successive calls to a Bellman-Ford algorithm in dimension 1. The proposition shows that, starting from a dependence graph with lexico-positive weight circuits, there always exists a way to shift the corresponding statements so that all dependence vectors become lexico-nonnegative. Furthermore, since there is no zero-weight circuit, the statements can be (textually) reordered so that all zero-weight dependences follow the textual order. In other words, a sequence of arbitrarily nested loops, with uniform dependences and such that all statements are surrounded by n loops, can always be transformed (by shifting) into a single set of n nested loops (possibly with a prologue and epilogue for each dimension). The real difficulty arises when we introduce loop parallelism, i.e., when we require some loops to be parallel. This question is the heart of the paper.

3.3   Internal Shifting

In this section, we deal with the case of loops that have to be grouped so as to get an outermost parallel loop (or a block of outermost parallel loops). For that, we seek a retiming for which all dependences have a zero weight after retiming. This problem is already known as *loop alignment* and the first solution was given by Peir [19, Lemma 4.1]. We recall it here because the characterization of its solutions is the basis for the more elaborated (multi-dimensional) techniques presented in the next section. We only detail here the case of simple loops (case $n = 1$). Extensions to $n > 1$ are mentioned at the end of the section. We define $w(c) = \sum_{e \in c} \mu_c(e) w(e)$ the weight of a cycle c, where $\mu_c(e) = 1$ if e is a forward arc of c, $\mu_c(e) = -1$ if e is a backward arc of c, and $\mu_c(e) = 0$ otherwise. It is straightforward to show that a retiming does not change the weight of a cycle (as for a circuit). The following proposition is a necessary and sufficient condition for the existence of a retiming achieving loop alignment (see Peir's thesis [19]).

Proposition 2. *Let $G = (V, E, w)$ be a graph. There exists a retiming r such that G_r has only zero-weight arcs if and only if G has only zero-weight cycles.*

In other words, the fact that the graph has only zero-weight circuits is not sufficient. It must have only zero-weight **cycles**, and this property makes everything more complicated when considering external retiming. The constructive proof suggests a $O(|V| + |E|)$ algorithm based on a graph traversal of G considered as an undirected graph. This algorithm is nothing but Algorithm 4.1 in [19], and it is actually more efficient than the algorithm suggested (later) in [24]. In the following algorithm, we assume that G is connected, otherwise we apply the procedure on each connected component.

Perfect Alignment$(G = (V, E), r)$

- for each $v \in V$, set $m(v) = 0$ (i.e., not yet visited).
- set $S = \{v\}$ where v is a vertex of G, set $m(v) = 1$ (i.e., visited) and $r(v) = 0$.
- while $S \neq \emptyset$, choose $v \in S$, and set $S = S - \{v\}$:
  - for each $e = (u, v) \in E$ with $m(u) = 0$,
    set $r(u) = r(v) + w(e)$, $m(u) = 1$, $S = S \cup \{u\}$.
  - for each $e = (v, u) \in E$ with $m(u) = 0$,
    set $r(u) = r(v) - w(e)$, $m(u) = 1$, $S = S \cup \{u\}$.
- if G_r has only zero-weight arcs, then return TRUE, else return FALSE.

This algorithm can be extended to the multi-dimensional case, that is to (fuse and) parallelize several loops of a succession of loop nests, either to transform a sequence of loop nests of same depth n into n nested loops, all parallel, or so that only the d outermost loops are parallel. Both problems can be solved with a similar algorithm, but the second one may need an extra retiming in the remaining dimensions in case lexico-negative weights appear. See details in [5].

4   External Shifting

We now address the case where internal shifting is not sufficient for parallelizing loops, i.e., when whatever the dimension, the dependence graph of the corresponding portion of code has a cycle of nonzero weight. The idea is to use external shifting to change the structure (i.e., the cycles) of the dependence graph to consider for the inner loops. Without loss of generality (thanks to Prop. 1 and a preliminary retiming), we can first restrict to legal graphs whose arc weights are all lexico-nonnegative. We now give an intermediate result – maybe surprising at first sight – that allows us to restrict to the case of a single outer sequential loop. Note: all missing proofs can be found in the extended version of this paper [5].

Theorem 1. *Let G be a multi-dimensional graph with only lexico-nonnegative weights. For any legal retiming r of G, there exists a legal retiming r' in the last dimension only such that all arcs loop carried in G_r are also loop carried in $G_{r'}$.*

This result shows that, to carry dependences by the d first loops, a retiming only in the dth dimension is sufficient. We can thus restrict to the case of one sequential loop surrounding a set of loops that have to be made parallel. Furthermore, as shown in Section 3.3, parallelizing a single loop or a block of nested loops is actually the same problem: the perfect alignment algorithm can be used, in any dimension. So, we can also restrict to the case of a single parallel inner loop. In other words, we can focus on the retiming problem in 2D, to make the outer loop sequential and the inner loop parallel. In 2D, the goal of the external retiming is to carry the "right" dependences so that the remaining dependences form a graph with only zero-weight cycles. In the graphs of Figure 12 (which corresponds to the example in Figure 9), dotted arcs are those carried by the first loop. In the first graph, the outer loop cannot be parallelized by retiming since there is a cycle with nonzero weight (in the first dimension). Now consider $\tilde{G}$ (the solid arcs): it has no circuit, but a nonzero-weight cycle, therefore the second loop cannot be parallelized by internal shifting. However, if we shift in the first dimension, we can change the structure of $\tilde{G}$ so that it now has no cycle at all (thus, no cycle of nonzero weight): an adequate shift in the second dimension makes the remaining dependence loop independent.

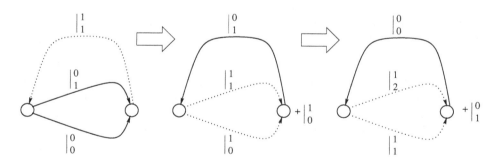

Fig. 12. Two-Step Inner Loop Parallelization.

Even if, for the previous example, the suitable external shifting was obvious, the problem is, in general, NP-complete in the strong sense.

Theorem 2. *Given a directed graph $G = (V, E, w)$ of dimension 2, the problem of deciding if there exists a legal retiming, in the first dimension, such that $\tilde{G}_r$ has only zero-weight cycles is strongly NP-complete.*

The proof is based on a polynomial reduction from 3SAT (Problem LO2 in [9, p. 259]). Note that, except if P=NP, this goes against the results of [14] where a polynomial algorithm is given to solve the problem. (A counter-example for this algorithm can easily be given, for example a graph with two vertices a and b, and two arcs from a to b with weight $(0,0)$ and $(0,1)$, and two arcs from b to a with weight $(1,0)$ and $(1,1)$.) Nevertheless, as many optimizations of this type, the problem can be formulated as a system of integer linear constraints. For the sake of simplicity, we again give only the formulation in 2D. Because of Theorem 1 and the remarks in Section 3.3, all other situations are similar and can be solved as variations of this problem.

Proposition 3. *Let $G = (V, E, w)$ be a graph of dimension 2 and define the quantity $M = \max(\sum_{e \in E} |w(e)_2|, 1)$. There exists a legal retiming such that the innermost loop is parallel if and only if there exists a retiming r such that, for each arc $e = (u, v)$, we have $M w_r(e)_1 + w_r(e)_2 \geq 0$ and $M w_r(e)_1 - w_r(e)_2 \geq 0$.*

We are thus able to find a solution – if it exists – with a relatively small problem, $n|V|$ variables (the retiming) and $2n|E|$ constraints. All problems mentioned before can be solved with a similar approach, with additional linear constraints. We now mention some particular cases that can be quickly solved in polynomial time. To make the discussion simpler, we again restrict to two loops that we want to transform, by shifting, into a sequential loop surrounding a parallel loop.

The simplest case is when all the dependences in $\tilde{G}$ (the dependences not carried by the first loop) can be transformed into zero-weight arcs, that is when what we called **internal shifting** alone is sufficient. This case can be quickly checked and solved by the perfect alignment algorithm (see Section 3.3). An example of this case is Peir and Cytron's code (see Figure 6). Now, consider **external shifting**. Since the general problem is NP-complete, we can hope that this result does not hold for particular graph structures. The first subproblem that comes in mind is to find a shift of the outer loop such that the dependence graph of the inner loop is a forest because, in this case, $\tilde{G}$ has no cycles and internal shifting will lead to a parallel loop. But the graph built in our NP-completeness proof is such that the only way to "eliminate" (by shifting) all the nonzero-weight cycles is also the only way to get a forest or, even simpler, a set of chains! Therefore, finding an external shift such that the dependence graph of the inner loop is a forest (or even a set of chains) is also NP-complete.

The simplest remaining subproblem is to find an external shift such that the dependence graph of the inner loop has only zero-weight arcs (or even no arcs at all). In this case, there is no need for an additional internal shifting because the inner loop is already parallel. These two cases can be easily identified by variants of the Bellman-Ford algorithm (see [5] for details). External shifting alone fails when there are too few arcs carried by the first dimension to "eliminate" all the dependences that have a nonzero weight in the inner dimension. This is the case for the graph of Figure 2. Formulated differently, we have thus classified the following list of problems: deciding if a retiming r in the first dimension exists such that 1) $\tilde{G}_r$ has no arc, 2) $\tilde{G}_r$ has only zero-weight arcs, 3) $\tilde{G}_r$ is a set of chains, 4) $\tilde{G}_r$ has no cycle, 5) $\tilde{G}_r$ has only zero-weight cycles. Problems 1 and 2 are polynomially solvable, while Problems 3, 4, and 5 are strongly NP-complete.

Let us push this study back in the perspective of classical parallelizing algorithms based on affine schedules. Their strategy is the most conservative one: they enforce constraints on the linear parts of the schedules so that $\tilde{G}_r$ has no arc at all (case 1). After scheduling, all dependences are loop carried (or between different loops) and the constraints are equivalent to the fact that, for all circuits c, $\lambda w(c) \geq l(c)$ where λ is the linear part of the schedule and $l(c)$ is the length of the circuit. These constraints are obviously too strong and this is why some codes cannot be found. And trying to fuse the loops back, after scheduling, can fail, due to a bad choice of linear parts. On the other hand, enforcing $\lambda w(c) \geq 1$ (as suggested in [7]) is indeed necessary to obtain a parallel inner loop, but not always sufficient. And finding the right trade-off (somewhere between cases 2 and 3) is NP-complete.

5   Conclusions and Extensions

The initial goal of this work was to be able to apply simple parallelization techniques such as loop fusion/distribution and loop shifting, when they are sufficient to reveal parallel loops, or to find a way to extend algorithms based on general affine transformations so that they can pick a simple one. Loop shifting has indeed a property similar to loop fusion/distribution: it does not change the iteration domains of statements, it just shifts them with respect to each other.

Shifting in one dimension so that all loops can be fused into a single parallel loop is well known: this is loop alignment. The main result of this paper concerns shifting in several dimensions. We showed that the simplicity of the problem in one dimension disappears in higher dimensions: determining if a shift exists for two nested loops such that the innermost loop is parallel is strongly NP-complete. This result is, at first sight, very surprising: indeed, all algorithms that detect parallel loops are polynomial (based on rational linear programming or simpler matrix computations) and most of them do capture loop shifting. But, they are not able to enforce a given linear part (i.e., to control the complexity of the solution), and they are not able to generate codes whose parallel loops contain loop independent dependences (they try to carry all dependences). Despite the complexity of the problem, for not too large problems, an exact solution can be found as an integral solution of a linear (in the number of dependences) number of integral linear constraints. Also, subcases can be solved with variants of the Bellman-Ford algorithm. A practical parallelization technique could be to check simple unimodular parts by enumeration as in [10] (indeed in practice, very few linear parts have to be checked, mainly the identity matrix, the matrices corresponding to interchange, and skews with small factors) and to find the adequate shift (which can be hard to find by enumeration) by first checking particular cases, then using our formulation with integer linear constraints. This technique has the advantage to give complete control on the chosen transformation.

References

1. J. Allen, D. Callahan, and K. Kennedy. Automatic decomposition of scientific programs for parallel execution. In *PoPL'87*, pages 63–76, Munich, Jan. 1987.
2. P. Boulet, A. Darte, G.-A. Silber, and F. Vivien. Loop parallelization algorithms: From parallelism extraction to code generation. *Journal of Parallel Computing*, 24(3), 1998. Special issue on Languages and Compilers for Parallel Computers.
3. J.-F. Collard, P. Feautrier, and T. Risset. Construction of DO loops from systems of affine constraints. *Parallel Processing Letters (PPL)*, 5(3):421–436, Sept. 1995.
4. A. Darte. On the complexity of loop fusion. In *PACT'99*, pages 149–157, Newport Beach, CA, Oct. 1999.
5. A. Darte and G. Huard. Loop shifting for loop parallelization. Technical Report RR2000-22, LIP, ENS-Lyon, France, May 2000.
6. A. Darte, Y. Robert, and F. Vivien. *Scheduling and Automatic Parallelization*. Birkhauser, 2000. ISBN 0-8176-4149-1.
7. A. Darte, G.-A. Silber, and F. Vivien. Combining retiming and scheduling techniques for loop parallelization and loop tiling. *PPL*, 7(4):379–392, 1997.
8. P. Feautrier. Some efficient solutions to the affine scheduling problem, Part II. *Internationl Journal of Parallel Programming*, 21(6):389–420, Dec. 1992.
9. M. R. Garey and D. S. Johnson. *Computers and Intractability: A Guide to the Theory of NP-Completeness*. W. H. Freeman and Company, 1979.
10. W. Kelly and B. Pugh. Selecting affine mappings based on performance estimation. *Parallel Processing Letters*, 4(3):205–219, Sept. 1994.
11. W. Kelly, W. Pugh, and E. Rosser. Code generation for multiple mappings. In *Frontiers of Massively Parallel Computation*, pages 332–341, McLean, 1995.
12. K. Kennedy and K. S. McKinley. Loop distribution with arbitrary control flow. In *Supercomputing'90*, Aug. 1990.
13. L. Lamport. The parallel execution of DO loops. *Communications of the ACM*, 17(2):83–93, Feb. 1974.
14. C. Lang, N. Passos, and E. Sha. Polynomial-time nested loop fusion with full parallelism. In *ICPP'96*, volume 3, pages 9–16, Bloomingdale, IL, 1996.
15. C. E. Leiserson and J. B. Saxe. Retiming synchronous circuitry. *Algorithmica*, 6(1):5–35, 1991.
16. A. W. Lim and M. S. Lam. Maximizing parallelism and minimizing synchronization with affine transforms. In *PoPL'97*. ACM Press, Jan. 1997.
17. K. S. McKinley and K. Kennedy. Maximizing loop parallelism and improving data locality via loop fusion and distribution. *LCPC'93*, volume 768 of *Lecture Notes in Computer Science*, pages 301–320. Springer-Verlag, 1993.
18. K. Okuda. Cycle shrinking by dependence reduction. In *Euro-Par'96*, volume 1123 of *Lecture Notes in Computer Science*, pages 398–401. Springer-Verlag, 1996.
19. J. K. Peir. *Program Partitioning and Synchronization on Multiprocessor Systems*. PhD thesis, University of Illinois at Urbana-Champaign, 1986.
20. J. K. Peir and R. Cytron. Minimum distance: A method for partitioning recurrences for multiprocessors. *IEEE Trans. on Computers*, 38(8):1203–1211, Aug. 1989.
21. F. Quilleré, S. Rajopadhye, and D. Wilde. Generation of efficient nested loops from polyhedra. *International Journal of Parallel Programming*, 28(5):469–498, 2000.
22. R. Schreiber, S. Aditya, B. R. Rau, V. Kathail, S. Mahlke, S. Abraham, and G. Snider. High-level synthesis of nonprogrammable hardware accelerators. In *Application-Specific Systems, Architectures, and Processors*, pages 113–124, 2000.

23. M. E. Wolf and M. S. Lam. A loop transformation theory and an algorithm to maximize parallelism. *IEEE TPDS*, 2(4):452–471, Oct. 1991.
24. M. Wolfe. *High Performance Compilers for Parallel Computing*. Addison-Wesley Publishing Company, 1996.
25. H. Zima and B. Chapman. *Supercompilers for Parallel and Vector Computers*. ACM Press, 1990.

A Probabilistic 3–SAT Algorithm Further Improved

Thomas Hofmeister[1], Uwe Schöning[*,2], Rainer Schuler[2], and Osamu Watanabe[**,3]

[1] Informatik 2, Universität Dortmund, 44221 Dortmund, Germany
[2] Abt. Theoretische Informatik, Universität Ulm, Oberer Eselsberg, 89069 Ulm, Germany
[3] Dept. of Mathematical and Computing Sciences, Tokyo Institute of Technology, Meguro-ku Ookayama, Tokyo 152-8552, Japan, : watanabe@is.titech.ac.jp

Abstract. In [Sch99], Schöning proposed a simple yet efficient randomized algorithm for solving the k-SAT problem. In the case of 3-SAT, the algorithm has an expected running time of $\mathrm{poly}(n) \cdot (4/3)^n = O(1.3334^n)$ when given a formula F on n variables. This was the up to now best running time known for an algorithm solving 3-SAT. Here, we describe an algorithm which improves upon this time bound by combining an improved version of the above randomized algorithm with other randomized algorithms. Our new expected time bound for 3-SAT is $O(1.3302^n)$.

1 Introduction and Preliminaries

In the k-satisfiability problem (k-SAT for short), a Boolean formula F is given by a set of clauses each of which has length at most k. The problem is to determine whether there is an assignment to the variables that satisfies the formula. This problem has been studied by many researchers, and various algorithms have been proposed. Some algorithms indeed have a worst-case time complexity much better than the trivial exhaustive search algorithm; see, e.g., [MS85,PPSZ98, Sch99,Kul99,DGHS00,Hir00a]. In particular, a random walk algorithm proposed by Schöning [Sch99], which can be used for the k-SAT problem in general, had so far the best worst-case time bound for 3-SAT. The algorithm consists of two phases. First, an initial assignment is chosen independently of the clauses, then, a random walk is started which is guided by the clauses.

Schöning's analysis showed that for 3-SAT, the probability of finding a satisfying assignment for a satisfiable formula F on n variables is at least $(3/4)^n/p(n)$, where $p(n)$ is some polynomial over n. Iterating the algorithm until a satisfying assignment is actually found yields a randomized algorithm with expected running time $p(n) \cdot (4/3)^n$.

In this paper, we improve upon Schöning's algorithm by modifying the first phase of the algorithm. The main idea is to make the choice of the initial assignment

---

[*] Supported in part by DFG grant Sch 302/5-2.
[**] Supported in part by JSPS/NSF cooperative research: Complexity Theory for Strategic Goals, 1998–2001.

H. Alt and A. Ferreira (Eds.): STACS 2002, LNCS 2285, pp. 192–202, 2002.
© Springer-Verlag Berlin Heidelberg 2002

depend on the input clauses. We have to combine the modified algorithm with another randomized algorithm to beat the time bound in all cases. Our new combined algorithm finds a satisfying assignment for a satisfiable formula in $O(1.3302^n)$ expected time. While this improvement in the time bound $O(c^n)$ to $c = 1.3302$ may seem minor, it does at least show that the more natural looking barrier $c = 4/3$ for 3-SAT can in fact be beaten.

First, we recall some basic definitions concerning k-SAT as well as Schöning's randomized algorithm for k-SAT.

Given n Boolean variables $x_1, \ldots, x_n$, an assignment to those variables is a vector $a = (a_1, \ldots, a_n) \in \{0, 1\}^n$. A *clause* C of length k is a disjunction $C = l_1 \vee l_2 \vee \cdots \vee l_k$ of *literals*, where a literal is either a variable x_i or its negation $\overline{x}_i$, for some $1 \le i \le n$. For some constant k, the problem k-SAT is defined as follows:

Input: A k-*SAT formula* $F = C_1 \wedge \cdots \wedge C_m$. Each C_i is a clause of length at most k.

Problem: Is there an assignment that makes F true, i.e., a *satisfying assignment* for F?

Throughout this paper, we will use n for the number of variables and m for the number of clauses in the given input formula. We will only be concerned with the problem 3-SAT, and the notion "formula" will always refer to a 3-SAT formula. It can be assumed without loss of generality (and we will do so in the rest of the paper) that each clause has length *exactly* 3 and that no variable appears twice in the same clause. Furthermore, we may assume that no clause appears twice in the input formula, hence, the number of clauses is bounded by $O(n^3)$ and it makes sense to estimate the running time in terms of n.

Schöning's algorithm is a randomized local search algorithm that is described as program Search in Figure 1.

> **Subroutine** RandomWalk(**assignment** a);
> $a' := a$;
>   **for** $3n$ times **do**
>   **if** a' is a satisfying assignment for F **then accept** (and halt);
> $C \leftarrow$ a clause of F that is not satisfied by a';
>   Modify a' as follows:
>   Select a literal of C uniformly at random and
>   flip the assignment to this literal;
>   **end-for**;
>
>
> **Program** Search(**3-SAT formula** F);
>   **repeat forever**
>   **begin**
>   Select an initial assignment a uniformly at random from $\{0, 1\}^n$;
>   RandomWalk(a);
>   **end**;

Fig. 1. Schöning's randomized local search algorithm

The algorithm, as stated in the figure, does not terminate until a satisfying assignment is found. Hence, it runs forever if an unsatisfiable formula is given as an input. Note, however, the following. The *success probability* of one repeat-iteration, i.e., the probability of finding a satisfying assignment (if one exists) during one execution of the repeat loop, is at least $1/T(n)$ for some $T(n)$. (Theorem 1 below gives a bound on $T(n)$.) This means that if an input formula is satisfiable, then the probability that none of $N \cdot T(n)$ repeat-iterations finds a satisfying assignment is at most $(1 - 1/T(n))^{N \cdot T(n)} < e^{-N}$. Therefore, one can modify the algorithm so that it halts and rejects an input formula after executing the repeat-iteration $N \cdot T(n)$ times, and its (one-sided) error probability is bounded by e^{-N}.

The following property of subroutine RandomWalk plays a key role in our discussion and was proved in [Sch99]. Here, and in the rest of this paper, let $d(a, b) = \#\{i \mid a_i \neq b_i\}$ denote the Hamming distance between two binary vectors a and b.

Lemma 1. *Let F be a satisfiable formula and a^* be a satisfying assignment for F. For each assignment a, the probability that a satisfying assignment is found by RandomWalk(a) is at least $(1/2)^{d(a,a^*)}/p(n)$, where p is a polynomial.*

In this paper, $p(n)$ will always denote the polynomial which is given by the bound in Lemma 1. It can also be assumed that $p(n) \geq 1$ for all n.

The following Theorem 1 states the bound for 3-SAT obtained in [Sch99]. We show here how Lemma 1 leads to Theorem 1, since we will later on apply Lemma 1 in a similar manner.

Theorem 1. *For any satisfiable formula F on n variables, the success probability of one repeat-iteration is at least $(3/4)^n/p(n)$ for some polynomial p. Thus, the expected number of repeat-iterations executed until some satisfying assignment is found is at most $(4/3)^n \cdot p(n)$.*

Proof. Let a^* be one of the satisfying assignments for F. We estimate the probability $\Pr\{\text{success}\}$ that some satisfying assignment is found in one repeat-iteration. The initial assignment a is chosen randomly, hence $X_i := d(a_i^*, a_i)$ is a random variable which is 0 if a^* and a agree in the i-th component and 1 otherwise. We use $E[Y]$ to denote the expected value of a random variable Y and obtain:

$$\Pr\{\text{success}\} \geq \sum_{a \in \{0,1\}^n} \Pr\{\ a \text{ is the initial assignment } \} \cdot \left(\frac{1}{2}\right)^{d(a,a^*)} /p(n)$$

$$= \frac{1}{p(n)} \cdot E[(\frac{1}{2})^{d(a,a^*)}] = \frac{1}{p(n)} \cdot E[(\frac{1}{2})^{X_1 + \cdots + X_n}]$$

$$= \frac{1}{p(n)} \cdot E[(\frac{1}{2})^{X_1}] \cdots E[(\frac{1}{2})^{X_n}].$$

Here, we have exploited that $E[Y \cdot Z] = E[Y] \cdot E[Z]$ for independent random variables Y and Z. In the initial assignment, each bit a_i is set to 1 with probability $1/2$, hence X_i is 1 or 0, each with probability $1/2$, hence we have

$$E[(\frac{1}{2})^{X_i}] = \frac{1}{2} \cdot (\frac{1}{2})^0 + \frac{1}{2} \cdot (\frac{1}{2})^1 = \frac{3}{4} .$$

We obtain $\Pr\{\text{success}\} \geq (\frac{3}{4})^n / p(n)$. The expected number of repeat-iterations is bounded by the inverse of the success probability $\Pr\{\text{success}\}$, hence it is bounded by $(4/3)^n \cdot p(n)$. □

2   Independent Clauses and the Basic Idea

Independent Clauses

We begin by introducing some notions. Two clauses C and C' are called *independent* if they have no variable in common. E.g., $C = x_1 \vee x_2 \vee x_3$ and $C' = \overline{x}_1 \vee x_5 \vee x_6$ are *not* independent.

For a formula F, a *maximal independent clause set* $\mathcal{C}$ is a subset of the clauses of F such that all clauses in $\mathcal{C}$ are (mutually) independent and no clause of F can be added to $\mathcal{C}$ without destroying this property.

If $\mathcal{C}$ is a maximal independent clause set for a formula F, then every clause C in F contains at least one variable that occurs in (some clause of) $\mathcal{C}$. This holds because otherwise, $\mathcal{C}$ would not be a *maximal* independent clause set.

This has the following consequence: If we assign constants to all the variables contained in the independent clauses, then – after the usual simplifications consisting of removing constants – we obtain a formula F^* which is a 2-SAT formula, since every clause in F^* has at most two literals. It is well known that there is a polynomial-time algorithm for checking the satisfiability of a 2-SAT formula [APT79] and finding a satisfying assignment if one exists.

Hence, given a satisfiable formula F and a maximal independent clause set with $\hat{m}$ independent clauses $C_1, \ldots, C_{\hat{m}}$, we are able to find a satisfying assignment in time $poly(n) \cdot 7^{\hat{m}}$. Namely, for each of the $\hat{m}$ independent clauses, we can run through all seven (of the eight) assignments that satisfy the clause. The remaining 2-SAT formula is tested in polynomial time for satisfiability.

For a better understanding of what follows, we remark that this algorithm could also be replaced by a randomized algorithm: Set each of the independent clauses with probability $1/7$ to one of the seven satisfying assignments. Then, solve the remaining 2-SAT problem. The success probability of this randomized algorithm is at least $(1/7)^{\hat{m}}$, which is large if $\hat{m}$ is small.

The Basic Idea

The basic idea behind our algorithm is as follows: If we are given an instance of 3-SAT where we find a maximal independent clause set of $\hat{m}$ clauses where $\hat{m}$ is small, we get a small running time (or large success probability) by the algorithm just described. On the other hand, $\hat{m}$ could be as large as $n/3$. But then we are able to improve the first phase of Schöning's algorithm. The idea is as follows: In the original algorithm, each variable is set to 0 with probability $1/2$. The information given by the clauses is ignored completely. Although a clause $C = x_1 \vee x_2 \vee x_3$ tells us that not all those three variables should be set to zero simultaneously, the original initialization selects such an assignment with probability $1/8$. The improvement is to initialize the variables in the independent clauses in groups of three such that an all-0-assignment is avoided in each group. The computation shows that the success probability of this modified version of Schöning's algorithm increases with $\hat{m}$.

3   The Underlying Algorithms

The algorithm starts by computing a maximal set of independent clauses. It should be clear that this can be done in polynomial time by a greedy algorithm which selects independent clauses until no more independent clause can be added. Let $C_1, \ldots, C_{\hat{m}}$ be the independent clauses thus chosen. By renaming variables and exchanging the roles of x_i and $\overline{x}_i$ if necessary, we may assume that $C_1 = x_1 \vee x_2 \vee x_3$, $C_2 = x_4 \vee x_5 \vee x_6$, etc. Hence, the variables $x_1, \ldots, x_{3\hat{m}}$ are those contained in the independent clauses and $x_{3\hat{m}+1}, \ldots, x_n$ are the remaining variables. For assigning constants to the variables in the independent clauses, we apply a randomized procedure called Ind-Clauses-Assign which depends on three parameters p_1, p_2, and p_3. They describe the probabilities with which we choose one of the seven assignments for the three variables of one clause. The details of the procedure, which is used by all our randomized algorithms, are explained in Figure 2. After Ind-Clauses-Assign is called, all variables $x_1, \ldots, x_{3\hat{m}}$ are assigned constants.

Subroutine Ind-Clauses-Assign(p_1, p_2, p_3)
\# p_i are probabilities with $3p_1 + 3p_2 + p_3 = 1$ \#
    for $C \in \{C_1, \ldots, C_{\hat{m}}\}$ **do**
        Assume that $C = x_i \vee x_j \vee x_k$.
        Set the variables x_i, x_j, x_k randomly in such a way
        that for $a = (x_i, x_j, x_k)$ the following holds:
            $\Pr\{a = (0,0,1)\} = \Pr\{a = (0,1,0)\} = \Pr\{a = (1,0,0)\} = p_1$
            $\Pr\{a = (0,1,1)\} = \Pr\{a = (1,0,1)\} = \Pr\{a = (1,1,0)\} = p_2$
            $\Pr\{a = (1,1,1)\} = p_3$.
    endfor;

Fig. 2. Subroutine Ind-Clauses-Assign

Assume in the following that a^* is an arbitrary, but fixed satisfying assignment to the input formula. For each independent clause C_i, we can count how many of the variables in C_i are set to 1 by a^*. Since a^* is a satisfying assignment for all clauses, we have that either one, two or three of the variables in C_i are set to 1, for each i. For our analysis, let m_1 (m_2 and m_3, respectively) be the number of clauses in $\{C_1, \ldots, C_{\hat{m}}\}$ in which exactly one variable is set to 1 by a^* (two and three variables, respectively). Let us also abbreviate $\alpha_i := m_i/\hat{m}$. We then have $\hat{m} = m_1 + m_2 + m_3$, hence $\alpha_1 + \alpha_2 + \alpha_3 = 1$.

We are now ready to describe the two randomized algorithms which are the ingredients of our 3-SAT algorithm.

Algorithm Red2 and Its Success Probability

Algorithm Red2 is the generalization of the algorithm which checks all $7^{\hat{m}}$ assignments to the variables in the maximal independent clause set. It reduces a 3-SAT formula to a 2-SAT formula and is successful if the 2-SAT formula is satisfiable. The algorithm is described in Figure 3.

> **Algorithm** Red2(q_1, q_2, q_3);
> # q_i are probabilities with $3q_1 + 3q_2 + q_3 = 1$ #
>   Ind-Clauses-Assign(q_1, q_2, q_3);
>   Simplify the resulting formula and start the
>   polynomial-time 2-SAT algorithm;

Fig. 3. Algorithm Red2

Algorithm Red2 finds a satisfying assignment when the partial assignment to the variables $x_1, \ldots, x_{3\hat{m}}$ (which satisfies the clauses $C_1, \ldots, C_{\hat{m}}$) can be extended to a complete satisfying assignment. This is the case, e.g., if the partial assignment agrees with a^* on $x_1, \ldots, x_{3\hat{m}}$. The probability of this event is exactly $q_1^{m_1} \cdot q_2^{m_2} \cdot q_3^{m_3}$, as we show now:

Let C_j be one of the $\hat{m}$ clauses and let a^* satisfy exactly i literals in C_j. The probability that algorithm Red2 assigns values to the three variables in C_j that agree with a^* is q_i. Multiplying over all clauses, we obtain the above probability. The following theorem states the bound with the parameters we need later on:

Theorem 2. *Algorithm Red2$(\alpha_1/3, \alpha_2/3, \alpha_3)$ has success probability at least*

$$(\frac{\alpha_1}{3})^{m_1} \cdot (\frac{\alpha_2}{3})^{m_2} \cdot \alpha_3^{m_3} = \left[(\frac{\alpha_1}{3})^{\alpha_1} \cdot (\frac{\alpha_2}{3})^{\alpha_2} \cdot \alpha_3^{\alpha_3} \right]^{\hat{m}} .$$

Algorithm RW and Its Success Probability

This algorithm describes the improvement on the first phase of Schöning's random walk (RW) algorithm in which an initial assignment is chosen.

Instead of initializing each variable x_i with probability $1/2$ to 1, the variables $x_1, \ldots, x_{3\hat{m}}$ in the independent clauses are set in groups of three. The parameters p_1, p_2 and p_3 describe the probability distribution according to which we set the variables. Algorithm RW is described in Figure 4 and its success probability is analyzed in the following theorem.

Algorithm RW(p_1, p_2, p_3)
\# p_i are probabilities with $3p_1 + 3p_2 + p_3 = 1$ \#
    Ind-Clauses-Assign(p_1, p_2, p_3);
    Set the variables $x_{3\hat{m}+1}, \ldots, x_n$ independently of each other
    to 0 or 1, each with probability 1/2.
    To the assignment a obtained in this way, apply RandomWalk(a).

Fig. 4. Algorithm RW

Theorem 3. *The success probability of algorithm RW is at least* $P_{RW} :=$

$$\left(\frac{3}{4}\right)^{n-3\hat{m}} \cdot \left(\frac{3p_1}{2} + \frac{9p_2}{8} + \frac{p_3}{4}\right)^{m_1} \cdot$$
$$\cdot \left(\frac{3p_2}{2} + \frac{p_3}{2} + \frac{9p_1}{8}\right)^{m_2} \cdot \left(p_3 + \frac{3p_2}{2} + \frac{3p_1}{4}\right)^{m_3} \cdot \frac{1}{p(n)} .$$

Proof. By Lemma 1, the success probability of RandomWalk(a), where a has Hamming distance $d(a, a^*)$ from the satisfying assignment $a^* = (a_1^*, \ldots, a_n^*)$ is at least $(\frac{1}{2})^{d(a,a^*)}/p(n)$. As in the proof of Theorem 1, we obtain for the success probability:

$$\Pr\{\text{success}\} \geq E[(\frac{1}{2})^{d(a,a^*)}]/p(n).$$

This time, it does not hold that all variables are fixed independently of each other. But: the only dependence is between variables which are in the same clause C_i, where $1 \leq i \leq \hat{m}$. Define $X_{1,2,3}$ to be the random variable which is the Hamming distance between (a_1, a_2, a_3) and (a_1^*, a_2^*, a_3^*). Define $X_{4,5,6}$ etc. similarly and let $X_i = d(a_i, a_i^*)$ for $i \geq 3\hat{m} + 1$. We have

$$d(a, a^*) = X_{1,2,3} + X_{4,5,6} + \cdots + X_{3\hat{m}-2,3\hat{m}-1,3\hat{m}} + X_{3\hat{m}+1} + X_{3\hat{m}+2} + \cdots + X_n,$$

hence

$$\Pr\{\text{success}\} \geq \frac{1}{p(n)} \cdot E[(\frac{1}{2})^{X_{1,2,3}+X_{4,5,6}+\cdots+X_{3\hat{m}-2,3\hat{m}-1,3\hat{m}}+X_{3\hat{m}+1}+X_{3\hat{m}+2}+\cdots+X_n}]$$
$$= \frac{1}{p(n)} \cdot E[(\frac{1}{2})^{X_{1,2,3}}] \cdots E[(\frac{1}{2})^{X_{3\hat{m}-2,3\hat{m}-1,3\hat{m}}}] \cdot \prod_{i=3\hat{m}+1}^{n} E[(\frac{1}{2})^{X_i}].$$

As in the proof of Theorem 1, we have $E[(\frac{1}{2})^{X_i}] = 3/4$ for $i \geq 3\hat{m} - 1$, hence

$$\prod_{i=3\hat{m}+1}^{n} E[(\frac{1}{2})^{X_i}] = (3/4)^{n-3\hat{m}}.$$

We show how to analyze $E[(\frac{1}{2})^{X_{1,2,3}}]$, the other terms $E[(\frac{1}{2})^{X_{4,5,6}}]$ etc. are analyzed in exactly the same way. It turns out that $E[(\frac{1}{2})^{X_{1,2,3}}]$ depends on how many ones (a_1^*, a_2^*, a_3^*) contains. We have to analyze the three possible cases:

Case $a_1^* + a_2^* + a_3^* = 3$:

Then $(a_1^*, a_2^*, a_3^*) = (1,1,1)$. The algorithm chooses $(x_1, x_2, x_3) = (1,1,1)$ with probability p_3, and the Hamming distance from (a_1^*, a_2^*, a_3^*) then is zero. Likewise, the algorithm sets (x_1, x_2, x_3) with probability $3p_2$ to one of $(0,1,1), (1,0,1)$ and $(1,1,0)$ which leads to Hamming distance 1, etc. Thus, by definition of the expected value, we obtain

$$E[(\frac{1}{2})^{X_{1,2,3}}] = (\frac{1}{2})^0 \cdot p_3 + (\frac{1}{2})^1 \cdot 3p_2 + (\frac{1}{2})^2 \cdot 3p_1 = p_3 + \frac{3}{2}p_2 + \frac{3}{4}p_1.$$

Likewise, we can analyze the other two cases:

Case $a_1^* + a_2^* + a_3^* = 1$:

$$E[(\frac{1}{2})^{X_{1,2,3}}] = (\frac{1}{2})^0 \cdot p_1 + (\frac{1}{2})^1 \cdot 2p_2 + (\frac{1}{2})^2 \cdot (2p_1 + p_3) + (\frac{1}{2})^3 \cdot p_2 = \frac{p_3}{4} + \frac{9}{8} \cdot p_2 + \frac{3}{2} \cdot p_1.$$

Case $a_1^* + a_2^* + a_3^* = 2$:

$$E[(\frac{1}{2})^{X_{1,2,3}}] = (\frac{1}{2})^0 \cdot p_2 + (\frac{1}{2})^1 \cdot (p_3 + 2p_1) + (\frac{1}{2})^2 \cdot 2p_2 + (\frac{1}{2})^3 \cdot p_1 = \frac{p_3}{2} + \frac{3}{2} \cdot p_2 + \frac{9}{8} \cdot p_1.$$

The values m_1, m_2, m_3 count for how many clauses which of the three cases holds. Hence, we obtain the bound on Pr{success} stated in the theorem.   □

4   Combining the Algorithms

The simplest way to improve upon the $O(poly(n) \cdot (4/3)^n)$ bound of Schöning's 3-SAT algorithm is as follows:
First call Red2(1/7, 1/7, 1/7). Then call RW(4/21, 2/21, 3/21).
The success probability of this combined algorithm is at least $\Omega(1.330258^{-n})$, which we will prove now: First, observe that the success probability of algorithm Red2(1/7, 1/7, 1/7) is at least $(1/7)^{\hat{m}} \geq (1/7)^{\hat{m}}/p(n)$. On the other hand, by Theorem 3, with the chosen parameters, the success probability of algorithm RW is at least

$$\left(\frac{3}{4}\right)^{n-3\hat{m}} \cdot \left(\frac{3}{7}\right)^{m_1} \cdot \left(\frac{3}{7}\right)^{m_2} \cdot \left(\frac{3}{7}\right)^{m_3} \cdot \frac{1}{p(n)} = \left(\frac{3}{4}\right)^{n-3\hat{m}} \cdot \left(\frac{3}{7}\right)^{\hat{m}} \cdot \frac{1}{p(n)}$$

$$= \left(\frac{3}{4}\right)^n \cdot \left(\frac{64}{63}\right)^{\hat{m}} \cdot \frac{1}{p(n)}.$$

The combined algorithm is successful if one of the two randomized algorithms is successful and thus the success probability of the combined algorithm is at least as large as the larger of the two success probabilities. We observe that the bound on the success probability of algorithm Red2 decreases with $\hat{m}$, while the bound on the success probability of algorithm RW increases with $\hat{m}$, hence it suffices to compute the $\hat{m}$ where both are equal. This is the case for

$$\frac{n}{\hat{m}} = \frac{\log 9/64}{\log 3/4} \approx 6.8188417, \text{i.e., } \hat{m} \approx 0.1466525 \cdot n.$$

This leads to a success probability of at least $\Omega(1.330258^{-n})$ and a randomized algorithm for 3-SAT with expected running time $O(1.330258^n)$.

In the following section, we will show how this running time can still be improved.

5   Refinements

The algorithms Red2 and RW can be fine-tuned if the values m_1, m_2, m_3 (or equivalently, $\alpha_1, \alpha_2, \alpha_3$) are known in advance.

E.g., if we knew that $(\alpha_1, \alpha_2, \alpha_3) = (1, 0, 0)$, then we could improve upon algorithm Red2 by just checking all possible $3^{\hat{m}}$ assignments to the variables in the independent clauses.

Of course, the values $\alpha_1, \alpha_2, \alpha_3$ are not known in advance. But since we know $\hat{m}$ and since $m_1 + m_2 + m_3 = \hat{m}$, there are only polynomially in n many possibilities for $\alpha_1, \alpha_2, \alpha_3$.

We proceed as follows: For each of the possible values for $\alpha_1, \alpha_2, \alpha_3$, we choose the parameters which maximize the success probabilities of algorithm Red2 and algorithm RW and call those two algorithms with the chosen parameters. The running time of this algorithm is still polynomial. The success probability of the whole algorithm is at least as large as the success probability of the largest success probability of all the algorithms we called.

The idea is simple, but the computations are a little tedious. We will sketch in the following that one can thus achieve a success probability of at least $\Omega(1.330193^{-n})$.

It is rather straightforward to see that one should call Red2($\alpha_1/3, \alpha_2/3, \alpha_3$) to maximize the success probability of algorithm Red2. Recall (Theorem 2) that the success probability then is at least:

$$[\underbrace{(\frac{\alpha_1}{3})^{\alpha_1} \cdot (\frac{\alpha_2}{3})^{\alpha_2} \cdot \alpha_3^{\alpha_3}}_{Z(\alpha_1, \alpha_2, \alpha_3)}]^{\hat{m}} = Z(\alpha_1, \alpha_2, \alpha_3)^{\hat{m}}.$$

Thus, $P_{Red2} := \frac{1}{p(n)} \cdot Z(\alpha_1, \alpha_2, \alpha_3)^{\hat{m}}$ is a lower bound for the success probability of algorithm Red2. Since $Z(\alpha_1, \alpha_2, \alpha_3) \leq 1$, this bound is decreasing with $\hat{m}$.

We recall that the success probability of algorithm RW is at least P_{RW} which is given by Theorem 3. Using that $p_3 + 3p_2 + 3p_1 = 1$, we rewrite $P_{RW} =$

$$\frac{1}{p(n)} \cdot \left[\left(\frac{3}{4}\right)^{\frac{n}{\hat{m}}-3} \cdot \underbrace{\left(\frac{3}{8} - \frac{p_3}{8} + \frac{3}{8} \cdot p_1\right)^{\alpha_1}}_{T_1(p_1, p_3)} \cdot \underbrace{\left(\frac{1}{2} - \frac{3}{8} \cdot p_1\right)^{\alpha_2}}_{T_2(p_1, p_3)} \cdot \underbrace{\left(\frac{1}{2} + \frac{p_3}{2} - \frac{3}{4} \cdot p_1\right)^{\alpha_3}}_{T_3(p_1, p_3)} \right]^{\hat{m}}.$$

Since we select the parameters p_1, p_2, and p_3 depending on $\alpha_1, \alpha_2, \alpha_3$, we can consider the following function as a function depending on the α_i only:

$$T(\alpha_1, \alpha_2, \alpha_3) := T_1(p_1, p_3)^{\alpha_1} \cdot T_2(p_1, p_3)^{\alpha_2} \cdot T_3(p_1, p_3)^{\alpha_3}.$$

We will choose p_1 and p_3 in such a way that

$$T(\alpha_1, \alpha_2, \alpha_3) \geq (\tfrac{3}{4})^3, \text{ i.e., } \log_{3/4} T(\alpha_1, \alpha_2, \alpha_3) \leq 3.$$

This guarantees that P_{RW} is increasing with $\hat{m}$.
Like we did in the last section, we have to analyze for which $\hat{m}$ both bounds on the success probabilities are equal. Equality of P_{Red2} and P_{RW} is achieved when

$$\left(\frac{3}{4}\right)^{\frac{n}{\hat{m}}-3} = \frac{Z(\alpha_1, \alpha_2, \alpha_3)}{T(\alpha_1, \alpha_2, \alpha_3)}, \text{ i.e., } \frac{n}{\hat{m}} = 3 + \log_{3/4} \frac{Z(\alpha_1, \alpha_2, \alpha_3)}{T(\alpha_1, \alpha_2, \alpha_3)}.$$

For a value of $\hat{m}$ where this equality holds, we have

$$P_{Red2} = \frac{1}{p(n)} \cdot Z(\alpha_1, \alpha_2, \alpha_3)^{n/[\log_{3/4}(Z(\alpha_1,\alpha_2,\alpha_3)/T(\alpha_1,\alpha_2,\alpha_3))+3]}.$$

Written differently:

$$\frac{1}{p(n)} \cdot \left(\frac{3}{4}\right)^{F(\alpha_1,\alpha_2,\alpha_3)\cdot n}, \text{ where } F(\alpha_1, \alpha_2, \alpha_3) = \frac{1}{1 + \frac{3-\log_{3/4} T(\alpha_1,\alpha_2,\alpha_3)}{\log_{3/4} Z(\alpha_1,\alpha_2,\alpha_3)}} \cdot \quad (1)$$

This lower bound for the success probability is increasing with $T(\alpha_1, \alpha_2, \alpha_3)$ and $Z(\alpha_1, \alpha_2, \alpha_3)$.
We now suggest a choice of p_1 and p_3 which only depends on α_1. We are aware that there are better choices which also take α_2 (and α_3) into account, but the gain in the success probability is only minor. For simplicity of exposition, we suggest the following choice:

$$p_3 := \begin{cases} 0, & \text{if } 0 \leq \alpha_1 \leq 0.5 \\ 2\alpha_1 - 1, & \text{if } 0.5 \leq \alpha_1 \leq 0.6 \\ \frac{1}{5}, & \text{else.} \end{cases} \qquad p_1 := (4/3) \cdot p_3.$$

This choice is valid since $3p_1 + p_3 \leq 1$, and it leads to $T_2(p_1, p_3) = T_3(p_1, p_3)$.
We have:

	for $0 \leq \alpha_1 \leq 0.5$	for $0.5 \leq \alpha_1 \leq 0.6$	for $0.6 \leq \alpha_1 \leq 1$
$T_1(p_1, p_3) =$	$3/8$	$3/4 \cdot \alpha_1$	$9/20$
$T_2 = T_3(p_1, p_3) =$	$1/2$	$1 - \alpha_1$	$2/5$
$T(\alpha_1, \alpha_2, \alpha_3) =$	$(\frac{3}{8})^{\alpha_1} \cdot (\frac{1}{2})^{1-\alpha_1}$	$(\frac{3}{4} \cdot \alpha_1)^{\alpha_1} \cdot (1 - \alpha_1)^{1-\alpha_1}$	$(\frac{9}{20})^{\alpha_1} \cdot (\frac{2}{5})^{1-\alpha_1}$

On the other hand, for fixed α_1, i.e., fixed $1 - \alpha_1 = \alpha_2 + \alpha_3$, the function Z is minimal for $\alpha_2 = (3/4) \cdot (1 - \alpha_1)$ and $\alpha_3 = (1/4) \cdot (1 - \alpha_1)$, hence

$$Z(\alpha_1, \alpha_2, \alpha_3) \geq (\frac{\alpha_1}{3})^{\alpha_1} \cdot (\frac{1 - \alpha_1}{4})^{1-\alpha_1}.$$

Given the above lower bounds for Z and T for all values of α_1, one can now verify by either a sequence of calculations or by numerically evaluating the function $F(\alpha_1, \alpha_2, \alpha_3)$ from equation (1) that the lower bound for the success probability is at least $\Omega(1.330193^{-n})$ for all $0 \leq \alpha_1 \leq 1$. We have the following theorem.

Theorem 4. *For any satisfiable formula F on n variables, the combined algorithm finds a satisfying assignment for F in expected time at most $O(1.3302^n)$.*

References

[APT79]   B. Aspvall, M.F. Plass, R.E. Tarjan, A linear-time algorithm for testing the truth of certain quantified Boolean formulas, *Information Processing Letters*, 8(3), 121–123, 1979.

[DGHS00]   E. Dantsin, A. Goerdt, E.A. Hirsch, and U. Schöning, Deterministic algorithms for k-SAT based on covering codes and local search, in *Proc. 27th International Colloquium on Automata, Languages and Programming*, ICALP'00, Lecture Notes in Comp. Sci. 1853, 236–247, 2000.

[Hir00a]   E.A. Hirsch, New worst-case upper bounds for SAT, *Journal of Automated Reasoning*, 24(4), 397–420, 2000.

[Kul99]   O. Kullmann, New methods for 3-SAT decision and worst-case analysis, *Theoretical Computer Science*, 223(1/2), 1–72, 1999.

[MS85]   B. Monien and E. Speckenmeyer, Solving satisfiability in less than 2^n steps, *Discrete Applied Mathematics*, 10, 287–295, 1985.

[PPSZ98]   R. Paturi, P. Pudlák, M.E. Saks, and F. Zane, An improved exponential-time algorithm for k-SAT, in *Proc. of the 39th Ann. IEEE Sympos. on Foundations of Comp. Sci.* (FOCS'98), IEEE, 628–637, 1998.

[Sch99]   U. Schöning, A probabilistic algorithm for k-SAT and constraint satisfaction problems, in *Proc. of the 40th Ann. IEEE Sympos. on Foundations of Comp. Sci.* (FOCS'99), IEEE, 410–414, 1999.

The Secret of Selective Game Tree Search, When Using Random-Error Evaluations*

U. Lorenz and B. Monien

Department of Mathematics and Computer Science
Paderborn University
Germany

Abstract. Game tree search deals with the problems that arise, when computers play two-person-zero-sum-games such as chess, checkers, othello etc. The greatest success of game tree search so far, was the victory of the chess machine 'Deep Blue' vs. G. Kasparov[14], the best human chess player in the world at that time. In spite of the enormous popularity of computer chess and in spite of the successes of game tree search in game playing programs, we do not know much about a useful theoretical background that could explain the usefulness of (selective) search in adversary games.

We introduce a combinatorial model, which allows us to model errors of a heuristic evaluation function, with the help of coin tosses. As a result, we can show that searching in a game tree will be 'useful' if, and only if, there are at least two leaf-disjoint strategies which prove the root value. In addition, we show that the number of leaf-disjoint strategies, contained in a game tree, determines the order of the quality of a heuristic minimax value. The model is integrated into the context of average-case analyses.

1 Introduction

When a game tree is so large that it is not possible to find a correct move, there are two standard approaches for computers to play games. In the first approach, the algorithms work in two phases. Initially, a subtree of the game tree is chosen for examination. This subtree may be a full width, fixed depth tree, or any other subtree rooted at the starting position. Thereafter, a search algorithm heuristically assigns evaluations to the leaves and propagates these numbers up the tree according to the minimax principle. Usually the chosen subtree is examined by the help of the $\alpha\beta$-algorithm [5] or one of its variants. As far as the error frequency is concerned, it does not make any difference whether the envelope is examined by the $\alpha\beta$-algorithm or by a pure minimax algorithm. In both cases, the result is the same. Only the effort to get the result differs drastically.

---

* Supported by the German Science Foundation (DFG) project *Efficient Algorithms For Discrete Problems And Their Applications*

H. Alt and A. Ferreira (Eds.): STACS 2002, LNCS 2285, pp. 203–214, 2002.
© Springer-Verlag Berlin Heidelberg 2002

The heuristic minimax value of such a static procedure already leads to high quality approximations of the root value. However, there are several improvements that form the selected subtree more individually. These lead us to a second class of algorithms which work in only one phase and which form the tree shape dynamically at execution time. Some of the techniques are domain independent such as Nullmoves [2], Fail High Reductions [3], or 'Conspiracy Number Search'. Conspiracy Number Search was introduced by D. McAllister [9]. J. Schaeffer [13] interpreted the idea and developed a search algorithm that behaves well on tactical chess positions. We presented an efficient flexible search algorithm which can deal with Conspiracy Numbers[6][8]. We implemented the algorithm in the chess program 'P.ConNerS', which was the first one that could win an official FIDE Grandmaster Tournament [7]. The success was widely recognized in the chess community.

In spite of the overwhelming successes of game tree search in practice, we do not know much about a theoretical background, which could explain how errors of the heuristic evaluation process are filtered out by the minimax-procedure, resp. why the flexible, adaptive algorithms work better in practice than static approaches do.

Pearl [12] examined game trees, assuming that WIN and LOSS are randomly, and indepenently from each other, assigned to the terminal positions. As a result the outcome at the root does only depend on the fraction of WIN-leaves. G. Schrüfer proposed a model in which he could explain the helpfulness of depth d search trees. Furthermore, he was able to distinguish between games where depth d-uniform searching is helpful and ones where it is detrimental [1]. The game trees in his model are random events themselves. A similar model, refined by the concept of quiescence, was used by Kaindl and Scheucher [4].

The results of this paper are applicable for all concrete (i.e. not randomly chosen, not vague) game trees, and they are the most general ones that we know of. We use a simple basic model which has been inspected before with non-encouraging results only [10,11]:

Model: An arbitrary but fixed finite game tree G is given. Each of its nodes has a value 1 or 0. These values follow the minimax principle, and we call them 'real' (=true) values. We perform coin tosses at the leaves of G. Thus, each leaf gets a second value, which we call a 'heuristic' one. With a probability of p a leaf gets the same heuristic value as its real one. With a probability of $1 - p$ it gets the complementary value. We assume p to be equal for all leaves. The heuristic value of G's root is the minimax value of the heuristic leaf values. The question of interest is the following: With which probability will the heuristic minimax value and the real value of G's root be equal to each other?

Insights: We show that the 'number of leave-disjoint strategies' is a key-term of error analyses in game trees.

The number of leaf-disjoint strategies (LDSs), which all prove the root value, determines to what degree the root value is approximated by a heuristic minimax value. If there are not at least two such LDSs, a minimax value will not lead to better approximations than a direct heuristic evaluation (with respect to a reasonable definition of 'better').

Note that our results are applicable for all game trees. They are neither limited to fixed-depth trees nor to trees with clustering leaf-values.

Example: In order to prove that the value of G's root (see Fig. 1) to be 0, we do only need to inspect the subtree S_1 := $\{v1, v2, v3, v5, v8, v9, v10, v12, v14\}$. We call this subtree (which proves the upper bound to be 0) a *strategy* which is contained in the game tree G. The nodes of S_2 := $\{v1, v2, v3, v5, v8, v9, v10, v11, v14\}$ build a strategy for the root value 0, as well. In contrast to S_3 := $\{v1, v2, v3, v5, v7, v9, v10, v11, v13\}$, however, S_2 is not *leaf-disjoint* to S_1.

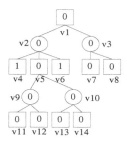

Fig. 1. An arbitrary Game Tree G

The paper is organized as follows: In Chapter 2 we introduce some general definitions, concerning game tree search. In Chapter 3 we analyze the model and discuss the results. We also link our results to Conspiracy Number Search. Chapter 4 contains the proofs, which profit from a new elegant technique of error analysis.

2   Definitions and Notations

2.1   General Game Tree Definitions

Definition 1. *In this paper $G = (T, h)$ will be a* game tree, *if $T = (V, E)$ is a tree and $h : L(G) \to \{0, 1\}$ is a function, $L(G)$ being the set of leaves of T. Remark: We identify the nodes of a game tree G with positions of the underlying game and the edges of T with moves from one position to the next.*

Definition 2. *There are two players MIN and MAX. MAX has to move on even levels of the game tree, MIN on the other levels. This defines a player function $p : V \to \{MAX, MIN\}$.*

Definition 3. *Let $G = (T, h)$ be a game tree and $v \in V$ a node of T. The function minimax $: V \to \{0, 1\}$ is inductively defined by*

$$minimax(v) := \begin{cases} h(v) & if\ v \in L(G) \\ \max\{minimax(v') \mid (v, v') \in E\} & if\ p(v) = MAX \\ \min\{minimax(v') \mid (v, v') \in E\} & if\ p(v) = MIN \end{cases}$$

We call minimax(v) the *minimax value* of v. The minimax value of the root of G is denoted by minimax(G).

Definition 4. *Let G be a game tree, $s \in \{MIN, MAX\}$, with root $v \in V$. Let $\Gamma(u)$ denote the set of $u's$ successors. A strategy for player s, $S_s = (V_s, E_s)$, is a subtree of G, inductively defined by*

- *$v \in V_s$*
- *If $u \in V_s$ is an internal node of T with $p(u) = s$ then there is exactly one $u' \in \Gamma(u)$ with $u' \in V_s$ and $(u, u') \in E_s$.*
- *If $u \in V_s$ is an internal node of T with $p(u) = \bar{s}$, then $\Gamma(u) \subset V_s$, and for all $u' \in \Gamma(u)$ is $(u, u') \in E_s$.*

As a consequence of that, the minimax-evaluation of a strategy S either provides us with a lower (in case S is a MAX-strategy) or with an upper (in case S is a MIN-strategy) bound of the root value of G. When a strategy S gives us 1 as a lower bound, resp. 0 as an upper bound, we say that S 'proves' the root value.

Definition 5. *Two strategies will be called leaf-disjoint if they have no leaf in common.*

Definition 6. *The depth of a game tree is the maximum distance between the root of G and its leaves.*

3   Error Analysis

When we agree on splitting an error analysis into a certain number of errors and the errors' positions, one can analyze how errors propagate in game trees from different points of view: Firstly, the errors may be positioned by a friend of ours (some kind of best-case scenario). Let S be the set of leaves of a smallest strategy, that proves the real value of the root. Let n be the number of leaves of G. Then we can make at least $n - |S|$ many errors at the leaves without the error reaching the root of G.

Secondly, the positioner of the errors may be our enemy (some kind of worst-case scenario). If x is the number of leaf-disjoint strategies in G, that all prove the root value, we will be able to make at least $x - 1$ errors without the error reaching the root. When there are exactly x faulty evaluations, they can be positioned in such a way that the root value gets faulty, too.

Thirdly, and much more interesting is the question of what happens, when we presume a certain error rate (in relation to n) and when these errors are arbitrarily or average-like positioned. For simplification, let us assume that errors occur randomly.

Let us perform coin tosses on all leaves of G. With a probability of p each leaf gets the same heuristic value as its real value. With a probability of $1 - p$ it gets the complementary value. We assume p to be equal for all leaves. For the inner nodes of G we build the minimax value that is based on the possibly incorrect heuristic values. The question of interest is, with which probability the heuristic value and the real value at the root of G will be the same.

Because of the tree structure the non-error probabilities of leaf-disjoint subtrees are independent of each other. The probability of making a correct evaluation at the root of a game tree G is a polynomial Q_G in the non-error probability p of a direct evaluation of a node. We will call $Q_G(p)$ the *polynomial of quality for* G.

Let v be a MAX-node. Then, in principle and without any loss of generality, we come to one of the following two situations:

1. v has got the real value 1:

    Let $v_1, \ldots, v_b$ be the successors of the node v, the root of the example tree (figure 2). Let $g_1(p), \ldots, g_b(p)$ be the probabilities for the event that the heuristic values $h_i, i \in \{1 \ldots b\}$ are equal to the real values $w_1 \ldots w_b$ of the nodes $v_1 \ldots v_b$. Now, we can compute the probability $Q_{G1}(p)$ that the heuristic minimax value of v is equal to the real value of v:

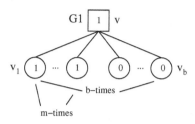

Fig. 2. MAX-node v with real value 1

$Q_{G1}(p)$ is equal to the probability that not all successors of v get the heuristic value 0:

$$Q_{G1}(p) = 1 - (\prod_{i=1}^{m}(1 - g_i(p)) \cdot \prod_{i=m+1}^{b} g_i(p))$$

2. v has got the real value 0 (figure 3):

    Let $v_1, \ldots, v_b$ be the successors of the node v again, v being the root of the example game tree. If we know the probabilities $g_1(p), \ldots, g_b(p)$ that a heuristic value $h_i, i \in \{1 \ldots b\}$ is equal to the real value $w_i = 0$ of the node v_i, we will compute the probability $Q_{G2}(p)$, that the heuristic

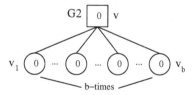

Fig. 3. MAX-node with real value 0

value h corresponds to the real value of the node v with a probability of $Q_{G2}(p) = \prod_{i=1}^{b} g_i(p)$.

Notion 1. Let $Q'_G(p)$ resp. $Q_G^{(1)}(p)$ denote the first derivative of $Q_G(p)$. We will call a game tree G *useful*, if $Q'_G(1) = 0$. (Because we will see that $Q'_G(1) = 0$ or $Q'_G(1) \geq 1$ holds for all polynomials of quality, this criterion will give us a clear distinction between 'good' and 'bad' game trees, if the evaluator is 'good enough'.)

Let $Q_G(p)$ be a polynomial of quality.

Lemma 1. *$Q'_G(1) = 0$ or $Q'_G(1) \geq 1$ for all game trees.*

Lemma 2. *If $Q'_G(1) > 0$, $Q'_G(1)$ will express the number of leaves, which are able to change the root value of G by a single flip of a leaf value.*

Theorem 1. *$Q'_G(1) = 0$ if, and only if, the game tree G contains at least 2 leaf-disjoint strategies, both proving the real value of the root of G.*

Lemma 1 reveals that there are two contrary types of game trees. Game trees which are clearly useful and such ones that are not useful at all. Theorem 1 specifies the useful game trees to be those that contain several leaf-disjoint strategies. Game trees with no two leaf-disjoint strategies have a damaging error behaviour.

Theorem 2. *$Q_G^{(n)}(1) = Q_G^{(n-1)}(1) = \ldots = Q_G^{(1)}(1) = 0 \Leftrightarrow$ there are $n + 1$ leaf-disjoint strategies below v that prove the real value of v.*

Let G be an arbitrary game tree, Q_G its polynomial of quality. Taylor's theorem, $f(p) = f(1) + f'(1)(p - 1) + \ldots + \frac{f^{(n)}(1)}{n!}(p - 1)^n + R_{n+1}(p)$, leads us to $|Q_G(p) - Q_G(1)| = O((1 - p)^{n+1})$ if, and only if there are $n + 1$-many leaf-disjoint strategies in the game tree G. Near the point $x = 1$ the number of leaf-disjoint strategies, contained in G, determines the order of the quality of a search.

Interpretation: Figure 4 presents us with three different courses of $Q(p)$ and the identity function. In the case of $Q(p) \leq p$, we do not see any motivation for performing any search at all. It is obviously more effective to directly evaluate the root. Thus, we see that the search tree is only useful, (in an intuitive sense, here) when $Q(p) > p$, since only then is the error rate of the computed minimax value of the root smaller than the error rate of a direct evaluation of the root. Thus, if $Q_G(p)$ and $Q_H(p)$ are polynomials of quality and $Q'_G(1) = 0$ and $Q'_H(1) \geq 1$,

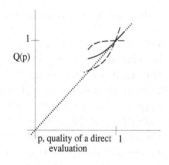

Fig. 4. Possible courses of $Q(p)$

there will be an $\epsilon > 0$ such that for all $p \in [1 - \epsilon, 1]$ it is $Q_G(p) > Q_H(p)$, and thus G is more useful than H.

Example: Let us re-inspect the game tree G in Figure 1. Let H arise from G by flipping the value of node v11 from 0 to 1. As a consequence, there are no two leaf-disjoint strategies in H for the root value.

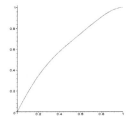 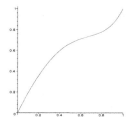

Fig. 5. $Q_{\epsilon(G)}(p)$ (on the left), $Q_{\epsilon(H)}(p)$ (on the right)

Figure 5 shows, how the non-error probabilities at the roots grow with increasing p at the leaves.

We derive the following conjectures:

- It is not the number of strategies, contained in G, that is the key to the quality of a search, but the number of leaf-disjoint ones. There may be thousands of different strategies, if there are not at least two leaf-disjoint ones, the game tree is not helpful.
- The branching factor (i.e. the degree of the inner nodes) of a game tree seems to have only little impact on the approximability of decisions taken in game trees. Indeed, it vanishes in our analysis. This gives us reason to hope that game tree search methods can also be used in games with quite a high branching factor, which has sometimes be doubted. Especially the branching factor argument has led to the general opinion that Go is not suited for game tree search.
- For chess, there are many heuristic-based ideas how to form a game tree. What is the effect of these heuristics such as Fail High Reductions or Singular Extensions? We guess that these heuristics do nothing but increase the number of leaf-disjoint strategies in game trees, concerning their real values. Conspiracy Number Search (CNS) can be interpreted as a meta-heuristic which tries to increase the number of LDSs, too. Indeed, with CNS you try to examine game trees which contain several LDSs, but seen from the side of the heuristic values.

Integration of the model: An objection to the model, which we use here, might be the fact that in practice (e.g. in chess programs) the evaluation-errors are not independent from each other, or that they do not occur randomly at all. Anyway, the used model is definitely a nice and elegant tool for analyses and it is not (as it is no model in the world) necessarily a one-to-one description of the reality. We can easily set the model in the context of average case analyses. Let us call an average-case error analysis, in which the number of occurring errors is weighted according to a binomial distribution, a *relaxed average-case analysis*. The term expresses that we are interested in the problem of how errors propagate in minimax trees when we presume an error rate of 'approximately x percent'.

Let G be an arbitrary game tree with n leaves, and let s be the string of 0s and 1s, consisting of the real values of the leaves of G in a left to right order. Furthermore, let $s' \in \{0,1\}^n$ be a further string of length n. If the value $s(i)$ and $s(i')$ have the same value at a position i, we say 'the i-th leaf of s' has been correctly classified'. When we put all strings of length n into clusters $C_1 \ldots C_n$, whith each cluster containing those strings which have i correctly classified items, there exists a certain number c_i for each cluster of strings, which tells us, in how many cases the root is correctly classified by a minimax-evaluation of G.

Since $Q_G(p)$ is indeed equal to $\sum_{i=0}^{n}$ Prob(the root value is correctly classified by the minimax-evaluation of G | there are exactly i correctly classified leaves) $\cdot$

Prob(exactly i leaves are correctly classified) $= \sum_{i=0}^{n} c_i/|C_i| \cdot \binom{n}{i} \cdot x^i \cdot (1-x)^{n-i}$,

whith x being equal to the probability p of our combinatorial model, the proposed model is nothing but a nice mathematical vehicle in which we perform a relaxed average-case error analysis.

4   Proofs

Let $Q_G(p)$ be a polynomial of quality.

Lemma 1: $Q'_G(1) = 0$ or $Q'_G(1) \geq 1$ for all game trees.

Theorem 1: $Q'_G(1) = 0$ if, and only if, the game tree G contains at least 2 leaf-disjoint strategies, both proving the real value of the root of G.

Without loss of generality we can recursively compose any game tree G with the help of the following three 'different' classes of depth-1 trees. They are different as far as their polynomials of quality are concerned.

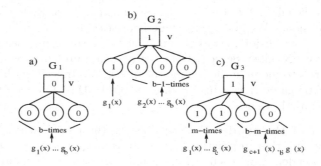

Fig. 6. Three quality-different (t=1)-trees

Figure 6a) shows a game tree rooted by a maxnode with value 0. Because of the minimax principle all successors are 0 as well. Figure 6b) is a game tree rooted by a maxnode with value 1. Exactly one successor of the root has a value of 1.

In figure 6c) there is more than one successor with a value of 1. By the help of section 4 we get polynomials of quality Q_1, Q_2 and Q_3 for the three types of game trees G_1, G_2 and G_3:

$$Q_1(x) = g_1(x) \cdots g_b(x)$$
$$Q_2(x) = 1 - (1 - g_1(x)) \cdot g_2(x) \cdots g_b(x)$$
$$Q_3(x) = 1 - (1 - g_1(x)) \cdots (1 - g_m(x)) \cdot g_{m+1} \cdots g_b(x)$$

The derivatives of these three polynomials are easily computable:
$$Q_1'(x) = \sum_{i=1}^{b} g_1(x) \cdots g_i'(x) \cdots g_b(x)$$
$$Q_2'(x) = -(1 - g_1(x))' \cdot g_2(x) \cdots g_b(x) + \sum_{i=2}^{b}(1 - g_1(x)) \cdot g_2(x) \cdots g_i'(x) \cdots g_b(x)$$
$$Q_3'(x) = -\sum_{i=1}^{c}(1 - g_1(x)) \cdots (1 - g_i(x)) \cdots (1 - g_c(x)) \cdot g_{c+1}(x) \cdots g_b(x) -$$
$$\sum_{i=c+1}^{b}(1 - g_1(x)) \cdots (1 - g_c(x)) \cdot g_{c+1}(x) \cdots g_i'(x) \cdots g_b(x)$$

Because $g_i(1) = 1$ for all i, the following holds:
$$Q_1'(1) = \underline{g_1'(1) + \ldots + g_b'(1)}$$
$$Q_2'(1) = \underline{g'(1)}$$
$$Q_3'(1) = \underline{0}$$

Any given game tree can be regarded as a recursive construction of $t = 1$-trees, as introduced above. By the fact that the polynomial of quality of any leaf is the identity function and by a simple implicit induction over the depth of game trees we see that Lemma 1, as well as Theorem 1 are correct. Example: Let Theorem 1 be proven for all game trees with a maximum depth of d. Let us now inspect a game G tree with a depth of $d + 1$, with real root value of e.g. 0. Therefore, all root successors $s_1 \ldots s_b$ of G have the real value 0. If there are two or more leaf-disjoint strategies contained in each subtree $\overline{G_1} \ldots \overline{G_b}$ below $s_1 \ldots s_b$, and if they all prove a value of 0, then $Q'_{G_i}(1) = 0$ for all $1 \leq i \leq b$. Therefore $Q'_G(1) = 0$, as well. If one or more of the subtrees $\overline{G_1} \ldots \overline{G_b}$ contains no two leaf-disjoint strategies, at least one of the addends is greater than 0 and thus $Q'_G(1) > 0$, as well.

Let G be an arbitrary game tree, and let v be its root. From now on, let $Q_G^{(n)}$ denote the n-th derivative of Q_G. By induction we will show

Theorem 2: $Q_G^{(n)}(1) = Q_G^{(n-1)}(1) = \ldots = Q_G^{(1)}(1) = 0 \Leftrightarrow$ there are $n + 1$ leaf-disjoint strategies below v that prove the real value of v.

Claim. The n-th derivative of a product of polynomials $h(x) := g_1(x) \cdots g_b(x)$ can be described as

$$\sum_{y_1 + \ldots y_b = n} a(y_1, \ldots, y_b) \cdot g_1^{(y_1)}(x) \cdots g_b^{(y_b)}(x)$$

with appropriate $a(y_1, \ldots, y_b) \in I\!N$ (natural numbers)

The Derivatives of Q_1, Q_2 and Q_3

Analogously to the previous subsection we will examine the n-th derivative of Q_1, Q_2 and Q_3. We will examine them on the following assumptions: **(i)** For any game tree G and for all $i \leq n$ there is valid $Q_G^{(i-1)}(1) = \ldots = Q_G^{(1)}(1) = 0 \Leftrightarrow$ there are i leaf-disjoint strategies below v (the root of G) that prove the real value of v. **(ii)** For all $G \in \{G_1, G_2, G_3\}$ it is $Q_G^{(n-1)}(1) = \ldots = Q_G^{(1)}(1) = 0$ and there are n leaf-disjoint strategies below v that prove the real value of v. Moreover **(iii)**, for all $i \in \{1 \ldots n-1\}$ the following staement is valid: the sign of $Q_G^{(i)}(1)$ is equal to $(-1)^{i-1}$.

Remark: When we will start the induction itself, (i) and (iii) will form the induction hypothesis. (ii) will be derived from one of the assumption that will be made in the induction step: For a given $G \in \{G_1, G_2, G_3\}$ it is either $Q_G^{(n)}(1) = \ldots = Q_G^{(1)}(1) = 0$, or, there are $n+1$ leaf-disjoint strategies below v (the root of G) that prove the real value of v.

We start describing the derivatives now:
$Q_1^{(n)}(x) = \sum_{y_1 + \ldots + y_b = n} a(y_1, \ldots, y_b) \cdot g_1^{(y_1)}(x) \cdots g_b^{(y_b)}(x)$ with appropriate $a(\ldots) \in I\!N$. With the help of assumption (b) we see that all addends that contain a derivative smaller than n are zero at the position $x = 1$. As $g_i(1) = 1$ for all i, we see that

$$Q_1^{(n)}(1) = \sum_1^b g_i^{(n)}(1)$$

$Q_2^{(n)}(x) = (-1) \cdot \sum_{y_1 + \ldots + y_b = n} a(y_1, \ldots, y_n) \cdot (1 - g_1(x))^{(y_1)} \cdot g_2^{(y_2)}(x) \cdots$
$g_b^{(y_b)}(x)$ with appro- priate $a(\ldots) \in I\!N$. With the help of assumption (ii) we see that at $x = 1$ only one addend of the sum is non-equal to zero. We conclude that

$$Q_2^{(n)}(1) = g_1^{(n)}(1).$$

The n-th derivative of $Q_3^{(n)}(x)$ is more complicated.
$Q_3^{(n)}(x) = (-1) \cdot \sum_{y_1 + \ldots + y_b = n} a(y_1, \ldots, y_b) \cdot (1 - g_1(x))^{(y_1)} \cdots (1 - g_c(x))^{(y_c)} \cdot g_{c+1}^{(y_{c+1})}(x) \cdots g_b^{(y_b)}(x)$ with appropriate $a(\ldots) \in I\!N$.

We distinguish between 3 cases:

1. $n < c$: $Q_3^{(n)}(1) = 0$, because one factor of each addend is zero, and there are n leaf-disjoint strategies below v because of the definition of a strategy.
2. $n = c$: Let $S_{y_1, \ldots, y_b}(x)$ be any addend of $Q_3^{(n)}(x)$, at $x = 1$.
   a) If there is an l with $l \leq c$ and ($y_l > 1$ or $y_l = 0$), it will follow $S_{y_1, \ldots, y_b}(1) = 0$, because $1 - g_l(1) = 0$.
   b) If there is an l with $l > c$ and $y_l \neq 0$ it will follow $S_{y_1, \ldots, y_b}(1) = 0$. It is zero because the assumption implies that there is an index i with $i \leq c$ and $y_i = 0$.
   c) Otherwise: $S_{y_1, \ldots, y_b}(1) = (-1)^c \cdot \prod_{i=1}^c g_i^{(1)}(1)$
   With $n = c$ we get $Q_3^{(n)}(1) = (-1) \cdot (-1)^n \cdot k \cdot \prod_{i=1}^n g_i^{(1)}(1)$, for a appropriate $k \in I\!N$. (In this case the sign of $Q_3^{(n)}(1)$ is (-1), if n is even, and $(+1)$ otherwise. That is because first derivatives of polynomials of quality are positive.)

3. $n > c$: Again, let $S_{y_1,\ldots,y_b}(x)$ be one of the addends of $Q_3^{(n)}(x)$.

   a) If there is an l with $l \le c$ and $y_l = 0$, we know that $S_{y_1,\ldots,y_b}(1) = 0$.

   b) If there is an l with $l > c$ and $y_l > 0$, we get $\sum_{i=1}^{c} y_i \le n - 1$. $S_{y_1,\ldots,y_b}(x)$ has the form $(1 - g_1(x))^{(y_1)} \cdots (1 - g_c(x))^{(y_c)} \cdot X, X \in \mathbb{R}(\text{ real numbers})$. From assumption (ii) we know that there are n leaf-disjoint strategies below v. By the help of the definition of strategies we know that the sum of leaf-disjoint strategies below $v_1 \ldots v_c$ is n, as well. As $\sum_{i=1}^{c} y_i \le n - 1$, we can conclude that there is one successor $v_i, i \in \{1 \ldots c\}$ that is supplied with more than y_i-many leaf-disjoint strategies. Thus, by the help of the assumption (i), we know that at least one of the $(1 - g_i^{(y_i)}(x))$ becomes zero at the position $x = 1$, for some $i \in \{1 \ldots c\}$.

   c) Last but not least there is the case of $\sum_{i=1}^{c} y_i \le n$ and $\prod_{i=1}^{c} y_i > 0$. Here we get $Q_3^{(n)}(1) = -\sum_{y_1+\ldots+y_c=n} a(y_1, \ldots, y_c, 0, \ldots, 0) \cdot (1 - g_1(x))^{(y_1)}(1) \cdots (1 - g_c(x))^{(y_c)}(1)$ with appropriate $a(y_1, \ldots y_c, 0, \ldots, 0) \in \mathbb{N}$.
   With some proper $a(y_1, \ldots, y_c, 0, \ldots, 0) \in \mathbb{N}$ we can conclude

$$Q_3^{(n)}(1) = (-1)^{c+1} \sum_{y_1+\ldots+y_c=n} a(y_1, \ldots, y_c, 0, \ldots, 0) g_1(x)^{(y_1)}(1) \cdots g_c(x)^{(y_c)}(1)$$

With the help of assumption (iii) the sign of $Q_3^{(n)}(1)$ is equal to $(-1) \cdot (-1)^c \cdot \prod_1^c \text{sign}(g_i^{(y_i)}(1))$, with $\sum_1^c y_i = n$: Let $k_i = y_i - 1, \forall i \in \{1 \ldots c\}$. Thus $(-1)^{k_i}$ is the sign of $g_i^{(y_i)}(1)$. Obviously, $\sum_1^c k_i = n - c$, and therefore, the sign of $Q_c^{(n)}(1) = \prod_{i=1}^c k_i \cdot (-1) \cdot (-1)^c = (-1)^{(n-1)}$.

Now, we can easily prove theorem 4-2 by induction. The induction hypothesis is:

For all $i \le n$ the following is valid: $Q_G^{(i-1)}(1) = \ldots = Q_G^{(1)}(1) = 0 \Leftrightarrow$ there are i leaf-disjoint strategies below v (the root of G) that prove the real value of v. Moreover, the sign of $Q_G^{(i)}(1)$ is equal to $(-1)^{i-1}$.

The induction start has already been done in the previous subsection.

Induction step $(n \to n + 1)$: We start with '$\Leftarrow$'. There are $n + 1$ leaf-disjoint strategies below v, the root of a game tree G. Thus, we know that there are n leaf-disjoint strategies, too. With the help of the induction hypothesis we know that the assumptions (i), (ii), and (iii) are fulfilled. By a simple, implicit induction over the depth of G, we see that the induction step is already done. We use the previously computed derivatives of Q_1, Q_2 and Q_3.

Now, we come to '$\Rightarrow$': Let Q_G be a polynomial of quality of a game tree G. Let $Q_G^{(n)} = \ldots = Q_G^{(1)} = 0$. Obviously, $Q_G^{(n-1)} = \ldots = Q_G^{(1)} = 0$, too. From the induction hypothesis we know that there are n leaf-disjoint strategies below the root of G, reasoning the real value of the root of G. Once again, we know that the assumptions (i), (ii), and (iii) are fulfilled. By a simple, implicit induction over the depth of G, we see that the induction step is already done. We must only consider the previously computed derivatives of Q_1, Q_2 and Q_3, and the fact that for all $i \in \{1 \ldots n\}$ the sign of $Q_G^{(i)}(1)$ (G a game tree) is $(-1)^{i-1}$.

5 Conclusion

We presented a combinatorial model, that allows us to model errors of a heuristic evaluation function with the help of coin tosses. The non-error probability of a heuristic minimax value at the root of any game tree G is a polynomial in the non-error probability of the heuristic evaluation function at the leaves of G. Let $Q_G(x)$ be that polynomial. We were able to prove a one to one relationship between the number of leaf-disjoint strategies that all prove the minimax value of G, and the derivatives of $Q_G(1)$. We showed that the number of leaf-disjoint strategies that are contained in a game tree, determines the order of the quality of a heuristic minimax value. We also got an easily understandable criterion for the usefulness of game tree searches with heuristic evaluations at all.

Acknowledgement. We would like to thank the unknown reviewers for their constructive and helpful comments.

References

1. I. Althöfer. Root evaluation errors: How they arise and propagate. *ICCA Journal*, 11(3):55–63, 1988.
2. C. Donninger. Null move and deep search. *ICCA Journal*, 16(3):137–143, 1993.
3. R. Feldmann. Fail high reductions. *Advances in Computer Chess 8 (ed. J. van den Herik)*, 1996.
4. H. Kaindl and A. Scheucher. The reason for the benefits of minmax search. In *Proc. of the 11 th IJCAI*, pages 322–327, Detroit, MI, 1989.
5. D.E. Knuth and R.W. Moore. An analysis of alpha-beta pruning. *Artificial Intelligence*, 6(4):293–326, 1975.
6. U. Lorenz. Controlled Conspiracy-2 Search. *Proceedings of the 17th Annual Symposium on Theoretical Aspects of Computer Science (STACS), (H. Reichel, S.Tison eds), Springer LNCS*, pages 466–478, 2000.
7. U. Lorenz. P.ConNers wins the 10th Grandmaster Tournament in Lippstadt. *ICCA Journal*, 23(3):188–192, 2000.
8. U. Lorenz. Parallel controlled conspiracy number search. *Proceedings of the 13th Annual ACM Symposium on Parallel Algorithms and Architectures (SPAA), ACM Press*, pages 320–321, 2001.
9. D.A. McAllester. Conspiracy Numbers for Min-Max searching. *Artificial Intelligence*, 35(1):287–310, 1988.
10. D.S. Nau. *Quality of Decision versus depth of search on game trees*. PhD thesis, Duke University, Durham, NC, 1979.
11. J. Pearl. On the nature of pathology in game searching. *Artificial Intelligence*, 20(4):427–453, 1983.
12. J. Pearl. *Heuristics – Intelligent Search Strategies for Computer Problem Solving*. Addison-Wesley Publishing Co., Reading, MA, 1984.
13. J. Schaeffer. Conspiracy numbers. *Artificial Intelligence*, 43(1):67–84, 1990.
14. J. Schaeffer and A. Plaat. Kasparov versus Deep Blue: The Rematch. *ICCA Journal*, 20(2):95–125, 1997.

Randomized Acceleration of Fundamental Matrix Computations⋆

Victor Y. Pan

Department of Mathematics and Computer Science
Lehman College of CUNY, Bronx, NY 10468, USA
vpan@lehman.cuny.edu

Abstract. We accentuate the power of several known effective methods by combining them together and adding some novel techniques. As a result, we substantially improve the known record randomized bit-operation complexity estimates for various fundamental computations with integer matrices. This includes the computation of the determinant, minimum and characteristic polynomials, Smith and Frobenius invariant factors, and the eigenvalues. Most of the algorithms can be further accelerated where the input matrix can be multiplied by a vector fast, they can be effectively parallelized and extended to matrix polynomials.

Keywords. Matrix determinant, minimum polynomial, characteristic polynomial, Smith invariant factors, Frobenius invariant factors, randomized algorithms, bit-operation complexity, block Wiedemann algorithm

2000 Math. Subject Classification: 15A23, 15A36, 68Q40, 68Q25, 15A15, 15A54

1 Introduction

The *Smith invariant factors* of an integer matrix $A = (a_{i,j})_{i,j=1}^n$ have important applications in computational number theory, abelian group theory, and homology theory (see [G95], [S96], [EGV00], [G01] on the bibliography). The Smith factors immediately define the *determinant* of A, whose computation is a classical problem with well-known applications to the solution of multivariate polynomial systems of equations and more recently to computational geometry (see, e.g., [PY01]). Furthermore the *minimum and characteristic polynomials* of a matrix A and its *Frobenius invariant factors* are defined by the Smith factors of the polynomial matrix $\lambda I - A$. Our goal is the accelerated randomized computation of the Smith and Frobenius factors, the determinant, and the minimum and characteristic polynomials of integer and polynomial matrices.

The best deterministic algorithm for the Smith factors in the integer case [S96] uses $\widetilde{O}(n^4 \log |A| + n^3 \log^2 |A|)$ bit-operations where $|A| = n \max_{i,j} |a_{i,j}|$. Here

⋆ Supported by NSF Grant CCR 9732206 and PSC CUNY Awards 61393-0030, 62435-0031, and 66383-0032

H. Alt and A. Ferreira (Eds.): STACS 2002, LNCS 2285, pp. 215–226, 2002.
© Springer-Verlag Berlin Heidelberg 2002

and hereafter, $\widetilde{O}(g(n))$ denotes $g(n)^{1+o(1)}$, and following the line of [EGV00] we ignore possible theoretical asymptotic acceleration based on fast square and rectangular matrix multiplication effective only for the inputs of very large size (cf. Remark 5.7 and [KV01]).

Faster algorithms use randomization. The randomized uncertified (Monte Carlo) algorithm of [G95], [G01] has roughly the same complexity as in [S96] but proposes distinct powerful techniques. Another approach originated in [P87, Appendix] and [P88], where it was proposed to combine Cramer's rule for linear systems of equations with Hensel's lifting and Hadamard's bound on $|\det A|$. A nontrival development of this approach in [ABM99], [EGV00], extending some techniques from [V00] and [G95], has resulted in an advanced randomized uncertified algorithm using $\widetilde{O}(n^{3.5}\log^{2.5}|A|)$ bit-operations. Here and hereafter, the estimates do not cover the cost of generating random parameters. On an average input matrix A, the algorithm uses $\widetilde{O}(n^3\log^2|A|)$ bit-operations.

In [KV01] a randomized certified (Las Vegas) algorithm computes $\det A$ in $\widetilde{O}(n^{10/3}\log|A|)$ bit-operations, yielding the current record for the worst case dense input matrices, the latter bound remains the same for an average matrix A. This important algorithm reduces the determinant problem for A to that for a polynomial matrix of a smaller size by combining Wiedemann's algorithm for the minimum polynomial of a matrix, its improvements [K95], and the minimum realization techniques from linear systems and control theory. The paper states the determinant computation as its main goal, but its fundamental techniques clearly lead to improving various other matrix computations. In the last section of the present paper, we explore this direction.

More generally, in this paper we accentuate the power of the effective advanced methods proposed and/or developed in the cited papers by combining them together, refining them, and adding some other old and new techniques. In section 4 (following some preliminaries in sections 2 and 3), we improve the worst case bit-operation complexity bound of [EGV00], preserving the average case bound. For the worst case input A, we compute the determinant and all Smith's factors of A by generating $O((n^{5/3}+\log|A|)\log n)$ random bits and performing $\widetilde{O}(n^{10/3}\log^2|A|)$ bit-operations. For an average case A the bounds decrease to $\widetilde{O}((n+\log|A|)\log n)$ and $\widetilde{O}(n^3\log^2|A|)$, respectively. Our progress is achieved by combining the approach of [P87], [P88], [ABM99], [EGV00] with the algorithms of [G95], [S96], [G01] and compressing the information carried in Smith's factors into much fewer bits that define the ratios of Smith's successive factors. In section 5 we first refine the randomized certified computation of $\det A$ in [KV01] to decrease the bit-operation cost bound to $\widetilde{O}(n^{16/5}\log|A|)$ using $O(n^{8/5}\log n+n\log|A|)$ random bits. (As in [KV01] and unlike [EGV00], the worst case bound is not improved for an average input A.) Technically, we just observe a simple transition to the computation with matrices of bounded displacement rank, for which a known algorithm (incorporating the compression of generators in [P92, Appendix] and Kaltofen-Saunders' random preconditioners) performs better than the respective algorithm in [KV01]. Further extensions of the same bit-cost bound to computing and certifying Smith and Frobenius factors and thus the minimum and characteristic polynomials enable dramatic progress versus the other known approaches. Compare the record bounds of $O(n^4 \log^d |A|)$ bit-operations for computing all Frobenius factors of A in [G94] for $d = 2$ (randomized, using $O(n \log |A|) \log n$ random bits) and in [S98] for $d = 1$ (deterministic). The extensions combine fundamental Theorem 1 in [KV01] with [G01], as is hinted to in the Conclusion in [KV01], but in addition incorporate the approaches in [P87]/[P88] with a modified stage of λ-adic lifting (see Remark 5.2) and [V00] and for every j exploit the bounds on the coefficients of the j-th Frobenius factor, in terms of n, j and $|A|$.

The proposed algorithms can be extended to the case of a polynomial input matrix A. Most of them can be accelerated where the input matrix is sparse and/or structured, that is, where it can be multiplied by a vector fast. The algorithms can be effectively parallelized, including their blocks of Toeplitz and Padé computations [P00] but with an exception of the stage of Hensel's p-adic lifting.

Our upcoming paper shows that the techniques supporting our present results can be extended to support substantially stronger bit-complexity estimates (see our section 6).

2   Some Definitions and Basic Facts

Hereafter $\mathbb{Z}$ and $\mathbb{Z}_q$ denote the rings of integers and of integers modulo q. I is the identity matrix. $\det A$, $c_A(\lambda) = \det(\lambda I - A)$ and $\mu_A(\lambda)$ are the determinant, characteristic and minimum polynomials of a matrix, respectively. M^T is the transpose of a matrix M.

Definition 2.1. *The greatest common divisor (gcd) $d_k = d_k(A)$ of all $k \times k$ minors (subdeterminants) of a matrix $A \in \mathbb{Z}^{n \times n}$ is called the k-th* determinantal divisor *of A, for $k = 1, \ldots, n$. We write $s_0 = d_0 = 1$ and define the k-th* Smith invariant factor *of A as*

$$s_k = s_k(A) = d_k/d_{k-1}, \quad k = 1, \ldots, n.$$

It is easily deduced that

$$s_1, \ldots, s_n \in \mathbb{Z}, \tag{2.1}$$

$$s_{i-1} | s_i \text{ for } i = 1, \ldots, n, \tag{2.2}$$

$$|\det A| = s_1 s_2 \cdots s_n. \tag{2.3}$$

Hereafter we write

$$s_i^{(q)} = s_i \bmod q, \quad i = 1, \ldots, n, \tag{2.4}$$

$$r_i = s_i/s_{i-1}, \quad r_i^{(q)} = r_i \bmod q, \quad i = 1, \ldots, n. \tag{2.5}$$

To estimate the *Smith ratios* r_i, we use (2.3) and the crude bound

$$|\det A| \leq |A|^n, \quad |A| = n \max_{i,j} |a_{i,j}| \tag{2.6}$$

where $A = (a_{i,j})$, $|a_{i,j}| = \|\mathbf{a}_{i,j}\|_1$ if $a_{i,j}$ are polynomials with the coefficient vectors $\mathbf{a}_{i,j}$. (2.3), (2.5) and (2.6) together imply that

$$s_i = \prod_{j=1}^{i} r_j, \qquad |\det A| = \prod_{g=1}^{n} r_g^{n-g+1} \le |A|^n, \tag{2.7}$$

$$r_g \le r_g^+ = |A|^{n/(n-g+1)}, \quad g = 1, \dots, n. \tag{2.8}$$

Having any lower bounds on r_g and/or s_i, we yield an improvement.

Lemma 2.2. Let $\tilde{r}_i \le r_i$, $i = 1, \dots, l$; $\tilde{s}_i \le s_i$ for $i = l+1, \dots, n$. Then

$$r_g \le \tilde{r}_g c_g, \quad c_g = \left(|A|^n \Big/ \left(\prod_{i=1}^{l} \tilde{r}_i^{l-i+1} \prod_{j=l+1}^{n} \tilde{s}_j \right) \right)^{1/(l-g+1)}, \quad g = 1, \dots, l.$$

The Smith factors $s_0(W) = 1$, $s_i(W) \in \mathbb{Z}(\lambda)$, $i = 1, \dots, n$, can be defined also for a matrix polynomial $W(\lambda) \in \mathbb{Z}^{n \times n}[\lambda]$. The properties (2.2) and (2.3) are extended. For $W(\lambda) = \lambda I - A$, we obtain the *Frobenius invariant factors* $f_i(A) = s_i(\lambda I - A) \in \mathbb{Z}[\lambda]$, $i = 0, \dots, n$, which define the characteristic and minimum polynomials,

$$\mu_A(\lambda) = f_n(A), \quad c_A(\lambda) = \det(\lambda I - A) = \prod_{i=1}^{n} f_i(A). \tag{2.9}$$

Hereafter $\mathbf{u}$ denotes the coefficient vector of a polynomial $u(\lambda)$, and $\delta(u)$ denotes its degree. Write $r_i(\lambda) = f_{n-i+1}(\lambda)/f_{n-i}(\lambda)$, $\delta_j = \delta(f_j)$.

Lemma 2.3. $|\mathbf{f}_j| \le (1 + |A|)^{\delta_j}$, $|\mathbf{r}_j| \le (1 + |A|)^{\delta(r_j)}$ for all j.

Proof. Each $r_j(\lambda)$ and $f_j(\lambda)$ is a polynomial $\prod_g (\lambda - \lambda_{i_g})$ where λ_{i_g} are the eigenvalues of A, so $|\lambda_{i_g}| \le |A|$ for all j. $\square$

Lemma 2.4. $\delta_{n-k+1} \le n/k$, $k = 1, \dots, n-1$.

Proof. We have $\sum_{i=0}^{n-1} \delta_{n-1} = n$, $\delta_{n-k+1} \le \delta_{n-k+i}$ for $i \ge 1$, so $\delta_{n-k+1} \le n - \sum_{i=0}^{k-2} \delta_{n-i} \le n - (k-1)\delta_{n-k+1}$, $\delta_{n-k+1} \le n/k$. $\square$

We recall that the number of bit-operations involved in an arithmetic operation in $\mathbb{Z}_s$ for an integer s, $\lceil \log_2 s \rceil = \xi$, is bounded by

$$\beta(\xi) = O((\xi \log \xi) \log \log \xi). \tag{2.10}$$

$M(m, n, p)$ denotes the number of arithmetic operations required to multiply an $m \times n$ and $n \times p$ matrices, $M(n) = M(n, n, n)$. In our estimates we rely on the bound supported by the straightforward algorithm (see Remark 5.7), for which

$$M(m, n, p) = (2n-1)mp, \quad M(n) = (2n-1)n^2. \tag{2.11}$$

We assume hereafter that v_A bit-operations are sufficient to multiply a matrix $A \in \mathbb{Z}^{n \times n}$ by a vector and that $v_A \ge n$. So $v_A \le 2n^2 - n$ for any A, whereas $v_A = o(n^2)$ for sparse and/or structured matrices A.

3   Three Basic Theorems

Theorem 3.1. *[EGV00]. For a positive ε, a nonsingular matrix $A \in \mathbb{Z}^{n \times n}$, and integers φ, $i(1), \ldots, i(\varphi)$, $n \geq i(1) > i(2) > \cdots > i(\varphi) \geq 1$, it is sufficient to generate $\rho = O((\varphi + n - i(\varphi))(n + \log |A|) \log n)$ random bits and then perform $\widetilde{O}((\varphi n^3 \log^2 |A|) \log(1/\epsilon))$ bit-operations in order to compute some divisors $s^*_{i(j)}$ of the Smith factors $s_{i(j)}$ of A for $j = 1, \ldots, \varphi$ such that* Probability$\{s_{i(j)} = s^*_{i(j)}, j = 1, \ldots, \varphi\} \geq 1 - \epsilon$.

Proof. In a very minor modification of the algorithm of [EGV00], choose random matrices $U_{i(\varphi),g}, V_{i(\varphi),g}$ for $g = 1, \ldots, c$, $c = O(1)$ and let all other matrices $U_{i(j),g}, V_{i(j),g} \in \mathbb{Z}^{n \times (n-i(j))}$, $j = 1, \ldots, \varphi$, be formed by the first $n - i(j)$ columns of $U_{i(\varphi),g}, V_{i(\varphi),g}$, respectively.   □

Theorem 3.2. *[G01] (cf. [G95], [S96]). For a matrix $A \in \mathbb{Z}^{n \times n}$, and a positive integer $k = n^{1+o(1)}$, sampling k random primes $q_1, \ldots, q_k$ from an interval $[a, b]$, $a, b \in n^{O(1)}$ and performing $\widetilde{O}(v_A n (\log q)(\log(1/\epsilon)) \log |A|)$ bit-operations, for $q = q_1 q_2 \cdots q_k$, are sufficient to compute all Smith factors $s_i^{(q)}$ and ratios $r_i^{(q)}$, $i = 1, \ldots, n$ (cf. (2.4), (2.5)), with an error probability of at most ε.*

Remark 3.3. *The overall computational cost in the algorithm [G01] supporting Theorem 3.2 is dominated by the stage of computing $\mu_B(\lambda)$ mod q where $B = D_1 T D_2 A$, $D_1, D_2 \in \mathbb{Z}^{n \times n}$ are diagonal matrices, $T_0 \in \mathbb{Z}^{n \times n}$ is a Toeplitz matrix, $\log |D_1 T_0 D_2| = O(n \log |A|)$, and the matrices D_1, D_2, T_0 are random. The reduction of the computation of the Smith factors of A to the computation of the minimum polynomial $\mu_B(\lambda)$ in [G01] can be extended from $A \in \mathbb{Z}_q^{n \times n}$ to $A \in \mathbb{Z}^{n \times n}$ and further to $A(\mathbf{t}) \in \mathbb{Z}^{n \times n}[\mathbf{t}]$, $\mathbf{t} = (t_i)_{i=1}^k$.*

Theorem 3.4. *[V00]. Let $A \in \mathbb{Z}^{n \times n}$, $U, V^T \in \mathbb{Z}_q^{n \times k}$ where $q = n^{O(1)}$ and U, V are random Toeplitz matrices with entries uniformly and independently of each other sampled from a finite set S. Let $a_i(\lambda)$ and $b_i(\lambda)$, $i = 1, \ldots, n$, be the Frobenius invariant factors of A and $B = A + UV$, respectively. Then $a_j(\lambda) = \gcd(a_n(\lambda), b_{j+k}(\lambda))$, $j = 1, \ldots, n - k$, with a probability of at least $1 - (nk + n + 1)/\operatorname{card}(S)$.*

4   An Algorithm for the Smith Factors and Determinant of an Integer Matrix

It is well-known [ABM99], [EGV00] that matrices $A \in \mathbb{Z}^{n \times n}$, with entries chosen in a fixed range, almost always have just a few Smith invariant factors exceeding 1, and all of them, besides $s_n(A)$, are small. If so, that is, for an average input matrix A, Theorem 3.1 can be applied for $\varphi = O(1)$, $n - i(\varphi) = O(1)$ to compute and certify (based on (2.3)) all Smith factors $s_i = s_i(A)$ by generating $O((n + \log |A|) \log n)$ random bits and performing $\widetilde{O}(n^3 \log^2 |A|)$ bit-operations. The next algorithm still performs within the latter cost bounds on an average input A but decreases the worst case bound of Theorem 3.1. To achieve this, we compute all $r_i^{(q)}$ and then all $s_i^{(q)}$ for smaller q based on Theorem 3.2. This gives us $s_i = s_i^{(q)}$ for all $i \leq \lambda_0$ provided $r_i < q$ for all $i \leq \lambda_0$ (cf. (2.8) and Lemma 2.2). The latter information is also enables us to decrease φ when we apply the algorithm supporting Theorem 3.1 to compute the remaining leading factors $s_i, i > \lambda_0$.

Algorithm 4.1. The Smith factors and determinant.

Input: $A \in \mathbb{Z}^{n \times n}$ and two bounds, $q^+ > 1$ (an upper bound on an integer parameter q) and $i_{exp}^+(\varphi) > 1$ (an expected upper bound on $i(\varphi)$ in Theorem 3.1); both bounds are specified later on.

Output: $s_1 = s_1(A), \ldots, s_n = s_n(A)$, $\det A$.

Initialization: $q \leftarrow \lceil (n \log |A|)^3 \rceil$ (our sample initial choice for the parameter q; in fact any choice of q, $1 < q = (n \log |A|)^{O(1)}$, will do).

Computations:

1. *Apply Theorem 3.2 to compute $s_i^{(q)}, r_i^{(q)}, i = 1, \ldots, n$.*
2. *Apply (2.8) or Lemma 2.2 to determine the largest integer κ such that*

$$r_i^+ < q \text{ for } i \leq \kappa. \tag{4.1}$$

If $\kappa + i_{exp}^+(\varphi) < n$ and if $q^2 \leq q^+$, replace q by q^2 and go to stage 1. Otherwise compute

$$r_i = r_i^{(q)}, \quad s_i = s_{i-1} r_i \tag{4.2}$$

for all $i \leq \kappa$. Output $s_1, \ldots, s_\kappa$ and write $\lambda = \kappa$, $\nu = n$.

3. *Apply Theorem 3.1 to compute and output s_ν. If $\kappa \geq \nu + 1$, go to stage 4. Otherwise if $\lambda \geq \nu + 1$, replace λ by κ, ν by λ and repeat stage 3.*

$$s_\nu / s_\lambda < q. \tag{4.3}$$

   - *If the pair (λ, ν) passes the test, compute and output $s_{i-1} = s_i / r_i^{(q)}$ for $i = \nu, \nu - 1, \ldots, \lambda + 1$. Update κ based on Lemma 2.2. Then, based on (4.2), compute and output $s_{\kappa(old)+1}, \ldots, s_{\kappa(new)}$. Replace ν by λ, λ by $\kappa(new)$ and repeat stage 3.*
   - *Otherwise keep ν but replace λ by $(\nu + \lambda)/2$ and repeat stage 3.*

4. *Extend the computation to computing $\det A$ as in [EGV00] and stop.*

Clearly, the algorithm applies Theorem 3.1 with φ not greater than the overall number of failures of test (4.3). Let us next estimate this number. By defining a binary tree with the passes of (4.3) as the leaves and the failures of (4.3) as the internal nodes (each node associated with a range (λ, ν)), we observe that within the factor of less than two, φ is the number of those failures whose child or both children pass (4.3). Let $(\lambda(1), \nu(1)), \ldots, (\lambda(\psi), \nu(\psi))$ denote all ranges (λ, ν) associated with such failures and enumerated in the order of increasing $\lambda(i)$ so that

$$\lambda(1) < \nu(1) \leq \lambda(2) < \nu(2) \leq \ldots \leq \lambda(\psi) < \nu(\psi). \tag{4.4}$$

(By construction, the ranges have no pairwise overlaps.) Let us write

$$h = \lceil \log_2 q \rceil - 1. \tag{4.5}$$

Lemma 4.2. $\varphi \le ((8n \log_2 |A|)/h)^{1/2}$ for h in (4.5).

Proof. The failure of test (4.3) at a node (λ, ν) implies the bound $s_\nu \ge q s_\lambda > 2^h s$. By combining this bound and (4.4), we obtain $|A|^n \ge |\det A| = \prod_{i=1}^n s_i \ge \prod_{i=1}^\psi s_{\lambda(i)} > \prod_{i=1}^\psi 2^{hi} = 2^{h \sum_{i=1}^\psi i} = 2^{(\psi+1)\psi h/2} > 2^{\psi^2 h/2}$. Therefore, $\psi^2 < 2\log_2(|A|^n)/h$, $\phi < 2\psi \le ((8n\log_2|A|)/h)^{1/2}$. $\square$

Let us next estimate (again in terms of h) the range where the Smith ratios exceed $q - 1$, that is, where Algorithm 4.1 invokes Theorem 3.1.

Lemma 4.3. *Let Algorithm 4.1 invoke Theorem 3.1 for $i(1), \ldots, i(\varphi)$ satisfying $n \ge i(1) > i(2) > \ldots > i(\varphi) \ge 1$. Then $n - i(\varphi) < (n/h)\log_2\|A\| - 1$ for h in (4.5).*

Proof. By construction, we have $r_{i(\varphi)} \ge q > 2^h$. On the other hand, by (2.8), $r_{i(\varphi)} \le |A|^{n/(n-i(\varphi)+1)}$. Therefore, $2^h < |A|^{n/(n-i(\varphi)+1)}$, $h < (n/(n-i(\varphi)+1))\log_2|A|$, $n - i(\varphi) + 1 < (n/h)\log|A|$. $\square$

Now, let Algorithm 4.1 be applied to the worst case input. Then we have $q \le q^+ \le q^2$ and we optimize the bit-operation cost bound by choosing q^+ such that $h = \Theta(n^{1/3}\log|A|)$. In this case $\varphi = \widetilde{O}(n^{1/3})$, and both stages, of computing a) smaller Smith's ratios and factors based on Theorem 3.2 and b) larger Smith's factors based on Theorem 3.1, are performed at roughly the same bit-operation cost, so we arrive at the overall bound $\widetilde{O}(n^{10/3}\log^2|A|)$. Furthermore, in this case we have $n - i(\varphi) = \widetilde{O}(n^{2/3})$, due to Lemma 4.3, and we may choose any $i^+_{exp}(\varphi) \ge n - n_0$, for $n_0 = \Theta(n^{2/3})$. (Our actual choice of $i^+_{exp}(\varphi)$ should rely on experimental statistics. One may start with $i^+_{exp}(\varphi) = n - x$ for $x = O(1)$ or $x = O(\log n)$ in the initial trials.) .

Theorem 4.4. *With a failure probability of at most ε, all Smith's factors of $A \in \mathbb{Z}^{n \times n}$ can be computed by generating $O((n^{5/3} + \log|A|)\log n)$ random bits and performing $\widetilde{O}((n^{10/3}\log^2|A|)\log(1/\varepsilon))$ bit-operations.*

Remark 4.5. *The bounds in Theorem 4.4 on the numbers of random bits and bit-operations are reached where Algorithm 4.1 is applied to such matrices as $A = \operatorname{diag}(s_i)_{i=1}^n$, $s_i = 1$ for $i < n - i(\varphi)$; $s_i = q$ for $i = i(\varphi), \ldots, n - \varphi$; $s_{i+1}/s_i = q$ for $i > n - \varphi$, where $n - i(\varphi) = \Theta(n^{2/3}\log|A|)$, $\varphi = \Theta(n^{1/3})$.*

Remark 4.6. *We may avoid binary search by slightly modifying Algorithm 4.1 (and accelerating it by roughly the factor of $\log^{2/3} n$). That is, let p and q be two products of distinct random primes in the range of $(n\log|A|)^{O(1)}$ each and let $\log p, \log q \in \Theta((n\log|A|)^{1/3})$. Apply Theorem 3.2 to compute $r_i^{(p)}$ and $r_i^{(q)}$ for all i and write $r_i = r_i^{(p)}$ if and only if $r_i^{(p)} = r_i^{(q)}$. This choice is correct with a high probability and enables a desired modification.*

Remark 4.7. *By choosing a larger q^+, one may decrease the upper bounds on $n - i(\varphi)$ and consequently on the number of random bits used in the algorithm at the expense of allowing more bit-operations.*

Remark 4.8. *The algorithm of [G95], [G01], supporting the basic Theorem 3.2, relies on the fast randomized computation of the content of a multivariate polynomial of bounded degree and norm, given with a black box for its evaluation. Suppose that this basic computation of [G95], [G01] can be effectively extended to the computation of the ratio of the contents of two polynomials provided that the ratio is an integer and we are given a black box subroutine for the evaluation of the ratio of these two polynomials. In our case, the ratios of the contents are the Smith factors s_i if the ratios of the polynomials are the i-th diagonal entries of the inverse of the preconditioned input matrix A (cf. [G95], [G01] on the preconditioning). By applying the p-adic lifting of [D82], we can probabilistically yield these entries for all i in $\widetilde{O}(n^2(n + \sum_{i=1}^{n} \log s_i))$ bit-operations. Due to (2.1)–(2.3), $\sum_{i=1}^{n} \log s_i = \log |\det A| \leq n \log |A|$. Thus having the desired extension of [G95], [G01], we would have computed all Smith's factors of A in $\widetilde{O}(n^3 \log^2 |A|)$ bit-operations.*

Remark 4.9. *By choosing structured (e.g., Toeplitz) auxiliary random matrices U_i and V_i in the algorithm of [EGV00], one may decrease the overall number of random parameters in Algorithm 4.1 provided that under this choice of the matrices U_i and V_i the algorithm still computes the Smith factors of A.*

5   Accelerated Fundamental Matrix Computations

Theorem 5.1. *Let $d_0 = t_0 = 1$, $\mathbf{t} = (t_i)_{i=1}^{k}$, $\mathbf{h} = (h_i)_{i=1}^{k}$, $\mathbf{d} = (d_i)_{i=1}^{k}$, $\mathbf{t^h} = t_1^{h_1}, \ldots, t_k^{h_k}$, $D = d_1 \cdots d_k$, $A \in \mathbb{Z}^{n \times n}[\mathbf{t}]/\mathbf{t^d}$. Let $1 \leq m \leq n$, $1 \leq r \leq n/m$; $m, n \in \mathbb{Z}$. Then*

I) det A *and all Smith's factors of A can be computed and certified by generating $O(mn \log n + n \log |A|)$ random bits and performing*

$$\beta_k = \widetilde{O}((\min\{v_A n^{k-1}/m^{k+1}, r^{k+1} + (n/m)^{k+1}/r\} + m^2 n^{k-1})n^3 D \log |A|) \tag{5.1}$$

bit-operations. β_k yields its minimums

$$\beta_k^- = \widetilde{O}(n^{k+2+2\gamma(k)} D \log |A|),$$

for $r = \Theta((n/m)^{(k+1)/(k+2)})$, $m = \Theta(n^{\gamma(k)})$, $\gamma(k) = (k+3)/(k^2 + 4k + 5)$, and

$$\beta_k^- = \widetilde{O}(n^{k+2} v_A^{2/(k+3)} D \log |A|),$$

for $m = \Theta(v_A^{1/(k+3)})$. In particular $\beta_0^- = \widetilde{O}(n^{16/5} \log |A|)$, for $r^2 = \Theta(n/m)$, $m = \Theta(n^{3/5})$; $\beta_1^- = \widetilde{O}(n^{19/5} D \log |A|)$, for $r = \Theta((n/m)^{2/3})$, $m = \Theta(n^{2/5})$; $\beta_0^- = \widetilde{O}(n^2 v_A^{2/3} \log |A|)$, for $m = \Theta(v_A^{1/3})$; $\beta_1^- = \widetilde{O}(v_A^{1/2} D n^3 \log |A|)$, for $m = \Theta(v_A^{1/4})$. Conversely, (5.1) enables us to decrease the bit-cost bound (although not below $O(n \log(n|A|))$) by decreasing m.

II) Furthermore, a factor of $\mu_A(\lambda)$, equal to $\mu_A(\lambda)$ with a high probability, can be computed within the same asymptotic cost bounds.

Proof. For $k = 0$, the algorithm in [KV01] reduces computing $\det A$ essentially to computing a) the $m \times m$ matrices $B^{[g]} = X^T A^g Y$, for $g = 0, \ldots, \alpha + \beta - 1$, random $X, Y \in \mathbb{Z}_t^{n \times m}$, $t = n^{O(1)}$; $\alpha + \beta = \Theta(n/m)$, and b) m linearly independent vectors $\mathbf{v}_1, \ldots, \mathbf{v}_m$ from the null space of the block Toeplitz matrix $T = (B^{[\beta+i-j]})_{i,j=0}^{\alpha-1,\beta-1}$. The algorithm in [KV01] performs stage a) within our cost bounds (to yield the bounds in terms v_A, choose $r = 1$, that is, trivialize the stage of baby steps/giant steps in this algorithm), but at stage b) we use $O(n \log n)$ random bits and $\widetilde{O}(m^2 n^2 \log |A|)$ bit-operations, versus $O(mn \log n)$ and $\widetilde{O}(m^3 n^2 \log |A|)$ in [KV01]. That is, instead of simply applying Levinson-Durbin's algorithm, as in [KV01], we first interchange the rows and columns of T to arrive at an $n \times n$ matrix with displacement rank m or less; then we apply the MBA divide-and-conquer algorithm [P01, chapter 5], including compression of generators [P92, Appendix] and Kaltofen-Saunders preconditioning. Summarizing the cost bounds, we obtain part a) for $\det A$ where $k = 0$.

To prove part II), apply Theorem 1 of [KV01], by which $\mu_A(\lambda)$ equals Smith's largest invariant factor of the matrix polynomial $F_X^{A,Y}(\lambda) \in \mathbb{Z}^{m \times m}[\lambda]/\lambda^\delta$, $\delta = \Theta(n/m)$. Assume that $\mu_A(0) \neq 0$ (otherwise shift to $F_X^{A,Y}(\lambda) - \rho I$ and $\mu_A(\lambda - \rho)$ for a random ρ), so $\det F_X^{A,Y}(0) \neq 0$. Then proceed along the line of [P87], [P88]. That is, first compute $Z_0 = (F_X^{A,Y}(0))^{-1}$ in m^3 operations in the field of rationals $\mathbb{Q}$; then apply two-stage λ-adic lifting (specified below) to obtain $\mathbf{w}(\lambda) = (F_X^{A,Y}(\lambda))^{-1}\mathbf{u} \bmod \lambda^{2n}$ for a random vector $\mathbf{u} \in \mathbb{Z}_q^m$, where $\log q = O(\log n)$; recover the m entries of the rational vector function $F_X^{A,Y}(\lambda)^{-1}\mathbf{u}$ as the $(n-1, n)$ Padé approximations to the entries of $\mathbf{w}(\lambda)$, and output $\mu_A(\lambda)$ as the least common multiple of the denominators of the m rational entries.

The dominating cost of performing $O(m^2 n)$ field operations in $\mathbb{Q}$ is at the lifting stage, partitioned into Newton's lifting of Z_0 to $Z_\delta = (F_X^{A,Y}(\lambda))^{-1} \bmod \lambda^\delta$ and Hensel's lifting [D82] of $\mathbf{w}(\lambda) \bmod \lambda^{i\delta}$ to $\mathbf{w}(\lambda) \bmod \lambda^{(i+1)\delta}$, for $i = 1, \ldots, \lceil 2n/\delta \rceil - 1$. At all stages we perform computations modulo sufficiently many primes in $(n \log |A|)^{O(1)}$ whose product exceeds $2|A|^n$, and we recover the integer coefficients of $\mu_A(\lambda)$ by applying the Chinese Remainder Algorithm. Summarizing, we extend the bit-cost bounds from part I) to part II). Finally, we apply this result to the computation of $\mu_B(\lambda)$ for B of Remark 3.3. Due to [G01], this enables us to extend the cost estimates to the computation of the Smith factors of A. The output is certified by testing (2.3). This proves Theorem 5.1 for $k = 0$. For $k \geq 1$ the proof is similar. $\square$

Remark 5.2. *In the Conclusion section of [KV01] it is suggested to use [G01] for an extension from determinant to Smith's factors, but we achieve our progress along this line only by applying the approach of [P87], [P88] adjusted with Newton/Hensel's partition of the λ-adic lifting.*

Next we compute the Frobenius factors and characteristic polynomial of A.

Then again we perform computations modulo sufficiently many random primes in $(n \log |A|)^{O(1)}$ and recover the integer coefficients of the factors $f_i(A)$

by means of the Chinese Remainder algorithm. To estimate the number of random primes required, we apply Lemmas 2.3 and 2.4 and the analysis in [G94], which enables us to identify and exclude those primes p_j for which $\deg f_i(A)$ drops in the transition from $\mathbb{Z}$ to $\mathbb{Z}_{p_j}$. Probabilistically for such primes p_j, the degree drops versus even its maximum in j. [G94] shows that these primes are sufficiently rare, so ignoring them does not affect our estimates.

Lemma 5.3. *The l leading Frobenius factors of $A \in \mathbb{Z}^{n\times n}$ can be computed with $O((m\log n + \log |A|)n)$ random bits and $\beta_0 \Gamma(l)$ bit-operations for $m \leq n$, β_0 of Theorem 5.1, and $\Gamma(l) = O(\sum_{h=1}^{l} 1/h) = O(\log l)$.*

Proof. Compute the l Frobenius factors $f_i(A)$, $i = n, n-1, \ldots, n-l+1$, by combining Theorems 3.4 and 5.1 II). Combine Lemmas 2.3 and 2.4 to compute an upper bound t_i on the magnitudes of the output coefficients of $f_i(\lambda)$ and compute $f_i(\lambda)$ in $\mathbb{Z}_{q_i}$ for $2t_i < q_i$, $q_i/t_i = (n\log |A|)^{O(1)}$. Reuse all random parameters for every new i. □

Theorem 5.4. *All Frobenius factors of $A \in \mathbb{Z}^{n\times n}$ (and consequently the characteristic and minimum polynomials of A) can be computed and verified by generating $O((n^{3/5}\log n + \log |A|)n)$ random bits and performing $\widetilde{O}(n^{16/5}\log |A|)$ bit-operations or alternatively with $O((v_A^{1/3}\log n + \log |A|)n)$ random bits and $\widetilde{O}(n^2 v_A^{2/3}\log |A|)$ bit-operations.*

Proof. Apply Lemma 5.3 for $l = n$ and either $m = \Theta(n^{3/5})$ or alternatively $m = \Theta(v_A^{1/3})$. Verify correctness of the output by checking if $\sum_{i=1}^{n} \deg f_i(A) = n$. □

Remark 5.5. *The number of random bits in the estimates of Theorem 5.4 can be decreased at the expense of using more bit-operations if a smaller value of m is chosen.*

Remark 5.6. *Typically, for a random $A \in \mathbb{Z}^{n\times n}$ with a fixed upper bound on $|A|$, the ratio $c_A(\lambda)/\mu_A(\lambda)$ is either 1 or a polynomial of a small degree, and $f_i(A) = 1$ for $i < n - O(1)$. Thus for an average input matrix A, the algorithm supporting Theorem 5.4 invokes Theorem 3.4 $O(1)$ times. Furthermore, even for the worst case input A, the number of these invocations can be decreased by extending our techniques in section 4. The asymptotic bit-cost, however, still remains bounded from below by the cost of computing $\mu_A(\lambda)$, estimated in Theorem 5.1 II).*

Remark 5.7. *The known fast algorithms for matrix multiplication enable a decrease of the upper bound (2.11) on $M(n)$ and $M(m,n,p)$ for very large m, n and p [HP98]. Based on incorporating these algorithms, the asymptotic bit-operation complexity estimates of Theorems 3.1, 3.4 and consequently 4.5, 5.1, and 5.4 can be theoretically decreased for very large n. We leave this as a challenge to the reader.*

Corollary 5.8. *All eigenvalues of a matrix $A \in \mathbb{Z}^{n\times n}$ can be approximated within 2^{-b} for $b \geq 2n\log(n|A|)$ with γ_A random bits and $\beta_A + O((n\log^2 n)(\log^2 n + \log b)\xi(bn))$ bit-operations for $\xi(q)$ of (2.10) and γ_A and β_A representing the respective cost bound in Theorem 5.1 II). The cost bound in Theorem 5.4 and Remarks 5.5–5.7 support computing also the algebraic multiplicities of the eigenvalues.*

The following lemma does not lead to any improvement of our cost estimates but is included to demonstrate another direction of extending the seminal approach of [KV01].

Lemma 5.9. *The Frobenius factors $f_{n-m+h+1}(A), \ldots, f_{n-m+1}(A)$, for $A \in \mathbb{Z}^{n \times n}$, $1 \leq h < m \leq n$, can be computed with $O(mn \log n + (n/(m-h)) \log |A|)$ random bits and $\widetilde{O}((\beta_0 + n^2 m^{14/5} \log |A|)/(m-h)^2)$ bit-operations.*

Proof. Theorem 1 in [KV01] reduces computing the m leading Frobenius factor of A to computing the m Smith factors of $F_X^{A,Y}(\lambda)$. By extending the algorithm of [G01], reduce the latter task to the computation of $\mu_{B(\lambda)}$ for $B(\lambda) = D_1 T_0 D_2 F_X^{A,Y}(\lambda)$ and for D_1, T_0, D_2 of Remark 3.3. Due to Lemmas 2.3 and 2.4, we may perform this computation in $\mathbb{Z}_v[\lambda]/\lambda^t$ where $t = \Theta(n/(m-h))$, $\log v = \Theta((n/(m-h)) \log |A|)$ provided we only seek $f_i(A)$ for $i < n - m + h$. It remains to apply Theorem 5.1 II) for $k = 1$, A replaced by $B(\lambda)$, n by m, t by λ, d by n/m, and $\log |B(\lambda)|$ by $n \log |A|$ and to decrease the cost bounds due to performing the computations in $\mathbb{Z}_v[\lambda]/\lambda^t$. □

6 Conclusion

In our upcoming paper we extend our present algorithms to compute the determinant and all Smith factors of an $n \times n$ integer matrix A by using $\widetilde{O}(n^{22/7} \log^2 |A|)$ bit-operations.

Acknowledgements. I am grateful to Mark Giesbrecht, Erich Kaltofen, Arne Storjohann, and Gilles Villard for (p)reprints and most valuable discussions and to the referees for many helpful comments.

References

[ABM99]   J. Abbott, M. Bronstein, T. Mulders, Fast Deterministic Computations of the Determinants of Dense Matrices, *Proceedings of International Symposium on Symbolic and Algebraic Computation (ISSAC'99)*, 197-204, ACM Press, New York, 1999.

[C94]   D. Coppersmith, Solving Homogeneous Linear Equations over $GF(2)$ via Block Wiedemann Algorithm, *Math. Comput.*, **62, 205**, 333-350, 1994.

[CEKSTVa]   L. Chen, W. Eberly, E. Kaltofen, B. D. Saunders, W. J. Turner, G. Villard, Efficient Matrix Preconditioners for Black Box Linear Algebra, *Linear Algebra and Its Applications*, in press.

[D82]   J. D. Dixon, Exact Solution of Linear Equations Using p-adic Expansions, *Numerische Math.*, **40**, 137-141, 1982.

[EGV00]   W. Eberly, M. Giesbrecht, G. Villard, On Computing the Determinant and Smith Form of an Integer Matrix, *Proc. 41st Annual Symposium on Foundations of Computer Science (FOCS' 2000)*, 675-685, IEEE Computer Society Press, Los Alamitos, California, 2000.

[G94]   M. Giesbrecht, Fast Algorithms for Rational Forms for Integer Matrices, *Proc. of International Symp. on Symbolic and Algebraic Computation (ISSAC'94)*, 305-311, ACM Press, New York, 1994.

[G95]   M. Giesbrecht, Fast Computation of the Smith Normal Forms of an Integer Matrix, *Proc. International Symposium on Symbolic and Algebraic Computation (ISSAC'95)*, 110-118, ACM Press, New York, 1995.

[G01]   M. Giesbrecht, Fast Computation of the Smith Form of a Sparse Integer Matrix, *Computational Complexity*, **10**, 41-69, 2001.

[GL96]   G. H. Golub, C. F. Van Loan, *Matrix Computations*, Johns Hopkins University Press, Baltimore, Maryland, 1996 (3rd edition).

[HP98]   X. Huang, V. Y. Pan, Fast Rectangular Matrix Multiplication and Applications, *Journal of Complexity*, **14**, 257-299, 1998.

[K92]   E. Kaltofen, On Computing Determinants of Matrices Without Divisions, *Proc. Intern. Symposium on Symbolic and Algebraic Computation (ISSAC'92)*, 342-349, ACM Press, New York, 1992.

[K95]   E. Kaltofen, Analysis of Coppersmith's Block Wiedemann Algorithm for the Parallel Solution of Sparse Linear Systems, *Math. Comput.*, **62, 210**, 777-806, 1995.

[KV01]   E. Kaltofen, G. Villard, On the Complexity of Computing Determinant, manuscript, August 2001, presented at *2001 AMS/IMS/SIAM Summer Research Conference on Fast Algorithms in Math., Computer Science and Engineering*, S. Hadley, Mass., August 2001.

[P87]   V. Y. Pan, Complexity of Parallel Matrix Computations, *Theoretical Computer Science*, **54**, 65-85, 1987.

[P88]   V. Y. Pan, Computing the Determinant and the Characteristic Polynomials of a Matrix via Solving Linear System of Equations, *Information Processing Letters*, **28**, 71-75, 1988.

[P92]   V. Y. Pan, Parametrization of Newton's Iteration for Computations with Structured Matrices and Applications, *Computers & Mathematics (with Applications)*, **24, 3**, 61–75, 1992.

[P00]   V. Y. Pan, Parallel Complexity of Computations with General and Toeplitz-like Matrices Filled with Integers and Extensions, *SIAM Journal on Computing*, **30**, 1080–1125, 2000.

[P01]   V. Y. Pan, *Structured Matrices and Polynomials: Unified Superfast Algorithms*, Birkhäuser/Springer, Boston/New York, 2001.

[P01a]   V. Y. Pan, Univariate Polynomials: Nearly Optimal Algorithms for Polynomial Factorization and Rootfinding, *Proc. International Symposium on Symbolic and Algebraic Computation (ISSAC'01)*, 253-266, ACM Press, New York, 2001.

[PY01]   V. Y. Pan, Y. Yu, Certification of Numerical Computation of the Sign of the Determinant of a Matrix, *Algorithmica*, **30**, 708-724, 2001.

[S96]   A. Storjohann, Near Optimal ALgorithms for Computing Smith Normal Forms of Integer Matrices, *Proc. International Symposium on Symbolic and Algebraic Computation (ISSAC'96)*, 267-274, ACM Press, New York, 1996.

[S98]   A. Storjohann, An $O(n^3)$ Algorithm for Frobenius Normal Form, *Proc. International Symposium on Symbolic and Algebraic Computation (ISSAC'98)*, 101-104, ACM Press, New York, 1998.

[V00]   G. Villard, Computing the Frobenius Normal Form of a Sparse Matrix, *Proc. 3rd Intern. Workshop on Computer Algebra in Scientific Computing*, 395-407, Springer, 2000.

Approximations for ATSP with Parametrized Triangle Inequality

L. Sunil Chandran and L. Shankar Ram*

Indian Institute of Science, Bangalore
{sunil,shankar}@csa.iisc.ernet.in

Abstract. We give a *constant factor* $(\frac{\gamma}{1-\gamma} + \gamma)$ approximation for the asymmetric traveling salesman problem in graphs with costs on the edges satisfying γ-parametrized triangle inequality (γ-Asymmetric graphs) for $\gamma \in [\frac{1}{2}, 1)$. We also give an improvement of the algorithm with approximation factor approaching $\frac{\gamma}{1-\gamma}$.
We also explore the $\frac{c_{max}}{c_{min}}$ ratio of edge costs in a general asymmetric graph. We show that for $\gamma \in [\frac{1}{2}, \frac{1}{\sqrt{3}})$, $\frac{c_{max}}{c_{min}} \leq \frac{2\gamma^3}{1-3\gamma^2}$, while for $\gamma \in [\frac{1}{\sqrt{3}}, 1)$, this ratio can be *arbitrarily large*. We make use of this result to give a better analysis to our main algorithm. We also observe that when $\frac{c_{max}}{c_{min}} > \frac{\gamma^2}{1-\gamma-\gamma^2}$ with $\gamma \in (\frac{1}{2}, \frac{\sqrt{5}-1}{2})$, the minimum cost and the maximum cost edges in the graph are *unique* and are *reverse* to each other.

1   Introduction

The Traveling Salesman problem(TSP) is perhaps one of the most well known NP-Complete Optimization problems. There are two main variations to the problem - *Symmetric(STSP)* and *Asymmetric(ATSP)*. The STSP is to find a minimum cost hamiltonian tour in a complete undirected graph $G = (V, E)$ with cost function, $c : E \to \Re^+$ associated with the edges. It is known that there does not exist any finite factor approximation algorithm for this problem in the most general case. But if the cost function satisfies the triangle inequality, i.e $c(u, v) \leq c(u, w) + c(w, v) \; \forall u, v, w \in V$, then an approximation algorithm due to Christofides [4] is available. This algorithm achieves a factor of $3/2$ which is so far the best one for the problem. In [1,2], STSP with cost function satisfying the γ-parametrized triangle inequality i.e $c(u, v) \leq \gamma(c(u, w) + c(w, v)) \; \forall u, v, w \in V$ has been studied where $\gamma \geq 1$. An approximation guarantee of $min\{\frac{3\gamma^2}{2}, 4\gamma\}$ is shown there. In [3], algorithms achieving $(1 + \frac{2\gamma-1}{3\gamma^2-2\gamma+1})$ and $\frac{2}{3} + \frac{\gamma}{3(1-\gamma)}$ factors are available with $\gamma \in [\frac{1}{2}, 1)$. In the same paper, the NP-hardness result for the problem is also provided.

The ATSP is to find a minimum cost directed hamiltonian tour in a given

---

* Dept. of Computer Science and Automation, Indian Institute of Science, Bangalore, India 560012. This research is partially supported by Infosys fellowship

H. Alt and A. Ferreira (Eds.): STACS 2002, LNCS 2285, pp. 227–237, 2002.
© Springer-Verlag Berlin Heidelberg 2002

complete directed graph $G = (V, E)$ with cost function $c : E \to \Re^+$. This problem also does not admit any finite factor approximation in the most general case. But when c satisfies the triangle inequality, a $\log(n)$ factor algorithm due to [7] is available. Getting a constant factor approximation for the problem has remained a challenge for researchers in combinatorial optimization.

In this paper, for the first time, we look at the ATSP with c satisfying γ-parametrized triangle inequality with $\gamma \in [\frac{1}{2}, 1)$. We are giving a constant factor approximation for this problem. The factor we obtain is $\frac{\gamma}{1-\gamma} + \gamma$. We also suggest an improvement of the algorithm with approximation factor approaching $\frac{\gamma}{1-\gamma}$. To design the algorithm, we make use of the fact that minimum cost cycle cover is a lower bound for the ATSP optimum as in [7].

The TSP (both ATSP and STSP) are studied in the literature with other kinds of restrictions on the edge costs also. For example [9], gives a $\frac{17}{12}$ - factor algorithm for the ATSP when the edge costs are restricted to either 1 or 2 (observe that a 2-factor algorithm is trivial in this case and a $\frac{3}{2}$-factor algorithm is also obtainable with slightly more effort). [8] study the STSP with the same restriction on the edge costs.

In [7], Frieze et al. consider two assumptions on the data for the ATSP and provide better approximation algorithms when these assumptions are valid.
1. $c(i, j) \leq \alpha c(j, i)$ $\alpha \geq 1$ and
2. $c(i, j) \leq \alpha(c(k, i) + c(k, j))$ $\alpha \geq \frac{1}{2}$.
We mention that the γ-parametrized triangle inequality is also a natural data dependent assumption but has not been considered in [7]. It is natural since we can find the γ for the given instance of the ATSP in polynomial time using the formula $\gamma = \max\{\frac{c(u,v)}{c(u,w)+c(w,v)}\}$ $\forall u, v, w \in V$. The $\frac{\gamma}{1-\gamma} + \gamma$ factor algorithm we give is better than the available $\log(n)$ factor algorithm for all $\gamma < 1 - \epsilon$ where $\epsilon = \frac{1}{2}(\sqrt{\log^2(n) + 4} - \log(n))$ (note that $\epsilon < 1$). Also we provide an improvement to the algorithm with factor tending to $\frac{\gamma}{1-\gamma}$. For $\gamma \in [\frac{1}{2}, \frac{4}{7})$, this algorithm will still be better even if the conjectured [10] $\frac{4}{3}$ factor algorithm materializes. But a $\frac{4}{3}$ factor algorithm even for STSP is considered a hard open problem which has been challenging researchers for more than two decades.

We also explore the $\frac{c_{max}}{c_{min}}$ ratio of a general asymmetric graph to give a better analysis for our main algorithm. We show that for $\gamma \in [\frac{1}{2}, \frac{1}{\sqrt{3}})$, $\frac{c_{max}}{c_{min}} \leq \frac{2\gamma^3}{1-3\gamma^2}$. We also show that for this range this upperbound is *tight*. Moreover when $\gamma \in [\frac{1}{\sqrt{3}}, 1)$, $\frac{c_{max}}{c_{min}}$ can be *arbitrarily large*. This is in contrast with the symmetric case (STSP) where $\frac{c_{max}}{c_{min}} \leq \frac{2\gamma^2}{1-\gamma}$ $\forall \gamma \in [\frac{1}{2}, 1)$, refer [3]. Thus we observe that the asymmetric graphs with γ triangle inequality for $\gamma \in [\frac{1}{\sqrt{3}}, 1)$ behave differently from their symmetric counterparts.

As mentioned earlier, in [3] it is shown that it is NP-hard to approximate STSP with γ-parametrized triangle inequality to within $\frac{7611+10\gamma^2+5\gamma}{7612+8\gamma^2+4\gamma} - \epsilon$, for $\gamma \in [1/2, 1)$ and any $\epsilon > 0$. So, since STSP is a special case of ATSP, ATSP with γ-parametrized triangle inequality also can not be approximated within the same bound.

We note that there is no loss in generality in assuming that the underlying graph is complete. If an edge (u, v) is not present, i.e $c(u, v) = \infty$ then there cannot be a hamiltonian tour of finite cost in the graph. This is because the cost of any edge is bounded above by the cost of the optimum tour. The assumption that $\gamma \geq 1/2$ also does not mean any loss in generality due to the following observation.

Lemma 1. *If in a complete directed graph $G = (V, E)$ with cost function $c : E \to \Re^+$, c satisfies γ-parametrized triangle inequality, then $\gamma \geq 1/2$.*

Proof :   Let (u, v) be the edge with maximum cost, c_m. Let w be a vertex other than u and v. Also let the costs $c(u, w)$ and $c(w, v)$ be a and b respectively. We have $a \leq c_m$ and $b \leq c_m$. Because of the parametrized triangle inequality, $c_m \leq \gamma(a + b)$ from which it follows that $\gamma \geq 1/2$. $\square$
It is easy to see that when $\gamma = 1/2$, all the edges will have the same cost.

2   The Main Algorithm

We are given a complete directed graph $G = (V, E)$ with cost function, $c : E \to \Re^+$, satisfying the γ-parametrized triangle inequality i.e. $c(u, v) \leq \gamma(c(u, w) + c(w, v))$ $\forall u, v, w \in V$ with $\gamma \in [\frac{1}{2}, 1)$. We define $c(X) = \sum_{e \in X} c(e)$ where $X \subseteq E$. Also assume that the cost of the optimal tour in G is OPT. Let $|V| = n$.

The following lemma captures the main observation which allows us to design the basic algorithm.

Lemma 2. *Let $C = \{u_1, u_2, \cdots u_k, u_{k+1} = u_1\}$ be a directed cycle in G with $2 \leq k \leq n$ and $\gamma \in [1/2, 1)$. Let $w \in V$. Then, $\exists(u_i, u_{i+1})$ where $1 \leq i \leq k$ such that*

$$c(u_i, w) \leq \tfrac{\gamma}{1-\gamma} c(u_i, u_{i+1})$$

Proof :   If $w \in C$ then say $w = u_{i+1}$. Clearly the edge (u_i, u_{i+1}) satisfies the required property since $\frac{\gamma}{1-\gamma} \geq 1$. If $w \notin C$ then let $c(u_j, w) = \max\{c(u_i, w) : i = 1, .., k\}$. We claim that the edge (u_j, u_{j+1}) satisfies the required property. By the γ-parametrized triangle inequality, we have

$$c(u_j, w) \leq \gamma(c(u_j, u_{j+1}) + c(u_{j+1}, w))$$

As $c(u_{j+1}, w) \leq c(u_j, w)$,

$$(\tfrac{1}{\gamma} - 1)c(u_j, w) \leq c(u_j, u_{j+1})$$

i.e

$$c(u_j, w) \leq \tfrac{\gamma}{1-\gamma} c(u_j, u_{j+1})$$

$\square$

Corollary   Let $\mathcal{C} = \{u_1, u_2, \cdots u_k, u_{k+1} = u_1\}$ be a directed cycle in G with $2 \leq k \leq n$ and $\gamma \in [1/2, 1)$. Let $w \in V \backslash \mathcal{C}$. If (u_m, w) is such that $c(u_m, w) = \max\{c(u_i, w) : i = 1 \cdots k\}$, then $c(u_m, w) \leq \frac{\gamma}{1-\gamma} c(u_m, u_{m+1})$.   □

We first present a subroutine which constructs a hamiltonian path ending at a specified vertex given a cycle cover and the vertex as input.

We start with a minimum cost cycle cover (Notice that there are efficient algorithms to find cycle covers [6]). If an edge is removed from a cycle, we are left with a path which spans the nodes of the cycle. If one edge is removed from each cycle, we get a collection of such paths. Now we can connect these paths *properly* so that we get a hamiltonian path. The trick is to carefully select the edge to be removed from each cycle.

ALGORITHM Ham_Path
Input : A cycle cover $\mathcal{C} = \{C_1, C_2 \cdots C_k\}$; a vertex $u_1 \in V$. Without loss of generality we can assume $u_1 \in C_1$.
Output : A hamiltonian path of G ending at vertex u_1.

1. Consider the edge $(u_1, v_1) \in C_1$ and remove it. Now, we have a path from v_1 to u_1.
2. For $i = 2 \cdots k$ do:
   (a) Get an edge (u_i, v_i) in C_i such that $c(u_i, v_{i-1}) = \max\{c(x, v_{i-1}) : \forall x \in C_i\}$ and remove it.
   (b) Add the edge (u_i, v_{i-1}). We have a path from v_i to u_1.
3. We have a hamiltonian path from v_k to u_1. Output this path.

Lemma 3. *If a cycle cover $\mathcal{C}$ is input to the algorithm* **Ham_Path**, *it outputs a hamiltonian path H of cost $\leq \frac{\gamma}{1-\gamma} c(\mathcal{C})$. In particular, if $\mathcal{C}$ is a minimum cost cycle cover, $c(H) \leq \frac{\gamma}{1-\gamma} OPT$.*

Proof : Let $e_2, \cdots, e_k$ be the edges removed from $C_2, \cdots, C_k$ respectively in Step2(a) and $f_2, \cdots, f_k$ be the edges included in Step2(b). From the corollary to Lemma 2, we see that $c(f_i) \leq \frac{\gamma}{1-\gamma} c(e_i)$. So the increase in cost by the replacements (ICR)

$$ICR = \sum_{i=2}^{k} [c(f_i) - c(e_i)]$$
$$\leq (\tfrac{\gamma}{1-\gamma} - 1) \sum_{i=2}^{k} c(e_i)$$
$$\leq (\tfrac{\gamma}{1-\gamma} - 1) c(\mathcal{C})$$

The total cost of the path H is

$$c(H) \leq c(\mathcal{C}) + ICR \leq \tfrac{\gamma}{1-\gamma} c(\mathcal{C})$$

If $\mathcal{C}$ is a minimum cost cycle cover, $c(H) \leq \frac{\gamma}{1-\gamma} OPT$, since the cost of any minimum cost cycle cover is a lower bound on OPT.   □

Now we present the main algorithm. We assume that the minimum cost cycle cover contains at least 2 cycles. Otherwise, we already have the optimum hamiltonian tour.

ALGORITHM γ-ATSP

1. $\tau = \phi$; $F = \phi$
2. Construct a minimum cost cycle cover $\mathcal{C} = \{C_1, C_2 \cdots C_k\}$; $k \geq 2$.
3. For each $t_i \in V$ do the following:
   (a) $H_i = \textbf{Ham\_Path}(\mathcal{C}, t_i)$. (Note that H_i is the hamiltonian path returned by the subroutine and its end vertex will be t_i.) . Suppose that s_i is the start vertex of H_i.
   (b) $T_i = H_i \bigcup \{(t_i, s_i)\}$
   (c) $F = F \bigcup \{(t_i, s_i)\}$, $\tau = \tau \bigcup \{T_i\}$
4. If the edges in F form a hamiltonian tour, let it be T_f. Add it to τ i.e $\tau = \tau \bigcup \{T_f\}$.
5. Output the minimum cost tour in τ.

Theorem 1. *The cost of the tour output by the algorithm γ-**ATSP** is at most $(\frac{\gamma}{1-\gamma} + \gamma)OPT$.*

Proof : Let T_o be an optimum hamiltonian tour in the graph. If $T_f = T_o$ then we are done - the algorithm will output the optimum tour itself . Otherwise, there exists an edge $(t_i, s_i) \in F$ such that $(t_i, s_i) \notin T_o$. Let the path from t_i to s_i in T_o be P. Obviously, $P \neq \{(t_i, s_i)\}$. Therefore, $c(t_i, s_i) \leq \gamma c(P) \leq \gamma OPT$. Thus $c(T_i) = c(H_i) + c(t_i, s_i) \leq (\frac{\gamma}{1-\gamma} + \gamma)OPT$ using Lemma 3. $\square$

We note that a better result exists for $\gamma \in [\frac{1}{2}, \frac{1}{\sqrt{3}})$ than implied by the above theorem.

Theorem 2. *For each tour $T_i \in \tau$, $c(T_i) \leq [\frac{\gamma}{1-\gamma} + \frac{2\gamma^3}{n(1-3\gamma^2)}]OPT$. (Note that the second term becomes negligible for large enough n).*

Proof : Let c_{min} and c_{max} be the cost of the minimum cost and maximum cost edges in the graph. Since the optimum tour should contain an edge of cost $\leq \frac{OPT}{n}$, it is obvious that $c_{min} \leq \frac{OPT}{n}$. Now, $c(T_i) = c(H_i) + c(t_i, s_i)$

$$\leq \frac{\gamma}{1-\gamma}OPT + rc_{min}$$
$$\leq [\frac{\gamma}{1-\gamma} + \frac{r}{n}]OPT$$

where $r = \frac{c_{max}}{c_{min}}$. The result follows since $r \leq \frac{2\gamma^3}{1-3\gamma^2}$ for $\gamma \in [\frac{1}{2}, \frac{1}{\sqrt{3}})$ from Theorem 4, which will be proved in the section 4. $\square$

The running time of the above algorithm is that of finding the minimum cost cycle cover. The minimum cost cycle cover is usually found by minimum cost bipartite perfect matching algorithm which can be done in $O(n^3)$. [5] The remaining operations run in $O(n^2)$ time.

3   Improvement to the Main Algorithm

Unfortunately, Theorem 2 does not work for $\gamma \in [\frac{1}{\sqrt{3}}, 1)$. Now, we give a modification of our basic algorithm, which has an approximation guarantee that *asymptotically approaches* $\frac{\gamma}{1-\gamma}$, for $\forall \gamma \in [\frac{1}{2}, 1)$.

We first observe that if the graph satisfies the γ - triangle inequality, then the edges that "short-circuit" *long paths* will be of much less cost than the path itself. We capture this idea in the following lemma.

Lemma 4. *Let* $P = \{u_0, u_1, \cdots, u_k\}$ *be a path of length* k. *Then* $c(u_0, u_k) \leq \gamma^{\lfloor \log(k) \rfloor} c(P)$.

Proof : First we show the result for $k = 2^i$, for some $i \geq 1$. When $i = 1$, this is obvious from the definition of the γ - triangle inequality. As the induction hypothesis, assume that the claim is true for all paths P' of length $k = 2^{i-1}$, $i > 1$. A path $P = \{u_0, u_1, \cdots, u_k\}$ where $k = 2^i$ can be seen as the concatenation of two subpaths $P_1 = \{u_0, u_1 \cdots, u_{\frac{k}{2}}\}$ and $P_2 = \{u_{\frac{k}{2}}, \cdots, u_k\}$ both having length 2^{i-1}. Therefore, $c(u_0, u_k) \leq \gamma(c(u_0, u_{\frac{k}{2}}) + c(u_{\frac{k}{2}}, u_k)) \leq \gamma(\gamma^{i-1}c(P_1) + \gamma^{i-1}c(P_2)) = \gamma^i c(P)$. To see the result when k is not a power of 2, consider the vertex u_j where $j = 2^{\lfloor \log(k) \rfloor} - 1$. We first short-circuit the subpath of P from u_j to u_k using the edge (u_j, u_k) to obtain a path $P' = \{u_0, u_1, \cdots, u_j, u_k\}$. Obviously, $c(P') \leq c(P)$ and P' has length $2^{\lfloor \log(k) \rfloor}$. It follows that $c(u_0, u_k) \leq \gamma^{\lfloor \log(k) \rfloor} c(P') \leq \gamma^{\lfloor \log(k) \rfloor} c(P)$.   □

The modified version of our algorithm makes use of a cycle of length $\leq \sqrt{n}$ in the minimum cost cycle cover. If such a cycle is not present, we obtain one, by splitting one of the cycles in such a way that the cost of the new cycle cover is not much more than the old one. The following subroutine achieves this.

ALGORITHM Modify_CycleCover
Input : A cycle cover $\mathcal{C} = \{C_1, C_2 \cdots C_k\}$; $k \geq 2$ such that $|C_i| > \sqrt{n}$ $\forall i$.
Output : A cycle cover $\mathcal{C}' = \{C'_1, \cdots C'_{k+1}\}$ such that $|C'_{k+1}| = 2$.

1. Let $C_r = \{u_1, u_2 \cdots u_l\}$ be a cycle in $\mathcal{C}$. Without loss of generality, assume that (u_1, u_2) is the minimum cost edge in C_r.
2. Now we split C_r into a 2-length cycle $C'_{k+1} = \{u_1, u_2\}$ and $C'_r = \{u_3, \cdots u_l\}$. (Note that C'_r is obtained by closing the path from u_3 to u_l by the edge (u_l, u_3)).
3. $C'_j = C_j$ for $j \neq r, 1 \leq j \leq k$.
4. Output $\mathcal{C}' = \{C'_1, \cdots C'_{k+1}\}$.

Lemma 5. $c(\mathcal{C}') \leq [1 + \frac{1}{\sqrt{n}} + \gamma^{\lfloor \log(\sqrt{n}-1) \rfloor}] c(\mathcal{C})$.

Proof : Since (u_1, u_2) is the minimum cost edge in C_r, $c(u_1, u_2) < \frac{1}{\sqrt{n}} c(\mathcal{C})$ since $|C_r| > \sqrt{n}$. We observe that (u_2, u_1) short-circuits the path $P = \{u_2, u_3, \cdots, u_l, u_1\}$ in C_r. Since $|C_r| > \sqrt{n}$, $|P| \geq \sqrt{n} - 1$. Hence by Lemma 4, $c(u_2, u_1) \leq \gamma^{\lfloor \log(\sqrt{n}-1) \rfloor} c(P) \leq \gamma^{\lfloor \log(\sqrt{n}-1) \rfloor} c(C_r)$. Thus,

$C'_{k+1} \leq (\frac{1}{\sqrt{n}} + \gamma^{\lfloor \log(\sqrt{n}-1) \rfloor})c(C_r)$. Also, $c(C'_r) \leq c(C_r)$ since C'_r is obtained just by short-circuiting C_r from u_l to u_3. Therefore, $c(C') \leq c(C) + c(C'_{k+1})$ and the lemma follows.   □

Now we present the algorithm. We assume that the minimum cost cycle cover is not already a hamiltonian cycle.

ALGORITHM Modified γ-ATSP

1. If there is a cycle C_i such that $|C_i| \leq \sqrt{n}$ in the minimum cost cycle cover C, let $C' = C$ else $C' = $ **Modify_CycleCover**(C).
2. Let $C' = \{C_1, \cdots, C_{last}\}$ and let C_{last} be the smallest length cycle in C'. Let $V' = V - \{u : u \in C_{last}\}$.
3. For each $t_i \in V'$ do the following:
   (a) Arrange the cycles of C' such that the cycle containing t_i is the first cycle and C_{last} is the last cycle.
   (b) $H_i = $ **Ham_Path**(C', t_i). Let s_i be the start vertex of H_i and remember that t_i will be the end vertex. Also note that $s_i \in C_{last}$.
   (c) $T_i = H_i \bigcup \{(t_i, s_i)\}$.
   (d) $\tau = \tau \bigcup \{T_i\}$.
4. Output the minimum cost tour in τ.

Lemma 6. *There exists a $t_i \in V'$ such that $c(t_i, s_i) \leq \gamma^{\lfloor \log(\sqrt{n}-1) \rfloor} OPT$, where (t_i, s_i) is the edge added to H_i in step 3(c) to make it a tour T_i.*

Proof : Consider an optimum hamiltonian tour T_o. Since $|C_{last}| \leq \sqrt{n}$, there exists two nodes $a, b \in C_{last}$ such that the length of the directed path $P = \{a, u_1, u_2, \cdots, b\}$ in T_o is at least $\sqrt{n}$ and $P \bigcap C_{last} = \{a, b\}$. Take $t_i = u_1$. Obviously, $t_i \in V'$. Consider the edge (t_i, s_i) added by step 3(c) of the algorithm. Since $s_i \in C_{last}$ the length of the directed path P' from t_i to s_i along the directed cycle T_o has length at least $\sqrt{n} - 1$. Otherwise, s_i has to be a vertex of P other than a and b which is not possible. Now by Lemma 4, $c(t_i, s_i) \leq \gamma^{\lfloor \log(\sqrt{n}-1) \rfloor} c(P') \leq \gamma^{\lfloor \log(\sqrt{n}-1) \rfloor} OPT$.   □

Theorem 3. *The tour output by the above algorithm achieves an approximation guarantee which asymptotically approaches $\frac{\gamma}{1-\gamma}$.*

Proof : Consider the tour T_i, corresponding to the vertex t_i indicated by Lemma 6. Let H_i be the hamiltonian path from which T_i was constructed. Now, $c(T_i) = c(H_i) + c(t_i, s_i)$. As $n \to \infty$, $c(T_i) \to c(H_i)$, since the second term becomes negligible from Lemma 6. $c(H_i) \leq \frac{\gamma}{1-\gamma} c(C')$ due to Lemma 3 where C' is the cycle cover input to the routine **Ham_Path**. But $c(C') \to c(C) \leq OPT$ as $n \to \infty$ since $c(C') \leq [1 + \frac{1}{\sqrt{n}} + \gamma^{\lfloor \log(\sqrt{n}-1) \rfloor}]c(C)$ by Lemma 5. Hence the result follows.   □

4 c_{max} vs. c_{min}

Let c_{max} be the cost of a maximum cost edge in the graph and c_{min} be the cost of a minimum cost edge. A study of the $\frac{c_{max}}{c_{min}}$ ratio can be motivated as follows : Consider an arbitrary ATSP tour in the graph. The cost can be at most nc_{max} , while the cost of the optimal tour is at least nc_{min}. Therefore, $\frac{c_{max}}{c_{min}}$ gives an upperbound for the ratio between the cost of an arbitrary and the optimal tour. In this section, we show that for $\gamma \in [\frac{1}{2}, \frac{1}{\sqrt{3}})$, $\frac{c_{max}}{c_{min}} \le \frac{2\gamma^3}{1-3\gamma^2}$. We made use of this result in section 2 to give a better analysis of our main algorithm, when $\gamma \in [\frac{1}{2}, \frac{1}{\sqrt{3}})$. We wish to mention that a weaker upperbound for $\frac{c_{max}}{c_{min}}$ for a narrower range of γ can be obtained by a careful, exhaustive case analysis. In comparison, the result we get in this section is *tight* both in value and range.

Lemma 7. *Let $u, v, w \in V$. Let (u, v) be the edge of minimum cost i.e $c(u, v) = \bar{c}_{min}$ in the induced subgraph on $\{u, v, w\}$. And let the maximum cost in this induced subgraph on $\{u, v, w\}$ be $\bar{c}_{max} = r\bar{c}_{min}$. Then*

$$c(u, w) = x \le \frac{r\gamma^2 + \gamma}{1-\gamma^2}\bar{c}_{min}, \quad c(v, w) = y \le \frac{r\gamma + \gamma^2}{1-\gamma^2}\bar{c}_{min}$$

and

$$c(w, v) = y' \le \frac{r\gamma^2 + \gamma}{1-\gamma^2}\bar{c}_{min}, \quad c(w, u) = x' \le \frac{r\gamma + \gamma^2}{1-\gamma^2}\bar{c}_{min}$$

Proof : Assume without loss of generality that $\bar{c}_{min} = 1$ (we can scale the costs such that $\bar{c}_{min} = 1$ and the ratios are not affected).

We define two sequences $\{x_i\} = x_0, x_1 \cdots$ and $\{y_i\} = y_0, y_1 \cdots$ as follows:

$$x_0 = r$$
$$x_i = \gamma + \gamma^2 r + \gamma^2 x_{i-1}$$
$$y_i = \gamma(r + x_i)$$

In the following, the reader can easily verify that if one substitutes y' with x and x' with y, the arguments will still hold good. So the bound we derive for x and y will be valid for y' and x' respectively also.

Now we show that $\{x_i\}$ and $\{y_i\}$ define sequences of upperbounds for x and y respectively.

We first note that if x_i is an upperbound for x, y_i has to be an upperbound for y, since $y \le \gamma(r + x) \le \gamma(r + x_i) = y_i$. Therefore we only need to prove that each term in the sequence $\{x_i\}$ is an upperbound for x. Obviously, $x_0 = r$ is an upperbound. Now assume that x_{i-1} is an upperbound where $i \ge 1$. Then y_{i-1} is an upperbound for y. Now

$$x \le \gamma(1 + y) \le \gamma(1 + y_{i-1}) = \gamma(1 + \gamma(r + x_{i-1})) = (\gamma + \gamma^2 r) + \gamma^2 x_{i-1} = x_i$$

Thus x_i is also an upperbound for x.

Now we prove that $\{x_i\}$ is a convergent sequence by showing that it is *bounded* and *monotone*.

Suppose that $r > \frac{\gamma}{1-2\gamma^2}$. Rearranging

$$r > \frac{\gamma^2 r + \gamma}{1-\gamma^2}$$

Thus we have $x_0 = r > \frac{\gamma^2 r + \gamma}{1-\gamma^2}$. Now we use induction on i to show that $\{x_i\}$ is bounded below. Suppose

$$x_{i-1} > \tfrac{\gamma^2 r + \gamma}{1 - \gamma^2} \text{ with } i \geq 1$$

But,

$$x_i = \gamma + \gamma^2 r + \gamma^2 x_{i-1} > \gamma + \gamma^2 r + \gamma^2 \left(\tfrac{\gamma^2 r + \gamma}{1 - \gamma^2}\right) = \tfrac{\gamma^2 r + \gamma}{1 - \gamma^2}$$

Thus we have

$$x_i > \tfrac{\gamma^2 r + \gamma}{1 - \gamma^2} \quad \forall i$$

Also we note that $\{x_i\}$ will be a strictly decreasing sequence, since

$$x_{i-1} > \tfrac{\gamma^2 r + \gamma}{1 - \gamma^2}$$

i.e

$$(1 - \gamma^2)x_{i-1} > \gamma^2 r + \gamma$$
$$x_{i-1} > \gamma + \gamma^2 r + \gamma^2 x_{i-1} = x_i$$

Similarly by replacing $\leq$ for $>$ in each inequality above, we find that if $r \leq \tfrac{\gamma}{1 - 2\gamma^2}$, then $\forall i, \quad x_i \leq \tfrac{\gamma^2 r + \gamma}{1 - \gamma^2}$ and $\{x_i\}$ is a nondecreasing sequence.

Therefore in both cases we have proved that $\{x_i\}$ is bounded and monotone and hence is a convergent sequence. Now as $i \to \infty$, $x_i = x_{i-1}$. Substituting this in $x_i = \gamma + \gamma^2 r + \gamma^2 x_{i-1}$,

$$\lim_{i \to \infty} x_i = \tfrac{\gamma^2 r + \gamma}{1 - \gamma^2}$$

Since, $\forall i, x \leq x_i$, we have $c(u, w) = x \leq \tfrac{\gamma^2 r + \gamma}{1 - \gamma^2}$ [1]
As $y_i = \gamma(r + x_i)$ and $\{x_i\}$ is a convergent sequence, $\{y_i\}$ is also convergent. Also, $\lim_{i \to \infty} y_i = \tfrac{\gamma^2 + \gamma r}{1 - \gamma^2}$. Therefore, $c(v, w) = y \leq \tfrac{\gamma^2 + \gamma r}{1 - \gamma^2}$.
As mentioned earlier we can also prove that

$$c(w, u) = x' \leq \tfrac{\gamma^2 + \gamma r}{1 - \gamma^2}, \quad c(w, v) = y' \leq \tfrac{\gamma^2 r + \gamma}{1 - \gamma^2}$$

□

Corollary   If $r > \tfrac{\gamma^2}{1 - \gamma - \gamma^2}$, then $x, y, x', y' < r$ and (v, u) is the unique maximum cost edge in the induced subgraph on $\{u, v, w\}$, where x, y, x', y', r are as defined in lemma 7.
Proof :   Now, $r > \tfrac{\gamma^2}{1 - \gamma - \gamma^2}$ i.e $(1 - \gamma^2)r - \gamma r > \gamma^2$ which implies that $r > \tfrac{\gamma^2 + \gamma r}{1 - \gamma^2} \geq y$ from the lemma 7. Hence $y < r$. Also, $r > 1$ implies that $\tfrac{\gamma^2 + \gamma r}{1 - \gamma^2} \geq \tfrac{\gamma^2 r + \gamma}{1 - \gamma^2} \geq x$. So, $x < r$. Similarly, we can prove that $x' < r$ and $y' < r$. Since every edge other than (v, u) has cost strictly less than r, it must be the case that (v, u) is the edge(unique) of cost r.   □

The upper bounds obtained in the above lemma help us to make an interesting observation about the maximum cost edge in a γ-asymmetric graph.

---

[1] To put it more rigorously, if $x = \tfrac{\gamma^2 r + \gamma}{1 - \gamma^2} + \epsilon$ for $\epsilon > 0$, then there exists an i such that $x_i < \tfrac{\gamma^2 r + \gamma}{1 - \gamma^2} + \epsilon$ by definition of convergence and this contradicts the fact that x_i is an upperbound for x.

Lemma 8. *Let c_{max} and c_{min} be the maximum cost and minimum cost edges in G and let $\frac{c_{max}}{c_{min}} = r > \frac{\gamma^2}{1-\gamma-\gamma^2}$; $\gamma \in (\frac{1}{2}, \frac{\sqrt{5}-1}{2})$. Then there will be a unique minimum cost edge (u, v) in the graph. Moreover, (v, u) will be the unique maximum cost edge.*

Proof :   Consider a minimum cost edge (u, v) in the graph. Let the cost of this edge be 1 without loss of generality (scale all the costs accordingly, otherwise). First we partition E into the two sets say $\overline{E}, F = E \backslash \overline{E}$. We define $\overline{E} = \{(x, y) \in E : x, y \in V \backslash \{u, v\}\}$ i.e $\overline{E}$ is the edge-set of the induced subgraph on $V \backslash \{u, v\}$. Now F, the remaining edges, forms a *flower like* structure with edges (u, v) and (v, u) at the centre. Consider a node $w \in V \backslash \{u, v\}$. The induced subgraph on $\{u, v, w\}$ forms a *petal* of the *flower*, say P_w.

Consider an edge $(a, b) \in \overline{E}$. Let $c(a, b) = z$. Now $z \le \gamma(c(a, u) + c(u, b))$. Now from Lemma 7, considering the induced graph on $\{u, v, a\}$ we know that $c(a, u) = x' \le \frac{\gamma r + \gamma^2}{1-\gamma^2}$ and similarly considering the induced subgraph on $\{u, v, b\}$, we have $c(u, b) = x \le \frac{\gamma^2 r + \gamma}{1-\gamma^2}$. Hence, $z \le \frac{\gamma^2}{1-\gamma}(1 + r)$. Now, since $r > \frac{\gamma^2}{1-\gamma-\gamma^2}$, we have $r + 1 > \frac{1-\gamma}{1-\gamma-\gamma^2}$ which gives by rearranging $r > \frac{\gamma^2}{1-\gamma}(1 + r) \ge z$. This means that the maximum cost edge of G occurs in F, the *flower*.

Now consider the *petal* P_w of the *flower* which contains the maximum cost edge. Since $r > \frac{\gamma^2}{1-\gamma-\gamma^2}$, we have from the corollary to Lemma 7, that the maximum cost edge must be (v, u) and its cost is r. But this edge is part of every *petal* $P_w, w \in V \backslash \{u, v\}$ and $r_w = \max\{c(e) : e \in P_w\} = r > \frac{\gamma^2}{1-\gamma-\gamma^2}$. Therefore, the corollary to the Lemma 7 is applicable for each *petal* P_w and hence (v, u) is the unique maximum cost edge in the graph. This also implies that (u, v) is the unique minimum cost edge also - otherwise we could have started with the other minimum cost edge and its reverse edge will turn out to be a maximum cost edge as well, which is contradictory.   □

Theorem 4. *Let G be a complete directed graph with costs on the edges satisfying γ-parametrized triangle inequality where $\gamma \in [\frac{1}{2}, \frac{1}{\sqrt{3}})$. Then $\frac{c_{max}}{c_{min}} = r \le \frac{2\gamma^3}{1-3\gamma^2}$.*

Proof :   Since $1 \le 2\gamma$ which implies $1 - \gamma - \gamma^2 \ge 1 - 3\gamma^2$, for $\gamma \in [\frac{1}{2}, \frac{1}{\sqrt{3}})$, we have $\frac{\gamma^2}{1-\gamma-\gamma^2} \le \frac{2\gamma^3}{1-3\gamma^2}$. So if $r \le \frac{\gamma^2}{1-\gamma-\gamma^2}$, we are done. Otherwise, let (v, u) be the unique maximum cost edge in the graph. (see Lemma 8). $c(v, u) \le \gamma(c(v, w) + c(w, u))$ for some $w \in V \backslash \{u, v\}$. By Lemma 7,

$$r \le \gamma(y + x') \le \frac{2\gamma(r\gamma + \gamma^2)}{1-\gamma^2}$$

rearranging, the lemma follows.   □

We illustrate the tightness of theorem 4. More precisely,

Theorem 5. *For $\gamma \in [\frac{1}{2}, \frac{1}{\sqrt{3}})$, there exist n-node γ-asymmetric graphs, for each n, such that $\frac{c_{max}}{c_{min}} = \frac{2\gamma^3}{1-3\gamma^2}$ and for $\gamma \in [\frac{1}{\sqrt{3}}, 1)$, $\frac{c_{max}}{c_{min}}$ can be arbitrarily large.*

Proof : The proof is omitted for want of space. □

References

1. T. Andreae, J. Bandelt, *Performance guarantees for approximation algorithms depending on parametrized triangle inequalities.* SIAM J. Discrete Math 8, 1995, pp1-16.
2. M.A. Bender and C. Chekuri, *Performance guarantees for the TSP with a parametrized triangle inequality.* Proc. WADS'99, LNCS 1663, Springer 1999, pp80-85.
3. J. Bockenhauer, J. Hromkovic, R. Klasing, S. Seibert and W. Unger, *An improved lower bound on the approximability of metric TSP and approximation algorithms for the TSP with sharpened triangle inequality.* Proc. STACS'00, LNCS, Springer, Berlin, 2000.
4. N. Christofides, *Worst-case analysis of a new heuristic for the travelling salesman problem.* Tech. Rep, GSIA, Carnegie Mellon University, 1976.
5. W. Cook et al., *Combinatorial Optimization.* John Wiley & Sons, Inc., 1998.
6. J. Edmonds, L. Johnson, *Matching: A well-solved class of integer linear programs.* Proc. Calgary International conference on combinatorial structures and their applications, Gordon and Breach, 1970, pp89-92.
7. A. Frieze, G. Galbiati and F. Maffioli, *On the worst-case performance of some algorithms for the asymmetric travelling salesman problem.* Networks 12, 23-39, 1982.
8. C. Papadimitriou and M. Yannakakis, *The traveling salesman problem with distances one and two.* Math. of Operations Research, Vol. 18, 1-11, 1993.
9. Sundar Vishwanathan, *An approximation algorithm for the asymmetric travelling salesman problem with distances one and two.* Information Processing Letters, 44(6), 297-302, 1992.
10. S. Vempala and R. Carr, *Towards a 4/3 approximation for the asymmetric TSP.* Proc. SODA'00.

A New Diagram from Disks in the Plane[*]

Joachim Giesen and Matthias John

Institute for Theoretical Computer Science, ETH Zürich, CH-8092 Zürich
{giesen,john}@inf.ethz.ch

Abstract. Voronoi diagrams and (weighted) Delaunay triangulations are well known diagrams associated with a set of disks in the plane. Here we introduce a new diagram associated with a set of disks and an efficient algorithm to compute it. The new diagram is closely related to the (weighted) Delaunay diagram.

Keywords. Computational geometry, computational topology

1 Introduction

Voronoi diagrams, also known as power diagrams, and (weighted) Delaunay triangulations of a set of disks in the plane are well studied objects [1]. They found applications in many areas of science and engineering.

Recently we introduced a new diagram associated with a set of disks in the plane [5]. This diagram is closely related to the Voronoi diagram of the disks. The way we constructed the new diagram suggests that there is another new diagram closely related to the Delaunay triangulation. The characterization and computation of the latter diagram is the topic of this paper.

The construction of the diagram in [5] goes as follows: The disks induce a dynamical system on the plane. We are going to study some of its mathematical properties in this paper. The fixpoints of the dynamical system are either minima, saddles or maxima. We are going to study stable- and unstable manifolds for the fixpoints. These are basically the sets of points that either flow into a fixpoint or that are contained in the flow of some point in a small neighborhood of a fixpoint. We are going to provide exact definitions and examples in the subsequent sections.

It turns out that for the dynamical system induced by the disks the minima are a subset of the disk centers and the maxima are a subset of the Voronoi vertices. In our recent paper [5] we characterized a smoothed version of the unstable manifolds of the minima. We call this smoothed version a Min region. The Min regions define a decomposition of the plane, called Min diagram, which is related to the Voronoi diagram. Though these diagrams are related there exist differences. In Voronoi diagrams there is a cell for every disk (except in degenerate

---

[*] Partly supported by the IST Programme of the EU as a Shared-cost RTD (FET Open) Project under Contract No IST-2000-26473 (ECG - Effective Computational Geometry for Curves and Surfaces).

© Springer-Verlag Berlin Heidelberg 2002

configurations), but in the new diagram there are only regions for certain disks, namely minima. A Voronoi region need not contain its corresponding disk, but the region assigned to a minimum always contains this minimum.

In this paper we want to characterize and compute smoothed versions of the stable manifolds of maxima. It turns out that this is more involved than computing the unstable manifolds of minima. Degeneracies can occur due to the fact that the dynamical system under investigation is not time reversible. We are going to deal with these degeneracies explicitly.

We call the new diagram Max diagram since it is composed out of smoothed stable manifolds of maxima which we call Max regions. The Max diagram is directly related to the definition of pockets in macromolecules, see [2], (here of course only their counterparts in the plane). It seems that pockets can be computed approximately via Min regions of a dual set of disks. That is of course a more indirect approach than the approach via Max regions. Both approaches coincide only under special conditions which we have discussed in [6]. Because of its close relationship to the Delaunay triangulation we are convinced that Max diagrams will find also other applications.

In the Max diagram cells are only assigned to some Voronoi vertices, namely maxima. This is in contrast to Delaunay triangulations which assign a cell to every Voronoi vertex. Another difference is that a Delaunay cell need not contain its dual Voronoi vertex, but the region assigned to a maximum always contains this maximum.

Please note that we had to omit many proofs due to the lack of space. We leave them for the full version of this paper.

2   Disks, Diagrams, and Critical Points

Disk. A disk is a pair $(z, r) \in \mathbb{R}^3$, where $z \in \mathbb{R}^2$ is the center of the disk and $\sqrt{r} \in \mathbb{C}$ its radius. Note that we also allow purely imaginary radii in which case the geometric intuition about disks is not helpful. We refer to r also as the power of the disk.

In the following we are going to consider finite sets of disks. Very often these disks are not in general position. That is, we do not assume general position unless stated differently.

Power distance. The power distance of a point $x \in \mathbb{R}^2$ from a disk (z, r) is $\pi(x) = \|x - z\|^2 - r$.

Power bisector. Given two disks in the plane. The power bisector of these two disks is the set of all points that have the same power distance to both disks. The power bisector is a straight line which is orthogonal to the line segment connecting the two disk centers.

Voronoi diagram. Given a set B of disks. The *Voronoi cell* of a disk b_i under the power distance is given as

$$V_i = \{x \in \mathbb{R}^2 \, : \, \forall b_j \in B, \, \pi_i(x) \leq \pi_j(x)\}.$$

The sets V_i are convex polygons or empty since the set of points that have the same power distance from two disks forms a straight line. Closed line segments shared by two Voronoi cells are called *Voronoi edges* and the points shared by three or more Voronoi cells are called *Voronoi vertices*. The term *Voronoi object* can denote either a Voronoi cell, edge or vertex. The *Voronoi diagram* of B is the collection of Voronoi cells, edges and vertices. It defines a cell decomposition of $\mathbb{R}^2$.

Delaunay diagram. The *Delaunay diagram* of a set of disks B is dual to the Voronoi diagram of B. The convex hull of three or more center points defines a *Delaunay cell* if the intersection of the corresponding Voronoi cells is not empty and there exists no superset of center points with the same property. Analogously, the convex hull of two center points defines a *Delaunay edge* if the intersection of their corresponding Voronoi cells is not empty. Every center point is called *Delaunay vertex*. The term *Delaunay object* can denote either a Delaunay cell, edge or vertex. The Delaunay diagram defines a decomposition of the convex hull of all center points. This decomposition is a triangulation if the disks are in general position.

We always refer to the interior and to the boundary of Voronoi-/Delaunay objects with respect to their dimension, e.g. the interior of a Delaunay edge contains all points in this edge besides the endpoints and the interior of a vertex and its boundary are the vertex itself.

Power height function. Let B be a set of disks in $\mathbb{R}^2$. The power height function is given as

$$h(x) = \min\{\pi_i(x) \, : \, b_i \in B\}. \tag{1}$$

Observe that the function h is continuous. It is smooth everywhere besides at points which have the same power distance from two or more disks, i.e. at points that lie on the boundary of Voronoi cells.

Regular and critical points. Let B be a set of disks. The critical points of the power height function h are the intersection points of Voronoi objects V and their dual Delaunay object σ. The index of a critical point is the dimension of σ. All points which are not critical are called regular.

Observe that in degenerate situations critical points can collapse, e.g. a minimum could also be a saddle. These situations have to be taken care of, but in the following we want to assume that such degenerate situations do not occur.

3   Disk Induced Flow

Disk induced flow. Given a set B of disks, the induced flow ϕ is given as follows: Since the Voronoi diagram of B is a decomposition of the plane any point $x \in \mathbb{R}^2$ lies in some Voronoi object. Let V be the lowest dimensional Voronoi object that contains x. Assume that x is the intersection point of V and its dual Delaunay object. In this case we set:

$$\phi(t, x) = x \, , \, t \in [0, \infty)$$

Otherwise let σ be the dual Delaunay object of V and $y = \mathrm{argmin}_{y' \in \sigma} \|x - y'\|$. Since σ is convex there is only one such y which we call a driver of the flow. Let R be the ray originating at x and shooting in the direction $x - y$. Let z be the first point on R for which $\mathrm{argmin}_{y' \in \tau} \|z - y'\|$ is different from y where τ denotes the dual Delaunay object of the lowest dimensional Voronoi object z lies in. Note that such a z need not exist in $\mathbb{R}^2$. In this case let z be the point at infinity. We set:

$$\phi(t, x) = x + t \frac{x - y}{\|x - y\|} \, , \, t \in [0, \|z - x\|]$$

For $t > \|z - x\|$ the flow is given as follows:

$$\phi(t, x) = \phi \left(t - \|z - x\| + \|z - x\|, \, x \right)$$
$$= \phi \left(t - \|z - x\|, \, \phi \left(\|z - x\|, \, x \right) \right)$$

It is not completely obvious but ϕ can be shown to be well defined on the whole of $[0, \infty) \times \mathbb{R}^2$. Furthermore, the following two properties of dynamical systems hold for disk induced flows,

(1) $\phi(0, x) = x$.
(2) $\phi(t, \phi(s, x)) = \phi(t + s, x)$.

The following three observations are helpful to get a better understanding of disk induced flows and their relationship to the power height function of the same set of disks.

(1) The fixpoints of ϕ are the critical points of the power height function.
(2) The orbits of ϕ are piecewise linear curves that are linear in Voronoi objects.
(3) The flow ϕ has no closed orbits.

Because of the first observation we want to refer to fixpoints of ϕ as minimum, saddle or maximum if the corresponding critical point of the power height function is a minimum, saddle or maximum, respectively.

Stable and Unstable Manifolds. Given a disk induced flow ϕ in the plane. The stable manifold $S(x)$ of a fixpoint $x \in M$ contains all points that flow into x, i.e.

$$S(x) = \{y \in \mathbb{R}^2 \, : \, \lim_{t \to \infty} \phi_y(t) = x\}.$$

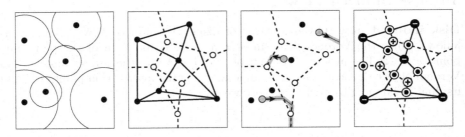

Fig. 1. A set of disk in the plane (on the left), its Voronoi- and Delaunay diagram (second from the left), some orbits of the flow induced by the disks (third form the left) and the minima $\ominus$, saddles $\odot$ and maxima $\oplus$ of this flow (on the right).

The unstable manifold $U(x)$ of a fixpoint x is a little bit more involved to define. Given a neighborhood U of x and let $V(U)$ be the set of all points which lie in an orbit that starts in U, i.e.

$$V(U) \; = \; \{y \in \mathbb{R}^2 \, : \, \exists z \in U, \, t \in [0, \infty) \text{ s.t. } \phi_z(t) = y \text{ or } \lim_{t \to \infty} \phi_z(t) = y\}.$$

Then $U(x)$ is given as the intersection of all such sets $V(U)$.

Lemma 1. *Let M be the set of minima of ϕ. The set $\mathcal{M} \; = \; \{U(m) \, : \, m \in M\}$ covers the whole $\mathbb{R}^2$, i.e. every $x \in \mathbb{R}^2$ is contained in at least one unstable manifold $U(m)$ of some minimum m of ϕ.* ◻

Lemma 2. *Let K be the set of all points that do not flow to infinity, i.e.*

$$K = \mathbb{R}^2 - \{x \in \mathbb{R}^2 \, : \, \forall n \in \mathbb{N} \exists t_n \geq 0 \text{ s.t. } \forall t > t_n \, \|\phi_x(t) - 0\| > n\}.$$

Let S bet the set of all saddles, M be the set of all maxima and M' be the set of all minima of ϕ. The set $\mathcal{C} \; = \; \{S(y) \, : \, y \in S \cup M\}$ covers $K - M'$, i.e. every $x \in K$ that is not a minimum is contained in exactly one stable manifold $S(y)$ of some saddle or maximum y of ϕ. ◻

The central theme of our discussion will be the stable manifolds of saddles. In contrast to the unstable manifolds the stable manifolds of saddles need not be one-dimensional curves.

Lemma 3. *Let s be a saddle of ϕ. If the stable manifold $S(s)$ of s does not contain a Voronoi vertex then* closure$(S(s))$ *is a piecewise linear curve.*

Proof. Assume that s is a saddle such that $S(s)$ does not contain a Voronoi vertex. According to the definition of critical points we have that s is the intersection point of a Delaunay edge E and its dual Voronoi edge E'. First we show that in a sufficiently small neighborhood U of s we have $S(s) \cap U \; = \; E \cap U$.

From the construction of ϕ we have that $E \cap U \subset S(s) \cap U$. It remains to show that $S(s) \cap U \subset E \cap U$.

Let $x \in U - E$. We have to show that $x \notin S(s)$. We distinguish two cases either $x \in E'$ or $x \notin E'$. In the first case x belongs to the unstable manifold $U(s)$ of s, provided U is sufficiently small. But the intersection $S(s) \cap U(s)$ contains only the point s, because otherwise one can construct an orbit that comes arbitrarily close to its starting point. This is not possible for the same reasons as there are no closed orbits. Thus $x \notin S(s)$.

In the second case the flow of x is driven by an endpoint of E. By the construction of ϕ we find that x flows through some point $x' \in E'$. As in the first case x' does not belong to the stable manifold of s. Thus $x \notin S(s)$.

Hence $S(s) \cap U \subset E \cap U$ and $S(s)$ is a piecewise linear curve in a small neighborhood of s.

Next we show that for all $x, y \in S(s)$ one of the following conditions holds:

(a) $\exists t \in [0, \infty)$ s.t. $\phi_x(t) = y$
(b) $\exists t \in [0, \infty)$ s.t. $\phi_y(t) = x$
(c) $\phi_x([0, \infty)) \cap \phi_y([0, \infty)) = \{s\}$

Observe that the first two conditions cannot hold together since ϕ has no closed orbits. Now assume that there exists $x, y \in S(s)$ such that none of the three conditions holds. Since both x and y belong to the stable manifold of s, the orbits ϕ_x and ϕ_y have to meet at some point $z \in S(s)$. The point z has to be either an interior point of a Voronoi edge or a Voronoi vertex. If z is an interior point of a Voronoi edge then z has to flow in one of the two Voronoi vertices incident to the Voronoi edge by the construction of ϕ. Hence in either case $S(s)$ has to contain a Voronoi vertex. That contradicts our assumption that $S(s)$ does not contain any Voronoi vertex.

Finally we show that the closure of $S(s)$ is a piecewise linear curve. Given $x \in S(s)$ such that $x \neq s$. We consider the following set $S(x) = \{y \in \mathbb{R}^2 : x \in \phi_y([0, \infty))\}$. This set has to be a semi open curve, because for every two points in $S(s)$ either condition (1) or (2) must hold. This curve is piecewise linear, because it is the union of orbits such that for every two such orbits one of them is completely contained in the other. Since all points in $S(x)$ have to flow into s via x none of them can be a minimum or saddle. We know from Lemma 1 that every point $x \in \mathbb{R}^2$ belongs to the unstable manifold of some minimum but it might also belong to the unstable manifold of a saddle. That is the one point m in the closure of $S(x)$ has to be a minimum or another saddle. Both x and s belong to the unstable manifold of m. Thus s is connected to m by a piecewise linear curve and every point on this curve besides m belongs to the stable manifold of s. Now let $x' \in S(s)$ be such that $x' \notin \phi_x([0, \infty))$ and repeat above construction for x'. Let m' be the minimum or saddle that we get from this construction. We cannot repeat this construction a third time with some point $x'' \notin \phi_x([0, \infty)) \cup \phi_{x'}([0, \infty))$, because we already know that $S(s)$ is a piecewise linear curve in a small neighborhood of s. This shows that the closure of $S(s)$ is the union of the two piecewise linear arcs that connect s with m and m'. Thus the closure of $S(s)$ is a piecewise linear curve. $\qquad \square$

Lemma 4. *Let v be a Voronoi vertex that is not contained in its dual Delaunay cell, i.e. which is not a maximum. Let U be a sufficiently small neighborhood of v and $S(v) = \{x \in U : v \in \phi_x([0,\infty)) \cap U\}$. The set $S(v)$ is an angular domain or a line segment that is completely contained in a closed halfspace which contains v in its boundary.*

The boundary segments of $S(v)$ are contained in segments that connect v with some Delaunay vertices from the boundary of the Delaunay cell dual to v.

Proof. From Lemma 1 we can conclude that $S(v)$ is not empty. Let σ be the dual Delaunay cell of v. Since v is not contained in σ there exists a line through v such that the smallest angular domain A centered at v that contains σ lies on one side of this line and touches the line only at v. Let $x \in U \cap (\mathbb{R}^2 - A)$. The flow of x is driven by points on the boundary of $\sigma \subset A$. Thus $\phi_x([0,\infty)) \cap U$ cannot contain any point from A. Furthermore the flow of x has to leave U in finite time since ϕ has no fixpoints in U for sufficiently small U. Hence $S(v)$ is contained in A. This implies altogether that there has to exist a smallest angular domain $A' \subset A$ centered at v that contains $S(v)$. We want to show that $S(v) = A' \cap U$. To follow the proof it might be helpful to look at Figure 2.

First we show that all points in the intersection of U with the two boundary rays of A' have to flow into v. Let R be a boundary ray of A'. By construction there exists a point $x \in R \cap U$ that flows into v. The flow of x is either controlled by a Delaunay edge or by a Delaunay vertex. In the first case x lies on a Voronoi edge and flows directly on this edge into v, i.e. the ray R contains this Voronoi edge and all points in U that lie on this edge flow into v. In the second case we know that the flow of x has to stay in A'. Thus the Delaunay vertex that drives the flow has to lie in $\mathbb{R}^2 - \mathrm{interior}(A')$ because otherwise the flow would leave A'. More specific this Delaunay vertex has to lie on R since otherwise the points that flow into x would lie outside of A' but also flow into v which is impossible. Thus all points on R in a small neighborhood of v flow into v.

Next we show that the first case in the above reasoning is not possible. That implies that the boundary rays of A' contain line segments that connect v with some Delaunay vertex. Assume that the flow in the intersection $R \cap U$ is controlled by a Delaunay edge. That is, the points in $R \cap U$ flow directly into v on a Voronoi edge E and the Delaunay vertices that correspond to E lie on opposite sides of the power bisector that contains E. Let $x \in R \cap U$. By the definition of ϕ all points in the intersection of a sufficiently small neighborhood of x and the two segments that connect x to the two Delaunay vertices that correspond to E have to flow through x, i.e. they have to flow through v. But that means that R cannot be a boundary ray of A' which is a contradiction.

Finally we show that all points in A' flow into v. By construction σ is contained in A and does not contain v. Thus there exists a line which separates v from σ which has non empty intersection with $A' \subset A$. Let L be such a line and consider the portion of A' that lies below L. For sufficiently small U this portion contains all points in $A' \cap U$. The flow of the points in $A' \cap U$ is driven by points in the boundary of σ. Hence the flow of these points cannot leave $A' \cap U$ by crossing L. Since ϕ has no fixpoints in U we know that all points in $A' \cap U$ have to leave $A' \cap U$ either through the boundary rays of A' or through v. We have already shown that all points in the intersection of U with the boundary rays of A' flow into v. That shows that all points in $A' \cap U$ flow through v before they leave U. That proves our claim.   □

Join. A join is a Voronoi vertex v that lies in the stable manifold of some saddle such that the intersection of every open neighborhood of v with the set of points that flow into v contains an open subset of $\mathbb{R}^2$.

Thinned stable manifold. At every join f of a saddle s of ϕ we modify the stable manifold of s as follows: Observe that the left- and right boundary rays of the angular domain A' that flows into f cannot coincide, see Lemma 4 and the definition of joins. Let R_l be the left- and R_r be the right boundary ray of A'. The thinned stable manifold $T(s)$ of s excludes all points from $S(s)$ that flow into f but whose flow does not contain any point from R_l or R_r different from f, i.e. at every join f we subtract the set $\{x \in S(s) : f \in \phi_x([0, \infty)), \phi_x([0, \infty)) \cap (R_l \cup R_r) = \{f\}\}$ from $S(s)$. Removing these sets from $S(s)$ leaves us with a one-dimensional set $T(s)$ that can be decomposed in piecewise linear arcs, see also Lemma 3.

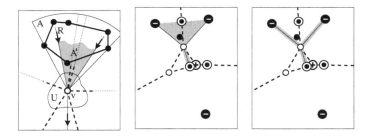

Fig. 2. The neighborhood of a join v and some of the sets that we need in the proof of Lemma 4 (on the left). The stable manifold of a saddle $\odot$ (in the middle) and its corresponding thinned stable manifold (on the right).

Stable decomposition graph. Let G be the following graph: Its vertex set consists of all minima, saddles and joins. Two vertices are connected by an edge if there exists a thinned stable manifold $T(s)$ of some saddle s such that the two vertices are both contained in closure$(T(s))$ and there is no other vertex on $T(s)$ between them. We have to take the closure of the thinned stable manifolds because otherwise the minima would not be contained in them. We refer to G as the stable decomposition graph associated with a disk induced flow. Let S be the set of saddles of ϕ and $T(S) = \bigcup_{s \in S} \text{closure}(T(s))$. We refer to $T(S)$ as the geometric realization of G.

Lemma 5. *The stable decomposition graph of ϕ is planar, i.e. it can be embedded crossing free in the plane.*   □

4   Max Diagram and Algorithm

Max and Join regions. If m is a maximum of a disk induced flow we call the closure of the stable manifold $S(m)$ the Max region of m, i.e. the Max region of m is a smoothed version of $S(m)$.

Let G be the stable decomposition graph associated with a disk induced flow ϕ. Since G was constructed via ϕ we can orient the edges of G in the direction of the flow along an edge. Every join has three incident edges in G. The geometric realization of G decomposes the plane into closed regions. Such a region is called a join region if it has a join f and two ingoing edges of f on its boundary.

Theorem 1. *Let ϕ be a disk induced flow. The Max- and join regions of ϕ are exactly the closures of the bounded regions of the geometric realization of the stable decomposition graph associated with ϕ.*   □

Corollary 1. *The following is true:*

(1) The Max- and join regions of ϕ are polygons.
(2) The stable decomposition graph associated with ϕ is connected.   □

Max diagram. We call the collection of all Max- and join regions the Max diagram associated with a set of disks.

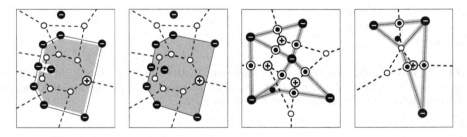

Fig. 3. A stable manifold of a maximum ⊕ (on the left) and its corresponding Max region (second from the left). The Max diagram of the set of disks from Figure 1 (second from the right) and the Max diagram of a degenerate situation (on the right).Note that here the boundaries of Max- and fork regions are emphasized.

The analysis in the previous sections suggests an algorithm to compute the Max diagram associated with with a set of disks in the plane.

MAXDIAGRAM(B)
1    $E, V, F := \emptyset$
2   compute Voronoi- and Delaunay diagram of B.
3   compute the saddles S of ϕ.
4   **for each** $s \in S$ **do**
5        $V := V \cup \{s\}$
6        $vw :=$ Delaunay edge which contains s.
7        $F := F \cup \{sv, sw\}$
8   **while** $F \neq \emptyset$ **do**
9   choose $xy \in F$
10           $F := F - \{xy\}$.
11           $x' :=$ first intersection point of the segment
                     from x to y with a Voronoi edge e that
                     intersects xy in only one point;
                     or y if such a point does not exist.
12           $V := V \cup \{x'\}$; $E := E \cup \{xx'\}$
13   **if** $x' \neq y$ **do**
14   **if** x' lies in the interior of e **do**
15                   $yy' :=$ Delaunay edge dual to e.
16                   $F := F \cup \{x'y'\}$
17   **else**
18                   $(z, z') := $ JOIN(x')
19                   $F := F \cup \{x'z, x'z'\}$
20   **end if**
21   **end if**
22   **end while**
23   **end for**
24   extract the regions of $G = (V, E)$.

The algorithm MAXDIAGRAM takes a set B of disks in the plane as argument and works as follows:

In the first line we initialize two arrays V, E and a stack F with the empty set. The array V is going to store the vertices of the geometric realization of the stable decomposition graph G, E is going to store its edges and F is only needed during the computation to store tuples of vertices of G and their drivers. In the second line we compute the Voronoi and Delaunay diagram of B and in the third line we compute the saddles of the flow ϕ induced by B.

Lines 4 and 23 enclose the loop in which we compute for every saddle s its thinned stable manifold $T(s)$. The saddle is included in the vertex set of G in line 5 and in line 7 the line segments from s to the endpoints v and w of the Delaunay edge that contains s are pushed on the stack F. The points v and w drive the flow of the points on the line segment vw in some neighborhood of s.

In the loop enclosed by lines 8 and 22 we compute the two components of $T(s)$ that flow into s. We compute a new vertex x' of G and insert it into V. The driver of x' is a Delaunay vertex, because one can show that the points on thinned stable manifolds are always driven by Delaunay vertices. The new vertex is either the Delaunay vertex y or it is the first intersection point starting from x of the segment xy with a Voronoi edge. One can construct the situation where xy contains part of a Voronoi edge. The flow does not change along this edge. Thus we demand that x' has to be an intersection point of xy with a Voronoi edge that intersects xy in only one point. The edge that connects x' with x is inserted into E. If $x' = y$ then x' has to be a minimum, because x' is connected to the Delaunay vertex y by a straight line segment whose interior points are all driven by y and that does not cross a Voronoi edge, i.e. y is contained in its own Voronoi cell which in turn means that it is a minimum. If y is a minimum we finished a component of $T(s)$ and no new segments have to be inserted into F.

Otherwise we compute in lines 14 to 20 the new segments and insert tuples of vertices of G and their drivers into F. If x' is not a join, i.e. x' is an interior point of a Voronoi edge, then only one tuple which is computed in line 15 is inserted into F. If x' is a join then possibly two tuples have to be inserted into F. These tuples are computed by calling the subroutine JOIN which returns the two Delaunay vertices that correspond to the boundary segments of the angular domain that flows through x', see Lemma 4. The two tuples given by x' and the two Delaunay vertices returned by JOIN are inserted into F in line 19. Note that not necessarily two tuples are inserted. We have to check explicitly if z and z' coincide to avoid multiple insertions into F.

Finally we extract in line 24 the interiors of the regions of G by removing recursively vertices of degree one and their incident edges from G.

The subroutine JOIN can be implemented as follows:

JOIN(v)
```
1   V := ∅
2   for all Voronoi cells c incident to v in
            counter clockwise order around v do
3       x := Delaunay vertex dual to c.
4       if vx crosses c do
5           V := V ∪ {x}
6       end if
7   end for
8   (y, y') := the tuple of consecutive points in the
            sequence V[1], . . . , V[|V|], V[1] that
            maximizes the smaller angle spanned
            by the segments vy and vy'.
9   return (y, y').
```

The subroutine JOIN takes as its argument a Voronoi vertex v that is contained in the stable manifold of some saddle. It returns a tuple (y, y') of Delaunay vertices incident to the Delaunay cell dual to v such that the segments vy and vy' cross the Voronoi cell dual to y and y', respectively. Among all such tuples (y, y') maximizes the smaller angle spanned by the segments vy and vy'. The segments vy and vy' contain the boundary segments of the set $S(v)$ character-ized in Lemma 4. If $S(v)$ degenerates to a line segment then y and y' coincide. In line 5 we insert the new element always at the end of V. Thus in line 8 we can exploit the ordering induced by the order in which the **for** loop circulates through the Voronoi cells incident to v.

Theorem 2. *The maximum geometric complexities of the geometric realization of the stable decomposition graph of a set of n disks is $\Theta(n^2)$ vertices, $\Theta(n^2)$ edges and $\Theta(n)$ regions.* □

Theorem 3. *The geometric realization of the stable decomposition graph of a set of n disks has a worst case algorithmic complexity of $\Theta(n^2)$.* □

5   Future Work

We are currently working on the extensions of our constructions to higher dimensions and on implementing the algorithm MAXDIAGRAM. We are also investigating the relationship between Max regions and Min regions of a dual set of disks. As we already said in the introduction many of the proofs had to be left for the full version of this paper.

References

1. F. Aurenhammer and R. Klein. Voronoi Diagrams. In *Handbook of Computational Geometry,* J.-R. Sack and J. Urrutia (eds.), pp. 201–290, Elsevier (2000)
2. H. Edelsbrunner, M. A. Facello and J. Liang. On the definition and the construction of pockets in macromolecules. *Discrete Apl. Math.* **88**, pp. 83–102, (1998)
3. H. Edelsbrunner. Triangulations and meshes in computational geometry. *Acta Numerica* **9**, pp. 133–213, (2000)
4. H. Edelsbrunner, J. Harer and A. Zomorodian. Hierarchy of Morse complexes for piecewise linear 2-manifolds. In *Proc. 17th Ann. Sympos. Comput. Geom.*, pp. 70–79, (2001)
5. J. Giesen and M. John. A Dynamical System from Disks in the Plane. Manuscript (2001)
6. J. Giesen and M. John. Duality in Disk Induced Flows. Manuscript (2001)

Computing the Maximum Detour and Spanning Ratio of Planar Paths, Trees, and Cycles*

Stefan Langerman[1], Pat Morin[1], and Michael Soss[2]

[1] School of Computer Science, McGill University
3480 University St., Suite 318
Montréal, Québec, CANADA, H3A 2A7
{sl,morin}@cgm.cs.mcgill.ca
[2] Chemical Computing Group, Inc.
1010 Sherbrooke Street West, Suite 910
Montréal, Québec, CANADA, H3A 2R7
soss@chemcomp.com

Abstract. The maximum detour and spanning ratio of an embedded graph G are values that measure how well G approximates Euclidean space and the complete Euclidean graph, respectively. In this paper we describe $O(n \log n)$ time algorithms for computing the maximum detour and spanning ratio of a planar polygonal path. These algorithms solve open problems posed in at least two previous works [5,10]. We also generalize these algorithms to obtain $O(n \log^2 n)$ time algorithms for computing the maximum detour and spanning ratio of planar trees and cycles.

1 Introduction

Let $G = (V, E)$ be an embedded connected graph with n vertices and m edges. Specifically, the vertex set V consists of points in R^2, and E consists of closed line segments whose endpoints are in V. Let s and t be two points in $\cup E$.[1] We denote by $\|st\|$ the Euclidean distance between s and t and by $\|st\|_G$ the length of the shortest path from s to t in G. The detour between two points $s, t \in \cup E$ is defined as

$$D(G, s, t) = \frac{\|st\|_G}{\|st\|} \ .$$

The *maximum detour* $D(G)$ of G is the maximum detour over all pairs of points in G, i.e.,

$$D(G) = \max\{D(G, s, t) : s, t \in \cup E, s \neq t\} \ .$$

The *spanning ratio* or *stretch factor* $S(G)$ of G is the maximum detour over all pairs of vertices of G, i.e.,

$$S(G) = \max\{D(G, s, t) : s, t \in V, s \neq t\} \ .$$

---

* This research was partly funded by CRM, FCAR, MITACS, and NSERC. This research was done while the third author was affiliated with SOCS, McGill University.
[1] Here and thoughout, $\cup S$ is shorthand for $\bigcup_{e \in S} e$.

H. Alt and A. Ferreira (Eds.): STACS 2002, LNCS 2285, pp. 250–261, 2002.
© Springer-Verlag Berlin Heidelberg 2002

The maximum detour and spanning ratio play important roles in the analysis of online routing algorithms [3,8] and the construction of spanners [6]. In the former case, the goal is to find paths that minimize maximum detour. In the latter, the goal is to construct graphs with few edges that minimize the spanning ratio.

1.1   Related Work

Recently, researchers have become interested in computing the maximum detour and spanning ratio of embedded graphs. The spanning ratio can be computed in $O(n(m + n \log n))$ time by computing the shortest paths between all pairs of vertices and then comparing these to the distances between all pairs of vertices. In R^2, the maximum detour is infinite if G is non planar, so the maximum detour can be computed in $O(n^2 \log n)$ time by computing shortest paths and using this information to find the maximum detour between each pair of edges. Surprisingly, these are the best known results for computing the maximum detour or spanning ratio. Even if the input graph G is a path, no sub-quadratic time algorithms are known, though fast approximation algorithms have been reported.

Narasimhan and Smid [10] study the problem of approximating the spanning ratio in a graph and give $O(n \log n)$ time algorithms that can $(1-\epsilon)$-approximate the spanning ratio when G is a path, a cycle or a tree. More generally, they show that, after $O(n \log n)$ preprocessing, the problem of approximating the spanning ratio can be reduced to $O(n)$ approximate shortest path queries on G. Their results hold even for graphs embedded in R^d. The authors also show that approximating the spanning ratio requires $\Omega(n \log n)$ time in the algebraic decision tree model of computation.

Ebbers-Baumann et al [5] study the problem of approximating the maximum detour of a polygonal path and give an $O(\frac{n}{\epsilon} \log n)$ time algorithm that finds a $(1 - \epsilon)$-approximation to the maximum detour.

1.2   New Results

In this paper we give randomized algorithms with $O(n \log n)$ expected running time that compute the exact spanning ratio or maximum detour of a polygonal path with n vertices. These are the first sub-quadratic time algorithms for finding the exact spanning ratio or maximum detour, and they solve open problems posed in at least two papers [5,10].[2]

We solve these problems by reducing the associated decision problem to computing the upper envelope of a set of identical cones in R^3. In the case of spanning ratio, the set of cones is finite, and the upper envelope that we compute is actually an additively-weighted Voronoi diagram of points in the plane. In the case of maximum detour, the set of cones is infinite, and corresponds to computing the additively-weighted Voronoi diagram of line segments in the plane, a diagram

---

[2] Subquadratic time algorithms were independently found by Agarwal et al [1], see below.

that seems not to have been considered previously. We then apply a general optimization technique of Chan [4] to convert the decision algorithm into an optimization algorithm.

We also show that more complicated structures can sometimes be treated by using multiple invocations of the above technique. As examples, we give $O(n \log^2 n)$ time algorithms for computing the maximum detour and spanning ratio of a planar tree and $O(n \log^2 n)$ time algorithms for computing the maximum detour and spanning ratio of a planar cycle.

Independently, Agarwal $et\ al$ [1] studied the problem of computing the maximum detour of paths, cycles and trees in two and three dimensions. For planar paths and cycles they give $O(n \log^4 n)$ time deterministic algorithms and $O(n \log^3 n)$ time randomized algorithms. For planar trees they give an $O(n \log^5 n)$ time deterministic algorithm and an $O(n \log^4 n)$ time randomized algorithm. For three-dimensional paths they give a randomized algorithm whose expected running time is $O(n^{16/9+\epsilon})$, where ϵ is any constant greater than zero.

The remainder of the paper is organized as follows: In Section 2 we show how to reduce the decision problems to an upper envelope computation. In Section 3 we give an algorithm for computing the upper envelope of a set of objects called bats that is required for solving the maximum detour decision problem. In Section 4 we describe how to use these decision algorithms to obtain optimization algorithms. In Section 5 we extend these algorithms to trees and cycles.

2   A Problem on Cones

For a point $p \in \mathrm{R}^3$, denote by p_z the z-coordinate of p and by p_{xy}, the projection of p onto the xy plane. Given a polygonal path C in R^2 whose vertices are $v_1, \ldots, v_n$ we lift it to a polygonal path C' in R^3 by assigning to each point p in C a z-coordinate equal to its distance along the path from v_1, i.e., for each point $p \in C$, C' has a point p' such that $p'_{xy} = p$ and $p'_z = \|v_1 p\|_C$.

In this way, we obtain a z-monotone polygonal path C' with vertices $u_1, \ldots, u_n$ such that for any two points $p', q' \in C'$, the unique path between the corresponding points $p, q \in C$ has length $\|pq\|_C = \|p'_z - q'_z\|$.

Consider the following construction. At each vertex u_i of C' we place the apex of a cone c_i that points downwards with its axis of rotation parallel to the z-axis and that spans an angle of $2 \arctan(1/d)$. Now, if some cone c_i contains some vertex u_j then $D(C, v_i, v_j) \geq d$ (see Fig. 1). Conversely, if there exists a pair of vertices v_i, v_j such that $D(C, v_i, v_j) \geq d$, then either c_i contains u_j or c_j contains u_i. Thus, the problem of determining whether the spanning ratio of C is greater than or equal to d is reducible to the problem of determining whether any cone c_i contains any vertex u_j.

The $upper\ envelope$ of the cones $c_1, \ldots, c_n$ is the bivariate function $f(x, y) = \max\{z : (x, y, z) \in c_i, \text{ for some } i\}$. From this definition it follows that u_i is contained in some cone if and only if u_i does not appear on the upper envelope, i.e., $f(u_i) \neq u_{i,z}$. The upper envelope of identical and identically oriented cones

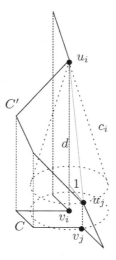

Fig. 1. If c_i contains u_j then $D(C, v_i, v_j) \geq d$.

has been given at least two other names: additively-weighted Voronoi diagram
[7] and Johnson-Mehl crystal growth diagram [9]. It is known that f consists of
$O(n)$ pieces and can be computed in $O(n \log n)$ time using a sweep line algorithm
[7]. Thus, the decision problem of determining whether the spanning ratio of C
is at least d can be solved in $O(n \log n)$ time.

Next we turn to the problem of determining whether the maximum detour
of G is at least d. For this problem we use the same construction except that
we place a cone with its apex on *every* point of C', not just the vertices. The
decision problem then reduces to the question: Does every point on C' appear on
the upper envelope of these (infinitely many) cones. Of course, computationally,
we do not compute the upper envelope of infinitely many cones. Instead, we
compute the upper envelope of n *bats*, where a *bat* is the convex hull of two
cones with apexes on the endpoints of an edge of C. We call the edge that
defines a bat the *core* of the bat.

Thus, for both maximum detour and spanning ratio, the associated decision
problem can be solved by computing the upper envelope of a suitably chosen
set of objects, either cones or bats. To the best of our knowledge, no algorithm
exists for computing the upper envelope of a set of bats. In the following section,
we derive such an algorithm.

3   The Upper Envelope of a Set of Bats

In this section we show how to compute the upper envelope of a set of bats
using a sweep line algorithm. This algorithm is essentially a modification of
Fortune's algorithm for computing additively-weighted Voronoi diagrams and

Voronoi diagrams of line segments [7]. We say that a point p in the core of some bat is *redundant* if p is contained in some other bat. We say that the input is *redundant* if some bat contains a redundant point and the input is *non-redundant* otherwise.

It is clear that solving the decision problem associated with detour is equivalent to determining whether the input is redundant or non-redundant. This is fortunate, since the upper envelope of n bats can have $\Omega(n^2)$ complexity (see Fig. 2), so any approach that requires computing the upper envelope is doomed to have quadratic running time in the worst case.

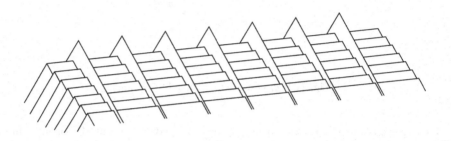

Fig. 2. The upper envelope of n bats can have $\Omega(n^2)$ complexity.

We describe an algorithm that takes as input a set of n bats which are the union of cones that span angles of $2\alpha \leq \pi/2$ and either reports that the input is redundant or correctly computes the upper envelope of the input. The algorithm sweeps a plane P through space and maintains, at all times, the intersection of P with the upper envelope E. The plane P is parallel to the x-axis and forms an angle of $\pi - \alpha$ with the xy plane (see Fig. 3). The reason we sweep with such a plane is that no bat b can contribute to $P \cap E$ until P has swept over some point in the core of b.

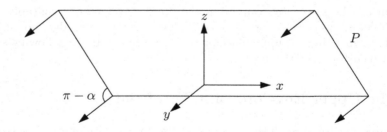

Fig. 3. The sweep plane makes an angle of $\pi - \alpha$ with the xy plane.

To understand the structure of $P \cap E$, it is helpful to note that the boundary of a bat consists of four pieces (see Fig. 4): two conic pieces and two planar pieces. It is easy to verify that the intersection $P \cap E$ is a weakly x-monotone curve consisting of pieces of parabolas and lines. Therefore, its pieces can be stored in a balanced binary tree sorted by x-coordinate. The intersection $P \cap E$ consists of parabolic arcs (where P intersects conic pieces) and straight line segments (where P intersects linear pieces).

Fig. 4. The boundary of a bat consists of two conic pieces and two planar pieces.

As P sweeps through space, $P \cap E$ changes continuously. Therefore, we store $P \cap E$ symbolically so that each arc and segment is represented by its equation as a function of the position of the plane P. The progress of the sweep plane is controlled by a priority queue Q. Initially, Q contains $2n$ events corresponding to the times at which the sweep plane passes over each endpoint of the core of a bat.

During the sweep, some arcs or segments of $P \cap E$ may disappear as they become obscured by neighbouring arcs. Since each arc and segment is parameterized by the position of the plane P, it is a constant time operation to determine the time (if any) that an arc will be obscured by its neighbours. In the following discussion, when we insert and delete arcs and segments from $P \cap E$, it is implicit that we recompute the times at which arcs in the neighbourhood of the inserted/deleted arc are obscured and insert these times into Q. For further details refer to Fortune's original paper [7].

During the sweep, we process three types of events:

1. *P sweeps over a point p that is the first endpoint of the core of some bat b.* Refer to Fig. 5.a. In this case, we first check if p is below $P \cap E$. If so, then p is contained in the bat that intersects P directly above p, so we quit and report that the input is redundant. Otherwise, we add four objects to $P \cap E$. These objects are two line segments representing the intersection of P with the two planar pieces of b and two parabolic arcs representing the intersection of P with the cone whose apex is at p.

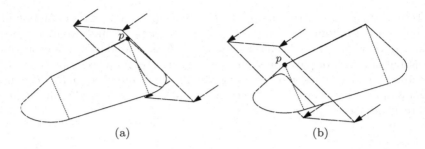

(a)   (b)

Fig. 5. Handling type 1 and type 2 events.

2. *P sweeps over a point p that is the last endpoint of the core of some bat b.* Refer to Fig. 5.b. In this case, we add two parabolic arcs to $P \cap E$ that correspond to the intersection P with the cone whose apex is at p.
3. *An arc or segment disappears from $P \cap E$.* In this case, we remove from Q any events associated with the arc or segment. In the case of a segment, we also check if one endpoint of the segment corresponds to a point in the core of a bat. If so, then that point is not part of E so we can quit and report that the input is redundant.

To see that the above algorithm runs in $O(n \log n)$ time, we observe that there are only $O(n)$ type 1 and 2 events. Each of these can be easily implemented in $O(\log n)$ time, and each such event adds $O(1)$ arcs or segments to $P \cap E$. Each type 3 event can also be implemented in $O(\log n)$ time, and deletes one element from $P \cap E$. Therefore, there are only $O(n)$ type 3 events and the entire algorithm runs in $O(n \log n)$ time.

To see that the algorithms is correct for non-redundant inputs we can use arguments which are standard by now [7]. In particular, we can show that there is a direct correspondence between the events processed by the algorithm and changes to the combinatorial structure of $P \cap E$. Suppose therefore that the input is redundant and let p be the first redundant point swept over by P. Either p is an endpoint of a core or it is in the interior of a core. In the former case, it will be handled as a type 1 event while in the latter case it will be handled as a type 3 event.

In either case, all input previously swept over by P is non-redundant, so the intersection $P \cap E$ has been correctly computed. If p is an endpoint of a core it will then be processed as a type 1 event and the algorithm will correctly detect that p is below $P \cap E$ and is therefore redundant. If p is in the interior of a core it will be processed as a type 3 event and the algorithm will correctly detect that the input is redundant. Therefore, either the algorithm correctly computes the upper envelope (if the input is non-redundant) or correctly reports that the input is redundant.

Lemma 1 *There exists an algorithm requiring $O(n \log n)$ time and $O(n)$ space that tests whether a set of n bats is redundant or non-redundant.*

4   Optimization

So far we have given all the tools required for solving the decision problems associated with finding the maximum detour and spanning ratio of a path. More specifically, we have solved the problem: Given a set of segments (possibly points) in 3 space, does there exist a cone c with angular radius $2\arctan(1/d)$, center of rotation parallel to the z-axis and apex on one of the segments such that c intersects another segment. The optimization problem is that of finding the largest value of d for which such a cone exists.

To solve the optimization problem we apply the randomized reduction of Chan [4], which requires only that we (1) be able to solve the decision problem in $f(n) = \Omega(n^\epsilon)$ time, for some constant $\epsilon > 0$, and (2) partition the problem into r subproblems, each of size at most αn, $\alpha < 1$ such that the optimal solution is the maximum of the solutions to the r subproblems. The reduction works by considering the subproblems in random order and recursively solving a subproblem only if its solution is larger than the current maximum (which can be tested by the decision algorithm). The resulting optimization algorithm has running time $O(f(n))$.

We have already shown how to do (1) in $f(n) = O(n \log n)$ time. To do (2), we simply note that we can partition our set of segments in to three sets A, B, and C, each of size $n/3$. The optimal solution is then the maximum of the solutions to $A \cup B$, $B \cup C$ and $A \cup C$. Since there are only $r = 3$ subproblems and each has size $\alpha n = \frac{2}{3}n$ we have satisfied the conditions required to use Chan's optimization technique.

Theorem 1 *The maximum detour and spanning ratio of a planar path with n vertices can be computed in $O(n \log n)$ expected time.*

5   Trees and Cycles

In this section we show how the tools developed for planar paths can be used for solving problems on more complicated types of objects.

5.1   Planar Trees

Let T be a tree embedded in the plane and assume T is rooted at a vertex v such that $T \setminus \{v\}$ has no component with more than $n/2$ vertices. Such a vertex is easily found in linear time. Refer to Fig. 6. We partition the children of v into two sets A and B. Let T_A, respectively T_B, denote the tree induced by v and all vertices having ancestors in A, respectively B. The partition A, B is chosen so that $\frac{1}{4}n \leq \|T_A\|, \|T_B\| \leq \frac{3}{4}n$. Since no descendent of v is the root of a subtree with size more than $\frac{n}{2}$, such a partition can be found with a greedy algorithm.

We lift T into a 3-dimensional tree T' in the following way. Each point $p \in T_A$ is assigned a z-coordinate equal to $\|vp\|_T$. Each point $p \in T_B$ is assigned a z-coordinate equal to $-\|vp\|_T$. This gives us a 3-dimensional tree T' consisting of points T'_A above the xy plane and points T'_B below the xy plane.

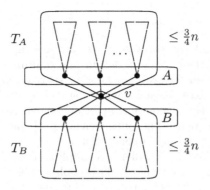

Fig. 6. Partitioning into subtrees T_A and T_B.

This lifting has the property that for any point $p' \in T_A'$ and any point $q' \in T_B'$ the distance between the corresponding points p and q in T is equal to the difference in z-coordinates of p' and q', i.e., $\|pq\|_T = \|p_z' - q_z'\|$. Furthermore, for two points p, q both in T_A or both in T_B, $\|pq\|_T \geq \|p_z' - q_z'\|$. It follows that if we run the maximum detour algorithm of the previous section on the tree T', the algorithm will respond with the correct answer $D(T)$ if there is a pair of points $p \in T_A$ and $q \in T_B$ such that $D(T) = D(T, p, q)$. If this is not the case, then the algorithm may report a value less than $D(T)$, but will never report a larger value.

Therefore, we can compute $D(T)$ using a recursive algorithm. We run the maximum detour algorithm on the tree T', make two recursive calls on T_A and T_B and output the maximum of the three values obtained. To see that this algorithm correctly computes $D(T)$, consider the pair $p, q \in T$ that maximizes $D(T, p, q)$. If $p \in A$ and $q \in B$ (or vice versa), the correct value of $D(T, p, q)$ is found when we run the maximum detour algorithm on T'. Otherwise, $p, q \in T_A$ or $p, q \in T_B$ and is correctly reported by one of the recursive calls.

The running time of the above algorithm is given by the recurrence $T(n) = T(n - k + 1) + T(k) + O(n \log n)$, with $\frac{1}{4}n \leq k \leq \frac{3}{4}n$, which solves to $O(n \log^2 n)$.

Theorem 2 *The maximum detour or spanning ratio of a planar tree with n vertices can be computed in $O(n \log^2 n)$ expected time.*

5.2 The Spanning Ratio of a Planar Cycle

To obtain an algorithm for computing the spanning ratio of a planar cycle, we study the following decision problem in $\mathbf{R}^3$: Given a set S of n points in $\mathbf{R}^3$, do there exist two points $p, q \in S$ such that $(p_z - q_z)/\|p_{xy}q_{xy}\| > d$ and $p_z - q_z < 1/2$? This problem is almost identical to our previous problem on cones except that now, instead of being infinite, the cones have height $1/2$.

To reduce the problem of computing the spanning ratio of a polygonal planar cycle C to the above problem we first normalize the cycle C so that it has length

1. We then remove an arbitrary edge of C so that it becomes a path and lift the vertices of this path into R^3 as described in Section 2. This gives us a set S_1 of n points in R^3. Next, we make a copy S_2 of S_1 and translate S_2 downwards (in the $-z$-direction) by a distance of 1. The union S' of S_1 and S_2 consists of $2n-1$ points. Then it is not hard to verify that the spanning ratio of C is larger than d if and only if there exists $p, q \in S'$ or such that $(p_z - q_z)/\|p_{xy}q_{xy}\| > d$ and $p_z - q_z \leq 1/2$.

So far we have reduced the problems of computing the spanning ratio of a planar cycle to a problem on finite cones. Unfortunately, this problem is very different from our previous problem that involved infinite cones, and can not be solved by computing the upper envelope of the cones. However, suppose we have a dynamic data structure that supports insertion and deletion of (infinite) cones and signals when the set of cones currently contained in the data structure is redundant.

Then we can solve our problem on finite cones by sweeping with a plane P that is parallel to the xy plane and maintaining the data structure so that it contains only the cones with apexes on points of S that are below P but at distance at most $1/2$ from P. If at any time the data structure reports that the input is redundant then we know that there are two points $p, q \in S$ such that $(p_z - q_z)/\|p_{xy}q_{xy}\| > d$ and $p_z - q_z \leq 1/2$. On the other hand, if two such points p and q exist, then they will both be in the data structure at some point in time and the data structure will report that the input is redundant. Since only $O(n)$ points are inserted and deleted in the data structure, this algorithm will run in $O(n \log n + nU(n))$ time, where $U(n)$ is the time it takes to perform an update operation on the data structure.

All that remains is to develop a data structure that supports insertion and deletion of cones and signals if, at any time, the set of cones currently contained in the data structure is redundant. In general, this can be treated as a decomposable search problem and a data structure with $O(n^{\frac{3}{2}} \log n)$ update time can be obtained using the technique of Bentley and Saxe [2]. However, in our application insertions and deletions are done in a FIFO manner, so that the ith point inserted will be the ith point deleted. We will use this property to obtain a data structure that supports updates in $O(\log^2 n)$ amortized time.

The data structure maintains a partition of the cones into $O(\log n)$ sets. These sets are denoted by $I_0, \ldots, (I_k = D_k), \ldots, D_0$ where I_i, respectively D_i, either contains exactly 2^i cones or is empty. At all times, the data structure maintains two invariants: (1) The set $I_k = D_k$ is non-empty, and (2) all the cones in I_{i+1} will be deleted before any of the cones in I_i, and all the cones in D_i will be deleted before any of the cones in D_{i+1}, for all $0 \leq i \leq k$.

When inserting a cone c into the data structure, we check if the apex of c (and hence all of c) is contained in any cone of the data structure. If it is, then the input is redundant and we can quit. We can perform this test efficiently by maintaining, for each I_i (respectively, D_i), the upper envelope of the cones in I_i stored in a point location structure that can test, in $O(\log n)$ time if a point $p \in R^3$ is above or below the envelope.

After performing this test, and assuming it is negative, we insert the cone c into the data structure by finding the smallest value of i such that I_i is empty. If no such value of i exists, we increase the value of k by 1 so that I_k is empty. We group together the cone c and all the cones contained in $I_0, \ldots, I_{i-1}$, place these in I_i and build the upper envelope for I_i. At the same time, we make the sets $I_0, \ldots, I_{i-1}$ empty. It is clear that this maintains invariants 1 and 2.

To delete a cone from the data structure we find the smallest value of i such that D_i is non-empty. Because of invariants 1 and 2, D_i must exist and contain the cone c that is being deleted. We then partition $D_i \setminus \{c\}$ into $D_{i-1}, \ldots, D_0$ in such a way that invariant 2 is maintained, build the data structures for $D_{i-1}, \ldots, D_0$ and make D_i empty. To maintain invariant 1 we check if $i = k$. If $i = k$ and I_{k-1} is empty then we decrease the value of k by 1. Otherwise if $i = k$ and I_{k-1} is not empty then we merge I_{k-1} and D_{k-1}, put the result into $I_k = D_k$ and make I_{k-1} and D_{k-1} empty.

A simple amortized analysis of this data structure, which we don't include because of space constraints, shows that the amortized cost of all operations is $O(\log^2 n)$. (Give each cone $2\log^2 n$ credits when it is inserted and make it pay $\log n$ credits every time it moves.) Combining this with the machinery of Chan's optimization technique, we have just proven Theorem 3.

Theorem 3 *The spanning ratio of a planar cycle with n vertices can be computed in $O(n \log^2 n)$ time.*

5.3   The Maximum Detour of a Planar Cycle

To compute the maximum detour of a planar cycle we make use of the following lemma, which is a generalization of a lemma by Ebbers-Baumann *et al* [5].

Lemma 2 *Let C be a planar cycle of length 1. Then there exist two points $s, t \in C$ such that $D(C, s, t)$ is maximal and*

1. *s is a vertex of C or*
2. *(2) $\|st\|_C = 1/2$.*

Proof. The proof is omitted due to space constraints.

Given a planar cycle C of length 1, we can find the maximum detour among all points $s, t \in C$ such that $\|st\|_C = 1/2$ in linear time by starting with any two points s and t and sweeping them around C while maintaining the invariant that $\|st\|_C = 1/2$. Once we have done this, Lemma 2 implies that we can limit our search to pairs $s, t \in C$ such that s is a vertex of C.

Given this result, we can proceed in the same manner as we did when computing the spanning ratio of a planar cycle, except that now our dynamic data structure maintains a set of bats and the upper and lower envelopes are constructed using the algorithm of Section 3. Although the basic approach is essentially the same, two technical difficulties occur. Due to space constraints, we only sketch their solution.

The first difficulty is that bats in our data structure change continuously as the sweep plane sweeps over the cores of bats. However, Lemma 2 allows us to discretize this change by inserting, for each vertex v of C, a *Steiner vertex* v' whose distance from v along C is exactly $1/2$. The second difficulty is that the plane sweep will only find the pair s, t with maximum detour if the shortest path from s to t is counterclockise around C. To handle this problem we perform two plane sweeps, one in the $+z$ direction and one in the $-z$ direction.

As before, if at any time during the maintenance of the data structure we find some subset of bats to be redundant then we can quit. When inserting a bat b into the data structure, Lemma 2 ensures that performing point-location queries on the endpoints of the core of b is enough to ensure the correctness of the algorithm. Leaving further details to the full version, we obtain:

Theorem 4 *The maximum detour of a planar cycle C with n vertices can be computed in $O(n \log^2 n)$ time.*

References

1. P. K. Agarwal, R. Klein, C. Knauer, and M. Sharir. Computing the detour of polygonal curves. Unpublished Manuscript, November 2001.
2. J. L. Bentley and J. B. Saxe. Decomposable searching problems. I. Static-to-dynamic transformation. *Journal of Algorithms*, 1(4):301–358, 1980.
3. P. Bose and P. Morin. Competitive online routing in geometric graphs. In *Proceedings of the VIII International Colloquium on Structural Information and Communication Complexity (SIROCCO 2001)*, 2001.
4. T. M. Chan. Geometric applications of a randomized optimization technique. *Discrete & Computational Geometry*, 22(4):547–567, 1999.
5. A. Ebbers-Baumann, R. Klein, E. Langetepe, and A. Lingas. A fast algorithm for approximating the detour of a polygonal chain. In *Proceedings of the 9th Annual European Symposium on Algorithms (ESA 2001)*, pages 321–332, 2001.
6. D. Eppstein. Spanning trees and spanners. In J.-R. Sack and J. Urrutia, editors, *Handbook of Computational Geometry*, pages 425–461. Elsevier, 1999.
7. S. J. Fortune. A sweepline algorithm for Voronoi diagrams. *Algorithmica*, 2:153–174, 1987.
8. C. Icking and R. Klein. Searching for the kernel of a polygon: A competitive strategy. In *Proceedings of the 11th Annual Symposium on Computational Geometry*, pages 258–266, 1995.
9. W. A. Johnson and R. F. Mehl. Reaction kinetics in processes of nucleation and growth. *Transactions of the Americal Institute of Mining and Metallurgy*, 135:416–458, 1939.
10. G. Narasimhan and M. Smid. Approximating the stretch factor of Euclidean graphs. *SIAM Journal on Computing*, 30(3):978–989, 2001.

On the Parameterized Intractability of Closest Substring and Related Problems

Michael R. Fellows[1], Jens Gramm[2*], and Rolf Niedermeier[2]

[1] Department of Computer Science and Software Engineering,
University of Newcastle, University Drive, Callaghan 2308, Australia
mfellows@cs.newcastle.edu.au
[2] Wilhelm-Schickard-Institut für Informatik, Universität Tübingen,
Sand 13, D-72076 Tübingen, Fed. Rep. of Germany
{gramm, niedermr}@informatik.uni-tuebingen.de

Abstract. We show that Closest Substring, one of the most important problems in the field of biological sequence analysis, is W[1]-hard with respect to the number k of input strings (even over a binary alphabet). This problem is therefore unlikely to be solvable in time $O(f(k)n^c)$ for any function f and constant c independent of k — effectively, the problem can be expected to be intractable, in any practical sense, for $k \geq 3$. Our result supports the intuition that Closest Substring is computationally much harder than the special case of Closest String, although both problems are NP-complete and both possess polynomial time approximation schemes. We also prove W[1]-hardness for other parameterizations in the case of unbounded alphabet size. Our main W[1]-hardness result generalizes to Consensus Patterns, a problem of similar significance in computational biology.

1 Introduction

According to Li *et al.* [11], Closest Substring "is a key open problem in the search for a potential generic drug sequence which is 'close' to some sequences (of harmful germs)..." (also see [6,10,12] and the references cited therein for other applications in computational biology). Closest Substring is defined as follows.

> **Input:** k strings $s_1, s_2, \ldots, s_k$ over alphabet Σ and integers d and L.
> **Question:** Is there a string s of length L such that, for all $i = 1, \ldots, k$, $d_H(s, s_i') \leq d$ where s_i' is a length L substring of s_i?

Herein, $d_H(s, s_i)$ denotes the Hamming distance between s and s_i. Closest Substring is NP-complete, and remains so for the special case of the Closest String problem, where the string s that we search for is of same length as the input strings. Closest String is NP-complete even for the further restriction

---

* Supported by the Deutsche Forschungsgemeinschaft (DFG), project OPAL (optimal solutions for hard problems in computational biology), NI 369/2-1.

H. Alt and A. Ferreira (Eds.): STACS 2002, LNCS 2285, pp. 262–273, 2002.
© Springer-Verlag Berlin Heidelberg 2002

to a binary alphabet [7,10]. On the positive side, both problems admit a polynomial time approximation scheme (PTAS) [11,12], although in both cases the
exponent of the polynomial bounding the running time depends on the goodness
of the approximation — they are not EPTASs in the sense of [4].

For CLOSEST STRING as well as for CLOSEST SUBSTRING it is natural and important to study their parameterized complexity [5]. Taking into account that
the number k of strings or the distance d are comparatively small in many practical situations (cf. [6,11]), it is important to know whether the problems are
fixed parameter tractable with respect to these parameters.[1] CLOSEST STRING
was recently shown to be linear time fixed parameter tractable for parameter d,
for a parameter function bounded by d^d, and also linear time fixed parameter
tractable for parameter k, but in this case the parametric complexity contribution is much less encouraging [8]. The parameterized complexity of CLOSEST
SUBSTRING has largely remained open.

Our main result is to show that CLOSEST SUBSTRING with parameter k is W[1]-
hard even for a binary alphabet. That is, the problem is fixed parameter intractable unless FPT = W[1], the parametric analog of P = NP.

For computational biologists, the fundamental message is that exact algorithms
for CLOSEST SUBSTRING, parameterized by the number of strings k, probably
cannot achieve the running time $f(k) \cdot n^{O(1)}$ (i.e., exponential *only* in k). For unbounded alphabet size, we show that the problem is W[1]-hard for the combined
parameters L, d, and k. In the case of constant alphabet size, the complexity
of the problem remains open when parameterized by d and k together, or by d
alone. We also show that the closely related CONSENSUS PATTERNS problem [11],
where one tries to minimize the *sum* of distances instead of the maximum distance, is W[1]-hard for a binary alphabet when parameterized by the number
of strings. Note that in the case of CONSENSUS PATTERNS our result gains particular importance, because here the distance parameter d usually is not small,
whereas assuming that k is small is still reasonable. Until now, it was known
only that if one additionally considers the substring length L as a parameter,
then running times exponential in L can be achieved [1,6,15].

Finally, it is worth noting that analogous parameterized complexity studies have
been performed for the famous LONGEST COMMON SUBSEQUENCE (LCS) problem [2,3]. There, however, parameterized hardness results could only be shown
in case of unbounded alphabet size and, in particular, it is a long-standing open
problem (cf. [5]) to determine the parameterized complexity of LCS with respect
to the parameter k for constant alphabet size. Constant alphabet size, however,
is the case of biological importance. In this sense, our results seem to contribute
some of the first "real" parameterized intractability results for core problems in
the field of biological sequence analysis. Eventually, our work gives strong theory-
based support for the common intuition that CLOSEST SUBSTRING (W[1]-hard)
seems to be a much harder problem than CLOSEST STRING (in FPT [8]). No-

---

[1] Many investigations have been made of the complexity of the similar LONGEST COM-
MON SUBSEQUENCE problem, even for $k = 2$ (see, e.g., [16]).

tably, this could *not* be expressed by "classical complexity measures," since both problems are NP-complete as well as both do have a PTAS.

Some proofs had to be omitted and are deferred to the full paper.

2   Preliminaries and Previous Work

In this section, we start with a brief introduction to parameterized complexity (more details to be found in [5]) and then continue with sketching some previous work on CLOSEST SUBSTRING and CONSENSUS PATTERNS and related problems.

2.1   A Crash Course in Parameterized Complexity

Given an undirected graph $G = (V, E)$ with vertex set V, edge set E and a positive integer k, the NP-complete VERTEX COVER problem is to determine whether there is a subset of vertices $C \subseteq V$ with k or fewer vertices such that each edge in E has at least one of its endpoints in C. VERTEX COVER is *fixed parameter tractable*: There now are algorithms solving it in time less than $O(kn + 1.3^k)$. The corresponding complexity class is called FPT. By way of contrast, consider the also NP-complete CLIQUE problem: Given an undirected graph $G = (V, E)$ and a positive integer k, CLIQUE asks whether there is a subset of vertices $C \subseteq V$ with at least k vertices such that C forms a clique by having all possible edges between the vertices in C. CLIQUE appears to be *fixed parameter intractable*: It is *not* known whether it can be solved in time $f(k)n^{O(1)}$, where f might be an arbitrarily fast growing function only depending on k. The best known algorithm solving CLIQUE runs in time $O(n^{ck/3})$, where c is the exponent on the time bound for multiplying two integer $n \times n$ matrices. The decisive point is that k appears in the exponent of n, and there seems to be no way "to shift the combinatorial explosion only into k," independent from n.

Downey and Fellows developed a completeness program for showing parameterized intractability. However, the completeness theory of parameterized intractability involves significantly more technical effort (as will also become clear when following the proofs presented in this paper). We very briefly sketch some integral parts of this theory in the following. Let $L, L' \subseteq \Sigma^* \times \mathbf{N}$ be two parameterized languages.[2] For example, in the case of CLIQUE, the first component is the input graph coded over some alphabet Σ and the second component is the positive integer k, that is, the parameter. We say that L *reduces to* L' *by a standard parameterized m-reduction* if there are functions $k \mapsto k'$ and $k \mapsto k''$ from $\mathbf{N}$ to $\mathbf{N}$ and a function $(x, k) \mapsto x'$ from $\Sigma^* \times \mathbf{N}$ to Σ^* such that

1. $(x, k) \mapsto x'$ is computable in time $k''|x|^c$ for some constant c and
2. $(x, k) \in L$ iff $(x', k') \in L'$.

---

[2] In general, the second component (representing the parameter) can also be drawn from Σ^*; for most cases, however, assuming the parameter to be a positive integer is sufficient.

Notably, most reductions from classical complexity turn out *not* to be parameterized ones. Now, the "lowest class of parameterized intractability," W[1], can be defined as the class of languages that reduce to the SHORT TURING MACHINE ACCEPTANCE problem (also known as the k-STEP HALTING problem). Here, we want to determine, for an input consisting of a nondeterministic Turing machine M (with unbounded nondeterminism and alphabet size), and a string x, whether M has a computation path accepting x in at most k steps. This can trivially be solved in time $O(n^{k+1})$ by exploring all k-step computation paths exhaustively, and we would be surprised if this can be much improved.

Therefore, this is the parameterized analogue of the TURING MACHINE ACCEPTANCE problem that is the basic generic NP-complete problem in classical complexity theory, and the conjecture that FPT $\neq$ W[1] is very much analogous to the conjecture that P $\neq$ NP. Other problems that are W[1]-*complete* (there are many) include CLIQUE and INDEPENDENT SET.

From a practical point of view, W[1]-hardness gives a concrete indication that a parameterized problem with parameter k problem is unlikely to allow for an algorithm with a running time of the form $f(k)n^{O(1)}$.

2.2   Motivation and Previous Results

Many biological problems with respect to DNA, RNA, or protein sequences can be solved based on consensus word analysis [13, Section 8.6]; CLOSEST SUBSTRING and CONSENSUS PATTERNS are central problems in this context. Applications include locating binding sites and finding conserved regions in unaligned sequences for genetic drug target identification, for designing genetic probes, and for universal PCR primer design. These problems can be regarded as various generalizations of the common substring problem, allowing errors (see [10, 11] and references there). This leads to CLOSEST SUBSTRING and CONSENSUS PATTERNS, where errors are modeled by the (Hamming) distance parameter d. There is a straightforward factor-2-approximation algorithm for CLOSEST SUBSTRING. The first better-than-2 approximation with factor $2 - 2/(2|\Sigma| + 1)$ was given by Li *et al.* [11] and this was further improved to factor $4/3 + \epsilon$ for any small constant $\epsilon > 0$ [10]. Finally, there is a PTAS for CONSENSUS PATTERNS [11] as well as for CLOSEST SUBSTRING [12], both of which, however, have impractical running times.

Concerning exact (parameterized) algorithms, we only briefly mention that, e.g., Sagot [15] studies motif discovery by solving CLOSEST SUBSTRING, Evans and Wareham [6] give FPT algorithms for the same problem, and Blanchette [1] developed a so-called phylogenetic footprinting method for a slightly more general version of CONSENSUS PATTERNS. All these results, however, make essential use of the parameter "substring length" L and show "exponential behavior with respect to L." To circumvent the computational limitations for larger values of L, many heuristics were proposed, e.g., Pevzner and Sze [14] present algorithms called WINNOWER (wrt. CLOSEST SUBSTRING) and SP-STAR (wrt. CONSENSUS PATTERNS). Our analysis makes a first step towards showing that, for exact solutions, we have to include L in the exponential growth; namely, we show that it is unlikely to find algorithms with a running time exponential *only* in k.

3   CLOSEST SUBSTRING: Unbounded Alphabet

We present a parameterized reduction from CLIQUE to CLOSEST SUBSTRING parameterized by the combination of L, d, and k. This shows that CLOSEST SUBSTRING is W[1]-hard for the aggregate parameter (L, d, k) in case of unbounded alphabet size.

3.1   Reduction from CLIQUE

A CLIQUE instance is given by a graph G, with a set $V = \{v_1, v_2, \ldots, v_n\}$ of n vertices, a set E of m edges, and a positive integer k denoting the clique size. We describe how to generate a set S of $\binom{k}{2}$ strings such that G has a clique of size k iff there is a string s of length $L := k + 1$ and every $s_i \in S$ has a substring s_i' of length L with $d_H(s, s_i') \leq d := k - 2$. If a string $s_i \in S$ has a substring s_i' of length L with $d_H(s, s_i') \leq d$, we call s_i' a *match*. We assume $k > 2$, because $k = 1, 2$ are trivial cases.

Alphabet. The alphabet of the produced instance is given by the disjoint union

$$\{\, \sigma_i \mid v_i \in V \,\} \dot\cup \{\, \varphi_j \mid j = 1, \ldots, \binom{k}{2} \,\} \dot\cup \{\#\}.$$

In words, we use a set of *encoding symbols*, i.e., an alphabet symbol for every vertex of the input graph, further we use a unique symbol for every of the $\binom{k}{2}$ produced strings, and we use a *synchronizing symbol* "$\#$," making a total of $n + \binom{k}{2} + 1$ alphabet symbols.

Choice strings. We generate a set of $\binom{k}{2}$ *choice strings* $S_c = \{c_{1,1}, c_{1,2}, \ldots, c_{1,k}, c_{2,3}, c_{2,4}, \ldots, c_{k-1,k}\}$ and assume that the strings in S_c are ordered as shown. Every choice string will encode the whole graph; it consists of m concatenated strings, each of length $k + 1$, called *blocks*; by this, we have one block for every edge of the graph. The blocks will be separated by *barriers*, which are length k substrings consisting of symbols that are uniquely determined for every choice string. A choice string $c_{i,j}$, which is the i'th choice string in S_c, is given by

$$c_{i,j} := \langle \text{block}(i, j, e_1) \rangle \, (\varphi_{i'})^k \, \langle \text{block}(i, j, e_2) \rangle \, (\varphi_{i'})^k \ldots (\varphi_{i'})^k \, \langle \text{block}(i, j, e_m) \rangle,$$

where $e_1, e_2, \ldots, e_m$ are the edges of G and $\langle \text{block}() \rangle$ will be defined below. The solution string s will have length $k + 1$, which is exactly the length of one block.

Block in a choice string. Every block is a string of length $k + 1$ and encodes an edge of the input graph. Every choice string contains a block for every edge of the input graph; different choice strings, however, encode the edges in different positions of their blocks: For a block in choice string $c_{i,j}$, positions i and j are called *active* and these positions encode the edge. Let e be the edge to be encoded and let e connect vertices v_r and v_s, $1 \leq r < s \leq n$. Then, the ith position of the block is σ_r in order to encode v_r and the jth position is σ_s in order to encode v_s.

The last position of a block is set to the synchronizing symbol "#." Let $c_{i,j}$ be the i'th choice string in S_c; then, all remaining positions in the block are set to character $\varphi_{i'}$, which is unique for the choice string. Thus, the block is given by

$$\langle \text{block}(i, j, (v_r, v_s)) \rangle :=$$
$$\varphi_{i',1}\, \varphi_{i',2} \cdots \varphi_{i',i-1}\, \sigma_r\, \varphi_{i',i+1} \cdots \varphi_{i',j-1}\, \sigma_s\, \varphi_{i',j+1} \cdots \varphi_{i',k}\, \#,$$

where all $\varphi_{i',1}, \varphi_{i',2}, \ldots, \varphi_{i',k}$ stand for $\varphi_{i'}$.

Values for L and d. We set $L := k + 1$ and $d := k - 2$.

3.2   Correctness of the Reduction

Proposition 1 *For a graph with a k-clique, the construction in Subsection 3.1 produces a* CLOSEST SUBSTRING *instance with a solution, i.e., there is a string s of length L such that every $c_{i,j} \in S_c$ has a substring $s_{i,j}$ with $d_H(s, s_{i,j}) \leq d$.*

Proof. Let the input graph have a clique of size k. Let $h_1, h_2, \ldots, h_k$ denote the indices of the clique's vertices, $1 \leq h_1 < h_2 < \ldots < h_k \leq n$. Then, we claim that a solution for the produced CLOSEST SUBSTRING instance is given by $s := \sigma_{h_1}\sigma_{h_2} \cdots \sigma_{h_k}\#$. Consider choice string $c_{i,j}$, $1 \leq i < j \leq k$. As the vertices $v_{h_1}, v_{h_2}, \ldots, v_{h_k}$ form a clique, we have an edge connecting v_{h_i} and v_{h_j}. Choice string $c_{i,j}$ contains a block $s_{i,j} := \langle \text{block}(i, j, (v_{h_i}, v_{h_j})) \rangle$ encoding this edge:

$$s_{i,j} := \varphi_{i',1}\, \varphi_{i',2} \cdots \varphi_{i',i-1}\, \sigma_{h_i}\, \varphi_{i',i+1} \cdots \varphi_{i',j-1}\, \sigma_{h_j}\, \varphi_{i',j+1} \cdots \varphi_{i',k}\, \#,$$

where i' is the number of the choice string in S_c. We have $d_H(s, s_{i,j}) = k - 2$, and we can find such a block for every $1 \leq i < j \leq k$.   □

For the reverse direction, we show in Proposition 2 that a solution in the produced CLOSEST SUBSTRING instance implies a k-clique in the input graph. To do this, we need the following lemma (proof omitted).

Lemma 1 *A solution s has the following properties:*

1. *s contains at least two encoding symbols.*
2. *s contains exactly one "#" symbol, at its last position.*
3. *s does not contain a symbol φ_i, $i = 1, \ldots, \binom{k}{2}$.*   □

Proposition 2 *The first k characters of a solution string correspond to k vertices of a clique in the input graph.*

Proof. By Lemma 1, a solution has encoding symbols at its first k positions and a synchronizing symbol at its last position. Consequently, the blocks are the only possible matches of a solution in the choice string. Now assume that a solution is given by $s = \sigma_{h_1}\sigma_{h_2} \cdots \sigma_{h_k}\#$ for $h_1, h_2, \ldots, h_k \in \{1, \ldots, n\}$. Consider any two h_i, h_j, $1 \leq i < j \leq k$, and choice string $c_{i,j}$. Recall that in this choice string, the blocks encode edges at their ith and jth position, have "#" at their last position, and all other positions are set to a symbol unique for this choice string. Thus, we can only find a block that is a match if there is a block with σ_{h_i} at its ith position and σ_{h_j} at its jth position. We have such a block only if there is an edge connecting v_{h_i} and v_{h_j}. Summarizing, the solution s implies that there is an edge between every pair of $\{v_{h_1}, v_{h_2}, \ldots, v_{h_k}\}$; these vertices form a k-clique in the input graph.   □

Propositions 1 and 2 establish the following hardness result. Note that hardness for the combination of all three parameters also implies hardness for each subset of the three.

Theorem 1 CLOSEST SUBSTRING *with unbounded alphabet is* W[1]-*hard for each of the single parameters L, d, and k.*

4   CLOSEST SUBSTRING: Binary Alphabet

We modify the reduction from Section 3 to achieve a CLOSEST SUBSTRING instance with binary alphabet. In contrast to there, we are not allowed to encode every vertex with its own symbol and we cannot use a unique symbol for every produced string. Also, we have to find new ways to "synchronize" the matches of our solution, a task previously done by the synchronizing symbol "#". To overcome these problems, we complement the set of produced strings by one string and lengthen the blocks in the produced choice strings considerably.

4.1   Reduction from CLIQUE

Number strings. To encode numbers between 1 and n, we introduce *number strings* $\langle \text{number}(pos) \rangle$, which have length n and have a "1" symbol at position pos and "0" symbols elsewhere: $0_1 0_2 \ldots 0_{pos-1} 1 0_{pos+1} \ldots 0_n$, [3] where all $0_1, 0_2, \ldots, 0_n$ stand for "0." In contrast to the reduction from Section 3, now we can use these number strings to encode the vertices of a graph.

Choice strings. As in Section 3, we generate a set of $\binom{k}{2}$ *choice strings* $S_c = \{c_{1,1}, c_{1,2}, \ldots, c_{k-1,k}\}$. Again, every choice string will consist of m *blocks*, one block for every edge of the graph. The choice string $c_{i,j}$ is given by

$$c_{i,j} := \langle \text{block}(i,j,e_1) \rangle \langle \text{block}(i,j,e_2) \rangle \ldots \langle \text{block}(i,j,e_m) \rangle,$$

where $e_1, e_2, \ldots, e_m$ are the edges of the input graph and $\langle \text{block}() \rangle$ will be defined below. The length of the solution string will be exactly the length of one block. **Block in a choice string.** Every block consists of a front tag, an encoding part, and a back tag. A block in choice string $c_{i,j}$ encodes an edge e; let e be an edge connecting vertices v_r and v_s, $1 \le r < s \le n$, and let $c_{i,j}$ be the i'th string in S_c. Then, the block is given by

$$\langle \text{block}(i,j,(v_r,v_s)) \rangle := \langle \text{front\_tag} \rangle \langle \text{encode}(i,j,(v_r,v_s)) \rangle \langle \text{back\_tag}(i') \rangle.$$

Front tags. We want to enforce that a solution string can only match substrings at certain positions in the produced choice strings, using front tags:

$$\langle \text{front\_tag} \rangle := (1^{3nk}0)^{2nk},$$

i.e., a front tag has length $(3nk+1) \cdot 2nk$. By this arrangement, the solution s and every match of s start (see Subsection 4.2) with the front tag.

Encoding part. The encoding part consists of k sections, each of length n. The encoding part corresponds to the blocks used in Section 3. As a consequence, in $\langle \text{block}(i,j,e) \rangle$ the ith and jth section are called *active* and encode edge $e = (v_r, v_s)$; section i encodes v_r by $\langle \text{number}(r) \rangle$ and section j encodes v_s by $\langle \text{number}(s) \rangle$. The other sections except for i and j are called *inactive* and are given by $\langle \text{inactive} \rangle := 0^n$. Thus,

$$\langle \text{encode}(i,j,(v_r,v_s)) \rangle := \langle \text{inactive}_1 \rangle \ldots \langle \text{inactive}_{i-1} \rangle \langle \text{number}(r) \rangle \langle \text{inactive}_{i+1} \rangle$$
$$\ldots \langle \text{inactive}_{j-1} \rangle \langle \text{number}(s) \rangle \langle \text{inactive}_{j+1} \rangle \ldots \langle \text{inactive}_k \rangle.$$

Back tag. The back tag of a block is intended to balance the Hamming distance of the solution string to a block, as will be explained later. The back tag consists of $\binom{k}{2}$ sections, each section has length $nk - 2k + 2$. The i'th section consists of "1" symbols, all other sections consist of "0" symbols:

$$\langle \text{back\_tag}(i') \rangle := 0^{(i'-1)(nk-2k+2)} 1^{nk-2k+2} 0^{(\binom{k}{2}-i')(nk-2k+2)}$$

Template string. The set of choice strings is complemented by one *template string*. It consists, in analogy to the blocks in the choice strings, of three parts: A front tag of length $(3nk+1) \cdot 2nk$, followed by a length nk string of "1" symbols, followed by a length $\binom{k}{2}(nk - 2k + 2)$ string of "0" symbols. Thus, the template string has the same length as a block in a choice string, i.e., $(3nk+1) \cdot 2nk + nk + \binom{k}{2}(nk - 2k + 2)$.

Values for d and L. We set $L := (3nk+1) \cdot 2nk + nk + \binom{k}{2}(nk - 2k + 2)$ and $d := nk - k$. As we will see in Subsection 4.2, the possible matches for a string of this length are the blocks in the choice strings, and, concerning the template string, the template string itself.

Notation. For a solution string s, we denote its first $(3nk+1) \cdot 2nk$ symbols (the front tag) by s', the following nk symbols (its encoding part) by s'', and the last $\binom{k}{2}(nk - 2k + 2)$ symbols (its back tag), by s'''. Analogously, the three parts of the template string t are denoted t', t'', and t'''. A particular block of a choice string $c_{i,j}$, is referred to by $s_{i,j}$; its three parts are called $s'_{i,j}$, $s''_{i,j}$, and $s'''_{i,j}$.

4.2   Correctness of the Reduction

Proposition 3 *For a graph with a k-clique, the construction in Subsection 4.1 produces a* CLOSEST SUBSTRING *instance with a solution, i.e., there is a string s of length L such that every $c_{i,j} \in S_c$ has a length L substring $s_{i,j}$ with $d_H(s, s_{i,j}) \leq d$ and $d_H(s, t) \leq d$.*

270 M.R. Fellows, J. Gramm, and R. Niedermeier

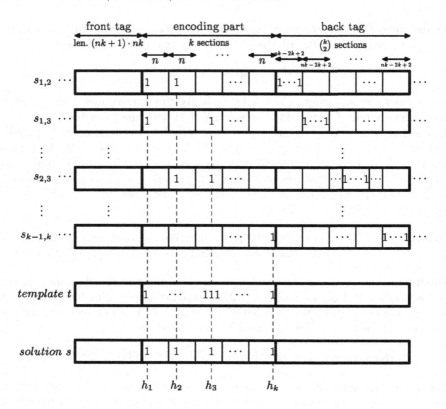

Fig. 1. Illustration of the solution for the CLOSEST SUBSTRING instance that is constructed from a graph having a k-clique. We display the template string t, the solution string s, and the matches $s_{i,j}$ in the choice strings. The front tags are equal in all strings and, therefore, are not displayed in detail. For the encoding part, we only indicate the positions of "1" symbols, where $h_1, h_2, \ldots, h_k$ name the position of a "1" symbol within its section. For the back tag part, we also only indicate the sections which consist of "1" symbols, all other sections consist of "0" symbols.

$d_H(\cdot, \cdot)$	s'	s''	s'''	s
match $s_{i,j}$ in choice string $c_{i,j}$	0	$k-2$	$nk-2k+2$	$nk-k$
template string t	0	$nk-k$	0	$nk-k$

Fig. 2. Hamming distance of solution s (with front tag s', encoding part s'', and back tag s''') to its matches in the choice strings and the template string.

Proof. Let the graph have a clique of size k. Let $h_1, h_2, \ldots, h_k$ denote the indices of the clique's vertices, $1 \le h_1 < h_2 < \ldots < h_k \le n$. Then, we can find a solution as outlined in Fig. 1. We display the template string, the solution string, and those blocks that are the matches in the choice strings. It follows from the construction that the choice strings have the indicated substrings: For every $1 \leq i < j \leq k$, we produced choice string $c_{i,j}$ with a block $s_{i,j}$ encoding the edge between vertices v_{h_i} and v_{h_j}. This block is a match for our solution. For this kind of block as well as for the template string, we report in Fig. 2 the distance they have to the solution string, separately for each of their three parts and in total. As is obvious from these distance values, the indicated substrings in the choice strings all have Hamming distance $nk - k$ to the solution string.   □

For the reverse direction, we assume that the CLOSEST SUBSTRING instance has a solution. We need the following statements (proof omitted):

Lemma 2 *A solution s has the following properties:*

1. *s and all its matches in the input instance start with the front tag.*
2. *The encoding part of s contains exactly k "1" symbols.*
3. *Every section of the encoding part of s contains exactly one "1" symbol.*   □

Lemma 2 directly implies that the only matches in a choice string $c_{i,j}$ are its blocks and that the solution's back tag s''' consists only of "0" symbols.

Proposition 4 *The k "1" symbols in the solution string's encoding part correspond to a k-clique in the graph.*

Proof. Let s be a solution for the CLOSEST SUBSTRING instance. Summarizing, we know by Lemma 2(1) that s can have as a match only one of the choice string's blocks. By Lemma 2(3), every section of the encoding part s'' contains exactly one "1" symbol; therefore, we can read this as an encoding of k vertices of the graph. Let $v_{h_1}, v_{h_2}, \ldots, v_{h_k}$ be these vertices. Further, we know that the back tag s''' consists only of "0" symbols. We have $d_H(s''', s'''_{i,j}) = nk - 2k + 2$ for *every* choice string match $s_{i,j}$ and, since every $s''_{i,j}$ contains only two "1" symbols, $d_H(s'', s''_{i,j}) \geq k - 2$. Now consider some $1 \leq i < j \leq k$ and the corresponding choice string $c_{i,j}$. Since s is a solution, we know that there is a block $s''_{i,j}$ with $d_H(s'', s''_{i,j}) = k - 2$. That means that the two "1" symbols in $s''_{i,j}$ have to match two "1" symbols in s''; this implies that the two vertices v_{h_i} and v_{h_j} are connected by an edge in the graph. Since this is true for all $1 \leq i < j \leq k$, vertices $v_{h_1}, \ldots, v_{h_k}$ are pairwisely interconnected by edges and form a k-clique.   □

Propositions 3 and 4 yield the following main theorem:

Theorem 2 CLOSEST SUBSTRING *is* W[1]-*hard for parameter k in the case of a binary alphabet.*

5 W[1]-Hardness for Consensus Patterns

Our ideas for showing hardness of Closest Substring, parameterized by the number k of input strings, also apply to Consensus Patterns. Because of the similarity to Closest Substring, we restrict ourselves to explaining the problem and pointing out new features in the hardness proof.

Given strings $s_1, s_2, \ldots, s_k$ over alphabet Σ and integers d and L, the Consensus Patterns problem asks whether there is a string s of length L such that $\sum_{i=1,\ldots,k} d_H(s, s_i') \le d$ where s_i' is a length L substring of s_i. Thus, Consensus Patterns aims for minimizing the *sum* of errors.[3] The problem is NP-complete and has a PTAS [11]. By reduction from Clique, we can show W[1]-hardness results as for Closest Substring given unbounded alphabet size. We omit the details here and focus on the case of binary input alphabet. We can apply basically the same ideas as were used in Section 4; however, some modifications are necessary.

As in Subsection 4.1, we generate a set of $\binom{k}{2}$ choice strings $c_{1,1}, \ldots, c_{k-1,k}$; each choice string consists of m blocks, one block for every edge of the input graph. The blocks, however, do consist of only two parts, the front tag and the encoding part. The back tag part is omitted. The encoding part is constructed as in Subsection 4.1. The front tag, however, is now given by $0^x(1^x0)^{x+1}0^x$ where $x := (\binom{k}{2}(k-1))nk$. We set parameter L to the block length: $L := x^2 + 4x + 1 + nk$. In contrast to Subsection 4.1, we produce not only one but $\binom{k}{2} - (k-1)$ many template strings. All template strings are equal, have length L, and are a concatenation of the front tag part (as given above) and an encoding part consisting of nk many "1" symbols. We set $d := (\binom{k}{2} - (k-1))nk$.[4]

For an overview, the front tag ensures that only the block of a choice string can be selected as a substring matching a solution. Regarding the distribution of mismatches, we note that a solution's front tag part will not cause any mismatches. In its encoding part, every of its nk positions causes at least $(\binom{k}{2} - (k-1))$ mismatches. It causes exactly $(\binom{k}{2} - (k-1))$ mismatches for every position iff the input graph contains a k-clique. Based on this reduction, we state the following theorem (proof omitted).

Theorem 3 Consensus Patterns *is* W[1]-*hard for parameter k in case of a binary alphabet.*

6 Conclusion

We have proved that Closest Substring and Consensus Patterns, parameterized by the number k of input strings and with alphabet size two, are W[1]-hard. This contrasts with related sequence analysis problems, such as Longest

---

[3] Here, the most significant parameterization seems to be the one by k, since errors are summed up over all strings and, therefore, d, usually, will not be a small value.

[4] We make sure that a match that is not a block is impossible by letting it, due to the front tags, imply a distance value larger than d. Since d has a higher value here compared to Section 4, we need a more involved front tag.

COMMON SUBSEQUENCE [2,3] and SHORTEST COMMON SUPERSEQUENCE [9], where parameterized hardness has so far only been established for the case of unbounded alphabet size. The parameterized complexity of CLOSEST SUBSTRING and CONSENSUS PATTERNS, parameterized by "distance parameter" d, remains open for alphabets of constant size. If these problems are also $W[1]$-hard, then no efficient and useful PTAS would be possible [4,5].

Finally, we leave it as an open problem whether CLOSEST SUBSTRING and CONSENSUS PATTERNS, parameterized by k, are in $W[1]$ and, thus, are $W[1]$-complete — it is easy to see that they are contained in the complexity class $W[P]$.

References

1. M. Blanchette. Algorithms for phylogenetic footprinting. In *Proc. of 5th ACM RECOMB*, pages 49–58, 2001, ACM Press.
2. H. L. Bodlaender, R. G. Downey, M. R. Fellows, and H. T. Wareham. The parameterized complexity of sequence alignment and consensus. *Theoretical Computer Science*, 147:31–54, 1995.
3. H. L. Bodlaender, R. G. Downey, M. R. Fellows, M. T. Hallett, and H. T. Wareham. Parameterized complexity analysis in computational biology. *Computer Applications in the Biosciences*, 11: 49–57, 1995.
4. M. Cesati and L. Trevisan. On the efficiency of polynomial time approximation schemes. *Information Processing Letters*, 64(4):165–171, 1997.
5. R. G. Downey and M. R. Fellows. *Parameterized Complexity*. Springer. 1999.
6. P. A. Evans and H. T. Wareham. Practical non-polynomial time algorithms for designing universal DNA oligonucleotides: a systematic approach. Manuscript, April 2001.
7. M. Frances and A. Litman. On covering problems of codes. *Theory of Computing Systems*, 30:113–119, 1997.
8. J. Gramm, R. Niedermeier, and P. Rossmanith. Exact solutions for Closest String and related problems. To appear in *Proc. of 12th ISAAC* (Christchurch, New Zealand), LNCS, December 2001. Springer.
9. M. T. Hallett. *An Integrated Complexity Analysis of Problems from Computational Biology*. PhD Thesis, University of Victoria, Canada, 1996.
10. J. K. Lanctot, M. Li, B. Ma, S. Wang, and L. Zhang. Distinguishing string selection problems. In *Proc. of 10th ACM-SIAM SODA*, pages 633–642, 1999, ACM Press. To appear in *Information and Computation*.
11. M. Li, B. Ma, and L. Wang. Finding similar regions in many strings. In *Proc. of 31st ACM STOC*, pages 473-482, 1999. ACM Press. To appear in *Journal of Computer and System Sciences*.
12. B. Ma. A polynomial time approximation scheme for the closest substring problem. In *Proc. of 11th CPM*, number 1848 in LNCS, pages 99–107, 2000. Springer.
13. P. A. Pevzner. *Computational Molecular Biology - An Algorithmic Approach*. MIT Press. 2000.
14. P. A. Pevzner and S.-H. Sze. Combinatorial approaches to finding subtle signals in DNA sequences. In *Proc. of 8th ISMB*, pages 269–278, 2000. AAAI Press.
15. M.-F. Sagot. Spelling approximate repeated or common motifs using a suffix tree. In *Proc. of 3rd LATIN*, number 1380 in LNCS, pages 111–127, 1998. Springer.
16. D. Sankoff and J. Kruskal (eds.). *Time Warps, String Edits, and Macromolecules*. Addison-Wesley. 1983. Reprinted in 1999 by CSLI Publications.

On the Complexity of Protein Similarity Search under mRNA Structure Constraints

Rolf Backofen*, N.S. Narayanaswamy**, and Firas Swidan

Institut für Informatik, Oettingenstrasse 67, 80538 Munich, Germany
{backofen,swamy,swidan}@informatik.uni-muenchen.de

Abstract. This paper addresses the complexity of a new problem associated with RNA secondary structure prediction. The issue is to identify an mRNA which has the secondary structure (=set of bonds) of a given mRNA, and the amino acid sequence it encodes has maximum similarity(defined below) to a given amino acid sequence. The problem is modeled as an optimization problem and a linear time algorithm is presented for the case when the given mRNA has a hairpin like secondary structure. Relevant extensions of this problem are shown to be NP-complete, and a factor 2 approximation algorithm is designed for the problem instances generated by the NP-completeness reduction.

1 Introduction

We start by introducing necessary biological terminology. *Nucleotides* are elements of the set $\{A, C, G, U\}$. (A, U) and (C, G) are known as complementary nucleotide pairs. An *mRNA* is a finite string of nucleotides. A *codon* is a string of three nucleotides. UAA,UAG, and UGA are called *stop* codons. Each of the remaining codons represents one of 20 objects called *amino acids*. A *protein* is a finite string of amino acids.

An mRNA is a blueprint for manufacturing proteins. Protein manufacture, known as mRNA translation mechanism, terminates whenever a stop codon is encountered in the mRNA, except in some special cases which form the central part of this paper. Translation of an mRNA is a blockwise (the mRNA can be visualized as a string of codons) replacement of codons by their corresponding amino acid until a stop codon is encountered. Every protein is output by the translation of some mRNA which is referred to as the mRNA of the protein.

While an mRNA is considered as a string, being a physical entity, it has a shape due to chemical interactions (bonds) between its nucleotides. The hairpin and pseudo-knot are two common mRNA shapes (see Figure 2). It helps to visualize an mRNA as a graph with edges between nucleotides that form a bond. For more details, the reader is referred to the book by Waterman [10].

Biological Motivation. *Selenocysteine* (Sec) is a rare amino acid, which was discovered as the 21st amino acid [2]. Proteins containing selenocysteine are called *selenoproteins*. The discovery of selenocysteine was another clue to the

---

* Partially supported by the DFG (national program SPP 1087 "Selenoprotein – Biochemische Grundlagen und klinische Bedeutung")
** Supported by DFG Grant No. Jo 291/2-1.

H. Alt and A. Ferreira (Eds.): STACS 2002, LNCS 2285, pp. 274–286, 2002.
© Springer-Verlag Berlin Heidelberg 2002

	U	C	A	G	
	Phe	Ser	Tyr	Cys	U
U	Phe	Ser	Tyr	Cys	C
	Leu	Ser	Stop	Stop	A
	Leu	Ser	Stop	Trp	G
	Leu	Pro	His	Arg	U
C	:	:	:	:	

Fig. 1. Translation of RNA. a) shows the 3-dimensional table of genetic code with the selenocysteine codon encircled, and b) shows the hairpin structure of the SECIS.

complexity and flexibility of the mRNA translation mechanism. Selenocysteine is encoded by UGA which normally causes the translation to stop. It has been shown [2] that in the case of selenocysteine, termination of translation is prevented due to the presence of a specific mRNA sequence in the region starting immediately after the UGA string. This mRNA sequence, called Selenocysteine Insertion Sequence (SECIS), forms a hairpin like three dimensional secondary structure. See Figure 1.

A motivating factor for research in the area of selenoproteins is that selenocysteines play an important role in increasing the catalytic activity of certain proteins. This is illustrated in the work of [9] on *E. coli*(a bacteria) where, the mRNA sequence of a protein was engineered (by hand) such that it formed a SECIS while preserving maximum similarity between the modified protein and the original protein (we refer to this as similarity at the amino acid level). The engineered protein then had a selenocysteine in place of a catalytic cysteine (at position 41) and showed a 4-fold enhanced catalytic activity.

Our Motivation. Our motivation here is to design algorithms that converts a given mRNA sequence into a SECIS while preserving the functionality of the original protein. In general, the problem is stated as follows: Find an mRNA sequence that has a required secondary structure (in this case the structure of a SECIS), is maximally similar to a given mRNA sequence, and the amino acid sequence encoded by it is maximally similar to a given amino acid sequence. This problem captures the essence of the experimental approach taken by [9] to insert selenocysteine in an *E. Coli* protein. Consequently, we believe that our study and results here will be of further use in designing mRNA sequences with different secondary structures.

A brute force search for a required mRNA sequence among the space of all mRNA sequences is not a feasible solution. The reason is that the typical target mRNA has at least 30 nucleotides in it. A brute force search would have to explore at least 4^{30} possible sequences. In particular, the SECIS in the *E. Coli* is 40 nucleotides long. Therefore, we should take into account input specific information like, the target mRNA's structure, and notion of the similarity between amino acids. We formally state the notions of similarity, the constraints imposed by the secondary structure, and the optimization issue in the description below.

Problem. Let $S = S_1 \ldots S_{3n}$ be the nucleotide sequence of an mRNA, and let $A = A_1 \ldots A_n$ be a given amino acid sequence. We have to find an appropriate mRNA sequence $N = N_1 \ldots \ldots N_{3n}$, which has the same secondary structure as S, has maximum similarity with S, and encodes an amino acid sequence

$A' = A'_1, \ldots, A'_n$ that has maximum similarity with A. The similarity between two amino acids is given by the PAM matrix which was introduced by Dayhoff et. al in [4]. For the purpose of this paper, the measure of similarity can be thought of as a function that assigns a value to each amino acid. Thus, we have the following picture:

$$
\begin{array}{ccccccc}
A = & A_1 & \cdots & A_i & \cdots & A_n \\
 & \wr & & \wr & & \wr \\
A' = & A'_1 & \cdots & A'_i & \cdots & A'_n \\
 & \uparrow & & \uparrow & & \uparrow \\
N = & \overbrace{N_1 N_2 N_3} & \cdots & \overbrace{N_{3i-2} N_{3i-1} N_{3i}} & \cdots & \overbrace{N_{3n-2} N_{3n-1} N_{3n}} \\
 & \wr \ \wr \ \wr & & \wr \ \ \wr \ \ \wr & & \wr \ \ \wr \ \ \wr \\
S = & S_1 \ S_2 \ S_3 & \cdots & S_{3i-2} S_{3i-1} S_{3i} & \cdots & S_{3n-2} S_{3n-1} S_{3n}.
\end{array}
$$

The similarities on both the amino acid ($A_i \sim A'_i$) and nucleotide level ($N_j \sim S_j$) will be measured by functions $F_{A_i}^{S_{3i-2} S_{3i-1} S_{3i}}(N_{3i-2} N_{3i-1} N_{3i})$, and we search an N that maximizes $\sum_{i=1}^n F_{A_i}^{S_{3i-2} S_{3i-1} S_{3i}}(N_{3i-2} N_{3i-1} N_{3i})$ (referred to as *cost*) and satisfies the constraints given by the secondary structure of S. The condition is that, if there is a bond between position q and r in S, then every valid N has to satisfy that N_q and N_r are complementary nucleotides. Note that, in S each nucleotide can be bonded to at most one other nucleotide.

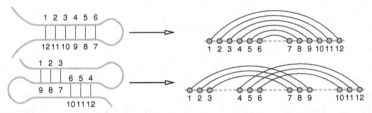

Fig. 2. Linear embedding of hairpin and pseudo-knot. We have numbered only the positions that are part of a bond.

Results. The hairpin shape of the SECIS has a natural outer-planar graph (defined below, see Figure 2) associated with it. We exploit this property and design a linear time algorithm to find the optimum cost. Here, the running time is linear in n, the length of the amino acid sequence.

We show that in the case when the graph associated with the mRNA shape is allowed to be arbitrary, and the similarity functions take only boolean values, the natural decision version of our optimization problem is NP-complete. The reason for addressing this issue is that the problem is not as nicely tractable when the mRNA is not hairpin shaped, but contains a pseudo-knot shape (see Figure 2). Like hair-pins, pseudo-knots have an extensive set of biological applications [5,7]. As an aside, some pseudo-knots have an associated graph which is outer-planar. Such pseudo-knots are known as simple pseudo-knots. Many known pseudo-knots are not simple and, recursive pseudo-knots are a set that do not have an associated outer-planar graph. For more on simple and recursive pseudo-knots and, other mRNA secondary structure problems, see [1].

Finally, we consider the case where similarity functions take only positive values. The reason for interest in this restriction is that, in the proof of NP-completeness the reduction outputs exactly such instances. For these instances, we design a factor 2 approximation algorithm running in polynomial time.

Plan of the Paper. We present the definitions that we use and the graph-theoretic definition of the problem in section 2. In section 3 we present our algorithm to solve the problem when the mRNA secondary structure is hairpin shaped as in the SECIS. We will then show in section 4 that the problem becomes NP-complete, if we allow arbitrary mRNA bond structures (e.g. pseudo-knot structures). The approximation algorithm is presented in section 5.

2   Preliminaries

We follow graph theoretic notation and definitions as presented in the standard text by Harary [8]. A vertex labeled graph $\Gamma(V, E)$ is an undirected graph. An undirected graph is said to be *outer-planar* if it can be drawn satisfying the following conditions: 1.) all the vertices lie on a line, 2) all the edges lie on one side of the line, and 3.) two edges that intersect do so only at their end points. Such a drawing is called an outer-planar embedding of the graph.

The complement of a variable taking nucleotide values, is denoted by the superscript C. We also use the set $\{0, 1, 2, 3\}$ to represent $\{A, C, G, U\}$. Recall that (C, G) and (A, U) are complementary pairs. We choose $0, 1, 2, 3$ to correspond to C, G, A, U, respectively.

Statement of the Problem
Input: A labeled graph $\Gamma = (V, E)$, $V = \{v_1, \ldots, v_{3n}\}$. n functions, $f_1, \ldots, f_n$ are also part of the input. f_i is associated with $\{v_{3i-2}, v_{3i-1}, v_{3i}\}, 1 \leq i \leq n$, and, is a function defined on $\{A, C, G, U\}^3$ and takes rational values.
Output: Find vector $(N_1, \ldots, N_{3n}) \in \{A, C, G, U\}^{3n}$ s.t. N_k is assigned to v_k and

1. *Complementarity Condition:* $\{v_k, v_l\} \in E(\Gamma)$ implies that $N_k = N_l^C$.
2. $\sum_{i=1}^{n} f_i(N_{3i-2}, N_{3i-1}, N_{3i})$ is maximized. This function is referred to as the *cost* of the assignment at which it is evaluated.

We denote this problem by $MRSO(\Gamma, f_1, \ldots, f_n)$. The edges of Γ are referred to as *bonds*. The decision version of MRSO is $MRSD(\Gamma, f_1, \ldots, f_n, c)$. In the decision version an input is accepted if and only if there is an assignment to the vertices that satisfies the complementarity condition and, the cost of the assignment is at least c.
About Graphs: Γ is referred to as *structure graph*. For a structure graph Γ, we define the *implied graph* Γ^{impl} to be a graph with $V(\Gamma^{\text{impl}}) = \{u_1, \ldots, u_n\}$ and $E(\Gamma^{\text{impl}}) = \left\{ \{u_i, u_j\} \middle| \begin{array}{l} \exists r \in \{3i - 2, 3i - 1, 3i\} : \\ \exists s \in \{3j - 2, 3j - 1, 3j\} : (v_r, v_s) \in E(\Gamma) \end{array} \right\}$. From the structure graph associated with a SECIS, it follows that the corresponding implied graph is outer-planar. We follow the convention that, independent of the subscript or superscript, u denotes a vertex from $V(\Gamma^{\text{impl}})$ and v denotes a vertex

from Γ. Given $I \subseteq \{1..n\}$, U_I denotes $\{u_j \mid j \in I\}$. $E(\Gamma^{\mathrm{impl}})|_I$ denotes the edge set of the induced subgraph of Γ^{impl} on U_I. $E(\Gamma)|_I$ denotes the edge set of the induced subgraph of Γ on $\{v_{3i-2}, v_{3i-1}, v_{3i} \mid i \in I\}$. For a vertex $u \in V(\Gamma^{\mathrm{impl}})$, we denote the corresponding vertices in $V(\Gamma)$ by $\mathrm{nuc}_i(u)$, $i = 1, 2, 3$. $\mathrm{nucs}(u)$ denotes the set $\{\mathrm{nuc}_1(u), \mathrm{nuc}_2(u), \mathrm{nuc}_3(u)\}$. f_u denotes the function associated with $\{\mathrm{nuc}_1(u), \mathrm{nuc}_2(u), \mathrm{nuc}_3(u)\}$ in the input.

About Codons: Given a sequence of codons $L_1 \ldots L_n$, let $N_1 \ldots N_{3n}$ be the corresponding nucleotide sequence obtained by replacing L_i by the corresponding nucleotide representation $N_{3i-2} N_{3i-1} N_{3i}$. $L_1 \ldots L_n$ is said to *satisfy* $E(\Gamma)$ iff the corresponding nucleotide representation $\{N_{3i-2} N_{3i-1} N_{3i}\}_{i=1}^n$ satisfies the complementarity conditions for Γ. L_i and L_j are said to be *valid for* $\{u_i, u_j\}$ *w.r.t.* $E(\Gamma)$ if the corresponding nucleotides $N_{3i-2} N_{3i-1} N_{3i}$ and $N_{3j-2} N_{3j-1} N_{3j}$ satisfy the complementarity condition imposed by $E(\Gamma)$.

3   When Γ^{impl} Is Outer-Planar

We present a linear time recursive algorithm to solve MRSO when Γ^{impl} is outer-planar. The hairpin shape of the SECIS is captured by the outer-planarity of Γ^{impl}. The algorithm is based on a recurrence relation that we prove in this section. Also, the algorithm solves MRSO even when the functions can take arbitrary values.

We fix $u_1, \ldots, u_n$ as the ordering of the vertices from left to right on the line in an outer-planar embedding of Γ^{impl}. For each $1 \leq i \leq n$, let f_i be the function associated with u_i. Having fixed an embedding of the graph on the line, we do not distinguish between a vertex and its index. That is, the interval $[i, \ldots, i+k]$ denotes the set of vertices $\{u_i, \ldots, u_{i+k}\}$.

We now define the notion of compatibility between codons L_i and L_j assigned to vertices u_i and u_j, respectively. We denote compatibility by $\equiv_{E(\Gamma)}$.[1] The three lines in $\equiv$ serve to indicate that the complementarity conditions dictated by $E(\Gamma)$ should be satisfied. For $1 \leq i, j \leq n$, define $(i, L_i) \equiv_{E(\Gamma)} (j, L_j)$ by

$$
(i, L_i) \equiv_{E(\Gamma)} (j, L_j) = \begin{cases} true & \text{if } \{u_i, u_j\} \notin E(\Gamma^{\mathrm{impl}}) \\ true & \text{if } \{u_i, u_j\} \in E(\Gamma^{\mathrm{impl}}) \text{ and} \\ & L_i \text{ and } L_j \text{ are } valid \text{ for } \{u_i, u_j\} \text{ w.r.t. } E(\Gamma) \\ false & otherwise, \end{cases}
$$

$(i, L_i) \not\equiv_{E(\Gamma)} (j, L_j)$ denotes that $(i, L_i) \equiv_{E(\Gamma)} (j, L_j)$ is false. We now define the following function:

$$
w(i, i+k, L_i, L_{i+k}) = \max_{L_{i+1}, \ldots, L_{i+k-1}} \left\{ \sum_{i \leq j \leq i+k} f_j(L_j) \ \middle| \ \begin{matrix} L_i, \ldots, L_{i+k} \\ \text{satisfies } \mathrm{E}(\Gamma)|_{[i..i+k]} \end{matrix} \right\}.
$$

---

[1] **Note:** To read, associate the symbol $\equiv$ with the word *compatible*

If the set over which the maximum is taken is empty, the value of the function is considered as $-\infty$. This function forms the central part of our algorithm to solve the problem because $\max\limits_{L_1,L_n} w(1,n,L_1,L_n)$ is the value that we are interested in computing. The base cases for w is the following:

$$w(i,i+1,L_i,L_{i+1}) = \begin{cases} f_i(L_i) + f_{i+1}(L_{i+1}) & \text{if } (i,L_i) \equiv_{E(\Gamma)} (i+1,L_{i+1}) \\ -\infty & \text{otherwise.} \end{cases}$$

We solve the problem on an interval $[i..i+k]$ by splitting $[i..i+k]$ into two parts and solving the resulting subproblems. If there is no edge between i and any vertex in the interval $[i..i+k-1]$, the interval is split into $[i..i+1]$ and $[i+1..i+k]$. Otherwise, we choose the farthest vertex p in $[i..i+k-1]$ which is adjacent to i. The edge (i,p) is called the *maximal* edge in $E(\Gamma^{\text{impl}})|_{[i..i+k-1]}$. Then, the interval is split into $[i..p]$ and $[p..i+k]$ (see Figure 3). Hence, define $\text{next}(i,i+k)$ for $k \geq 2$ by

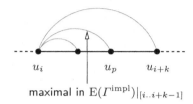

maximal in $\text{E}(\Gamma^{\text{impl}})|_{[i..i+k-1]}$

Fig. 3. Def. of $\text{next}(i,i+k)$

$$\text{next}(i,i+k) = \begin{cases} i+r & \text{if } \{i,i+r\} \text{ is maximal in } \text{E}(\Gamma^{\text{impl}})|_{[i..i+k-1]} \\ i+1 & \text{otherwise.} \end{cases}$$

Theorem 1 (Recurrence). *Let* $(\Gamma, f_1, \dots, f_n)$ *be instance of the MRSO. For* $k \geq 2, 1 \leq i \leq n-k$, *let* $p = \text{next}(i,i+k)$. *Then*

$$w(i,i+k,L_i,L_{i+k}) =$$
$$\begin{cases} -\infty & (i,L_i) \not\equiv_{E(\Gamma)} (i+k,L_{i+k}) \\ \max\limits_{L_p} \begin{pmatrix} w(i,p,L_i,L_p) \\ + w(p,i+k,L_p,L_{i+k}) \\ - f_p(L_p) \end{pmatrix} & \text{if } (i,L_i) \equiv_{E(\Gamma)} (i+k,L_{i+k}), \end{cases}$$

Proof. For the case when $(i,L_i) \not\equiv_{E(\Gamma)} (i+k,L_{i+k})$, the result is true because we follow the convention that the maximum over an empty set is $-\infty$. So we consider the case when $(i,L_i) \equiv_{E(\Gamma)} (i+k,L_{i+k})$. Let $k \geq 2$. We wish to evaluate,

$$\max\limits_{L_p} \big(w(i,p,L_i,L_p) + w(p,i+k,L_p,L_{i+k}) - f_p(L_p). \big)$$

The above term is equal to

$$= \max\limits_{L_p} \left\{ \begin{array}{l} \max\limits_{L_{i+1},\dots,L_{p-1}} (\sum\limits_{i \leq j \leq p} f_j(L_j) \mid L_i, \dots, L_p \text{ satisfies } \text{E}(\Gamma)|_{[i..p]}) \\ + \max\limits_{L_{p+1},\dots,L_{i+k-1}} (\sum\limits_{p \leq j \leq i+k} f_j(L_j) \mid L_p, \dots, L_{i+k} \text{ satisfies } \text{E}(\Gamma)|_{[p..i+k]}) \\ - f_p(L_p) \end{array} \right\}$$

Now we show that this equals

$$\max_{L_{i+1},\ldots,L_{i+k-1}} \left\{ \sum_{i \leq j \leq i+k} f_j(L_j) \mid \begin{array}{c} L_i,\ldots,L_{i+k} \\ \text{satisfies } E(\Gamma)|_{[i..i+k]}) \end{array} \right\}$$

We first observe that there are no edges between vertices in the interval $[i + 1..p - 1]$ to vertices in the interval $[p + 1, i + k]$. This is because we have assumed as input an outer-planar drawing of Γ^{impl}. Therefore, the above two quantities are equal because the maximum for the problem in the region $[i..i + k]$ for a fixed L_i, L_p, L_{i+k} can be obtaining by finding $w(i, p, L_i, L_p)$ and $w(p, i + k, L_p, L_{i+k})$. We make this statement precise now. Let $L'_{i+1},\ldots,L'_{p-1}$ be some codon sequences such that $L_i, L'_{i+1},\ldots,L'_{p-1}, L_p$ satisfies $E(\Gamma)|_{[i..p]}$, and $L''_{p+1},\ldots,L''_{i+k-1}$ be a codon sequence such that $L_p, L''_{p+1},\ldots,L''_{i+k-1}, L_{i+k}$ satisfies $E(\Gamma)|_{[p..i+k]}$. We have to show that $L_i, L'_{i+1},\ldots,L'_{p-1}, L_p, L''_{p+1},\ldots,L''_{i+k-1}, L_{i+k}$ satisfies $E(\Gamma)|_{[i..i+k]}$. L_i and L_{i+k} are compatible since we have assumed $(i, L_i) \equiv_{E(\Gamma)} (i + k, L_{i+k})$. The compatibility within regions $[i..p]$ and $[p..i + k]$ hold by assumption. The compatibility for vertex pairs (u_r, u_s) with $r \in [i..p - 1]$, and the second point is in $s \in [p + 1..i + k]$ holds since (i, p) is a maximal edge, and there are no edges in $E(\Gamma^{impl})$ between $r \in [i + 1..p - 1]$ and $s \in [p + 1..i + k]$.

Algorithm Based on the Recurrence. This theorem gives us a recursive algorithm to find the optimum value and an assignment of codons attaining that value for an input $(\Gamma, f_1,\ldots,f_n)$. We present the properties of the algorithm at the top-most level of the recursion. Recall that our goal is to compute $\max_{L_1,L_n} w(1, n, L_1, L_n)$ and a corresponding assignment of codons. Let $p = \text{next}(1, n)$. If $w(1, p, L_1, L_p)$ and $w(p, n, L_p, L_n)$ is known for all choices of L_1, L_p, L_n, then we can compute $\max_{L_1,L_n} w(1, n, L_1, L_n)$. Observe that we can find a codon L_p that achieves this maximum value. This computation takes constant time (64^3 to be exact). In particular, we can compute $w(1, n, L_1, L_n)$ for all choices of L_1, L_n and a corresponding assignment to L_p. Since Γ^{impl} is an outer-planar, we get only n subproblems and each one takes only 64^3 time. Therefore, we can compute $\max_{L_1,L_n} w(1, n, L_1, L_n)$ and an assignment of codons attaining the optimum value in $64^3 n$ time.

4 NP-Completeness of MRSD

In the previous section we have designed an efficient algorithm when the implied graph is outer-planar. While this captures the essence of the SECIS and simple pseudo-knots, it does not address the case of arbitrary shapes for the mRNA secondary structure. We now show that in its generality the MRSD is NP-complete. We first show that MRSD is in NP and then, we present a polynomial time many-one reduction of 3SAT to MRSD. As 3SAT is an NP-complete problem, it follows that MRSD is also NP-complete. The 3SAT problem is defined as follows in [6]:

3-Satisfiability (3SAT)

Instance: Collection $\psi = \{c_1, \dots, c_m\}$ of clauses on a finite set U of variables such that $|c_i| = 3$ for $1 \leq i \leq m$.

Question: Is there a truth assignment to U which satisfies all the clauses in ψ?

MRSD is in NP because, given an instance $(\Gamma, f_1, \dots, f_n, c)$, we guess an assignment whose cost is at least c and then verify in polynomial time that it is indeed valid. The verification involves checking if the assignment satisfies the complementarity conditions and if its cost is at least c.

Reduction. We present a polynomial time reduction that takes an instance ψ of 3SAT as input and outputs an instance M of MRSD such that ψ is satisfiable if and only if M is a true instance of MRSD. For ease of presentation, with out loss of generality, we consider instances of 3SAT in which every variable occurs more than once.

Outline: Given a set of m clauses $\{c_1, \dots, c_m\}$, each with 3 literals from a set of n variables $\{x_1, \dots, x_n\}$, we construct, a graph on $12m$ vertices and, a set of $4m$ functions, each one a function of an ordered set of 3 variables which take values from the set $\{0, 1, 2, 3\}$.

Variables: Let $\{y_1, \dots, y_{3m}, z_1, \dots, z_{3m}\}$ be a set of variables. Each vertex of the graph is labeled by one of these variables or its complement such that each variable is assigned to exactly two vertices.

The functions: Let $r_1, \dots, r_n$ be positive integers where r_i denotes the number occurrences of variable x_i in the set of m clauses. In the set of functions we construct, there are r_i functions associated with $x_i, 1 \leq i \leq n$, and, one function for each clause. The functions associated with $x_i, 1 \leq i \leq n$, are denoted by $Vf_{ij}, 1 \leq j \leq r_i$, and, the function associated with clause $c_i, 1 \leq i \leq m$, is denoted by Cf_i. Each function is associated with 3 distinct variables (association is described below) from the set $\{y_1, \dots, y_{3m}, z_1, \dots, z_{3m}\}$.

Variables associated with the functions: The variables associated with function Cf_i are $(y_{3i-2}, y_{3i-1}, y_{3i})$. For example, if x_1, x_2, x_3 were variables occurring in c_i, then x_1 is replaced by y_{3i-2}, x_2 by y_{3i-1} and, x_3 by y_{3i}. This substitution naturally defines the set of y_js associated with a variable x_i. In the above example, y_{3i-2} is in the set associated with x_1. Clearly, variable x_i has r_i variables from the set $\{y_1, \dots, y_{3m}\}$ associated with it. To this set of variables associated with x_i, we add r_i more variables from the set $\{z_1, \dots, z_{3m}\}$ such that a variable from this set is not associated with two different variables x_i and x_j. Now what is left is to describe the variables associated with Vf_{ij} for x_i, $1 \leq i \leq n$ and $1 \leq j \leq r_i$. Without loss of generality, let $\{y_{i_1}, \dots, y_{i_{r_i}}, z_{i_1}, \dots, z_{i_{r_i}}\}$ be the set of variables associated with x_i. For every $j, 1 \leq j \leq r_i$, the literals associated with Vf_{ij} are $z_{i_j}^{\mathsf{C}}, y_{i_j}^{\mathsf{C}}, z_{i_{(j+1) \bmod r_i}}$.

Values taken by the functions: The functions evaluate to 0 on all the non-boolean settings to its variables. For each $1 \leq i \leq m$, on a boolean assignment to its variables, Cf_i takes the value taken by c_i under the same boolean assignment to its (c_i's) variables. For this definition to make sense, the above mentioned substitution correspondence between the variables of Cf_i and c_i must be respected. For $1 \leq i \leq n, 1 \leq j \leq r_i$, Vf_{ij} is the same function. It evaluates to 1 on the two settings $100, 011$ to its variables and, 0 elsewhere.

For a fixed i, the variables associated with Vf_{ij} along with the definition of Vf_{ij} guarantees that, in the solution (for MRSD) corresponding to a satisfying assignment, the variables associated with x_i get the same value. It is easy to observe that each variable from $\{y_1, \ldots, y_{3m}, z_1, \ldots, z_{3m}\}$ is associated with exactly two functions.

The Graph Γ: The graph Γ has $12m$ vertices arranged on a line. The first $3m$ are assigned unique labels from $y_1, \ldots, y_{3m}$. The remaining $9m$ vertices are partitioned into n parts such that the i-th part has $3r_i$ vertices and, this part is associated with the variable x_i. The $3r_i$ vertices are partitioned into r_i consecutive blocks of 3 vertices each. The 3 vertices in the j-th block are labeled $z_{i_j}^{\mathbb{C}}, y_{i_j}^{\mathbb{C}}, z_{i_{(j+1) \mod r_i}}$. There is an edge between two vertices that have complementary labels. Having described Γ, to get an instance of $MRSD$ we must describe the associated function. For this, the $12m$ vertices are partitioned into $4m$ consecutive blocks of 3 vertices each and a function is associated with each block. The function associated with the i-th block for $1 \leq i \leq m$ is Cf_i. For the remaining blocks: the j-th block in the part associated with the variable x_i gets the function Vf_{ij}. The graph Γ with the set of associated functions is an instance of the MRSO problem. The natural decision version that captures the satisfiability of the input $3CNF$ is, $MRSD(\Gamma, f_1, \ldots, f_{4m}, 4m)$. We make the following claim whose proof follows easily from the construction.

Theorem 2. *Let ϕ be a 3-CNF and let $\Gamma, f_1, \ldots, f_{4m}$ be the graph and similarity functions output by the reduction described above. Then ϕ is satisfiable if and only if $MRSD(\Gamma, f_1, \ldots, f_{4m}, 4m)$ is true.*

Proof. Let $\phi = c_1 \wedge \ldots \wedge c_m$ be a 3-SAT formula on variables $\{x_1, \ldots, x_n\}$. Let $r_1, \ldots, r_n$ be non negative integers such that r_i denotes the number of occurrences of x_i in ϕ. Let $S = \{y_1, \ldots, y_{3m}, z_1, \ldots, z_{3m}\}$ be a set of new variables. Let $S_i \subseteq S, 1 \leq i \leq n$ be associated with $x_i, 1 \leq i \leq n$, respectively. From the construction above, we know that S_i has $2r_i$ elements in it, $1 \leq i \leq n$. Let $y_i, y_i^{\mathbb{C}}, z_i, z_i^{\mathbb{C}}, 1 \leq i \leq 3m$ be the labels associated with the $12m$ vertices in the graph Γ. Let ϕ be satisfiable and let $a_1, \ldots, a_n$ be a satisfying assignment to the variables $x_1, \ldots, x_n$, respectively. Furthermore, let $b_1, \ldots, b_{6m}$ be the assignment to the variables $y_1, \ldots, y_{3m}, z_1 \ldots, z_{3m}$, respectively, obtained by assigning the variables in S_i the value a_i. Each of the $4m$ similarity functions output by the reduction evaluate to 1 under $b_1 \ldots, b_{6m}$. Consequently, $MRSD(\Gamma, f_1, \ldots, f_{4m}, 4m)$ is true.

To prove the other direction, let $MRSD(\Gamma, f_1, \ldots, f_{4m}, 4m)$ be true. Since, there are $4m$ similarity functions, there exists an assignment $b_1, \ldots, b_{6m}$ to the variables $y_1, \ldots, y_{3m}, z_1, \ldots, z_{3m}$, respectively, such that each of the similarity functions evaluates to 1. Consequently, each of the variables in the sets $S_i, 1 \leq i \leq n$, gets the same value. Let $a_1, \ldots, a_n$ be the assignment to $x_1, \ldots, x_n$, respectively, by assigning to x_i the value of the variables in S_i in the assignment $b_1, \ldots, b_{6m}$. $a_1, \ldots, a_n$ is a satisfying assignment to ϕ because, the similarity functions associated with the clauses evaluate to 1 under $b_1, \ldots, b_{6m}$ and the remaining similarity functions ensure that by evaluating to 1, the variables in $S_i, 1 \leq i \leq n$, get the same value. Therefore, ϕ is satisfiable.

5   Constant Factor Approximation Algorithms for MRSO

We design a polynomial time constant approximation algorithm for instances of $MRSO$ in which every vertex has degree at most 1 and the functions take non-negative values. Such instances are exactly those that are output by the reduction from $3SAT$. Clearly, the maximum vertex degree in the implied graph of such instances is at most 3.

Outline. An approximate solution for the problem is obtained by partitioning the vertex set of the graph, into two parts, such that the induced graph on each part is outer-planar. We solve the problem separately for each part, using our algorithm for outer-planar graph, and then combine the two solutions to yield a solution whose cost is at least a constant fraction of the optimum cost. To present the approximation algorithm we need the following notation.

Notation. Let $W \subseteq V(\Gamma)$ such that for each $v \in W$, if $\{v, v'\} \in E(\Gamma)$, then $v' \in W$. Unless stated explicitly, the sets denoted by W satisfy this property. Let $\phi : W \to \{A, C, G, U\}$ be an assignment that satisfies the complementarity conditions. Given Γ, ϕ, W, let $\Gamma_{(\phi, W)}$ denote the graph obtained from Γ by removing the edges whose end points are in W. $(\Gamma_{(\phi, W)}, f_1, \ldots, f_n)$ is a new instance of the $MRSO$ obtained from $(\Gamma, f_1, \ldots, f_n)$ and ϕ. The optimization problem on $(\Gamma_{(\phi, W)}, f_1, \ldots, f_n)$ is to find the best assignment to vertices in $V(\Gamma) - W$ given an assignment to the vertices in W. It is easy to see, due the degree of every vertex in Γ being at most 1, that there is an assignment such that the complementarity conditions are satisfied. Furthermore, if $\Gamma_{(\phi, W)}^{impl}$ is outer-planar then we can find an optimal assignment to vertices in $V - W$ using the algorithm for outer-planar graphs with a simple modification. The modification relates to the choice of codons for $u \in \Gamma_{(\phi, W)}^{impl}$ for which some vertices in $\mathrm{nucs}(u)$ have been assigned a value by ϕ. We illustrate this modification by an example whose generalization is straightforward. Let $\mathrm{nuc}_1(u)$ be assigned A under ϕ. Given this, the codons that are to be assigned to u are to be chosen only from the set $\{A\} \times \{A, C, G, U\} \times \{A, C, G, U\}$.

The Algorithm. We use the algorithm for the outer-planar case (see Section 3) as a subroutine. Let M be a maximal independent set of Γ^{impl} and $O = V(\Gamma^{impl}) - M$. An independent set is a maximal independent set if it is not the subset of a larger independent set. As M is a maximal independent set, every vertex in O is adjacent to some vertex in M. Let G_o and G_m denote the induced subgraphs of Γ^{impl} on O and M, respectively. Each vertex in O has degree at most 2 in G_o (recall, maximum degree in Γ^{impl} is 3). Therefore, G_o is outer-planar. Let Γ_o and Γ_m be the induced subgraphs of Γ on the vertex sets $\{\mathrm{nucs}(u) : u \in O\}$ and $\{\mathrm{nucs}(u) : u \in M\}$, respectively.

We first solve MRSO on the instances $(\Gamma_o, (f_u)_{u \in G_o})$ and $(\Gamma_m, (f_u)_{u \in G_m})$. Let ϕ_o and ϕ_m be assignments achieving the optimum costs C_o and C_m, respectively. Recall that ϕ_o (resp. ϕ_m) are assignments to the nucleotides corresponding to the vertices in G_o (resp. G_m). Of the two assignments, we consider the assignment that achieves the larger cost. Let us consider the case when it is ϕ_m. The case when C_o is larger is addressed by replacing m by o in the presentation below. Let $V_m \subseteq V(\Gamma)$ be the set of vertices that are assigned values by ϕ_m. Let $W = V_m \bigcup \{$ vertices of V_o with a neighbour in $V_m\}$. ϕ_m is extended on $W - V_m$ by assigning to each $v \in W - V_m$ a value so that the complementarity conditions are satisfied. Let ϕ denote the resulting assignment to vertices in W. Now we solve the $MRSO$, $(\Gamma_{(\phi,W)}, f_1, \ldots, f_n)$. The edges in $\Gamma_{(\phi,W)}$ are exactly the edges of Γ_o. Since G_o is outer-planar, it follows that $\Gamma_{(\phi,W)}^{\text{impl}}$ is outer-planar. Consequently, we can use our algorithm for outer-planar graphs to solve $(\Gamma_{(\phi,W)}, f_1, \ldots, f_n)$. Let A be the assignment to the vertices of $V(\Gamma)$ obtained as above. Clearly, A satisfies the complementarity conditions. Furthermore,

Theorem 3. *Cost of $A \geq$ half the optimum cost.*

Proof. Assuming that the table has only non-negative entries, it follows that the optimum cost for Γ cannot exceed $2C_m$. The reason is that C_m is the larger of C_m and C_o. Consequently, all the other assignments to G_o have cost at most C_m. By the algorithm described above, the cost of the assignment obtained is at least C_m. This is because, in A, ϕ_m is the assigned to the vertices in G_m. Therefore, the above algorithm guarantees a solution whose cost is with in half the optimum cost.

The running time of the algorithm is dominated by the time to solve the MRSO on the graph G_o, G_m and $\Gamma_{(\phi,W)}$. Since the graphs are all outer-planar, it follows that the algorithm has only a linear running time. Also, for implementation G_o is chosen as a spanning forest of Γ^{impl} and G_m is the induced graph on $V(\Gamma^{\text{impl}}) - V(G_o)$. Due to the maximum degree in Γ^{impl} being at most 3, G_m and G_o are outer-planar.

6   Discussion and Concluding Remarks

We have addressed the problem of finding an mRNA sequence that has a required secondary structure, is maximally similar to a given mRNA sequence, and the amino acid sequence it encodes by it is maximally similar to a given amino acid sequence. This problem is motivated by an experiment on an *E. coli* amino acid sequence, where a selenocysteine was added in place of a cysteine, and the functionality of the original amino acid sequence was retained. The requirements of our problem captures exactly the effects of the experiment on *E. coli*. Our results show that when the given mRNA sequence is an SECIS, the problem can be solved by an efficient computer program. The reason is that the hairpin shape of the SECIS secondary structure allows the problem to be split recursively into two smaller subproblems. Our algorithm also works when the shape of the given mRNA is an outer-planar graph. Outer-planarity captures certain mRNA shapes which are known as simple pseudo-knots [1]. We then consider the complexity of the problem when arbitrary shapes are allowed. In that case, we observe that the problem becomes intractable as shown by the NP-completeness result. We then design a factor 2 approximation algorithm for this case. The exact algorithm for outer-planar graphs has been implemented. In a first version, only one optimum assignment was calculated. Since any mathematical modeling of a biological problem (including ours) cannot fully capture the biological function, it is inadequate to output just one optimum assignment. An optimum assignment output by any program may not be the right biological answer. Therefore, it is important to be able to inspect all possible optima. Towards this end the program has been updated to output all assignments attaining the optimum. Then, we pick the required sequence by inspection.

The NP-completeness is proved by reducing 3-SAT to our problem. The fact that our problem can be solved when the shape of the input mRNA is outer-planar suggests an open direction of research toward algorithms for SAT. Our results show that when a certain graph associated with a 3-SAT formula is outer-planar, then satisfiability can be decided in linear time. The question is, for what other graph structures can 3-SAT be solved in polynomial time? These questions remain interesting with respect to sub-exponential algorithms for 3-SAT. For the current state of research on sub-exponential time algorithms for SAT see [3] and the references therein.

The approximation algorithm is able to guarantee a factor 2 approximation due to the assumption that the functions take only non-negative values. While this addresses the inputs obtained from the NP-completeness reduction, it is not a practical assumption considering that the input functions can take negative values in the biological setting. Another problem faced is that the output mRNA sequence should not contain any stop codon. We have observed that this property is not satisfied by the algorithm and an attempt to satisfy this yields in a highly technical factor 14 approximation algorithm. An interesting question here is to design a better approximation algorithm that would satisfy the stop codon constraint.

Acknowledgment. The first author thanks Prof. Böck from the Institute of Microbiology for explaining the biochemical background of selenocysteine, for many suggestions and the fruitful cooperation on the problems discussed in this paper. The authors thank Sebastian Will for many discussions, and for reading draft versions of this paper. The second author thanks Jan Johannsen for his support of this research.

References

1. T. Akutsu. Dynamic programming algorithms for rna secondary structure prediction with pseudoknots. In *Discrete Applied Mathematics, 104*, pages 45–62. 2000.
2. A. Böck, K. Forchhammer, J. Heider, and C. Baron. Selenoprotein synthesis: an expansion of the genetic code. *Trends Biochem Sci*, 16(12):463–467, 1991.

3. E. Dantsin, E.A. Hirsch, S. Ivanov, and M. Vsemirnov. Algorithms for sat and upper bounds on their compleixty. In *Electronic Colloquium on Computational Compleixty, Report 12.* 2001.
4. M. O. Dayhoff, R. M. Schwartz, and B. C. Orcutt. A model of evolutionary change in proteins. In M. O. Dayhoff, editor, *Atlas of Protein Sequence and Structure,* volume 5, supplement 3, pages 345–352. National Biomedical Research Foundation, Washington, DC, 1978.
5. P. J. Farabaugh. Programmed translational frameshifting. *Microbiology and Molecular Biology Reviews,* 60(1):103–134, 1996.
6. Michael R. Garey and David S. Johnson. Computers and intractability: A guide to the theory of NP-completeness. *W.H.Freeman,* 1979.
7. D. P. Giedroc, C. A. Theimer, and P. L. Nixon. Structure, stability and function of rna pseudoknots involved in stimulating ribosomal frameshifting. *Journal of Molecular Biology,* 298(2):167–186, 2000.
8. Frank Harary. Graph theory. *Addison Wesley,* 1972.
9. Stephane Hazebrouck, Luc Camoin, Zehava Faltin, Arthur Donny Strosberg, and Yuval Eshdat. Substituting selenocysteine for catalytic cysteine 41 enhances enzymatic activity of plant phospholipid hydroperoxide glutathione peroxidase expressed in *escherichia coli. Journal of Biological Chemistry,* 275(37):28715–28721, 2000.
10. M.S. Waterman. Introduction to computational biology. Chapman and Hall, London, 1995.

Pure Dominance Constraints

Manuel Bodirsky[1] and Martin Kutz[2]

[1] Humboldt Universität zu Berlin, Germany
bodirsky@informatik.hu-berlin.de
[2] Freie Universität Berlin, Germany
kutz@math.fu-berlin.de

Abstract. We present an efficient algorithm that checks the satisfiability of *pure dominance constraints*, which is a tree description language contained in several constraint languages studied in computational linguistics. Pure dominance constraints partially describe unlabeled rooted trees. For arbitrary pairs of nodes they specify sets of admissible relative positions in a tree. The task is to find a tree structure satisfying these constraints.
Our algorithm constructs such a solution in time $\mathcal{O}(m^2)$ where m is the number of constraints. This solves an essential part of an open problem posed by Cornell.

Keywords. Efficient algorithms, tree descriptions, constraint satisfaction

1 Introduction

Tree description languages became an important tool in computational linguistics over the last twenty years. Grammar formalisms have been proposed which derive logical descriptions of trees representing the syntax of a string [18, 22, 8]. Acceptance of a string is then equivalent to the satisfiability of the corresponding logical formula.

In computational semantics, the paradigm of *underspecification* aims at manipulating the partial description of tree-structured semantic representations of a sentence rather than at manipulating the representations themselves [21, 9]. So the key issue in both, constraint based grammar and semantic formalisms, is to collect partial descriptions of trees and to *solve* them, i.e., to find a tree structure that satisfies all constraints.

Cornell [5] introduced a simple but powerful tree description language. It contains literals for dominance, precedence and equality between nodes, and disjunctive combinations of the these. He also gave a saturation algorithm based on local propagations but the completeness proof for the algorithm turned out to be wrong [6]. Recently Bodirsky, Duchier and Niehren [personal communication] in fact found a counterexample to the completeness of this algorithm.

In this paper, we will always talk about finite rooted unlabeled trees without any order on the children of a node. The constraint language that we are considering contains variables denoting the nodes in a tree together with dominance,

H. Alt and A. Ferreira (Eds.): STACS 2002, LNCS 2285, pp. 287–298, 2002.
© Springer-Verlag Berlin Heidelberg 2002

disjointness, and equality constraints between them. Moreover, we can combine the above constraints disjunctively. For example, we can state that some node x must either lie above some other node y or must be disjoint to it, i.e., lie in a different branch of the tree. We call the resulting tree description language *pure dominance constraints*.

In contrast to pure dominance constraints, the constraint language introduced by Cornell is based on trees with an order on the children of the tree nodes. His *precedence constraints* are stronger than our disjointness constraints. They allow to state that some node x in the tree lies to the left of some other node y. Interestingly, the counterexample to the completeness of Cornell's algorithm does not make use of this distinction. It can be formulated using disjointness only instead of precedence.

Pure dominance constraints are a fragment of *dominance constraints* with set operators [18, 3, 7] which have applications both in syntax and in semantics of computational linguistics [22, 12, 9]. Those provide literals for *labeling*, *arity*, *parent-child relations*, and *dominance* to partially describe finite labeled trees. Checking satisfiability of dominance constraints is NP-complete [15]. For the restrictive, though linguistically still relevant fragment of *normal dominance constraints*, there is a polynomial time satisfiability algorithm [2, 14]. Dominance constraints with set operators have been studied in [7] and are important for formulating algorithms for extensions of dominance constraints with more expressive constraints [10, 4]. Pure dominance constraints are a strict subset of dominance constraints with set operators since they lack the possibility to specify labeling and arity of nodes in the described tree.

Hilfinger, Lawler and Rote investigated a similar problem. They gave an efficient algorithm that minimizes the depth of a given tree satisfying additional tree constraints [13], which has applications in compiling block structured computer languages.

If nodes in a tree are interpreted as intervals over the real line, pure dominance constraints translate into a fragment of Allen's interval algebra [1] where the intervals are a laminar family, i.e., they are overlap-free. Dominance between nodes then corresponds to containment of intervals. Allen's full interval constraint logic has many applications in temporal reasoning but is NP-complete in its unrestricted form. Nebel and Bürckert [20] presented an algorithm for the largest tractable subclass "ORD-Horn" of Allen's interval algebra, which decides satisfiability in time $\mathcal{O}(n^3)$ and constructs a solution in time $\mathcal{O}(n^5)$, where n is the number of intervals. Note that this algorithm cannot be used to solve pure dominance constraints since the translation into interval constraints may have non-laminar solutions, which do not retranslate into trees.

In this paper, we will present an efficient and complete algorithm that tests satisfiability of pure dominance constraints by directly constructing a solution to the problem instance. It runs in time $\mathcal{O}(m^2)$, where m is the number of constraints in the input. This is considerably faster than the known algorithms for the related problems on interval constraints. The performance is achieved by

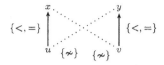

Fig. 1. A visualization of the unsatisfiable pure dominance constraint $\{\ x\{\approx\}v, u\{<,=\}$ $x, v\,\{<,=\}\,y, u\,\{\approx\}\,y\ \}$.

a recursive greedy strategy that works directly on the constraint graph avoiding the consistent path technique and saturation algorithms of, for example, [17].

2 Pure Dominance Constraints

Pure dominance constraints are tree specifications. So let us first fix some conventions for trees. We will always consider finite rooted directed trees, the arcs pointing upward towards the root. Normally the nodes of a tree will be denoted by $a, b, c, \ldots$. We write $a \le b$ and say that b dominates a if there is a path from a to b. We write $a < b$ for $a \le b$ and $a \ne b$. The expression $a \sim b$ denotes comparability of a and b, that is, $a \le b$ or $a \ge b$. Two incomparable nodes $a \not\sim b$ are called *disjoint*. Note that for every pair a, b of nodes in a tree, exactly one of the following cases holds:

$$a < b, \quad a > b, \quad a \approx b, \quad a = b.$$

Pure dominance constraints allow to partially describe the structure of a tree by use of arbitrary disjunctions of these four cases.

Definition 1. Let V be a set of variable symbols ranged over by $\{x, y, z, \ldots\}$, and $\mathcal{R} := \{<, >, \approx, =\}$ a set of binary relation symbols. Then a *pure dominance constraint (PDC)* is a set Φ of literals of the form $x\ \Omega\ y$ where $\Omega \subseteq \mathcal{R}$ and $x, y \in V$. We write V_Φ for the variable symbols occurring in Φ.

A *Solution* (T, α) of a PDC Φ is a tree T together with a surjective mapping $\alpha : V_\Phi \to T$ from the used variables onto the nodes of this tree *satisfying* all literals in Φ. A literal $x\ \Omega\ y$ is *satisfied* by a solution (T, α) if $\alpha(x)\ \rho\ \alpha(y)$ holds in the tree T for some relation $\rho \in \Omega$.

We will omit the subscript of V_Φ whenever the reference to a specific pure dominance constraint Φ is clear. For readability we will often write $\bar{x}$ for $\alpha(x)$.

If a PDC has a solution we call it *satisfyable*. Note that we are not only interested in the *constraint satisfaction problem* for PDCs, that is, the question whether a given PDC is satisfiable.[1] We also want to *solve* a given constraint Φ efficiently, i.e., compute a solution (T, α).

---

[1] Note that the constraint satisfaction problem considered here does not fall into the class of *classical constraint satisfaction problems* since we do not map the variables into a *finite* template. Hence the classification of tractable constraint satisfaction problems of, e.g., [11] does not apply to our problem.

Figure 1 shows an unsatisfiable constraint. It has no solution since there is no candidate for a root and the mapping α from variables into the tree must be surjective.

The requirement that α be onto is natural since we are looking for a tree structure on V_Φ and not just any homomorphism into a tree. Nevertheless, dropping this restriction only changes the problem marginally. If (T, α) is a solution with nonsurjective α, we can remove every nonroot node in T that does not have a preimage, making all children of α children of α's parent. It remains the problem to map some $x \in V_\Phi$ to the root of T. As shown in Figure 1 this may not be possible although a nonsurjective solution might exist. But the general case can easily be transformed into the surjective case: Given a PDC Φ, we introduce a new variable x_0 and constraints $x_0 \left\{ > \right\} x$ for each $x \in V_\Phi$. This forces x_0 to map to the root of T and therefore, finding an arbitrary solution for Φ is equivalent to finding a surjective one for the extended constraint.

3   Restricted Constraints

As a first step towards solving pure dominance constraints, we show that each PDC can be expressed with the following three types of literals only:

$$x \left\{ <, = \right\} y, \quad x \left\{ \approx, = \right\} y, \quad x \left\{ <, >, \approx \right\} y. \tag{1}$$

The singletons $\{>\}$, $\{<\}$, $\{\approx\}$ and $\{=\}$ can be written as intersections of these; and $\{>, =\}$ is simply the first literal of (1) flipped.

If we show how to express the literals

$$x \left\{ <, >, = \right\} y \quad \text{and} \quad x \left\{ <, \approx, = \right\} y \tag{2}$$

with those in (1) we are finished, since $\{<, \approx\}$, $\{>, \approx\}$ and $\{<, >\}$ are representable as intersections of literals from (1) and (2). The two extremal sets $\{<, >, \approx, =\}$ and $\emptyset$ are not needed since the former imposes no restrictions on the tree and the latter is, by definition, unsatisfyable. We can construct the literals in (2) from those in (1) by means of auxiliary variables:

For each literal $\phi = x \left\{ <, >, = \right\} y$, we introduce a new variable z and replace ϕ by the two literals $z \left\{ <, = \right\} x$ and $z \left\{ <, = \right\} y$. By tree shape, these two imply $\overline{x} \sim \overline{y}$, just as ϕ does.

For each literal $\phi = x \left\{ <, \approx, = \right\} y$, we introduce a new variable z and replace ϕ by the two literals $x \left\{ <, = \right\} z$ and $z \left\{ \approx, = \right\} y$. Now either $\overline{z} = \overline{y}$, which implies $\overline{x} \leq \overline{y}$, or $\overline{z} \not\sim \overline{y}$, and therefore $\overline{x} \not\sim \overline{y}$ by tree shape. Since the transformed constraint implies the original one, a solution (T, α) for the former is also a solution for the latter if we restrict α to the original variables.

The above modifications also maintain solvability. Let (T, α) be a solution for the original constraint Φ containing $x \left\{ <, >, = \right\} y$. Then we have $\overline{x} \sim \overline{y}$. Assume wlog. that $\overline{x} \leq \overline{y}$. Letting $\overline{z} := \overline{x}$ then satisfies the new constraints $x \left\{ >, = \right\} z$ and $z \left\{ <, = \right\} y$. If Φ contained a literal $x \left\{ <, \approx, = \right\} y$, we can extend the solution (T, α) by $\overline{z} := \overline{y}$ if $\overline{x} < \overline{y}$, and $\overline{z} := \overline{x}$ otherwise. In both cases $x \left\{ <, = \right\} z$ and $z \left\{ \approx, = \right\} y$ are satisfied in the modified constraint.

Thus we can express any PDC with the three basic literals from (1). We give them special names: The literal $x\,\{\approx, =\}\,y$ is called *parallelity* literal—in analogy to a pair of parallel lines in the plane which are either disjoint or coincide. The literal $x\,\{<, >, \approx\}\,y$ is called a *distinctness* and $x\,\{<, =\}\,y$ a *dominance* literal. Our algorithm works on PDCs containing these three kinds of literals only. Note that the above transformations may introduce a new node for each constraint in Φ. Hence, when dealing with running times, we will have to state whether we consider the general problem or the reduced one.

4   Preparations

We assume that our PDC has already been transformed so that it only contains dominance, parallelity and distinctness literals. We define the *dominance graph* $G = (V, E)$ of a PDC Φ on the variables $V := V_\Phi$ by letting $(x, y) \in E$ iff $x\,\{<, =\}\,y$ is in the PDC. Parallelity and distinctness literals are represented by two symmetric irreflexive binary relations P and D, respectively. We let $x\,P\,y$ iff $x\,\{\approx, =\}\,y$ and $x\,D\,y$ iff $x\,\{<, >, \approx\}\,y$. Because our algorithm and its correctness proofs are given in terms of graph theory, we will normally call elements $x \in V$ *nodes*. This should cause no confusion; it will always be clear whether we talk about nodes of V or T.

It turns out useful to define the relation $\leq$ on the set V, letting $x \leq y$ if and only if there is a directed path from x to y in the graph G. Obviously $\leq$ is a quasi order (that is, it is transitive, reflexive, but not necessarily antisymmetric) and we may use the symbols $\geq, <, >$ as usual. As with trees, we use the expression $x \sim y$, which is a short for $x \leq y$ or $x \geq y$. Observe that a binary relation $\leq$ is now defined on the tree T and on the variable set V. Again, the reference will always be clear.

The problem can now be restated as follows: Given an instance (V, E, P, D) of a pure dominance constraint, find a tree T and a surjection $\alpha : V \to T$ so that the following conditions are satisfied:

$$
\begin{array}{ll}
x \leq y \Rightarrow \overline{x} \leq \overline{y} & \text{dominance} \\
x\,P\,y \Rightarrow \overline{x} \not\sim \overline{y} \text{ or } \overline{x} = \overline{y} & \text{parallelity} \\
x\,D\,y \Rightarrow \overline{x} \neq \overline{y} & \text{distinctness}
\end{array}
$$

The critical parts of the problem are pairs of parallel elements that may not become equivalent in the solution. If D is empty, we can simply map all nodes to the root. And it is also easy to see how to construct a solution if P is empty. So let us consider pairs of nodes that are mapped to incomparable nodes in the tree.

Lemma 1. *Let (T, α) be a solution to the instance (V, E, P, D). Consider any sequence $(a_0, a_1, a_2, \ldots, a_r)$ in V with $a_{i-1} \sim a_i$ for $1 \leq i \leq r$ and with $\overline{a_0} \not\sim \overline{a_r}$. Then there exists an index $j \in \{1, \ldots, r-1\}$ with $\overline{a_j} > \overline{a_0}$ and $\overline{a_j} > \overline{a_r}$.*

Proof. By induction on the length r. For $r = 1$ there is nothing to show since the combination $a_0 \sim a_1$, $a_0 \ P \ a_1$, and $a_0 \ D \ a_1$ has no solution. If $r = 2$, uncomparability of a_0 and a_2 directly implies $\overline{a_1} > \overline{a_0}, \overline{a_2}$.

So let $r > 2$. If $a_{i-1} \sim a_{i+1}$ for some $0 < i < r$, we may remove a_i from the sequence and our claim follows by induction. Otherwise, the relations in the sequence alternate and hence there exists an $i \in \{1, \ldots, r-1\}$ with $a_i < a_{i-1}$ and $a_i < a_{i+1}$. By tree shape, $\overline{a_{i-1}}$ and $\overline{a_{i+1}}$ have to be comparable. Assume wlog. $\overline{a_{i-1}} \leq \overline{a_{i+1}}$. If we let $E' := E \cup \{(a_{i-1}, a_{i+1})\}$ then (T, α) is also a solution to the instance (V, E', P, D). Now $(a_0, \ldots, a_{i-1}, a_{i+1}, \ldots, a_r)$ is a sequence of length $r - 1$ with any two neighboring elements related in (V, E') and again the lemma follows by induction. $\qquad\square$

As an example, let us consider the constraint graph in Figure 2. In any solution, we must have $\overline{x_1} \not\sim \overline{x_3}$. Thus by Lemma 1, we get $\overline{x_2} > \overline{x_1}, \overline{x_3}$. Now $\overline{x_2} \not\sim \overline{x_4}$ would contradict Lemma 1; hence $\overline{x_2} = \overline{x_4}$ by **parallelity**. The positions of u, v, and w are not of interest.

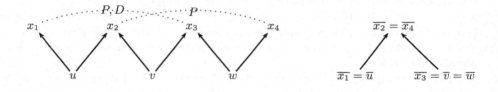

Fig. 2. A constraint graph and a solution

5   The Algorithm

Our algorithm is based on the recursive function `treeify`. Given a set $W \subseteq V$ of nodes, `treeify` creates a new tree T and defines the mapping α on W, yielding $\alpha(W) = T$. Finally the tree T is returned. More specifically, `treeify` first creates an empty tree T containing only a root r and maps a certain subset $S \subseteq W$ to r. Afterwards it calls itself recursively on subsets $W_i \subseteq W \setminus S$. The trees returned by these recursive calls are then linked to the root r.

Before we come to the crucial part of this function, namely the choice of the sets S and W_i, let us recollect some basic notions from graph theory: An *undirected* path in a directed graph may use arcs in any direction, ignoring their orientation. A strongly (weakly) connected component of G is a maximally induced subgraph U of G with a directed (an undirected) path from a to b for any two nodes $a, b \in U$. We introduce the expression $(V, E)|_W := (W, E \cap W^2)$ for the subgraph of (V, E) induced by a node set $W \subseteq V$.

So let us find out how to choose the set S. By construction we get $\overline{x} > \overline{y}$ for all $x \in S$ and $y \in W \setminus S$. Thus **dominance** demands that S be closed under

```
treeify(W):

    compute the strongly connected components of the graph (V, E ∪ P)|_W;
    if no free component exists then exit "problem has no solution"
    else pick a free component S;
    create new tree T with root r;
    for each x ∈ S do α[x] := r od;
    compute the weakly connected components W_1, ..., W_k of the graph (V, E)|_{W\S};
    for i = 1 to k do
        T_i := treeify(W_i);
        link T_i directly under r;
    od;
    return T;
```

Fig. 3. The function treeify

directed reachability in (V, E), i.e., no E-edges may leave S. Suppose now that a pair of parallel elements $a \ P \ b$ gets split by S, say $a \in S$ but $b \notin S$. This implies $\bar{a} > \bar{b}$, in contradiction to parallelity. Hence either $a, b \in S$ or $a, b \notin S$ for $a \ P \ b$. And finally, S may obviously not contain a pair of distinct elements $u \ D \ v$. These observations lead to the following definition.

Definition 2. A strongly connected component C of the graph $(V, E \cup P)|_W$, $W \subseteq V$, is called a *terminal* component (of W) if its outdegree in $(V, E \cup P)|_W$ is zero. That is, $(E \cup P) \cap (C \times (W \setminus C)) = \emptyset$. We call a terminal component $C \subseteq W$ *free* if $x \ \not{D} \ y$ for all $x, y \in C$.

Note that the graph $(V, E \cup P)|_W$ contains—in addition to the arcs from E—a bidirectional edge between each pair $x \ P \ y$ of parallel nodes in W.

For the set S, the function treeify picks an arbitrary free component of W. We will show later that, given the whole problem has a solution, such a free component always exists. It remains to choose the partition $W_1, ..., W_k$ of $W \setminus S$. By construction, elements in different sets W_i of this partition will become incomparable in the tree. So we may simply choose the W_i to be the weakly connected components of the graph $(V, E)|_{W\setminus S}$. This way parallelity and distinctness constraints between different components are automaticly resolved. The proofs will be given below.

Figure 3 gives a detailed description of the function treeify. The whole algorithm for an input (V, E, P, D) just consists of the function call treeify(V). Note that although the tree structure T is created recursively, the mapping α is not passed at function calls but is accessed globally. We assume that at the beginning we have $\alpha(x) = \text{undef}$ for all $x \in V$.

5.1   Correctness Proof

Soundness. We first show that the algorithm only returns correct solutions, i.e., it satisfies the conditions dominance, parallelity and distinctness of Section 4.

To verify **dominance**, we need to show that reachability in the induced graphs $(V, E)|_W$ is equivalent to reachability in the input graph (V, E), which defines the relation $\leq$ on V.

Lemma 2. *Let $W \subseteq V$ be a node set that occurs as an argument to the function* `treeify` *throughout the execution of the algorithm. For any pair $x, y \in W$, each directed path p in (V, E) from x to y is also a path in $(V, E)|_W$.*

Proof. By induction over the recursion. For the initial case $W = V$ there is nothing to do. So assume that the statement is true for some call `treeify`(W), and that we have just chosen a free component S. We show that it remains true for all recursive calls of `treeify` on weakly connected components W' of $(V, E)|_{W \setminus S}$. Let x, y be nodes in W'. No path p from x to y contains a node from S since S is terminal. Hence, p lies completely in W' because W' contains all nodes $z \in W \setminus S$ that are reachable from x. $\qquad\square$

Now **dominance** follows easily by induction. Consider a call `treeify`(W) and a pair $x, y \in W$ with $x \leq y$. If $x \in S$, we also have $y \in S$ and thus $\overline{x} = \overline{y}$. If $x \notin S$ and $y \in S$, we get $\overline{y} = r$ while x is passed on through a recursive call and hence $\overline{x}$ will lie in a subtree below r. Otherwise, x and y both fall into the same weakly connected component W_i and we may conclude **dominance** by induction.

To verify **parallelity**, we again consider an arbitrary call `treeify`(W). Let x, y be nodes in W. If $x, y \in S$ we have $\overline{x} = \overline{y}$. Otherwise, none of x and y lies in S since these nodes are strongly connected in $(V, E \cup P)|_W$. If they lie in different weakly connected components of $(V, E)|_{W \setminus S}$, their images $\overline{x}$ and $\overline{y}$ will fall into different subtrees of r. Hence $\overline{x} \not\sim \overline{y}$. If they lie in the same component, our statement again follows by induction.

Condition **distinctness** is satisfied since any two nodes x, y with $\overline{x} = \overline{y}$ lie in the same free component S of some instance of `treeify`. But since S is free such pairs satisfy $x \not\!\!D y$.

The constructed mapping $\alpha : V \to T$ is obviously a surjection.

Completeness. So we know that the algorithm only returns correct solutions. We now show that it also always finds one if the PDC has a solution. To this end, we assume that the input (V, E, P, D) has a solution (T, α). Then we consider an arbitrary instance of `treeify`(W) that is executed by the algorithm and prove that the set W must contain a free component.

The basic idea for the proof is to construct iteratively a sequence $x_0, x_1, x_2, \ldots$ in W with $\overline{x_0} \leq \overline{x_1} \leq \ldots$. Some of these inequalities will be strict. For the others with $\overline{x_i} = \overline{x_{i+1}}$ we will guaranty the existence of a path from x_i to x_{i-1} in $(V, E \cup P)|_W$. From these paths we will be able to conclude that our sequence eventually enters a free component. For preparation we prove the following extension of Lemma 1.

Lemma 3. *Let (T, α) be a solution to the input (V, E, P, D) and let W be the argument of a call to* `treeify`. *Then for any pair $u, v \in W$ with $\overline{u} \not\sim \overline{v}$, there exists a node $y \in W$ with $\overline{y} > \overline{u}, \overline{v}$.*

Proof. We distinguish two cases. If $W = V$, α maps neither u nor v to the root r of the tree T since otherwise $\overline{u} \sim \overline{v}$. Hence any $y \in \alpha^{-1}(r)$ has the desired property.

If $W \neq V$, the graph $(V, E)|_W$ is weakly connected. So there is an undirected path q in this graph connecting u and v. By Lemma 1 there exists a node $y \in W$ with $\overline{y} > \overline{u}$ and $\overline{y} > \overline{v}$.   □

The next lemma is the basis for our construction of the sequence $x_0, x_1, x_2, \ldots$.

Lemma 4. *Let (T, α) be a solution to the input (V, E, P, D) and let W be the argument of a call to* `treeify`. *Let C be a strongly connected component of $(V, E \cup P)|_W$ that is not free and let $x \in C$. Then there either exists*

1. *a node $y \in W$ with $\overline{y} > \overline{x}$ or*
2. *a node $w \in W \setminus C$ satisfying $\overline{w} \geq \overline{x}$, together with a directed path q in $(V, E \cup P)|_W$ from x to w.*

Proof. If C is terminal, there exist two nodes $a, b \in C$ with $a \, D \, b$ since C is not free. Choose two directed paths $p_1 = (x = u_1, \ldots, u_k = a)$ and $p_2 = (a = u_k, \ldots, u_l = b)$ in $(V, E \cup P)|_C$. By distinctness, we have $\overline{u_k} \neq \overline{u_l}$. Hence there must be an index $j < l$ with $\overline{x} = \overline{u_j} \neq \overline{u_{j+1}}$.

Since $(u_j, u_{j+1}) \in E \cup P$, we have $u_j \leq u_{j+1}$ or $u_j \, P \, u_{j+1}$. In the former case let $y := u_{j+1}$; then $\overline{y} > \overline{u_j} = \overline{x}$. In the latter case we get $\overline{u_j} \not\sim \overline{u_{j+1}}$ and Lemma 3 yields the desired $y \in W$.

If C is not terminal there exist two nodes $z \in C$ and $w \in W \setminus C$ with $(z, w) \in E \cup P$. Since $(z, w) \in P$ would imply $w \in C$, we even have $z \leq w$. Let $p = (x = u_0, \ldots, u_k = z)$ be a directed path in $(V, E \cup P)|_C$. Consider the index set $I = \{i \mid 0 \leq i < k, \ \overline{u_i} \not\leq \overline{u_{i+1}}\}$.

If $I \neq \emptyset$, let $j := \min I$. We have $u_j \not\leq u_{j+1}$ by dominance and thus $u_j \, P \, u_{j+1}$ by the definition of p. Therefore $\overline{x} \leq \overline{u_j} \not\leq \overline{u_{j+1}}$; and by parallelity even $\overline{u_j} \not\sim \overline{u_{j+1}}$. Now Lemma 3 again yields the sought-after $y \in W$.

If $I = \emptyset$, we have $\overline{x} \leq \overline{z} \leq \overline{w}$ and the path $q = (u_0, \ldots, u_k, w)$ connects x and w in $(V, E \cup P)|_W$.   □

Theorem 1. *If the instance (V, E, P, D) has a solution then for each function call* `treeify(W)` *throughout the execution of the algorithm, the respective graph $(V, E \cup P)|_W$ contains a free component.*

Proof. We pick an arbitrary node $x_0 \in W$. If it lies in a free component we are finished. If not, we apply Lemma 4 to x_0 yielding some $x_1 \in W$ with $\overline{x_1} > \overline{x_0}$ or only $\overline{x_1} \geq \overline{x_0}$. But in the latter case, x_1 is reachable from x_0 via a directed path in $(V, E \cup P)|_W$. We may repeat this step as long as no free component is found. This yields a sequence $x_0, x_1, x_2, \ldots$ in W.

We claim that this sequence is finite, i.e., some x_i will eventually lie in a free component. Assume for contradiction that it is infinite. Then we have $x_k = x_l$ for some pair of indices $k < l$. Since $\overline{x_{i+1}} > \overline{x_i}$ for any index $i \in \{k, \ldots, l-1\}$ would imply the contradiction $\overline{x_k} < \overline{x_k}$, all pairs x_i, x_{i+1}, $k \le i < l$ of neighbouring nodes are connected by a directed path. But then $x_k, x_{k+1}, \ldots, x_{l-1}$ all lie in the same strongly connected component of $(V, E \cup P)_W$, in contradiction to Lemma 4.   □

5.2   Running Time

We measure the running time of our algorithm in terms of the number of nodes $n := \#V$ and constraints $m := \#E + \#P + \#D$. For our estimates, we assume $m \ge n - 1$. Otherwise, the graph $(V, E \cup P \cup D)$ is not connected and we can simply treat its weakly connected components independently and then link the resulting trees in any arbitrary order.

Theorem 2. *The algorithm runs in time $\mathcal{O}(nm)$.*

Proof. Computing the strongly connected components of $(V, E \cup P)|_W$ takes $\mathcal{O}(m)$ steps [16, Sec. 3]. Checking each components outdegree and internal distinctness constraints also takes $\mathcal{O}(m)$ steps. Creating the tree and linking the subtrees can be done in $\mathcal{O}(n)$ steps. Hence, each instance of `treeify` performs $\mathcal{O}(m)$ operations, not counting the time needed for the recursive calls.

Since in each instance of `treeify` at least one node is mapped to the tree, there happen no more than n calls to `treeify` throughout the whole execution of the algorithm.   □

Observe that this result applies to the reduced constraint set only. Since the reduction of pure dominance constraints to their restricted form might introduce a new vertex for each constraint, constructing a solution for a general pure dominance constraint might take time quadratic in the number of constraints.

Note that we cannot improve upon this bound of Theorem 2 by arguing that the sets $W \subseteq V$ in calls to `treeify` become smaller and smaller throughout the recursion. As a counterexample, consider a dominance graph that consists of a single directed path with all neighbouring nodes distinct. Then in each instance `treeify($W$)`, the free component S contains only a single element and the remaining graph $(V, E)|_{W \setminus S}$ remains weakly connected. On this input, our algorithm actually performs $\Omega(nm)$ steps.

Possible improvements. We want to hint at a part in our algorithm that appears inefficient and which might be subject to future improvements. The strongly connected components of subgraphs of $(V, E \cup P)$ and the weakly connected components of subgraphs of (V, E) are computed over and over again. The algorithm does not reuse any connectivity information.

Mapping the component S to the tree in any instance of `treeify` only removes nodes from the respective graph, and splitting the remaining set $W \setminus S$ into its weakly connected components only deletes edges. Thus we never introduce nodes or edges. Therefore, it might be possible to reuse information by means of the decremental dynamic connectivity techniques presented in [23]. But those results only apply to undirected graphs and it is not clear how to extend them to strongly connected components. And even if we could speed up the connectivity computations, we would still have to find the free components.

6 Conclusion

We introduced the tree description language *pure dominance constraints* which forms a natural fragment of constraint languages used in computational linguistics. It is expressive enough to contain a counterexample to the completeness of Cornell's algorithm.

We presented an algorithm that constructs a solution of pure dominance constraints in quadratic time. It is mainly based on weak and strong graph connectivity. We think that the running time can still be improved using decremental dynamic connectivity techniques.

A main contribution of this paper is the method of using greedy algorithms and graph theoretic concepts for constraint satisfaction problems instead of the consistent path method and saturation algorithms. We may ask if it is possible to avoid the consistent path method for Nebel and Bürckert's maximal tractable subclass of Allen's interval algebra, too. A faster algorithm for this subclass would be of practical importance since it could be used to bound the branching in backtracking algorithms for the important full interval algebra [19, 24]. A first step in this direction would be the extension of our algorithm such that it can also deal with precedence constraints as defined in the original language by Cornell.

Acknowledgements. We want to thank Denys Duchier and Joachim Niehren for pointing us to the constraint problem, and an anonymous referee for his usefull remarks.

References

1. J. F. Allen. Maintaining knowledge about temporal intervals. *Communications of the ACM*, 26(11):832–843, 1983.
2. E. Althaus, D. Duchier, A. Koller, K. Mehlhorn, J. Niehren, and S. Thiel. An efficient algorithm for the configuration problem of dominance graphs. In *Proceedings of the 12th ACM-SIAM Symposium on Discrete Algorithms*, pages 815–824, Washington, DC, Jan. 2001.
3. R. Backofen, J. Rogers, and K. Vijay-Shanker. A first-order axiomatization of the theory of finite trees. *Journal of Logic, Language, and Information*, 4:5–39, 1995.
4. M. Bodirsky, K. Erk, A. Koller, and J. Niehren. Beta Reduction Constraints. In *Proceedings of the 12th International Conference on Rewriting Techniques and Applications*, Utrecht, 2001.
5. T. Cornell. On determining the consistency of partial descriptions of trees. In *32nd Annual Meeting of the Association for Computational Linguistics*, pages 163–170, 1994.
6. T. Cornell. On determining the consistency of partial descriptions of trees. http://tcl.sfs.nphil.uni-tuebingen.de/~cornell/mss.html, 1996.

7. D. Duchier and J. Niehren. Dominance constraints with set operators. In *First International Conference on Computational Logic (CL2000)*, LNCS, July 2000.
8. D. Duchier and S. Thater. Parsing with tree descriptions: a constraint-based approach. In *Sixth International Workshop on Natural Language Understanding and Logic Programming (NLULP'99)*, pages 17–32, Las Cruces, New Mexico, Dec. 1999.
9. M. Egg, A. Koller, and J. Niehren. The Constraint Language for Lambda Structures. *Journal of Logic, Language, and Information*, 2001. To appear.
10. K. Erk, A. Koller, and J. Niehren. Processing underspecified semantic representations in the constraint language for lambda structures. *Journal of Language and Computation*, 2001. To appear.
11. T. Feder and M. Vardi. The computational structure of monotone monadic SNP and constraint satisfaction: a study through datalog and group theory. *SIAM J. Comput.*, (28):57–104, 1999.
12. C. Gardent and B. Webber. Describing discourse semantics. In *Proceedings of the 4th TAG+ Workshop*, Philadelphia, 1998. University of Pennsylvania.
13. P. Hilfinger, E. L. Lawler, and G. Rote. Flattening a rooted tree. *Applied Geometry and Discrete Mathematics*, 4:335–340, 1991.
14. A. Koller, K. Mehlhorn, and J. Niehren. A polynomial-time fragment of dominance constraints. In *Proceedings of the 38th Annual Meeting of the Association of Computational Linguistics*, Hong Kong, Oct. 2000.
15. A. Koller, J. Niehren, and R. Treinen. Dominance constraints: Algorithms and complexity. In *Third International Conference on Logical Aspects of Computational Linguistics (LACL '98)*, Grenoble, France, Dec. 1998.
16. T. Lengauer. *Combinatorial algorithms for integrated circuit layout*. Wiley – Teubner, 1990.
17. A. K. Mackworth. Consistency in networks of relations. *Artificial Intelligence*, 8:99–118, 1977.
18. M. P. Marcus, D. Hindle, and M. M. Fleck. D-theory: Talking about talking about trees. In *Proceedings of the 21st ACL*, pages 129–136, 1983.
19. B. Nebel. Solving hard qualitative temporal reasoning problems: Evaluating the efficiency of using the ORD-Horn class. 1(3):175–190, 1997.
20. B. Nebel and H.-J. Bürckert. Reasoning about temporal relations: A maximal tractable subclass of Allen's interval algebra. *Journal of the ACM*, 42(1):43–66, 1995.
21. M. Pinkal. Radical Underspecification. In *Proc. 10th Amsterdam Colloquium*, 1996.
22. J. Rogers and V. Shanker. Reasoning with descriptions of trees. In *Proceedings of the 30th Meeting of the Association for Computational Linguistics*, 1992.
23. M. Thorup. Decremental dynamic connectivity. *J. Algorithms*, 33(2):229–243, Nov. 1999.
24. P. van Beek. Exact and approximate reasoning about temporal relations. *Computational Intelligence*, 6:132–144, 1990.

Improved Quantum Communication Complexity Bounds for Disjointness and Equality

Peter Høyer[1],[*] and Ronald de Wolf[2],[**]

[1] Dept. of Comp. Sci., Univ. of Calgary, AB, Canada. hoyer@cpsc.ucalgary.ca
[2] UC Berkeley. 583 Soda Hall, Berkeley CA 94720, USA. rdewolf@cs.berkeley.edu

Abstract. We prove new bounds on the quantum communication complexity of the disjointness and equality problems. For the case of exact and non-deterministic protocols we show that these complexities are all equal to $n+1$, the previous best lower bound being $n/2$. We show this by improving a general bound for non-deterministic protocols of de Wolf. We also give an $O(\sqrt{n} \cdot c^{\log^* n})$-qubit bounded-error protocol for disjointness, modifying and improving the earlier $O(\sqrt{n} \log n)$ protocol of Buhrman, Cleve, and Wigderson, and prove an $\Omega(\sqrt{n})$ lower bound for a class of protocols that includes the BCW-protocol as well as our new protocol.

1 Introduction

The area of *communication complexity* deals with abstracted models of distributed computing, where one only cares about minimizing the amount of communication between the parties and not about the amount of computation done by the individual parties. The standard setting is the following. Two parties, Alice and Bob, want to compute some function $f : \{0,1\}^n \times \{0,1\}^n \to \{0,1\}$. Alice receives input $x \in \{0,1\}^n$, Bob receives $y \in \{0,1\}^n$, and they want to compute $f(x,y)$. For example, they may want to find out whether $x = y$ (the *equality* problem) or whether x and y are characteristic vectors of disjoint sets (the *disjointness* problem). A communication protocol is a distributed algorithm where Alice first does some computation on her side, then sends a message to Bob, who does some computation on his side, sends a message back, etc. The *cost* of the protocol is measured by the number of bits (or qubits, in the quantum case) communicated on a worst-case input (x,y).

As in many other branches of complexity theory, we can distinguish between various different "modes" of computation. Letting $P(x,y)$ denote the acceptance probability of the protocol (the probability of outputting 1), we consider four different kinds of protocols for computing f,

- An *exact* protocol has $P(x,y) = f(x,y)$, for all x, y

[*] Supported in part by Canada's NSERC and the Pacific Institute for the Mathematical Sciences.
[**] Supported by Talent grant S 62–565 from the Netherlands Organization for Scientific Research. Work conducted while at CWI, Amsterdam, partially supported by EU fifth framework project QAIP, IST–1999–11234.

H. Alt and A. Ferreira (Eds.): STACS 2002, LNCS 2285, pp. 299–310, 2002.
© Springer-Verlag Berlin Heidelberg 2002

- A *non-deterministic* protocol has $P(x,y) > 0$ if and only if $f(x,y) = 1$, for all x, y
- A *one-sided error* protocol has $P(x,y) \geq 1/2$ if $f(x,y) = 1$, and $P(x,y) = 0$ if $f(x,y) = 0$
- A *two-sided error* protocol has $|P(x,y) - f(x,y)| \leq 1/3$, for all x, y.

These four modes of computation correspond to those of the computational complexity classes P, NP, RP, and BPP, respectively.

Protocols may be *classical* (send and process classical bits) or *quantum* (send and process quantum bits). Classical communication complexity was introduced by Yao [35], and has been studied extensively. It is well motivated by its intrinsic interest as well as by its applications in lower bounds on circuits, VLSI, data structures, etc. We refer to the book of Kushilevitz and Nisan [26] for definitions and results. We use $D(f)$, $N(f)$, $R_1(f)$, and $R_2(f)$ to denote the minimal cost of classical protocols for f in the exact, non-deterministic, one-sided error, and two-sided error settings, respectively.[1] Note that $R_2(f) \leq R_1(f) \leq D(f) \leq n+1$ and $N(f) \leq R_1(f) \leq D(f) \leq n+1$ for all f. Similarly we define $Q_E(f)$, $NQ(f)$, $Q_1(f)$, and $Q_2(f)$ for the quantum versions of these communication complexities (we will be a bit more precise about the notion of a quantum protocol in the next section). For all of these complexities, we assume Alice and Bob start out without any shared randomness or entanglement.

Quantum communication complexity was introduced by (again) Yao [36] and the first examples of functions where quantum communication complexity is less than classical communication complexity were given in [14,10,11,15,9]. In particular, Buhrman, Cleve, and Wigderson [9] showed for a specific *promise version* of the equality problem that $Q_E(f) \in O(\log n)$ while $D(f) \in \Omega(n)$. They also showed for the *intersection* problem (the negation of the disjointness problem) that $Q_1(\mathrm{INT}_n) \in O(\sqrt{n} \log n)$, whereas $R_2(\mathrm{INT}_n) \in \Omega(n)$ is a well known and non-trivial result from classical communication complexity [20,31]. Later, Raz [30] exhibited a promise problem with an exponential quantum-classical gap even in the bounded-error setting: $Q_2(f) \in O(\log n)$ versus $R_2(f) \in \Omega(n^{1/4}/\log n)$. Other results on quantum communication complexity may be found in [25,2,28,13,21,34,24,23].

The aim of this paper is to sharpen the bounds on the quantum communication complexities of the equality and disjointness (or intersection) problems, in the four modes we distinguished above. We summarize what was known prior to this paper,

- $n/2 \leq Q_1(\mathrm{EQ}_n), Q_E(\mathrm{EQ}_n) \leq n+1$ [25,13]
  $n/2 \leq NQ(\mathrm{EQ}_n) \leq n+1$ [34]
  $Q_2(\mathrm{EQ}_n) \in \Theta(\log n)$ [25]
- $n/2 \leq Q_1(\mathrm{DISJ}_n), Q_E(\mathrm{DISJ}_n) \leq n+1$ [25,13]
  $n/2 \leq NQ(\mathrm{DISJ}_n) \leq n+1$ [34]
  $\log n \leq Q_1(\mathrm{INT}_n), Q_2(\mathrm{DISJ}_n) \in O(\sqrt{n} \log n)$ [9].

---

[1] Kushilevitz and Nisan [26] use $N^1(f)$ for our $N(f)$, $R^1(f)$ for our $R_1(f)$ and $R(f)$ for our $R_2(f)$.

In Section 3 we first sharpen the non-deterministic bounds, by proving a general algebraic characterization of $NQ(f)$. In [34] it was shown for all functions f that

$$\frac{\log nrank(f)}{2} \leq NQ(f) \leq \log(nrank(f)) + 1,$$

where $nrank(f)$ denotes the rank of a "non-deterministic matrix" for f (to be defined more precisely below). It is interesting to note that in many places in quantum computing one sees factors of $\frac{1}{2}$ appearing that are essential, for example in the query complexity of parity [4,17], in the bounded-error query complexity of all functions [16], in superdense coding [5], and in lower bounds for entanglement-enhanced quantum communication complexity [13,28]. In contrast, we show here that the $\frac{1}{2}$ in the above lower bound can be dispensed with, and the upper bound is tight,[2]

$$NQ(f) = \log(nrank(f)) + 1.$$

Equality and disjointness both have non-deterministic rank 2^n, so their non-deterministic complexities are maximal: $NQ(\text{EQ}_n) = NQ(\text{DISJ}_n) = n + 1$. (This contrasts with their complements: $NQ(\text{NEQ}_n) = 2$ [27] and $NQ(\text{INT}_n) \leq N(\text{INT}_n) = \log n + 1$.) Since $NQ(f)$ lower bounds $Q_1(f)$ and $Q_E(f)$, we also obtain optimal bounds for the one-sided and exact quantum communication complexities of equality and disjointness. In particular, $Q_E(\text{EQ}_n) = n + 1$, which answers a question posed to one of us (RdW) by Gilles Brassard in December 2000.

The two-sided error bound $Q_2(\text{EQ}_n) \in \Theta(\log n)$ is easy to show, whereas the two-sided error complexity of disjointness is still wide open. In Section 4 we give a one-sided error protocol for the intersection problem that improves the $O(\sqrt{n} \log n)$ protocol of Buhrman, Cleve, and Wigderson by nearly a log-factor,

$$Q_1(\text{INT}_n) \in O(\sqrt{n} \cdot c^{\log^\star n}),$$

where c is a (small) constant. The function $\log^\star n$ is defined as the minimum number of iterated applications of the logarithm function necessary to obtain a number less than or equal to 1: $\log^\star n = \min\{r \geq 0 \mid \log^{(r)} n \leq 1\}$, where $\log^{(0)}$ is the identity function and $\log^{(r)} = \log \circ \log^{(r-1)}$. Even though $c^{\log^\star n}$ is exponential in $\log^\star n$, it is still very small in n, in particular $c^{\log^\star n} \in o(\log^{(r)} n)$ for every constant $r \geq 1$. It should be noted that our protocol is asymptotically somewhat more efficient than the BCW-protocol ($\sqrt{n} c^{\log^\star n}$ versus $\sqrt{n} \log n$), but is also more complicated to describe; it is based on a recursive modification of the BCW-protocol, an idea that previously has been used for claw-finding by Buhrman *et al.* [12, Section 5].

Proving good *lower* bounds on the Q_2-complexity of the disjointness and intersection problems is one of the main open problems in quantum communication complexity. Only logarithmic lower bounds are known so far for general

---

[2] Similarly we can improve the query complexity result $ndeg(f)/2 \leq NQ_q(f) \leq ndeg(f)$ of [34] to the optimal $NQ_q(f) = ndeg(f)$.

protocols [25,2,13]. A lower bound of $\Omega(n^{1/k})$ is shown in [24] for protocols exchanging at most $k \in O(1)$ messages. In Section 4.1 we prove a nearly tight lower bound of $\Omega(\sqrt{n})$ qubits of communication for all protocols that satisfy the constraint that their acceptance probability is a function of $x \wedge y$ (the n-bit AND of Alice's x and Bob's y), rather than of x and y "separately." Since DISJ_n itself is also a function only of $x \wedge y$, this does not seem to be an extremely strong constraint. The constraint is satisfied by a class of protocols that includes the BCW-protocol and our new protocol. It seems plausible that the general bound is $Q_2(\mathrm{DISJ}_n) \in \Omega(\sqrt{n})$ as well, but we have so far not been able to weaken the constraint that the acceptance probability is a function of $x \wedge y$.

2   Preliminaries

2.1   Quantum Computing

Here we briefly sketch the setting of quantum computation, referring to the book of Nielsen and Chuang [29] for more details. An m-qubit quantum *state* $|\phi\rangle$ is a superposition or linear combination over all classical m-bit states,

$$|\phi\rangle = \sum_{i \in \{0,1\}^m} \alpha_i |i\rangle,$$

with the constraint that $\sum_i |\alpha_i|^2 = 1$. Equivalently, $|\phi\rangle$ is a unit vector in $\mathbb{C}^{2^m}$. Quantum mechanics allows us to change this state by means of unitary (i.e., norm-preserving) operations: $|\phi_{\mathrm{new}}\rangle = U|\phi\rangle$, where U is a $2^m \times 2^m$ unitary matrix. A *measurement* of $|\phi\rangle$ produces the outcome i with probability $|\alpha_i|^2$, and then leaves the system in the state $|i\rangle$.

The two main examples of quantum algorithms so far, are Shor's algorithm for factoring n-bit numbers using a polynomial number (in n) of elementary unitary transformations [32] and Grover's algorithm for searching an unordered n-element space using $O(\sqrt{n})$ "look-ups" or queries in the space [18]. Below we use a technique called *amplitude amplification*, which generalizes Grover's algorithm.

Theorem 1 (Amplitude amplification [7]). *There exists a quantum algorithm* **QSearch** *with the following property. Let $\mathcal{A}$ be any quantum algorithm that uses no measurements, and let $\chi : \{1, \ldots, n\} \to \{0,1\}$ be any Boolean function. Let a denote the initial success probability of $\mathcal{A}$ of finding a solution (i.e., the probability of outputting some $i \in \{1, \ldots, n\}$ so that $\chi(i) = 1$). Algorithm* **QSearch** *finds a solution using an expected number of $O\left(\frac{1}{\sqrt{a}}\right)$ applications of $\mathcal{A}$, $\mathcal{A}^{-1}$, and χ if $a > 0$, and it runs forever if $a = 0$.*

Consider the problem of searching an unordered n-element space. An algorithm $\mathcal{A}$ that creates a uniform superposition over all $i \in \{1, \ldots, n\}$ has success probability $a \geq 1/n$, so plugging this into the above theorem and terminating after $O(\sqrt{n})$ applications gives us an algorithm that finds a solution with probability at least $1/2$ provided there is one, and otherwise outputs 'no solution'.

2.2   Communication Complexity

For classical communication protocols we refer to [26]. Here we briefly define quantum communication protocols, referring to the surveys [33,8,22,6] for more details. The space in which the quantum protocol works, consists of three parts: Alice's part, the communication channel, and Bob's part (we do not write the dimensions of these spaces explicitly). Initially these three parts contain only 0-qubits,

$$|0\rangle|0\rangle|0\rangle.$$

We assume Alice starts the protocol. She applies a unitary transformation $U_1^A(x)$ to her part and the channel. This corresponds to her initial computation and her first message. The length of this message is the number of channel qubits affected. The state is now

$$(U_1^A(x) \otimes I^B)|0\rangle|0\rangle|0\rangle,$$

where $\otimes$ denotes tensor product, and I^B denotes the identity transformation on Bob's part. Then Bob applies a unitary transformation $U_2^B(y)$ to his part and the channel. This operation corresponds to Bob reading Alice's message, doing some computation, and putting a return-message on the channel. This process goes back and forth for some k messages, so the final state of the protocol on input (x, y) will be (in case Alice goes last)

$$(U_k^A(x) \otimes I^B)(I^A \otimes U_{k-1}^B(y)) \cdots (I^A \otimes U_2^B(y))(U_1^A(x) \otimes I^B)|0\rangle|0\rangle|0\rangle.$$

The total *cost* of the protocol is the total length of all messages sent, on a worst-case input (x, y). For technical convenience, we assume that at the end of the protocol the output bit is the first qubit on the channel. Thus the acceptance probability $P(x, y)$ of the protocol is the probability that a measurement of the final state gives a '1' in the first channel-qubit. Note that we do not allow intermediate measurements during the protocol. This is without loss of generality: it is well known that such measurements can be postponed until the end of the protocol at no extra communication cost. As mentioned in the introduction, we use $Q_E(f)$, $NQ(f)$, $Q_1(f)$, and $Q_2(f)$ to denote the cost of optimal exact, non-deterministic, one-sided error, and two-sided error protocols for f, respectively.

The following lemma was stated summarily without proof by Yao [36] and in more detail by Kremer [25]. It is key to many of the earlier lower bounds on quantum communication complexity as well as to ours, and is easily proven by induction on ℓ.

Lemma 2 (Yao [36]; Kremer [25]). *The final state of an ℓ-qubit protocol on input (x, y) can be written as*

$$\sum_{i \in \{0,1\}^\ell} |A_i(x)\rangle|i_\ell\rangle|B_i(y)\rangle,$$

where the $A_i(x), B_i(y)$ are vectors (not necessarily of norm 1), and i_ℓ denotes the last bit of the ℓ-bit string i (the output bit).

The acceptance probability $P(x, y)$ of the protocol is the squared norm of the part of the final state that has $i_\ell = 1$. Letting a_{ij} be the 2^n-dimensional complex column vector with the inner products $\langle A_i(x)|A_j(x)\rangle$ as entries, and b_{ij} the 2^n-dimensional column vector with entries $\langle B_i(y)|B_j(y)\rangle$, we can write P (viewed as a $2^n \times 2^n$ matrix) as the sum $\sum_{i,j:i_\ell=j_\ell=1} a_{ij} b_{ij}^T$ of $2^{2\ell-2}$ rank 1 matrices, so the rank of P is at most $2^{2\ell-2}$. For example, for exact protocols this gives immediately that ℓ is lower bounded by $\frac{1}{2}$ times the logarithm of the rank of the communication matrix, and for non-deterministic protocols ℓ is lower bounded by $\frac{1}{2}$ times the logarithm of the non-deterministic rank (defined below). In the next section we show how we can get rid of the factor $\frac{1}{2}$ in the non-deterministic case.

We use $x \wedge y$ to denote the bitwise-AND of n-bit strings x and y, and similarly $x \oplus y$ denotes the bitwise-XOR. Let OR denote the n-bit function which is 1 if at least one of its n input bits is 1, and NOR be its negation. We consider the following three communication complexity problems,

- *Equality*: $\text{EQ}_n(x, y) = \text{NOR}(x \oplus y)$
- *Intersection*: $\text{INT}_n(x, y) = \text{OR}(x \wedge y)$
- *Disjointness*: $\text{DISJ}_n(x, y) = \text{NOR}(x \wedge y)$.

3 Optimal Non-deterministic Bounds

Let $f : \{0, 1\}^n \times \{0, 1\}^n \to \{0, 1\}$. A $2^n \times 2^n$ complex matrix M is called a *non-deterministic matrix* for f if it has the property that $M_{xy} \neq 0$ if and only if $f(x, y) = 1$ (equivalently, $M_{xy} = 0$ if and only if $f(x, y) = 0$). We use $nrank(f)$ to denote the *non-deterministic rank* of f, which is the minimal rank among all non-deterministic matrices for f. In [34] it was shown that

$$\frac{\log nrank(f)}{2} \leq NQ(f) \leq \log(nrank(f)) + 1.$$

In this section we show that the upper bound is the true bound. The proof uses the following technical lemma.

Lemma 3. *If there exist two families of vectors $\{A_1(x), \dots, A_m(x)\} \subseteq \mathbb{C}^d$ and $\{B_1(y), \dots, B_m(y)\} \subseteq \mathbb{C}^d$ such that for all $x \in \{0, 1\}^n$ and $y \in \{0, 1\}^n$, we have*

$$\sum_{i=1}^m A_i(x) \otimes B_i(y) = 0 \text{ if and only if } f(x, y) = 0,$$

then $nrank(f) \leq m$.

Proof. Assume there exist two such families of vectors. Let $A_i(x)_j$ denote the jth entry of vector $A_i(x)$, and let similarly $B_i(y)_k$ denote the kth entry of vector $B_i(y)$. We use pairs $(j, k) \in \{1, \dots, d\}^2$ to index entries of vectors in the d^2-dimensional tensor space. Note that

if $f(x,y) = 0$ then $\sum_{i=1}^{m} A_i(x)_j B_i(y)_k = 0$ for all (j,k),
if $f(x,y) = 1$ then $\sum_{i=1}^{m} A_i(x)_j B_i(y)_k \neq 0$ for some (j,k).

As a first step, we want to replace the vectors $A_i(x)$ and $B_i(y)$ by numbers $a_i(x)$ and $b_i(y)$ that have similar properties. We use the probabilistic method [1] to show that this can be done.

Let I be an arbitrary set of 2^{2n+1} numbers. Choose coefficients $\alpha_1, \dots, \alpha_d$ and $\beta_1, \dots, \beta_d$, each coefficient picked uniformly at random from I. For every x, define $a_i(x) = \sum_{j=1}^{d} \alpha_j A_i(x)_j$, and for every y define $b_i(y) = \sum_{k=1}^{d} \beta_k B_i(y)_k$. Consider the number

$$v(x,y) = \sum_{i=1}^{m} a_i(x) b_i(y) = \sum_{j,k=1}^{d} \alpha_j \beta_k \left(\sum_{i=1}^{m} A_i(x)_j B_i(y)_k \right).$$

If $f(x,y) = 0$, then $v(x,y) = 0$ for all choices of the α_j, β_k.

Now consider some (x,y) with $f(x,y) = 1$. There is a pair (j', k') for which $\sum_{i=1}^{m} A_i(x)_{j'} B_i(y)_{k'} \neq 0$. We want to prove that $v(x,y) = 0$ happens only with very small probability. In order to do this, fix the random choices of all α_j, $j \neq j'$, and β_k, $k \neq k'$, and view $v(x,y)$ as a function of the two remaining not-yet-chosen coefficients $\alpha = \alpha_{j'}$ and $\beta = \beta_{k'}$,

$$v(x,y) = c_0 \alpha \beta + c_1 \alpha + c_2 \beta + c_3.$$

Here we know that $c_0 = \sum_{i=1}^{m} A_i(x)_{j'} B_i(y)_{k'} \neq 0$. There is at most one value of α for which $c_0 \alpha + c_2 = 0$. All other values of α turn $v(x,y)$ into a linear equation in β, so for those α there is at most one choice of β that gives $v(x,y) = 0$. Hence out of the $(2^{2n+1})^2$ different ways of choosing (α, β), at most $2^{2n+1} + (2^{2n+1} - 1) \cdot 1 < 2^{2n+2}$ choices give $v(x,y) = 0$. Therefore

$$\Pr[v(x,y) = 0] < \frac{2^{2n+2}}{(2^{2n+1})^2} = 2^{-2n}.$$

Using the union bound, we now have

$$\Pr\left[\text{there is an } (x,y) \in f^{-1}(1) \text{ for which } v(x,y) = 0\right]$$
$$\leq \sum_{(x,y) \in f^{-1}(1)} \Pr[v(x,y) = 0] < 2^{2n} \cdot 2^{-2n} = 1.$$

This probability is strictly less than 1, so there exist sets $\{a_1(x), \dots, a_m(x)\}$ and $\{b_1(y), \dots, b_m(y)\}$ that make $v(x,y) \neq 0$ for every $(x,y) \in f^{-1}(1)$. We thus have that

$$\sum_{i=1}^{m} a_i(x) b_i(y) = 0 \text{ if and only if } f(x,y) = 0.$$

View the a_i and b_i as 2^n-dimensional vectors, let A be the $2^n \times m$ matrix having the a_i as columns, and B be the $m \times 2^n$ matrix having the b_i as rows. Then $(AB)_{xy} = \sum_{i=1}^{m} a_i(x) b_i(y)$, which is 0 if and only if $f(x,y) = 0$. Thus AB is a non-deterministic matrix for f, and $nrank(f) \leq rank(AB) \leq rank(A) \leq m$. $\square$

Lemma 3 allows us to prove tight bounds for non-deterministic quantum protocols.

Theorem 4. $NQ(f) = \log(nrank(f)) + 1$.

Proof. The upper bound $NQ(f) \leq \log(nrank(f)) + 1$ was shown in [34] (actually, the upper bound shown there was $\log(nrank(f))$ for protocols where only Bob has to know the output value). For the sake of completeness we repeat that proof here. Let $r = nrank(f)$ and M be a rank-r non-deterministic matrix for f. Let $M^T = U\Sigma V$ be the singular value decomposition of the transpose of M [19], so U and V are unitary, and Σ is a diagonal matrix whose first r diagonal entries are positive real numbers and whose other diagonal entries are 0. Below we describe a one-round non-deterministic protocol for f, using $\log(r) + 1$ qubits. First Alice prepares the state $|\phi_x\rangle = c_x \Sigma V |x\rangle$, where $c_x > 0$ is a normalizing real number that depends on x. Because only the first r diagonal entries of Σ are non-zero, only the first r amplitudes of $|\phi_x\rangle$ are non-zero, so $|\phi_x\rangle$ can be compressed into $\log r$ qubits. Alice sends these qubits to Bob. Bob then applies U to $|\phi_x\rangle$ and measures the resulting state. If he observes $|y\rangle$, then he puts 1 on the channel and otherwise he puts 0 there. The acceptance probability of this protocol is

$$P(x,y) = |\langle y|U|\phi_x\rangle|^2 = c_x^2 |\langle y|U\Sigma V|x\rangle|^2 = c_x^2 |M_{yx}^T|^2 = c_x^2 |M_{xy}|^2.$$

Since M_{xy} is non-zero if and only if $f(x,y) = 1$, $P(x,y)$ will be positive if and only if $f(x,y) = 1$. Thus we have a non-deterministic quantum protocol for f with $\log(r) + 1$ qubits of communication.

For the lower bound, consider a non-deterministic ℓ-qubit protocol for f. By Lemma 2, its final state on input (x, y) can be written as

$$\sum_{i \in \{0,1\}^\ell} |A_i(x)\rangle |i_\ell\rangle |B_i(y)\rangle.$$

Without loss of generality we assume the vectors $A_i(x)$ and $B_i(y)$ all have the same dimension d. Let $S = \{i \in \{0,1\}^\ell \mid i_\ell = 1\}$ and consider the part of the state that corresponds to output 1 (we drop the $i_\ell = 1$ and the $|\cdot\rangle$-notation here),

$$\phi(x,y) = \sum_{i \in S} A_i(x) \otimes B_i(y).$$

Because the protocol has acceptance probability 0 if and only if $f(x,y) = 0$, this vector $\phi(x,y)$ will be the zero vector if and only if $f(x,y) = 0$. The previous lemma gives $nrank(f) \leq |S| = 2^{\ell-1}$, and hence that $\log(nrank(f)) + 1 \leq NQ(f)$.
□

Note that any non-deterministic matrix for the equality function has non-zeroes on its diagonal and zeroes off-diagonal, and hence has full rank. Thus $NQ(EQ_n) = n + 1$, which contrasts sharply with the non-deterministic complexity of its complement (inequality), which is only 2 [27]. Similarly, a non-deterministic matrix for disjointness has full rank, because reversing the ordering of the columns gives an upper triangular matrix with non-zero elements on the diagonal. This gives tight bounds for the exact, one-sided error, and non-deterministic settings.

Corollary 5. *We have that* $Q_E(EQ_n) = Q_1(EQ_n) = NQ(EQ_n) = n + 1$ *and that* $Q_E(DISJ_n) = Q_1(DISJ_n) = NQ(DISJ_n) = n + 1$.

4   On the Bounded-Error Complexity of Disjointness

4.1   Improved Upper Bound

Here we show that we can take off most of the $\log n$ factor from the $O(\sqrt{n} \log n)$ protocol for the intersection problem (the complement of disjointness) that was given by Buhrman, Cleve, and Wigderson in [9].

Theorem 6. *There exists a constant c such that $Q_1(INT_n) \in O(\sqrt{n} \cdot c^{\log^\star n})$.*

Proof. We recursively build a one-sided error protocol that can find an index i such that $x_i = y_i = 1$, provided such an i exists (call such an i a 'solution'). Clearly this suffices for computing $INT_n(x, y)$. Let C_n denote the cost of our protocol on n-bit inputs.

Alice and Bob divide the n indices $\{1, \dots, n\}$ into $n/(\log n)^2$ blocks of $(\log n)^2$ indices each. Alice picks a random number $j \in \{1, \dots, n/(\log n)^2\}$ and sends the number j to Bob. Now they recursively run our protocol on the jth block, at a cost of $C_{(\log n)^2}$ qubits of communication. Alice then measures her part of the state, and they verify whether the measured i is indeed a solution. If there is a solution in the jth block, then Alice finds one with probability at least $1/2$, so the overall probability of finding a solution (if there is one) is at least $(\log n)^2/2n$. By using a superposition over all j we can push all intermediate measurements to the end without affecting the success probability. Therefore, applying $O(\sqrt{n}/\log n)$ rounds of amplitude amplification (Theorem 1) boosts this protocol to having error probability at most $1/2$. We thus have the recurrence

$$C_n \leq O(1) \frac{\sqrt{n}}{\log n} \left(C_{(\log n)^2} + O(\log n) \right).$$

Since $C_1 = 2$, this recursion unfolds to the bound $C_n \in O(\sqrt{n} \cdot c^{\log^\star n})$ for some constant c. Careful inspection of the protocol gives that the constant c is reasonably small. $\qquad\square$

4.2   Lower Bound for a Specific Class of Protocols

We give a lower bound for two-sided error quantum protocols for disjointness. The lower bound applies to all protocols whose acceptance probability $P(x, y)$ is a function just of $x \wedge y$, rather than of x and y "separately." In particular, the protocols of [9] and of our previous section fall in this class.

The lower bound basically follows by combining various results from [13].

Theorem 7. *Any two-sided error quantum protocol for DISJ$_n$ whose acceptance probability is a function of $x \wedge y$, has to communicate $\Omega(\sqrt{n})$ qubits.*

Proof. Consider an ℓ-qubit protocol with error probability at most $1/3$. By the comment following Lemma 2, we can write its acceptance probability $P(x, y)$ as a $2^n \times 2^n$ matrix P of rank $r \leq 2^{2\ell-2}$.

We now invoke a relation between the rank of the matrix P and properties of the $2n$-variate multilinear polynomial that equals $P(x, y)$.[3] There is an n-variate function g such that $P(x, y) = g(x \wedge y)$. Let $g(z) = \sum_S a_S z_S$ be the polynomial representation of g. Then $P(x, y) = g(x \wedge y) = \sum_S a_S (x \wedge y)_S = \sum_S a_S x_S y_S$, so the $2n$-variate multilinear polynomial P only contains monomials in which the set of x-variables is the same as the set of y-variables. For polynomials of this form (called "even"), [13, Lemmas 2 and 3] imply that the number of monomials in $P(x, y)$ equals the rank r of the matrix P.

Setting $y = x$ in $P(x, y)$ gives a polynomial $p(x) = \sum_S a_S x_S$ that has r monomials and that approximates the n-bit function NOR, since $|p(x) - \text{NOR}(x)| = |P(x, x) - \text{DISJ}(x, x)| \leq 1/3$. But [13, Theorem 8] implies that every polynomial that approximates NOR must have at least $2^{\sqrt{n/12}}$ monomials. Hence $2^{\sqrt{n/12}} \leq r \leq 2^{2\ell-2}$, which gives $\ell \geq \sqrt{n/48} + 1$. $\square$

5   Open Problems

This paper fits in a sequence of papers that (slowly) extend what is known for quantum communication complexity, e.g. [9,2,30,13,21,34,24,23]. The main open question is still the bounded-error complexity of disjointness. Of interest is whether it is possible to prove an $O(\sqrt{n})$ upper bound for disjointness, thus getting rid of the factor of $c^{\log^* n}$ in our upper bound of Theorem 6, and whether it is possible to extend the lower bound of Theorem 7 to broader classes of protocols. Since disjointness is coNP-complete for communication complexity problems [3], strong lower bounds on the disjointness problem imply a host of other lower bounds.

A second question is whether qubit communication can be significantly reduced in case Alice and Bob can make use of prior entanglement (shared EPR-pairs). Giving Alice and Bob n shared EPR-pairs trivializes the non-deterministic complexity (use the EPR-pairs as a public coin to randomly guess some n-bit z, Alice then sends Bob 1 bit indicating whether $x = z$, if $x = z$ then Bob can compute the answer $f(x, y)$ and send it to Alice, if $x \neq z$ then they output 0), but for the exact and bounded-error models it is open whether prior entanglement can make a significant difference.

---

[3] For $S \subseteq [n] = \{1, \ldots, n\}$, we use x_S for the monomial $\prod_{i \in S} x_i$. An n-variate multilinear polynomial $p(x) = \sum_{S \subseteq [n]} a_S x_S$, $a_S \in \mathbb{R}$, is a weighted sum of such monomials. The number of monomials in p is the number of S for which $a_S \neq 0$. One can show that for every function $g : \{0, 1\}^n \to \mathbb{R}$ there is a unique n-variate multilinear polynomial p such that $g(x) = p(x)$ for all $x \in \{0, 1\}^n$.

Acknowledgments. We thank Harry Buhrman and Hartmut Klauck for helpful discussions concerning the proof of Lemma 3, and Richard Cleve for helpful discussions on the protocol for disjointness.

References

1. N. Alon and J. H. Spencer. *The Probabilistic Method*. Wiley-Interscience, 1992.
2. A. Ambainis, L. Schulman, A. Ta-Shma, U. Vazirani, and A. Wigderson. The quantum communication complexity of sampling. In *Proceedings of 39th IEEE FOCS*, pages 342–351, 1998.
3. L. Babai, P. Frankl, and J. Simon. Complexity classes in communication complexity theory. In *Proceedings of 27th IEEE FOCS*, pages 337–347, 1986.
4. R. Beals, H. Buhrman, R. Cleve, M. Mosca, and R. de Wolf. Quantum lower bounds by polynomials. In *Proceedings of 39th IEEE FOCS*, pages 352–361, 1998. quant-ph/9802049.
5. C. Bennett and S. Wiesner. Communication via one- and two-particle operators on Einstein-Podolsky-Rosen states. *Physical Review Letters*, 69:2881–2884, 1992.
6. G. Brassard. Quantum communication complexity (a survey). quant-ph/0101005, 1 Jan 2001.
7. G. Brassard, P. Høyer, M. Mosca, and A. Tapp. Quantum amplitude amplification and estimation. quant-ph/0005055. To appear in Quantum Computation and Quantum Information: A Millennium Volume, AMS Contemporary Mathematics Series, 15 May 2000.
8. H. Buhrman. Quantum computing and communication complexity. *EATCS Bulletin*, pages 131–141, February 2000.
9. H. Buhrman, R. Cleve, and A. Wigderson. Quantum vs. classical communication and computation. In *Proceedings of 30th ACM STOC*, pages 63–68, 1998. quant-ph/9802040.
10. H. Buhrman, W. van Dam, and R. Cleve. Quantum entanglement and communication complexity. *SIAM Journal on Computing*, 30(6):1829–1841, 2001. quant-ph/9705033.
11. H. Buhrman, W. van Dam, P. Høyer, and A. Tapp. Multiparty quantum communication complexity. *Physical Review A*, 60(4):2737–2741, 1999. quant-ph/9710054.
12. H. Buhrman, Ch. Dürr, M. Heiligman, P. Høyer, F. Magniez, M. Santha, and R. de Wolf. Quantum algorithms for element distinctness. In *Proceedings of 16th IEEE Conference on Computational Complexity*, pages 131–137, 2001. quant-ph/0007016.
13. H. Buhrman and R. de Wolf. Communication complexity lower bounds by polynomials. In *Proceedings of 16th IEEE Conference on Computational Complexity*, pages 120–130, 2001. cs.CC/9910010.
14. R. Cleve and H. Buhrman. Substituting quantum entanglement for communication. *Physical Review A*, 56(2):1201–1204, 1997. quant-ph/9704026.
15. R. Cleve, W. van Dam, M. Nielsen, and A. Tapp. Quantum entanglement and the communication complexity of the inner product function. In *Proceedings of 1st NASA QCQC conference*, volume 1509 of *Lecture Notes in Computer Science*, pages 61–74. Springer, 1998. quant-ph/9708019.
16. W. van Dam. Quantum oracle interrogation: Getting all information for almost half the price. In *Proceedings of 39th IEEE FOCS*, pages 362–367, 1998. quant-ph/9805006.

17. E. Farhi, J. Goldstone, S. Gutmann, and M. Sipser. A limit on the speed of quantum computation in determining parity. *Physical Review Letters*, 81:5442–5444, 1998. quant-ph/9802045.
18. L. K. Grover. A fast quantum mechanical algorithm for database search. In *Proceedings of 28th ACM STOC*, pages 212–219, 1996. quant-ph/9605043.
19. R. A. Horn and C. R. Johnson. *Matrix Analysis*. Cambridge University Press, 1985.
20. B. Kalyanasundaram and G. Schnitger. The probabilistic communication complexity of set intersection. *SIAM Journal on Computing*, 5(4):545–557, 1992.
21. H. Klauck. On quantum and probabilistic communication: Las Vegas and one-way protocols. In *Proceedings of 32nd ACM STOC*, pages 644–651, 2000.
22. H. Klauck. Quantum communication complexity. In *Proceedings of Workshop on Boolean Functions and Applications at 27th ICALP*, pages 241–252, 2000. quant-ph/0005032.
23. H. Klauck. Lower bounds for quantum communication complexity. In *Proceedings of 42nd IEEE FOCS*, pages 288–297, 2001. quant-ph/0106160.
24. H. Klauck, A. Nayak, A. Ta-Shma, and D. Zuckerman. Interaction in quantum communication and the complexity of set disjointness. In *Proceedings of 33rd ACM STOC*, pages 124–133, 2001.
25. I. Kremer. Quantum communication. Master's thesis, Hebrew University, Computer Science Department, 1995.
26. E. Kushilevitz and N. Nisan. *Communication Complexity*. Cambridge University Press, 1997.
27. S. Massar, D. Bacon, N. Cerf, and R. Cleve. Classical simulation of quantum entanglement without local hidden variables. *Physical Review A*, 63(5):052305, 2001. quant-ph/0009088.
28. M. A. Nielsen. *Quantum Information Theory*. PhD thesis, University of New Mexico, Albuquerque, 1998. quant-ph/0011036.
29. M. A. Nielsen and I. L. Chuang. *Quantum Computation and Quantum Information*. Cambridge University Press, 2000.
30. R. Raz. Exponential separation of quantum and classical communication complexity. In *Proceedings of 31st ACM STOC*, pages 358–367, 1999.
31. A. Razborov. On the distributional complexity of disjointness. *Theoretical Computer Science*, 106(2):385–390, 1992.
32. P. W. Shor. Polynomial-time algorithms for prime factorization and discrete logarithms on a quantum computer. *SIAM Journal on Computing*, 26(5):1484–1509, 1997. quant-ph/9508027.
33. A. Ta-Shma. Classical versus quantum communication complexity. *ACM SIGACT News (Complexity Column 23)*, pages 25–34, 1999.
34. R. de Wolf. Characterization of non-deterministic quantum query and quantum communication complexity. In *Proceedings of 15th IEEE Conference on Computational Complexity*, pages 271–278, 2000. cs.CC/0001014.
35. A. C.-C. Yao. Some complexity questions related to distributive computing. In *Proceedings of 11th ACM STOC*, pages 209–213, 1979.
36. A. C.-C. Yao. Quantum circuit complexity. In *Proceedings of 34th IEEE FOCS*, pages 352–360, 1993.

On Quantum Computation with Some Restricted Amplitudes

Harumichi Nishimura[1,2]

[1] CREST, Japan Science and Technology
[2] Center for Quantum Computation, Clarendon Laboratory
Parks Road, Oxford, OX1 3PU, UK
H.Nishimura@qubit.org

Abstract. In this paper we explore the power of quantum computers with restricted amplitudes. Adleman, DeMarrais and Huang showed that quantum Turing machines (QTMs) with the amplitudes from $\mathcal{A} = \{0, \pm\frac{3}{5}, \pm\frac{4}{5}, \pm 1\}$ are equivalent to ones with the polynomial-time computable amplitudes as machines implementing bounded-error polynomial time algorithms. We show that QTMs with the amplitudes from $\mathcal{A}$ is polynomial-time equivalent to deterministic Turing machines as machines implementing exact algorithms. Extending this result, it is shown that exact computers with rational biased 'quantum coins' are equivalent to classical computers. We also show that from the viewpoint of zero-error algorithms $\mathcal{A}$ is not more useful than $\mathcal{B} = \{0, \pm\frac{1}{\sqrt{2}}, \pm 1\}$ but sufficient for the factoring problem as the set of amplitudes taken by QTMs.

1 Introduction

In the late 1980's, Deutsch proposed quantum Turing machines (QTMs) [9] and quantum circuits [10] as models of quantum computers. Using these models, some researchers provided results suggesting that quantum computers are more powerful than classical ones [3,5,11,22]. In 1994, Shor [23] found out a polynomial-time quantum algorithm solving the factoring problem and the discrete logarithm problem. Since then, many researchers have studied the computational power of quantum computers.

Bernstein and Vazirani [5] introduced the complexity classes **BQP** and **EQP** based on QTMs, which are respectively the sets of languages recognized by bounded-error and exact polynomial-time QTMs. Afterward, in [6] they restricted their classes to the classes defined by QTMs with the amplitudes from the polynomial-time computable complex numbers. Adleman, DeMarrais and Huang [1] introduced more general notions $\mathbf{BQP}_K$, $\mathbf{EQP}_K$ and $\mathbf{NQP}_K$, which are respectively the sets of languages recognized by bounded-error, exact and 'non-deterministic' polynomial-time QTMs with the amplitudes from a subset K of the complex number field $\mathbb{C}$. In recent years, complexity classes of QTMs with variously restricted amplitudes have been investigated [1,12,13,28], not only from theoretical interest but also in order to explore the origin of the

H. Alt and A. Ferreira (Eds.): STACS 2002, LNCS 2285, pp. 311–322, 2002.
© Springer-Verlag Berlin Heidelberg 2002

power of quantum computers. In particular, QTMs with amplitudes from the sets $\mathcal{A} = \{0, \pm\frac{3}{5}, \pm\frac{4}{5}, \pm 1\}$ and $\mathcal{B} = \{0, \pm\frac{1}{\sqrt{2}}, \pm 1\}$ have been often used for the investigation of quantum complexity classes or quantum algorithms.

Adleman, DeMarrais and Huang [1] showed that although $\mathbf{BQP}_{\mathbb{C}}$ is uncountable, $\mathbf{BQP} = \mathbf{BQP}_{\mathcal{A}}$. Shor [24] claimed that the following set of three gates, rotations around the y-axis and the z-axis by $\frac{\pi}{2}$, and Toffoli gates, are universal for quantum computation. (This universality was also proved by Kitaev [15], who independently gave a similar universal set). This implies that $\mathbf{BQP} = \mathbf{BQP}_{\mathcal{B}}$ by the equivalence between bounded-error uniform quantum circuits and QTMs, which was implicitly mentioned by Yao [29] and proved in [17] with an argument on the uniformity of quantum circuit families. Fenner, Green, Homer and Pruim [12] showed $\mathbf{NQP}_{\overline{\mathbb{Q}}} = \mathbf{NQP}_{\mathcal{A}} = \mathbf{NQP}_{\mathcal{B}} = \text{co-}\mathbf{C}_{=}\mathbf{P}$, where $\overline{\mathbb{Q}}$ denotes the set of algebraic numbers and $\mathbf{C}_{=}\mathbf{P}$ is a counting class defined by Wagner [26]. Afterward, Yamakami and Yao [28] showed $\mathbf{NQP}_{\mathbb{C}} = \text{co-}\mathbf{C}_{=}\mathbf{P}$. From these results, we can see that bounded-error and nondeterministic quantum complexity classes are stable for the choice of the amplitudes taken by QTMs. However, as shown later, exact quantum complexity classes are so sensitive to the choice of the amplitudes. In this paper we explore exact and zero-error (or Las-Vegas) quantum complexity classes with some restricted amplitudes.

Firstly, we investigate the power of quantum computers with restricted amplitudes from the exact algorithms. Deutsch and Jozsa [11] and Brassard and Høyer [7] provided exact quantum algorithms suggesting that exact quantum computers are superior to classical ones. On the other hand, Adleman, DeMarrais and Huang showed that for any $\theta \in [0, 2\pi)$ such that $\cos\theta$ is transcendental, $\mathbf{EQP}_{\{0, \pm\cos\theta, \pm\sin\theta, \pm 1\}} = \mathbf{P}$ and left the following question. *Does it hold* $\mathbf{EQP}_{\{0, \pm\frac{3}{5}, \pm\frac{4}{5}, \pm 1\}} = \mathbf{EQP}_{\{0, \pm\frac{5}{13}, \pm\frac{12}{13}, \pm 1\}}$? In this paper, we give a general positive answer for this open question; For any $\theta \in [0, 2\pi)$ such that $\cos\theta$ and $\sin\theta$ are rational, $\mathbf{EQP}_{\{0, \pm\cos\theta, \pm\sin\theta, \pm 1\}} = \mathbf{P}$. This result shows that the set $\mathcal{A}$ of amplitudes is useless for exact quantum computation, while it is sufficient for bounded-error quantum computation. Therefore, the set $\mathcal{A}$ seems to be so different from the set $\mathcal{B}$ for exact quantum algorithms, since the exact quantum algorithm of Deutsch and Jozsa uses the amplitudes from $\mathcal{B}$. In fact, extending the above result, we show that if the bias of 'quantum coins' is rational, exact quantum computers with such biased coins have only equivalent power to deterministic computers.

Secondly, we investigate the power of quantum computers with restricted amplitudes from the zero-error algorithms. Combining Preskill's quantum circuit which produces the angle θ with $\cos\theta = \frac{3}{5}$ probabilistically [21] and the construction of universal QTMs [6], we show that $\mathbf{ZQP}_{\mathcal{A}} \subseteq \mathbf{ZQP}_{\mathcal{B}}$, where for a set $K \subseteq \mathbb{C}$ the class $\mathbf{ZQP}_K$ is the set of the languages recognized by zero-error polynomial-time QTMs with the amplitudes from K. This means that QTMs with the amplitudes from $\mathcal{A}$ are not more powerful than ones with the amplitudes from $\mathcal{B}$ for zero-error algorithms. On the other hand, we show that the factoring problem can be solved by a zero-error polynomial-time QTM with the amplitudes from $\mathcal{B}$. This result implies that in case of zero-error algorithms $\mathcal{A}$ is useful as the set of amplitudes taken by QTMs, different from exact case.

2   Preliminaries

Let $\mathbb{N}, \mathbb{Z}, \mathbb{Q}, \mathbb{R}$, and $\mathbb{C}$ denote the set of natural numbers, integers, rational numbers, reals, and complex numbers. Let $P\mathbb{C}$ denote the set of polynomial-time computable complex numbers [14]. For any $n \in \mathbb{Z}$, let $\mathbb{Z}_{\geq n}$ denote the set $\{n, n+1, \cdots\}$. For any $n, m \in \mathbb{N}$, let $\mathrm{GCD}(n, m)$ denote the greatest common divisor of n and m. Let $\#A$ denote the number of the elements of a set A. Let $|x|$ denote the length of a string x. Let denote $\mathcal{A} = \{0, \pm\frac{3}{5}, \pm\frac{4}{5}, \pm 1\}$ and $\mathcal{B} = \{0, \pm\frac{1}{\sqrt{2}}, \pm 1\}$. For classical complexity theory, for example, see the textbook by Papadimitriou [20].

A quantum Turing machine (QTM) is a quantum system consisting of a processor, a bilateral infinite tape consisting of cells numbered by the integers, and a head to read and write the symbols on the cells. The formal definition of a QTM is given as follows. A *processor configuration set* is a finite set with two specific elements q_0 and q_f, where q_0 represents the *initial processor configuration* and q_f represents the *final processor configuration*. A *symbol set* is a finite set with the specific element B called the *blank*. A *Turing frame* is a pair (Q, Σ) of a processor configuration set Q and a symbol set Σ. In what follows, let (Q, Σ) be a Turing frame. A *tape configuration* from a symbol set Σ is a function T from $\mathbb{Z}$ to Σ such that $T(m) = B$ except for finitely many $m \in \mathbb{Z}$. The set of all the possible tape configurations is denoted by $\Sigma^{\#}$. The *configuration space* of (Q, Σ) is the product set $\mathcal{C}(Q, \Sigma) = Q \times \Sigma^{\#} \times \mathbb{Z}$. A *configuration* of (Q, Σ) is an element $C = (q, T, \xi)$ of $\mathcal{C}(Q, \Sigma)$. Specifically, if $q = q_0$ and $\xi = 0$ then C is called an *initial configuration* of (Q, Σ), and if $q = q_f$ then C is called a *final configuration* of (Q, Σ). The *quantum state space* of (Q, Σ) is the Hilbert space $\mathcal{H}(Q, \Sigma)$ spanned by $\mathcal{C}(Q, \Sigma)$ with the canonical basis $\{|C\rangle \,|\, C \in \mathcal{C}(Q, \Sigma)\}$ called the *computational basis*. A *quantum transition function* for (Q, Σ) is a function from $Q \times \Sigma \times Q \times \Sigma \times \{0, \pm 1\}$ into $\mathbb{C}$. A *prequantum Turing machine* is defined to be a triple $M = (Q, \Sigma, \delta)$ consisting of a Turing frame (Q, Σ) and a quantum transition function δ for (Q, Σ).

Let $M = (Q, \Sigma, \delta)$ be a prequantum Turing machine. An element of Q is called a *processor configuration* of M, a tape configuration from Σ is called a *tape configuration* of M, the function δ is called the *quantum transition function* of M, and an (initial or final) configuration of (Q, Σ) is called an *(initial or final) configuration* of M. A unit vector in $\mathcal{H}(Q, \Sigma)$ is called a *state* of M. The *evolution operator* of M is a linear operator M_δ on $\mathcal{H}(Q, \Sigma)$ such that

$$M_\delta |q, T, \xi\rangle = \sum_{p \in Q, \tau \in \Sigma, d \in \{0, \pm 1\}} \delta(q, T(\xi), p, \tau, d) |p, T_\xi^\tau, \xi + d\rangle$$

for all $(q, T, \xi) \in \mathcal{C}(Q, \Sigma)$, where T_ξ^τ is a tape configuration such that $T_\xi^\tau(m) = \tau$ if $m = \xi$, and $T(m)$ if $m \neq \xi$. The above equation uniquely defines the bounded operator M_δ on the space $\mathcal{H}(Q, \Sigma)$ [19]. A prequantum Turing machine is said to be a *quantum Turing machine (QTM)* if the evolution operator is unitary. A QTM $M = (Q, \Sigma, \delta)$ satisfying range(δ) $\subseteq K$ is called a *QTM with the amplitudes from K*. The condition that δ should satisfy for M_δ to be unitary is given in [19, 27], including the multi-tape case. Those conditions extend ones given by Bernstein and Vazirani [5], who defined a QTM whose head moves either to the left or to the right at each step (a *two-way* QTM).

A QTM $M = (Q, \Sigma, \delta)$ is called an *m-track* QTM if Σ can be factorized as $\Sigma = \Sigma_1 \times \Sigma_2 \times \cdots \times \Sigma_m$. Then, $T \in \Sigma^{\#}$ can be written in the form $(T^1, T^2, \ldots, T^m)$, where $T^i \in \Sigma_i^{\#}$ for $i = 1, \ldots, m$. The function T^i is called an *i-th track configuration*. For a string $x = x_0 x_1 \cdots x_{k-1}$ of length k, we denote by $\mathrm{T}[x]$ a tape (or track) configuration such that $\mathrm{T}[x](i) = x_i$ if $0 \le i \le k-1$, and B otherwise. For any tape configuration T, we will write $T = (T^1, \ldots, T^j)$ if $T = (T^1, \ldots, T^j, \mathrm{T}[\epsilon], \ldots, \mathrm{T}[\epsilon])$, where ϵ denotes the empty string. Let $E(\hat{\xi} = j)$, $E(\hat{q} = p)$, and $E(\hat{T}^i = T_0)$ be respectively projections on span$\{|q, T, j\rangle|\ q \in Q,\ T \in \Sigma^{\#}\}$, span$\{|p, T, \xi\rangle|\ T \in \Sigma^{\#},\ \xi \in \mathbb{Z}\}$, and span$\{|q, T, \xi\rangle|\ q \in Q,\ T = (T^1, \cdots, T_0, \cdots, T^m) \in \Sigma^{\#},\ \xi \in \mathbb{Z}\}$.

We introduce some of the restricted QTMs given in [1,5,6]. A QTM $M = (Q, \Sigma, \delta)$ is said to be *unidirectional*, if we have $d = d'$ whenever $\delta(q, \sigma, p, \tau, d)$ and $\delta(q', \sigma', p, \tau', d')$ are both non-zero. Let $M = (Q, \Sigma, \delta)$ be a unidirectional QTM. For any $p \in Q$, we denote by $D(p)$ the unique value d determined from the set $\{(q, \sigma, \tau) \in Q \times \Sigma^2 |\ \delta(q, \sigma, p, \tau, d) \ne 0\}$. We denote by L_δ the $\#(Q \times \Sigma)$-dimensional unitary operator on the Hilbert space span$\{|q, \sigma\rangle|\ (q, \sigma) \in Q \times \Sigma\}$ such that $\langle p, \tau | L_\delta | q, \sigma \rangle = \delta(q, \sigma, p, \tau, D(p))$ for any $(q, \sigma, p, \tau) \in (Q \times \Sigma)^2$. Unidirectional QTMs can efficiently simulate two-way QTMs [6] and multi-tape QTMs [17,27]. A QTM $M = (Q, \Sigma, \delta)$ is called a *binary QTM with angle θ*, if the matrix representing L_δ in the computational basis is block diagonal (up to permutations of rows and columns) with each block either 1, -1, or 2 by 2 of the form

$$\cos\theta \quad -\sin\theta$$
$$\sin\theta \quad \cos\theta.$$

A QTM $M = (Q, \Sigma, \delta)$ is said to be *stationary*, if given an initial configuration C, there exists some $t \in \mathbb{N}$ satisfying $||E(\hat{\xi} = 0)E(\hat{q} = q_f)M_\delta^t|C\rangle||^2 = 1$ and for all $s < t$ we have $||E(\hat{q} = q_f)M_\delta^s|C\rangle||^2 = 0$. Its positive integer t is called the *computation time* of M for input state $|C\rangle$, and $M_\delta^t|C\rangle$ is called the *final state* of M for $|C\rangle$. Specifically, if $|C\rangle = |q_0, \mathrm{T}[x], 0\rangle$, the integer t is called the computation time on input x. A *polynomial-time* QTM is a stationary QTM such that the computation time on every input is a polynomial in the length of the input. We say that $M = (Q, \Sigma, \delta)$ is in *normal form* if $\delta(q_f, \sigma, q_0, \sigma, 1) = 1$ for any $\sigma \in \Sigma$. Henceforth, we consider only unidirectional stationary normal from QTMs and simply call them 'QTMs'. See Ozawa [18] for QTMs which are not stationary.

We say that a QTM M *accepts* (resp. *rejects*) a bit string x *with probability* p if the final state $|\psi\rangle$ of M for the initial state $|q_0, \mathrm{T}[x], 0\rangle$ satisfies $||E(\hat{T}^1 = \mathrm{T}[x])E(\hat{T}^2 = \mathrm{T}[1])|\psi\rangle||^2 = p$ (resp. $||E(\hat{T}^1 = \mathrm{T}[x])E(\hat{T}^2 = \mathrm{T}[0])|\psi\rangle||^2 = p$).

We say that M *recognizes* a language L with probability p if M accepts x with probability at least p for any $x \in L$ and rejects x with probability at least p for any $x \notin L$. Let K be a subset of $\mathbb{C}$. A language L is in $\mathbf{EQP}_K$ if there is a polynomial-time QTM M with the amplitudes from K such that M recognizes L with probability 1. A language L is in $\mathbf{ZQP}_K$ if there is a polynomial-time QTM M with the amplitudes from K satisfying the following conditions: (1) M recognizes L with probability at least $\frac{1}{2}$; (2) If M accepts (resp. rejects) x with a positive probability, M rejects (resp. accepts) x with probability 0. Then, M is called a *QTM recognizing L with zero-error*. In particular, if $K = \mathbb{PC}$, we denote $\mathbf{EQP}_K$ and $\mathbf{ZQP}_K$ by $\mathbf{EQP}$ and $\mathbf{ZQP}$, respectively.

3   Exact Quantum Computation

A computation by a QTM can be represented by a tree such that each edge has a non-zero complex number as its weight. Let $M = (Q, \Sigma, \delta)$ be a QTM. Let $p = p(|x|)$ be the computation time of M on input x. Then, the computation of M on input x can be represented by the following tree of depth p. The root (the node of depth 0) corresponds to an initial configuration $C_0 = (q_0, \mathrm{T}[x], 0)$ of M, each leaves (the nodes of depth p) correspond to final configurations of M, and each internal nodes except the root correspond to configurations of M which are neither initial nor final. Henceforth, we identify an node of the tree with the corresponding configuration. An internal node C has $\#\{C' | \langle C' | M_\delta | C \rangle \neq 0\}$ outgoing edges and the edge connecting C and $C'' \in \{C' | \langle C' | M_\delta | C \rangle \neq 0\}$ has weight $\langle C'' | M_\delta | C \rangle$. We denote such a tree by $\mathrm{Tree}(M, x)$ and call it *a computation tree of M on input x*. Then, a path $P = (C_0, C_1, \ldots, C_p)$ of $\mathrm{Tree}(M, x)$ has non-zero weight $\langle C_1 | M_\delta | C_0 \rangle \times \cdots \times \langle C_p | M_\delta | C_{p-1} \rangle$. We denote the weight of P by $W(P)$. The probability that we obtain a final configuration C by measuring M at time p is $|\sum_P W(P)|^2$, where the sum is taken over the set of paths $P = (C_0, \ldots, C_p)$ of $\mathrm{Tree}(M, x)$ satisfying $C_p = C$. The following fact can be easily seen.

Lemma 1. (1) *For any binary QTM M, $\mathrm{Tree}(M, x)$ is a tree of depth $p(|x|)$ whose nodes have at most 2 outgoing edges, where $p(|x|)$ is the computation time of M on input x.*
(2) *Let $\theta \in [0, 2\pi)$ be such that $\cos\theta$ and $\sin\theta$ are rational, and let M be a QTM with the amplitudes from $\{0, \pm\cos\theta, \pm\sin\theta, \pm 1\}$. Then, $\mathrm{Tree}(M, x)$ is a tree of depth $p(|x|)$ whose nodes have at most 2 outgoing edges, where $p(|x|)$ is the computation time of M on input x.*

By Lemma 1 (1), computations by binary QTMs with angle θ can be viewed as computations with 'quantum coins' of bias $|\frac{1}{2} - \cos^2\theta|$. Now, we show that 'rational biased quantum coins' are useless for exact quantum computation.

Theorem 1. *Assume that $\theta \in [0, 2\pi)$ satisfy $\cos\theta = \sqrt{a}$, where $a \in \mathbb{Q} \setminus \{\frac{1}{2}\}$. If a function f can be exactly computed by a binary QTM with angle θ in polynomial time, there is a deterministic Turing machine which computes f in polynomial time.*

Proof. In what follows, let denote by $\mathbb{Q}(\alpha)$ the field generated by α. The case $a = 0$ or 1 is trivial, so we consider the case $0 < a < 1$. Let $a = \frac{n}{m}$, where $\mathrm{GCD}(n, m) = 1$. Then, $\mathrm{GCD}(m - n, m) = 1$ and $\mathrm{GCD}(n, m - n) = 1$. Thus, n, $m - n$ and m can be represented as $n = p_1^{e_1} \cdots p_s^{e_s}$, $m - n = q_1^{f_1} \cdots q_t^{f_t}$ and $m = r_1^{g_1} \cdots r_u^{g_u}$, where $s, t \in \mathbb{Z}_{\geq 0}$, u, $e_1, \ldots, e_s$, $f_1, \ldots, f_t$, $g_1, \ldots, g_u \in \mathbb{Z}_{\geq 1}$ and $p_1, \ldots, p_s$, $q_1, \ldots, q_t$ and $r_1, \ldots, r_u$ are different primes. When $s = 0$ (resp. $t = 0$), let $n = 1$ (resp. $m - n = 1$). Note that s or t is not 0 since $a \neq \frac{1}{2}$. We shall show the theorem for the case $t \neq 0$. Changing $\cos \theta$ with $\sin \theta$, we can also prove the theorem for the case $s \neq 0$ similarly according to the following argument. Let $M = (Q, \Sigma, \delta)$ be a binary QTM which computes f with probability 1. Let x the input of M. Let $p = p(|x|)$ be the computation time of M on input x. We assume without loss of generality that p is even. By Lemma 1 (1), the sum of the weights of all computation paths leading to a configuration C is expressed as $S_0 + S_1 + \ldots + S_p$, where $S_0 = n_0 (\cos \theta)^{m_0}$ and $S_j = (\sin \theta)^j \left(\sum_{k=0}^{p-1} n_{jk} (\cos \theta)^k \right)$. Here, $n_0 \in \{0, \pm 1\}$, $0 \leq m_0 \leq p$ and $n_{jk} \in \mathbb{Z}$ for $j = 1, \ldots, p$. Using the equations

$$\cos \theta = \left(\frac{p_1^{e_1} \cdots p_s^{e_s}}{r_1^{g_1} \cdots r_u^{g_u}} \right)^{\frac{1}{2}} \quad \text{and} \quad \sin \theta = \left(\frac{q_1^{f_1} \cdots q_t^{f_t}}{r_1^{g_1} \cdots r_u^{g_u}} \right)^{\frac{1}{2}}, \tag{1}$$

$S_0, \ldots, S_p$ can be rewritten in the form

$$S_0 = n_0 \frac{(r_1^{g_1} \cdots r_u^{g_u})^{k_0} (p_1^{e_1} \cdots p_s^{e_s})^{h_0}}{(r_1^{g_1} \cdots r_u^{g_u})^p} (\cos \theta)^{l_0} \tag{2}$$

$$S_{2j} = \frac{(q_1^{f_1} \cdots q_t^{f_t})^j (N_{2j}^1 + N_{2j}^2 \cos \theta)}{(r_1^{g_1} \cdots r_u^{g_u})^p} \tag{3}$$

$$S_{2j-1} = \frac{(q_1^{f_1} \cdots q_t^{f_t})^{j-1} \sin \theta (N_{2j-1}^1 + N_{2j-1}^2 \cos \theta)}{(r_1^{g_1} \cdots r_u^{g_u})^p}, \tag{4}$$

where $k_0, h_0 \in \mathbb{Z}_{\geq 0}$, $l_0 \in \{0, 1\}$ and $N_{2j}^1, N_{2j}^2, N_{2j-1}^1, N_{2j-1}^2 \in \mathbb{Z}$ for $j = 1, \ldots, \frac{p}{2}$. If C represents a wrong answer, the sum $S_0 + \ldots + S_p$ should vanish, since M produces a correct answer with probability 1. Then, we show that S_0 should also vanish, which implies that the only configuration C satisfying $S_0 \neq 0$ should represent a correct answer. This means that the only path such that all the egdes have weight ± 1 or $\pm \cos \theta$, which has weight S_0, leads to a correct answer. Then, this path can be deterministically found as follows: If the present configuration of M is $C_1 = (q, T, \xi)$, then using the $\#(Q \times \Sigma)$-dimensional matrix corresponding to the unitary operator L_δ, search the only pair (p, τ) satisfying $\langle p, \tau | L_\delta | q, T(\xi) \rangle = \pm 1$ or $\pm \cos \theta$ (Note that M is a binary QTM), and replace C_1 with $C_2 = (p, T_\xi^\tau, \xi + D(p))$. Thus, we can deterministically compute f by calculating the sequence of configurations corresponding to its only path. In the rest of this proof, we show that if $S_0 + \ldots + S_p = 0$ then $S_0 = 0$. We can consider the following five cases: (1) $\cos \theta \notin \mathbb{Q}$ and $\sin \theta \notin \mathbb{Q}(\cos \theta)$; (2) $\cos \theta \notin \mathbb{Q}$ and $\sin \theta \in \mathbb{Q}(\cos \theta) \setminus \mathbb{Q}$; (3) $\cos \theta \notin \mathbb{Q}$ and $\sin \theta \in \mathbb{Q}$; (4) $\cos \theta \in \mathbb{Q}$ and $\sin \theta \notin \mathbb{Q}$; (5) $\cos \theta \in \mathbb{Q}$ and $\sin \theta \in \mathbb{Q}$. For space restriction, we show $S_0 = 0$ for case (2) only, which is the most difficult case. In case (2), we first show that $\sin\theta = B\cos\theta$ with $B \in \mathbb{Q}$. From $\sin\theta \in \mathbb{Q}(\cos\theta) \setminus \mathbb{Q}$, we can write $\sin\theta = A + B\cos\theta$, where $A, B \in \mathbb{Q}$ and $B \neq 0$. Substituting $\cos\theta = \sqrt{a}$ and $\sin\theta = A + B\sqrt{a}$ into the equation $\sin^2\theta + \cos^2\theta = 1$, we have the equation

$$(A^2 + B^2 a + a - 1) + 2AB\sqrt{a} = 0.$$

Since $\sqrt{a} = \cos\theta \notin \mathbb{Q}$, we have $AB = 0$. Thus, we have $A = 0$ since $B \neq 0$. Now, by Eqs.(1) we have

$$B = \left(\frac{q_1^{f_1} \cdots q_t^{f_t}}{p_1^{e_1} \cdots p_s^{e_s}}\right)^{\frac{1}{2}}. \tag{5}$$

Since $B \in \mathbb{Q}$ and $p_1, \ldots, p_s, q_1, \ldots, q_t$ are different primes, the numbers $p_1^{e_1} \cdots p_s^{e_s}$ and $q_1^{f_1} \cdots q_t^{f_t}$ should be square. Thus, $e_1, \ldots, e_s, f_1, \ldots, f_t$ are even. For each of the cases $l_0 = 0$ and $l_0 = 1$, let us rewrite $S_0 + \ldots + S_p = 0$ in the form $\alpha_{l_0} + \beta_{l_0}\cos\theta = 0$ with $\alpha_{l_0}, \beta_{l_0} \in \mathbb{Q}$ by using Eqs.(1)–(5). Then, $\alpha_{l_0} = \beta_{l_0} = 0$ from $\cos\theta \notin \mathbb{Q}$. In case $l_0 = 0$, from $\alpha_0 = 0$ we have

$$(r_1^{g_1} \cdots r_u^{g_u})^{k_0+1}(p_1^{e_1} \cdots p_s^{e_s})^{h_0} n_0$$
$$+ \sum_{j=1}^{\frac{p}{2}} (q_1^{\frac{f_1}{2}} \cdots q_t^{\frac{f_t}{2}})^{2j-1} [(r_1^{g_1} \cdots r_u^{g_u})(q_1^{\frac{f_1}{2}} \cdots q_t^{\frac{f_t}{2}})N_{2j}^1 + (p_1^{\frac{e_1}{2}} \cdots p_s^{\frac{e_s}{2}})N_{2j-1}^2] = 0,$$

and in case $l_0 = 1$, from $\beta_1 = 0$ we have

$$(r_1^{g_1} \cdots r_u^{g_u})^{k_0}(p_1^{e_1} \cdots p_s^{e_s})^{h_0+\frac{1}{2}} n_0$$
$$+ \sum_{j=1}^{\frac{p}{2}} (q_1^{\frac{f_1}{2}} \cdots q_t^{\frac{f_t}{2}})^{2j-1} [(p_1^{\frac{e_1}{2}} \cdots p_s^{\frac{e_s}{2}})(q_1^{\frac{f_1}{2}} \cdots q_t^{\frac{f_t}{2}})N_{2j}^2 + N_{2j-1}^1] = 0.$$

In both cases, since $t \geq 1$, all the terms except the first term are 0 modulo q_1. Since $q_1, p_1, \ldots, p_s, r_1, \ldots, r_u$ are different primes and $n_0 \in \{0, \pm 1\}$, the first term should be 0 modulo q_1. Thus, we have $n_0 = 0$, which implies $S_0 = 0$. In other cases, we can also show that $S_0 + \ldots + S_p = 0$ implies $S_0 = 0$ by algebraic arguments similar to case (2). Therefore, the proof is completed.

From Lemma 1 (2) and Theorem 1 we can now show that QTMs with the amplitudes from $\mathcal{A}$ are equivalent to deterministic Turing machines as machines implementing exact polynomial-time algorithms.

Theorem 2. *For any $\theta \in [0, 2\pi)$ such that $\cos\theta$ and $\sin\theta$ are rational, it holds the relation* $\mathbf{EQP}_{\{0, \pm\cos\theta, \pm\sin\theta, \pm 1\}} = \mathbf{P}$. *In particular,* $\mathbf{EQP}_{\{0, \pm\frac{3}{5}, \pm\frac{4}{5}, \pm 1\}} = \mathbf{EQP}_{\{0, \pm\frac{5}{13}, \pm\frac{12}{13}, \pm 1\}} = \mathbf{P}$.

We can define a tree representing a computation by a quantum circuit and show similar results for quantum circuits. Let $M_3(N)$ denote a Toffoli gate, which takes $M_3(N)|b_1, b_2, b_3\rangle = |b_1, b_2, (b_1 \wedge b_2) \oplus b_3\rangle$, where $b_1, b_2, b_3 \in \{0, 1\}$, the symbol $\wedge$ denotes AND, and the symbol $\oplus$ denotes exclusive-or. Let R_θ

denote a 1-qubit gate which implements a rotation around the y-axis by an angle 2θ, which takes $R_\theta|0\rangle = \cos\theta|0\rangle + \sin\theta|1\rangle$ and $R_\theta|1\rangle = -\sin\theta|0\rangle + \cos\theta|1\rangle$. Let P_θ denote a 1-qubit gate which implements a rotation around the z-axis by an angle θ, which takes $P_\theta|0\rangle = |0\rangle$ and $P_\theta|1\rangle = e^{i\theta}|1\rangle$. Then, we can see that if $\cos^2\theta$ is a rational number except $\frac{1}{2}$ the exact computational power of quantum circuits constructed from the set $\{M_3(N), R_\theta, P_\theta\}$ of quantum gates are polynomially equivalent to Boolean circuits, using a similar technique to Theorem 1 and a technique that simulates QTMs with non-real amplitudes by ones with real amplitudes [5].

Berthiaume and Brassard [3] constructed an oracle language L such that $\mathbf{EQP}^L \not\subseteq \mathbf{NP}^L \cup \mathrm{co-}\mathbf{NP}^L$, recasting the promise problem of Deutsch and Jozsa [11] (See [8] for the definition of oracle QTMs). The exact quantum algorithm of Deutsch and Jozsa, which solves their promise problem, uses only the amplitudes from $\mathcal{B}$. Therefore, from the result of [3] and the fact that Theorem 1 can be relativized, we have the following corollary.

Corollary 1. *Assume that $\theta \in [0, 2\pi)$ satisfy $\cos\theta = \sqrt{a}$, where $a \in \mathbb{Q} \setminus \{\frac{1}{2}\}$. Then, there is a language L such that $\mathbf{P}^L = \mathbf{EQP}^L_\theta \subset \mathbf{EQP}^L_\mathcal{B}$, where $\mathbf{EQP}_\theta$ is the class of the languages which are exactly recognized by binary QTMs with angle θ in polynomial time, and $A \subset B$ means that A is a proper subset of B.*

According to Corollary 1 and the result by Adleman, DeMarrais and Huang [1], it seems that from the viewpoint of exact algorithms, computers with 'unbiased' quantum coins are more powerful than classical ones, different from computers with rational or transcendental biased quantum coins. Now one may ask the following question. *Are all computers with biased quantum coins useless for exact quantum computation?* The answer is no. In fact, from the fact $R_{\frac{\pi}{4}} = R_{\frac{\pi}{8}}^2$, we can see that there exists a computer with an *algebraic* biased quantum coin which exactly simulates one with an unbiased quantum coin.

Proposition 1. *For any QTM M with the amplitudes from $\mathcal{B}$, there is a binary QTM with angle $\frac{\pi}{8}$ which exactly simulates M.*

4   Zero-Error Quantum Computation

In this section we show that $\mathbf{ZQP}_\mathcal{A}$ is included in $\mathbf{ZQP}_\mathcal{B}$. This relation can be essentially obtained from the construction of a universal set given in the theory of fault-tolerant quantum computation. Shor [24] claimed that the set $\{M_3(N), R_{\frac{\pi}{4}}, P_{\frac{\pi}{2}}\}$ is universal for quantum computation. Afterward, Preskill [21] concretely gave a quantum circuit implementing rotations of irrational multiples of π based on the set $\{M_3(N), H, P_{\frac{\pi}{2}}\}$ of gates, where H denotes an Hadamard gate. In fact, his circuit implements P_θ with probability $\frac{5}{8}$ and the identity transformation with probability $\frac{3}{8}$, where $\cos\theta = \frac{3}{5}$. The excellent point is that the success probability can be amplified by repetitions, since even if we fail to apply P_θ to the input state, we still retain its input state. In what follows, when there are a QTM M and a constant $a > 0$ such that M implements a

unitary transformation U with probability a and the identity transformation
with probability $1 - a$, the operation implemented by M is called the *repeatable*
implementation of U.

Now, using Preskill's circuit and the construction of universal QTMs[1] by
Bernstein and Vazirani [6], we show that from the viewpoint of zero-error algorithms $\mathcal{A}$ is not more useful than $\mathcal{B}$ as the set of amplitudes taken by QTMs.

Theorem 3. $\mathbf{ZQP}_{\mathcal{A}} \subseteq \mathbf{ZQP}_{\mathcal{B}}$.

Proof. Assume that $L \in \mathbf{ZQP}_{\mathcal{A}}$. Let $M = (Q, \Sigma, \delta)$ be a QTM recognizing
L with zero-error. Let $r = r(n)$ be the computation time of M on input x of
length n. We first consider the following QTM M' simulating M. Given input
$x = x_0 x_1 \cdots x_{n-1}$, the QTM M' writes on its first track the sequence of $2r + 1$
k_1-bit strings

$$(b, B) \cdots (b, B)(q_0, x_0)(b, x_1) \cdots (b, x_{n-1})(b, B) \cdots (b, B)$$

representing that the current state of M is $|q_0, \mathrm{T}[x], 0\rangle$. Here, $k_1 = \lceil \log(\#Q+1) \rceil$
$+ \lceil \log(\#\Sigma) \rceil$. Now we assume that when M' completes the simulation of t steps
of M, the first track configuration is $\mathrm{T}[y]$, where $t < r$ and y denotes

$$(b, \sigma_{-r}) \cdots (b, \sigma_{l-1})(q, \sigma)(b, \sigma_{l+1}) \cdots (b, \sigma_r).$$

Then, M' simulates one step of M in the following three stages: (1) M' searches
a pair (q, σ) whose first component is not b; (2) M' transforms (q, σ) into (p, τ)
with amplitude $\langle p, \tau | L_\delta | q, \sigma \rangle$; (3) according to the value $D(p) \in \{0, \pm 1\}$, M'
implements the operation corresponding to the movement of the head of M. For
example, if $D(p) = 1$, M' transforms y into

$$(b, \sigma_{-r}) \cdots (b, \sigma_{l-1})(b, \tau)(p, \sigma_{l+1}) \cdots (b, \sigma_r)$$

with amplitude $\langle p, \tau | L_\delta | q, \sigma \rangle$. We can see easily that M' simulates M. All the
operation of M' are deterministic except stage (2), which can be viewed as the
implementation of the k_1-qubit gate L_δ. Now, we give a QTM M_0 implement-
ing L_δ with zero-error. Note that all the components $\langle p, \tau | L_\delta | q, \sigma \rangle$ of L_δ are in
$\mathcal{A}$. By the method given in [4], we can decompose L_δ into the sequence of k_2
Toffoli gates, k_3 P_π's and k_4 controlled-R_θ's with $\cos \theta = \frac{3}{5}$, where k_2, k_3, k_4 are
independent of the length of the input of M. We can easily construct QTMs
M_1 and M_2 with the amplitudes $\mathcal{B}$ implementing a Toffoli gate and P_π. We use
Preskill's circuit (and its modification) in order to implement a controlled-R_θ
with zero-error. Now let R_a and R_b be the two 1-qubit gates represented by the
matrices

$$R_a = e^{\frac{i\theta}{2}} \begin{pmatrix} \cos \frac{\theta}{2} & \sin \frac{\theta}{2} \\ \sin \frac{\theta}{2} & -\cos \frac{\theta}{2} \end{pmatrix} \quad \text{and} \quad R_b = e^{-\frac{i\theta}{2}} \begin{pmatrix} \cos \frac{\theta}{2} & -\sin \frac{\theta}{2} \\ -\sin \frac{\theta}{2} & -\cos \frac{\theta}{2} \end{pmatrix}$$

in the computational basis. Note that $R_a R_b = R_\theta$. The implementation of a
controlled-R_a with zero-error and high probability is given by the following algorithm defined in Fig 1. In this algorithm, Step 3.2.2 corresponds to the operation

---

[1] The construction of more general universal QTM simulating even multi-tape QTMs
is given in [17] based on the claim in [29].

(Input) 2-qubit given in Registers 1 and 2.

1. Set $C = 0$ and $A = 0$, and prepare the state $|0\rangle \otimes \cdots \otimes |0\rangle$ in Register 3 which consists of $2s(n) + 1$ cells.
2. While $A = 0$, iterate the following Steps 3–4.
3. (Repeatable implementation of a controlled-R_a) Implement Steps 3.1–3.4.
   3.1. Implement $S = P_{\frac{\pi}{2}}$ and H on Register 2 in this order.
   3.2. (Repeatable implementation of a controlled-P_θ) Implement Steps 3.2.1–3.2.3.
      3.2.1. If the bit 0 is written in Register 1, swap the contents of Register 2 and the $(2s(n) + 1)$-th cell of Register 3.
      3.2.2. (Repeatable implementation of P_θ) Implement Steps (a)–(e).
         (a). Implement H on the $(C + 1)$-th and $(C + 2)$-th cells in Register 3.
         (b). Implement a Toffoli gate on Register 2 and the $(C+1)$-th and $(C+2)$-th cells of Register 3, where the former and the latters work as the target qubit and the control qubits.
         (c). Implement S on Register 2.
         (d). Implement the same operations as Step (b) and Step (a) in this order.
         (e). If the bit in either the $(C + 1)$-th or the $(C + 2)$-th cell of Register 3 is 1, implement $Z = P_\pi$ on Register 2.
      3.2.3. Implement the same operation as Step 3.2.1.
   3.3. Implement H and S on Register 2 in this order.
   3.4. If the 3 bits in Register 1, the $(C + 1)$-th and $(C + 2)$-th cells of Register 3 is not 100, implement Z on Register 2.
4. If the 2 bits 00 is written in the $(C + 1)$-th and $(C + 2)$-th cells of Register 3, then set $A = 1$. Otherwise, set $C = C + 2$. If $C = 2s(n)$, output the special symbol '?' and set $A = 1$.

Fig. 1. Algorithm implementing a controlled-R_a with zero-error and probability $1 - (\frac{3}{8})^{s(n)}$, where Registers 1 and 2 respectively correspond to controlled and target qubits.

implemented by Preskill's circuit. Step 3.2 and Step 3 respectively corresponds to the repeatable implementations of a controlled-P_θ and a controlled-R_a. Now, using the programming lemmas of QTMs [6], we can give a QTM M_3 implementing the algorithm in Fig.1, i.e., implementing a controlled-R_a with zero-error and probability $1 - (\frac{3}{8})^{s(n)}$. We can also construct a QTM M_4 implementing a controlled-R_b with zero-error and probability $1 - (\frac{3}{8})^{s(n)}$ by replacing $S = P_{\frac{\pi}{2}}$ in Step 3.2.2(c) of Fig.1 with $S' = \imath P_{-\frac{\pi}{2}}$. From M_3 and M_4 we can construct a QTM M_5 which implements a controlled-R_θ with zero-error and probability $(1 - (\frac{3}{8})^{s(n)})^2$. We can now construct M_0 by combining M_1, M_2 and M_5. Let M'' be a QTM obtained by replacing stage (2) of M' with M_0. Then, we can see that M'' simulates M with probability $(1 - (\frac{3}{8})^{s(n)})^{2k_4 r(n)}$ and M'' outputs '?' if M'' fails in the simulation. The quantum transition function of M'' has the range in $\mathcal{B}$. Thus by taking $s(n) = cr(n)$ for sufficiently large $c \in \mathbb{N}$, we have $L \in \mathbf{ZQP}_{\{0, \pm \frac{1}{\sqrt{2}}, \pm 1, \imath\}}$. Using a technique that simulates QTMs with non-real amplitudes by ones with real amplitudes [5], we have $L \in \mathbf{ZQP}_\mathcal{B}$.

However, different from the exact case, it seems that zero-error quantum computation with the amplitudes from $\mathcal{A}$ is more powerful than zero-error classical computation, since we can construct a QTM with the amplitudes from $\mathcal{A}$ solving the factoring problem in zero-error polynomial-time. To this end, it is sufficient to show that the language

FACTOR $= \{\langle N, k\rangle|$ N has a non-trivial prime factor larger than k $\}$

is in $\mathbf{ZQP}_{\mathcal{A}}$, since the factoring problem is polynomial-time Turing reducible to FACTOR and the class of problems solved by zero-error algorithms is closed under polynomial-time Turing reductions. In fact, we can show more general statement by combining Shor's factoring algorithm [23], a zero-error primality testing [2], Solovay-Kitaev theorem [15,16], and the fact that probabilistic Turing machines with coins of bias $0 < \alpha < \frac{1}{2}$ are polynomially equivalent to ones with unbiased coins [25].

Theorem 4. *If a real θ is an irrational multiple of π, it holds the relation* FACTOR $\in \mathbf{ZQP}_{\{0,\pm\cos\theta,\pm\sin\theta,\pm1\}}$. *In particular,* FACTOR $\in \mathbf{ZQP}_{\mathcal{A}}$.

Acknowledgements. The author would like to thank Artur Ekert, Hartmut Klauck, and Masanao Ozawa for helpful comments. He is also grateful to anonymous referees for their critical comments on an early draft.

References

1. L. M. Adleman, J. DeMarrais, and M. A. Huang: Quantum computability. SIAM J. Comput. **26** (1997) 1524–1540.
2. L. Adleman and M. Huang: Recognizing primes in random polynomial time. in Proc. of 19th ACM Symposium on Theory of Computing, ACM (1987) 462–470.
3. A. Berthiaume and G. Brassard: Oracle quantum computing. J. Modern Opt. **41** (1994) 2521–2535.
4. A. Barenco, C. H. Bennett, R. Cleve, D. DiVicenzo, N. Margolus, P. Shor, T. Sleator, J. Smolin and H. Weinfurter: Elementary gates for quantum computation. Phys. Rev. A, **52** (1995) 3457–3467.
5. E. Bernstein and U. Vazirani: Quantum complexity theory (Preliminary abstract). in Proc. of 25th ACM Symposium on Theory of Computing, ACM (1993) 11–20.
6. E. Bernstein and U. Vazirani: Quantum complexity theory. SIAM J. Comput. **26** (1997) 1411–1473.
7. G. Brassard and P. Høyer: An exact quantum polynomial-time algorithm for Simon's problem. in Proc. of 5th Israeli Symposium on Theory of Computing and Systems, IEEE (1997) 12–23.
8. C. H. Bennett, E. Bernstein, G. Brassard, and U. Vazirani: Strengths and weaknesses of quantum computing. SIAM J. Comput. **26** (1997) 1510–1523.
9. D. Deutsch: Quantum theory, the Church-Turing principle and the universal quantum computer. Proc. Roy. Soc. London Ser. A, **400** (1985) 96–117.
10. D. Deutsch: Quantum computational networks. Proc. Roy. Soc. London Ser. A, **425** (1989) 73–90.

11. D. Deutsch and R. Jozsa: Rapid solution of problems by quantum computation. Proc. Roy. Soc. London Ser. A, **439** (1992) 553–558.
12. S. Fenner, F. Green, S. Homer and R. Pruim: Determining acceptance probability for a quantum computation is hard for **PH**. in Proc. 6th Italian Conference on Theoretical Computer Science, World-Scientific (1998) 241–252
13. L. Fortnow and J. Rogers: Complexity limitations on quantum computation. in Proc. 13th Conference on Computational Complexity, IEEE (1998) 202–209.
14. Ker-I. Ko and H. Friedman: Computational complexity of real functions. Theoret. Comput. Sci. **20** (1982) 323–352.
15. A. Kitaev: Quantum computations: algorithms and error correction. Russian Math. Surveys **52** (1997) 1191-1249.
16. M. A. Nielsen and I. L. Chuang: Quantum Computation and Quantum Information, Cambridge (2000).
17. H. Nishimura and M. Ozawa: Computational complexity of uniform quantum circuit families and quantum Turing machines. Theoret. Comput. Sci. (to appear). Also `quant-ph/9906095` [2].
18. M. Ozawa: Quantum nondemolition monitoring of universal quantum computers. Phys. Rev. Lett. **80** (1998) 631–634.
19. M. Ozawa and H. Nishimura: Local transition functions of quantum Turing machines. RAIRO Theoret. Informatics Appl. **34** (2000) 379-402.
20. C. H. Papadimitriou: Computational Complexity, Addison-Wesley (1994).
21. J. Preskill: Reliable quantum computers. Proc. Roy. Soc. London ser A, **454** (1998) 385–410.
22. D. Simon: On the power of quantum computation. in Proc. 35th IEEE Symposium on Foundations of Computer Science, IEEE (1994) 116–123; SIAM J. Comput. **26** (1997) 1474–1483.
23. P. W. Shor: Algorithms for quantum computations: Discrete log and factoring. in Proc. 35th IEEE Symposium on Foundations of Computer Science, IEEE (1994) 124–134. ; Polynomial-time algorithms for prime factorization and discrete logarithms on a quantum computer. SIAM J. Comput. **26** (1997) 1484-1509.
24. P. W. Shor: Fault-tolerant quantum computation. in Proc. 37th IEEE Symposium on Foundations of Computer Science, IEEE (1996) 56–65.
25. U. V. Vazirani and V. V. Vazirani: Random polynomial time equals semi-random polynomial time. in Proc. 26th IEEE Symposium on Foundations of Computer Science, IEEE (1985) 417–428.
26. K. Wagner: The complexity of combinatorial problems with succinct input representation. Acta Inf. **23** (1986) 325–361.
27. T. Yamakami: A foundation of programming a multi-tape quantum Turing machine. in Proc. 24th International Symposium on Mathmatical Foundations of Computer Science, Lecture Notes in Comput. Sci. **1672** (1999) 430–441.
28. T. Yamakami and A. Yao: $\mathbf{NQP}_\mathbb{C} = $ co-$\mathbf{C_=P}$. Inform. Process. Lett. **71** (1999) 63–69.
29. A. Yao: Quantum circuit complexity. in Proc. 34th Annual IEEE Symposium on Foundations of Computer Science, IEEE (1993) 352–361.

---

[2] quant-ph preprints are available at `http://xxx.lanl.gov/abs/quant-ph/number`.

A Quantum Goldreich-Levin Theorem with Cryptographic Applications

Mark Adcock and Richard Cleve*

Dept. of Computer Science, University of Calgary, Calgary, Alberta, Canada T2N 1N4
mark.adcock@cdcgy.com, cleve@cpsc.ucalgary.ca

Abstract. We investigate the Goldreich-Levin Theorem in the context of quantum information. This result is a reduction from the problem of inverting a one-way function to the problem of predicting a particular bit associated with that function. We show that the quantum version of the reduction is quantitatively more efficient than the known classical version. If the one-way function acts on n-bit strings then the overhead in the reduction is by a factor of $O(n/\varepsilon^2)$ in the classical case but only by a factor of $O(1/\varepsilon)$ in the quantum case, where $\frac{1}{2} + \varepsilon$ is the probability of predicting the hard-predicate. We also show that, using the Goldreich-Levin Theorem, a quantum bit (or qubit) commitment scheme that is perfectly binding and computationally concealing can be obtained from any quantum one-way permutation.

1 Introduction

Fast quantum algorithms are potentially useful in that, if quantum computers that can run them are built, they can then be used to solve computational problems quickly. Algorithms can also be the basis of *reductions* between computational problems in instances where the underlying goals are different from fast computations. For example, reductions are often used as indicators that certain problems are computationally hard, as in the theory of NP-completeness (see [12] and references therein). Another domain where reductions play an important role is in complexity-based cryptography, where a reduction can show that breaking a particular cryptosystem is as difficult (or almost as difficult) as solving a computational problem that is presumed to be hard.

We investigate such a cryptographic setting where quantum algorithms yield different reductions than are possible in the classical case: the so-called Goldreich-Levin Theorem [13]. This result is a reduction from the computational problem of inverting a one-way function to the problem of predicting a particular hard-predicate associated with that function. Roughly speaking, a *one-way function* is a function that can be efficiently computed in the forward direction but is hard to compute in the reverse direction, and a *hard-predicate* of a function is a bit that can be efficiently computed from the input to the function but is hard to estimate from the output of the function. We show that

---

* Partially supported by Canada's NSERC and the Killam Trust.

© Springer-Verlag Berlin Heidelberg 2002

the quantum version of the reduction is quantitatively more efficient than the classical version.

Goldreich and Levin essentially showed that, for a problem instance of size n bits, if their hard-predicate can be predicted with probability $\frac{1}{2} + \varepsilon$ with computational cost T then the one-way function can be inverted with computational cost $O(T\,D(n, \varepsilon))$, where $D(n, \varepsilon)$ is polynomial in n/ε. Taken in its contrapositive form, this means that, if inverting the one-way function requires a computational cost of $\Omega(T)$, then predicting the hard-predicate with probability $\frac{1}{2} + \varepsilon$ requires a computational cost of $\Omega(T/D(n, \varepsilon))$. Note that if we start with a specific lower bound of $\Omega(T)$ for inverting the function then we end up with a weaker lower bound—by a *dilution factor* of $D(n, \varepsilon)$—for breaking the hard-predicate. In [14], it is shown that the dilution factor can be as small as $O(n/\varepsilon^2)$.

We show that there is a quantum implementation of the reduction where the dilution factor is only $O(1/\varepsilon)$. We also show that $\Omega(n/\varepsilon^2)$ is a lower bound on the dilution factor for classical versions of the reduction in a suitable black-box framework. In the standard parameterization of interest in cryptography, T is assumed to be superpolynomial in n and $\varepsilon \in 1/n^{O(1)}$. In this case, although $1/\varepsilon$ is smaller than n/ε^2, the diluted computational cost, $T/D(n, \varepsilon)$, remains superpolynomial in both cases. However, there are other parameterizations where the difference between the achievable quantum reduction and best possible classical reduction is more pronounced. One example is the case where $T = n^3$ and $\varepsilon = 1/n$. If we start with a classical one-way function that requires a computational cost of $\Omega(n^3)$ to invert and apply the Goldreich-Levin Theorem to construct a classical hard-predicate then the reduction implies only that the computational cost of predicting the predicate with probability $\frac{1}{2} + \frac{1}{n}$ is lower bounded only by a *constant*. However, if we start with a *quantum* one-way function that requires a computational cost of $\Omega(n^3)$ to invert and apply our quantum version of the Goldreich-Levin Theorem then the computational cost of predicting the predicate with probability $\frac{1}{2} + \frac{1}{n}$ is lower bounded by $\Omega(n^2)$.

A particular application of hard-predicates is for bit commitment. It is well-known that an information theoretically secure bit commitment scheme cannot be based on the information-theoretic properties of quantum devices alone [16, 17]. Of course, this is also the case with *classical* devices, though *computationally secure* bit commitment schemes have been widely proposed, investigated, and applied. Such schemes can be based on the existence of one-way permutations. Most of these proposed one-way permutations are hard to invert only if problems such as factoring or the discrete logarithm are hard, and are insecure against quantum computers, which can efficiently solve such problems [19]. Recently, Dumais, Mayers and Salvail considered the possibility of *quantum* one-way permutations [11], and showed how to base quantum bit commitment on them (see also [10]). Their scheme is perfectly concealing and *computationally binding*, in the sense that changing a commitment is computationally hard if inverting the permutation is hard. We exhibit a complementary quantum bit commitment scheme that is perfectly binding and computationally concealing. As with hard-predicates, the dilution factor in the measure of computational security is lower than possible with the corresponding classical construction. Furthermore, a possible advantage of our protocol is that the information that must be communicated and stored between the parties consists of $O(n)$ classical bits for bit commitment (and $O(n)$ classical bits plus one qubit for qubit commitment), whereas the scheme in [11] employs $O(n)$ qubits.

The organization of this paper is as follows. In Section 2, we investigate a simple black-box problem associated to the Goldreich-Levin Theorem. In Section 3, we give definitions pertaining to one-way permutations and hard-predicates (classical and quantum versions) and investigate the complexity of reductions from the former to the latter. In Section 4, we show how to use the Goldreich-Levin Theorem to construct a perfectly binding and computationally concealing quantum bit commitment scheme from a quantum one-way permutation.

2   A Black-Box Problem

Our results about the Goldreich-Levin Theorem are based on the query complexity of the following black-box problem, which we refer to as the *GL problem* (see, e.g., [2]). Let n be a positive integer and $\varepsilon > 0$. Let $a \in \{0,1\}^n$ and let information about a be available only from inner product and equivalence queries.

Definition 1. A *classical inner product (IP)* query (with bias ε) has input $x \in \{0,1\}^n$ and outputs a bit that is correlated with $a \cdot x$ (the inner product of a and x modulo two) in the sense that, with respect to the uniform distribution,

$$\Pr_x[IP(x) = a \cdot x] \geq \tfrac{1}{2} + \varepsilon. \tag{1}$$

Definition 2. An *classical equivalence (EQ)* query has input $x \in \{0,1\}^n$, and the output is 1 if $x = a$ and 0 otherwise.

The goal is to determine a with a minimum number of IP and EQ queries. A secondary resource under consideration is the number of auxiliary bit/qubit operations. When $\varepsilon = \tfrac{1}{2}$, this is essentially a problem that Bernstein and Vazirani [3] considered, where IP queries return $a \cdot x$ on input x. For this problem, n IP queries are necessary and sufficient to solve it classically; however, it can be solved with a single (appropriately defined) quantum IP query. (See also [20].) When ε is small—say, $\varepsilon \in 1/n^{O(1)}$—an efficient classical solution to this problem is nontrivial. The correctness probability of an IP query for a particular x cannot readily be amplified by simple techniques such as repeating queries; for some x, $IP(x)$ may always be wrong. Goldreich and Levin [13] were the first to (implicitly) solve this problem with a number of queries and auxiliary operations that is polynomial in n/ε—and this is the basis of their cryptographic reduction in Theorem 11.

We show that any classical algorithm solving the GL problem with constant probability must make $\Omega(n/\varepsilon^2)$ queries (for a reasonable range of values of ε), whereas there is a quantum algorithm that solves the GL problem with $O(1/\varepsilon)$ queries.

Theorem 3. *Any classical probabilistic algorithm solving the GL problem with success probability $\delta > 0$ requires either more than $2^{n/2}$ EQ queries or $\Omega(\delta n/\varepsilon^2)$ IP queries when $\varepsilon \geq \sqrt{n}2^{-n/3}$.*

Proof. The proof uses classical information theory, bounding the conditional mutual information about an unknown string that is revealed by each *IP* query, in conjunction with an analysis of the effect of *EQ* queries. Consider an algorithm to be *successful* on a particular input if and only if it performs an *EQ* query whose output is 1.

We begin by showing that it is sufficient to consider algorithms (formally, decision trees) that are in a convenient simple form. First, by a basic game-theoretic argument [21], it suffices to consider deterministic algorithms, where their input data—embodied in the black-boxes for *IP* and *EQ* queries—may be generated in a probabilistic manner. Second, it can be assumed that all *EQ* queries occur only after all *IP* queries have been completed. To see why this is so, start with an algorithm that interleaves *IP* and *EQ* queries, and modify it as follows. Whenever an *EQ* query occurs before the end of the *IP* queries, the modified algorithm stores the value of the input to the query and proceeds as if the result were 0. Then, at the end of the *IP* queries, each such deferred *EQ* query is applied. The modified algorithm will behave consistently whenever the actual output of a deferred *EQ* query is 0, and also it will perform (albeit later) any *EQ* query where the output is 1. Henceforth, we consider only algorithms with the above simplifications.

Now we describe a probabilistic procedure for constructing the black boxes that perform *IP* and *EQ* queries. First, $a \in \{0,1\}^n$ is chosen randomly, uniformly. Then a set $S \subseteq \{0,1\}^n$ is chosen randomly, uniformly subject to the condition that $|S| = (\frac{1}{2} + \varepsilon)2^n$ (assuming that $\varepsilon 2^n$ is an integer). Then

$$IP(x) = \begin{cases} a \cdot x & \text{if } x \in S \\ \overline{a \cdot x} & \text{if } x \notin S \end{cases} \tag{2}$$

$$EQ(x) = \begin{cases} 1 & \text{if } x = a \\ 0 & \text{if } x \neq a. \end{cases} \tag{3}$$

Consider an algorithm that makes m *IP* queries. If $m \geq \delta n/\varepsilon^2$ then the theorem is proven. Otherwise, since $\varepsilon \geq \sqrt{n}2^{-n/3}$, we have

$$m < \frac{\delta n}{\varepsilon^2} \leq \delta 2^{2n/3}. \tag{4}$$

We proceed by determining the amount of information about a that is conveyed by the application of m *IP* queries. Let A be the $\{0,1\}^n$-valued random variable corresponding to the probabilistic choice of $a \in \{0,1\}^n$, and let $Y_1, Y_2, \ldots, Y_m$ be the $\{0,1\}$-valued random variables corresponding to the respective outputs of the m *IP* queries. Let H be the Shannon entropy function (see, e.g., [9]). Then, for each $i \in \{1, 2, \ldots, m\}$,

$$H(A|Y_1, \ldots, Y_i) = H(A|Y_1, \ldots, Y_{i-1}) - H(Y_i|Y_1, \ldots, Y_{i-1}) + H(Y_i|A, Y_1, \ldots, Y_{i-1}).$$

Combining the above equations yields

$$H(A|Y_1, \ldots, Y_m) = H(A) + \sum_{i=1}^{m} (H(Y_i|A, Y_1, \ldots, Y_{i-1}) - H(Y_i|Y_1, \ldots, Y_{i-1})).$$

$$(5)$$

We shall bound each term on the right side of Eq. 5. Since the *a priori* distribution of A is uniform, $H(A) = n$. Since Y_i is $\{0, 1\}$-valued, $H(Y_i|Y_1, \ldots, Y_{i-1}) \leq 1$ for all $i \in \{1, 2, \ldots, m\}$. Next, we show that, for all $i \in \{1, 2, \ldots, m\}$,

$$H(Y_i|A, Y_1, \ldots, Y_{i-1}) \geq 1 - (16/\ln 2)\varepsilon^2. \tag{6}$$

It is useful to view the set S as being generated during the execution of the *IP* queries as follows. Initially S is empty, and when the first *IP* query is performed on some input x, x is placed in S with probability $\frac{1}{2} + \varepsilon$ and in $\overline{S}$ with probability $\frac{1}{2} - \varepsilon$. The inputs to subsequent *IP* queries are placed in S or $\overline{S}$ with appropriate probabilities: the input to the ith query is placed in S with probability

$$\frac{(\frac{1}{2} + \varepsilon)2^n - j}{2^n - (i-1)}, \tag{7}$$

where $j \in \{0, 1, \ldots, i-1\}$ is the number of previous inputs to queries that have been placed in S. Using Eq. 4, the above probability can be shown to lie between $\frac{1}{2} - 2\varepsilon$ and $\frac{1}{2} + 2\varepsilon$. It follows that

$$\begin{aligned} H(Y_i|A, Y_1, \ldots, Y_{i-1}) &\geq H(\tfrac{1}{2} + 2\varepsilon, \tfrac{1}{2} - 2\varepsilon) \\ &= -(\tfrac{1}{2} + 2\varepsilon)\log(\tfrac{1}{2} + 2\varepsilon) - (\tfrac{1}{2} - 2\varepsilon)\log(\tfrac{1}{2} - 2\varepsilon) \\ &\geq 1 - (16/\ln 2)\varepsilon^2, \end{aligned} \tag{8}$$

establishing Eq. 6. Substituting the preceding inequalities into Eq. 5, we obtain

$$H(A|Y_1, \ldots, Y_m) \geq n - (16/\ln 2)m\varepsilon^2. \tag{9}$$

Intuitively, the *IP* queries yield information about the value of A in terms of their effect on the probability distribution of A conditioned on the values of $Y_1, \ldots, Y_m$. Eq. 9 lower bounds the decrease in entropy possible.

From the conditions of the theorem, it can be assumed that, after the *IP* queries, $2^{n/2}$ *EQ* are performed. The algorithm succeeds with probability at least δ only if there exist $2^{n/2}$ elements of $\{0, 1\}^n$ whose total probability is at least δ. The maximum entropy that a distribution with this property can have is for a bi-level distribution, where $2^{n/2}$ elements of $\{0, 1\}^n$ each have probability $\delta/2^{n/2}$ and $2^n - 2^{n/2}$ elements each have probability $(1-\delta)/(2^n - 2^{n/2})$. Therefore,

$$\begin{aligned} H(A|Y_1, \ldots, Y_m) &\leq H\big(\underbrace{\tfrac{\delta}{2^{n/2}}, \ldots, \tfrac{\delta}{2^{n/2}}}_{2^{n/2}}, \underbrace{\tfrac{1-\delta}{2^n - 2^{n/2}}, \ldots, \tfrac{1-\delta}{2^n - 2^{n/2}}}_{2^n - 2^{n/2}}\big) \\ &= H(\delta, 1-\delta) + \delta \log(2^{n/2}) + (1-\delta)\log(2^n - 2^{n/2}) \\ &< 1 + \delta n/2 + (1-\delta)n \\ &= n - \delta n/2 + 1. \end{aligned} \tag{10}$$

Combining Eq. 9 with Eq. 10, yields $m > (\ln 2)(\delta n - 2)/(32\varepsilon^2) \in \Omega(\delta n/\varepsilon^2)$. $\square$

Definition 4. A *quantum inner product* query (with bias ε) is a unitary transformation U_{IP} on $n + m$ qubits, or its inverse $U_{IP}^\dagger$, such that U_{IP} satisfies the following two properties:

1. If $x \in \{0,1\}^n$ is chosen uniformly at random and the last qubit of $U_{IP}|x\rangle|0^m\rangle$ is measured, yielding $w \in \{0,1\}$, then $\Pr[w = a \cdot x] \geq \frac{1}{2} + \varepsilon$.
2. For any $x \in \{0,1\}^n$ and $y \in \{0,1\}^m$, the state of the first n qubits of $U_{IP}|x\rangle|y\rangle$ is $|x\rangle$.

The first property captures the fact that, taking a query to be a suitable application of U_{IP} followed by a measurement of the last qubit, Eq. 1 is satisfied. Any implementation of a quantum circuit that produces an output that is $a \cdot x$ with probability on average $\frac{1}{2} + \varepsilon$ can be modified to consist of a unitary stage U_{IP} followed by a measurement of one qubit. The second property is for technical convenience and does not restrict generality. Moreover, given a circuit implementing U_{IP}, it is easy to construct a circuit implementing $U_{IP}^\dagger$.

Definition 5. A *quantum equivalence query* is the unitary operation U_{EQ} such that, for all $x \in \{0,1\}^n$ and $b \in \{0,1\}$,

$$U_{EQ}|x\rangle|b\rangle = \begin{cases} |x\rangle|\bar{b}\rangle & \text{if } x = a \\ |x\rangle|b\rangle & \text{if } x \neq a. \end{cases} \tag{11}$$

For the quantum GL problem, $a \in \{0,1\}^n$ and information about a in available only from quantum *IP* and *EQ* queries and the goal is to determine a. We can now state and prove the result about quantum algorithms for the GL problem (which is similar to a result in [8] in a different context).

Theorem 6. *There exists a quantum algorithm solving the GL problem with constant probability using $O(1/\varepsilon)$ U_{IP}, $U_{IP}^\dagger$ and U_{EQ} queries in total. Also, the number of auxiliary qubit operations used by the procedure is $O(n/\varepsilon)$.*

Proof. The proof is by a combination of two techniques: the algorithm in [3] for the exact case (i.e., when $\varepsilon = \frac{1}{2}$), which is shown to be adaptable to "noisy" data in [8] (with a slightly different noise model than the one that arises here); and amplitude amplification [5,15,6].

Since U_{IP} applied to $|x\rangle|y\rangle$ has no net effect on its first n input qubits, for each $x \in \{0,1\}^n$, $U_{IP}|x\rangle|0^m\rangle = |x\rangle \, (\alpha_x|v_x\rangle|a \cdot x\rangle + \beta_x|w_x\rangle|\overline{a \cdot x}\rangle)$, where α_x and β_x are nonnegative real numbers, and $|v_x\rangle$ and $|w_x\rangle$ are $m - 1$ qubit quantum states. If the last qubit of $U_{IP}|x\rangle|0^m\rangle$ is measured then the result is: $a \cdot x$ with probability α_x^2, and $\overline{a \cdot x}$ with probability β_x^2. Therefore, since, for a random uniformly distributed $x \in \{0,1\}^n$, measuring the last qubit of $U_{IP}|x\rangle|0^m\rangle$ yields $a \cdot x$ with probability at least $\frac{1}{2} + \varepsilon$, it follows that

$$\frac{1}{2^n} \sum_{x\in\{0,1\}^n} \alpha_x^2 \geq \frac{1}{2} + \varepsilon \qquad \text{and} \qquad \frac{1}{2^n} \sum_{x\in\{0,1\}^n} \beta_x^2 \leq \frac{1}{2} - \varepsilon. \tag{12}$$

Now, consider the quantum circuit C in Figure 1 (for information about circuit notation, see, e.g., [18]).

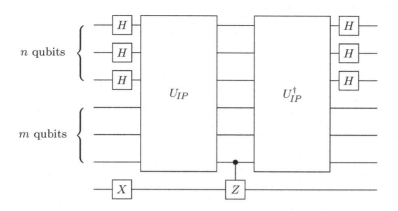

Fig. 1. Quantum circuit C.

We will begin by showing that $\langle a, 0^m, 1 | C | 0^n, 0^m, 0 \rangle$ is real-valued and

$$\langle a, 0^m, 1 | C | 0^n, 0^m, 0 \rangle \geq 2\varepsilon. \tag{13}$$

Note that C can be decomposed into the following five operations:

C_1: Apply H to each of the first n qubits, and NOT to the last qubit.
C_2: Apply U_{IP} to the first $n + m$ qubits.
C_3: Apply a controlled-Z to the last two qubits.
C_4: Apply $U_{IP}^\dagger$ to the first $n + m$ qubits.
C_5: Apply H to each of the first n qubits.

The value $\langle a, 0^m, 1 | C | 0^n, 0^m, 0 \rangle$ is the inner product of $C_3 C_2 C_1 | 0^n \rangle | 0^m \rangle | 0 \rangle$ and $C_4^\dagger C_5^\dagger | a \rangle | 0^m \rangle | 1 \rangle$. These states are (omitting normalization factors of $\frac{1}{\sqrt{2^n}}$)

$$
\begin{aligned}
C_3 C_2 C_1 | 0^n \rangle | 0^m \rangle | 0 \rangle &= C_3 C_2 \sum_{x \in \{0,1\}^n} |x\rangle | 0^m \rangle | 1 \rangle \\
&= C_3 \sum_{x \in \{0,1\}^n} |x\rangle \left(\alpha_x |v_x\rangle |a \cdot x\rangle + \beta_x |w_x\rangle |\overline{a \cdot x}\rangle \right) |1\rangle \\
&= \sum_{x \in \{0,1\}^n} |x\rangle \left(\alpha_x (-1)^{a \cdot x} |v_x\rangle |a \cdot x\rangle + \beta_x (-1)^{\overline{a \cdot x}} |w_x\rangle |\overline{a \cdot x}\rangle \right) |1\rangle \\
&= \sum_{x \in \{0,1\}^n} (-1)^{a \cdot x} |x\rangle \left(\alpha_x |v_x\rangle |a \cdot x\rangle - \beta_x |w_x\rangle |\overline{a \cdot x}\rangle \right) |1\rangle
\end{aligned}
$$

and

$$
\begin{aligned}
C_4^\dagger C_5^\dagger | a \rangle | 0^m \rangle | 1 \rangle &= C_4^\dagger \sum_{x \in \{0,1\}^n} (-1)^{a \cdot x} |x\rangle | 0^m \rangle | 1 \rangle \\
&= \sum_{x \in \{0,1\}^n} (-1)^{a \cdot x} |x\rangle \left(\alpha_x |v_x\rangle |a \cdot x\rangle + \beta_x |w_x\rangle |\overline{a \cdot x}\rangle \right) |1\rangle.
\end{aligned}
$$

From these equations and the fact that $\langle x | y \rangle = 0$ whenever $x \neq y$, we obtain

$$\langle a, 0^m, 1 | C | 0^n, 0^m, 0 \rangle = \frac{1}{2^n} \sum_{x \in \{0,1\}^n} (\alpha_x^2 - \beta_x^2) \geq (\tfrac{1}{2} + \varepsilon) - (\tfrac{1}{2} - \varepsilon) = 2\varepsilon,$$

which establishes Eq. 13.

Note that Eq. 13 implies that, if C is executed on input $|0^n, 0^m, 0\rangle$ and the result is measured in the classical basis, then the first n bits of the result will be a with probability at least $4\varepsilon^2$. Therefore, if this process is repeated $O(1/\varepsilon^2)$ times, checking each result with an EQ query, then a will be found with constant probability. A more efficient way of finding the value of a is to use *amplitude amplification* [5,15,6] with C and $C^\dagger$ in combination with EQ queries. As shown in [6], an expected number of $O(1/\varepsilon)$ executions of C, $C^\dagger$, and U_{EQ} are sufficient to obtain a. This implies that $O(1/\varepsilon)$ U_{IP}, $U_{IP}^\dagger$, and U_{EQ} are sufficient to succeed with constant probability. Also, it can be shown that the number of auxiliary operations required is $O(n/\varepsilon)$. $\square$

3   Hard-Predicates from One-Way Permutations

In this section, we give definitions pertaining to one-way permutations and hard-predicates (classical and quantum versions) and investigate the complexity of the reduction of Goldreich and Levin [13] from the former to the latter.

In the definitions below, the *size* of a classical [quantum] circuit, is understood to be relative to a suitable set of gates on one and two bits [qubits]. For a quantum circuit C acting on m qubits, and $x \in \{0,1\}^n$ (for $n \leq m$), let $C_k(x)$ ($k \in \{1, \dots, m\}$), denote the result of measuring the first k qubits of $C|x\rangle|0^{m-n}\rangle$ in the classical basis. The subscript k may be omitted when the value of k is clear from the context.

Intuitively, a quantum one-way permutation f on n bits is easy to compute in the forward direction but is hard to invert[1]. For the former property, the standard requirement is that f be computable by a uniform circuit of size $n^{O(1)}$ (though it is also possible to impose other upper bounds on the uniform circuit size).

Definition 7. A permutation $f : \{0,1\}^n \to \{0,1\}^n$ *is classically [quantumly] (δ, T)-hard to invert* if there is no classical [quantum] circuit C of size T such that $\Pr_a[C(f(a)) = a] \geq \delta$.

Now the standard requirement for the hard-to-invert condition is that f is (δ, T)-hard to invert for all $\delta \in 1/n^{O(1)}$ and $T \in n^{O(1)}$ (again, other bounds can be imposed). The idea behind a hard-predicate [4] is to concentrate the information that a one-way function "hides" about its input into a single bit. Intuitively, $h : \{0,1\}^n \to \{0,1\}$ is a hard-predicate of f if, given $a \in \{0,1\}^n$, it is easy to compute $h(a)$; whereas, given $f(a)$ for randomly chosen $a \in \{0,1\}^n$, it is hard to predict the value of the bit $h(a)$ with probability significantly better than $\frac{1}{2}$.

---

[1] The reversibility of quantum computations does not exclude this possibility [7].

The hard-predicate defined in [13] is

$$h(y, x) = y \cdot x, \tag{14}$$

(the inner product modulo two of x and y), for $(y, x) \in \{0,1\}^n \times \{0,1\}^n$. This is not a hard-predicate of f, but for a slightly modified version of f, as follows.

Definition 8. For a permutation $f : \{0,1\}^n \to \{0,1\}^n$, let $\tilde{f}$ denote the permutation $\tilde{f} : \{0,1\}^n \times \{0,1\}^n \to \{0,1\}^n \times \{0,1\}^n$ defined as $\tilde{f}(y, x) = (f(y), x)$.

Goldreich and Levin showed that if f is one-way then h is hard to predict from $\tilde{f}$. Instead of quantifying how well a circuit predicts h from $\tilde{f}$ as the amount by which $\Pr_{y,x}[C(\tilde{f}(y,x)) = h(y,x)]$ exceeds $\frac{1}{2}$, we adopt a related but slightly more complicated definition.

Definition 9. A circuit C (δ, ε)-predicts h from $\tilde{f}$ if

$$\Pr_y[\Pr_x[C(\tilde{f}(y,x)) = h(y,x)] \geq \tfrac{1}{2} + \varepsilon] \geq \delta. \tag{15}$$

To explain Eq. 15 in words, call $y \in \{0,1\}^n$ ε-good if $\Pr_x[C(\tilde{f}(y,x)) = h(y,x)] \geq \frac{1}{2} + \varepsilon$ for that value of y. Then then Eq. 15 means $\Pr_y[y$ is ε-good$] \geq \delta$.

The following lemma, which relates the two measures of prediction, is straightforward to prove by an averaging argument.

Lemma 10. If $\Pr_{y,x}[G(\tilde{f}(y,x)) = h(y,x)] \geq \frac{1}{2} + \varepsilon$ then G $(\varepsilon/(1-\varepsilon), \varepsilon/2)$-predicts h from $\tilde{f}$.

Proof. Let $p = \Pr[y$ is $\frac{\varepsilon}{2}$-good$]$. Then

$$\begin{aligned}
\tfrac{1}{2} + \varepsilon &\leq \Pr[G(\tilde{f}(y,x)) = h(y,x)|y \text{ is } \tfrac{\varepsilon}{2}\text{-good}] \Pr[y \text{ is } \tfrac{\varepsilon}{2}\text{-good}] \\
&+ \Pr[G(\tilde{f}(y,x)) = h(y,x)|y \text{ is not } \tfrac{\varepsilon}{2}\text{-good}] \Pr[y \text{ is not } \tfrac{\varepsilon}{2}\text{-good}] \\
&< p + (\tfrac{1}{2} + \tfrac{\varepsilon}{2})(1 - p),
\end{aligned}$$

which implies $p > \varepsilon/(1 - \varepsilon)$. $\qquad\square$

Note in particular that, if $\Pr_{y,x}[G(\tilde{f}(y,x)) = h(y,x)] \geq \frac{1}{2} + 1/n^{O(1)}$ then G $(1/n^{O(1)}, 1/n^{O(1)})$-predicts h from $\tilde{f}$.

Theorem 11 ([13,14]). *If $f : \{0,1\}^n \to \{0,1\}^n$ is classically $(\delta/2, T)$-hard to invert then any classical circuit that (δ, ε)-predicts h from $\tilde{f}$ must have size $\Omega(T\varepsilon^2/n)$.*

The proof of this theorem is essentially a reduction from the problem of inverting f to the problem of (δ, ε)-predicting h. One begins by assuming that a circuit G of size $o(T\varepsilon^2/n)$ (δ, ε)-predicts h from $\tilde{f}$ and then shows that, by making $O(n/\varepsilon^2)$ calls to both G and f (plus some additional computations), f can be inverted with probability $\delta/2$ [14]. The total running time of the inversion procedure is $o((n/\varepsilon^2)(T\varepsilon^2/n)) = o(T)$, contradicting the fact that f is $(\delta/2, T)$-hard to invert.

Our quantum version of the Goldreich-Levin Theorem is the following.

Theorem 12. *If $f : \{0,1\}^n \to \{0,1\}^n$ is quantumly $(\delta/2, T)$-hard to invert then any quantum circuit that (δ, ε)-predicts h from $\tilde{f}$ must have size $\Omega(T\varepsilon)$.*

Proof. As in the classical case, the proof is essentially a reduction from the problem of inverting f to the problem of (δ, ε)-predicting h. Let $b = f(a)$ be an input instance—the goal is to determine a from b. We will show how to simulate EQ and IP queries in this setting and then apply the bounds in Theorem 6. It is easy to simulate an EQ query (relative to a) by making one call to f and checking if the result is b. Suppose that there exists a circuit G of size $o(T\varepsilon)$ that (δ, ε)-predicts h from $\tilde{f}$. Thus, $\Pr_y[\Pr_x[G(\tilde{f}(y,x)) = h(y,x)] \geq \frac{1}{2} + \varepsilon] \geq \delta$. Note that, with probability at least δ, a is ε-good, in the sense that $\Pr_x[G(\tilde{f}(a,x)) = h(a,x)] \geq \frac{1}{2} + \varepsilon$. When a is ε-good, computing $G(\tilde{f}(a,x)) = G(b,x)$ is simulating an IP query for x (relative to a). It follows from Theorem 6 that a can be computed with circuit-size $o((1/\varepsilon)(T\varepsilon)) = o(T)$ with success probability at least $\delta/2$ (where $1/2$ is the success probability of the algorithm that finds a when a is ε-good and δ is the probability that a is ε-good to begin with). This contradicts the $(\delta/2, T)$-hardness of inverting f, thus such a G cannot exist. □

To conclude this section, we note that it can be shown that the classical reduction in Theorem 11 cannot be improved asymptotically. Using Theorem 3, it can be shown that a lower bound for the size of $\tilde{f}$ that is larger than $\Omega(T\varepsilon^2/n)$ cannot be obtained classically by a black-box reduction from the problem of inverting f. The details of this are omitted in this version of the paper.

4   Bit Commitment from Quantum One-Way Permutations

In this section, we show how to use the Goldreich-Levin Theorem to construct a quantum bit commitment scheme from a quantum one-way permutation.

Definition 13. *A permutation $f : \{0,1\}^n \to \{0,1\}^n$ is a quantum one-way permutation if:*

1. *There is a uniform quantum circuit of size $n^{O(1)}$ computing $f(x)$ from x.*
2. *f is quantumly (δ, T)-hard to invert for any $\delta \in 1/n^{O(1)}$ and $T \in n^{O(1)}$.*

Theorem 14. *If there exists a quantum one-way permutation $f : \{0,1\}^n \to \{0,1\}^n$ then there exists a bit [or qubit] commitment scheme that is perfectly binding and computationally concealing, in the sense that the committed bit cannot be predicted with probability $\frac{1}{2} + 1/n^{O(1)}$ by a circuit of size $n^{O(1)}$.*

Proof. From Theorem 12, it is straightforward to construct a quantum bit commitment scheme from Alice to Bob based on a one-way permutation f as follows (where $h(y,x) = y \cdot x$).

Bit-commit. Let z be the bit to commit to. Alice chooses $a, x \in \{0,1\}^n$ randomly, sets $c = z \oplus h(a,x)$, computes $b = f(a)$, and sends (b,x,c) to Bob.

Bit-decommit. Alice sends a to Bob. Bob checks if $f(a) = b$ and rejects if this is not the case. Otherwise, Bob accepts and computes $c \oplus h(a, x)$ as the bit.

Since f is a permutation there is at most one *classical* value of a that is an acceptable decommitment of Alice's bit. This implies that the scheme is perfectly binding to Alice.

Theorem 12 implies that the scheme is also computationally concealing, since any $n^{O(1)}$-size circuit that enables Bob to guess z from (b, x, c) with probability $\frac{1}{2} + 1/n^{O(1)}$ can be converted to a $n^{O(1)}$-size circuit that inverts f with probability $1/n^{O(1)}$, violating the fact that f is one-way.

Finally, we explain how a *qubit* commitment scheme can be constructed using some of the ideas discussed in [1].

Qubit-commit. Let $|\psi\rangle$ be the qubit to commit to. Alice chooses $a_1, a_2, x_1, x_2 \in \{0,1\}^n$ randomly, constructs the state $|\psi'\rangle = X^{h(a_1,x_1)} Z^{h(a_2,x_2)} |\psi\rangle$, computes $b_1 = f(a_1)$ and $b_2 = f(a_2)$, and sends $(|\psi'\rangle, b_1, b_2, x_1, x_2)$ to Bob.

Qubit-decommit. Alice sends a_1, a_2 to Bob. Bob checks if $f(a_1) = b_1$ and $f(a_2) = b_2$, rejecting if this is not the case. Otherwise, Bob accepts and computes $Z^{h(a_2,x_2)} X^{h(a_1,x_1)} |\psi'\rangle$ as the qubit.

Clearly, the scheme is perfectly binding. Intuitively, the scheme is computationally concealing, because $h(a_1, x_1)$ and $h(a_2, x_2)$ "look random" to Bob. If Bob can use his information to efficiently significantly distinguish between the qubit that he receives from Alice in the commitment stage and a totally mixed state (density matrix $\frac{1}{2}I$) then this procedure can be adapted to distinguish between the pair of bits $r_1 = h(a_1, x_1)$ and $r_2 = h(a_2, x_2)$ and a pair of truly random bits, which would violate the result proven in Theorem 12. $\qquad\square$

Acknowledgments. We would like to thank Lynn Burroughs, Paul Dumais, Peter Høyer, Dominic Mayers, and Louis Salvail for helpful discussions.

References

1. A. Ambainis, M. Mosca, A. Tapp, R. de Wolf, "Private quantum channels", *Proc. 41st Ann. IEEE Symp. on Foundations of Computer Science (FOCS '00)*, pp. 547–553, 2000.
2. M. Bellare, "The Goldreich-Levin Theorem", Manuscript, 1999. (Available at http://www-cse.ucsd.edu/users/mihir/.)
3. E. Bernstein and U. V. Vazirani, "Quantum complexity theory", *SIAM J. on Comput.*, Vol. 26, No. 5, pp. 1411–1473, 1997.
4. M. Blum and S. Micali, "How to generate cryptographically strong sequences of pseudo-random bits", *SIAM J. on Comput.*, Vol. 13, No. 4, pp. 850–864, 1984.
5. G. Brassard and P. Høyer, "An exact quantum polynomial-time algorithm for Simon's problem", *Proc. Fifth Israeli Symp. on Theory of Computing and Systems*, pp. 12–23, 1997.

6. G. Brassard, P. Høyer, M. Mosca, A. Tapp, "Quantum amplitude amplification and estimation", To appear in *Quantum Computation and Quantum Information: A Millennium Volume*, AMS Contemporary Mathematics Volume. Available on the LANL preprint archive as quant-ph/0005055, 2000.

7. H. F. Chau and H.-K. Lo, "One way functions in reversible computations", *Cryptologia*, Vol. 21, No. 2, pp. 139–148, 1997.

8. R. Cleve, W. van Dam, M. Nielsen, and A. Tapp, "Quantum entanglement and the communication complexity of the inner product function", *Proc. of the First NASA International Conf. on Quantum Computing and Quantum Communications*, C. P. Williams (Ed.), Lecture Notes in Computer Science 1509, Springer-Verlag, pp. 61-74, 1999.

9. T. M. Cover and J. A. Thomas, *Elements of Information Theory*, John Wiley and Sons, 1991.

10. C. Crépeau, F. Légaré and L. Salvail, "How to convert the flavor of a quantum bit commitment", to appear in *Advances in Cryptology — EUROCRYPT 2001*, B. Pfitzmann (Ed.), Lecture Notes in Computer Science 2045, Springer-Verlag, pp. 60-77, 2001.

11. P. Dumais, D. Mayers, and L. Salvail, "Perfectly concealing quantum bit commitment from any one-way permutation", *Advances in Cryptology — EUROCRYPT 2000*, B. Preneel (Ed.), Lecture Notes in Computer Science 1807, Springer-Verlag, pp. 300–315, 2000.

12. M. R. Garey and D. S. Johnson, *Computers and Intractability: A Guide to the Theory of NP-Completeness*, W. H. Freeman & Co., 1979.

13. O. Goldreich and L. Levin, "Hard-core predicates for any one-way function", *Proc. 21th Ann. ACM Symp. on Theory of Computing (STOC '89)*, pp. 25–32, 1989.

14. O. Goldreich, *Modern Cryptography, Probabilistic Proofs and Pseudo-Randomness*, Springer, 1999.

15. L. K. Grover, "A fast quantum mechanical algorithm for database search", *Proc. 28th Ann. ACM Symp. on Theory of Computing (STOC '96)*, pp. 212–219, 1996.

16. H.-K. Lo and H. F. Chau, "Is quantum bit commitment really possible?", *Phys. Rev. Lett.*, Vol. 78, No. 17, pp. 3410–3413, 1997.

17. D. Mayers, "Unconditionally secure bit commitment is impossible", *Phys. Rev. Lett.*, Vol. 78, No. 17, pp. 3414–3417, 1997.

18. M. A. Nielsen and I. L. Chuang, *Quantum Computation and Quantum Information*, Cambridge, 2000.

19. P. W. Shor, "Polynomial-time algorithms for prime factorization and discrete logarithms on a quantum computer", *SIAM J. on Computing*, Vol. 26, No. 5, pp. 1484–1509, 1997.

20. B. M. Terhal and J. A. Smolin, "Single quantum querying of a database", *Phys. Rev. A*, Vol. 58, No. 3, pp. 1822–1826, 1998.

21. A. C.-C. Yao, "Lower bounds by probabilistic arguments", *Proc. 24th Ann. IEEE Symp. on Foundations of Computer Science (FOCS '83)*, pp. 420–428, 1983.

On Quantum and Approximate Privacy

(Extended Abstract*)

Hartmut Klauck**

CWI
P.O. Box 94079, 1090 GB Amsterdam
The Netherlands
klauck@cwi.nl

Abstract. This paper studies privacy and secure function evaluation in communication complexity. The focus is on quantum versions of the model and on protocols with only approximate privacy against honest players. We show that the privacy loss (the minimum divulged information) in computing a function can be decreased exponentially by using quantum protocols, while the class of privately computable functions (i.e., those with privacy loss 0) is not increased by quantum protocols. Quantum communication combined with small information leakage on the other hand makes certain functions computable (almost) privately which are not computable using quantum communication without leakage or using classical communication with leakage. We also give an example of an exponential reduction of the communication complexity of a function by allowing a privacy loss of $o(1)$ instead of privacy loss 0.

1 Introduction

Mafiosi Al and Bob, both honest men, claim rights to protect a subset of the citizens of their hometown. To find out about possible collisions of interest they decide to communicate and find out whether there is a citizen they both intend to protect. Of course they would like to do this in a way that gives each other as little information as possible on the subset they think of. In other words, they want to compute a function with as much privacy as possible. This problem is one of the kind studied in the theory of private computation resp. secure function evaluation, initiated by Yao [19]. Another example is the two millionaires' problem, in which Al and Bob try to determine who is richer without revealing more about their actual wealth.

Informally a protocol for the computation of some function on inputs distributed to several players is private, if all information that can be deduced by one player during a run of the protocol can also be deduced from the player's input and the function value. A function has privacy loss k, if the minimum information divulged to the other player is k. Furthermore the leakage of a protocol

---

* For the full version of the paper see http://www.arXiv.org/abs/quant-ph/0110038.
** Supported by the EU 5th framework program QAIP IST-1999-11234 and by NWO grant 612.055.001.

H. Alt and A. Ferreira (Eds.): STACS 2002, LNCS 2285, pp. 335–346, 2002.
© Springer-Verlag Berlin Heidelberg 2002

may be measured as a distance between "not to be distinguishable" message states. This model generalizes several cryptographic settings, see [19].

There are several variants of this basic model. One can distinguish computationally secure and information theoretically secure protocols. The first variant is studied e.g. in [19,12]. Multiparty protocols in the information theoretically secure setting are given in [5,9]. A second kind of variations concerns the type of players. Basically "honest but curious" and "malicious" (or "byzantine") players have been considered in the literature. The first type of players sticks to the protocol but tries to get information by running some extra program on the messages he has received. The second type of players deviates arbitrarily from the protocol to get as much information as possible.

The two millionaires' problem has a computationally secure solution [19] relying on the existence of one-way functions, but it cannot be solved in the information theoretic sense [10] (among honest players), i.e., some cryptographic hardness assumption has to be used. A variant of the millionaires' problem that can actually be solved with information theoretic privacy (for honest players) is the identified minimum problem, in which the wealth of the less rich player and his identity is revealed, but no other information [14].

While considering only honest players seems to strongly restrict the model, it is important for several reasons. First, understanding honest players is a prerequisite to understanding actively cheating players. Secondly, these players capture "passive" attacks that cannot be detected in any way. Furthermore [12] gives a quite general reduction from multiparty protocols with honest majority to protocols with only honest players (in the computationally secure setting). Other motivations for considering this model include close connections to complexity measures like circuit size [16].

We focus on the following aspects of private computing. Al and Bob have heard that quantum computers can break cryptographic schemes classically assumed to be secure, so they do not want to rely on computational solutions. They are interested in whether quantum communication increases the set of privately computable functions or substantially decreases the privacy loss of functions. Furthermore they are interested in whether it is possible to decrease the communication cost of a protocol by allowing leakage of a small amount of information. We concentrate on the two player model in this paper, though some of the results have implications for the multiparty setting, which we mention in the conclusions. We investigate the model with honest players. In the quantum case a major point will also be whether the players might be trusted to not quit the protocol before the end. Our main model will measure the maximum information obtainable over all the rounds, not only the information obtainable at the end of the protocol. This corresponds to players that might quit the protocol before its end[1]. The other model of nonpreemptive players will be investigated also, but here every function turns out to be computable almost privately and at the same time efficiently in the quantum case.

---

[1] Al has the habit of shooting his guests after dessert, which may be well before the end of the protocol.

The functions we mainly consider in this paper are the disjointness problem $DISJ_n$, in which Al and Bob each receive a subset of a size n universe, and have to decide whether their subsets are disjoint or not, and the identified minimum problem $IdMin_n$, in which Al and Bob receive numbers x, y from 1 to 2^n, and the output is $2x + 1$, if $x \leq y$, and $2y$ otherwise.

Our main results are the following. We show that the quantum protocol for disjointness with communication complexity $O(\sqrt{n} \log n)$ given in [7] can be adapted to have privacy loss $O(\log^2 n)$. We proceed to show that any classical bounded error protocol for disjointness divulges $\Omega(\sqrt{n}/ \log n)$ bits of information. Thus Al and Bob are highly motivated to use the quantum protocol for privacy reasons.

We then show that the class of privately computable functions is not increased by using quantum computers, i.e., every function that can be computed privately using a quantum protocol can also be computed privately by a deterministic protocol. This result leads to the same characterization of privately computable functions as in the classical case. We furthermore show that allowing a small leakage combined with quantum communication allows to compute Boolean functions which are nonprivate. This does not hold for both quantum communication without leakage and classical communication with leakage. We also analyze a tradeoff between the number of communication rounds and the leakage required to quantum compute a nonprivate function.

We then turn to the question, whether leakage can decrease the communication complexity and show that $IdMin_n$ can be computed with leakage $1/poly(n)$ and communication $poly(n)$, while any perfectly private protocol is known to need $\Omega(2^n)$ rounds and communication. Thus a tiny leakage reduces the communication cost exponentially. It has been known previously [3] that one bit of privacy loss in the "hint sense" can decrease the communication complexity exponentially, but in our result the privacy loss is much smaller, and the function we consider is more natural.

Note that most proofs are omitted from this extended abstract.

2   Preliminaries

2.1   The Communication Complexity Model

In the quantum communication complexity model [20], two parties Al and Bob hold qubits. When the game starts Al holds a superposition $|x\rangle$ and Bob holds $|y\rangle$ (representing the input to the two players), and so the initial joint state is simply $|x\rangle \otimes |y\rangle$. Furthermore each player has an arbitrarily large supply of private qubits in some fixed basis state. The two parties then play in *rounds*. Suppose it is Al's turn to play. Al can do an arbitrary unitary transformation on his qubits and then send one or more qubits to Bob. Sending qubits does not change the overall superposition, but rather changes the ownership of the qubits, allowing Bob to apply his next unitary transformation on the newly received qubits. Al may also (partially) measure his qubits during his turn. At the end of the protocol, one player makes a measurement and sends the result of the protocol to the other player. The overall number of message exchanges is called the number of rounds. In a *classical* probabilistic protocol the players may only exchange messages composed of classical bits.

The complexity of a quantum (or classical) protocol is the number of qubits (respectively, bits) exchanged between the two players. We say a protocol *computes* a function $f : \mathcal{X} \times \mathcal{Y} \mapsto \mathcal{Z}$ with $\epsilon \geq 0$ error if, for any input $x \in \mathcal{X}, y \in \mathcal{Y}$, the probability that the two players compute $f(x, y)$ is at least $1 - \epsilon$.

We say a protocol $\mathcal{P}$ *computes* f with ϵ error with respect to a distribution μ on $\mathcal{X} \times \mathcal{Y}$, if

$$\text{Prob}_{(x,y) \in \mu, \mathcal{P}}(\mathcal{P}(x,y) = f(x,y)) \geq 1 - \epsilon.$$

A randomized classical or a quantum protocol has access to a *public coin*, if the players can flip a classical coin and both read the result without communication. If not explicitly mentioned we do not consider this model.

The communication matrix of a function $f(x, y)$ has rows labeled by the x's, columns labeled by the y's, and contains $f(x, y)$ at position x, y.

2.2   Information Theory Background

The quantum mechanical analogue of a random variable is a probability distribution over superpositions, also called a *mixed state*. For the mixed state $X = \{p_i, |\phi_i\rangle\}$, where $|\phi_i\rangle$ has probability p_i, the *density matrix* is defined as $\rho_X = \sum_i p_i |\phi_i\rangle\langle\phi_i|$. Density matrices are Hermitian, positive semidefinite, and have trace 1.

The *trace norm* of a matrix A is defined as $\| A \|_t = \text{Tr} \sqrt{A^\dagger A}$, which is the sum of the magnitudes of the singular values of A. A useful theorem states that for two mixed states ρ_1, ρ_2 their distinguishability is reflected in $\| \rho_1 - \rho_2 \|_t$ [1]:

Fact 1 *Let ρ_1, ρ_2 be two density matrices on the same space $\mathcal{H}$. Then for any measurement $\mathcal{O}$,*

$$\| \rho_1^{\mathcal{O}} - \rho_2^{\mathcal{O}} \|_1 \leq \| \rho_1 - \rho_2 \|_t ,$$

where $\rho^{\mathcal{O}}$ denotes the classical distribution on outcomes resulting from the measurement of ρ, and $\| \cdot \|_1$ is the ℓ_1 norm. Furthermore, there is a measurement $\mathcal{O}$, for which the above is an equality.

The *Shannon entropy* $H(X)$ of a classical random variable X and *mutual information* $I(X : Y)$ of a pair of random variables X, Y are defined as usual (see, e.g., [11]). The *von Neumann entropy* $S(\rho)$ of a density matrix ρ is defined as $S(\rho) = -\text{Tr} \rho \log \rho = -\sum_i \lambda_i \log \lambda_i$, where $\{\lambda_i\}$ is the multi-set of all the eigenvalues of ρ. Notice that the eigenvalues of a density matrix form a probability distribution. For properties of this function see [18]. We define the "mutual information" $I(X : Y)$ of two disjoint quantum systems X, Y as $I(X : Y) = S(X) + S(Y) - S(XY)$, where XY is density matrix of the system that includes the qubits of both systems.

We will employ the following fact from [13].

Fact 2 *Let* ρ_1, ρ_2 *be two mixed states with support in a Hilbert space* $\mathcal{H}$, $\mathcal{K}$ *any Hilbert space of dimension at least* $\dim(\mathcal{H})$, *and* $|\phi_i\rangle$ *any purifications of* ρ_i *in* $\mathcal{H} \otimes \mathcal{K}$. *Then, there is a local unitary transformation* U *on* $\mathcal{K}$ *that maps* $|\phi_2\rangle$ *to* $|\phi_2'\rangle = I \otimes U |\phi_2\rangle$ *such that*

$$\| \, |\phi_1\rangle\langle\phi_1| - |\phi_2'\rangle\langle\phi_2'| \, \|_t \; \leq \; 2 \, \| \, \rho_1 - \rho_2 \, \|_t^{\frac{1}{2}} \, .$$

2.3 Privacy

Given a protocol a player is *honest* if for each input and all messages he receives he sends exactly the messages prescribed by the protocol. All operations are allowed as long as this requirement is met. It is e.g. allowed to copy a classical message to some additional storage (if it is known that the message is classical). Copying general unknown quantum states is impossible, see [18].

We make this requirement a bit more formal in the following way. In a quantum protocol as defined above the actions of the players are defined as a series of unitary transformations plus sending a certain choice of qubits. For player Al to be honest we demand that for all rounds t of the protocol, for all inputs x, and for all sequences of pure state messages he may have received in the previous rounds, the density matrix of the message in the next round equals the density matrix defined by the protocol and the input. Note that in a run of the protocol the player might actually receive mixed state messages, but the behaviour of the player on these is defined by his behaviour on pure state messages.

We define the privacy loss of a protocol as follows. Let ρ_{ABXY} denote a state of the registers containing Al's private qubits in A, Bob's private qubits in B, Al's input in X, Bob's input in Y. We assume that the (classical) inputs are never erased by the players.

For a distribution μ on the inputs to a protocol computing f the information divulged from Al to Bob at time t is $L(t, B, \mu) = I(B : X|Y, f(X, Y))$, for the state $\rho_{ABXY}^{(t)}$ of the protocol at time t induced by the distribution μ on the inputs. Symmetrically we define Al's loss $L(t, A, \mu)$. The privacy loss of a protocol is the supremum of $L(t, \cdot, \mu)$ over all t and A, B and all μ.

The *privacy loss* $L_\epsilon(f)$ of a function f is the infimum privacy loss over all quantum protocols computing f with error ϵ. The classical privacy loss $CL_\epsilon(f)$ is defined analogously, with the infimum over all classical randomized protocols.

A function f is said to be *private*, if $CL_0(f) = 0$. It is known that $CL_\epsilon(f) = 0$ with $\epsilon < 1/2$ holds only for private functions [14].

Note that in the above definition we have assumed that the information available to a player is small in all rounds. Thus even if one player decides to quit the protocol at some point the privacy loss is guaranteed.

If we consider only the final state of the protocol in our definition we call the players *honest and nonpreemptive*. For a classical protocol there is no difference between these two possibilities, since the information available only increases with time. In the quantum case, however, this is not true.

The information divulged by a nonprivate protocol can also be measured in a different way, namely via distinguishability, see [10]. Let ρ_{AB}^{xy} denote the state of Al's and Bob's qubits in some round for inputs x, y. A protocol is said to leak at most δ from Al to Bob, if for all x, x' and y with $f(x, y) = f(x', y)$ it is true that $\left\| \rho_B^{xy} - \rho_B^{x'y} \right\|_{\mathrm{t}} \leq \delta$. This means that no quantum operation on Bob's qubits can distinguish the two states better than δ. An analogous definition is made for Al. We say a protocol *leaks* at most δ, if the maximum leakage to a player in any round is at most δ.

Two other definitions of privacy loss are also considered in [3]. In the first variant there is not one protocol which is good against all distributions on the inputs, but for each distribution there may be one specialized protocol for which the privacy loss is measured. The second definition is privacy loss in the hint sense, which we do not discuss here (but briefly in the full version). Let us show that our standard definition of privacy loss and the first alternative definition mentioned above are equivalent for randomized and quantum protocols. The following lemma is a simple consequence of the Yao principle.

Lemma 3. *The following statements are equivalent:*

There is a randomized [quantum] public coin protocol for a function f with communication c, error ϵ, privacy loss δ against all distributions μ on inputs.

For every distribution μ on inputs there is a deterministic [quantum] protocol for f with communication c, error ϵ, and privacy loss δ on that distribution.

Our definition of communication complexity allows only private coins, however. If we are only interested in the privacy loss, one of the players may simply flip enough coins and communicate them to the other player, then they simulate the public coin protocol. This increases the communication, but none of the other parameters. We need another result to get rid of the public coin at a lower cost. First consider the following lemma concerning leakage, proven like [17].

Lemma 4. *Let $f : \{0,1\}^n \times \{0,1\}^n \to \mathcal{Z}$ be computable by a randomized protocol with error ϵ, public randomness, c bits of communication, and leakage δ.*

Then for all $\gamma > 0$ there is a (private coin) randomized protocol for f with error $(1 + \gamma)\epsilon$, leakage $(1 + \gamma)\delta$, and communication $O(c + \log n + \log(1/\delta) + \log(1/\gamma) + \log(1/\epsilon))$.

If leakage is small a bound on privacy loss is implicit.

Lemma 5. *Let $f : \{0,1\}^n \times \{0,1\}^n \to \mathcal{Z}$ be computable by a randomized protocol with error ϵ (using private randomness), c bits of communication, and leakage δ. Then the same protocol has privacy loss $O(n \cdot \delta - \delta \log \delta)$.*

Now we can say something about privacy loss and private coins.

Lemma 6. *Let $f : \{0,1\}^n \times \{0,1\}^n \to \mathcal{Z}$ be computable by a randomized protocol with error ϵ (using public randomness) and c bits of communication, that has privacy loss $\delta \geq 1/2^{2n}$.*

Then there is a (private coin) randomized protocol for f with error 2ϵ, privacy loss $O(n \cdot \sqrt{\delta})$, and communication $O(c + \log n + \log(1/\delta) + \log(1/\epsilon))$.

Actually the above lemmas easily generalize to the quantum case.

3   An Exponential Decrease in Privacy Loss

In this section we give an example of a function that can be computed with an exponentially smaller privacy loss in the quantum case than in the classical case. This function is the disjointness problem, and the quantum protocol is derived from [7]. In fact we describe a general way to protect a certain type of protocols against large privacy loss.

We now roughly sketch how the protocol works, and then how to make it secure. The protocol is based on a general simulation of black-box algorithms given in [7]. A black-box algorithm for a function g is turned into a communication protocol for a function $g(x \wedge y)$ for the bitwise defined operation $\wedge$. The black-box algorithm for OR is the famous search algorithm by Grover or rather its variant in [6]. The important feature of the protocol for us is that the players exchange a set of $\log n + O(1)$ qubits in each round and apart from that no further qubits or classical storage depending on the inputs are used. Also the protocol runs in $O(\log n)$ stages, each concluded by a measurement. If this measurement yields an index i with $x_i = y_i = 1$, then the protocol stops, else it continues. The qubits contain a superposition over indices i from 1 to n plus the values of x_i and $x_i \wedge y_i$. So an honest player that does not attempt to get more information learns $O(\log n)$ times the measurement result for $O(\log n)$ qubits.

The main tool to show that the privacy loss is small against players trying to get more information is the following generalization of the famous no-cloning theorem, see [18]. While the no-cloning theorem says that we cannot make a perfect copy of an unknown quantum state (which would enable us to find out some information about the state without changing the original by measuring the copy), this lemma says that no transformation leaving two nonorthogonal originals both unchanged gives us any information about those states.

Lemma 7. *Let $|\phi_1\rangle$ and $|\phi_2\rangle$ be two states that are nonorthogonal. Assume a unitary U sends $|\phi_1\rangle \otimes |0\rangle$ to $|\phi_1\rangle \otimes |a\rangle$ and $|\phi_2\rangle \otimes |0\rangle$ to $|\phi_2\rangle \otimes |b\rangle$. Then $|a\rangle = |b\rangle$.*

Now assume a protocol sends k qubits (in a pure state) back and forth without using any private storage depending on the input and without measuring. If we manage to change the messages in a way so that for no inputs x, x', y the message sent in round t for input x, y is orthogonal to the message for input x', y, then there is no transformation for Bob that leaves the message unchanged, yet extracts some information. In other words, honest players are forced to follow the protocol without getting further information. The only information is revealed at the end of a stage, when one player is left with the qubits from the last message, resp. at the time when one player decides to quit the protocol. We now describe how to make the messages nonorthogonal.

Lemma 8. *For all $\epsilon > 0$ and for any finite set of l-dimensional unit vectors $\{v_i\}$ there is a set of $l+1$ dimensional unit vectors $\{v'_i\}$ such that $||v_i - v'_i|| < \epsilon$, and $v'_i \perp v'_j$ for no i, j.*

Theorem 1. *$DISJ_n$ can be computed by a quantum protocol with error $1/3$, communication $O(\sqrt{n} \log n)$, and privacy loss $O(\log^2 n)$.*

PROOF: In [7] a quantum protocol with error $1/4$, communication $O(\sqrt{n}\log n)$ is described, in which Al and Bob exchange pure state messages of length $\log n + O(1)$, but use no further storage depending on the input. The protocol consists of $O(\log n)$ stages each of which ends with a measurement of the qubits in the standard basis. No further measurements are used.

We modify the protocol. We add one more qubit to the messages. Then we change the first message to be sent (prepared by Al) as described in lemma 8. The error introduced by this change is arbitrarily small. Then the protocol is used as before, ignoring the new qubit in all transformations, but always sending that qubit with the other qubits back and forth between the players. This can be done in a way ensuring that no message sent for any pair of inputs x, y in any round will be orthogonal to another such message.

Assume Bob wants to get more information than he can get from the $O(\log n)$ classical strings of length $\log n + O(1)$ obtained from the measurements. In some round he will first start an attack on the message. He has to map the message received to another message he must send back. The second message is the result of a fixed unitary transformation (depending on y) on the first. He has to combine the attack with that unitary transformation. So we may assume that he first attacks the message and then applies the transformation to get the next message. The attack transformation maps the message and some empty qubits to the tensor product of the same message and another state, that depends on the other player's input. Lemma 7 ensures that this is impossible. So Bob has to stick to the protocol without getting more information than allowed.   □

Now we turn to the lower bound. Every classical deterministic protocol partitions the communication matrix into rectangles labeled with the output of the protocol. Let μ be a distribution on the inputs. A labeled rectangle is $1 - \epsilon$-correct, if according to μ at least $1 - \epsilon$ of the weight of the rectangle is on correctly labeled inputs. The *width* of a rectangle is $\min\{|A|, |B|\}$. Let $r_\epsilon(f)$ denote the largest width of any $1 - \epsilon$-correct rectangle. [3] proves:

Fact 9 $CL_0(f) \geq (n - \log r_0(f))/2 - 1$ for all $f : \{0,1\}^n \times \{0,1\}^n \to \{0,1\}$.

We now describe a new bound. The maximum size of a $1 - \epsilon$-correct rectangle according to μ is called $s_\epsilon(f, \mu)$.

Lemma 10. *All randomized protocols with error $1/3$ computing $f : \{0,1\}^n \times \{0,1\}^n \to \mathcal{Z}$ have privacy loss*

$$\Omega\left(\frac{-\log(s_{1/n^2}(f, unif)) - \log|Z| - 1}{\log n}\right).$$

The following is proved in [2]

Fact 11 *Let μ be the uniform distribution on pairs of sets of size $\sqrt{n}$ from a size n universe. Then the largest $1 - \epsilon$-correct rectangle for disjointness has size $1/2^{\Omega(\sqrt{n})}$ for some constant ϵ.*

Corollary 1. $CL_{1/3}(DISJ_n) = \Omega(\sqrt{n}/\log n)$.

4   The Class of Private Functions

4.1   Players That Do Not Preempt

First let us take a look at the model of honest players, in which only the information retrievable at the end of the protocol is counted.

Theorem 2. *For every function f with deterministic communication complexity c there is a quantum protocol with communication $O(c^2)$, where the final state obtainable by every honest player has an arbitrarily small distance to the final state of a player that knows only his input and the function value at the end.*

Thus we get an arbitrarily close approximation of privacy against honest players if we consider only the information available at the end. In other words if we trust the other player not only to play honest, but also to not quit before the end, every function can be computed securely.

4.2   The Characterization of Quantum Privacy

Now we return to our regular definition of privacy. The set of classically private functions has been characterized in [14] and [4].

Definition 1. *Let $M = C \times D$ be a matrix. A relation $\equiv$ is defined as follows: rows x and x' satisfy $x \equiv' x'$, if there is a column y with $M_{x,y} = M_{x',y}$. $\equiv$ is the transitive closure of $\equiv'$. Similar relations are defined for columns.*

A matrix is called forbidden, if all its rows are equivalent, all its columns are equivalent, and the matrix is not monochromatic.

Theorem 3. *If the communication matrix of f contains a forbidden submatrix then f cannot be computed by a quantum protocol with error smaller than $1/2$ and no privacy loss.*

PROOF: A quantum protocol with error smaller than $1/2$ for f must also solve the (nontrivial) problem g corresponding to the forbidden submatrix. We will show that one round after the other can be shaved off the protocol, eventually yielding a protocol for g with one round. Such a protocol cannot compute g with error better than $1/2$, thus we reach a contradiction.

We show that the first message (w.l.o.g. sent by Al) does not depend on the input, and can thus be computed by Bob, whereupon the first round of communication can be skipped. Let $x_1, \ldots, x_l$ denote the rows of the forbidden submatrix, enumerated in such a way that $x_i \equiv' x_j$ for some $j < i$ for all $i > 1$. If $x_i \equiv' x_j$ then there is a y, so that $g(x_i, y) = g(x_j, y)$. Since it is possible that Bob holds y, it is not allowed for Al to send different messages on x_i and x_j, since otherwise Bob may obtain information about x_i and x_j not deducable from the function value alone. So for all x_i the same message is sent. Let ρ_{AM}^{xy} denote the state of Al's qubits right before the first message is sent (on inputs x, y), with M containing the message. ρ_M^{xy} is the same for all x, y. Also ρ_{AM}^{xy} purifies such a state. Due to fact 2 Al has a unitary operation on A switching between those states without introducing any error. Thus Bob may prepare $\rho_{AM}^{x,y}$ for some fixed x, y, send the part of the state in A to Al and keep the M part. Al can then change the received state to the one for the correct x. Furthermore Bob can send the message for round 2 together with the first message, saving one round.

Eventually we arrive at a protocol with one round only, in which, say, a message is sent from Al to Bob. Thus if the error is smaller than $1/2$ the output does not depend on y. Then the communication matrix of g consists of a set of monochromatic rows. Such a matrix is clearly not a forbidden submatrix, and we arrive at a contradiction to our assumptions.   □

Now that we know a forbidden submatrix excludes a private quantum protocol, the other piece for a characterization is as follows, see [14].

Fact 12 *If the communication matrix of f contains no forbidden submatrix, then f can be computed by a deterministic private protocol.*

The class of private functions is unchanged by quantum communication.

4.3   Boolean Functions and Leakage

Next we consider the case of Boolean functions. It is known [10] that the class of private Boolean functions is the class of functions $f_A(x) \oplus f_B(y)$, even if one considers protocols that leak δ and have error ϵ with $\epsilon + \delta < 1/2$. These functions are combinatorially characterized by the so-called "corners lemma" [10]. As a corollary of theorem 3 we get a result for the quantum case.

Corollary 2. *If the communication matrix of a function f contains a 2×2 rectangle with exactly 3 times the same entry, then no private quantum protocol with error smaller than $1/2$ can compute f.*

Corollary 3. *The class of Boolean functions computable by private quantum protocols is the class of functions $f_A(x) \oplus f_B(y)$.*

Is corollary 2 valid in the quantum case with leakage? The answer is no. There is, however, a leakage-communication tradeoff.

Theorem 4. *There is a quantum protocol computing the AND function on two bits with error $1/3$ that has leakage δ and uses $O(1/\delta^2)$ communication.*

Theorem 5. *If the communication matrix of a function f contains a 2×2 rectangle with exactly 3 times the same entry, then no quantum protocol with error $1/3$, leakage δ, and at most $1/(12\sqrt{\delta})$ rounds can compute f.*

5   Trading Privacy Loss against Complexity

In this section we show that allowing a privacy loss of much less than one bit (instead of privacy loss 0) can reduce the communication complexity of a certain function, namely the identified minimum problem, exponentially.

Theorem 6. *The function $IdMin_n$ can be computed by a randomized protocol with privacy loss δ, error δ, and communication $O(n^4/\delta)$.*

Every randomized protocol computing $IdMin_n$ with error $\epsilon < 1/2$ and with privacy loss 0 needs communication $\Omega(2^n)$.

The proof proceeds as follows. First we describe for each distribution on the inputs a deterministic protocol that has low *expected* leakage on that distribution, as well as low error. Then we use a Yao-type result to find a single public coin randomized protocol with low expected leakage for every distribution. Such a protocol has low leakage in our ordinary definition. Then we turn this into a private coin protocol with lemma 4 and finally into a protocol having small privacy loss with lemma 5. The lower bound is from [14].

6   Conclusions and Open Problems

In this paper we have discussed privacy with respect to honest players with a focus on the themes quantum communication and protocols with privacy loss or leakage. We have given an example of a function that can be computed with exponentially smaller privacy loss using quantum communication than in the case of classical communication. The set of functions with privacy loss 0 is, however, not increased by quantum communication. For Boolean functions we were able to give a simple characterization of the quantum private functions as $f_A(x) \oplus f_B(y)$. The characterization for Boolean functions can be extended to the case of multiparty private computation. As in [10] it can be shown that only functions of the form $f_1(x_1) \oplus \cdots \oplus f_k(x_k)$ can be quantum computed in a way so that every set of $\lceil k/2 \rceil$ players learns nothing more about the other players' inputs than what is deducible from their inputs and the output alone. Since every function can be computed classically so that no coalition of less than $k/2$ players learns more than allowed [5,9], and the aforementioned functions are private against coalitions of even $k - 1$ players, there are only 2 levels in this hierarchy of privacy for quantum computation, as in the classical case, see [10].

We now give some open problems. A more realistic type of adversary is an adversary that has two objectives: with highest priority he wants the output to be correct with high probability. But then he also wants to learn as much as possible under this constraint. A restricted form of this adversary is an almost honest adversary that follows the protocol, but only sends messages that are in distance ϵ from the "correct" messages. This allows in the quantum case e.g. to use approximate cloning as in [8]. A study of privacy against such adversaries would be interesting. In particular can a quantum protocol for disjointness have small privacy loss against them? Note that the set of Boolean private functions is robust against such players, since those are of the form $f_A(x) \oplus f_B(y)$. About non-Boolean functions no results seem to be known even in the classical case.

Another open problem is whether we can extend the characterization of (non-Boolean) classically private functions to the case of small leakage.

Finally, how can one prove lower bounds on the privacy loss of quantum protocols?

References

1. D. Aharonov, A. Kitaev, and N. Nisan. Quantum circuits with mixed states. *30th ACM Symposium on Theory of Computing*, pp. 20–30, 1998.
2. L. Babai, P. Frankl, and J. Simon. Complexity classes in communication complexity theory. *27th IEEE Symposium on Foundations of Computer Science*, pp.303–312, 1986.
3. R. Bar-Yehuda, B. Chor, E. Kushilevitz, and A. Orlitsky. Privacy, Additional Information, and Communication. *IEEE Transactions on Information Theory*, vol.39, pp.1930–1943, 1993.
4. D. Beaver. Perfect privacy for two party protocols. Technical Report TR-11-89, Harvard University, 1989.
5. M. Ben-Or, S. Goldwasser, and A. Wigderson. Completeness Theorems for Non-Cryptographic Fault-Tolerant Distributed Computation. *20th ACM Symposium on Theory of Computing*, pp. 1–10, 1988.
6. M. Boyer, G. Brassard, P. Høyer, and A. Tapp. Tight bounds on quantum searching. *4th Workshop on Physics and Computation*, pp.36-43, 1996.
7. H. Buhrman, R. Cleve, and A. Wigderson. Quantum vs. classical communication and computation. *30th ACM Symposium on Theory of Computing*, pp. 63–68, 1998.
8. V. Buzek and M. Hillery. Quantum Copying: Beyond the No-Cloning Theorem. *Phys. Rev. A*, vol.54, pp.1844-1852, 1996.
9. D. Chaum, C. Crepeau, and I. Damgard. Multiparty Unconditionally Secure Protocols. *20th ACM Symposium on Theory of Computing*, pp. 11–19, 1988.
10. B. Chor and E. Kushilevitz. A zero-one law for Boolean privacy. *SIAM Journal Discrete Math.*, vol.4, pp.36–47, 1991.
11. T.M. Cover and J.A. Thomas. *Elements of Information Theory*. Wiley Series in Telecommunications. John Wiley & Sons, 1991.
12. O. Goldreich, S. Micali, and A. Wigderson. How to Play Any Mental Game. *19th ACM Symposium on Theory of Computing*, pp.218–229, 1987.
13. H. Klauck, A. Nayak, A. Ta-Shma, and D. Zuckerman. Interaction in Quantum Communication and the Complexity of Set Disjointness. *33rd ACM Symposium on Theory of Computing*, pp. 124–133, 2001.
14. E. Kushilevitz. Privacy and Communication Complexity. *SIAM Journal Discrete Math.*, vol.5, pp.273–284, 1992.
15. E. Kushilevitz and N. Nisan. *Communication Complexity*. Cambridge University Press, 1997.
16. E. Kushilevitz, R. Ostrovsky, and A. Rosen. Characterizing Linear Size Circuits in Terms of Privacy. *Journal of Computer and System Sciences*, vol.58, pp. 129–136, 1999.
17. I. Newman. Private vs. Common Random Bits in Communication Complexity. *Information Processing Letters*, vol.39, pp.67–71, 1991.
18. M.A. Nielsen and I.L. Chuang. Quantum Computation and Quantum Information. Cambridge University Press, 2000.
19. A.C.-C. Yao. Protocols for Secure Computations. *23rd IEEE Symposium on Foundations of Computer Science*, pp. 160–164, 1982.
20. A.C.-C. Yao. Quantum circuit complexity. *34th IEEE Symposium on Foundations of Computer Science*, pp. 352–361, 1993.

On Quantum Versions of the Yao Principle

Mart de Graaf[1][*] and Ronald de Wolf[2][**]

[1] CWI. P.O. Box 94079, 1090 GB Amsterdam, The Netherlands. mgdgraaf@cwi.nl
[2] UC Berkeley. 583 Soda Hall, Berkeley CA 94720, USA. rdewolf@cs.berkeley.edu

Abstract. The classical Yao principle states that the complexity $R_\epsilon(f)$ of an optimal *randomized* algorithm for a function f with success probability $1 - \epsilon$ equals the complexity $\max_\mu D_\epsilon^\mu(f)$ of an optimal *deterministic* algorithm for f that is correct on a fraction $1 - \epsilon$ of the inputs, weighed according to the hardest distribution μ over the inputs. In this paper we investigate to what extent such a principle holds for quantum algorithms. We propose two natural candidate quantum Yao principles, a "weak" and a "strong" one. For both principles, we prove that the quantum bounded-error complexity is a lower bound on the quantum analogues of $\max_\mu D_\epsilon^\mu(f)$. We then prove that equality cannot be obtained for the "strong" version, by exhibiting an exponential gap. On the other hand, as a positive result we prove that the "weak" version holds up to a constant factor for the query complexity of all symmetric Boolean functions.

1 Introduction

1.1 Motivation

In classical computing, the *Yao principle* [17] gives an equivalence between two kinds of randomness in algorithms: randomness inside the algorithm itself, and randomness on the inputs. Let us fix some model of computation for computing a Boolean function f, like query complexity, communication complexity, etc. Let $R_\epsilon(f)$ be the minimal complexity among all *randomized* algorithms that compute $f(x)$ with success probability at least $1 - \epsilon$, for all inputs x. Let $D_\epsilon^\mu(f)$ be the minimal complexity among all *deterministic* algorithms that compute f correctly on a fraction of at least $1 - \epsilon$ of all inputs, weighed according to a distribution μ on the inputs. The Yao principle now states that these complexities are equal if we look at the "hardest" input distribution μ:

$$R_\epsilon(f) = \max_\mu D_\epsilon^\mu(f).$$

It is a special case of Von Neumann's minimax theorem in game theory [10].

[*] Partially supported by EU fifth framework project QAIP, IST–1999–11234, and grant 612.055.001 from the Netherlands Organization for Scientific Research (NWO).
[**] Supported by Talent grant S 62–565 from NWO. Work done while at CWI.

Since its introduction, the Yao principle has been an extremely useful tool for deriving lower bounds on randomized algorithms from lower bounds on deterministic algorithms: choose some "hard" input distribution μ, prove a lower bound on deterministic algorithms that compute f correctly for "most" inputs, weighed according to μ, and then use $R_\epsilon(f) \geq D_\epsilon^\mu(f)$ to get a lower bound on $R_\epsilon(f)$. This method is used very often, because it is usually much easier to analyze deterministic algorithms than to analyze randomized ones.

In recent years *quantum* computation received a lot of attention. Here quantum mechanical principles are employed to realize more efficient computation than is possible with a classical computer. Famous examples are Shor's efficient quantum factoring algorithm [15] and Grover's search algorithm [8]. However, the field is still young and open questions abound. In particular, there has been a search for good techniques to provide lower bounds on quantum algorithms, particularly in the query model of computation. Two general methods in this direction are the *polynomial method* introduced by Beals, Buhrman, Cleve, Mosca, and de Wolf [3] and the *quantum adversary method* of Ambainis [2]. In this paper we investigate the possibility of a third method, a *quantum Yao principle*. It is our hope that such a principle will prove itself useful as a link between techniques for lower bounds on exact and bounded-error quantum algorithms.

The first difficulty one runs into when investigating a quantum version of the Yao principle, is the question what the proper quantum counterparts of $R_\epsilon(f)$ and $D_\epsilon^\mu(f)$ are. Let us fix the error probability at $\epsilon = \frac{1}{3}$ here (any other value in $(0, \frac{1}{2})$ would do as well). The quantum analogue of $R_{1/3}(f)$ is straightforward: let $Q_2(f)$ denote the minimal complexity among all *quantum* algorithms that compute $f(x)$ with probability at least $\frac{2}{3}$, for all inputs x. However, the inherently "random" nature of quantum algorithms prohibits a straightforward definition of "deterministic" quantum algorithms in analogy of deterministic classical algorithms. We therefore propose two different definitions, a weak and a strong one. In the following, let $f : D \to \{0,1\}$ be some function that we want to compute, with $D \subseteq \{0,1\}^N$. If $D = \{0,1\}^N$ then f is a *total* function, otherwise f is a *promise* function. Let A be a quantum algorithm, $P_A(x)$ the acceptance probability of A on input x (the probability of outputting 1 on input x), and $\mu : D \to [0,1]$ a probability distribution over the inputs.

Definition 1. *A is weakly $\frac{2}{3}$-exact for f with respect to μ iff $\mu(\{x \mid P_A(x) = f(x)\}) \geq \frac{2}{3}$.*

Definition 2. *A is strongly $\frac{2}{3}$-exact for f with respect to μ iff A is weakly $\frac{2}{3}$-exact for f with respect to μ and $P_A(x) \in \{0,1\}$ for all inputs $x \in \{0,1\}^N$.*

The second definition most closely mimics the behavior of a classical deterministic algorithm: the input x fully determines the output bit (even on $x \notin D$) and the algorithm gives correct output $f(x)$ for "most" x. The first definition is more liberal: here we only require this "input-determines-output" behavior to occur for a μ-fraction of at least $\frac{2}{3}$ of the inputs where the algorithm gives the correct output $f(x)$. Note that a strongly $\frac{2}{3}$-exact algorithm for f with respect to μ actually computes some total function $g : \{0,1\}^N \to \{0,1\}$ with success probability 1, namely the function $g(x) = P_A(x)$.

These two definitions lead to a weak and a strong quantum counterpart to the classical distributional complexity $D_{1/3}^\mu(f)$: let $Q_{WE}^\mu(f)$ and $Q_{SE}^\mu(f)$ denote the minimal complexity among all weakly and strongly $\frac{2}{3}$-exact algorithms for f with respect to μ, respectively. Note that $Q_{WE}^\mu(f) \leq Q_{SE}^\mu(f)$ for all f and μ. We can now state two potential quantum versions of the Yao principle:

- Weak quantum Yao principle: $Q_2(f) \overset{?}{=} \max_\mu Q_{WE}^\mu(f)$
- Strong quantum Yao principle: $Q_2(f) \overset{?}{=} \max_\mu Q_{SE}^\mu(f)$

In this paper we investigate to what extent these two principles hold.

1.2 Results

Our results are threefold. First, we prove that both principles hold in the '$\leq$'-direction, for all f:

- $Q_2(f) \leq \max_\mu Q_{WE}^\mu(f) \leq \max_\mu Q_{SE}^\mu(f)$

The proof of the first inequality is analogous to the classical game-theoretic proof. We emphasize that this result is perfectly general, and applies to all computational models to which the classical Yao principle applies.

In order to investigate to what extent the '$\geq$'-directions of these two quantum Yao principles hold, we instantiate our complexity measures to the query complexity setting. Our second result is an exponential gap between $Q_2(f)$ and $Q_{SE}^\mu(f)$ for the query complexity of Simon's problem [16]:

- There exist f and μ such that $Q_2(f)$ is exponentially smaller than $Q_{SE}^\mu(f)$.

This shows that the strong quantum Yao principle is false.

Thirdly, we prove that the weak quantum Yao principle holds up to a constant factor for the query complexity of all *symmetric* functions:

- $Q_2(f) = \Theta \left(\max_\mu Q_{WE}^\mu(f) \right)$ for all symmetric f

For this result we first construct a quantum algorithm that can determine the N-bit input x *with certainty* in $O(\sqrt{kN})$ queries if k is a known upper bound on the Hamming weight of x. We then use that algorithm to construct, for every symmetric function f and distribution μ, a quantum algorithm that computes $f(x)$ with certainty for "most" inputs x. In addition to this result for symmetric functions, we also show that for a particular *monotone* non-symmetric function f (the AND-OR tree), the $\max_\mu Q_{WE}^\mu(f)$ complexity lies in between the best known bounds for $Q_2(f)$. The gist of this third batch of results is that most known quantum algorithms that are somehow based on Grover's algorithm can be made weakly $\frac{2}{3}$-exact. This may actually be the main contribution of this paper.

2   Preliminaries

In this section we formalize the notion of query complexity, define several complexity measures, and state Von Neumann's minimax theorem.

2.1   Query Complexity

We assume familiarity with classical computation theory and briefly sketch the basics of quantum computation; an extensive introduction may be found in the book by Nielsen and Chuang [12]. Quantum algorithms operate on *qubits* as opposed to bits in classical computers. The state of an m-qubit quantum system can be written as $|\phi\rangle = \sum_{i \in \{0,1\}^m} \alpha_i |i\rangle$, where $|i\rangle$ denotes the basis state i, which is a classical m-bit string. The α_i's are complex numbers known as the *amplitudes* of the basis states $|i\rangle$ and we require $\sum_{i \in \{0,1\}^m} |\alpha_i|^2 = 1$. Mathematically, the state of a system is thus described by a 2^m-dimensional complex unit vector. If we measure the value of $|\phi\rangle$, then we will see the basis state $|i\rangle$ with probability $|\alpha_i|^2$, after which the system collapses to $|i\rangle$. Operations that are not measurements correspond to *unitary transformations* on the vector of amplitudes.

In the *query model* of computation, the goal is to compute some function $f : D \to \{0,1\}$ on an input $x \in D \subseteq \{0,1\}^N$, using as few accesses ("queries") to the N input bits as possible. It is by now standard to formalize a quantum query as an application of a unitary transformation O that acts as $O|i, b, z\rangle = |i, b \oplus x_i, z\rangle$. Here $i \in \{1, \ldots, N\}$, $b \in \{0,1\}$, $\oplus$ denotes the exclusive-or function, and z denotes the workspace of the algorithm, which is not affected by O. A T-query quantum algorithm A then has the form $A = U_T O U_{T-1} O \cdots U_1 O U_0$, with each U_i a fixed unitary transformation independent of the input x. Algorithm A is assumed to start in the all-zero state $|0 \ldots 0\rangle$, and its output (0 or 1) is obtained by measuring the rightmost bit of its final state $A|0 \ldots 0\rangle$. The *acceptance probability* $P_A(x)$ of a quantum algorithm A is defined as the probability of getting output 1 on input x. Its *success probability* $S_A(x)$ is the probability of getting the correct output $f(x)$ on input x.

A quantum algorithm A computes a function $f : D \to \{0,1\}$ *exactly* if $S_A(x) = 1$ for all inputs $x \in D$. Algorithm A computes f with *bounded-error* if $S_A(x) \geq \frac{2}{3}$ for all $x \in D$. We use $Q_E(f)$ and $Q_2(f)$ to denote the minimal number of queries required by exact and bounded-error quantum algorithms for f, respectively. These complexities are the quantum versions of the classical deterministic and bounded-error decision tree complexities $D(f)$ and $R_2(f)$, respectively. For completeness, we repeat our two alternative quantum versions of the classical distributional complexity $D^\mu(f)$ from the introduction. Let μ be a probability distribution on the set of all possible inputs. An algorithm A is *weakly $\frac{2}{3}$-exact for f with respect to μ* if $\mu(\{x \mid P_A(x) = f(x)\}) \geq \frac{2}{3}$, and A is *strongly $\frac{2}{3}$-exact for f with respect to μ* if A is weakly $\frac{2}{3}$-exact for f with respect to μ and $P_A(x) \in \{0,1\}$ for all $x \in \{0,1\}^N$. By $Q_{SE}^\mu(f)$ and $Q_{WE}^\mu(f)$ we denote the minimal number of queries needed by strongly and weakly $\frac{2}{3}$-exact quantum algorithms for f with respect to μ, respectively. Note that $Q_{WE}^\mu(f) \leq Q_{SE}^\mu(f)$ for all f and μ, hence in particular $\max_\mu Q_{WE}^\mu(f) \leq \max_\mu Q_{SE}^\mu(f)$.

One of the first quantum algorithms operating in the query model is Grover's search algorithm [8,4]. Let $|x|$ denote the Hamming weight (number of 1's) in the input x, and let x_i denote the ith bit of x. If $t = |x| > 0$ then Grover's algorithm uses $\frac{\pi}{4}\sqrt{N/t}$ queries and with high probability outputs an i such that $x_i = 1$. If $|x| = 0$ then the algorithm always outputs 'no solutions'. Brassard, Høyer, Mosca, and Tapp [4] gave an exact version of Grover's algorithm that accomplishes the same task with probability 1 if $|x|$ is known.

A function $f : \{0,1\}^N \to \{0,1\}$ is *symmetric* if its value $f(x)$ depends only on $|x|$. For such f, define $f_k = f(x)$ where $|x| = k$. In [3] it is proven that $Q_2(f) = \Theta(\sqrt{N(N-\Gamma(f))})$, where $\Gamma(f) = \min\{|2k - N - 1| \mid f_k \neq f_{k+1}$ and $0 \leq k \leq N - 1\}$. Informally, the quantity $\Gamma(f)$ (introduced by Paturi [14]) measures the length of the interval around Hamming weight $\frac{N}{2}$ where f is constant. A symmetric function f is a *threshold* function if there is a $0 < t \leq N$, such that $f(x) = 1$ iff $|x| \geq t$. Note that for $t \leq N/2$ we have $Q_2(f) = \Theta(\sqrt{tN})$ as a direct consequence of the bound for symmetric functions. A function $f : \{0,1\}^N \to \{0,1\}$ is *monotone* if $(\forall i\ x_i \leq y_i) \Rightarrow f(x) \leq f(y)$.

2.2 Von Neumann's Minimax Theorem

The book by Owen [13] provides an excellent introduction to game theory. Here we only state Von Neumann's famous minimax theorem [10]. Consider a two-player, zero-sum game with payoff matrix P. Player 1 wants to maximize the payoff, player 2 wants to minimize. Both players have available a finite set of *pure* strategies. If player 1 plays pure strategy i and player 2 plays pure strategy j, then the payoff is $P_{ij} = e_i^T P e_j$, where e_i and e_j are the appropriate unit vectors and superscript-T denotes vector transposition. In addition, they may also use a *mixed* strategy. This is a probability distribution over the set of pure strategies, modeled by a vector of non-negative reals that sum to 1. If player 1 plays mixed strategy ρ and player 2 plays mixed strategy μ, then the expected payoff of the game is $\rho^T P \mu$. The minimax theorem states that the maximal payoff that player 1 can assure if he can base ρ on μ, equals the minimal payoff that player 2 can assure if he can base μ on ρ:

$$\min_{\mu} \max_{\rho} \rho^T P \mu = \max_{\rho} \min_{\mu} \rho^T P \mu.$$

Without loss of generality the "inner" choices can be assumed to be pure strategies, hence

$$\min_{\mu} \max_{i} e_i^T P \mu = \max_{\rho} \min_{j} \rho^T P e_j.$$

As mentioned in the introduction, the classical Yao principle is an easy consequence of this theorem. In the next section we use it to prove one half of the *quantum* Yao principle.

3   Proof of One Half of the Quantum Yao Principle

Here we prove $Q_2(f) \leq \max_{\mu} Q_{WE}^{\mu}(f)$. The proof is similar to the derivation of the classical Yao principle, but the details are a bit more messy.

Theorem 1. *For all $f : D \to \{0,1\}$, with D finite, $Q_2(f) \leq \max_{\mu} Q_{WE}^{\mu}(f)$.*

Proof. Consider the (infinite) set of all quantum algorithms of complexity $\leq$ $\max_{\mu} Q_{WE}^{\mu}(f)$. Let i be any algorithm from this set, and $x \in D$ an input. Consider the quantity $\lfloor S_i(x) \rfloor$, which is 1 if algorithm i computes $f(x)$ with success probability 1, and which is 0 otherwise. Call algorithms i and j *similar* if $\lfloor S_i(x) \rfloor = \lfloor S_j(x) \rfloor$ for all $x \in D$. In this way, similarity is an equivalence relation on the set of all quantum algorithms of complexity $\leq \max_{\mu} Q_{WE}^{\mu}(f)$. Note that similarity partitions this set into at most $2^{|D|}$ equivalence classes. From each equivalence class, we choose as a representative an algorithm from that class with the least complexity.

Now consider the game in which player 1 wants to compute f, and as pure strategies he has available the (finite) set of representatives of the equivalence classes. Player 2 is an adversary that chooses hard inputs $x \in D$ to f. Let S be the matrix of success probabilities ($S_{ix} = S_i(x)$). Define the payoff matrix as $P_{ix} = \lfloor S_{ix} \rfloor$. Now consider the quantity $\max_i e_i^T P \mu$. This represents the μ-fraction of inputs on which the best weakly $\frac{2}{3}$-exact quantum algorithm for f with respect to that μ is correct. This quantity is at least $\frac{2}{3}$ for all μ, since we've been considering all quantum algorithms of complexity up to $\max_{\mu} Q_{WE}^{\mu}(f)$. From the minimax theorem we now obtain

$$\frac{2}{3} \leq \min_{\mu} \max_i e_i^T P \mu = \max_{\rho} \min_x \rho^T P e_x \leq \max_{\rho} \min_x \rho^T S e_x.$$

Here the last term can be interpreted as the success probability of a quantum algorithm formed by a probability distribution ρ over the set of representatives of the equivalence classes (such a distribution can be easily realized in a quantum algorithm using a superposition). By the above inequality, this algorithm has success probability $\geq \frac{2}{3}$ for all inputs $x \in D$. Since it is a probability distribution over algorithms of complexity $\leq \max_{\mu} Q_{WE}^{\mu}(f)$, its complexity is at most $\max_{\mu} Q_{WE}^{\mu}(f)$. Hence $Q_2(f) \leq \max_{\mu} Q_{WE}^{\mu}(f)$. $\square$

Corollary 1. *For all $f : D \to \{0,1\}$, with D finite, $Q_2(f) \leq \max_{\mu} Q_{SE}^{\mu}(f)$.*

We again emphasize that this result applies to all computational models where the classical Yao principle applies.

4 A Counterexample for the Strong Quantum Yao Principle

From here on, we will instantiate our complexity measures to the query complexity setting. Ambainis [1] has proven that for almost all Boolean functions f we have $Q_2(f) = \Omega(N)$. This result immediately implies that both the strong and weak quantum Yao principle hold up to a constant factor for almost all Boolean functions in the query complexity setting.

However, the strong quantum Yao principle does not hold in general. Below we exhibit a function f and distribution μ where $Q_2(f)$ is exponentially less than $Q_{SE}^\mu(f)$. The function is Simon's problem [16], and our separation is based on Simon's classical lower bound combined with the result that classical and quantum query complexity are polynomially related for all total functions [3].

Theorem 2. *There exist a problem f on $N = n2^n$ bits and a distribution μ such that $Q_2(f) = O(n^2)$ and $Q_{SE}^\mu(f) = \Omega(2^{\frac{n}{8}})$.*

Proof. Consider Simon's problem: given a function $\phi : \{0,1\}^n \rightarrow \{0,1\}^n$ with the promise that there is an $s \in \{0,1\}^n$ such that $\phi(a) = \phi(b)$ iff $a \oplus b = s$, decide whether $s = 0$ or not. This function ϕ is given as an input x of $N = n2^n$ bits, using n 1-bit entries for each function value $\phi(\cdot)$. The input bits can be queried in the usual way. Using Simon's bounded-error quantum algorithm, this problem can be solved in $O(n^2)$ queries, and hence $Q_2(Simon) = O(n^2)$. Now define a distribution μ which uniformly places half the total weight on inputs with $s = 0$ and half the total weight on inputs with $s \neq 0$. Simon proved that under this distribution, any classical algorithm that is correct on a fraction $\geq \frac{2}{3}$ requires $\Omega(\sqrt{2^n})$ queries. Now take any strongly $\frac{2}{3}$-exact T-query quantum algorithm A for this problem, then A computes some *total* function g. Since $D(g) = O(Q_E(g)^4)$ [3], there exists a deterministic classical algorithm that computes g using $O(T^4)$ queries. But this classical algorithm is then correct on a μ-fraction $\frac{2}{3}$ of all Simon inputs. Simon's lower bound on classical algorithms now implies that $O(T^4) = \Omega(\sqrt{2^n})$, and hence $Q_{SE}^\mu(Simon) = \Omega(2^{\frac{n}{8}})$. $\square$

5   A Positive Result for the Weak Quantum Yao Principle

In this section we show that the weak quantum Yao principle holds for all symmetric functions. We start with the special case of threshold functions.

5.1   Equality up to a Constant Factor for Threshold Functions

Consider a threshold function with threshold $t \leq N/2$. For every distribution μ, we will exhibit a weakly $\frac{2}{3}$-exact quantum algorithm for f with respect to μ with $O(\sqrt{tN})$ queries. This, together with Theorem 1 and the known fact that $Q_2(f) = \Theta(\sqrt{tN})$ for threshold functions f [3], gives the desired result.

Note that given a threshold function $f : \{0,1\}^N \rightarrow \{0,1\}$ with threshold t, in order to be sure that $f(x) = 1$, it suffices to find at least t 1's in the input. The crucial idea behind our algorithm is that if the number of 1's in the input is large enough, then for each distribution μ over the inputs, we can pick a substantially smaller part of the input such that there are between t and $100t$ 1's in this sub-part for a large μ-fraction of the inputs. This idea is formally stated in the following technical lemma.[1]

---
[1] We need the condition $i \geq 10$ in this lemma in order to be able to approximate the hypergeometric distribution by a binomial distribution with sufficient accuracy.

Lemma 1. *Let t be a threshold, μ a probability distribution over the $x \in \{0,1\}^N$, and i an integer such that $10 \leq i \leq \log N - \log t - 1$. Denote the event $t2^i \leq |x| \leq t2^{i+1}$ by I, and let $x \wedge y$ denote the bitwise AND of x and y. There is a $y \in \{0,1\}^N$ with $|y| = \min\{\frac{10N}{2^i}, N\}$, such that $\Pr_\mu[t \leq |x \wedge y| \leq 100t \mid I] > 0.7$.*

Proof. We assume $\frac{10N}{2^i} \leq N$, for otherwise the lemma trivially holds. Consider any $x \in \{0,1\}^N$ with $t2^i \leq |x| \leq t2^{i+1}$. We claim that if we pick a $y \in \{0,1\}^N$ with $|y| = \frac{10N}{2^i}$ uniformly at random, then $\Pr[t \leq |x \wedge y| \leq 100t] > 0.7$. To prove this claim, note that $|x \wedge y|$ is hypergeometrically distributed, with expected value $E(|x \wedge y|) = \frac{|x||y|}{N} \in [10t, 20t]$. By Markov's inequality it follows directly that $\Pr[|x \wedge y| > 100t] \leq 0.2$.

We can approximate the above distribution with a binomial distribution since the number of draws is small compared to the size of the sample space, see e.g. [11], and we shall henceforth treat $|x \wedge y|$ as if it were binomially distributed, with success probability $\theta = \frac{|x|}{N}$ and number of draws $n = |y|$. To bound $\Pr[|x \wedge y| < t]$, we use the Chernoff bound as explained in [9, pp.67-73]:

$$\Pr[|x \wedge y| < (1-\delta)E(|x \wedge y|)] < e^{\frac{-\delta^2 E(|x \wedge y|)}{2}}.$$

Choosing $\delta = \frac{9}{10}$, we obtain $\Pr[|x \wedge y| < t] < e^{-\frac{810t}{200}} < 0.1$. Combining the previous two inequalities, it then follows that $\Pr[t \leq |x \wedge y| \leq 100t] > 0.7$, which proves the above claim.

Now imagine a matrix whose rows are indexed by the x satisfying $t2^i \leq |x| \leq t2^{i+1}$ and whose columns are indexed by the $M = \binom{N}{|y|}$ different y of weight $|y| = \frac{10N}{2^i}$. We give the (x, y) entry of this matrix value $\mu(x|I)$ if $t \leq |x \wedge y| \leq 100t$ and value 0 otherwise. By the above claim, each row will contain at least 70% non zero entries, so the sum of the entries of the x-row is at least $0.7M\mu(x|I)$. Hence, the sum of all entries in the matrix is equal to $\sum_x 0.7M\mu(x|I) = 0.7M$. But then there must be a column with $\mu(\cdot \mid I)$-weight at least 0.7. The y corresponding to this column is the y we are looking for in this lemma. $\square$

We will use the fact stated in the previous lemma to successively search for t 1's in exponentially smaller parts of the inputs, assuming the presence of increasingly more 1's in the original input. The following lemma states that this searching can be done efficiently:

Lemma 2. *There exists a quantum algorithm that can find all the 1's in an input x of size N with probability 1, using at most $\frac{\pi}{2}\sqrt{kN}$ queries, if k is a known upper bound on the number of 1's in x.*

Proof. Assume an upper bound k on the number of 1's in x. Suppose we run the exact version of Grover's algorithm assuming $|x| = k$. Either we find a solution, in which case we can remove that solution from the search space, lower our upper bound k by 1 and continue; or we do not find a solution, in which case we know that $|x|$ must be less than k, so we can safely lower our upper bound k by 1 and continue. Accordingly, it easily follows by induction on k that Algorithm 1 below finds all $|x|$ solutions with certainty. The number of queries it uses is

$$\sum_{i=1}^{k} \frac{\pi}{4} \sqrt{\frac{N}{i}} \leq \frac{\pi}{4} \sqrt{N} \int_0^k \frac{\mathrm{d}i}{\sqrt{i}} = \frac{\pi}{2} \sqrt{kN}.$$

$\square$

Algorithm 1

---

for $i = k$ down to 1 **do**
    Apply the exact version of Grover's algorithm, assuming
        there are i solutions.
    if a solution has been found **then**
        mark its index as a zero in the search space
    end if
end for
output the positions of all solutions found

We are now ready to prove an upper bound on $Q_{WE}^{\mu}(f)$:

Lemma 3. *For threshold function f with threshold t, and for every distribution μ, we have $Q_{WE}^{\mu}(f) = O(\sqrt{tN})$.*

Proof. Fix a distribution μ. Invoking Lemmas 1 and 2, our algorithm (Algorithm 2 below) is as follows. First we count the number of 1's in the input using Algorithm 1, assuming an upper bound of $2^{10}t$ 1's. If after that we haven't found at least t 1's yet, then we successively assume that there are between $t2^i$ and $t2^{i+1}$ 1's in the input, with i going up from 10 to $\log N - \log t - 1$. For each of these assumptions, we search a smaller part of the input. If we have reached the i for which $t2^i \leq |x| \leq t2^{i+1}$, then Lemma 1 guarantees that for a large μ-fraction of those inputs we can find a small sub-part containing between t and $100t$ 1's. We then count the number of 1's in this sub-part using Algorithm 1. This algorithm will be correct on all inputs x with $|x| < t$ and will produce a correct answer on at least a μ-fraction 0.7 of all inputs x with $|x| \geq t$ as guaranteed by Lemma 1. Hence it will be correct on a μ-fraction at least $\mu(\{x \mid |x| < t\}) + 0.7(1 - \mu(\{x \mid |x| < t\})) \geq 0.7$. Furthermore, its query complexity is

$$O(\sqrt{tN}) + \sum_{i=10}^{\log N - \log t - 1} O\left(\sqrt{\frac{tN}{2^i}}\right) = O(\sqrt{tN}),$$

where the first term corresponds to the cost of searching the entire space once with a small upper bound, and the summation corresponds to searching consecutively smaller sub-parts $y^{(i)}$. $\square$

Algorithm 2

Count the number of 1's in the input using Algorithm 1, assuming an upper bound
of $2^{10}t$ 1's
if at least t 1's are found **then**
    output 1
end if
for $i = 10$ to $\log N - \log t - 1$ **do**
    Let $y^{(i)} \in \{0,1\}^N$ be a string of weight $\min\{N, \frac{10N}{2^i}\}$ satisfying Lemma 1
    Using Algorithm 1, count the number of solutions in the sub-part
        of the input induced by $y^{(i)}$, assuming an upper bound of $100t$ 1's.
    if at least t 1's are found **then**
        output 1
    end if
end for
output 0

Recall that for threshold functions $f : \{0,1\}^N \to \{0,1\}$ with threshold
$t \leq N/2$, we have $Q_2(f) = \Theta(\sqrt{tN})$. By Theorem 1 it then follows that
$\max_\mu Q^\mu_{WE}(f) = \Omega(\sqrt{tN})$. In combination with Lemma 3, this yields:

Lemma 4. *For all threshold functions* $f : \{0,1\}^N \to \{0,1\}$ *with* $t \leq N/2$

$$Q_2(f) = \Theta \left(\max_\mu Q^\mu_{WE}(f) \right) = \Theta \left(\sqrt{tN} \right).$$

5.2   Equality up to a Constant Factor for Symmetric Functions

With the result about threshold functions in mind, we can easily prove that the
quantum Yao principle holds for all symmetric functions as well.

Theorem 3. *For all symmetric functions* $f : \{0,1\}^N \to \{0,1\}$

$$Q_2(f) = \Theta \left(\max_\mu Q^\mu_{WE}(f) \right) = \Theta \left(\sqrt{N(N - \Gamma(f))} \right).$$

We give an informal sketch of the proof whose details are straightforward.
Firstly, note that $\Gamma(f)$ measures the length of the interval around Hamming
weight $\frac{N}{2}$ where f is constant, so in order to compute $f(x)$ it suffices to know
$|x|$ exactly if $|x| \in [0, \frac{N-\Gamma(f)}{2})$ or $|x| \in (\frac{N+\Gamma(f)-2}{2}, N]$, or to know that $|x| \in$
$[\frac{N-\Gamma(f)}{2}, \frac{N+\Gamma(f)-2}{2}]$ otherwise. Using the threshold algorithm from Section 5.1
twice, we can, at a cost of $O(\sqrt{N(N - \Gamma(f))})$ queries, compute which of three
intervals $|x|$ is in. If $|x|$ is in the interval of length $\Gamma(f)$ around $\frac{N}{2}$ where f is
constant we are done. In both other cases we now in effect have an upper bound
on the number of 1's in the input, and we can use Algorithm 1 to exactly count
the number of 1's, again using $O(\sqrt{N(N - \Gamma(f))})$ queries.

5.3   A Result for the AND-OR Tree

Above we proved that the weak quantum Yao principle holds (up to a constant factor) for all *symmetric* functions. A similar result might be provable for all *monotone* functions. In this section we state a preliminary result in this direction, namely that the known upper and lower bounds on the $Q_2(f)$-complexity of the 2-level *AND-OR tree* carry over to weakly $\frac{2}{3}$-exact quantum algorithms. This monotone but non-symmetric function is the AND of $\sqrt{N}$ independent ORs of $\sqrt{N}$ variables each. In the sequel, we use AO to denote this N-bit AND-OR tree.

No tight characterization of $Q_2(AO)$ is known, but Buhrman, Cleve, and Widgerson [5] proved $Q_2(AO) = O(\sqrt{N}\log N)$ via a recursive application of Grover's algorithm. Using a result about efficient error-reduction in quantum search from [6], this can be improved to $Q_2(AO) = O(\sqrt{N\log N})$. This nearly matches Ambainis' lower bound of $\Omega(\sqrt{N})$ [2]. Note that Ambainis' bound together with our Theorem 1 immediately gives the lower bound $\max_\mu Q^\mu_{WE}(AO) = \Omega(\sqrt{N})$. Using the same techniques as in the previous section one can show that the best known *upper* bound carries over to weakly $\frac{2}{3}$-exact algorithms. Due to space constraints we omit the proof, which may be found at the Los Alamos preprint server at http://xxx.lanl.gov/abs/quant-ph/0109070.

Theorem 4. *For every distribution μ we have $Q^\mu_{WE}(AO) = O(\sqrt{N\log N})$.*

6   Summary and Open Problems

In this paper we investigated to what extent quantum versions of the classical Yao principle hold. We formulated a strong and a weak version of the quantum Yao principle, showed that both hold in one direction, falsified the other direction for the strong version, and proved the weak version for the query complexity of all symmetric functions.

The main question left open by this research is the general validity of the weak quantum Yao principle. On the one hand, we may be able to find a counterexample to the weak principle as well, perhaps based on the query complexity of the *order-finding problem*. Shor showed that the order-finding problem can be solved by a bounded-error quantum algorithm using $O(\log N)$ queries [15]. Using Cleve's $\Omega(N^{1/3}/\log N)$ lower bound on classical algorithms for order-finding [7], we might be able to exhibit a μ such that any weakly $\frac{2}{3}$-exact quantum algorithm for f with respect to μ requires $N^{\Omega(1)}$ queries, as it seems hard to construct weakly $\frac{2}{3}$-exact quantum algorithms for this problem.

On the other hand, we may try to extend the class of functions for which we know the weak quantum Yao principle *does* hold. A good starting point here might be the class of all *monotone* functions. We discussed one such function, the 2-level AND-OR tree, in Section 5.3. Unfortunately, at the time of writing no general characterization of the $Q_2(f)$ complexity of monotone functions is known.

Acknowledgments. We thank Harry Buhrman for initiating this research, for coming up with the counterexample of Theorem 2, and for useful comments on a preliminary version of this paper. We thank him and Peter Høyer for their contributions to an initial proof of the weak quantum Yao principle for the OR function, which forms the basis for the current proof of Theorem 3. We also thank Leen Torenvliet and Chris Klaassen for useful discussions.

References

1. A. Ambainis. A note on quantum black box complexity of almost all Boolean functions. In *Information Processing Letters 71*, pages 5–7, 1999. quant-ph/9811080.
2. A. Ambainis. Quantum lower bounds by quantum arguments. In *Proceedings of 32nd ACM STOC*, pages 636–643, 2000. quant-ph/0002066.
3. R. Beals, H. Buhrman, R. Cleve, M. Mosca, and R. de Wolf. Quantum lower bounds by polynomials. In *Proceedings of 39th IEEE FOCS*, pages 352–361, 1998. quant-ph/9802049.
4. G. Brassard, P. Høyer, M. Mosca, and A. Tapp. Quantum amplitude amplification and estimation. quant-ph/0005055. To appear in Quantum Computation and Quantum Information: A Millennium Volume, AMS Contemporary Mathematics Series, 15 May 2000.
5. H. Buhrman, R. Cleve, and A. Wigderson. Quantum vs. classical communication and computation. In *Proceedings of 30th ACM STOC*, pages 63–68, 1998. quant-ph/9802040.
6. H. Buhrman, R. Cleve, R. de Wolf, and Ch. Zalka. Bounds for small-error and zero-error quantum algorithms. In *Proceedings of 40th IEEE FOCS*, pages 358–368, 1999. cs.CC/9904019.
7. R. Cleve. The query complexity of order-finding. In *Proceedings of 15th IEEE Conference on Computational Complexity*, pages 54–59, 2000. quant-ph/9911124.
8. L. K. Grover. A fast quantum mechanical algorithm for database search. In *Proceedings of 28th ACM STOC*, pages 212–219, 1996. quant-ph/9605043.
9. R. Motwani and P. Raghavan. *Randomized Algorithms*. Cambridge University Press, 1995.
10. J. von Neumann and O. Morgenstern. *Theory of Games and Economic Behavior*. Princeton University Press, 1947.
11. W. L. Nicholson. On the normal approximation to the hypergeometric distribution. *Annals of Mathematical Statistics*, 27:471–483, 1956.
12. M. A. Nielsen and I. L. Chuang. *Quantum Computation and Quantum Information*. Cambridge University Press, 2000.
13. G. Owen. *Game Theory*. Academic Press, second edition, 1982.
14. R. Paturi. On the degree of polynomials that approximate symmetric Boolean functions. In *Proceedings of 24th STOC*, pages 468–474, 1992.
15. P. W. Shor. Polynomial-time algorithms for prime factorization and discrete logarithms on a quantum computer. *SIAM Journal on Computing*, 26(5):1484–1509, 1997. Earlier version in FOCS'94. quant-ph/9508027.
16. D. Simon. On the power of quantum computation. *SIAM Journal on Computing*, 26(5):1474–1483, 1997. Earlier version in FOCS'94.
17. A. C-C. Yao. Probabilistic computations: Toward a unified measure of complexity. In *Proceedings of 18th IEEE FOCS*, pages 222–227, 1977.

Describing Parameterized Complexity Classes

Jörg Flum and Martin Grohe

[1] Institut für Mathematische Logik, Universität Freiburg, Eckerstr. 1, 79104 Freiburg, Germany.
flum@uni-freiburg.de
[2] Laboratory for Foundations of Computer Science, University of Edinburgh,
Edinburgh EH9 3JZ, Scotland, UK.
grohe@dcs.ed.ac.uk

Abstract. We describe parameterized complexity classes by means of classical complexity theory and descriptive complexity theory. For every classical complexity class we introduce a parameterized analogue in a natural way. In particular, the analogue of polynomial time is the class of all fixed-parameter tractable problems. We develop a basic complexity theory for the parameterized analogues of classical complexity classes and give, among other things, complete problems and logical descriptions. We then show that most of the well-known intractable parameterized complexity classes are not analogues of classical classes. Nevertheless, for all these classes we can provide natural logical descriptions.

1 Introduction

Parameterized complexity theory provides a framework for a refined complexity analysis, which in recent years has found its way into various areas of computer science, such as database theory [15,21], artificial intelligence [14], and computational biology [3,22]. Central to the theory is the notion of *fixed-parameter tractability*, which relaxes the classical notion of tractability, polynomial time computability, by admitting algorithms whose runtime is exponential, but only in terms of some *parameter* that is usually expected to be small. As a complexity theoretic counterpart, a theory of *parameterized intractability* has been developed. Remarkably, complexity theoretic assumptions from this theory of parameterized intractability have recently been used to prove results in a purely classical setting [2,16].

Unfortunately, the landscape of fixed-parameter intractable problems is not as nice and simple as the landscape of classical intractable problems provided by the theory of NP-completeness. Instead, there is a huge variety of seemingly different classes of fixed-parameter intractable problems. Moreover, the definitions of these classes are quite intractable themselves, at least for somebody who is not an expert in the area. One reason for this is that the classes are not defined in terms of a simple machine model, but as the closure of particular problems under parameterized reductions. Indeed, for the most interesting of these classes there seem to be no nice descriptions in terms of, say, Turing machines. However, in this paper we give simple and uniform logical descriptions of the classes using variants of logics that are also used in classical descriptive complexity theory. A related issue we address here is the relation between parameterized and classical complexity classes. In particular when proving classical results under assumptions from

H. Alt and A. Ferreira (Eds.): STACS 2002, LNCS 2285, pp. 359–371, 2002.
© Springer-Verlag Berlin Heidelberg 2002

parameterized complexity theory, an understanding of these assumptions in classical terms would be desirable.

What is fixed-parameter tractability? An instance of a *parameterized problem* is a pair (x, k) consisting of the actual input x and a parameter k. Thus formally a parameterized problem is a subset of $\Sigma^* \times \mathbb{N}$ for some finite alphabet Σ. Often, parameterized problems P are derived from classical problems $X \subseteq \Sigma^*$ by choosing a suitable parameterization $p : \Sigma^* \to \mathbb{N}$ and letting $P = \{(x, p(x)) \mid x \in X\}$. The idea is to choose the parameterization in such a way that for those instances that most often occur in practice the parameter value is small. An illustrative example is the problem of evaluating a database query. Its input consists of a relational database and a query, and a reasonable parameterization of this problem is by the length of the input query, which can be expected to be small compared to the size of the input database. It is important, however, to keep in mind that every classical problem has different parameterizations that may lead to parameterized problems of drastically different (parameterized) complexities.

Let $P \subseteq \Sigma^* \times \mathbb{N}$ be a parameterized problem. We always denote the parameter by k and the length $|x|$ of the input string x by n. The problem P is *fixed-parameter tractable* if there is a computable function $f : \mathbb{N} \to \mathbb{N}$, a constant c, and an algorithm deciding P in time $O(f(k) \cdot n^c)$. Our starting point is the observation that this notion of fixed-parameter tractability is very robust and has other natural interpretations. The first of these is inspired by the database query evaluation example. Instead of requiring a query to be evaluated in polynomial time, we may allow an algorithm (a "pre-computation") to first optimise the input query, that is, to transform it to an equivalent query that is easier to evaluate, and then evaluate the optimised query in polynomial time (see [15]). In our abstract setting, we can show that a parameterized problem P is fixed-parameter tractable if, and only if, it is in polynomial time after such a pre-computation. The following characterisation of fixed-parameter tractability is more technical, but it contributed a lot to our intuitive understanding of fixed-parameter tractability: We say that $P \subseteq \Sigma^* \times \mathbb{N}$ is *eventually in polynomial time* if there is a computable function f and a polynomial time algorithm that, given an instance (x, k) of P with $n \geq f(k)$, decides if $(x, k) \in P$. Then P is fixed-parameter tractable if, and only if, it is computable and eventually in polynomial time.

Parameterized classes derived from classical classes. If the class FPT is a parameterized version of PTIME, can we understand other parameterized complexity classes in a similar way as being derived from classical complexity classes? This is the first problem we shall study in this paper. For every complexity class K we introduce a parameterized complexity class para-K, taking FPT = para-PTIME as a prototype. Our mechanism of deriving a parameterized complexity class para-K from a classical class K is closely related to a similar mechanism introduced by Cai, Chen, Downey, and Fellows [4]. We show that the classes para-K are quite robust; for all reasonable complexity classes K, the class para-K can equivalently be characterised as the class of all problems being in K after a pre-computation, the class of all computable problems being eventually in K, and by a resource bound similar to that in the original definition of fixed-parameter tractability. We develop a basic complexity theory of these classes. It turns out that the classes para-K are closely related to the corresponding K: para-K is contained in para-K' if, and only if, K is contained in K'. Complete problems for the parameterized classes para-K under parameterized reductions are essentially unparameterized problems that are also complete for the corresponding K under standard reductions. Moreover, the classes para-K admit logical descriptions closely related to such descriptions of the corresponding K.

Overall, we get a fairly accurate picture of the structure of these parameterized complexity classes; it simply reflects the structure of the corresponding classical classes.

The W-hierarchy and other "inherently parametric" classes. The most important intractable parameterized complexity classes are the classes of the W-hierarchy [5]. Other classes we shall study here are AW[∗] [1] and the classes of the A-hierarchy [12]. Many natural parameterized problems that are not known to be fixed-parameter tractable have been shown to be complete for one of these classes, most of them for the first-level W[1] of the W-hierarchy. Therefore, the main goal of our research in structural parameterized complexity theory is a better understanding of these classes. Remembering the previous paragraph, we may hope that the intractable parameterized complexity classes are derived from intractable classical classes such as NP and the classes of the polynomial hierarchy. Unfortunately, the situation is not so simple; we show that none of the classes Q mentioned above is of the form para-K for a classical class K (unless Q = FPT). The classes turn out to be interwoven with para-NP, the parameterized classes para-Σ_t^P derived from the levels of the polynomial hierarchy, and para-PSPACE in a non-trivial way.

So we cannot really use our structure theory for the classes para-K to understand the classes of W-hierarchy, the A-hierarchy, and AW[∗]. However, we can extend our descriptive complexity theoretic approach and give logical descriptions of these classes. Descriptive characterisations of classical complexity classes have been criticised for being merely simple translations of the original Turing-machine based definitions of the classes into a logical formalism and therefore not providing new insights. Although we do not entirely agree with this criticism, it is certainly true that the logical characterisations are often very close to the machine descriptions. However, in our context this is rather an advantage, because there are no known simple and natural machine descriptions of the parameterized classes we characterise here.

Organisation and Preliminaries. We have organised the paper following the outline given in this introduction. Due to space limitations, we defer the proofs to the full version of the paper [13]. We assume that the reader is familiar with the fundamentals of complexity theory (see, e.g., [20]) and logic (see, e.g., [10]). We do not assume familiarity with parameterized complexity theory (for background see [6]).

2 Parameterized Classes Derived from Classical Classes

Classical problems, which we usually denote by X, Y, are subsets of Σ^* for some alphabet Σ. *Parameterized problems*, which we denote by P, Q, are subsets of $\Sigma^* \times \mathbb{N}$. Of course, we may view any parameterized problem $P \subseteq \Sigma^* \times \mathbb{N}$ as a classical problem in some larger alphabet, say, the alphabet obtained from Σ by adding new symbols '(', ',', ')', and '1'. [1] Conversely, with every classical problem $X \subseteq \Sigma^*$ we can associate

---

[1] Parameters are usually encoded in unary, although this is inessential.

its *trivial parameterization* $X \times \{0\}$. A classical or parameterized complexity class is simply a set of classical problems or parameterized problems, respectively.

In this section, we associate a parameterized complexity class para-K with every classical complexity class K and then discuss basic properties of these classes and their relation to the classical counterparts.

Definition 1. Let K be a classical complexity class. Then *para-K* is the class of all parameterized problems P, say with $P \subseteq \Sigma^* \times \mathbb{N}$, for which there exists an alphabet Π, a computable function $f : \mathbb{N} \to \Pi^*$, and a problem $X \subseteq \Sigma^* \times \Pi^*$ such that $X \in$ K and for all instances $(x, k) \in \Sigma^* \times \mathbb{N}$ of P we have

$$(x, k) \in P \iff (x, f(k)) \in X.$$

Intuitively, para-K consists of all problems that are in K *after a pre-computation* that only involves the parameter. This means that a problem in para-K can be solved by two algorithms $\mathbb{P}$ and $\mathbb{A}$, where $\mathbb{P}$ is arbitrary and $\mathbb{A}$ has resources constrained by K. The pre-computation $\mathbb{P}$ only involves the parameter; it transforms k into some string $f(k)$, presumably one that makes it easier to solve the problem. Then the algorithm $\mathbb{A}$ solves the problem, given the original input x and the result $f(k)$ of the pre-computation, with resources constrained by the complexity class K.

Example 2. We want to evaluate a sentence of monadic second-order logic MSO on a string. The input to this problem consists of a string x and a sentence φ of MSO. We assume that we have fixed a reasonable encoding $[\cdot]$ of MSO-sentences by natural numbers. We parameterize the problem by the encoding $[\varphi] \in \mathbb{N}$ of the input sentence φ. The pre-computation translates $[\varphi]$ into an equivalent finite automaton $\mathfrak{A}(\varphi)$. Then the main algorithm runs $\mathfrak{A}(\varphi)$ on the string x; this only requires linear time.

We have just proved that the problem of evaluating a sentence of MSO on a string, parameterized by an encoding of the input sentence, is in para-LINTIME.

Our definition of para-K is more or less the same as the definition of the class *uniform K + advice*, which Cai et al. [4] introduced as a parameterized version of the complexity class K. Instead of considering the result $f(k)$ of the pre-computation as part of the input of the main computation, Cai et al. treat it as an "advice string", which is accessed as an "oracle". Our model is slightly more general, because by treating it as part of the input we do not have to worry about how the result of the pre-computation is accessed, and we have to make no reference to a particular machine model.

We have mentioned in the introduction that para-PTIME is precisely the class FPT of all fixed-parameter tractable problems (formally, this is an immediate consequence of Theorem 3 below and also of [4]). It seems somewhat arbitrary to use the pre-computation view on fixed-parameter tractability as the basis of our generalisation of FPT = para-PTIME to other classes K. However, it turns out that the equivalent characterisations of FPT given in the introduction have an analogue for any class para-K derived from a "reasonable" complexity class K. We want to have a notion of "reasonable" class that includes all the standard classes such as LOGSPACE, NLOGSPACE, PTIME, NP, Σ_i^P (the ith level of the polynomial hierarchy) for $i \geq 1$, PSPACE, et cetera. We first fix a machine model $\mathcal{M}$, for our purposes it seems best to use alternating Turing machines with some restriction put on the alternations that are allowed. Our Turing machines have an input tape, if necessary an output tape, and an arbitrary finite number of work tapes. For example, $\mathcal{M}$ may just be the class of deterministic Turing machines, or the class of non-deterministic Turing machines, or maybe the class of alternating Turing machines with at most 5 quantifier alternations starting with an existential state. Then we fix a resource $\mathcal{R}$, either time or space. For a function $\gamma : \mathbb{N} \to \mathbb{N}$, we let $K(\mathcal{M}, \mathcal{R}, \gamma)$ be the class of all problems accepted by a machine in $\mathcal{M}$ that on all inputs of size n uses an amount of at most $\gamma(n)$ of $\mathcal{R}$, for all $n \in \mathbb{N}$. For a class Γ of functions we let $K(\mathcal{M}, \mathcal{R}, \Gamma) = \bigcup_{\gamma \in \Gamma} K(\mathcal{M}, \mathcal{R}, \gamma)$.

We call a class Γ of functions on the natural numbers *regular*, if for all $\gamma \in \Gamma$:

– γ is computable.
– γ is monotone growing with $\lim_{m \to \infty} \gamma(m) = \infty$.
– For all $c \in \mathbb{N}$ there are functions $\gamma_1, \gamma_2, \gamma_3 \in \Gamma$ such that for all $m \in \mathbb{N}$ we have $c + \gamma(m) \le \gamma_1(m), c \cdot \gamma(m) \le \gamma_2(m)$, and $\gamma(c \cdot m) \le \gamma_3(m)$.

A *regular complexity class* is a class $K(\mathcal{M}, \mathcal{R}, \Gamma)$ for a regular class Γ of functions. It is easy to see that all the complexity classes mentioned above are regular.

Recall that n always denotes the length of the input x of a parameterized problem and k denotes the parameter.

Theorem 3. *Let* $K = K(\mathcal{M}, \mathcal{R}, \Gamma)$ *be a regular complexity class. Then for every parameterized problem* $P \subseteq \Sigma^* \times \mathbb{N}$ *the following three statements are equivalent:*

(1) P *is in para-K.*
(2) P *is computable and **eventually in K**, i.e., there is a computable* $f : \mathbb{N} \to \mathbb{N}$ *and a problem* $X \in K$ *such that for all* $(x, k) \in \Sigma^* \times \mathbb{N}$ *with* $n \ge f(k)$ *we have*

$$(x, k) \in P \iff (x, k) \in X.$$

(3) There is a computable f *and a* $\gamma \in \Gamma$ *such that* $P \in K(\mathcal{M}, \mathcal{R}, f(k) + \gamma(n))$.

Remark 4. For certain complexity classes $K = K(\mathcal{M}, \mathcal{R}, \Gamma)$, in particular for PTIME and PSPACE, Cai et al. [4] have proved that for every parameterized problem P, P is in para-K if, and only if, there is a computable function f and a function $\gamma \in \Gamma$ such that $P \in K(\mathcal{M}, \mathcal{R}, f(k) \cdot \gamma(n))$. However, there are interesting classes such as LOGSPACE for which this equivalence does not seem to hold.

Example 5. Cai et al. [4] have proved that the natural parameterization of the vertex cover problem (by the size of the vertex cover) is in para-LOGSPACE. In this example we shall prove that many interesting properties of graphs of bounded degree are in para-LOGSPACE.

For a class D of structures and a class Φ of sentences of some logic, the *Φ-model-checking problem on* D, denoted by $MC(D, \Phi)$, is the problem of deciding whether a given structure $\mathcal{A} \in D$ satisfies a given sentence $\varphi \in \Phi$. This problem is naturally parameterized by the length of the input sentence. We choose a slightly different parameterization (which can easily be seen to have the same parameterized complexity). We fix some encoding $[\cdot] : \Phi \to \mathbb{N}$ and parameterize the model-checking problem by $[\varphi]$. Formally, we let

$$p\text{-}MC(D, \Phi) = \{(\mathcal{A}, [\varphi]) \mid \mathcal{A} \in D, \varphi \in \Phi\}.$$

Denoting the class of all first-order sentences by FO and, for any $g \geq 1$, the class of all graphs of degree at most g by $\mathrm{DEG}(g)$, we claim:

$$\mathrm{p\text{-}MC}(\mathrm{DEG}(g), \mathrm{FO}) \in \mathrm{para\text{-}LOGSPACE}.$$

This implies that the standard parameterized versions of problems like dominating set or subgraph isomorphism restricted to graphs of bounded degree are in para-LOGSPACE.

There is another natural way of associating a parameterized complexity class with a classical class. For $k \geq 1$, the *kth slice* of a parameterized problem $P \subseteq \Sigma^* \times \mathbb{N}$ is the problem $P_k = \{x \mid (x, k) \in P\} \subseteq \Sigma^*$.

Definition 6. Let K be a classical complexity class. Then XK is the class of all parameterized problems P all of whose slices are in K.

To compare para-K and XK and to obtain further basic facts about these classes, we only require the complexity class K to be *closed*, i.e., to satisfy the following two conditions (for all alphabets Σ and Π):

- For every problem $X \subseteq \Sigma^*$ with $X \in$ K and every word $y \in \Pi^*$: $X \times \{y\} \in$ K.
- For every problem $X \subseteq \Sigma^* \times \Pi^*$ with $X \in$ K and every word $y \in \Pi^*$: $X_y = \{x \in \Sigma^* \mid (x, y) \in X\} \in$ K.

Clearly, every regular class is closed. Moreover, for every closed complexity class K and every problem $X \in$ K, the trivial parameterization $X \times \{0\}$ of X is in para-K. And para-K $\subseteq$ XK holds for every closed complexity class K.

Proposition 7. *For closed complexity classes* K *and* K' *the following are equivalent:*

(1) K $\subseteq$ K'.   *(2)* para-K $\subseteq$ para-K'.   *(3)* para-K $\subseteq$ XK'.

In particular, there is no closed K_0 *such that* para-K_0 *is between* para-K *and* XK.

In Section 3, we shall use Proposition 7 to show that the classes of the W-hierarchy and other important parameterized complexity classes, which are between FPT and XPTIME, are "inherently parametric", that is, not of the form para-K.

Remark 8. For regular complexity classes K the containment of para-K in XK is always strict just because we defined XK in a highly non-uniform way (so that it contains problems that are not computable).

For the standard, well-behaved complexity classes K, which are all of the form $K(\mathcal{M}, \mathcal{R}, \Gamma)$ for a simple class Γ, we can define a uniform version of XK (see [6]). There are classes K for which this uniform-XK coincides with para-K. An example is the class K of all linear time solvable problems. For other classes K, such as LOGSPACE, PTIME, NP, or PSPACE, a simple diagonalisation shows that para-K $\neq$ uniform-XK.

2.1 Reductions and Completeness

The following definition introduces the basic type of reduction on which much of the theory of parameterized intractability is built:

Definition 9. Let $P \subseteq \Sigma^* \times \mathbb{N}$ and $P' \subseteq (\Sigma')^* \times \mathbb{N}$ be parameterized problems.

A *para-PTIME-reduction*, or *FPT-reduction*, from P to P' is a mapping $R : \Sigma^* \times \mathbb{N} \to (\Sigma')^* \times \mathbb{N}$ such that:

(1) For all $(x, k) \in \Sigma^* \times \mathbb{N}$ we have $(x, k) \in P \iff R(x, k) \in P'$.

(2) There exists a computable function $g : \mathbb{N} \to \mathbb{N}$ such that for all $(x, k) \in \Sigma^* \times \mathbb{N}$, say with $R(x, k) = (x', k')$, we have $k' \leq g(k)$.

(3) There exists a computable function $f : \mathbb{N} \to \mathbb{N}$ and a constant $c \in \mathbb{N}$ such that R is computable in time $f(k) \cdot n^c$.

We write $P \leq^{\mathrm{FPT}} P'$ if there is an FPT-reduction from P to P'.

FPT-reductions are parameterized versions of polynomial-time many one reductions. Similarly, we define the analogue of LOGSPACE-reductions:

Definition 10. Let $P \subseteq \Sigma^* \times \mathbb{N}$ and $P' \subseteq (\Sigma')^* \times \mathbb{N}$ be parameterized problems.

A *para-LOGSPACE-reduction*, or *PL-reduction*, from P to P' is a mapping $R : \Sigma^* \times \mathbb{N} \to (\Sigma')^* \times \mathbb{N}$ that satisfies conditions (1) and (2) of Definition 9 and

(3') There exists a computable function $f : \mathbb{N} \to \mathbb{N}$ and a constant $c \in \mathbb{N}$ such that R is computable in space $f(k) + c \cdot \log n$.

We write $P \leq^{\mathrm{PL}} P'$ if there is a PL-reduction from P to P'.

Clearly, if $P \leq^{\mathrm{FPT}} P'$ and $P' \in \mathrm{FPT}$, then $P \in \mathrm{FPT}$, and the analogous result holds for $\leq^{\mathrm{PL}}$ and para-LOGSPACE. We define *hardness* and *completeness* in the usual way.

What are complete problems for our classes para-K? One might guess that suitable parameterizations of complete problems for K. It is not obvious which parameterizations are suitable. It came as a surprise to us that essentially we do not need any parameterization: the trivial parameterization of a complete problem for K is complete for para-K. So para-K has a complete problem that is essentially unparameterized.

Proposition 11. *Let* K *be a closed complexity class, and let* X *be complete for* K *under PTIME-reductions (LOGSPACE-reductions, respectively). Then* $X \times \{0\}$ *is complete for* para-K *under FPT-reductions (PL-reductions, respectively).*

The following proposition shows that completeness for a class para-K is always based on a somewhat trivial parameterization.

Proposition 12. *Let* K *be a closed complexity class, and let* $P \subseteq \Sigma^* \times \mathbb{N}$ *be complete for* para-K *under FPT-reductions (PL-reductions, respectively).*

Then there is an integer $\ell \in \mathbb{N}$ *such that the problem* $P \cap (\Sigma^* \times \{0, \dots, \ell\})$ *is complete for* para-K *under FPT-reductions (PL-reductions, respectively).*

Remark 13. The proof shows that if para-K has a complete problem P then K also has a complete problem. Moreover, $P_0 := P \cap (\Sigma^* \times \{0, \dots, \ell\})$, considered as a classical problem, is hard for K under PTIME-reductions (LOGSPACE-reductions, respectively).

At first sight, Propositions 11 and 12 seem to indicate that the classes para-K are trivial from the perspective of parameterized complexity theory. Of course this is not the case, these classes contain problems with interesting parameterizations. After all, the most important parameterized complexity class FPT is of this form. As another example, we have seen that p-MC(DEG(g), FO) $\in$ para-LOGSPACE, for every $g \geq 1$. It is worthwhile to note that the trivial parameterization of MC(DEG(g), FO) is complete for para-PSPACE under PL-reductions (because MC(DEG(g), FO) is complete for PSPACE under LOGSPACE-reductions [23]). Since LOGSPACE $\subset$ PSPACE and thus para-LOGSPACE $\subset$ para-PSPACE, this implies that there is no PL-reduction from MC(DEG(g), FO) $\times \{0\}$ to p-MC(DEG(g), FO).

2.2   Logical Descriptions

In this section, we show how to derive logical descriptions of the classes para-K from such descriptions of the corresponding K.

Preliminaries from logic and descriptive complexity theory. A *vocabulary* τ is a finite set of relation, function, and constant symbols. Each relation and function symbol has an *arity*. A τ-structure $\mathcal{A}$, consists of a set A called the universe, and an interpretation $T^{\mathcal{A}}$ of each symbol $T \in \tau$: Relation symbols and function symbols are interpreted by relations and functions on A of the appropriate arity, and constant symbols are interpreted by elements of A. *We only consider structures whose universe is finite.* An *ordered structure* is a structure $\mathcal{A}$ whose vocabulary contains the binary relation symbol $\leq$, the unary function symbol S, and constant symbols min and max, such that $\leq^{\mathcal{A}}$ is a linear order of A, $\min^{\mathcal{A}}$ and $\max^{\mathcal{A}}$ are the minimum and maximum element of $\leq^{\mathcal{A}}$, and $S^{\mathcal{A}}$ is the successor function associated with $\leq^{\mathcal{A}}$, where we let $S^{\mathcal{A}}(\max^{\mathcal{A}}) = \max^{\mathcal{A}}$. To avoid notational overkill, we often omit superscripts $^{\mathcal{A}}$. By $S^k(\min)$ we denote the term $\underbrace{S(S(\ldots S(\min) \ldots))}_{k \text{ times}}$, which is interpreted by the k-th element of the ordering. By ORD we denote the class of all ordered structures, and by ORD$[\tau]$ we denote the class of all ordered $\tau^{\mathrm{ord}} := \tau \cup \{\leq, S, \min, \max\}$-structures. *In the following, we assume that vocabularies do not contain any function symbols besides the successor function S.*

Until the end of this section, and also in Subsection 3.1, classical problems are subclasses of ORD$[\tau]$, for a vocabulary τ, that are closed under isomorphism. Similarly, parameterized problems are subclasses P of ORD$[\tau] \times \mathbb{N}$ for a vocabulary τ, where for each $k \in \mathbb{N}$ the class $P_k = \{\mathcal{A} \in \mathrm{ORD}[\tau] \mid (\mathcal{A}, k) \in P\}$ is closed under isomorphism.

The class of all formulas of first-order logic is denoted by FO.– If $\varphi(x_1, \ldots, x_k)$ is a formula with free variables among $x_1, \ldots, x_k$, $\mathcal{A}$ is a structure, and $a_1, \ldots, a_k \in A$, then we write $\mathcal{A} \models \varphi(a_1, \ldots, a_k)$ to denote that $\mathcal{A}$ satisfies φ if $x_1, \ldots, x_k$ are interpreted by $a_1, \ldots, a_k$, respectively. We let $\varphi(\mathcal{A}) = \{(a_1, \ldots, a_k) \in A^k \mid \mathcal{A} \models \varphi(a_1, \ldots, a_k)\}$. Recall that a *sentence* is a formula without free variables. Sentences define classes of structures. In particular, a sentence φ whose vocabulary is contained in τ^{ord} defines the class $X(\varphi) = \{\mathcal{A} \in \mathrm{ORD}[\tau] \mid \mathcal{A} \models \varphi\} \subseteq \mathrm{ORD}[\tau]$ of ordered structures. We say that a logic L *captures* a complexity class K, and write K $=$ L, if for every vocabulary τ and every problem $X \subseteq \mathrm{ORD}[\tau]$ we have

$$X \in \mathrm{K} \iff \text{There exists a sentence } \varphi \in \mathrm{L}[\tau^{\mathrm{ord}}] \text{ such that } X = X(\varphi).$$

Here a *logic* L is simply a class of formulas (together with a semantics). We denote by $\mathrm{L}[\tau]$ the class of formulas of L whose vocabulary is contained in τ.

We summarise the most important results from descriptive complexity theory (compare [9] or [19] for the definitions of the corresponding logics):

Theorem 14 (Fagin [11], Immerman [17,18], Vardi [24]). *(1) Deterministic transitive closure logic DTC captures LOGSPACE.*

(2) Transitive closure logic TC captures NLOGSPACE.

(3) Least fixed-point logic LFP captures PTIME.

(4) Existential second-order logic Σ_1^1 captures NP. Moreover, for every $t \geq 2$, Σ_t^1 captures Σ_t^{P}.

(5) Partial fixed-point logic PFP *captures* PSPACE.

Capturing parameterized complexity classes.

Definition 15. Let L be a logic.
(1) A parameterized problem $P \subseteq \text{ORD}[\tau] \times \mathbb{N}$ is *slicewise* L-*definable*, if there is a computable $\delta : \mathbb{N} \to \text{L}[\tau^{\text{ord}}]$ such that for all $\mathcal{A} \in \text{ORD}[\tau]$ and $k \in \mathbb{N}$ we have

$$(\mathcal{A}, k) \in P \iff \mathcal{A} \models \delta(k).$$

(2) L *captures* a parameterized complexity class Q, written Q $=$ slicewise-L, if for every vocabulary τ and every parameterized problem $P \subseteq \text{ORD}[\tau] \times \mathbb{N}$ we have

$$P \in \text{Q} \iff P \text{ is slicewise L-definable.}$$

For well-behaved complexity classes K and logics L with K $=$ L this equality holds "effectively"; therefore, in view of Remark 8,

$$\text{K} = \text{L} \iff \text{uniform-XK} = \text{slicewise-L.}$$

Proposition 16. *Let* K *be a classical complexity class and* Q *a parameterized complexity class. If* para-K $\subseteq$ Q $\subset$ uniform-XK *then there is no logic* L *with* Q $=$ *slicewise-*L.

Remark 8 shows that most of the interesting parameterized complexity classes are not of the form uniform-XK. Thus, by Proposition 16, if we want to describe any of these by logics, we need a more refined notion of capturing.

Definition 17. A family $(\text{L}_s)_{s \in \mathbb{N}}$ of logics *captures* a parameterized complexity class Q, written Q $= \bigcup_{s \geq 1}$ slicewise-L_s, if for every vocabulary τ and every parameterized problem $P \subseteq \text{ORD}[\tau] \times \mathbb{N}$ we have

$$P \in \text{Q} \iff \text{There is an } s \geq 1 \text{ such that } P \text{ is slicewise } \text{L}_s\text{-definable.}$$

To state our capturing results, we need some additional notation. For $s \in \mathbb{N}$, we denote by FO_s the set of first-order formulas of quantifier rank $\leq s$ and by FO^s the set of first-order formulas that contain at most s variables.

It is well known that every formula of least fixed point logic LFP is equivalent to one of the form

$$[\text{LFP}_{\bar{x}, X}\, \varphi]\, \bar{t}, \tag{1}$$

where φ is a first-order formula and $\bar{t}$ a tuple of terms [17]. LFP^s and LFP_s denote the fragments of LFP consisting of all formulas of the form (1) where $\varphi \in \text{FO}^s$ and $\varphi \in \text{FO}_s$, respectively, and where $\bar{t}$ has length at most s.

On ordered structures, every formula of transitive closure logic TC is equivalent to one of the form

$$[\text{TC}_{\bar{u}, \bar{v}}\, \varphi]\, \bar{t}_1, \bar{t}_2 \tag{2}$$

where φ is a first-order formula and $\bar{t}_1, \bar{t}_2$ are tuples of terms [18]. TC_s denotes the fragment of TC consisting of all formulas of the form (2) where $\varphi \in \text{FO}_s$ and $\bar{t}_1, \bar{t}_2$ are tuples of length at most s.

$(\Sigma_1^1)^s$ and $(\Sigma_1^1)_s$ are the sets of formulas of the form $\exists X_1 \ldots \exists X_m\, \varphi$ with $\varphi \in \text{FO}^s$ or $\varphi \in \text{FO}_s$, respectively. Similarly, we can define fragments DTC_s, $(\Sigma_t^1)^s$ and $(\Sigma_t^1)_s$, PFP^s and PFP_s of the respective logics.

We proved a first parameterized capturing result in [12]. The second equality in the following theorem has not been stated explicitly in [12], but follows from the proof.

Theorem 18 (Flum, Grohe [12]). $\text{FPT} = \bigcup_s \text{slicewise-LFP}^s = \bigcup_s \text{slicewise-LFP}_s$.

The following theorem lifts the other statements of Theorem 14:

Theorem 19. *(1)* $\text{para-LOGSPACE} = \bigcup_{s \geq 1} \text{slicewise-DTC}_s$.
(2) $\text{para-NLOGSPACE} = \bigcup_{s \geq 1} \text{slicewise-TC}_s$.
(3) $\text{para-NP} = \bigcup_{s \geq 1} \text{slicewise-}(\Sigma_1^1)^s = \bigcup_{s \geq 1} \text{slicewise-}(\Sigma_1^1)_s$,
and for every $t \geq 2$, $\text{para-}\Sigma_t^P = \bigcup_{s \geq 1} \text{slicewise-}(\Sigma_t^1)^s = \bigcup_{s \geq 1} \text{slicewise-}(\Sigma_t^1)_s$.
(4) $\text{para-PSPACE} = \bigcup_{s \geq 1} \text{slicewise-PFP}^s = \bigcup_{s \geq 1} \text{slicewise-PFP}_s$.

3   The W-Hierarchy and Other "Inherently Parametric" Classes

In this section, we shall study the classes $\text{W}[t]$, for $t \geq 1$, forming the so-called *W-hierarchy*, the classes $\text{A}[t]$, for $t \geq 1$, forming the *A-hierarchy*, and the class $\text{AW}[*]$. Each of these classes is defined as the closure of a particular problem or family of problems under FPT-reductions. Instead of giving these definitions, we characterise all the classes in a uniform way by complete model-checking problems for fragments of first-order logic. We need some notation: For $t \geq 1$, we let Σ_t be the class of all first-order formulas of the form

$$\exists x_{11} \ldots \exists x_{1\ell_1} \, \forall x_{21} \ldots \forall x_{2\ell_2} \, \exists x_{31} \ldots \exists x_{3\ell_3} \, \ldots \, Qx_{t1} \ldots Qx_{t\ell_t} \, \theta, \qquad (3)$$

where θ is a quantifier-free formula and $Q = \exists$ if t is odd, $Q = \forall$ if t is even. For $u \geq 1$, we let $\Sigma_{t,u}$ be the class of all Σ_t formulas of the form (3) where $\ell_2, \ell_3, \ldots, \ell_t \leq u$. Note that $\Sigma_1 = \Sigma_{1,1}$, but $\Sigma_t \neq \Sigma_{t,u}$ for all $t \geq 2, u \geq 1$.

Recall from Example 5 that for a class D of structures and a class Φ of formulas, $\text{p-MC}(D, \Phi)$ denotes the model-checking problem for structures from D and sentences from Φ parameterized by (a natural encoding of) the input formula. If D is the class of all structures, we denote $\text{p-MC}(D, \Phi)$ by $\text{p-MC}(\Phi)$.

Theorem 20 (Downey et al. [7], Flum, Grohe [12], Downey et al. [8]). *Let τ be a vocabulary that contains at least one binary relation symbol.*
(1) For all $t \geq 1$ and every parameterized problem P we have

$$P \in \text{W}[t] \iff \text{There exists a } u \geq 1 \text{ such that } P \leq^{\text{FPT}} \text{p-MC}(\Sigma_{t,u}[\tau]).$$

(2) For all $t \geq 1$, $\text{p-MC}(\Sigma_t[\tau])$ is complete for $\text{A}[t]$ under FPT-reductions.
(3) $\text{p-MC}(\text{FO}[\tau])$ is complete for $\text{AW}[*]$ under FPT-reductions.

Since for every fixed $\varphi \in \text{FO}$ there is a polynomial time algorithm deciding whether a given structure satisfies φ, we have $\text{p-MC}(\text{FO}) \in \text{XPTIME}$. Thus, the classes $\text{W}[t]$, $\text{A}[t]$, and $\text{AW}[*]$ are contained in XPTIME. On the other hand, since they are closed under FPT-reductions, they contain FPT. Thus by Proposition 7, none of these classes Q is of the form para-K for a closed classical complexity class K, unless $Q = \text{FPT}$.

It is easy to see that for $t, u \geq 1$ the problem $MC(\Sigma_{t,u})$ is in NP. Thus $p\text{-}MC(\Sigma_{t,u}) \in$ para-NP and therefore $W[t] \subseteq$ para-NP. Unless PTIME $=$ NP, by the above considerations we actually have $W[t] \subset$ para-NP. Similarly, the class $A[t]$ is easily seen to be contained in Σ_t^P, the tth level of the polynomial hierarchy, and unless PTIME $=$ NP, this containment is strict. We do not know if $A[t] \subseteq \Sigma_{t-1}^P$ for any $t \geq 2$, but we conjecture that this is not the case. Finally, we have $AW[*] \subseteq$ para-PSPACE, and unless PTIME $=$ PSPACE this containment is strict.

Since $\Sigma_1 = \Sigma_{1,u}$ for every $u \geq 1$, we have $A[1] = W[1]$. It is not known whether $A[t] = W[t]$ for any $t \geq 2$. The following example may help to understand the difference between $A[2]$ and $W[2]$ on an intuitive level. Downey and Fellows [5] proved that the natural parameterization of the dominating set problem is complete for $W[2]$. We give a similar graph theoretic problem that is complete for $A[2]$.

Example 21. We assume that a computable bijective pairing function $\langle , \rangle : \mathbb{N} \times \mathbb{N} \to \mathbb{N}$ is given. Then, the *parameterized dominating clique problem* DC is defined by

$$(\mathcal{G}, \langle k, l \rangle) \in DC \iff \mathcal{G} \text{ is a graph that contains a set of at most } k \text{ vertices which}$$
$$\text{dominates every clique of size } l.$$

Let X and Y be sets of vertices of the graph $\mathcal{G} = (G, E^\mathcal{G})$. Y is a *clique*, if $E^\mathcal{G}ab$ holds for all $a, b \in Y$, $a \neq b$. X *dominates* Y, if there are $a \in X$ and $b \in Y$ such that $E^\mathcal{G}ab$. A vertex a *dominates* Y, if $\{a\}$ dominates Y. Then DC is complete for $A[2]$ under FPT-reductions. In comparison, for every fixed $u \geq 1$ the problem of deciding whether a graph contains a set of k vertices that dominates every clique of size u, parameterized by k, is in $W[2]$.

3.1   Logical Descriptions

In this section, we give logical descriptions of the classes introduced above. Let $BOOL(LFP^s)$ be the class of Boolean combinations φ of formulas in LFP^s. Note that φ itself may contain more than s variables. For $t \geq 1$, we let $\Sigma_t\text{-}BOOL(LFP^s)$ be the class of all first-order formulas of the form

$$\exists x_{11} \ldots \exists x_{1\ell_1} \forall x_{21} \ldots \forall x_{2\ell_2} \exists x_{31} \ldots \exists x_{3\ell_3} \ldots Qx_{t1} \ldots Qx_{t\ell_t} \chi, \qquad (4)$$

where $\chi \in BOOL(LFP^s)$ and $Q = \exists$ if t is odd, $Q = \forall$ if t is even. For $u \geq 1$, we let $\Sigma_{t,u}\text{-}BOOL(LFP^s)$ be the class of all $\Sigma_t\text{-}BOOL(LFP^s)$ formulas of the form (4) where $\ell_2, \ell_3, \ldots, \ell_t \leq u$. Finally, we let $FO(LFP^s)$ be the closure of LFP^s under Boolean combinations and first-order existential and universal quantification.

Theorem 22. *(1) For all $t \geq 1$, $W[t] = \bigcup_{u \geq 1} \bigcup_{s \geq 1}$ slicewise-$\Sigma_{t,u}\text{-}BOOL(LFP^s)$.*
(2) For all $t \geq 1$, $A[t] = \bigcup_{s \geq 1}$ slicewise-$\Sigma_t\text{-}BOOL(LFP^s)$.
(3) $AW[*] = \bigcup_{s \geq 1}$ slicewise-$FO(LFP^s)$.

It is not known if the analogue for the W-hierarchy of the following result holds.

Corollary 23. *Let $t \geq 1$ and $A[t] = A[t+1]$. Then for all $k \geq 1$, $A[t] = A[t+k]$.*

The following corollary contains descriptive characterizations of open complexity-theoretic problems.

Corollary 24. *(1)* $\text{FPT} = W[1] \iff \forall \tau \, \forall s \geq 1 \, \exists r \geq 1:$
$\Sigma_1\text{-BOOL}(\text{LFP}^s)[\tau^{\text{ord}}] \subseteq \text{LFP}^r[\tau^{\text{ord}}].^2$

(2) $W[t] = A[t] \iff \forall \tau \, \forall s \geq 1 \, \exists u, r \geq 1:$
$\Sigma_t\text{-BOOL}(\text{LFP}^s)[\tau^{\text{ord}}] \subseteq \Sigma_{t,u}\text{-BOOL}(\text{LFP}^r)[\tau^{\text{ord}}].$

References

1. K.A. Abrahamson, R.G. Downey, and M.R. Fellows. Fixed-parameter tractability and completeness IV: On completeness for W[P] and PSPACE analogs. *Annals of pure and applied logic*, 73:235–276, 1995.

2. M. Alekhnovich and A. Razborov. Resolution is not automatizable unless W[P] is tractable. In *Proceedings of the 41st Annual IEEE Symposium on Foundations of Computer Science*, 2001. To appear.

3. H.L. Bodlaender, R.G. Downey, M.R. Fellows, M.T. Hallett, and H.T. Wareham. Parameterized complexity analysis in computational biology. *Computer Applications in the Biosciences*, 11:49–57, 1995.

4. L. Cai, J. Chen, R.G. Downey, and M.R. Fellows. Advice classes of parameterized tractability. *Annals of pure and applied logic*, 84:119–138, 1997.

5. R.G. Downey and M.R. Fellows. Fixed-parameter tractability and completeness I: Basic results. *SIAM Journal on Computing*, 24:873–921, 1995.

6. R.G. Downey and M.R. Fellows. *Parameterized Complexity*. Springer-Verlag, 1999.

7. R.G. Downey, M.R. Fellows, and K. Regan. Descriptive complexity and the W-hierarchy. In P. Beame and S. Buss, editors, *Proof Complexity and Feasible Arithmetic*, volume 39 of *AMS-DIMACS Volume Series*, pages 119–134. AMS, 1998.

8. R.G. Downey, M.R. Fellows, and U. Taylor. The parameterized complexity of relational database queries and an improved characterization of $W[1]$. In Bridges et al., editors, *Combinatorics, Complexity, and Logic – Proceedings of DMTCS '96*, pages 194–213. Springer-Verlag, 1996.

9. H.-D. Ebbinghaus and J. Flum. *Finite Model Theory*. Springer-Verlag, 2nd edition, 1999.

10. H.-D. Ebbinghaus, J. Flum, and W. Thomas. *Mathematical Logic*. Springer-Verlag, 2nd edition, 1994.

11. R. Fagin. Generalized first–order spectra and polynomial–time recognizable sets. In R. M. Karp, editor, *Complexity of Computation, SIAM-AMS Proceedings, Vol. 7*, pages 43–73, 1974.

12. J. Flum and M. Grohe. Fixed-parameter tractability, definability, and model checking. *SIAM Journal on Computing*, 31(1):113–145, 2001.

13. J. Flum and M. Grohe. Describing parameterized complexty classes. Currently available at http://www.dcs.ed.ac.uk/home/grohe/pub.html.

14. G. Gottlob, N. Leone, and M. Sideri. Fixed-parameter complexity in AI and nonmonotonic reasoning. In M. Gelfond et al., editors, *Logic Programming and Nonmonotonic Reasoning, 5th International Conference, LPNMR'99*, volume 1730 of *Lecture Notes in Computer Science*, pages 1–18. Springer-Verlag, 1999.

15. M. Grohe. The parameterized complexity of database queries. In *Proceedings of the 20th ACM Symposium on Principles of Database Systems*, pages 82–92, 2001.

16. M. Grohe, T. Schwentick, and L. Segoufin. When is the evaluation of conjunctive queries tractable. In *Proceedings of the 33rd ACM Symposium on Theory of Computing*, pages 657–666, 2001.

---

^{2} For logics L and L', $\text{L}'[\tau^{\text{ord}}] \subseteq \text{L}[\tau^{\text{ord}}]$ means that every $\text{L}'[\tau^{\text{ord}}]$ is equivalent to an $\text{L}[\tau^{\text{ord}}]$-sentence on ordered structures.

17. N. Immerman. Relational queries computable in polynomial time. *Information and Control*, 68:86–104, 1986.
18. N. Immerman. Languages that capture complexity classes. *SIAM Journal on Computing*, 16:760–778, 1987.
19. N. Immerman. *Descriptive Complexity*. Springer-Verlag, 1999.
20. C.H. Papadimitriou. *Computational Complexity*. Addison-Wesley, 1994.
21. C.H. Papadimitriou and M. Yannakakis. On the complexity of database queries. In *Proceedings of the 17th ACM Symposium on Principles of Database Systems*, pages 12–19, 1997.
22. U. Stege. *Resolving Conflicts in Problems from Computational Biology*. PhD thesis, ETH Zuerich, 2000. PhD Thesis No.13364.
23. L.J. Stockmeyer. *The Complexity of Decision Problems in Automata Theory*. PhD thesis, Department of Electrical Engineering, MIT, 1974.
24. M.Y. Vardi. The complexity of relational query languages. In *Proceedings of the 14th ACM Symposium on Theory of Computing*, pages 137–146, 1982.

On the Computational Power of Boolean Decision Lists

Matthias Krause*

Theoretische Informatik, Univ. Mannheim, 68131 Mannheim, Germany
krause@informatik.uni-mannheim.de

Abstract. We study the computational power of decision lists over AND-functions versus threshold-$\oplus$ circuits. AND-decision lists are a natural generalization of formulas in disjunctive or conjunctive normal form. We show that, in contrast to CNF- and DNF-formulas, there are functions with small AND-decision lists which need exponential size unbounded weight threshold-$\oplus$ circuits. This implies that *Jackson's* polynomial learning algorithm for DNFs [7] which is based on the efficient simulation of DNFs by polynomial weight threshold-$\oplus$ circuits [8], cannot be applied to AND-decision lists. A further result is that for all $k \geq 1$ the complexity class defined by polynomial length AC_k^0-decision lists lies *strictly* between AC_{k+1}^0 and AC_{k+2}^0.

Keywords: Boolean Complexity Theory, Learnability, Lower Bounds

1 Introduction

A decision list L of input size n and length m for computing a function $f \in B_n$ is a sequence of m instructions of the form **if** $f^i(x) = a^i$ **then output** $f(x) = b^i$ **and stop**, followed by the single instruction **output** $f(x) = \neg b^m$ **and stop**. For $i = 1, \ldots, m$ the functions $f^i \in B_n$ are called query functions, and a^i and b^i are Boolean constants. B_n denotes the set of all Boolean functions in n variables. If the query functions are defined to belong to a function basis $S \subseteq B_n$ then L is called S-decision list.

Decision lists are a special kind of decision trees and form a basic and natural model of computation. The standard models of decision lists and decision trees, where the query functions ask for variables, have been studied in numerous papers. With the results given here we intend to initiate a more systematic study of decision lists and decision trees which are defined over different sets of query functions like AND-functions, AC_k^0-functions, MOD-functions and threshold functions. We think that this could be a promising way to extend the known reservoir of lower bound arguments, learning rules for Boolean concept classes and efficient data structures and algorithms for implementing and manipulating Boolean functions.

---

* Supported by DFG grant Kr 1521/3-2.

H. Alt and A. Ferreira (Eds.): STACS 2002, LNCS 2285, pp. 372–383, 2002.
© Springer-Verlag Berlin Heidelberg 2002

Our results concern the computational power of AND-decision lists (i.e., decision lists which query AND-functions), k-decision lists (i.e., decision lists for which all query functions depend on at most k variables, k a constant), and AC_k^0-decision lists (i.e., decision lists which query AC_k^0-functions).

The first group of results is devoted to comparing the computational power of k-decision lists and AND-decision lists with the power of threshold-$\oplus$ circuits. AND-decision lists are a natural generalization of formulas in disjunctive or conjunctive normal form (for short, DNF- and CNF-formulas) which can be considered as monotone AND-decision lists. The comparision function $COMP_n(x,y)$, which outputs 1 iff the n-bit number x is not smaller than the n-bit number y, is a natural witness for the fact that AND-decision lists are strictly more powerful than DNF- and CNF-formulas. $COMP_n$ does not belong to AC_2^0 but it has linear length AND-decision lists, it has even linear length 2-decision lists.

Threshold-$\oplus$ circuits (i.e., unbounded fan-in depth-2 circuits with $\oplus$-gates at the bottom level and a threshold output gate at the top) were extensively studied during the last decade (see, e.g., [1],[2],[3],[4],[5],[7],[8],[9]). One reason for the importance of this computational model is that representing a Boolean function f by threshold-$\oplus$ circuits is equivalent to representing f as the sign of a polynomial with integer coefficients. This allows to characterize the complexity of f with respect to threshold-$\oplus$ circuits in terms of the spectral coefficients of f [3],[4],[10]. Another nice property is that majority-$\oplus$ circuits (i.e., threshold-$\oplus$ circuits with an unweighted threshold top gate) are polynomially PAC-learnable with respect to the uniform distribution (*Jackson* 1994 [7]). As polynomial size CNF- and DNF-formulas can be efficiently simulated by polynomial size majority-$\oplus$ circuits [8], *Jackson*'s learning algorithm yields the only known polynomial learning algorithm for CNF- and DNF-formulas. Note that polynomial size majority-$\oplus$ circuits are equivalent to polynomial weight threshold-$\oplus$ circuits. Let, as usual, $\hat{PT}_1$ and PT_1 denote the complexity class containing those sequences of Boolean functions which have polynomial size majority-$\oplus$ circuits, and polynomial size threshold-$\oplus$ circuits, respectively.

The main question which stimulated this research was to find out whether even polynomial length AND-decision lists can be simulated by polynomial size majority-$\oplus$ circuits. This would yield that *Jackson*'s learning algorithm polynomially learns even functions with polynomial length AND-decision lists. One main result of this paper is a negative answer to this question. There are functions with polynomial length AND-decision lists for which even weighted threshold-$\oplus$ circuits have exponentially many nodes (Theorem 1). Theorem 1 is based on analyzing the computational power of k-decision lists versus majority-$\oplus$ circuits. The main result here is Theorem 3 saying that there are even functions with 2-decision lists which need exponential size majority-$\oplus$ circuits (i.e., exponential weight threshold-$\oplus$ circuits). As each decision list of length m can be simulated by an m-ary exponential weight threshold function over the query functions (see relation (F) in section 2), k-decision lists can always be simulated by polynomial size (but exponential weight) threshold-$\oplus$ circuits. Note that due to a result of *Bruck* in [3], the function $COMP_n$, a function with a 2-decision lists, *has* polynomial weight threshold-$\oplus$ circuits.

Theorem 3 is based on our main technical contribution, the proof that there is a Boolean function f with 1-decision list, which cannot be computed by majority-$\oplus$ circuits with subexponential size and sublinear bottom fan-in (Theorem 2).

Another part of this paper concerns the computational power of AC_k^0-decision lists. For all integers $k \geq 1$ let DL^k denote the class of all Boolean functions having polynomial length AC_k^0-decision lists. Due to property (G) in section 2 it follows that $AC_k^0 \subseteq DL^k \subseteq AC_{k+1}^0$. Our second main result is that all these inclusions are strikt (Theorem 5). The proof is done via induction on k. The induction step is based on *Håstads* Switching lemma [6]. Note that DL^2 coincides with the set of all Boolean functions having polynomial length AND-decision lists. Consequently, our result $DL^2 \not\subseteq PT_1$ strengthens the main results in [8] saying that $AC_2^0 \subseteq PT_1$ but $AC_3^0 \not\subseteq PT_1$.

The paper is organized as follows. In section 2 we introduce some more denotations and discuss some basic properties of decision lists and threshold-$\oplus$ circuits. Section 3 contains all about AND-decision lists versus threshold-$\oplus$ circuits. In section 4 we give the proof of Theorem 5.

2   Some More Basics

Let us use expressions of the kind $L = (f^1, a^1, b^1), \ldots, (f^m, a^m, b^m)$ as a compressed notation of decision lists, where for $i = 1, \ldots, m$, f^i, a^i, b^i are defined as at the beginning of section 1. Observe that the 2-decision list

$$(x_{n-1} > y_{n-1}?, 1, 1), (x_{n-1} < y_{n-1}?, 1, 0), \ldots, (x_0 > y_0?, 1, 1)(x_0 < y_0?, 1, 0),$$

computes the decision whether the n-bit number $x = (x_{n-1}, \ldots, x_0)$ is not smaller than the n-bit number $y = (y_{n-1}, \ldots, y_0)$, i.e. $COMP_n$. Observe

(A) If L computes the function f then $(\neg f^1, \neg a^1, b^1), \ldots, (\neg f^m, \neg a^m, b^m)$ also does.

(B) If L computes f then $(f^1, a^1, \neg b^1), \ldots, (f^m, a^m, \neg b^m)$ computes $\neg f$.

(C) $\bigvee_{k=1}^m f^k$ can be computed by $(f^1, 1, 1), \ldots, (f^m, 1, 1)$.

(D) $\bigwedge_{k=1}^m f^k$ can be computed by $(f^1, 0, 0), \ldots, (f^m, 0, 0)$.

(E) Each k-decision list of length m can be simulated by an AND-decision list of length at most $2^{k-1}m$.

(F) If $L = (f^1, a^1, b^1), \ldots, (f^m, a^m, b^m)$ computes f then

$$f(x) = 1 \quad \text{iff} \quad 2\sum_{i=1}^m 2^{m-i}(-1)^{b^i}(f^i(x) = a^i) + (-1)^{1-b^m} \leq 0,$$

i.e., f can be written as exponential weight threshold function over $f^1, \ldots, f^m$.

(G) If $L = (f^1, a^1, b^1), \ldots, (f^m, a^m, b^m)$ computes f then

$$f(x) = \bigvee_{i, b^i = 1} \left(\bigwedge_{j=1}^{i-1} (f^j(x) \neq a^j) \wedge (f^i(x) = a^i) \right) \vee (\neg b_m) \bigwedge_{j=1}^{m} (f^j(x) \neq a^j)$$

$$= \bigwedge_{i, b^i = 0} \left(\bigvee_{j=1}^{i-1} (f^j(x) = a^j) \vee (f^i(x) \neq a^i) \right) \wedge \left(\bigvee_{j=1}^{m} (f^j = a^j) \vee \neg b^m \right).$$

Properties (A)–(D) follow directly from the definitions and the *DeMorgan*-laws. Property (E) can be obtained by replacing all query functions f^k by a monotone sublist simulating a DNF-formula for f^k according to (C). The coefficients of the threshold representation in (F) are constructed in such a way that the computational mode of a decision list is simulated. In particular the sign is $(-1)^{b^i}$ for the smallest i for which $f^i(x) = a^i$, or $(-1)^{1-b^m}$, if $f^i(x) \neq a^i$ for all i. The formulas in (G) reflect the fact that L accept x iff x reaches an accepting sink iff there is an i, $1 \leq i \leq m$ with $b^i = 1$ such that $f^i(x) = a^i$ and $f^j(x) \neq a^j$ for all $1 \leq j < i$, or if $f^j(x) \neq a^j$ for all $1 \leq j \leq m$ and $b^m = 0$. Conversely, L accept x iff x does not reach any rejecting sink iff for all i with $b^i = 0$ it holds that $f^i(x) \neq a^i$ or there is an $j < i$ such that $f^j(x) = a^j$, and $b_m = 0$ or there is an $j \leq m$ with $f^j(x) = a^j$.

Let us denote for all constants k by $WIDTH^k DL$ the set of all functions having k-decision lists. As for each decision list it makes no sense to pose a query twice, k-decision lists can always be supposed to have polynomial length.

Typical representatives of these classes are functions $d_{k,n} : \{0,1\}^{kn} \longrightarrow \{0,1\}$ defined by $d_{k,n}(x_{1,1}, \ldots, x_{1,k}, \ldots, x_{n,1}, \ldots, x_{n,k}) = 1$ iff the maximal i for which $\bigwedge_{j=1}^{k} x_{i,j} = 1$ is odd. An AND-decision list for $d_{k,n}$, n odd, is

$$\left((\bigwedge_{i=1}^{k} x_{i,n}, 1, 1), (\bigwedge_{i=1}^{k} x_{i,n-1}, 1, 0), \ldots, (\bigwedge_{i=1}^{k} x_{i,2}, 1, 0), (\bigwedge_{i=1}^{k} x_{i,1}, 1, 1) \right)$$

Note that $(d_{k,n})_{n \in \mathbb{N}}$ is $WIDTH^k DL$-complete and that $(d_{n,n})_{n \in \mathbb{N}}$ is DL^2-complete with respect to projections.

A Boolean function $f \in B_n$ is called a threshold function if it can be computed by a single threshold gate, i.e., if there are integers $a_0, \ldots, a_n$ such that $f(x) = 1$ iff $a_1 x_1 + \ldots + a_n x_n \geq a_0$.

The numbers $a_0, \ldots, a_n$ are called the edge weights of the gate and the value $|a_0| + \ldots + |a_n|$ is called the weight gate. The systematic study of threshold functions startet with the work on *perceptrons* by *Minsky* and *Papert* [11]. It was shown there, for instance, that all n-ary threshold functions can be written down as threshold functions of weight $\exp(n \log n)$. There are threshold functions which cannot be written as polynomial weight threshold functions, see, for instance, $COMP_n$. A threshold gate is called majority-gate if all edge weights are from $\{0, 1, -1\}$. Note that polynomial size majority-$\oplus$ circuits are equivalent to polynomial weight threshold-$\oplus$ circuits. It is known that unbounded weight polynomial size threshold-$\oplus$ circuits can do more than polynomial weight threshold-$\oplus$ circuits, i.e. $\hat{PT}_1 \subset PT_1$, [5].

Computing a Boolean function $f \in B_n$ by a threshold-$\oplus$ circuit containing s $\oplus$-gates of fan-in at most d and one threshold top gate of weight w is the same as representing f as the signum of a *voting polynomial*, i.e., a degree-d integer polynomial in n variables with s monomials, where the sum of the absolute values of the coefficients is w. As voting polynomials allow a more elegant presentation of our proof techniques we will use them in the following.

For $b \in \{0,1\}$ we denote $b^* = (-1)^b$. For $x = (x_1, \ldots, x_n) \in \{0,1\}^n$ we denote by $x^* \in \{1,-1\}^n$ the vector $x^* = (x_1^*, \ldots, x_n^*)$, and for a function $f : X \longrightarrow \{0,1\}$ by $f^* : X \longrightarrow \{1,-1\}$ the function defined by $f^*(x) = (f(x))^*$. For a nonzero integer z we denote by $sgn(z) \in \{1,-1\}$ the signum of z which is -1 if z is negative and 1 if positive. Observe that a Boolean function $f \in B_n$ is a treshold function iff there are integers $b_0, \ldots, b_n$ such that $f^*(x) = sgn(b_0 + b_1 x_1 + \ldots b_n x_n)$ iff there are integers $c_0, \ldots, c_n$ such that $f^*(x) = sgn(c_0 + c_1 x_1^* + \ldots c_n x_n^*)$. For each subset $I \subseteq [n]$ we denote by m_I the monomial $m_I = \prod_{i \in I} x_i$, where $m_\emptyset$ correspond to the constant-1 function.

We call p a $\{0,1\}$-voting polynomial for f if for all inputs $x \in \{0,1\}^n$ it holds $f^*(x) = sgn(p(x))$ and a $\{1,-1\}$-voting polynomial for f if for all $x \in \{0,1\}^n$ $f^*(x) = sgn(p(x^*))$.

As the monomial $2m_I - 1$ is a $\{0,1\}$-voting polynomial for $\bigwedge_I$, and as m_I is a $\{1,-1\}$-voting polynomial for $\bigoplus_I$, $\{0,1\}$-voting polynomials correspond to threshold-AND circuits and $\{1,-1\}$-voting polynomials correspond to threshold-$\oplus$ circuits in a straightforward way. The relevant cost measures of integer polynomials $p = \sum_{I \subseteq [n]} a_I m_I$ are the degree (the maximal length of a monomial occuring in p), the weight (the sum of the absolute values of the coefficients), and the length (the number of monomials occuring in p).

We get the complexity measures $\deg(f)$, $weight_\wedge(f)$, $weight_\oplus(f)$, $length_\wedge(f)$, $length_\oplus(f)$ for Boolean functions f, defined as the minimal degree, the minimal weight and the minimal length of a $\{0,1\}$-voting polynomial or a $\{1,-1\}$-voting polynomial for f, respectively.

Moreover we need to consider the measures $weight_\wedge^d(f)$ and $weight_\oplus^d(f)$ defined as the minimal weight of a degree-d $\{0,1\}$-voting polynomial, or a degree-d $\{1,-1\}$-voting polynomial for f, respectively.

The following relations between $weight_\wedge^d(f)$ and $weight_\oplus^d(f)$ can be proved quite straightforwardly by replacing each $\oplus$-gate (resp. $\wedge$-gate) by an integer polynomial which *exactly* simulates the gate over $\{0,1\}^n$ (resp. over $\{1,-1\}^n$).

Lemma 1. *For all n, $f \in B_n$ and $d < n$ it holds that $weight_\oplus^d(f) \leq 2^d weight_\wedge^d(f)$ and $weight_\wedge^d(f) \leq 3^d weight_\oplus^d(f)$.* $\square$

Note that due to (F) $d_{1,n}$ is a threshold function for all n. The following technical result on writing $d_{1,n}$ as a threshold function can be verified quite straightforwardly.

Lemma 2. $d_{1,n}^* = sgn\left(1 + b\sum_{i=1}^n (-a)^i x_i\right)$ *for all natural $a \geq 2$ and $b \geq 1$.* $\square$

It follows that for all $f \in WIDTH^k DL$ it holds $\deg(f) \leq k$, and thus $WIDTH^k DL \subseteq PT_1$.

3   On AND-Decision Lists versus Threshold-$\oplus$ Circuits

This section is devoted to the proof of

Theorem 1. *It holds $length_{\oplus}(d_{n,n}) \in 2^{n^{\Omega(1)}}$, i.e. $DL^2 \not\subseteq PT_1$.*

The proof of Theorem 1 is performed in two stages. The first and most complicated step is to show that for sublinear d it holds that $weight_{\oplus}^d(d_{1,n})$ is exponential (Theorem 2). The second step is to conclude Theorem 1 from Theorem 2. As a byproduct we obtain that AC_3^0 and $WIDTH^3 DL$ contain *heavy* threshold functions, i.e. threshold functions which do not belong to $\hat{PT_1}$. The first example of a heavy threshold functions (which is not an AC^0-function) was given in [5].

Using a spectral theoretic upper bound technique from [4] one can show that $d_{1,n} \in PT_1$, i.e. $d_{1,n}$ has polynomial weight $\{1, -1\}$-voting polynomials. On the other hand, $\deg(d_{1,n}) = 1$. The following result says that it is impossible to achieve polynomial weight and small degree simultaneously.

Theorem 2. *For all $d < n$ it holds that $weight_{\wedge}^d(d_{1,n}) \geq 2^{\lfloor \frac{n}{8d} \rfloor}$ and $weight_{\oplus}^d(d_{1,n}) \geq 2^{\lfloor \frac{n}{8d} \rfloor - \log_2 3d}$.*

The second relation follows from the first one and Lemma 1.

Corollary 1. *For $d \leq \frac{\sqrt{n}}{10}$ it holds that $weight_{\oplus}^d(d_{1,n}) \geq 2^{\sqrt{n}}$.*

Proof. It is sufficient to show that for $d \leq \frac{\sqrt{n}}{10}$ it holds $\frac{n}{8d} - 1 - \log_2 3d \geq \sqrt{n}$, which can be done in a straightforward way. $\square$

The proof of Theorem 2 will be given below. Let us first prove Theorem 1 using Theorem 2. This will be done in several steps.

Theorem 3. *$weight_{\oplus}(d_{2,n}) \in 2^{n^{\Omega(1)}}$, i.e., $WIDTH^2 DL \not\subseteq \hat{PT_1}$.*

Proof. We use the following construction which was introduced in [8]. For all $f \in B_n$ let the function $f^{op} \in B_{3n}$ be defined by

$$f^{op}(x_1, \ldots, x_n, y_1, \ldots, y_n, z_1, \ldots, z_n) = f(u_1, \ldots, u_n),$$

where for all i, $1 \leq i \leq n$, $u_i = \bar{z}_i x_i \vee z_i y_i$. Theorem 3 can be derived from the following technical lemma which is proved with the same probabilistic argument as Proposition 2.1 in [8].

Lemma 3. *For all $f \in B_n$ it holds that $weight_{\oplus}(f^{op}) \geq weight_{\oplus}^d(f)$ for all d with $2^d \geq weight_{\oplus}(f^{op})$.*

Proof of Theorem 3 using Lemma 3. Suppose that $weight_\oplus(d_{1,n}^{op}) < 2^{\lfloor \frac{\sqrt{n}}{10} \rfloor}$. Then, for $d = \lfloor \frac{\sqrt{n}}{10} \rfloor$ we have by Lemma 3 and Corollary 1 that $weight_\oplus(d_{1,n}^{op}) \geq weight_\oplus^d(d_{1,n}^{op}) \geq 2^{\sqrt{n}}$ which gives a contradiction. $\square$

Proof of Lemma 3. Fix an arbitrary $f \in B_n$, a $\{1,-1\}$-voting polynomial $p = \sum_{I \subseteq X \cup Y \cup Z} a_I m_I = \sum_{I \in M} a_I m_I$ of weight W for f^{op}, and let $d' = d+1$, where d denotes the minimal natural number for which $2^d > W$. (M denotes the set of all I for which $a_I \neq 0$. Observe that $|M| \leq W$.) Let f^{op} depend on the sets of variables $X = \{x_1, \ldots, x_n\}$, $Y = \{y_1, \ldots, y_n\}$ and $Z = \{z_1, \ldots, z_n\}$.

Now observe that each assignment c of the Z-variables defines a partition $X \cup Y = U(c) \cup V(c)$, $|U(c)| = |V(c)| = n$, such that $(f^{op})^c$ depends only on the variables of $U(c)$. Consequently, for all assignments $u \in \{0,1\}^{U(c)}$ it holds

$$f^*(u) = \mathbf{E}_{v \in_U \{0,1\}^{V(c)}}[(f^{op*})^c(u,v)] = \mathbf{E}_{v \in_U \{0,1\}^{V(c)}}\left[\sum_{I \in M} a_I m_I^c(u^*, v^*)\right]$$

$$= \sum_{I \in M} a_I \mathbf{E}_{v_U \in \{0,1\}^{V(c)}}[m_I^c(u^*, v^*)].$$

$\mathbf{E}_{v \in_U \{0,1\}^{V(c)}}$ denotes the expected value with respect to the uniform distribution on $\{0,1\}^{V(c)}$.

Note that $\mathbf{E}_{v \in_U \{0,1\}^{V(c)}}[m_I^c(u^*, v^*)] = 0$ if $I \cap V(c) \neq \emptyset$. Corresponding to this fact we say that c destroys a monomial m_I of p if $I \cap V(c) \neq \emptyset$. Each $c \in \{0,1\}^Z$ defines a $\{1,-1\}$-voting polynomial $p|_c$ of weight $\leq W$ for f by removing all those monomials which are destroyed by c.

Let us call a monomial m_I large if I contains at least d' variables from $X \cup Y$. Clearly, p contains at most W large monomials. The probability (with respect to $c \in_U \{0,1\}^Z$) that a large monomial is not destroyed by c is at most $2^{-d'}$. Consequently, the probability that there is an assignment $c \in \{0,1\}^Z$ that destroys all large gates is at least $1 - W2^{-d'}$ which is positive because of the choice of d'. The resulting $\{1,-1\}$-voting polynomial for f contains only monomials of length at most d, we get that $weight_\oplus^d(f) \leq W$. $\square$

Theorem 4. $WIDTH^3 DL$ contains a heavy threshold function, i.e., a threshold functions t_n with $weight_\oplus^*(t_n) \in 2^{n^{\Omega(1)}}$.

Proof. We have seen in Theorem 3 that $weight_\oplus^*(d_{2,n}) \in 2^{n^{\Omega(1)}}$. Observe that for all $a \geq 2$, by Lemma 2,

$$d_{2,n}^* = sgn\left(1 + 2\sum_{i=1}^n (-a)^i x_{i,1} x_{i,2}\right)$$

$$= sgn\left(1 + 2\sum_{i=1}^n (-a)^i \frac{1}{2}(x_{i,1} + x_{i,2} - (x_{i,1} \oplus x_{i,2}))\right).$$

I.e., $T_n^*(x, y, z) = sgn\left(1 + \sum_{i=1}^n (-a)^i (x_i + y_i - z_i)\right)$ defines a threshold function $T_n \in B_{3n}$ with $weight_{\oplus}^*(T_n) \in 2^{n^{\Omega(1)}}$. It is not hard to show that $T_n \in WIDTH^3 DL$. $\square$

Proof of Theorem 1 using Theorem 4. The proof uses the main result from [9] saying that for all threshold functions $t_n \in B_n$ it holds that if $weight_{\oplus}(t_n) \in 2^{n^{\Omega(1)}}$ then $length_{\oplus}(t_n \circ AND_{n,n}) \in 2^{n^{\Omega(1)}}$, where $t_n \circ AND_{n,m} \in B_{nm}$ is defined as

$$f_n \circ AND_{n,m}(x_{1,1}, \ldots, x_{1,m}, \ldots, x_{n,1}, \ldots, x_{n,m}) = f_n\left(\bigwedge_{i=1}^m x_{1,i}, \ldots, \bigwedge_{i=1}^m x_{n,i}\right).$$

As a consequence we obtain that $length_{\oplus}(T_n \circ AND_{n,n}) \in 2^{n^{\Omega(1)}}$, where T_n denotes the heavy threshold function considered in Lemma 4. As obviously $T_n \circ AND_{n,n} \in DL^2$ the Theorem follows. $\square$

Proof of Theorem 2. We have to show that for all $d \geq 10$ and $n > 8d$ it holds $weight_{\wedge}^d(d_{1,n}) \geq 2^{\lfloor \frac{n}{8d} \rfloor}$.

We will sometimes identify subsets $I \subseteq [n]$ with its characteristic vector in $\{0,1\}^n$ which is defined to have ones at exactly those positions i for which $i \in I$. For $I \subseteq [n]$ we denote $\max(I) = \max\{i, i \in I\}$.

Observe at first that $1 + \sum_{i=1}^n (-1)^i 2^i x_i$ is a $\{0,1\}$-voting polynomial for $d_{1,n}$.

Let us fix some d, n as above. For all $I \subseteq [n]$ let $Dom(I)$ denote the dominator set of I which is defined as

$$Dom(I) = \{\max(I) + 1, \max(I) + 3, \max(I) + 5, \ldots, N\},$$

where $N = n$ if $\max(I) + 1$ and n are both even or both odd, and $N = n - 1$ otherwise. Further let for $k \leq \lceil \frac{n - \max(I)}{2} \rceil$

$$Dom_k(I) = \{\max(I) + 1, \max(I) + 3, \ldots, \max(I) + 2k - 1\}.$$

Observe that for all $I \subseteq [n]$, $\max(I) < n$, all $I' \subseteq I$ and all nonempty $J \subseteq Dom(I)$ it holds that

$$d_{1,n}^*(I) = -d_{1,n}^*(I' \cup J). \tag{1}$$

Now let us fix a $\{0,1\}$-voting polynomial $p = \sum_{I \subseteq [n], |I| \leq d} a_I m_I$ for $d_{1,n}$ of degree d. We will prove the following

Claim 1. Let $I \subseteq [n]$ be fixed with $\max(I) \leq n - 8d$. Then there exists some $J \subseteq I \cup Dom_{4d}(I)$ with $|p(J)| \geq 2|p(I)|$.

Proof of Theorem 2 with Claim 1. Let $I_0 = \emptyset$ and observe that $p(I_0) = a_{\emptyset} \geq 1$. By Claim 1 we get a sequence of subsets $I_0, I_1, \ldots, I_k, \ldots$ where $I_k \subseteq [8dk]$ such that $|p(I_k)| \geq 2|p(I_{k-1})|$ for all $k, 1 \leq k \leq \lfloor \frac{n}{10d} \rfloor$. As $weight(p) \geq |p(I)|$ for all $I \subseteq [n]$ we get $weight(p) \geq 2^{\lfloor \frac{n}{8d} \rfloor}$ which proves the theorem.

Proof of Claim 1. We have to introduce some denotations. For all $J \subseteq Dom(I)$ let

$$l(J) = \sum_{I' \subseteq I} a_{I' \cup J}$$

denote the link-weight induced by J relative to I. For all j, $1 \leq j \leq \min\{|J|, d\}$, let

$$L^j(J) = \sum_{J' \subseteq J, |J'| = j} l(J')$$

and

$$L^{>j}(J) = \sum_{r=j+1}^{\min\{|J|,d\}} L^r(J).$$

Now observe that for all $J \subseteq Dom(I)$

$$p(I \cup J) = p(I) + L^1(J) + L^{>1}(J). \tag{2}$$

Further observe that for all $j \in Dom(I)$ it holds that $p(I)$ and $p(I \cup \{j\})$ have to have different sign. This implies by (2) that $|l(\{j\})| \geq |p(I)| + 1$ and, consequently,

$$|L^1(J)| \geq |J|(|p(I)| + 1) \tag{3}$$

for all $J \subseteq Dom(I)$.

Let us now fix a number $k > d$ in such a way that

$$|p(I \cup J)| < 2|p(I)| \tag{4}$$

for all $J \subseteq Dom_k(I)$. We will show a small upper bound for k. At first observe that $l(\{j\}) < 3|p(I)|$, otherwise $J = \{j\}$ would contradict assumption (4). We obtain that

$$|L^1(J)| < 3|J||p(I)| \tag{5}$$

for all $J \subseteq Dom_k(I)$. Next observe that, by (3), if $|J| \geq 3$ then $|p(I) + L^1(J)| > 2|p(I)|$, i.e., $L^1(J)$ and $L^{>1}(J)$ have to have different sign for all $J \subseteq Dom_k(I)$, $|J| \geq 3$, and it holds

$$|L^{>1}(J)| > |J|(|p(I)| + 1) - 3|p(I)| > (|J| - 3)|p(I)|. \tag{6}$$

On the other hand, as $L^{>1}(J)$ and $p(I)$ have the same sign, which equals $-d^*_{1,n}(I \cup J)$, we get, using (5), that

$$|L^{>1}(J)| < |L^1(J)| - |p(I)| < (3|J| - 1)|p(I)| < 3|J||p(I)| \tag{7}$$

for all $J \subseteq Dom_k(I)$, $|J| \geq 3$. Otherwise, p would produce the wrong signum on $I \cup J$. Summarizing all these observations we obtain that

$$\binom{k}{d}(d-3)|p(I)| < \sum_{J \subseteq Dom_k(I), |J|=d} |L^{>1}(J)| = \left| \sum_{J \subseteq Dom_k(I), |J|=d} L^{>1}(J) \right|.$$

This is due to (6) and as all $L^{>1}(J)$ have the same sign. I.e., due to (7)) Now observe that

$$\left| \sum_{J \subseteq Dom_k(I), |J|=d} L^{>1}(J) \right| = \left| \sum_{i=2}^{d} \binom{k-i}{d-i} L^i(Dom_k(I)) \right|$$

$$\leq \binom{k-2}{d-2} \left| \sum_{i=2}^{d} L^i(Dom_k(I)) \right| = \binom{k-2}{d-2} |L^{>1}(Dom_k(I))| < \binom{k-2}{d-2} 3k|p(I)|.$$

We obtain the relation $\binom{k}{d}(d-3) < \binom{k-2}{d-2}3k$ which is equivalent to $(k-1)(d-3) < 3d(d-1)$, and $k \leq \frac{3d(d-1)}{d-3}$. It can be easily verified that this relation does not hold for all $d \geq 10$ and $k \geq 4d$. I.e., if $d \geq 10$, $k \geq 4d$ and p is correct then there is some $J \subseteq Dom_k(I)$ fulfilling $|p(I \cup J)| \geq 2|p(I)|$. $\square$

4   The Decision List Hierarchy and the AC^0 Hierarchy

In this section we prove

Theorem 5. *For all $k \geq 1$ it holds $AC_k^0 \subset DL^k \subset AC_{k+1}^0$.*

Proof. The proof is done via induction on k. The induction step makes use of an argument from Theorem 5 (p.17) of Håstads Switching Lemma paper [6]. At page 15 (figure 7) in [6] there is defined a Boolean function g_m^k. We define a function h_m^k in a slightly modified way by giving the definition of a circuit H_m^k computing h_m^k. For $k \geq 3$ let H_m^k consists of $k-1$ layers consisting of gates of fan-in $4.4m^2, 4.4m^3, \ldots, 4.4m^{k-2}, 1.1m^{k-1}, m^{k-1}$. The top gate is a gate computing $d_{1,4.4m^2}$ (resp. $d_{1,1.1m^2}$ for $k = 3$), followed by a level of AND-gates, followed by a level of OR-gates, followed by a level of AND-gates and so on. The bottom gates are ANDs, resp. ORs, over pairwise disjoint blocks $B_j^{k,m}$, $j = 1, 2, \ldots$, containing m^{k-1} variables each. Observe that h_m^k belongs to DL^{k-1}. We show that $h_m^k \notin AC_0^{k-1}$ by the following modification of Theorem 5 in [6].

Lemma 4. *There are no depth k AND, OR circuits computing h_m^k with bottom fan-in $\leq \frac{1}{10}m$ and size $\leq 2^{\frac{1}{10}m}$ for $m > m_0$ some absolute constant.*

This lemma is proved by induction on k. It is quite straightforward to check that the induction step can be performed in a similar way as in the proof of Theorem 5 in [6]. In particular, consider the probability distribution $R^k(m)$ on the set of partial assignments to the variables of h_m^k. For each block $B_j^{k,m}$ of the input variables of h_m^k do independently the following experiment.

With probability $\frac{1}{m^{k-1}}$ set all $B_j^{k,m}$ variables to one. With probability $1 - \frac{1}{m^{k-1}}$ choose a random $x \in B_j^{k,m}$, set all $x' \in B_j^{k,m} \setminus \{x\}$ to 1, and with probability $\frac{1}{m}$ set x to * and with probability $1 - \frac{1}{m} - \frac{1}{m^{k-1}}$ set x to 0.

The induction step follows from the two observations that

(1) Applying a random restriction, distributed according to $R^k(m)$, to an AND,OR circuit of depth $k \geq 3$, size $2^{\frac{m}{10}}$, and bottom fan-in $\frac{m}{10}$ computing h_m^k leads to a circuit of depth $k-1$ and the same bottom fan-in with probability $1 - 2^{\frac{1}{10}m} \alpha^{\frac{1}{10}m}$, where $\alpha \leq 0.42$ for m large enough (Lemma 5 in [6]).

(2) Applying a random restriction, distributed according to $R^k(m)$, to h_m^k, $k \geq 4$, yields a function which contains h_m^{k-1} as a subfunction with probability at least $\frac{2}{3}$. This can be proved in the same manner as Lemma 8 in [6].

Consequently, for m large enough and $k \geq 4$ the existence of an AND,OR circuit of depth $k \geq 3$, size $2^{\frac{m}{10}}$ and bottom fan-in $\frac{m}{10}$, computing h_m^k implies the existence of an AND,OR circuit of depth $k-1$ of at most the same size and bottom fan-in for h_m^{k-1}. We have to show the base case $k = 3$. Observe that we can write

$$h_m^3(B_1, \ldots, B_{1.1m^2}) = d_{1,1.1m^2}\left(\bigwedge_{x_i \in B_1} x_i, \ldots, \bigwedge_{x_i \in B_{1.1m^2}} x_i\right).$$

Let us assume that there is a depth-3 AND,OR circuit C of size $\leq 2^{\frac{1}{10}m}$ and bottom fan-in $\leq \frac{1}{10}m$ computing h_m^3. Observe that, under $R^3(m)$, with probability $(1 - m^{-2})^{1.1m^2} \approx e^{-1.1} > 0.3$ no bottom AND of H_m^3 is set to constant 1 and that the expected number of bottom ANDs which are set to * is $1.1m$. Consequently, if m is large enough then with probability at least 0.3 at least m bottom ANDs are set to * and no bottom AND is set to 1. Thus, if m is large enough our assumption implies that there is a depth two AND,OR circuit of bottom fan-in $\frac{1}{10}m$ computing $d_{1,m}$. This would imply that either $d_{1,m}^{-1}(1)$ or $d_{1,m}^{-1}(0)$ can be covered by prime implicants of length at most $\frac{1}{10}m$. But this can be disproved by observing that

$$d_{1,m} = \bigwedge_{r \leq m \text{ odd}} \bar{x}_1 \ldots \bar{x}_{r-1} x_r, \text{ resp. } \neg d_{1,m} = \bigwedge_{r \leq m \text{ even}} \bar{x}_1 \ldots \bar{x}_{r-1} x_r$$

are the only possibilities to cover $d_{1,m}^{-1}(1)$ (resp. $d_{1,m}^{-1}(0)$) by prime implicants.

For showing that $DL^k \subset AC_0^{k+1}$ we prove a more general result. Let us define a $Th(k)$-circuit to be a circuit consisting of an unbounded weight threshold gate on the top followed by $k-1$ levels of AND, OR gates. As $d_{1,n}$ is a threshold function it holds that for all $k \geq 2$ all problems in DL^k have polynomial size $Th(k)$-circuits. Let us consider a function f_m^k which is defined via a circuit F_n as follows. F_n is a tree consisting of k levels, which contain alternatingly AND and OR gates. For $k > 2$ the top gate is an AND gate of fan-in $4.4m^4$ followed by levels of gates of fan-in $4.4m^5, \ldots, 4.4m^{k+1}, 1.1m^{k+2}, m^{k+2}$. The gates at the bottom level are defined over pairwise disjoint blocks of m^{k+2} variables each. The function f_m^2 is an AND over $1.1m^4$ OR-gates of fan-in m^4. We complete the proof of the Theorem by proving the following

Lemma 5. *There are no $Th(k)$-circuit computing f_m^k with bottom fan-in $\leq \frac{1}{10}m$ and size $\leq 2^{\frac{1}{10}m}$ for $m > m_0$ some absolute constant.*

This lemma is proved by induction on k again. The induction step can be performed exactly in the same way as described above, resp. as in [6]. We consider the base case $k = 2$ and suppose that there is a Th(3)-circuit $C = t(C_1, \ldots, C_r)$ of size $\leq 2^{\frac{1}{10}m}$ and bottom fan-in $\leq \frac{1}{10}m$ computing f_m^3. Here, t denotes a threshold function corresponding to the top gate of C. By lemma 5 from [6], a random restriction from $R^5(m)$ with probability $1 - 2^{\frac{1}{10}m} \alpha^{\frac{1}{10}m} \geq 1 - 0.84^{m/10}$ converts all functions C_i, $i = 1, \ldots, r$, into functions for which the function itself and its negation have prime implicants of size at most $\frac{1}{10}m$. This implies that all these restricted functions have decision trees of depth $(\frac{1}{10}m)^2$ (see, e.g., [12], Theorem 2.5.11) and, thus, can be written as rational polynomials of degree at most $(\frac{1}{10}m)^2$. Consequently, with probability at least $1 - 0.84^{m/10}$ it holds that $\deg(C|_\rho) \leq (1/100)m^2$, where ρ is $R^5(m)$-distributed. On the other hand, $f_m^3|_\rho$ has f_m^2 as a subfunction with probability at least $\frac{2}{3}$. This contradicts a result of [11] saying that $\deg\left(\bigwedge_{i=1}^{n} \bigvee_{j=1}^{4n^2} x_{i,j}\right) = n$, i.e., $\deg(f_m^2) \geq \frac{1}{2}m^2$. $\qquad\square$

References

1. N. Alon, J. Bruck. Explicite constructions of depth two majority circuits for comparison and addition. Techn. Report RJ 8300 (75661) IBM San Jose, 1991.
2. J. Aspnes, R. Beigel, M. Furst, S. Rudich. The expressive power of Voting Polynomials. Proc. STOC'91, 402-409.
3. J. Bruck. Harmonic Analysis of polynomial threshold functions. SIAM Journal of Discrete Mathematics. 3:22, 1990, pp. 168-177.
4. J. Bruck, R. Smolensky. Polynomial Threshold functions, AC^0-functions, and spectral norms. Proc. of FOCS'90, 632-641.
5. M. Goldmann, J. Håstad, A. A. Razborov. Majority gates versus general weighted Threshold gates. J. Computational Complexity 2, 1992, 277-300.
6. J. Håstad. Almost optimal lower bounds for small depth circuits. STOC'86, pp. 6-20.
7. J. Jackson. An efficient membership-query algorithm for learning DNF with respect to the uniform distribution, FOCS'94, pp. 42-53.
8. M. Krause, P. Pudlak. On the computational power of depth-2 circuits with threshold and modulo gates. J. Theoretical Computer Science 174, 1997, pp. 137-156. Prel. version in STOC'94, pp. 49-59.
9. M. Krause, P. Pudlak. Computing Boolean functions by polynomials and threshold circuits. J. Comput. complex. 7 (1998), pp. 346-370. Prel. version in FOCS'95, pp. 682-691.
10. N. Linial, Y. Mansour, N. Nisan. Constant depth circuits, Fourier transform, and learnability. J. of the ACM, vol. 40(3), 1993, pp. 607-620. Prel. version in FOCS'89, pp. 574-579.
11. M. Minsky, S. Papert. Perceptrons. MIT Press, Cambridge 1988 (expanded edition, first edition 1968).
12. I. Wegener. Branching Programs and Binary Decision Diagrams. SIAM Monographs on Discrete Mathematics and Applications. Philadelphia 2000.

How Many Missing Answers Can Be Tolerated by Query Learners?[*]

Hans Ulrich Simon

Fakultät für Mathematik, Ruhr-Universität Bochum, D-44780 Bochum, Germany
simon@lmi.ruhr-uni-bochum.de

Abstract. We consider the model of exact learning using an equivalence oracle and an incomplete membership oracle. In this model, a random subset of the learners membership queries is left unanswered. Our results are as follows. First, we analyze the obvious method for coping with missing answers: search exhaustively through all possible "answer patterns" associated with the unanswered queries. Thereafter, we present two specific concept classes that are efficiently learnable using an equivalence oracle and a (completely reliable) membership oracle, but are provably not polynomially learnable if the membership oracle becomes slightly incomplete. The first class will demonstrate that the aforementioned method of exhaustively searching through all possible answer patterns cannot be substantially improved in general (despite its apparent simplicity). The second class will demonstrate that the incomplete membership oracle can be rendered useless even if it leaves only a fraction $1/\text{poly}(n)$ of all queries unanswered. Finally, we present a learning algorithm for monotone DNF formulas that can cope with a relatively large fraction of missing answers (more than sixty percent), but is as efficient (in terms of run-time and number of queries) as the classical algorithm whose questions are always answered reliably.

1 Introduction

Certain classes of concepts, such as monotone DNF formulas and deterministic finite automata, have been shown to be polynomially learnable with equivalence and membership queries [1,9,2], but they are provably not polynomially learnable with equivalence queries alone [3]. Algorithms that rely upon a membership oracle are often of little pactical value because their questions cannot be answered reliably (even by human experts).[1] It seems natural to ask how many missing answers to membership queries can be tolerated by a (properly designed) learning algorithm. Angluin and Slonim [4] introduced the so-called "incomplete

---

[*] This work has been supported in part by the ESPRIT Working Group in Neural and Computational Learning II, NeuroCOLT2, No. 27150. The author was also supported by the Deutsche Forschungsgemeinschaft Grant SI 498/4-1.
[1] The analogous problem for the equivalence oracle can be circumvented by applying standard transformations in either the mistake bound or the pac-learning model [7, 2] (augmented by incomplete membership queries).

H. Alt and A. Ferreira (Eds.): STACS 2002, LNCS 2285, pp. 384–395, 2002.
© Springer-Verlag Berlin Heidelberg 2002

membership oracle", which behaves as follows. First, it is "persistent" in the sense that a question that is posed many times will always be answered (or left unanswered) in the same way. Second, it is "p-fallible" (for some fixed $0 \leq p \leq 1$) in the sense that a new question is left unanswered with probability p. Note that this model allows a smooth transition from learning with equivalence and membership queries (the case $p = 0$) to learning with equivalence queries alone (the case $p = 1$). A class being polynomially learnable in the former model but not in the latter must exhibit a "phase transition" at some "critical value" of p. For classes that are learnable in a robust sense, this phase transition will not happen before p approaches 1.

Example 1. Consider the class of monotone DNF formulas with n Boolean variables and m terms. There is a classical algorithm [2] that learns this class with n equivalence queries and at most mn membership queries. Angluin and Slonim [4] presented an algorithm that learns this class using an equivalence oracle and a p-fallible membership oracle. The expected number of queries and the run-time of their algorithm is polynomial in n and m for each constant $0 < p < 1$ (but superpolynomial in $1/(1-p)$). If $p = 1/2$, for instance, then the expected number of queries is $O(mn^2)$. Bshouty and Eiron [5] presented another learning algorithm for monotone DNF formulas whose expected total number of queries is $O(m^2 n^2/(1-p)^2)$. Thus, their algorithm remains polynomial if p is a function of the form $1 - 1/\text{poly}(n, m)$. Note that a $(1 - 1/\text{poly}(n, m))$-fallible membership oracle may leave almost all queries unanswered (except for a polynomial fraction). Since membership oracles that answer only a superpolynomially small fraction of the queries are easily shown to be useless for polynomial learners, it follows that the class of monotone DNF formulas exhibits the largest possible robustness against incomplete membership oracles.[2]

In this paper, we explore to which extent query learners can be made robust against missing answers. After the formal definition of the learning model in Section 2, we present three main results:
In Section 3, we describe and analyze a general transformation of a given polynomial learner (expecting a completely reliable membership oracle) into a new polynomial learner that can cope with an expected number of $O(\log n)$ missing answers. The new learner performs an exhaustive search through all possible answer patterns associated with unanswered membership queries. The analysis has to deal with the issue of exhaustively searching through a space of random (and exponentially fast growing) size.
In Section 4, we present two specific concept classes that are efficiently learnable using an equivalence oracle and a (completely reliable) membership oracle, but are provably not polynomially learnable if the membership oracle becomes slightly incomplete. The first class will demonstrate that the aforementioned method of exhaustively searching through all possible answer patterns cannot be substantially improved in general (despite its apparent simplicity). The second

---

[2] Bshouty and Owshanko [6] have shown a similar result for the class of deterministic finite automata.

class will demonstrate that the incomplete membership oracle can be rendered useless even if it leaves only a fraction $1/\text{poly}(n)$ of all queries unanswered. The derivation of the lower bounds on the query complexity of these classes must cope with some subtle issues because the learner can extract random bits from the answers of the incomplete membership oracle. For randomized learners, the standard adversary arguments do not readily apply.

In Section 5, we present a learning algorithm for monotone DNF formulas that can cope with a relatively large fraction of missing answers (more than sixty percent), but is as efficient as the classical algorithm whose questions are always answered reliably. In particular, the expected number of queries issued by this algorithm is $O(mn)$ for an arbitrary constant $p < (\sqrt{5} - 1)/2 \approx 0.62$. Our algorithm results from a slight modification of the original algorithm of Angluin and Slonim [4]. In order to bound the expected number of queries by $O(mn)$, we have to apply a considerably refined analysis. It uses a method of "stripping-off" dependencies between random variables, which may be interesting in its own right.

2   The Learning Model

A *concept class* C over domain X is a set of functions of the form $f : X \to \{0,1\}$. In the query-learning model, a learner has to identify an unknown function $f_* \in C$, called *target concept*, by means of queries that are honestly answered by an oracle. In this paper, we focus on the following types of queries:

Equivalence Queries (EQs). The learner passes a function $h \in C$ (functions outside C are not allowed!), called its *hypothesis*, to the equivalence oracle (EQ-oracle). The oracle answers either YES, signifying that $h \equiv f_*$, or returns a so-called *counterexample* $x \in X$ such that $h(x) \neq f_*(x)$.

Membership Queries (MQs). The learner passes an element $x \in X$ to the membership oracle (MQ-oracle). The oracle returns $f_*(x)$.

Incomplete Membership Queries (IMQs). The learner passes an element $x \in X$ to the incomplete membership oracle (IMQ-oracle). If an IMQ at query point x had already been issued before, then the oracle returns the same answer again. Otherwise, it flips a coin showing "heads" with some fixed probability p. If the coin shows "heads", the oracle returns "$*$" ($=$ no answer). Otherwise, it returns $f_*(x)$. In order to indicate the probability parameter associated with the IMQ-oracle, we say that the IMQ-oracle is *p-fallible*.

Note that the EQ-oracle has, in general, many possibilities to answer honestly: it may return any counterexample x. Since we will expose the learning algorithms to a worstcase analysis, we may assume that the EQ-oracle pursues an adversary strategy such as to slow down the learning progress as much as possible. In contrast to the EQ-oracle, the answer of the MQ- oracle (or IMQ-oracle) is determined by the query (and the outcome of the coin flip). Note that the IMQ-oracle flips its coins in an on-line fashion. Adversary strategies for the EQ-oracle can use the outcomes of past coin flips, but are ignorant to outcomes of future coin flips.

Formally, a *learner* for concept class C is an oracle algorithm, say A. In the EQ-model of learning, it is equipped with an EQ-oracle. An analogous remark is valid for the (EQ,MQ)-model and the (EQ,IMQ)-model of learning. An *adversary strategy* consists of rules that specify the target concept and the answers of the oracles, which may depend on the past interaction between the learner and the oracles (including the actual query). Learning then proceeds as an interactive process of the following kind:

1. First a target concept $f_* \in C$ (unknown to A) is fixed.
2. Then the program of A is executed. To each (allowed) query, the corresponding oracle returns an honest answer.

In query-learning models, it is usually assumed that the learner A is deterministic. In the (EQ,IMQ)-model however, even a deterministic program for A can extract random bits from the answers of the IMQ-oracle. For this reason, it is natural to allow randomized learners in this model. This has some subtle consequences. In a recent paper [10], it has been shown that the typical adversary arguments are no longer valid if A has access to random bits that cannot be predicted by the adversary.[3] Since the adversary will compete with a randomized learner, it will be convenient (for proof technical reasons) to consider randomized adversary strategies as well. The following definition takes these complications of the classical situation (of purely deterministic interactions) into account.

Definition 1. *We say that A (a potentially randomized oracle algorithm) learns C with an expected total number of q queries (of the allowed types) if, for any (potentially randomized) adversary strategy (concerning the choice of the target concept and the answers of the oracles), the following holds:*

1. *A issues at most q queries on the average (where the average is taken over all coin-flips of the IMQ-oracle, all internal coin-flips of the adversary strategy, and all internal coin-flips of A).*
2. *When A stops, target concept f_* is the only function from C that is consistent with all answers that were returned to A.*

Since asymptotic bounds will be an issue in this paper, we will consider parameterized concept classes and domains. (C_n) denotes a parameterized family of concept classes and concepts from C_n are functions from X_n to $\{0,1\}$. A learning algorithm for (C_n) must be able to learn C_n for each $n \geq 1$. Its time bound and its (expected) total number of queries are then considered as functions depending on n (and sometimes depending also on an additional parameter representing the complexity of the target concept). A is called a *polynomial learner*

---

[3] For deterministic learners, it can be theoretically justified to evaluate the learner in a scenario where the adversary does *not* have to make the initial commitment to a target concept. Oracle answers are considered as honest as long they do not rule out any consistent explanation in C. For probabilistic learners however, the initial commitment is essential. See [10] for more information on this issue.

if its (expected) run-time is polynomially bounded and if it learns its concept class with an (expected) polynomially bounded number of queries. We will typically parameterize C such that $|X_n| = 2^{n(1+o(1))}$. For instance, for Boolean concept classes with $X_n = \{0,1\}^n$, n will denote the number of Boolean variables. This normalization condition rules out "dirty tricks" (such as realizing a small query bound in dependence on n by using a "crazy" parameterization). We will also apply the parameterization to the probability associated with the IMQ-oracle, i.e., we consider $p(n)$-fallible IMQ-oracles. Non-constant choices of p are particularly interesting when $p(n)$ approaches either zero or one. In the former case, the (EQ,IMQ)-model approaches the (EQ,MQ)-model. In the latter case, it approaches the EQ-model.

3   Coping with Few Missing Answers

In this section, we show that an algorithm which learns a concept class (C_n) with EQs and MQs can be made robust (to some extent) against a $p(n)$-fallible IMQ-oracle: as long as $p(n)$ does not exceed a critical threshold, there will be only a polynomially bounded loss in efficiency. In order to equip an algorithm with robustness to missing answers, we will apply the straightforward method of exhaustively searching through all possible "answer patterns". Note, however, that S missing answers lead to 2^S possible patterns. In other words, we will discuss an exhaustive search through a space of random (and exponentially fast growing) size.

Clearly, the expectation of 2^S is, in general, much different from $2^{E[S]}$. As for a binomially distributed variable, we obtain the following result:

Lemma 1. *Let $S = S_{m,p}$ be the (binomially distributed) random variable that counts the number of successes in m independent Bernoulli trials, where each trial has probability p of success. Then $E\left[2^S\right] = (1+p)^m$.*

The proof (given in the full paper) is a simple application of the binomial theorem.

We now get back to the main subject of this section. We aim to transform a fault-intolerant learner (expecting completely reliable oracles) into a more fault-tolerant learner (being able to cope with an incomplete membership oracle to some extent). Let A be an algorithm with time bound $t(n)$ that learns (C_n) with $l(n)$ EQs and $m(n)$ MQs. Assume $t(n), l(n), m(n)$ are time-constructable. We will design an algorithm A' that learns (C_n) with EQs and IMQs. The basic idea is quite simple:

A' proceeds like A except that MQs of A are passed to an IMQ-oracle; if an IMQ is left unanswered then both possible answers are returned (which increases the number of A-simulations that have to be pursued further).

As more and more IMQs are left unanswered, more and more A-simulations are run in parallel. In order to proceed with all these A-simulations in a well-organized fashion, A' uses an administrative tool: the so-called "simulation-tree" $\mathcal{T}$ (consisting only of a single root in the beginning). At any time, the leaves in $\mathcal{T}$ decompose into "active" and "inactive" leaves. Each "active" leaf v in $\mathcal{T}$ represents an active A-simulation $\mathcal{S}(v)$. Each "inactive" leaf v corresponds to an aborted simulation. Intuitively, A' will abort simulations that cannot be correct. Technically, A' checks whether an A-simulation respects the resource bounds $t(n), l(n), m(n)$. If not, it can be aborted without damage. Furthermore, A' checks a final hypothesis of an A-simulation for correctness. If it does not coincide with the target concept, then again the A-simulation can be aborted without damage. In order to control the resource bounds, A' keeps track of the following quantities: the number t_v of simulated steps within A-simulation $\mathcal{S}(v)$, the number l_v of EQs that A has issued in simulation $\mathcal{S}(v)$, and the number m_v of MQs that A has issued in simulation $\mathcal{S}(v)$. A high level description of A' (based on the simplifying assumption that A never issues the same query twice) reads as follows:

Initialization. Create a simulation-tree $\mathcal{T}$ consisting of a single "active" root-node r. Initialize the quantities t_r, l_r, m_r to value 0.

Main Loop. Select an active leaf v of lowest depth in $\mathcal{T}$ and proceed with the A-simulation $\mathcal{S}(v)$ — on the way, the quantities t_v, l_v, m_v are properly updated — until one of the following conditions holds:

  a) *Resource-Bound-Condition:* $t_v > t(n)$, $l_v > l(n)$, or $m_v > m(n)$.
     Corresponding Action: Abort $\mathcal{S}(v)$ and declare v as "inactive".

  b) *Termination-Condition:* A-simulation $\mathcal{S}(v)$ terminates with final hypothesis h_*.
     Corresponding Action: Invest an EQ to check whether h_* coincides with the target concept f_*. If $h_* \equiv f_*$, then return h_* and stop, else abort $\mathcal{S}(v)$ and declare v as "inactive".

  c) *Blind-Spot-Condition:* An IMQ, say at query point x, is left unanswered.
     Corresponding Action: Grow $\mathcal{T}$ by creating two "active" children of v: a left child v_0 and a right child v_1. For A-simulation $\mathcal{S}(v_0)$, classification label 0 is returned to the IMQ on x. Symmetrically, classification label 1 is returned in A-simulation $\mathcal{S}(v_1)$. Initialize the quantities $t_{v_0}, m_{v_0}, l_{v_0}, t_{v_1}, m_{v_1}, l_{v_1}$ properly.

We move on to the analysis of A'. Note that $\mathcal{T}$ contains exactly one designated path, starting from the root, that represents the "correct" A-simulation whose IMQs are always answered correctly. Thus, if the target concept is not accidentally identified within a wrong A-simulation, this path must finally lead to a leaf that satisfies the termination-condition with a correct final hypothesis. Note that the length of this path is bounded by the number S of unanswered IMQs within the correct A-simulation. Since, in the beginning of the main loop, A' always selects an "active" leaf v of lowest depth, the depth of $\mathcal{T}$ is also bounded by S, and the total number of leaves in $\mathcal{T}$ (= number of A-simulations that are run in parallel) is bounded by 2^S. Note that S is binomially distributed. From these observations and from Lemma 1, we can easily derive the following result (whose proof is found in the full paper):

Theorem 1. *Let A be an algorithm with time bound $t(n)$ that learns (C_n) with $l(n)$ EQs and $m(n)$ MQs. Assume $t(n), l(n), m(n)$ are time-constructable. Let A' be the corresponding robust algorithm (as described above). Assume the IMQ-oracle is $p(n)$-fallible. Then A' is an algorithm with expected-time bound $O((1 + p(n))^{m(n)} \cdot t(n))$ that learns (C_n) with an expected number of at most $(1 + p(n))^{m(n)} \cdot (l(n) + 1)$ EQs and with an expected number of at most $(1 + p(n))^{m(n)} \cdot m(n)$ IMQs. If $p(n) \leq k \cdot \log(n)/m(n)$ for some constant k, then the "blow-up" factor $(1 + p(n))^{m(n)}$ is upper-bounded by the polynomial $n^{k \log(e)} \approx n^{1.44k}$.*

In the sequel, we will refer to A' as the *exhaustive search simulation* of A.

4 High Inherent Vulnerability to Missing Answers

In this section, we demonstrate the existence of concept classes that are efficiently learnable with EQs and MQs, but are provably not polynomially learnable if the membership oracle becomes slightly incomplete. In other words, each learning strategy is doomed to fail (in polynomial time) even if only few answers are missing. The first class that we employ for this purpose is called ADDRESSING (first considered by Maass and Turan [8]):

Example 2. For each fixed n, the domain $X_n = A_n \cup D_n$ of ADDRESS-ING consists of n "address points" $A_n = \{a_1, \ldots, a_n\}$ and 2^n "data points" $D_n = \{d_\alpha : \alpha \in \{0,1\}^n\}$. With each $\alpha \in \{0,1\}^n$, we associate the function f_α that maps a_i to α_i, $i = 1, \ldots, n$, d_α to 1, and all remaining data points to 0. ADDRESSING$_n = \{f_\alpha : \alpha \in \{0,1\}^n\}$.[4]

It has been observed by Maass and Turan [8] that (ADDRESSING$_n$) can be learned with n MQs, but at least $2^n - 1$ queries are needed if one learns with equivalence queries alone. The following result demonstrates that the two models are still separated by a superpolynomial gap if the EQ-oracle is augmented by an $\omega(\log(n)/n)$-fallible IMQ-oracle. If the IMQ-oracle is p-fallible for some constant $p > 0$, the gap is still exponential:

Lemma 2. *The expected total number of queries needed to learn the concept class (ADDRESSING$_n$) with EQs and $p(n)$-fallible IMQs is at least $(1 + p(n))^n/2$.*

Proof. The expected number of queries needed for learning ADDRESSING$_n$ cannot increase if we change the learning model in favour of the learner as follows:

- In a first phase, the learner issues n IMQs on all the adress points for free. Afterwards (phase 2), we start counting the number of queries that are still needed until the target concept is exactly identified.

---

[4] The name ADDRESSING is used because α is a binary address of length n that uniquely determines 1-out-of-2^n data points.

Consider the following adversary strategy against any learner:

- Pick $\alpha \in \{0,1\}^n$ uniformly at random and commit yourself to target concept f_α.
- Upon query $\text{EQ}(f_\beta)$, $\beta \neq \alpha$, return d_β as counterexample.

Let $S = S_{n,p(n)}$ be the binomially distributed random variable that counts the number of unanswered IMQs in the first phase of the learner. Let Q_u be the random variable that counts the number of queries in the second phase conditioned to the event that $S = u$. Finally, Q denotes the unconditioned random variable that counts the number of queries in the second phase. We claim that $E[Q_u] \geq 2^u/2$. Let us first show how the proof can be completed using this claim and Lemma 1:

$$E[Q] = \sum_{u=0}^{n} \Pr[S = u] \cdot E[Q_u] \geq \frac{1}{2} \cdot \sum_{u=0}^{m} \Pr[S = u] \cdot 2^u = \frac{1}{2} \cdot E[2^S] = \frac{1}{2} \cdot (1 + p(n))^n$$

We finally sketch the proof for $E[Q_u] \geq 2^u/2$. The main argument is as follows. Since u answers are missing after phase 1, there are still 2^u functions in ADDRESSING_n that are possible target functions. From the perspective of the learner, all of them are equally likely. Furthermore, each query in phase 2 — except for query $\text{EQ}(f_\alpha)$ — will rule out at most one of these functions, and the remaining ones are still equally likely.[5] It is not hard to show that this implies that, on the average, it takes (at least) $2^u/2$ queries until the learner accidentally finds the target concept (by issuing query $\text{EQ}(f_\alpha)$).[6]   •

Consider the exhaustive search simulation of the algorithm that learns the class (ADDRESSING_n) with n MQs. According to Theorem 1, its espected total number of queries is $(1+p(n))^n \cdot n$.[7] A comparison to the lower bound in Lemma 2 shows that no learner can perform substantially better. In particular, exhaustive search simulation is a polynomial learning method for (ADDRESSING_n) if the expected number of missing answers, which is actually $p(n)n$, satisfies $p(n)n = O(\log(n))$, or equivalently, if $p(n) = O(\log(n)/n)$. The lower bound in Lemma 2 shows that no algorithm can polynomially learn (C_n) if $p(n)$ goes beyond this barrier.

In the full paper, we present another concept class with a sort of highest possible inherent vulnerability to missing answers. The basic idea is to use a variant of ADDRESSING where each address bit is computed as a parity of many other bits. If one of these many bits is not returned by the IMQ-oracle, then the parity function cannot be evaluated and the address bit remains unknown.

---

[5] Formally, one can show that all functions that are still possible target functions have the same à-posteriori probability in the sense of Bayesean decision theory.

[6] If the learner issues only (irredundant) equivalence queries in phase 2, then the bound $(1 + p(n))^n)/2$ is actually tight for phase 2.

[7] In fact, if the simulation never issues the same query twice, it can be shown that it learns (ADDRESSING_n) with $O(n + (1 + p(n))^n)$ queries on the average.

By this construction (formally given in the full paper), we will establish an exponential gap between the (EQ,MQ)- and the (EQ,IMQ)-model even if only a polynomially small fraction of the answers is missing on the average.[8]

5   Monotone DNF Learning Revisited

We assume some familiarity with the theory of Boolean formulas. Recall that a monotone DNF formula is a disjunction of monotone terms (term = Boolean monomial). The class of monotone DNF formulas with at most m terms and n Boolean variables is denoted as $\text{MDNF}_{m,n}$. We impose the following lattice structure on the Boolean cube $\{0,1\}^n$:

$$x \leq y :\Leftrightarrow \forall i \in \{1, \ldots, n\} : x_i \leq y_i \ .$$

For each Boolean point x, let x^i denote the point obtained from x by flipping the i-th coordinate (and keeping the other coordinates fixed). A point $x \in \{0,1\}^n$ is called *minimal point* of $f \in \text{MDNF}_{m,n}$ if x is a minimal point in the Boolean lattice such that $f(x) = 1$. Clearly, the minimal points of f are in one-to-one correspondence with the monotone terms of f. In what follows, we identify f with its set of minimal points. Reversely, each $h \subseteq \{0,1\}^n$ is identified with the function from $\text{MDNF}_{m,n}$ that maps x to 1 iff there exists a point $p \in h$ such that $x \geq p$, i.e., x must be located above a point from h in the Boolean lattice. Here is the main result of this sectio:

Theorem 2. *For each constant $p < c_0 := (\sqrt{5} - 1)/2 \approx 0.62$, MDNF-LEARNER (described below) learns (MDNF$_{m,n}$) with EQs and p-fallible IMQs such that the expected total number of queries is bounded by $O(mn)$.*

The proof of Theorem 2 makes use of the following technical lemma:

Lemma 3. *Let X be a random variable with positive integer values. P denotes another random variable with (possibly multidimensional) "values" in some finite set $\mathcal{P}$.[9] For each $j \geq 1$, let Y_j denote a random variable with non-negative real values that satisfies $E[Y_j | X < j] = 0$ and*

$$\forall j \geq 1, \forall p \in \mathcal{P} : E[Y_j | X \geq j, P = p] \leq B \ . \tag{1}$$

for some bound $B \geq 0$. Then: $E\left[\sum_{j \geq 1} Y_j\right] \leq B \cdot E[X]$.

Proof. $E\left[\sum_{j \geq 1} Y_j\right] = \sum_{j \geq 1} E[Y_j] = \sum_{j \geq 1} E[Y_j | X \geq j] \cdot \Pr[X \geq j] = \sum_{j \geq 1} \sum_{p \in \mathcal{P}} E[Y_j | P = p, X \geq j] \cdot \Pr[P = p | X \geq j] \cdot \Pr[X \geq j] \leq B \cdot \sum_{j \geq 1} \Pr[X \geq j] \cdot \left(\sum_{p \in \mathcal{P}} \Pr[P = p | X \geq j]\right) = B \cdot \sum_{j \geq 1} \Pr[X \geq j] = B \cdot \sum_{j \geq 1} \Pr[X = j] \cdot j = B \cdot E[X]$ •

---

[8] Note that superpolynomially small fractions of missing answers will necessarily blur the distinction between the two models.

[9] The type of this set will be irrelevant.

In our application of Lemma 3, X will count how many times a probabilistic procedure must be called until it is (for the first time) "succesful". Y_j will count the number of queries during the j-th call of the procedure (which is zero by default if there is no j-th call because $X < j$). Random variable P will be chosen such that $E[Y_j | X \geq j, P = p] = E[Y_j | P = p]$. Loosely speaking, P allows to strip-off the statistical dependencies between Y_j and X. This will enable us derive a suitable upper bound B on $E[Y_j | X \geq j, P = p]$ and to apply Lemma 3.

We move on and present the algorithm MDNF-LEARNER (a slight modification of the original algorithm of Angluin and Slonim [4]):

algorithm MDNF-LEARNER
begin
  $h \leftarrow \emptyset$;
  while EQ(h) returns
      a counterexample x
  do if $h(x) = 0$
      then add newpoints(x) to h
      else remove wrongpoints(x)
         from h
    fi
  od
end

procedure wrongpoints(x)
begin
  return $\{p \in h : p \leq x\}$
end

procedure newpoints(x)
begin
  $Q \leftarrow \{i \in \{1, \ldots, n\} : x_i = 1\}$;
  $B \leftarrow \emptyset$; BS$\leftarrow \emptyset$;
  while $Q \neq \emptyset$
  do select and remove
    an element i from Q;
    $b \leftarrow$ IMQ(x^i);
    case of
      $b = 0$: do nothing
      $b = *$: add i to B
         and x^i to BS;
      $b = 1$: $x \leftarrow x^i$; $Q \leftarrow Q \cup B$;
          $B = \emptyset$
    esac
  od
  return $\{x\}\cup$BS
end

Whenever MDNF-LEARNER gets a positive counterexample, say x, to its current hypothesis h, it calls the procedure newpoints(x). The points returned by this procedure are added to h. Procedure newpoints, when called on a positive counterexample x, tries to find a minimal point of the target concept f_* that is located below x in the lattice. It can work towards such a minimal point by flipping x_i from 1 to 0 whenever $h(x^i) = 1$. Condition $h(x^i) = 1$ can be checked by means of an IMQ at x^i (which, however, is not always answered). Technically, the algorithm keeps track of a set Q (initially containing the 1-coordinates of x) and (initially empty) sets B and BS (BS = Blind Spots). In general, a coordinate i belongs to Q if $x_i = 1$ and the IMQ on query point x^i has not yet been issued; i belongs to B if $x_i = 1$ and the IMQ on query point x^i has already been issued but was left unanswered. In the latter case, x^i (with unknown classification label) is added as "blind spot" to BS. Whenever the IMQ on x^i is answered 1, the algorithm replaces x by x^i (thereby moving one step down in the lattice) and updates Q and B properly. When the walk down the lattice is getting stuck (which happens when $Q = \emptyset$), the current point x and all "blind spots" in BS are returned as new points (and then included in the current hypothesis h by the main program). We say that a call of `newpoints` at a positive counterexample x is *succesful* if the set `newpoints`(x) contains a minimal point of f_*. Note that the inclusion of blind spots in `newpoints`(x) increases the success probability, but it may insert negative examples of f_* into h. This is corrected by removing points from h that are located below a negative counterexample. (Compare with procedure `wrongpoints`.) Our analysis will use the following central notion. A *configuration* is a mapping $K : \{0,1\}^n \to \{0,1,*,-\}$ such that $K(x) = 0 \Rightarrow f_*(x) = 0$ and $K(x) = 1 \Rightarrow f_*(x) = 1$. $\mathcal{K}$ denotes the set of all configurations. Intuitively, $K(x) = -$ signifies that the IMQ at query point x has not yet been issued. The other labels indicate that this IMQ has already been issued, and that the label $0, 1$, or $*$ (no answer), respectively, had been returned. MDNF-LEARNER is started in *initial configuration* K_0 that maps each Boolean point to $-$. Note that the current configuration K and the coin flips of the IMQ-oracle completely determine the labels that are returned by the oracle. When the label $b \in \{0, 1, *\}$ is returned upon an IMQ at query point x, the current configuration K is updated by setting $K(x) = b$. We are now in the position to analyze MDNF-LEARNER. The claims of the theorem become obvious from the following observations (whose proof is sketched only):

1. Since all blind spots are added to h (and removed only if they are negative examples of f_*), no positive counterexample is ever located above a blind spot. This motivates the following definitions. The pair (K, x) is called *legal* if $K(p) \neq *$ for all $p \leq x$. A call `newpoints`(x) in configuration K is called a (K, x)-call of `newpoints`. It is called *legal* if (K, x) is a legal pair. It follows that MDNF-LEARNER initiates only legal calls of `newpoints`.

2. For each legal pair (K, x), the (K, x)-call of `newpoints` is not succesful with probability smaller than $p^2 + p^3 + p^4 + \cdots < p^2/(1-p)$. Thus, the probability of being succesful exceeds $q = 1 - p^2/(1-p)$. Note that q is a constant satisfying $q > 0$ because p is a constant satisfying $p < c_0$. In case of success, a new minimal point of f_* is added to h. Thus, the expected number of positive counterexamples (= the number of times procedure `newpoints` is called) is smaller than m/q.

3. For each legal pair (K, x) the following holds. During the execution of the (K, x)-call of `newpoints`, at least every $1/(1 - p)$-th IMQ (on the average) decrements the size of $Q \cup B$ by 1. Thus, the expected number of IMQs per legal call of `newpoints` is bounded by $n/(1 - p)$.

4. The legal pairs will play the role of $\mathcal{P}$ in Lemma 3. We denote the random variable whose values are legal pairs by P. The preceding observation can now be restated as follows. If $Y_{K,x}$ denotes the random variable that counts the number of IMQs during the execution of a (K, x)-call of `newpoints`, then $E[Y_{K,x}] \leq n/(1 - p)$ for each legal pair (K, x).

5. We decompose the run of MDNF-LEARNER into m phases. Each phase ends after a succesful call of `newpoints` (when a new minimal point of the target concept f_* has been inserted into the hypothesis h). Consider an arbitrary but fixed phase i. Let X be the random variable that counts the

number of calls of `newpoints` in phase i. Clearly, $E[X] = 1/q$. Let P_j be the random variable (with legal pairs as values) such that the j-th call of `newpoints` in phase i is a P_j-call. Let Y_j be the random variable that counts the number of IMQs during the execution of this call (with default $Y_j = 0$ if `newpoints` is called less than j times in phase i). Thus, Y_j conditioned to $X < j$ is always zero. Note that Y_j conditioned to $X \geq j$ depends on X only through P_j. Furthermore Y_j conditioned to $P_j = (K(j), x(j))$ is distributed like $Y_{K(j),x(j)}$. We conclude that

$$E\left[Y_j \mid X \geq j, P_j = (K(j), x(j))\right] = E\left[Y_j \mid P_j = (K(j), x(j))\right] \leq n/(1 - p) \ .$$

Lemma 3 implies that the expected total number of IMQs in phase i is at most $E[X] \cdot n/(1 - p) = O(n)$. Summing over all m phases, we get that the expected total number of IMQs is bounded by $O(mn)$.

6. Finally note that the total number of negative counterexamples is upper-bounded by the total number of blind-spots, which, in turn, is upper-bounded by the total number of IMQs.   •

It would be interesting to know more robust learners for specific classes, and to know more natural examples of classes with high inherent vulnerability. Furthermore, we would like to know whether the barrier $(\sqrt{5} - 1)/2$ from Theorem 2 can be pushed closer to 1.

References

1. Dana Angluin. Learning regular sets from queries and counterexamples. *Information and Computation*, 75:87–106, 1987.
2. Dana Angluin. Queries and concept learning. *Machine Learning*, 2(4):319–342, 1988.
3. Dana Angluin. Negative results for equivalence queries. *Machine Learning*, 5:121–150, 1990.
4. Dana Angluin and Donna K. Slonim. Randomly fallible teachers: Learning monotone DNF with an incomplete membership oracle. *Machine Learning*, 14:7–26, 1994.
5. Nader H. Bshouty and Nadav Eiron. Learning monotone DNF from a teacher that almost does not answer membership queries. In *Proceedings of the 14th Annual Workshop on Computational Learning Theory*, pages 546–557. Springer Verlag, 2001.
6. Nader H. Bshouty and Avi Owshanko. Learning regular sets with an incomplete membership oracle. In *Proceedings of the 14th Annual Workshop on Computational Learning Theory*, pages 574–588. Springer Verlag, 2001.
7. Nick Littlestone. Learning quickly when irrelevant attributes abound: a new linear threshold algorithm. *Machine Learning*, 2(4):245–318, 1988.
8. Wolfgang Maass and György Turán. Lower bound methods and separation results for on-line learning models. *Machine Learning*, 9:107–145, 1992.
9. Ronald L. Rivest and Robert E. Schapire. Inference of finite automata using homing sequences. *Information and Computation*, 103(2):299–347, 1993.
10. Hans U. Simon. How many queries are needed to learn one bit of information? In *Proceedings of the 14th Annual Workshop on Computational Learning Theory*, pages 1–13. Springer Verlag, 2001.

Games with a Uniqueness Property

Shin Aida[1], Marcel Crasmaru[2], Kenneth Regan[3], and Osamu Watanabe[2] [*]

[1] Graduate School of Human Informatics, Nagoya Univ.
`aida@info.human.nagoya-u.ac.jp`
[2] Dept. of Mathematical and Computing Sciences, Tokyo Institute of Technology
`{marcel,watanabe}@is.titech.ac.jp`
[3] Dept. of Computer Science, State Univ. of New York at Buffalo
`regan@cse.buffalo.edu`

Abstract. For two-player games of perfect information such as Chess, we introduce "uniqueness" properties to describe game positions in which a winning strategy (should one exist) is forced to be unique. Depending on how uniqueness is forced, and whether it applies to both players, the uniqueness property is classified as (*bi-*) *weak*, (*bi-*) *strong*, or *global*. We prove that any reasonable two-player game G is extendable to a game G^* with the bi-strong uniqueness property, so that e.g., QBF remains PSPACE-complete under this restriction. For global uniqueness, we introduce a simple game GUPQBF over Boolean formulas with this property, and prove that any reasonable two-player game with global uniqueness is reducible to this game. On the other hand, we also show that GUPQBF resides in "small" counting classes believed properly contained in PSPACE. Our results give a new characterization to some complexity classes such as PSPACE and EXPTIME.

1 Introduction

Is it easier to analyze a position in a game to determine which player stands to win if one is told that the winning strategy for that player is unique? A question of this type corresponds, in complexity theory, to the one asking whether problems in UP are easier than NP, or more precisely, whether the satisfiability of a given formula F is easier to solve under the promise that it has at most one satisfying assignment. (For the case of NP problems, it is known [8] that the promise does not reduce the complexity of NP problems so much.)

This paper analyzes theoretically whether the same intuition holds for two-player games. Since the notion of unique winning strategy is not so clear for a two-player game we first explain uniqueness properties we will discuss in this paper by using some intuitive examples. In the game Chess, a problem of the "White to move and win" variety is called a *study*. For a study to be "publishable" it is generally an absolute requirement that the first winning move — called the "key" — be unique. Moreover, the study should have a "main line"

---

[*] Supported in part by JSPS/NSF cooperative research: Complexity Theory for Strategic Goals, 1998–2001, # INT-9726724.

H. Alt and A. Ferreira (Eds.): STACS 2002, LNCS 2285, pp. 396–407, 2002.
© Springer-Verlag Berlin Heidelberg 2002

in which the defender forces the attacker to find unique winning moves at every turn (possibly excepting trivial junctures). The existence of a main line with unique play by the winner corresponds to our condition called *weak uniqueness*. When *all* options by the defender force unique replies at every move, then the study position has the *strong uniqueness* property. However, it is desirable that certain tempting mis-steps by the attacker — called "tries" — be punishable by unique defenses. When this applies not only to all poor first moves by the attacker but to every position reachable from the start, then we speak of the *global uniqueness* property of the study.

If all positions of a game have this global uniqueness property, then we simply say that the game has the global uniqueness property. An example of a game with the global uniqueness property is the basic form of Nim in which each player may remove from 1 to m counters from a single pile initially with N counters, and the player taking the last counter wins. If N is not a multiple of $m + 1$, then the first player wins uniquely by removing $N \bmod (m + 1)$ counters, while if not so, then the first player loses — with a unique rejoinder to every move.

Consider any game and any position P of the game that does not have the global uniqueness property. Suppose furthermore that there is no winning strategy of the current side. Since we assume no "draw" in the game, this means that there is a winning strategy of the opponent side for any of P's succeeding positions. Here if some (resp., any) succeeding position of P has the weak (resp., strong) uniqueness property, then we say that P has the *weak* (resp., *strong*) *co-uniqueness* property. For simplify our statement, let us also regard a position with no winning strategy (of the current side) to have the weak and strong uniqueness property. Similarly, we regard a position with no winning strategy (of the opponent side) to have the weak and strong co-uniqueness property. Then it is possible that a position has the weak (resp., strong) uniqueness and co-uniqueness properties, in which case we say that this position has the *weak* (resp., *strong*) *bi-uniqueness* property.

In general, finding a winning move for a given position is computationally hard. In complexity theory, the hardness of a game is usually classified in terms of its *Winner Decision Problem*, namely *deciding* whether a given position indeed has a winning move, i.e., a winning strategy. For example, it has been shown [6] that Winner Decision Problem is EXPTIME-complete for certain liberal generalizations to $n \times n$ boards of Chess and Go and several other games and becomes PSPACE-complete when those games are constrained to limit all plays to length polynomial in n. We would like to investigate whether this decision problem becomes easier if we can assume some of the uniqueness properties mentioned above. For example, we consider some subset of positions of Go that has the strong uniqueness property, and investigate the complexity of Winner Decision Problem. (Recall that a position without winning strategy is also regarded to have the strong uniqueness property.)

There have been some preliminary results on our question. Crasmaru and Tromp [3] proved that Winner Decision Problem for Go is still PSPACE-complete even instances are restricted to very simple positions called "ladder".

In their paper, they also showed that the problem remains PSPACE-complete even if we consider only ladders with the strong uniqueness and the weak co-uniqueness. Aida and Tsukiji [1] also discussed a way to define a game with some uniqueness property simulating a given alternating Turing machine.

In this paper, we consider our question in the most general case, and we first show the following result.

(1) For every reasonable game G, we can define its extension G^* in such a way that every position P of G has the strong bi-uniqueness property in G^*. Furthermore, the (maximal) depth of P, i.e., the depth of the game tree rooted by P, becomes at most three times of the original depth in G.

Here, by an "extension of G", we mean a game that includes all positions of G without changing their winners. Thus, this result illustrates that the hardness of the problem of deciding which side of a given position wins does not change even if we may assume that the position has the strong bi-uniqueness property. In fact, by using this result, we can define a game G^{PS} (resp., G^{EXP}) and a set P^{PS} (resp., P^{EXP}) of strong bi-unique game positions such that it is PSPACE- (resp., EXPTIME-) complete to determine a given position in P^{PS} (resp., P^{EXP}) has a winning strategy.

In contrast to the case of the strong bi-uniqueness property, it seems that the Winner Decision Problem becomes easier when the global uniqueness property can be assumed. Formally we define the following *promise problem* GUQBF:

> INSTANCE: A quantified Boolean formula ψ.
> PROMISE: The logic game on ψ has the global uniqueness property.
> QUESTION: Is ψ true?

For this problem GUQBF, we show the following complexity-theoretic results.

(2) GUQBF has solutions in $\mathrm{PP} \cap \oplus\mathrm{P}$ and in $\mathrm{C}_=\mathrm{P} \cap \mathrm{co}\text{-}\mathrm{C}_=\mathrm{P}$.

In particular, it follows from (2) that GUQBF cannot be PSPACE-hard unless, for example, $\mathrm{PSPACE} = \oplus\mathrm{P}$. However we can also prove the following result, which shows that GUQBF is not trivial because it is at least harder than the class UP.

(3) For every reasonable game of polynomial depth, if it has the global uniqueness property, then its Winner Decision Problem is poly-time reducible to GUQBF.

This implies, for example, the problem of factoring integers reduces to GUQBF. (Due to the space limit, we omit some proofs. See the technical report version, available from www.is.titech.ac.jp/research/research-report/C/, for all proof details.)

2   Formal Definition of Game and Uniqueness Properties

Here we define the notion of game and several uniqueness properties and prepare some notations for our discussion.

For the notion of two-player game, we adopt the definition given in [2].

Definition 1. *A two person perfect-information game* G *is a triple* (P_0^G, P_1^G, R^G), *where* P_0^G *and* P_1^G *are sets such that* $P_0^G \cap P_1^G = \emptyset$, *and* R^G *is a subset of* $P_0^G \times P_1^G \cup P_1^G \times P_0^G$.

Where no ambiguity arises, we will abbreviate P_0^G as P_0, etc. The set P_0 (resp., P_1) stands for the set of positions in which *player 0* (resp., 1) is to move, and R stands for the set of legal moves. We assume some reasonable coding for elements of $P_0 \cup P_1$; in the following, we simply identify elements of $P_0 \cup P_1$ and their description under this coding. We assume that P_0 and P_1 are disjoint; furthermore, for any position π in $P_0 \cup P_1$, we can easily compute the player that is to move at π (which we denote by $x(\pi)$). More precisely, the value $x(\pi)$ is computable within polynomial-time w.r.t. the length $|\pi|$ of the representation of π. For a generic player x, where $x \in \{0,1\}$, we denote by $\bar{x}$ its opponent, i.e., $\bar{x} = 1 - x$, and subsequently we denote by P_x (resp., $P_{\bar{x}}$) the set of moves of the player x (resp., $\bar{x}$). A *move* for player x in the current position $\pi \in P_x$ consists in choosing of a pair $(\pi, \pi') \in R$, π' becoming the next position. For any π, we define $R(\pi) = \{\,\pi' : (\pi, \pi') \in R\,\}$. For any $d \geq 1$, we inductively define R_d as follows.

$$R_1(\pi) = R(\pi), \quad R_{d+1}(\pi) = \bigcup_{\pi' \in R_d} R(\pi'), \quad \text{and} \quad R_\infty(\pi) = \bigcup_{d \geq 1} R_d(\pi).$$

For the requirements for "reasonable game", we consider the following polynomial-time computability. First, for any $\pi \in P_0 \cup P_1$ and any $\pi' \in R_\infty(\pi)$, it must hold that $|\pi'| \leq q(|\pi|)$ for some polynomial q. Note that this condition implies that the set $R_\infty(\pi)$ is finite for any π. Secondly, there must be some algorithm that determines $R(\pi, \pi')$ within polynomial-time w.r.t. $|\pi| + |\pi'|$. Notice that the polynomial-time computability of $R(\pi)$ is not required. It may be the case that $|R(\pi)|$ becomes exponential w.r.t. $|\pi|$. Thirdly, there is no loop; that is, for any $\pi, \pi' \in P_0 \cup P_1$, it must not be the case that both $\pi' \in R_\infty(\pi)$ and $\pi \in R_\infty(\pi')$ hold. The games satisfying these conditions are called *polynomially definable game*.

By convention, a player who is unable to move is declared the loser. Now, for any player $x \in \{0,1\}$, we would like to define a set W_x^G of winning positions of x in the game G. (Again we omit the superscript and write it as W_x.) Intuitively, a player x *wins* at position π (or, π is a *winning position* of x) if
1. the opponent $\bar{x}$ is to move at π but $\bar{x}$ is unable to move, or
2. x is to move at π and there exists a move to a winning position for x, or
3. $\bar{x}$ is to move at π and any available move drives to a winning position for x.

Formally, we define $W_{x,d}$ inductively as follows, and define $W_x = \cup_{d \geq 0} W_{x,d}$.

$$W_{x,0} = \{\,\pi \in P_{\bar{x}} : R(\pi) = \emptyset\,\},$$
$$W_{x,d+1} = W_{x,d} \cup \{\,\pi \in P_x : R(\pi) \cap W_{x,d} \neq \emptyset\,\} \cup \{\,\pi \in P_{\bar{x}} : R(\pi) \subseteq W_{x,d}\,\}.$$

By definition, we clearly have $W_0 \cap W_1 = \emptyset$. Notice also that it holds that $W_0 \cup W_1 = P_0 \cup P_1$; that is, all positions are winning positions of either player 0 or player 1. This is due to the condition prohibiting the existence of a loop. For any $\pi \in P_0 \cup P_1$, the *maximal depth* of π (denoted as $d_{max}(\pi)$) is the length of the longest path from π to a leaf of the game tree. Precisely, $d_{max}(\pi)$ is defined inductively as follows.

$$d_{max}(\pi) = \begin{cases} 0, & \text{if } R(\pi) = \emptyset, \\ 1 + \max_{\pi' \in R(\pi)}(d_{max}(\pi')), & \text{otherwise.} \end{cases}$$

Since we assume polynomially definable game, $d_{max}(\pi)$ is bounded by $2^{q(|\pi|)}$ for some polynomial q.

Now we are ready to introduce formally the concept of positions having a unique solution property. For any game G and any player $x \in \{0, 1\}$, we define WU_x^G (abbrev. WU_x) and SU_x^G (abbrev. SU_x) inductively in the following way.

$$\begin{aligned} WU_{x,d+1} &= \{\, \pi \in P_x : \pi \in W_x \Rightarrow |R(\pi) \cap W_x| = 1 \,\} \\ &\quad \cup \{\, \pi \in P_{\overline{x}} : \pi \in W_x \Rightarrow R(\pi) \cap WU_{x,d} \neq \emptyset \,\}, \quad \text{and} \\ SU_{x,d+1} &= \{\, \pi \in P_x : \pi \in W_x \Rightarrow |R(\pi) \cap W_x| = 1 \,\} \\ &\quad \cup \{\, \pi \in P_{\overline{x}} : \pi \in W_x \Rightarrow R(\pi) \subseteq SU_{x,d} \,\}. \end{aligned}$$

Define $WU_x = \bigcup_{d \geq 1} WU_{x,d}$ and $SU_x = \bigcup_{d \geq 1} SU_{x,d}$. We say that a position π has the *weak* (resp., *strong*) *uniqueness property* for player x if π belongs to the set WU_x (resp., SU_x). We say that a position π has the *weak* (resp., *strong*) *bi-uniqueness property* for player x if π belongs to the set $WU_x \cap WU_{\overline{x}}$ (resp., $SU_x \cap SU_{\overline{x}}$). Finally, we say that a position π has the *global uniqueness property* if every position π of $R_\infty(\pi)$ has the strong bi-uniqueness property, and that a game itself has the *global uniqueness property* if every position of the game has the global uniqueness property.

One can observe that a position π that is not a winning position for x has by definition the weak/strong uniqueness property for x.

It is easy to see that polynomially definable games can be used as a model of polynomially space bounded alternating computation. Furthermore, the uniqueness properties considered here correspond the uniqueness of accepting path at each alternation. In particular, for any UP-machine, a game simulating it has the global unique solution property.

3   Game with the Strong Uniqueness Property

Here we show how every polynomially definable game G can be converted into a game G^* in such a way that the following holds for each position π of G, (i) the conversion does not change π and its winner in G^*, and (ii) π has the strong uniqueness property in G^*. Furthermore, the conversion increases $d_{max}^*(\pi)$, the maximal depth of π in G^*, by at most three times from $d_{max}(\pi)$.

For any game $G = (P_0, P_1, R)$, we define the dual game of G as being $\overline{G} = (\overline{P}_0, \overline{P}_1, \overline{R})$, where roughly speaking, $\overline{P}_0 = P_1$, $\overline{P}_1 = P_0$, and $\overline{R} = R$. That is, the dual of G is simply the game in which the players "exchanged" only their places. Recall that we assume that the player to move is encoded in a position. Thus, precisely speaking, $\overline{P}_0$ is not the same as P_1, for example; instead, every position

$\pi \in P_1$, its dual $\overline{\pi} \in \overline{P}_0$ is obtained from π by replacing the part encoding "player 0" with "player 1". From this, we may assume that $(P_0 \cup P_1) \cap (\overline{P}_0 \cup \overline{P}_1) = \emptyset$. For the dual game $\overline{G}$, we can define $W_x^{\overline{G}}$ in the same way as G. To simplify our notation, we will denote $W_x^{\overline{G}}$ by $\overline{W}_x$.

One can observe that the dual of the dual of a game is the game itself, i.e., $\overline{\overline{G}} = G$, and that player x wins in a position π of G if and only if x loses in the dual position $\overline{\pi}$ of $\overline{G}$. Moreover, one can easily show that for any position π of G, we have $d_{\max}(\pi) = d_{\max}(\overline{\pi})$.

Theorem 1. *For any game $G = (P_0, P_1, R)$, we can define a game $G^* = (P_0^*, P_1^*, R^*)$ such that, for each $x \in \{0, 1\}$, we have (i) $P_x \subseteq P_x^*$ and $W_x \subseteq W_x^*$, and (ii) $W_x \subseteq SU_x^*$. (Here we abbreviate $P_0^{G^*}$ as P^*, etc.) Furthermore, if G is polynomially definable, then so is G^*, and for every $\pi \in P_0 \cup P_1$, we have $d_{\max}^*(\pi) \leq 3d_{\max}(\pi)$.*

Here we explain our conversion idea intuitively, the formal proof being obtained by induction on the maximal depth of a position. Let G and G^* denote an original game and its conversion, and let π be any position of G for a player x. Suppose that $R(\pi) = \{\pi_1, \pi_2, \ldots, \pi_k\}$; that is, $\{\pi_1, \pi_2, \ldots, \pi_k\}$ is the set of the possible next move from π in G. (See Figure 1.) Assume that there are some winning moves of x in $\{\pi_1, \pi_2, \ldots, \pi_k\}$. That is, x can win the game by choosing one of these winning moves. On the other hand, in our new game G^*, player x has to choose the *leftmost* winning move, i.e., the winning move with the smallest index. Furthermore, x has to verify (if it is requested by the opponent player $\overline{x}$) that his choice is indeed the leftmost.

Assume for example, π_3 is the leftmost winning move. Then to win in the new game G^*, x has to choose (π, π_3) as a next position of π, where (π, π_3) is a newly introduced position. Intuitively, choosing this position is to declare that the third position (i.e., π_3) in the set of the next moves from π is the leftmost winning move. Then in the next move, the opponent player $\overline{x}$ can choose $\overline{\pi}_1$, $\overline{\pi}_2$, or $[\pi, \pi_3]$. Intuitively, choosing $[\pi, \pi_3]$ means to continue the game as in the original game; that is, $\overline{x}$ accepts π_3 as a next position. Then x continues the game by choosing π_3, and hence, the next move of x from $[\pi, \pi_3]$ is always π_3. (See Figure 1.) On the other hand, choosing, e.g., $\overline{\pi}_1$, intuitively means that player x has to show that π_1 is not the winning move in G; that is, he would have lost the game by choosing π_1 in G (unless the opponent makes any mistake). This is because x loses from π_1 if and only if x wins from $\overline{\pi}_1$.

We keep this conversion inductively throughout the game tree. (The star symbol in Figure 1 means that the same conversion is made in the corresponding subtrees.) Since the leftmost winning move is unique, it is intuitively clear that some unique solution property is guaranteed in the new game, and indeed, it is not so difficult to prove that the strong bi-unique solution property holds. Also it is clear that the maximal depth increases at most three times from the original maximal depth.

When using polynomially definable games as a computation model, we may usually assume that the depth is the same as the maximal depth. Thus, from the

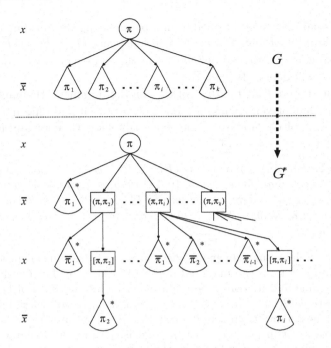

Fig. 1. Conversion from G to G^*

above result, we can show that any alternating computation can be simulated by some alternating computation with (i) the same order of alternations, and (ii) the strong bi-uniqueness property. From this observation, we can show, for example, that QBF remains PSPACE-complete even when restricted to formulas with the strong bi-uniqueness property.

4 Global Uniqueness Property and Boolean Formulas

In this section, we consider games with the global uniqueness property. But for simplifying our discussion, we will concentrate on games with polynomially bounded depth. Such games can be formulated as a natural logic game on the set of quantified Boolean formulas (QBFs), explained below. Consider any Boolean formula F with labeled variables $x_1, \ldots, x_d$; this formula induces the following quantified Boolean formula $\phi = \exists x_d \forall x_{d-1} \exists x_{d-2} \cdots Q_d x_1 F$, where Q_d is $\forall$ if d is even, $\exists$ if d is odd. Then we can define a natural logic game on such a Boolean formula. That is, the game is played by two players, E-player and U-player. E-player's goal is to assign 0 or 1 to variables with an existential quantifier to make the final value of the formula becomes 1 (i.e., true), whereas U-player's goal is to prevent this (i.e., making the value of the formula false) by assigning 0 or 1 to variables with a universal quantifier. Here we consider this logic game.

For simplifying for our discussion, for QBFs representing game positions, let us consider for a while only QBFs of a prenex form like the above. That is, existential and universal quantifiers appear alternatively starting with an existential quantifier. (Thus, the game always starts with E-player.) We use PQBF to denote the set of QBFs of this form, and let GUPQBF denote the promise problem corresponding to GUQBF where input formulas are restricted to those in PQBF. It will be shown later that this restriction is not essential for discussing the complexity of deciding a winner or evaluating a QBF. In the following, we use, e.g., $F, G, \ldots$ for denoting unquantified Boolean formulas, and use, e.g., $\phi, \psi, \ldots$ for denoting quantified Boolean formulas. Observe first that if F and G are equivalent as Boolean formulas, then the QBFs induced from F and G define the same logic game. Here we characterize game positions with global uniqueness. Let TRUE (resp., FALSE) denote a constant Boolean formula whose value is 1 (resp., 0). We define inductively the following sets of Boolean formulas:

$$A_0 = \{\text{FALSE}\}, \quad B_0 = \{\text{TRUE}\}, \quad \text{and for } d \geq 1,$$
$$A_d = \{(x_d \wedge \neg F_1) \vee (\bar{x}_d \wedge \neg F_0) : F_0, F_1 \in B_{d-1}\},$$
$$B_d = \{(x_d \wedge \neg F_1) \vee (\bar{x}_d \wedge \neg F_0) :$$
$$(F_0 \in A_{d-1} \text{ and } F_1 \in B_{d-1}) \text{ or } (F_0 \in B_{d-1} \text{ and } F_1 \in A_{d-1})\}.$$

Now syntactically, $A_1 = \{(x_1 \wedge \text{FALSE}) \vee (\bar{x}_1 \wedge \text{FALSE})\}$, which is equivalent to $A_1 = \{\text{FALSE}\}$. For optical appeal we may also use $A_1 = \{x_1 \wedge \bar{x}_1\}$. Also

$$B_1 = \{(x_1 \wedge \text{TRUE}) \vee (\bar{x}_1 \wedge \text{FALSE}), (x_1 \wedge \text{FALSE}) \vee (\bar{x}_1 \wedge \text{TRUE})\},$$

which reduces to $B_1 = \{x_1, \bar{x}_1\}$. For simplicity, let us use $\pm x$ to denote either x or $\bar{x}$. Then syntactically, A_2 consists of the four formulas $(x_2 \wedge \pm x_1) \vee (\bar{x}_2 \wedge \pm x_1)$ over the four possible combinations of $\pm x_1$ in both places. Semantically this reduces to $\{x_1, \bar{x}_1\}$, which is the same as B_1. Also

$$B_2 = \{(x_2 \wedge \text{TRUE}) \vee (\bar{x}_2 \wedge x_1), (x_2 \wedge \text{TRUE}) \vee (\bar{x}_2 \wedge \bar{x}_1),$$
$$(x_2 \wedge x_1) \vee (\bar{x}_2 \wedge \text{TRUE}), (x_2 \wedge \bar{x}_1) \vee (\bar{x}_2 \wedge \text{TRUE})\},$$

which is equivalent to the set of four formulas $\pm x_1 \vee \pm x_2$. The corresponding ϕ in each case gives a game position where E-player wins for exactly one choice of value for x_2, and for the other value, gives U-player exactly one winning response. Thus each of these ϕ has the global uniqueness property with E-player as a winner. Likewise for a QBF ϕ induced from any formula in A_2, E-player's value for x_2 is immaterial and U-player has the same unique reply; so ϕ has the global uniqueness property with E-player as a loser. The same reasoning applies to the recursive cases in general. We obtain the following characterization.

Lemma 1. *For every Boolean formula F on $d \geq 1$ variables, let ϕ be a QBF induced from F.*

(a) *ϕ is a position with global uniqueness such that U-player wins if and only if F is equivalent to a Boolean formula in A_d.*

(b) ϕ is a position with global uniqueness such that E-player wins if and only if F is equivalent to a Boolean formula in B_d.

With this lemma, we can characterize completely true/false PQBFs with global uniqueness. Unfortunately, however, each formula in $A_d \cup B_d$ has exponential length. Thus, this characterization is not so helpful for giving a nontrivial upper bound to the complexity of evaluating a given PQBF with global uniqueness nor that of deciding whether a given PQBF has global uniqueness.

On the other hand, for the complexity of the decision problem GUPQBF, we have give some upper bound in terms of counting classes PP, etc. PP is known to be polynomial-time Turing equivalent to Valiant's class #P of functions f such that for some polynomial-time bounded NTM N and all x, $f(x)$ equals the number of accepting computations of N on input x. A language L belongs to $C_=P$ iff there is a #P function f and a polynomial-time computable function $g : \Sigma^* \longrightarrow \mathbf{N}$ such that for all x, $x \in L \Leftrightarrow f(x) = g(x)$. $\oplus P$ is the class defined by stipulating $x \in L \Leftrightarrow f(x)$ is odd.

Theorem 2. *The promise problem GUPQBF has solutions in* $PP \cap \oplus P$ *and in* $C_=P \cap co\text{-}C_=P$.

The above theorem is a direct consequence of the following fact[1].

Fact 1 *For even d, every formula in A_d has $N_d = (2/3)(2^d - 1)$ satisfying assignments, while for odd d, every formula in A_d has $N_d = N_{d-1}$ satisfying assignments. On the other hand, for all d, every formula in B_d has $N_d + 1$ satisfying assignments.*

Since PP is believed as a proper subclass of PSPACE, it follows from this theorem that it is unlikely that GUQBF (or more precisely, GUPQBF) is as hard as PSPACE. Now does it really mean that the global uniqueness property makes games easy? To answer this question and also to give some nontrivial lower bound for GUPQBF, we would like to reduce any game with global uniqueness to GUPQBF. For this, we begin with considering the conversion of any (prenex) QBF to some in PQBF; for example, we would like to convert, e.g., $\exists x_3 \exists x_2 \forall x_1 F(x_1, x_2, x_3)$ to an equivalent one $\exists x_4' \forall x_3' \exists x_2' \forall x_1' F'(x_1', x_2', x_3', x_4')$ in PQBF, while keeping its global uniqueness property. For this, we make use of the following set of formulas, which we call *trivial unique game positions*. (Here we write the $\wedge$ operation like an invisible tighter-binding multiplication for better visual impact.)

$$t_0 = \text{TRUE}$$
$$t_2 = x_1 \vee x_2$$
$$t_4 = x_1 \vee x_2(x_3 \vee x_4)$$
$$t_6 = x_1 \vee x_2(x_3 \vee x_4(x_5 \vee x_6))$$
$$\vdots$$

---

[1] This actually places GUPQBF into a "small" subclass of SPP whose exact identity we have not yet worked out from unpublished sequels to paper [5].

and so on. Also for odd $d \geq 1$, define t_d by substituting $x_{d+1} = \text{FALSE}$ in t_{d+1}. Let τ_d denote the QBF induced from t_d. Notice that these formulas are much simpler than those in $A_d \cup B_d$; in fact, the size of the formula τ_d is linear in d. The main property of these formulae is captured by the following lemma.

Lemma 2. *For even d, the formula τ_d is a game position from which E-player wins with global uniqueness, while for odd d, U-player wins from τ_d with global uniqueness.*

Now we see how these formulas fill an essential padding role.

Lemma 3. *There is a polynomial-time procedure that, given any prenex form QBF ϕ' in d' variables with global uniqueness, produces a Boolean formula F in $d \leq 2d'$ variables that induces a formula ϕ in PQBF with global uniqueness that is equivalent to ϕ'. (In fact, it holds that F is equivalent to some formula in B_d if ϕ' is true and equivalent to some formula in A_d if ϕ' is false.)*

Proof. By renumbering variables in ϕ', we may assume that the innermost variable is numbered x_1 (resp., x_2) if it is universally (resp., existentially) quantified, and that variables are numbered going outward so that every universally (resp., existentially) quantified variable has odd (resp., even) index. Let x_d (resp., x_{d-1}) be the last index if x_d is existentially (resp., universally) quantified. (Thus, we assume that d is even.) Now we define a recursive procedure $\mathcal{R}$ that works on prenex QBFs with free variables allowed. $\mathcal{R}$ does not change on single-quantifier QBFs; otherwise, it works as follows according to the first two quantifiers.

$$
\begin{aligned}
\mathcal{R}(\exists x_d \exists x_{d-2} \psi) &= \exists x_d \forall x_{d-1} [(\bar{x}_{d-1} \wedge \tau_{d-2}) \vee (x_{d-1} \wedge \mathcal{R}(\exists x_{d-2} \psi))], \\
\mathcal{R}(\exists x_d \forall x_{d-1} \psi) &= \exists x_d \, \mathcal{R}(\forall x_{d-1} \psi), \\
\mathcal{R}(\forall x_{d-1} \exists x_{d-2} \psi) &= \forall x_{d-1} \, \mathcal{R}(\exists x_{d-2} \psi), \quad \text{and} \\
\mathcal{R}(\forall x_{d-1} \forall x_{d-3}) &= \forall x_{d-1} \exists x_{d-2} [(\bar{x}_{d-2} \wedge \tau_{d-3}) \vee (x_{d-2} \wedge \mathcal{R}(\forall x_{d-3} \psi))].
\end{aligned}
$$

It is easy to see that the logic game on $\mathcal{R}(\eta)$ is equivalent to that on η and retains global uniqueness.

Now we must transform $\mathcal{R}(\phi)$ into a prenex formula. We claim that we can simply move all quantifiers in $\mathcal{R}(\phi)$ to the far left without changing the semantics! This is because the invariant that existentially quantified variables have even index and the others have odd index is maintained throughout the computation of $\mathcal{R}(\phi)$, and because the inserted variables (such as x_{d-1} in the $\exists\exists$ case) enforce a separation that offsets the change in scope.

This facilitates our proof of one of the main results of this section.

Theorem 3. *Consider any polynomially definable game $G = (P_0, P_1, R)$ with the global uniqueness property. Suppose that its maximal depth function $d_{\max}$ is polynomially bounded; that is, with some polynomial p, we have $d_{\max}(\pi) \leq p(|\pi|)$ for any $\pi \in P_0 \cup P_1$. Then we can define a polynomial-time computable procedure $\mathcal{RED}$ such that given position $\pi \in P_0 \cup P_1$, the output $\mathcal{RED}(\pi)$ is a formula ϕ in PQBF with global uniqueness such that ϕ is true if and only if π is a winning position for player 0.*

Proof. We can modify the game G, while keeping global uniqueness, to a new polynomially definable game $G' = (P_0', P_1', R')$ so that (i) every original game position $\pi \in P_0 \cup P_1$ has some corresponding position (which is denoted by $[\pi]$) in G', and (ii) the game starting from any such position $[\pi]$ always terminates at the $2p(|\pi|)$th step with some special constant positions either π_0 or π_1, where this final position is π_0 (resp., π_1) if and only if π is a winning position for player 0 (resp., player 1) in the original game. Furthermore, we may assume that every position appearing in the game tree under $[\pi]$ is represented by a binary string of length $p(|\pi|)$ for some polynomial q. (Since the modified game (starting from $[\pi]$) terminates at even steps, *by definition* player 1 always wins from $[\pi]$. But here we want to keep the original win/loss criteria, and so, we will determine the win/loss by the final position. The global uniqueness property is defined under this win/loss criteria.)

Let us consider some game position π in G, and let $n = q(\pi)$ and $d = 2p(\pi)$. We use a binary vector $\boldsymbol{x} = x_1, ..., x_n$ to represent a game position in the game tree rooted $[\pi]$. Since R' is polynomial-time computable, by the standard technique, we can define a Boolean formula $DELTA$ such that $\exists \boldsymbol{w} DELTA(\boldsymbol{y}, \boldsymbol{z}, \boldsymbol{w})$ holds if and only if a position $\boldsymbol{z}$ follows from $\boldsymbol{y}$ in G'. Furthermore, we may assume that the choice of $\boldsymbol{w}$ for having $DELTA(\boldsymbol{y}, \boldsymbol{z}, \boldsymbol{w}) = 1$ is unique if it indeed exists. Also we define a Boolean formula WIN_0 such that $WIN_0(\boldsymbol{x})$ holds if and only if $\boldsymbol{x}$ is π_0.

Now we define, for our given π, a Boolean formula F and its quantified version ψ. Variables of F are

$$x_1^m, ..., x_n^m, x_1^{m-1}, ..., x_n^{m-1}, ..., x_1^1, ..., x_n^1, \text{ and}$$
$$w_1^m, ..., w_n^m, w_1^{m-1}, ..., w_n^{m-1}, ..., w_1^1, ..., w_n^1.$$

Intuitively, variables $\boldsymbol{x}^{m-i+1} = x_1^{m-i+1}, ..., x_n^{m-i+1}$ are used to encode a game position at the ith steps. For any $j \le m - 1$, we use a formula $\boldsymbol{t}_j$ that is defined in the same way as t_j by substituting every x_i, $1 \le i \le j$, with $x_1^i \vee \cdots \vee x_n^i \vee w_1^i \vee \cdots \vee w_n^i$ if i is odd, and with $x_1^i \wedge \cdots \wedge x_n^i \wedge w_1^i \wedge \cdots \wedge w_n^i$ if i is even. For example,

$$\boldsymbol{t}_3 = (x_1^1 \vee \cdots \vee x_n^1 \vee w_1^1 \vee \cdots \vee w_n^1)$$
$$\vee (x_1^2 \wedge \cdots \wedge x_n^2 \wedge w_1^2 \wedge \cdots \wedge w_n^2)(x_1^3 \vee \cdots \vee x_n^3 \vee w_1^3 \vee \cdots \vee w_n^3).$$

Then our F is defined as follows.

$$F = [\neg DELTA(\boldsymbol{x}^{m+1}, \boldsymbol{x}^m) \wedge \boldsymbol{t}_{m-1}]$$
$$\vee [DELTA(\boldsymbol{x}^{m+1}, \boldsymbol{x}^m) \wedge \neg DELTA(\boldsymbol{x}^m, \boldsymbol{x}^{m-1}) \wedge \boldsymbol{t}_{m-2}]$$
$$\vee [DELTA(\boldsymbol{x}^{m+1}, \boldsymbol{x}^m) \wedge DELTA(\boldsymbol{x}^{m-1}, \boldsymbol{x}^{m-2}) \wedge \neg DELTA(\boldsymbol{x}^{m-2}, \boldsymbol{x}^{m-3}) \wedge \boldsymbol{t}_{m-3}]$$
$$\vdots$$
$$\vee \left[\left(\bigwedge_{i=m}^{1} DELTA(\boldsymbol{x}^{i+1}, \boldsymbol{x}^i) \right) \wedge WIN_0(\boldsymbol{x}^1) \right].$$

Here we use the symbol x^{m+1} for simplifying the expression, it is a binary representation of $[\pi]$ that is a constant vector in F. A quantified formula ψ is obtained from F by simply quantifying variables with even superscripts existentially and variables with odd superscripts universally in the order of their superscripts. That is, $\phi = \exists x_1^m \exists x_2^m \cdots F$.

Our definition of F is based on the following interpretation: If E-player tries to cheat by setting the values of the existentially-quantified variables x^m to form a non-legal game position or he fails to give a unique witness for w^m, then E-player loses uniquely according to the trivial game t_{m-1} on the remaining variables. If E-player give correct assignments to x^m and w^m, then U-player must give correct assignments to x^{m-1} and w^{m-1} on pain of otherwise losing the trivial global unique game t_{m-2}. This compulsion holds all the way down until the end.

As in the last proof, we apply the padding of Lemma 3 between adjacent like quantifiers in ψ to obtain a final Boolean formula ϕ in PQBF. Clearly, the whole process can be done within polynomial-time in $|\pi|$.

From this theorem, we can show that the upper bound of Theorem 2 holds not only GUPQBF but also any polynomial-time game with the global uniqueness property. This reduction also us a meaningful lower bound. Recall that any UP and co-UP computation can be regarded as a one alternation game with the global uniqueness property.

Theorem 4. *The promise problem* GUPQBF *(and thus* GUQBF*) is hard for* UP *and for* co-UP *under polynomial-time many-one reductions.*

Acknowledgements. The authors wish to thank R. Schuler, J. Tromp, and T. Tsukiji for stimulating discussion on this topic.

References

1. S. Aida and T. Tsukiji, Complexity of two person games with unique winning strategy, Technical Report of IEICE COMP2000-87, 2001.
2. A.K. Chandra and L.J. Stockmeyer, Provably difficult combinatorial games, *SIAM J. Comput.*, 8, 151–174, 1979.
3. M. Crasmaru and J. Tromp, Ladders are PSPACE-complete, in *Proc. the 2nd Int'l Conference on Computers and Games*, Lecture Notes in Computer Science 2063, 253–263, 2000.
4. M. Crasmaru, *PSPACE versus EXP-TIME and the game of GO*, Master Thesis, Tokyo Institute of Technology, 2001.
5. S. Fenner, L. Fortnow, and S. Kurtz, Gap-definable counting classes, *J. Comp. Sys. Sci.*, 48 116–184, 1994.
6. J.M. Robson, Combinatorial games with exponential space complete decision problems, in *Proc. of Mathematical Foundations of Computer Science*, Lecture Notes in Computer Science 176, 498–506, 1984.
7. W.J. Savitch, Relationship between nondeterministic and deterministic tape classes, *J. Comput. Syst. Sci.*, 4, 177-192, 1970.
8. L.G. Valiant and V.V. Vazirani, NP is as easy as detecting unique solutions, *Theoret. Comput. Sci.*, 47, 85–93, 1986.

Bi-Immunity Separates Strong
NP-Completeness Notions

A. Pavan[*1] and Alan L. Selman[2]

[1] NEC Research Institute, 4 Independence way, Princeton, NJ 08540.
apavan@research.nj.nec.com
[2] Department of Computer Science and Engineering, University at Buffalo, Buffalo,
NY 14260. selman@cse.buffalo.edu

Abstract. We prove that if for some $\epsilon > 0$, NP contains a set that is
DTIME(2^{n^ϵ})-bi-immune, then NP contains a set that is 2-Turing complete
for NP (hence 3-truth-table complete) but not 1-truth-table complete for NP. Thus this hypothesis implies a strong separation of completeness notions for NP. Lutz and Mayordomo [LM96] and Ambos-Spies and Bentzien [ASB00] previously obtained the same consequence using strong hypotheses involving resource-bounded measure and/or category theory. Our hypothesis is weaker and involves no assumptions about stochastic properties of NP.

1   Introduction

We obtain a strong separation of polynomial-time completeness notions under the hypothesis that for some $\epsilon > 0$, NP contains a set that is DTIME(2^{n^ϵ})-bi-immune. We prove under this hypothesis that NP contains a set that is $\leq^P_{2-T}$-complete (hence $\leq^P_{3-tt}$-complete) for NP but not $\leq^P_{1-tt}$-complete for NP. In addition, we prove that if for some $\epsilon > 0$, NP $\cap$ co-NP contains a set that is DTIME(2^{n^ϵ})-bi-immune, then NP contains a set that is $\leq^P_{2-tt}$-complete for NP but not $\leq^P_{1-tt}$-complete for NP. (We review common notation for polynomial-time reducibilities in the next section.)

The question of whether various completeness notions for NP are distinct has a very long history [LLS75], and has always been of interest because of the surprising phenomenon that no natural NP-complete problem has ever been discovered that requires anything other than many-one reducibility for proving its completeness. This is in contrast to the situation for NP-hard problems. There exist natural, combinatorial problems that are hard for NP using Turing reductions that have not been shown to be hard using nonadaptive reductions [JK76]. The common belief is that NP-hardness requires Turing reductions, and this intuition is confirmed by the well-known result that if P $\neq$ NP, then there are sets that are hard for NP using Turing reductions that are not hard for NP using many-one reductions [SG77].

There have been few results comparing reducibilities within NP, and we have known very little concerning various notions of NP-completeness. The first result to distinguish reducibilities within NP is an observation of Wilson in one

---

[*] Work done while the author was at University at Buffalo.

H. Alt and A. Ferreira (Eds.): STACS 2002, LNCS 2285, pp. 408–418, 2002.
© Springer-Verlag Berlin Heidelberg 2002

of Selman's papers on p-selective sets [Sel82]. It is a corollary of results there that if NE $\cap$ co-NE $\neq$ E, then there exist sets A and B belonging to NP such that $A \leq^P_{postt} B$, $B \leq^P_{tt} A$, and $B \not\leq^P_{postt} A$, where $\leq^P_{postt}$ denotes positive truth-table reducibility. Regarding completeness, Longpré and Young [LY90] proved that there are $\leq^P_m$-complete sets for NP for which $\leq^P_T$-reductions to these sets are *faster*, but they did not prove that the completeness notions differ. Lutz and Mayordomo [LM96] were the first to give technical evidence that $\leq^P_T$- and $\leq^P_m$-completeness for NP differ. They proved that if the p-measure of NP is not zero, then there exists a $\leq^P_{2-T}$-complete language for NP that is not $\leq^P_m$-complete. Ambos-Spies and Bentzien [ASB00] extended this result significantly. They used an hypothesis of resource-bounded category theory that asserts that "NP has a p-generic language, " which is weaker than the hypothesis of Lutz and Mayordomo, to separate nearly all NP-completeness notions for the bounded truth-table reducibilities, including the consequence obtained by Lutz and Mayordomo.

Here we prove that the consequence of Lutz and Mayordomo follows from the hypothesis that NP contains a DTIME(2^{n^ϵ})-bi-immune language. This hypothesis is weaker than the genericity hypothesis in the sense that the genericity hypothesis implies the existence of a 2^{n^ϵ}-bi-immune language in NP. Indeed, there exists a DTIME(2^{n^ϵ})-bi-immune language, in EXP, that is not p-generic [PS01]. Notably, our hypothesis, unlike either the measure or genericity hypotheses, involves no stochastic assumptions about NP.

Pavan and Selman [PS01] proved that if for some $\epsilon > 0$, NP $\cap$ co-NP contains a set that is DTIME(2^{n^ϵ})-bi-immune, then there exists a $\leq^P_T$-complete set for NP that is not $\leq^P_m$-complete. The results that we present here are significantly sharper. Also, they introduced an Hypothesis H from which it follows that there exists a $\leq^P_T$-complete set for NP that is not $\leq^P_{tt}$-complete. We do not need to state this hypothesis here. Suffice it to say that if for some $\epsilon > 0$, UP $\cap$ co-UP contains a DTIME(2^{n^ϵ})-bi-immune set, then Hypothesis H is true. Thus, we may partially summarize the results of the two papers as follows:

1. If for some $\epsilon > 0$, NP contains a DTIME(2^{n^ϵ})-bi-immune set, then NP contains a set that is $\leq^P_{2-T}$-complete (hence $\leq^P_{3-tt}$-complete) that is not $\leq^P_{1-tt}$-complete.
2. If for some $\epsilon > 0$, NP $\cap$ co-NP contains a DTIME(2^{n^ϵ})-bi-immune set, then NP contains a set that is $\leq^P_{2-tt}$-complete that is not $\leq^P_{1-tt}$-complete.
3. If for some $\epsilon > 0$, UP $\cap$ co-UP contains a DTIME(2^{n^ϵ})-bi-immune set, then NP contains a set that is $\leq^P_T$-complete that is not $\leq^P_{tt}$-complete.

2   Preliminaries

We use standard notation for polynomial-time reductions [LLS75] and we assume that readers are familiar with Turing, $\leq^P_T$, and many-one, $\leq^P_m$, reducibilities. Given any positive integer $k > 0$, a *k-Turing* reduction ($\leq^P_{k-T}$) is a Turing reduction that on each input word makes at most k queries to the oracle. A set

A is *truth-table* reducible to a set B $(A \leq_{tt}^{P} B)$ if there exist polynomial-time computable functions g and h such that on input x, $g(x)$, for some $m \geq 0$, is (an encoding of) a set of queries $Q = \{q_1, q_2, \cdots, q_m\}$, and $x \in A$ if and only if $h(x, B(q_1), \cdots, B(q_m)) = 1$. For a constant $k > 0$, A is *k-truth-table* reducible to B $(A \leq_{k\text{-}tt}^{P} B)$ if for all x, $\|Q\| = k$. Given a polynomial-time reducibility $\leq_r^P$, recall that a set S is $\leq_r^P$-complete for NP if $S \in$ NP and every set in NP is $\leq_r^P$-reducible to S.

A language is DTIME($T(n)$)-*complex* if L does not belong to DTIME($T(n)$) almost everywhere; that is, every Turing machine M that accepts L runs in time greater than $T(|x|)$, for all but finitely many words x. A language L is *immune* to a complexity class $\mathcal{C}$, or $\mathcal{C}$-*immune*, if L is infinite and no infinite subset of L belongs to $\mathcal{C}$. A language L is *bi-immune* to a complexity class $\mathcal{C}$, or $\mathcal{C}$-*bi-immune*, if both L and $\overline{L}$ are $\mathcal{C}$-*immune*. Balcázar and Schöning [BS85] proved that for every time-constructible function T, L is DTIME($T(n)$)-complex if and only if L is bi-immune to DTIME($T(n)$). We will use the following property of bi-immune sets. See Balcázar *et al.* [BDG90] for a proof.

Proposition 1. *Let L be a* DTIME($T(n)$)-*bi-immune language and A be an infinite set in* DTIME($T(n)$). *Then both $A \cap L$ and $A \cap \overline{L}$ are infinite.*

3   Separation Results

Our first goal is to separate $\leq_{2\text{-}T}^{P}$-completeness from $\leq_m^P$-completeness under the assumption that NP contains a DTIME(2^{2n})-bi-immune language.

Theorem 1. *If NP contains a* DTIME(2^{2n})-*bi-immune language, then NP contains a $\leq_{2\text{-}T}^{P}$-complete set S that is not $\leq_m^P$-complete.*

Proof. Let L be a DTIME(2^{2n})-bi-immune language in NP. Let $k > 0$ be a positive integer such that $L \in$ DTIME(2^{n^k}). Let M decide L in 2^{n^k} time. Define

$$t_1 = 2^k, \text{ and, for } i \geq 1,$$
$$t_{i+1} = (t_i)^{k^2},$$

and, for each $i \geq 1$, define

$$I_i = \{x \mid t_i^{1/k} \leq |x| < t_i^k\}.$$

Observe that $\{I_i\}_{i \geq 1}$ partitions $\Sigma^* - \{x \mid |x| < 2\}$. Define the following sets:

$$E = \cup_{i \text{ even}} I_i,$$
$$O = \cup_{i \text{ odd}} I_i,$$
$$L_e = L \cap E,$$
$$L_o = L \cap O,$$
$$\text{PadSAT} = \text{SAT} \cap E.$$

Since L belongs to NP, L_e and L_o also belong to NP. We can easily see that PadSAT is NP-complete.

We now define our $\leq_{2-T}^P$-complete set S. To simplify the notation we use a three letter alphabet.

$$S = 0(L_e \cup \text{PadSAT}) \cup 1(L_e \cap \text{PadSAT}) \cup 2L_e.$$

It is easy to see that S is $\leq_{2-T}^P$-complete: To determine whether a string x belongs to PadSAT, first query whether $x \in L_e$. If $x \in L_e$, then $x \in \text{PadSAT}$ if and only if $x \in (L_e \cap \text{PadSAT})$, and, if $x \notin L_e$, then $x \in \text{PadSAT}$ if and only if $x \in (L_e \cup \text{PadSAT})$. The same reduction, since it consists of three distinct queries, demonstrates also that S is $\leq_{3-tt}^P$-complete for NP.

The rest of the proof is to show that S is not $\leq_m^P$-complete for NP. So assume otherwise and let f be a polynomial-time computable many-one reduction of L_o to S. We will show this contradicts the hypothesis that L is DTIME(2^{2n})-bi-immune.

We need the following lemmas about L_o. Note that $L_o \subseteq O$.

Lemma 1. *Let A be an infinite subset of O that can be decided in 2^{2n} time. Then both the sets $A \cap L_o$ and $A \cap \overline{L_o}$ are infinite.*

Proof. Since A is a subset of O, a string x in A belongs to L_o if and only if it belongs to L. Thus $A \cap L_o$ is infinite if and only if $A \cap L$ is infinite. Similarly, $A \cap \overline{L_o}$ is infinite if and only if $A \cap \overline{L}$ is infinite. Since A can be decided in 2^{2n} time, and L is 2^{2n}-bi-immune, by Proposition 1, both the sets $A \cap L$ and $A \cap \overline{L}$ are infinite. Thus, $A \cap L_o$ and $A \cap \overline{L_o}$ are infinite. ∎

Lemma 2. *Let A belong to DTIME(2^{n^k}), and suppose that g is a $\leq_m^P$-reduction from L_o to A. Then the set*

$$T = \{x \in O \mid |g(x)| < |x|^{1/k}\}$$

is finite.

Proof. It is clear that $T \in P$. Recall that M is a deterministic algorithm that correctly decides L. Let N decide A in 2^{n^k} time. The following algorithm correctly decides L and runs in 2^n time on all strings belonging to T: On input x, if x does not belong to T, then run M on x. If $x \in T$, then $x \in L$ if and only if $x \in L_o$, so run N on $g(x)$ and accept if and only if N accepts $g(x)$. N takes $2^{|g(x)|^k}$ steps on $g(x)$. Since $|g(x)| < |x|^{1/k}$, N runs in $2^{|x|}$ time. Thus, the algorithm runs in 2^n steps on all strings belonging to T. Unless T is finite, this contradicts the fact that L is DTIME(2^{2n})-bi-immune. ∎

Next we show that the reduction should map almost all the strings of O to strings of form by, where $y \in E$ and $b \in \{0, 1, 2\}$.

Lemma 3. *Let*

$$A = \{x \mid x \in O, f(x) = by, \text{ and } y \in O\}.$$

Then A is finite.

Proof. It is easy to see that A belongs to P. Both PadSAT and L_e are subsets of E. Thus if a string by belongs to S, where $b \in \{0, 1, 2\}$, then $y \in E$. For every string x in A, $f(x) = by$ and $y \in O$. Thus $by \notin S$, which implies, since f is a many-one reduction from L_o to S, that $x \notin L_o$. Thus $A \cap L_o$ is empty. Since $A \subseteq O$, if A were infinite, then this would contradict Lemma 1, so A is finite. ∎

Thus, for all but finitely many x, if $x \in O$ and $f(x) = by$, then $y \in E$. Now we consider the following set B,

$$B = \{x \mid |x| = t_i \text{ and } i \text{ is odd}\}.$$

Observe that $B \in$ P and that B is an infinite subset of O. Thus, by Lemma 1, $B \cap L_o$ is an infinite set. Since, for all strings x, $x \in L_o \Leftrightarrow f(x) \in S$, it follows that f maps infinitely many of the strings in B into S. The rest of the proof is dedicated to showing a contradiction to this fact. Exactly, we define the sets

$$B_0 = \{x \in B \mid f(x) = 0y\},$$
$$B_1 = \{x \in B \mid f(x) = 1y\}, \text{ and}$$
$$B_2 = \{x \in B \mid f(x) = 2y\},$$

and we prove that each of these sets is finite.

Lemma 4. B_0 *is finite.*

Proof. Assume B_0 is infinite. Let

$$C = \{x \in B_0 \mid f(x) = 0y \text{ and } y \in E\}.$$

Since B_0 is a subset of O, by Lemma 3, for all but finitely strings in B_0, if $f(x) = 0y$, then $y \in E$. Thus B_0 is infinite if and only if C is infinite.

Consider the following partition of C.

$$C_1 = \{x \in C \mid f(x) = 0y, |y| < |x|^{1/k}\},$$
$$C_2 = \{x \in C \mid f(x) = 0y, |x|^{1/k} \leq |y| < |x|^k\},$$
$$C_3 = \{x \in C \mid f(x) = 0y, |y| \geq |x|^k\}.$$

We will show that each of the sets C_1, C_2, and C_3 is finite.

Claim 1 C_1 *is finite.*

Proof. Since $S \in \text{DTIME}(2^{n^k})$, the claim follows from Lemma 2. ∎

Claim 2 C_2 *is the empty set.*

Proof. Assume that $x \in C_2$. Since $C_2 \subseteq C \subseteq B$, $|x| = t_i$, for some odd i. So, $|x|^{1/k} \leq |y| < |x|^k$ implies that $t_i^{1/k} \leq |y| < t_i^k$, which implies $y \in I_i$. Since i is odd, $y \in O$. However, by definition of C, $y \in E$. Thus, $C_2 = \emptyset$. ∎

Claim 3 C_3 *is finite.*

Proof. Observe that $C_3 \in P$. Suppose C_3 is infinite. Define $C_4 = C_3 - L_o$. We first show, under the assumption C_3 is infinite, that C_4 is infinite. Suppose C_4 is finite. Then the set $C_5 = C_3 \cap L_o$ differs from C_3 by a finite set. Thus, since $C_3 \in P$, $C_5 \in P$ also. At this point, we know that C_5 is an infinite subset of O that belongs to P, and that C_5 is a subset of L_o. Thus, $C_5 \cap \overline{L_o}$ is empty, which contradicts Lemma 1. Thus, C_4 is an infinite subset of C_3.

Let

$$F = \{y \in E \mid \exists x \, [x \in O, x \notin L_o, f(x) = 0y, \text{ and } |y| \geq |x|^k]\}.$$

The following implications show that F is infinite:

$$C_4 \text{ is infinite}$$
$$\Rightarrow$$
$$\exists^\infty x \, [x \in O, x \notin L_o, f(x) = 0y, |y| \geq |x|^k, y \in E]$$
$$\Rightarrow$$
$$\exists^\infty y \in E \, [\exists x \, x \in O, x \notin L_o, f(x) = 0y, |y| \geq |x|^k].$$

For each string y in F, there exists a string $x \in O - L_o$ such that $f(x) = 0y$. Since f is a many-one reduction from L_o to S, $f(x) = 0y \notin S$. Thus $y \notin L_e \cup \text{PadSAT}$, and so $y \notin L_e$. However, since $y \in E$, we conclude that $y \notin L$. Thus, F is an infinite subset of $\overline{L}$.

Now we contradict the fact that L is DTIME(2^{2n})-bi-immune by showing that F is decidable in time 2^{2n}. Let y be an input string. First decide, in polynomial time, whether y belongs to E. If $y \notin E$, then $y \notin F$. If $y \in E$, compute the set of all x such that $|x| \leq |y|^{1/k}$, $x \in O$, and $f(x) = 0y$. Run M on every string x in this set until M rejects one of them. Since $x \in O$, M rejects a string x only if $x \notin L_o$. If such a string is found, then $y \in F$, and otherwise $y \notin F$. There are at most $2 \times 2^{|y|^{1/k}}$ many x's such that $|x| \leq |y|^{1/k}$ and $f(x) = 0y$. The time taken to run M on each such x is at most $2^{|x|^k} \leq 2^{|y|}$. Thus, the total time to decide whether $y \in F$ is at most $2^{|y|} \times 2^{|y|^{1/k}} \times 2 \leq 2^{2|y|}$. Thus, F is decidable in time 2^{2n}.

We conclude that F must be a finite set. Therefore, C_4 is finite, from which it follows that C_3 is finite. ∎

Each of the claims is established. Thus, $C = C_1 \cup C_2 \cup C_3$ is a finite set, and this proves that B_0 is a finite set. ∎

Lemma 5. B_1 *is a finite set.*

Proof. Much of the proof is similar to the proof of Lemma 4. Assume that B_1 is infinite. This time, define

$$C = \{x \in B_1 \mid f(x) = 1y \text{ and } y \in E\}.$$

By Lemma 3, C is infinite if and only if B_1 is infinite. Thus, by our assumption, C is infinite. Partition C as follows.

$$C_1 = \{x \in C \mid f(x) = 1y, |y| < |x|^{1/k}\}$$
$$C_2 = \{x \in C \mid f(x) = 1y, |x|^{1/k} \le |y| < |x|^k\}$$
$$C_3 = \{x \in C \mid f(x) = 1y, |y| \ge |x|^k\}$$

As in the proof of Lemma 4, we can show that C_1 is a finite set and C_2 is empty. Now we proceed to show that C_3 is also a finite set.

Claim 4 C_3 *is finite.*

Proof. Assume C_3 is infinite and observe that $C_3 \in$ P. Define $C_4 = C_3 \cap L_o$. Now we show that C_4 is infinite. If C_4 is finite, then $C_5 = C_3 - L_o$ contains all but finitely many strings of C_3. Thus, since C_3 belongs to P, C_5 also belongs to P. Thus C_5 is an infinite subset of O that belongs to P, for which $C_5 \cap L_o$ is empty. That contradicts Lemma 1. Thus, C_4 is infinite.

Consider the following set:

$$F = \{y \in E \mid \exists x[x \in L_o, f(x) = 1y, |y| \ge |x|^k]\}$$

The following implications show that F is infinite.

$$C_4 \text{ is infinite}$$
$$\Rightarrow$$
$$\exists^\infty x \ [x \in L_o, f(x) = 1y, |y| \ge |x|^k, y \in E]$$
$$\Rightarrow$$
$$\exists^\infty y[\exists x \ f(x) = 1y, |y| \ge |x|^k, x \in L_o, y \in E].$$

For each string $y \in F$, there exists a string $x \in L_o$ such that $f(x) = 1y$. Since f is a $\le_m^P$-reduction from L_o to S, $f(x) = 1y \in S$, so $y \in L_e \cap$ PadSAT. In particular, $y \in L_e \subseteq L$. Therefore, F is an infinite subset of L. However, as in the proof of Claim 3, we can decide whether $y \in F$ in $2^{2|y|}$ steps, which contradicts the fact that L is DTIME(2^{2n})-bi-immune: Let y be an input string. First decide whether $y \in E$, and if not, then reject. If $y \in E$, then search all strings x such that $|x| \le |y|^{1/k}$, $x \in O$, and $f(x) = 1y$. For each such x, run M on x to determine whether $x \in L \cap O = L_o$. If an $x \in L_o$ is found, then $y \in F$, and otherwise $y \notin F$. The proof that this algorithm runs in 2^{2n} steps is identical to the argument in the proof of Claim 3.

Therefore, F is finite, from which it follows that C_4 is finite, and so C_3 must be finite. ∎

Now we know that C is finite. This proves that B_1 is finite, which completes the proof of Lemma 5. ∎

Lemma 6. B_2 *is a finite set.*

Proof. Assume B_2 is infinite. Then

$$C = \{x \in B \mid f(x) = 2y, \text{ and } y \in E\}$$

is infinite. We partition C into

$$C_1 = \{x \in C \mid f(x) = 2y, |y| < |x|^{1/k}\}$$
$$C_2 = \{x \in C \mid f(x) = 2y, |x|^{1/k} \leq |y| < |x|^k\}$$
$$C_3 = \{x \in C \mid f(x) = 2y, |y| \geq |x|^k\}$$

The proofs that C_1, C_2, and C_3 are finite are identical to the arguments in the proof of Lemma 5. (In particular, it suffices to define F as in the proof of Lemma 5.) ∎

Now we have achieved our contradiction, for we have shown that the each of the sets B_1, B_2, and B_3 are finite. Therefore, f cannot map infinitely many of the strings in B into S, which proves that f cannot be a $\leq_m^P$-reduction from L_o to S. Therefore, S is not $\leq_m^P$-complete. ∎

Next we show that NP has a $\mathrm{DTIME}(2^{n^\epsilon})$-bi-immune set if and only if NP has a $\mathrm{DTIME}(2^{n^k})$-bi-immune set using a reverse padding trick [ASTZ97].

Theorem 2. *Let* $0 < \epsilon < 1$ *and* k *be any positive integer.* NP *has a* $\mathrm{DTIME}(2^{n^\epsilon})$-*bi-immune set if and only if* NP *has a* $\mathrm{DTIME}(2^{n^k})$-*bi-immune set.*

Proof. The implication from right to left is obvious. Let $L \in$ NP be a $\mathrm{DTIME}(2^{n^\epsilon})$-bi-immune set. Define

$$L' = \{x \mid 0^{n^{k/\epsilon}} x \in L, |x| = n\}$$

and observe that $L' \in$ NP. We claim that L' is $\mathrm{DTIME}(2^{n^k})$-bi-immune. Suppose otherwise. Then there exists an algorithm M that decides L' and M runs in 2^{n^k} steps on infinitely many strings. Consider the following algorithm for L:

input y;
if $y = 0^{n^{k/\epsilon}} x$ $(|x| = n)$
    then run M on x
        and accept y if and only if M accepts x
    else run a machine that decides L;

Since M runs in 2^{n^k} time on infinitely many x, the above algorithm runs in time $2^{|x|^k}$ steps on infinitely many strings of the form $y = 0^{|x|^{k/\epsilon}} x$. Observe that $|y| \geq |x|^{n^{k/\epsilon}}$. Thus, the above algorithm runs in $2^{|y|^\epsilon}$ steps on infinitely many y. This contradicts the $\mathrm{DTIME}(2^{n^\epsilon})$-bi-immunity of L. ∎

Corollary 1. *If* NP *contains a* 2^{n^ϵ}-*bi-immune language, then* NP *contains a* $\leq_{2-T}^P$-*complete set* S *that is not* $\leq_m^P$-*complete.*

The proof of the next theorem shows that we can extend the proof of Theorem 1 to show that the set S defined there is not $\leq^P_{1-tt}$-complete. Thus, we arrive at our main result.

Theorem 3. *If* NP *contains a* 2^{n^ϵ}*-bi-immune language, then* NP *contains a* $\leq^P_{2-T}$*-complete set* S *that is not* $\leq^P_{1-tt}$*-complete.*

Proof. The proof is a variation of the proof of Theorem 1, and we demonstrate the interesting case only. Assume that the set S defined there is $\leq^P_{1-tt}$-complete and let (g, h) be a 1-truth-table reduction from L_o to S. Recall that, for each string x, $g(x)$ is a query to S and that

$$x \in L_o \Leftrightarrow h(x, S(g(x))) = 1.$$

The function h on input x implicitly defines four possible truth-tables. Let us define the sets

$$T = \{x \mid h(x, 1) = 1 \text{ and } h(x, 0) = 1\},$$
$$F = \{x \mid h(x, 1) = 0 \text{ and } h(x, 0) = 0\},$$
$$Y = \{x \mid h(x, 1) = 1 \text{ and } h(x, 0) = 0\},$$
$$N = \{x \mid h(x, 1) = 0 \text{ and } h(x, 0) = 1\}.$$

Each of the sets T, F, Y, and N belongs to P. Also, $T \subseteq L_o$, $F \subseteq \overline{L_o}$, for all strings $x \in Y$,

$$x \in L_o \Leftrightarrow x \in S,$$

and for all strings $x \in N$,

$$x \in L_o \Leftrightarrow x \in \overline{S}.$$

It follows immediately that T and F are finite sets. Now, as we did in the proof of Theorem 1, we consider the set $B = \{x \mid |x| = t_i \text{ and } i \text{ is odd}\}$. Recall that $B \in$ P and that B is an infinite subset of O. For all but finitely many strings $x \in B$, either $x \in Y$ or $x \in N$. In order to illustrate the interesting case, let us assume that $B^N = B \cap N$ is infinite. Note that $B^N \in$ P and that B^N is an infinite subset of O. By Lemma 1, $B^N \cap \overline{L_o}$ is infinite. For all $x \in B^N$, $x \in \overline{L_o} \Leftrightarrow x \in S$. Thus, g maps infinitely many of the strings in B^N into S. Similar to our earlier analysis, we contradict this by showing that each of the following sets is finite:

$$B_0 = \{x \in B^N \mid g(x) = 0y\},$$
$$B_1 = \{x \in B^N \mid g(x) = 1y\},$$
$$B_2 = \{x \in B^N \mid g(x) = 2y\}.$$

Here we will demonstrate that B_0 is finite. The other cases will follow similarly.

Define $A = \{x \in B_0 \mid g(x) = by, \text{ and } y \in O\}$. Again we need to show that A is a finite set, but we need a slightly different proof from that for Lemma 3. Note that $A \in P$. If $g(x) = 0y \in S$, then $y \in E$. Thus, $x \in A \Rightarrow g(x) \notin S \Rightarrow x \in L_o$. Thus $A \subseteq L_o$, from which it follows that A is finite. Hence, the set

$$C = \{x \in B_0 \mid g(x) = 0y \text{ and } y \in E\}$$

is an infinite set. As earlier, we partition C into the sets

$$C_1 = \{x \in C \mid f(x) = 0y, |y| < |x|^{1/k}\},$$
$$C_2 = \{x \in C \mid f(x) = 0y, |x|^{1/k} \leq |y| < |x|^k\},$$
$$C_3 = \{x \in C \mid f(x) = 0y, |y| \geq |x|^k\},$$

and we show that each of these sets is finite. To show that C_1 is finite, we show more generally, as in the proof of Lemma 2, that $V = \{x \in B^N \mid |g(x)| < |x|^{1/k}\}$ is a finite set. (The critical fact is that for $x \in V$, $x \in S \Leftrightarrow x \in \overline{L_o} \Leftrightarrow x \notin L$, because $V \subseteq O$.) Also, it is easy to see that $C_2 = \emptyset$.

We need to show that C_3 is finite. Assume that C_3 is infinite. Noting that $C_3 \in P$, the proof of Claim 4 (not Claim 3!) shows that the set $C_4 = C_3 \cap L_o$ is infinite. Then,

$$\exists^\infty x[x \in C_4, g(x) = 0y, |y| < |x|^{1/k}]$$
$$\Rightarrow$$
$$\exists^\infty x[x \in B^N, x \in L_o, y \in E, g(x) = 0y, |y| < |x|^{1/k}]$$
$$\Rightarrow$$
$$\exists^\infty y \exists x[x \in B^N, x \in L_o, y \in E, g(x) = 0y, |y| < |x|^{1/k}].$$

Thus, the set

$$U = \{y \mid \exists x[x \in B^N, x \in L_o, y \in E, g(x) = 0y, |y| < |x|^{1/k}]\}$$

is infinite. For each string $y \in U$, there exists $x \in B^N \cap L_o$ such that $g(x) = 0y$. For each such x, $g(x) = 0y \in \overline{S}$. Thus, $y \notin L_e \cup \text{PadSAT}$, so, in particular, $y \notin L_e$. However, $y \in E$, so $y \in L$. Thus, U is an infinite subset of $\overline{L}$.

Now we know that C is finite, from which it follows that B_0 is a finite set. In a similar manner we can prove that B_1 and B_2 are finite, which completes the proof of the case that B^N is infinite. The other possibility, that $B^Y = B \cap Y$ is infinite can be handled similarly. ∎

There is no previous work that indicates a separation of $\leq^P_{2-tt}$-completeness from $\leq^P_{1-tt}$-completeness. Our next result accomplishes this, but with a stronger hypothesis.

Theorem 4. *If* $\mathrm{NP} \cap \mathrm{co}\text{-}\mathrm{NP}$ *contains a* 2^{n^ϵ}*-bi-immune set, then* NP *contains a* $\leq^P_{2-tt}$*-complete set that is not* $\leq^P_{1-tt}$*-complete.*

Proof. The hypothesis implies the existence of a 2^{n^k}-bi-immune language L in $\mathrm{NP} \cap \mathrm{co}\text{-}\mathrm{NP}$. Let

$$S = 0(L_e \cap \text{PadSAT}) \cup 1((E - L_e) \cap \text{PadSAT}).$$

Since L belongs to NP $\cap$ co-NP, S belongs to NP. Since both PadSAT and L_e are subsets of E, for any string x

$$x \in \text{PadSAT} \Leftrightarrow (x \in L_e \cap \text{PadSAT}) \vee (x \in (E - L_e) \cap \text{PadSAT}).$$

Thus S is 2-tt-complete for NP. The rest of the proof is similar to the proof of Theorem 3.   ∎

Acknowledgements. The authors benefitted from conversations about this work with Osamu Watanabe, and with Lance Fortnow and Jack Lutz.

References

[ASB00]   K. Ambos-Spies and L. Bentzien. Separating NP-completeness under strong hypotheses. *Journal of Computer and System Sciences*, 61(3):335–361, 2000.

[ASTZ97]   K. Ambos-Spies, A. Terwijn, and X. Zheng. Resource bounded randomness and weakly complete problems. *Theoretical Computer Science*, 172(1):195–207, 1997.

[BDG90]   J. Balcázar, J. Diaz, and J. Gabarró. *Structural Complexity II*. Springer-Verlag, Berlin, 1990.

[BS85]   J. Balcázar and U. Schöning. Bi-immune sets for complexity classes. *Mathematical Systems Theory*, 18(1):1–18, June 1985.

[JK76]   D. Johnson and S. Kashdan. Lower bounds for selection in $x + y$ and other multisets. Technical Report 183, Pennsylvania State Univ., University Park, PA, 1976.

[LLS75]   R. Ladner, N. Lynch, and A. Selman. A comparison of polynomial time reducibilities. *Theoretical Computer Science*, 1:103–123, 1975.

[LM96]   J. Lutz and E. Mayordomo. Cook versus Karp-Levin: Separating completeness notions if NP is not small. *Theoretical Computer Science*, 164:141–163, 1996.

[LY90]   L. Longpré and P. Young. Cook reducibility is faster than Karp reducibility. *Journal of Computer and System Sciences*, 41:389–401, 1990.

[PS01]   A. Pavan and A. Selman. Separation of NP-completeness notions. In *16th Annual IEEE Conference on Computational Complexity*, pages 78–89, 2001.

[Sel82]   A. Selman. Reductions on NP and P-selective sets. *Theoretical Computer Science*, 19:287–304, 1982.

[SG77]   I. Simon and J. Gill. Polynomial reducibilities and upward diagonalizations. *Proceedings of the Ninth Annual ACM Symposium on Theory of Computing*, pages 186–194, 1977.

Complexity of Semi-algebraic Proofs[*]

Dima Grigoriev[1], Edward A. Hirsch[2,**], and Dmitrii V. Pasechnik[3]

[1] IRMAR, Université de Rennes, Campus de Beaulieu, 35042 Rennes, cedex France.
dima@maths.univ-rennes1.fr.
http://www.maths.univ-rennes1.fr/~dima/
[2] Steklov Institute of Mathematics at St.Petersburg, 27 Fontanka, 191011
St.Petersburg, Russia.
hirsch@pdmi.ras.ru.   http://logic.pdmi.ras.ru/~hirsch/
[3] Department of Technical Mathematics and Informatics, Faculty ITS, Delft
University of Technology, Mekelweg 4, 2628 CD Delft, The Netherlands.
d.pasechnik@its.tudelft.nl.
http://ssor.twi.tudelft.nl/~dima/

Abstract. Proof systems for polynomial inequalities in 0-1 variables include the well-studied *Cutting Planes* proof system (CP) and the *Lovász-Schrijver calculi* (LS) utilizing linear, respectively, quadratic, inequalities. We introduce generalizations LS^d of LS involving polynomial inequalities of degree at most d.

Surprisingly, the systems LS^d turn out to be very strong. We construct polynomial-size bounded degree LS^d proofs of the *clique-coloring tautologies* (which have no polynomial-size CP proofs), the *symmetric knapsack problem* (which has no bounded degree Positivstellensatz Calculus (PC) proofs), and *Tseitin's tautologies* (hard for many known proof systems). Extending our systems with a division rule yields a polynomial simulation of *CP with polynomially bounded coefficients*, while other extra rules further reduce the proof degrees for the aforementioned examples.

Finally, we prove lower bounds on Lovász-Schrijver *ranks*, demonstrating, in particular, their rather limited applicability for proof complexity.

1 Introduction

An observation that a propositional formula can be written as a system of polynomial equations has lead to considering, in particular, the Nullstellensatz (NS) and the Polynomial Calculus (PC) proof systems, see Sect. 2.1 below (we do not dwell much here on the history of this rich area, several nice historical overviews one could find in e.g., [2,3,4,5,6,7]).

---

[*] Due to the space considerations we had to shorten this text rather drastically. For the full version, that also contains new results not mentioned here, the reader is referred to [1].

[**] A part of this work was completed while visiting Delft University of Technology. Partially supported by grant #1 of the 6th RAS contest-expertise of young scientists projects (1999) and a grant from NATO.

H. Alt and A. Ferreira (Eds.): STACS 2002, LNCS 2285, pp. 419–430, 2002.
© Springer-Verlag Berlin Heidelberg 2002

For these proof systems several interesting complexity lower bounds on the degrees of the derived polynomials were obtained [4,5,7]. When the degree is close enough to linear (in fact, greater than the square root), these bounds imply exponential lower bounds on the proof complexity (more precisely, on the number of monomials in the derived polynomials) [5]. If polynomials are given by formulas rather than by sums of monomials as in NS or in PC, then the complexity could decrease significantly. Several gaps between these two kinds of proof systems were demonstrated in [8].

Systems of polynomial *inequalities* yield much more powerful proof systems than these operating with equations only, such as NS or PC. Historically first such system is Cutting Planes (CP) [9,10,11,12], see also Sect. 2.2. This system uses linear inequalities (with integer coefficients). Exponential lower bounds on proof size were established for CP with polynomially bounded coefficients [13] as well as for the general case [14].

Another family of well-studied proof systems are so-called Lovász-Schrijver calculi (LS) [15,16], see also [17] and Sect. 2.2 below. In these systems one is allowed to deal with quadratic inequalities. No non-trivial complexity lower bounds are known for them so far. Moreover, generalizing LS to systems LS^d that use inequalities of degree at most d (rather than 2 as in $LS=LS^2$) yields a very powerful proof system. In particular, there exists a short LS^4 proof of the clique-coloring tautologies (see Sect. 4). On the other hand, for these tautologies an exponential lower bound on the complexity of CP proofs was obtained in [14], relying on the lower bound for the monotone complexity [18]. Furthermore, we construct a short proof for the clique-coloring tautologies in the proof system $LS + CP^2$ (see Sect. 4) that manipulates just quadratic inequalities, endowed with the rounding rule (it generalizes directly the rounding rule for linear inequalities in CP). These results mean, in particular, that neither LS^4 nor $LS + CP^2$ have monotone effective interpolation, while for a system $LS + CP^1$ where the use of rounding rule is limited to linear inequalities, a (non-monotone) effective interpolation is known [17].

An analogue of (already mentioned) non-trivial lower bounds on the degree of derived polynomials in PC would fail in LS^d as we show in Sect. 3, namely, every system of inequalities of degree at most d having no real solutions posseses an LS^{2d} refutation.

A proof system manipulating polynomial inequalities called the Positivstellensatz Calculus was introduced in [19]. Lower bounds on the degree in this system were established for the parity principle, for Tseitin's tautologies [20] and for the knapsack problem [21]. Lower bounds on the Positivstellensatz Calculus degree are possible because its "dynamic" part is restricted to an ideal and an element of a cone is obtained from an element of ideal by adding the sum of squares to it. On the contrary, LS is a completely "dynamic" proof system. (The discussion on static and dynamic proof systems can be found in [19]. Briefly, the difference is that in LS a derivation constructs gradually an element of the cone generated by the input system of inequalities, while in the Positivstellensatz Calculus the sum of squares is given explicitly.)

Also the lower bound on the Positivstellensatz Calculus degree of the knapsack problem [21] entails (see Theorem 6) a lower bound on the so-called LS-rank [15,16]. Roughly speaking, the LS-rank counts the *depth* of multiplications invoked in a derivation. A series of lower bounds for various versions of the LS-rank were obtained in the context of optimization theory [22,23,24,25]. For a counterpart notion in CP, the so-called Chvátal rank [10], lower bounds were established in [11,12]. To the best of our knowledge, the connection between the Chvátal rank and CP proof complexity is not very well understood, despite a number of interesting recent results [26,27]. As a rule, however, diverse versions of the rank grow at most linearly, while we are looking for non-linear (exponential as a dream) lower bounds on the proof complexity. It turns out that for the latter purpose the rank is a too weak invariant. Indeed, there are short proofs for the pigeon-hole principle (PHP) in CP [11] and in LS [14], while we exhibit in Theorem 7 a linear lower bound on the LS-rank of the PHP. Another example of this sort is supplied by the symmetric knapsack problem for which in Sect. 5 we give a short LS^3-proof.

The above-mentioned LS^3-proof of the symmetric knapsack follows from a general fact that LS^d systems allow to reason about integers. In Sect. 6 we extend this technique to Tseitin's tautologies (which have no polynomial-size proofs in resolution [28], Polynomial Calculus [7] and bounded-depth Frege systems [29]). In Sect. 5 we also consider a certain extended version $LS_{*,split}$ of LS that, apart from the issue with integers, allows one to perform a trial of cases with respect to whether $f > 0$, $f < 0$, $f = 0$ for a linear function f (similar sorts of an extension of CP were introduced by Chvátal [unpublished] [17] and Krajíček [30]) and allows also to multiply inequalities. We show that $LS_{*,split}$ polynomially simulates CP with small coefficients. The same effect can be achieved by replacing the multiplication and the trial of cases by the *division rule* that derives $g \geq 0$ from $fg \geq 0$ and $f > 0$.

Finally, we formulate numerous open questions in Sect. 8.

2 Definitions

A *proof system* [31] for a language L is a polynomial-time computable function mapping words (proof candidates) onto L (whose elements are considered as theorems). A *propositional proof system* is a proof system for any fixed co-NP-complete language of Boolean tautologies (e.g., tautologies in DNF). When we have two proof systems Π_1 and Π_2 for the same language L, we can compare them. We say that Π_1 *polynomially simulates* Π_2, if there is a function g mapping proof candidates of Π_2 to proof candidates of Π_1 so that for every proof candidate π for Π_2, one has $\Pi_1(g(\pi)) = \Pi_2(\pi)$ and $g(\pi)$ is at most polynomially longer than π. Proof system Π_1 is *exponentially separated* from Π_2, if there is an infinite sequence of words $t_1, t_2, \ldots \in L$ such that the length of the shortest Π_1-proof of t_i is polynomial in the length of t_i, and the length of the shortest Π_2-proof of t_i is exponential. Proof system Π_1 is *exponentially stronger* than Π_2, if Π_1 polynomially simulates Π_2 and is exponentially separated from it.

When we have two proof systems for different languages L_1 and L_2, we can also compare them if we fix a reduction between these languages. However, it can be the case that the result of the comparison is more due to the reduction than to the systems themselves. Therefore, if we have propositional proof systems for languages L_1 and L_2, and the intersection $L = L_1 \cap L_2$ of these languages is co-NP-complete, we will compare these systems as systems[1] for L.

2.1   Proof Systems Manipulating with Polynomial Equations

There is a series of proof systems for languages consisting of unsolvable systems of polynomial equations. To transform such a proof system into a propositional proof system, one needs to translate Boolean tautologies into systems of polynomial equations.

To translate a formula F in k-DNF, we take its negation $\neg F$ in k-CNF and translate each clause of $\neg F$ into a polynomial equation. A clause containing variables $v_{j_1}, \ldots, v_{j_t}$ $(t \leq k)$ is translated into an equation

$$(1 - l_1) \cdot \ldots \cdot (1 - l_t) = 0, \tag{1}$$

where $l_i = v_{j_i}$ if variable v_{j_i} occurs positively in the clause, and $l_i = (1 - v_{j_i})$ if it occurs negatively. For each variable v_i, we also add the equation $v_i^2 - v_i = 0$ to this system.

Note that F is a tautology if and only if the obtained system S of polynomial equations $f_1 = 0$, $f_2 = 0$, $\ldots$, $f_m = 0$ has no solutions. Therefore, to prove F it suffices to derive a contradiction from S.

Nullstellensatz (NS) [2]. A proof in this system is a collection of polynomials $g_1, \ldots, g_m$ such that

$$\sum_i f_i g_i = 1.$$

Polynomial Calculus (PC) [6]. This system has two derivation rules:

$$\frac{p_1 = 0; \; p_2 = 0}{p_1 + p_2 = 0} \quad \text{and} \quad \frac{p = 0}{p \cdot q = 0}. \tag{2}$$

I.e., one can take a sum[2] of two already derived equations $p_1 = 0$ and $p_2 = 0$, or multiply an already derived equation $p = 0$ by an arbitrary polynomial q. The proof in this system is a derivation of $1 = 0$ from S using these rules.

---

[1] If one can decide in polynomial time for $x \in L_1$, whether $x \in L$, then any proof system for L_1 can be restricted to $L \subseteq L_1$ by mapping proofs of elements of $L_1 \setminus L$ into any fixed element of L. For example, this is the case for L_1 consisting of all tautologies in DNF and L consisting of all tautologies in k-DNF.

[2] Usually, an arbitrary linear combination is allowed, but clearly it can be replaced by two multiplications and one addition.

Positivstellensatz [19]. A proof in this system consists of polynomials $g_1, \ldots, g_m$ and $h_1, \ldots, h_l$ such that

$$\sum_i f_i g_i = 1 + \sum_j h_j^2 \tag{3}$$

Positivstellensatz Calculus [19]. A proof in this system consists of polynomials $h_1, \ldots, h_l$ and a derivation of $1 + \sum_j h_j^2 = 0$ from S using the rules (2).

2.2   Proof Systems Manipulating with Inequalities

To define a propositional proof system manipulating with inequalities, we again translate each formula $\neg F$ in CNF into a system S of linear inequalities, such that F is a tautology if and only if S has no 0-1 solutions. Given a Boolean formula in CNF, we translate each its clause containing variables $v_{j_1}, \ldots, v_{j_t}$ into the inequality

$$l_1 + \ldots + l_t \geq 1, \tag{4}$$

where $l_i = v_{j_i}$ if the variable v_{j_i} occurs positively in the clause, and $l_i = 1 - v_{j_i}$ if v_{j_i} occurs negatively. We also add to S the inequalities $x \geq 0$ and $x \leq 1$ for every variable x.

Cutting Planes (CP) [9,10,11,12], cf. also [17]. In this proof system, the system S defined above must be refuted (i.e., the contradiction $0 \geq 1$ must be obtained) using the following two derivation rules:

$$\frac{f_1 \geq 0; \; \ldots; \; f_t \geq 0}{\sum_{i=1}^t \lambda_i f_i \geq 0} \quad \text{(where } \lambda_i \geq 0\text{)}, \tag{5}$$

$$\frac{\sum_i a_i x_i \geq c}{\sum_i a_i x_i \geq \lceil c \rceil} \quad \text{(where } a_i \in \mathbb{Z}, \text{ and } x_i \text{ is a variable)}. \tag{6}$$

We restrict the intermediate inequalities in a CP derivation to the ones having integer coefficients (except the constant term).

Lovász-Schrijver calculus (LS) [15,16], cf. also [17]. In the weakest of Lovász-Schrijver proof systems, the contradiction must be obtained using the rule (5) applied to linear or quadratic f_i's and the rules

$$\frac{f \geq 0}{fx \geq 0}; \quad \frac{f \geq 0}{f(1-x) \geq 0} \quad \text{(where } f \text{ is linear, } x \text{ is a variable)}. \tag{7}$$

Also, the system S is extended by the axioms

$$x^2 - x \geq 0, \quad x - x^2 \geq 0 \tag{8}$$

for every variable x.

LS$_+$ [15,16,17]. This system has the same axioms and derivation rules as LS, and also has the axiom

$$l^2 \geq 0 \tag{9}$$

for every linear l.

LS$_*$ [15,16,17]. This system has the same axioms and derivation rules as LS, and also the derivation rule

$$\frac{f \geq 0;\ g \geq 0}{fg \geq 0} \quad (f, g \text{ are linear}). \tag{10}$$

LS$_{+,*}$. This system unites LS$_+$ and LS$_*$.

LS + CP^1 [17]. It has the same axioms and derivation rules as LS and also the rounding rule (6) of CP which can be applied only to linear inequalities.

Note that all Lovász-Schrijver systems described in this subsection deal either with linear or quadratic inequalities.

2.3   New Systems

In this paper we consider several extensions of Lovász and Schrijver proof systems. First, we define system LS + CP^2 which is slightly stronger than Pudlák's LS + CP^1.

LS + CP^2. It has the same axioms and rules as LS and also the extension of rounding rule (6) of CP to quadratic inequalities:

$$\frac{\sum_{i,j} a_{ij} x_i x_j + \sum_i a_i x_i \geq c}{\sum_{i,j} a_{ij} x_i x_j + \sum_i a_i x_i \geq \lceil c \rceil} \quad (\text{where } a_i, a_{ij} \in \mathbb{Z}, \text{ and } x_i \text{ is a variable}). \tag{11}$$

We then consider extensions of Lovász-Schrijver proof systems allowing monomials of degree up to d.

LS^d. This system is an extension of LS. The difference is that rule (7) is now restricted to f of degree at most $d - 1$ rather than to linear inequalities. Rule (5) can be applied to any inequality of degree at most d.

Similarly, we consider **LS$_*^d$**, transforming in (10) the condition "f, g are linear" into "$\deg(fg) \leq d$".

LS$_{\text{split}}^d$. This system allows not only inequalities of the form $f \geq 0$, but also of the form $f > 0$. The derivation rules (5) and (7) are extended in a clear way to handle both types of inequalities, and $f > 0$ can be always relaxed to $f \geq 0$. The axiom $1 > 0$ is added. Also if under each of the three assumptions $f > 0$, $f < 0$ and $f = 0$ (a shorthand for the two inequalities $f \geq 0$ and $f \leq 0$) there is an LS$_{\text{split}}^d$ derivation of inequality $h \geq 0$, then we say that $h \geq 0$ is derived in LS$_{\text{split}}^d$.

LS$_{*,\text{split}}^d$ is defined similarly. Note that the version of (10) for strict inequalities is

$$\frac{f > 0;\ g > 0}{fg > 0}.$$

LS$^d_{0/1\text{-split}}$ is a restricted version of $\text{LS}^d_{\text{split}}$ where the splitting is made for the assumptions $x = 0$, $x = 1$ only (x is a variable).

LS$^d_/$ is an extension of LS^d with strict inequalities by another useful rule:

$$\frac{fg \geq 0;\ f > 0}{g \geq 0}.$$

LS$_{\text{split}}$, **LS**$_{*,\text{split}}$, **etc.** are shorthands for the corresponding systems restricted to $d = 2$.

3 Encodings of Formulas in LS^d and Upper Bounds on the Refutation Degree

In LS^d, Boolean formulas are encoded as linear inequalities. However, this is not the only possible way to encode them, since in LS^d we can operate with polynomials of degree up to d. In particular, for formulas in k-CNF, one can use the same encoding as in Polynomial Calculus (1).

Consider system $\overline{\text{LS}^d}$ that has the same derivation rules as LS^d, but uses the encoding (1) instead of (4). It is clear that when $d = n$ is the number of variables, $\overline{\text{LS}^n}$ polynomially simulates Polynomial Calculus. Does LS^n polynomially simulate $\overline{\text{LS}^n}$ (and Polynomial Calculus)? To give the positive answer, it suffices to show that there is a polynomial-size derivation of the encoding by polynomial equations from the encoding by linear inequalities.

Lemma 1. *There is a polynomial-size LS^t derivation of (1) from (4), (8) and $0 \leq x \leq 1$.*

Corollary 1. *LS^d polynomially simulates $\overline{LS^d}$ (and, hence, LS^n polynomially simulates Polynomial Calculus).*

Corollary 2. *LS^n_+ polynomially simulates Positivstellensatz Calculus.*

It turns out that the converse of Lemma 1 is also true. In particular, due to [15, Theorem 1.4] that means that there is an LS^k refutation of every formula in k-CNF. Below, we also show (Theorem 1) that there is an LS^{2k} refutation of any system of polynomial inequalities of degree at most k.

Lemma 2. *There is a polynomial-size LS^t derivation of (4) from (1), (8), and $0 \leq x \leq 1$.*

Corollary 3. *$\overline{LS^d}$ polynomially simulates LS^d.*

Corollary 4. *There is an $\overline{LS^k}$ refutation of every formula in k-CNF.*

Theorem 1. *There is a polynomial-size LS^{2k} refutation of any unsolvable system of polynomial inequalities of degree at most k.*

4   Short LS + CP^2 and LS^4 Proofs of the Clique-Coloring Tautologies

Theorem 2. *There is a set of inequalities that has polynomial-size refutations in LS^4 and LS + CP^2, but has only exponential-size refutations in CP.*

The set of inequalities we use is close to the one used by Pudlák for proving an exponential lower bound for CP [14]. Pudlák's bound remains valid for this system. Therefore, to achieve the result, we show that this set of inequalities has polynomial-size refutations in LS^4 and LS + CP^2.

Clique-coloring tautologies. Given a graph G with n vertices, we try to colour it with $m - 1$ colours, while assuming the existense of a clique of size m in G. Each edge (i, j) is represented by a (0-1) variable p_{ij}. Variables q_{ki} encode a (possibly multivalued) function from the integers $\{1 \ldots m\}$ denoting the vertices of a m-clique to the set $\{1 \ldots n\}$ of the vertices of G. Namely, q_{ki} represents the i-th vertex of G being the k-th vertex of the clique. Variables $r_{i\ell}$ encode a (possibly multivalued) coloring of vertices by $m - 1$ colors. The assignment of the colour ℓ to the node i is represented by a variable $r_{i\ell}$. The inequalities (1)–(5) in [14] (see also [1]) state that G has an m-clique and is $(m-1)$-colorable. We now add to them one more family of inequalities, without affecting applicability of [14, Corollary 7], that is, any CP refutation of the new system will still require at least $2^{\Omega((n/\log n)^{1/3})}$ steps. Namely, we add

$$\sum_{i=1}^{n} q_{ki} \leq 1, \quad k = 1 \ldots m. \tag{12}$$

PHP interpretation of weak clique-coloring tautologies. The fact that the i-th vertex of G is the k-th vertex of the clique and is coloured with the colour ℓ is encoded as $q_{ki}r_{i\ell} \geq 1$. Then the fact that the k-th vertex of the clique has colour ℓ is encoded as

$$\sum_{i=1}^{n} q_{ki}r_{i\ell} \geq 1.$$

Let us denote this sum by $x_{k\ell}$. Note that $x_{k\ell}$'s define an injective (possibly multivalued) mapping from $\{1 \ldots m\}$ to $\{1 \ldots m - 1\}$.

For $x_{k\ell}$'s, the inequalities of another well-known unsolvable system of inequalities hold. Namely, these are inequalities that express the negation of the propositional pigeonhole principle:

$$\sum_{\ell=1}^{m-1} x_{k\ell} \geq 1; \qquad 1 \leq k \leq m; \tag{13}$$

$$x_{k\ell} + x_{k'\ell} \leq 1; \qquad 1 \leq k < k' \leq m; \ 1 \leq \ell \leq m - 1. \tag{14}$$

(That says that the k-th pigeon must get into a hole, while two pigeons k and k' cannot share the same hole ℓ.) For short, we denote this system of inequalities by PHP.

There is a polynomial-size CP refutation for PHP [11]. In our notation (note that x_{kl} denotes a quadratic polynomial) such refutation translates into an LS + CP^2 refutation. Alternatively, Pudlák shows that PHP also has polynomial-size refutation in LS. In our notation, this translates into an LS^4 refutation. We show that there is a polynomial-size derivation of (13)–(14) from (1)–(5) of [14] and (12) in LS^4 as well as in LS + CP^2.

5   Reasoning about Integers

Versions of Lovász-Schrijver calculi can be used for reasoning about integers. In the following lemma the basic primitive for the latter, the family of quadratic inequalities $f_d(Y) \geq 0$, is introduced. The lemma shows that there are short proofs of the fact that an integer linear combination of variables is either at most $d - 1$ or at least d for any integer d. It follows then that there are short LS^3 (as well as $LS_{0/1\text{-split}}$) proofs of the symmetric knapsack problem, and that CP with polynomially bounded coefficients can be simulated in $LS^3_/$ (as well as in $LS_{*,\text{split}}$).

Lemma 3. *Let $Y = \sum_{i=1}^n a_i x_i$ and $f_d(Y) = (Y - (d - 1))(Y - d)$, where a_i are integers and x_i are variables. Then the inequality $f_d(Y) \geq 0$ has a derivation in LS^3 and $LS_{0/1\text{-split}}$ of size polynomial in d, n and $\max_i |a_i|$.*

Theorem 3. *For $m \notin \mathbb{Z}$, there is a polynomial-size LS^3 (as well as $LS_{0/1\text{-split}}$) refutation of*

$$m - x_1 - x_2 - \ldots - x_n = 0. \tag{15}$$

CP with polynomially bounded coefficients are defined in [13].

Theorem 4. *The systems $LS_{*,\text{split}}$ and $LS^3_/$ polynomially simulate CP with polynomially bounded coefficients.*

6   Short Proof of Tseitin's Tautologies in LS^d

We recall the construction of Tseitin's tautologies. Let $G = (V, E)$ be a graph with an odd number n of vertices. Attach to each edge $e \in E$ a Boolean variable x_e, i.e. $x_e^2 = x_e$. The negation $T = T_G$ of Tseitin's tautologies with respect to G (see e.g., [7,8]) is a family of formulas meaning that for each vertex v of G the sum $\sum_{e \ni v} x_e$ ranging over the edges incident to v is odd. Clearly, T is contradictory.

In the applications to the proof theory [7,28] the construction of G is usually based on an expander. In particularly, G is d-regular, i.e., each vertex has degree d, where d is a constant. The respective negation $T = T_G$ of Tseitin's tautologies is given by the following equalities (due to Lemmas 1 and 2 we give them directly in PC translation):

$$\prod_{e \in S'_v} x_e \cdot \prod_{e \notin S'_v} (1 - x_e) = 0 \tag{16}$$

(for each vertex v and each subset S'_v of even cardinality of the set S_v of edges incident to v). There are 2^{d-1} equalities of degree d for each vertex of G.

Theorem 5. *For every constant $d \geq 1$ and every d-regular graph G, there is a polynomial-size refutation of (16) in LS^{d+2}.*

7   Lower Bounds on Lovász-Schrijver Rank

There is a series of lower bounds on Lovász-Schrijver rank in the literature (see e.g. [23,25] and the references there). However, these bounds are not suitable for the use in the propositional proof theory, because these are either bounds for *solvable* systems of inequalities, or bounds for systems with *exponentially many* inequalities.

The standard geometric setting for the Lovász-Schrijver procedures LS and LS$_+$ [15] involves, given a system $Ax \leq b$ of m linear inequalities in variables $x_1, \ldots, x_n$, homogenization by adding an extra variable $x_0 \geq 0$ and considering the cone K (resp., K_I) of (resp. 0-1) solutions to the homogenized system. Lift-and-project operators $N^r(K)$ and $N^r_+(K)$ are introduced, and the inclusions

$$K_I \subseteq N^n_{(+)}(K) \subseteq N^{n-1}_{(+)}(K) \subseteq \ldots \subseteq N^k_{(+)}(K) \subseteq \ldots \subseteq N_{(+)}(K) \subseteq K \quad (17)$$

are shown in [15]. The LS-*rank* (respectively, LS$_+$-*rank*) of a system of linear inequalities $Ax \leq b$ is the minimal k in (17) such that $N^k(K) = K_I$ (respectively, $N^k_+(K) = K_I$), where $K = K(A, b)$, as above.

Alternative definitions of Lovász-Schrijver ranks in proof systems terms are as follows. A proof in Lovász-Schrijver proof system is a directed acyclic graph whose vertices correspond to the derived inequalities, and there is an edge between $f \geq 0$ and $g \geq 0$ iff g is derived from f (and maybe something else) in one step. We now drop the edges corresponding to the rule (5). The *rank of a refutation* is the length of the longest path from an axiom to the contradiction in this graph. The *LS-rank* of a system is the smallest rank of an LS-refutation for it. The *LS$_+$-rank* is the smallest rank of an LS$_+$-refutation. Similarly, one can define LS$_*$- and LS$_{+,*}$-ranks. Note that this definition generalizes smoothly to LS^d, LS$^d_+$, LS$^d_*$ and LS$^d_{+,*}$.

The following Theorem 6 gives a linear lower bound on the LS$_+$-rank (and a logarithmic lower bound on the LS$_{+,*}$-rank) of symmetric knapsack problem by reducing it to a lower bound on the degree of Positivstellensatz Calculus refutation [21]. However, this system of inequalities is not obtained as a translation of a propositional formula, and thus lower bounds for it cannot be directly used in the propositional proof theory.

The system of inequalities for the symmetric knapsack problem is given by (15) and usual axioms $0 \leq x \leq 1$, (8). We restrict our attention to system K obtained by setting $m = \lfloor \frac{n}{2} \rfloor + \frac{1}{2}$.

Theorem 6. *The LS$_+$-rank of K is at least $n/4$, while the LS$_{+,*}$-rank of K is at least $\log_2 n - 1$.*

The following Theorem 7 gives an $O(m)$ lower bound on the LS-rank of PHP. Note that the LS_+-rank of PHP is a constant.

Theorem 7. *At least $m-2$ iterations of the N-operator are needed to prove the integer infeasibility of* (13)-(14), *that is, its LS-rank is at least $m-2$.*

8   Open Questions

1. What is the proof complexity of the knapsack problem in LS (cf. Sect. 5 and 7)? We conjecture it as a candidate for a lower bound.
2. Prove an exponential lower bound on the size of Positivstellensatz refutations.
3. Suggest a candidate for a lower bound in LS^d for (arbitrarily large) constant d.
4. How precise is the logarithmic lower bound on the LS_*-rank for the knapsack problem from Theorem 6?
5. Can one relax in Theorem 4 the condition on the polynomial growth of the coefficients?
6. Is it possible to simulate LS by means of a suitable version of CP (e.g. by the R(CP) introduced in [30])? In other words, is there an inverse to Theorem 4?

Acknowledgment. The authors are grateful to Arist Kojevnikov, Alexander Razborov and Hans van Maaren for useful discussions.

References

1. Grigoriev, D., Hirsch, E.A., Pasechnik, D.V.: Complexity of semi-algebraic proofs. Technical report, Electronic Colloquim on Computational Complexity (2001 (submitted)) http://eccc.uni-trier.de/eccc-local/Lists/TR-2001.html.
2. Beame, P., Impagliazzo, R., Krajíček, J., Pitassi, T., Pudlák, P.: Lower bounds on Hilbert's Nullstellensatz and propositional proofs. Proc. London Math. Soc. **73** (1996) 1–26
3. Beame, P., Impagliazzo, R., Krajíček, J., Pudlák, P., Razborov, A.A., Sgall, J.: Proof complexity in algebraic systems and bounded depth Frege systems with modular counting. Computational Complexity **6** (1996/97) 256–298
4. Razborov, A.A.: Lower bounds for the polynomial calculus. Computational Complexity **7** (1998) 291–324
5. Impagliazzo, R., Pudlák, P., Sgall, J.: Lower bounds for the polynomial calculus. Computational Complexity **8** (1999) 127–144
6. Clegg, M., Edmonds, J., Impagliazzo, R.: Using the Groebner basis algorithm to find proofs of unsatisfiability. In: Proceedings of the 28th Annual ACM Symposium on Theory of Computing, STOC'96, ACM (1996) 174–183
7. Buss, S., Grigoriev, D., Impagliazzo, R., Pitassi, T.: Linear gaps between degrees for the polynomial calculus modulo distinct primes. Journal of Computer and System Sciences **62** (2001) 267–289

8. Grigoriev, D., Hirsch, E.A.: Algebraic proof systems over formulas. Technical Report 01-011, Electronic Colloquim on Computational Complexity (2001) ftp://ftp.eccc.uni-trier.de/pub/eccc/reports/2001/TR01-011/index.html.
9. Gomory, R.E.: An algorithm for integer solutions of linear programs. In: Recent Advances in Mathematical Programming. McGraw-Hill (1963) 269–302
10. Chvátal, V.: Edmonds polytopes and a hierarchy of combinatorial problems. Discrete Math. **4** (1973) 305–337
11. Cook, W., Coullard, C.R., Turán, G.: On the complexity of cutting-plane proofs. Discrete Appl. Math. **18** (1987) 25–38
12. Chvátal, V., Cook, W., Hartmann, M.: On cutting-plane proofs in combinatorial optimization. Linear Algebra Appl. **114/115** (1989) 455–499
13. Bonet, M., Pitassi, T., Raz, R.: Lower bounds for Cutting Planes proofs with small coefficients. In: Proceedings of the 27th Annual ACM Symposium on Theory of Computing, STOC'95, ACM (1995) 575–584
14. Pudlák, P.: Lower bounds for resolution and cutting plane proofs and monotone computations. J. Symbolic Logic **62** (1997) 981–998
15. Lovász, L., Schrijver, A.: Cones of matrices and set-functions and 0–1 optimization. SIAM Journal on Optimization **1** (1991) 166–190
16. Lovász, L.: Stable sets and polynomials. Discrete Mathematics **124** (1994) 137–153
17. Pudlák, P.: On the complexity of propositional calculus. In: Sets and Proofs: Invited papers from Logic Colloquium'97. Cambridge University Press (1999) 197–218
18. Razborov, A.: Lower bounds on the monotone complexity of boolean functions. Dokl. AN USSR **22** (1985) 1033–1037
19. Grigoriev, D., Vorobjov, N.: Complexity of Null- and Positivstellensatz proofs. Annals of Pure and Applied Logic **113** (2001) 153–160
20. Grigoriev, D.: Linear lower bound on degrees of Positivstellensatz calculus proofs for the parity. Theoretical Computer Science **259** (2001) 613–622
21. Grigoriev, D.: Complexity of Positivstellensatz proofs for the knapsack. Computational Complexity **10** (2001) 139–154
22. Stephen, T., Tunçel, L.: On a representation of the matching polytope via semidefinite liftings. Math. Oper. Res. **24** (1999) 1–7
23. Cook, W., Dash, S.: On the matrix-cut rank of polyhedra. Math. Oper. Res. **26** (2001) 19–30
24. Dash, S.: On the Matrix Cuts of Lovász and Schrijver and their use in Integer Programming. Technical report tr01-08, Rice University (2001) http://www.caam.rice.edu/caam/trs/2001/TR01-08.ps.
25. Goemans, M.X., Tunçel, L.: When does the positive semidefiniteness constraint help in lifting procedures. Mathematics of Operations Research (2001) to appear.
26. Bockmayr, A., Eisenbrand, F., Hartmann, M., Schulz, A.S.: On the Chvátal rank of polytopes in the 0/1 cube. Discrete Applied Mathematics **98** (1999) 21–27
27. Eisenbrand, F., Schulz, A.S.: Bounds on the Chvátal rank of polytopes in the 0/1-cube. In G. Cornuéjos, R.E. Burkard, G., ed.: IPCO'99. Volume 1610 of Lecture Notes in Computer Science., Berlin Heidelberg, Springer-Verlag (1999) 137–150
28. Urquhart, A.: Hard examples for resolution. JACM **34** (1987)
29. Ben-Sasson, E.: Hard examples for bounded depth Frege. Manuscript (2001)
30. Krajíček, J.: Discretely ordered modules as a first-order extension of the cutting planes proof system. Journal of Symbolic Logic **63** (1998) 1582–1596
31. Cook, S.A., Reckhow, A.R.: The relative efficiency of propositional proof systems. Journal of Symbolic Logic **44** (1979) 36–50

A Lower Bound Technique for Restricted Branching Programs and Applications

(Extended Abstract)

Philipp Woelfel[*]

FB Informatik, LS2, Univ. Dortmund, 44221 Dortmund, Germany
woelfel@Ls2.cs.uni-dortmund.de

Abstract. We present a new lower bound technique for two types of restricted Branching Programs (BPs), namely for read-once BPs (BP1s) with restricted amount of nondeterminism and for $(1, +k)$-BPs. For this technique, we introduce the notion of *(strictly) k-wise l-mixed* Boolean functions, which generalizes the concept of l-mixedness defined by Jukna in 1988 [3]. We prove that if a Boolean function $f \in B_n$ is (strictly) k-wise l-mixed, then any nondeterministic BP1 with at most $k - 1$ nondeterministic nodes and any $(1, +k)$-BP representing f has a size of at least $2^{\Omega(l)}$. While leading to new exponential lower bounds of well-studied functions (e.g. linear codes), the lower bound technique also shows that the polynomial size hierarchy for BP1s with respect to the available amount of nondeterminism is strict. More precisely, we present a class of functions $g_n^k \in B_n$ which can be represented by polynomial size BP1s with k nondeterministic nodes, but require superpolynomial size if only $k - 1$ nondeterministic nodes are available (for $k = o(n^{1/3}/\log^{2/3} n)$). This is the first hierarchy result of this kind where the BP1 does not obey any further restrictions. We also obtain a hierarchy result with respect to k for $(1, +k)$-BPs as long as $k = o(\sqrt{n/\log n})$. This extends the hierarchy result of Savický and Žák [9], where k was bounded above by $\frac{1}{2} n^{1/6}/\log^{1/3} n$.

1 Introduction and Results

Branching Programs (BPs) or equivalently Binary Decision Diagrams (BDDs) belong to the most important nonuniform models of computation. Deterministic and nondeterministic BPs can be simulated by the corresponding Turing machines, and the BP complexity of a Boolean function is a measure for the space complexity of the corresponding model of sequential computation. Therefore, one is interested in large lower bounds for BPs.

Definition 1. *A (deterministic) Branching Program (short: BP) on the variable set $X_n = \{x_1, \ldots, x_n\}$ is a directed acyclic graph with one source and two sinks. The internal nodes are marked with variables in X_n and the sinks are*

---

[*] Supported in part by DFG grant We 1066/10-1.

H. Alt and A. Ferreira (Eds.): STACS 2002, LNCS 2285, pp. 431–442, 2002.
© Springer-Verlag Berlin Heidelberg 2002

labeled with the Boolean constants 0 and 1. Further, each internal node has two outgoing edges, marked with 0 and 1, respectively. A nondeterministic *(short: n.d.) Branching Program is a BP with some additional unmarked nodes with out-degree two, called* nondeterministic *nodes. The* size *of a (possibly n.d.) Branching Program G is the number of its nodes, and ist denoted by* $|G|$.

Let G *be a (possibly n.d.) BP on* X_n *and* $a = (a_1, \ldots, a_n) \in \{0, 1\}^n$ *an assignment to the variables in* X_n. *A source-to-sink path in G is called* computation path *of* a, *if it leaves any node marked with* x_i *over the edge labeled with* a_i. *Note that an input may have multiple computation paths if G is n.d.*

Let B_n *denote the set of Boolean functions* $\{0, 1\}^n \to \{0, 1\}$. *The BP G represents the function* $f \in B_n$ *for which* $f(a) = 1$ *if and only if there exists a computation path of* a *leading to the 1-sink.*

Until today, no superpolynomial lower bounds for general BPs representing an explicitly defined function are known. Therefore, various types of restricted BPs have been investigated, and one is interested in refining the proof techniques in order to obtain lower bounds for less restricted BPs. For this paper, the following two common types of restricted BPs are most important (for an in-depth discussion of other restricted BP models we refer to [13]).

Definition 2. (i) *A* (n.d.) read-k-times BP *(short: BPk) is a (n.d.) BP where each variable appears on each computation path at most k times.*

(ii) *A* (n.d.) $(1, +k)$-BP *is a (n.d.) BP where for each computation path p there exist at most k variables appearing on p more than once.*

Especially deterministic BP1s have been studied to a great extent. The first exponential lower bounds date back to the 80s [12,14], and today, lower bounds for explicitly defined functions in P are as large as $2^{n-O(\log^2 n)}$ [1].

If one considers BPs which allow multiple tests of the same variable during a computation, then one has to distinguish between *syntactic* and *semantic* restrictions. The restrictions given in the definition above are *semantic*, because they have to hold on each computation path. But since graph theoretical paths may be inconsistent, one may obtain BPs with less computational power if the restriction has to hold even on each graph theoretical path. Such restrictions are called *syntactic*, and the BPs corresponding to Definition 2 but with syntactic restrictions are called *syntactic BPks* and *syntactic $(1, +k)$-BPs*, respectively.

Besides the general interest in finding exponential lower bounds for less and less restricted BP-models, much of the research on Branching Programs has focused on separating the power of different types of Branching Programs. Similarly, it has been of considerable interest, how the computational power of BPs is influenced by e.g. the available amount of nondeterminism or the multiplicity of variable tests (i.e. the term k in BPks or $(1, +k)$-BPs).

Results on the influence of the available amount of nondeterminism have so far been obtained only for n.d. BP1s with additional restrictions.

Definition 3. (i) *An* $(\vee, k)$-BP1 *is a family of k deterministic BP1s and represents the function $f_1 \vee \ldots \vee f_k$, where f_i is the function represented by the ith BP1.*

(ii) *A BP1 is called* Ordered Binary Decision Diagram *(short: OBDD), if the nodes can be partitioned into levels such that all edges point only from lower to higher levels and all internal nodes of one level are marked with the same variable. A k-Partitioned Binary Decision Diagram (short: k-PBDD) is an $(\vee, k)$-BP1 whose BP1s are in fact OBDDs.*

Note that we can regard an $(\vee, k)$-BP1 as a n.d. BP1 having a binary tree of exactly $k - 1$ n.d. nodes at the top such that the outgoing edges of the leaves lead to the sources of k disjoint BP1s. Hence, the set of functions which can be represented in polynomial size by $(\vee, k)$-BP1s is a subset of the functions which can be represented in polynomial size by BP1s with at most $k - 1$ n.d. nodes. Although not yet proven, the results of [5] indicate that $(\vee, k)$-BP1s might be in fact less powerful than n.d. BP1s with $k - 1$ nondeterministic nodes.

Bollig and Wegener [2] have proven the first hierarchy result for k-PBDDs with respect to k, which has been extended later by Sauerhoff [6]. He presented functions being representable by polynomial size $(k + 1)$-PBDDs but requiring superpolynomial size k-PBDDs if $k = O\big((n/\log^{1+\epsilon} n)^{1/4}\big)$ for arbitrary $\epsilon > 0$. This means that for these k, the polynomial size hierarchy of k-PBDDs with respect to k is strict. A generalization of this result was obtained by Savický and Sieling [7], who proved that the polynomial size hierarchy of $(\vee, k)$-BP1s with respect to k ist strict for $k \le 2/3\sqrt{\log n}$.

One of the main contributions of this paper is a hierarchy result for n.d. BP1s without any restrictions except on the number of n.d. nodes. We present a class of multipointer functions $g_n^k \in B_n$, which can be represented by polynomial size n.d. BP1s having k n.d. nodes, but require superpolynomial size if at most $k - 1$ n.d. nodes are available (for $k = o(n^{1/3}/\log^{2/3} n)$).

The other main contribution of this paper is an improved hierarchy result for $(1, +k)$-BPs with respect to k. For syntactic $(1, +k)$-BPs, the first hierarchy result was obtained by Sieling in 1996 [10] and later improved as well as generalized for the semantic restriction by Savický and Žák [9]. They showed that the polynomial size hierarchy with respect to k is strict for $(1, +k)$-BPs if $k \le \frac{1}{2} n^{1/6} / \log^{1/3} n$ and for syntactic $(1, +k)$-BPs if $k \le \frac{1}{2} \sqrt{n} / \log n$. We extend their result for both types of restrictions by presenting a class of functions for which polynomial size syntactic $(1, +k)$-BPs but no polynomial size (semantic) $(1, +(k - 1))$-BPs exist if $k = o(\sqrt{n/\log n})$.

The hierarchy results for n.d. BP1s and $(1, +k)$-BPs are possible because of a new lower bound technique. Interestingly enough, this technique can be equivalently applied for both types of BPs. It mainly consists of introducing the notion of strictly k-wise l-mixed functions, and of proving for such functions a lower bound of $2^l + 1$ for n.d. BP1s with at most $k - 1$ n.d. nodes and a lower bound of $2^{l/2}$ for $(1, +k)$-BPs. This will be done in Section 3. In Section 4, we will show how to prove the k-wise l-mixedness of functions. As an easy example, we show that d-rare m-dense functions, investigated e.g. by Jukna and Razborov [4], are in fact k-wise l-mixed for $l < \min\{d, m/k\}$. We obtain as a corollary exponential lower bounds for linear codes in the $(1, +k)$-BP model, which have already been proven in [4], and new exponential lower bounds in the n.d. BP1 model with restricted amount of nondeterminism. We also show how to construct easily from a 1-wise l-mixed function in B_n a k-wise l/k-mixed function in B_{kn}. This construction helps us in Section 5 to obtain the hierarchy result for $(1, +k)$-BPs. The hierarchy result for n.d. BP1s will finally be stated in Section 6.

2   Notation

In the following text, we consider functions defined on the n Boolean variables in $X_n = \{x_1, \ldots, x_n\}$. A *partial input* is an element $\alpha = (\alpha^1, \ldots, \alpha^n) \in \{0, 1, *\}^n$. While a position α^i with value 0 or 1 means that the input variable x_i is fixed to the corresponding constant, a value of $*$ means that the input variable remains free. If $f \in B_n$ is a Boolean function and α is a partial input, then $f|_\alpha$ means the subfunction of f obtained by restricting all inputs to α. For a partial input $\alpha \in \{0, 1, *\}^n$, we denote the *support* of α by $S(\alpha) := \{x_i \mid \alpha^i \neq *\}$. The empty partial input, i.e. the partial input with support $\emptyset$, is written as ε. For two partial inputs α, β with the same support, we let $D(\alpha, \beta)$ be the set which contains all variables $x_i \in X_n$ for which $\alpha^i \neq \beta^i$. If α and β are partial inputs with disjoint supports, then we denote by $\alpha\beta$ the partial input where $(\alpha\beta)^i$ equals α^i or β^i if x_i is an element of $S(\alpha)$ or $S(\beta)$, respectively, and equals $*$ if $x_i \notin S(\alpha) \cup S(\beta)$.

3   The Lower Bound Technique

Our lower bound technique relies mainly on a generalization of the following property of Boolean functions, which was defined by Jukna in 1988 [3].

Definition 4. *Let $l \in \mathbb{N}$. A function $f \in B_n$ is called l-mixed, if for all $V \subseteq X_n$ such that $|V| = l$, any two distinct partial inputs α, β with support V yield different subfunctions, i.e. $f|_\alpha \neq f|_\beta$.*

Many exponential lower bound proofs for BP1s use the following folklore fact.

Proposition 1. *The size of any BP1 for an l-mixed function is at least $2^l + 1$.*

The following definition generalizes the above definition of l-mixed functions and is fundamental for our lower bound technique.

Definition 5. *Let $kl \leq n$. In the following formula, we restrict the choices of $W_1, \ldots, W_k, V_1, \ldots, V_k$ to disjoint subsets of X_n such that $|V_j| = l$ ($1 \leq j \leq k$). The choices of the partial inputs $\lambda_j, \alpha_j, \beta_j$ are restricted in such a way that $S(\lambda_j) = W_j$ and $S(\alpha_j) = S(\beta_j) = V_j$. Below, we allow for each $1 \leq j \leq k$ the choice of a partial assignment $\gamma_j \in \{\alpha_j, \beta_j\}$. We then denote by γ_j^* the element in $\{\alpha_j, \beta_j\} \setminus \gamma_j$ and let $c = \lambda_1 \gamma_1 \ldots \lambda_k \gamma_k$ and c_j be the partial assignment obtained from c by replacing γ_j with γ_j^*.*

A Boolean function $f \in B_n$ is called k-wise l-mixed if

$$\exists W_1, \lambda_1 \, \forall V_1, \alpha_1 \neq \beta_1 \, \exists \gamma_1 \in \{\alpha_1, \beta_1\} \, \ldots \, \exists W_k, \lambda_k \, \forall V_k, \alpha_k \neq \beta_k \, \exists \gamma_k \in \{\alpha_k, \beta_k\} :$$

$$\sum_{i=1}^k |W_i| \leq n - kl \quad \wedge \quad \exists x^* \, \forall 1 \leq j \leq k : \, f|_c(x^*) \neq f|_{c_j}(x^*),$$

where x^* is an input for the subfunction $f|_c$ (and $f|_{c_j}$), i.e. $S(x^*) = X_n \setminus (W_1 \cup V_1 \cup \ldots \cup W_k \cup V_k)$. If in the above formula we even have $f|_c(x^*) > f|_{c_j}(x^*)$ (instead of "$\neq$"), then f is called strictly k-wise l-mixed.

Remark 1. Any strictly k-wise l-mixed function is k-wise l-mixed. Furthermore, any (strictly) $(k+1)$-wise $(l+1)$-mixed function is (strictly) k-wise $(l+1)$-mixed and also (strictly) $(k+1)$-wise l-mixed. Finally, a function f is 1-wise l-mixed if and only if there exists an l-mixed subfunction of f.

The property k-wise l-mixed of a Boolean function implies lower bounds for $(1, +k)$-BPs as well as for n.d. BP1s with limited amount of nondeterminism. In order to measure the amount of nondeterminism in BPs, we need an appropriate measurement. Here, we choose the number of n.d. nodes, but another possibility would be to count the maximum number of n.d. nodes on any computation path. The proof of the following theorem though, can be adapted in order to obtain similar results for other measures of nondeterminism.

Theorem 1. (a) *If $f \in B_n$ is a strictly k-wise l-mixed function and G is a n.d. BP1 with at most $k - 1$ n.d. nodes representing f, then $|G| \geq 2^l + 1$.*
(b) *If $f \in B_n$ is a k-wise l-mixed function and G is a $\left(1, +(k-1)\right)$-BP representing f, then $|G| \geq 2^{l/2}$.*

Note that the $(1, +k)$-restriction is semantic in this case.

Due to space restrictions, we only prove part (a) of the theorem. In order to do so, we make use of the notion of filters and of the idea behind a lower bound technique of Simon and Szegedy [11] for deterministic BP1s. A *filter* of a set X is a closed upward subset of 2^X (i.e. if $S \in \mathcal{F}$, then all supersets of S are in $\mathcal{F}$). Let $\mathcal{F}$ be a filter of X_n. A subset $B \subseteq X_n$ is said to be in the *boundary* of $\mathcal{F}$ if $B \notin \mathcal{F}$ but $B \cup \{x_i\} \in \mathcal{F}$ for some $x_i \in X_n$.

Let p be a path starting at the source of a n.d. BP1 and leading to an arbitrary edge $e = (v, w)$. We say that a partial input α *induces* the path p, if no variable in $S(\alpha)$ is tested on any path from w to a sink and if α is consistent with p (i.e. for any c-edge of p, $c \in \{0, 1\}$, leaving a node marked with x_i, either $\alpha^i = c$ or $\alpha^i = *$).

Lemma 1. *Let G be a (possibly n.d.) BP1 on X_n. For each filter $\mathcal{F}$ of X_n there exists a set B in the boundary of $\mathcal{F}$ for which at least $\lceil 2^{|\overline{B}|}/(|G| - 1) \rceil$ different partial assignments with support $\overline{B}$ ($= X_n \setminus B$) induce paths leading to the same edge.*

The proof of this lemma is omitted, but it mainly adapts an idea of [11] to the nondeterministic model. We use it in the following proof of Theorem 1 (a).

Proof (of Theorem 1 (a)). In the following, we write $f \leq g$ for two functions f, g defined on the same domain, if $f(x) \leq g(x)$ for all inputs x.

The proof is by induction on k. If $k = 1$, then G contains no n.d. nodes and is deterministic. Then, upon choosing W_1 and λ_1 appropriately, $f|_{\lambda_1}$ is an l-mixed subfunction of f (see Remark 1) and the claim follows from Proposition 1.

Let now $k > 1$ and G be a n.d. BP1 consisting of $L \leq 2^l$ nodes, of which at most $k - 1$ are nondeterministic. We assume w.l.o.g. that each n.d. node in G has two different successors (if this is not the case, we may replace the n.d. node with its successor). We show that if G computes the function f, then

$$\forall W_1, \lambda_1 \, \exists V_1, \alpha_1 \neq \beta_1 \, \forall \gamma_1 \in \{\alpha_1, \beta_1\} \ldots \, \forall W_k, \lambda_k \, \exists V_k, \alpha_k \neq \beta_k \, \forall \gamma_k \in \{\alpha_k, \beta_k\} :$$

$$\sum_{i=1}^{k} |W_i| > n - kl \quad \vee \quad f|_c \leq \bigvee_{j=1}^{k} f|_{c_j}, \quad (1)$$

where c and c_j are defined as in Definition 5. Hence, f is not strictly k-wise l-mixed.

Let first $W_1 \subseteq X_n$ as well as a partial input λ_1 with support W_1 be chosen arbitrarily. We may assume that $|W_1| \leq n - kl$, since otherwise there is nothing to prove. Consider the restricted n.d. BP1 $G|_{\lambda_1}$ representing the function $f|_{\lambda_1}$ on the variables in $X' = X_n \setminus W_1$ and note that $n' := |X'| \geq kl$. We define a filter $\mathcal{F}$ on X'.

$$\mathcal{F} := \{V \subseteq X' \mid |V| > n' - l\}.$$

Each set B in the boundary of $\mathcal{F}$ has a cardinality of $n' - l$, and hence $|\overline{B}| = l$ (note that $\overline{B} = X' \setminus B$ in this case). Because of Lemma 1, there exists a set B in the boundary of $\mathcal{F}$ such that at least $\lceil 2^{|\overline{B}|}/(L-1) \rceil \geq \lceil 2^l/(2^l-1) \rceil = 2$ distinct partial assignments with support $\overline{B}$ induce paths leading to the same edge (v, w). We let $V_1 = \overline{B}$ and $\alpha_1 \neq \beta_1$ be two such partial assignments with support V_1. Note that $|V_1| = l$. Finally, let γ_1 be chosen arbitrarily among α_1 and β_1. Let f_w be the subfunction defined by the (possibly n.d.) BP1 rooted at node w. All 1-inputs for f_w obviously are also 1-inputs for $f_{\lambda_1 \gamma_1}$ and $f_{\lambda_1 \gamma_1^*}$, because the inputs γ_1 and γ_1^* both induce paths leading to the edge (v, w) in $G|_{\lambda_1}$. Hence,

$$f_w \leq f|_{\lambda_1 \gamma_1}, f|_{\lambda_1 \gamma_1^*}. \quad (2)$$

Now let p be the path induced by the partial input γ_1, leading from the source of $G|_{\lambda_1}$ to the edge (v, w). If there is no n.d. node on p, then $f|_{\lambda_1 \gamma_1} = f_w$ and thus $f|_{\lambda_1 \gamma_1} \leq f|_{\lambda_1 \gamma_1^*}$ by (2). No matter how the choice of the remaining $W_i, \lambda_i, V_i, \alpha_i, \beta_i, \gamma_i$ will be, statement (1) is fulfilled.

Therefore, we assume that there is at least one n.d. node on the path p. Let u be the last n.d. node on this path and let (u, u_0) and (u, u_1) be the two outgoing edges of u, where (u, u_0) is the edge on p. We replace the n.d. node u with the node u_1 by redirecting all edges pointing to u in such a way that they point to u_1. Let f' $(= f'|_{\lambda_1})$ be the function computed by the resulting Branching Program. Obviously

$$f'|_{\lambda_1} \leq f|_{\lambda_1}, \quad (3)$$

because following the (u, u_0)-edge in the original BP1 $G|_{\lambda_1}$ might only allow additional inputs to lead to the 1-sink. Furthermore, since u was the last n.d. node on the path from the source to (v, w) induced by the partial input γ_1,

$$f|_{\lambda_1 \gamma_1} = f'|_{\lambda_1 \gamma_1} \vee f_w \overset{(2)}{\leq} f'|_{\lambda_1 \gamma_1} \vee f|_{\lambda_1 \gamma_1^*}. \quad (4)$$

As a last step, we restrict the so obtained n.d. BP1 for the function $f'|_{\lambda_1}$ to a n.d. BP1 G' for the function $f'|_{\lambda_1\gamma_1}$. Note that G' is of size smaller than 2^l and that $f'|_{\lambda_1\gamma_1}$ is a function on $n'' = n - l - |W_1| \geq (k-1)l$ variables. Furthermore, since we have removed the n.d. node u from $G|_{\lambda_1}$, G' contains at most $k-2$ n.d. nodes. This means by the induction hypothesis that the subfunction $f'|_{\lambda_1\gamma_1}$ is not strictly $(k-1)$-wise l-mixed. In other words

$$\forall W_2, \lambda_2 \,\exists V_2, \alpha_2 \neq \beta_2 \,\forall \gamma_2 \in \{\alpha_2, \beta_2\} \ldots \;\forall W_k, \lambda_k \,\exists V_k, \alpha_k \neq \beta_k \,\forall \gamma_k \in \{\alpha_k, \beta_k\} :$$

$$\sum_{i=2}^{k} |W_i| > n'' - (k-1)l \quad \vee \quad (f'|_{\lambda_1\gamma_1})|_{c'} \leq \bigvee_{j=2}^{k} (f'|_{\lambda_1\gamma_1})|_{c'_j},$$

where c' is the partial input $\lambda_2\gamma_2 \ldots \lambda_k\gamma_k$ and c'_j is obtained from c' by replacing γ_j with γ_j^*.

Assume first that $\sum_{i=2}^{k} |W_i| > n'' - (k-1)l$. Because n'' equals $n - l - |W_1|$ it follows that $\sum_{i=1}^{k} |W_i| > n - kl$, and property (1) is fulfilled. Therefore, we assume $\sum_{i=2}^{k} |W_i| \leq n'' - (k-1)l$ and hence

$$(f'|_{\lambda_1\gamma_1})|_{c'} \leq \bigvee_{j=2}^{k} (f'|_{\lambda_1\gamma_1})|_{c'_j}. \tag{5}$$

Altogether we obtain

$$
\begin{aligned}
f|_c = f|_{\lambda_1\gamma_1 \ldots \lambda_k\gamma_k} &= (f|_{\lambda_1\gamma_1})|_{\lambda_2\gamma_2 \ldots \lambda_k\gamma_k} \\
&\overset{(4)}{\leq} (f'|_{\lambda_1\gamma_1})|_{\lambda_2\gamma_2 \ldots \lambda_k\gamma_k} \vee (f|_{\lambda_1\gamma_1^*})|_{\lambda_2\gamma_2 \ldots \lambda_k\gamma_k} \\
&\overset{(5)}{\leq} \left(\bigvee_{j=2}^{k} (f'|_{\lambda_1,\gamma_1})|_{\lambda_2 \ldots \lambda_k\gamma_2 \ldots \gamma_{j-1}\gamma_j^*\gamma_{j+1} \ldots \gamma_k} \right) \vee (f|_{\lambda_1\gamma_1^*})|_{\lambda_2\gamma_2 \ldots \lambda_k\gamma_k} \\
&\overset{(3)}{\leq} \left(\bigvee_{j=2}^{k} (f|_{\lambda_1\gamma_1})|_{\lambda_2 \ldots \lambda_k\gamma_2 \ldots \gamma_{j-1}\gamma_j^*\gamma_{j+1} \ldots \gamma_k} \right) \vee (f|_{\lambda_1\gamma_1^*})|_{\lambda_2\gamma_2 \ldots \lambda_k\gamma_k} \\
&= \bigvee_{j=1}^{k} f|_{\lambda_1 \ldots \lambda_k\gamma_1 \ldots \gamma_{j-1}\gamma_j^*\gamma_{j+1} \ldots \gamma_k} = \bigvee_{j=1}^{k} f|_{c_j}.
\end{aligned}
$$

Hence, we have proven (1). $\qquad\square$

4 Applications

4.1 Linear Codes and d-Rare m-Dense Functions

As a first application of our lower bound technique, we consider d-rare m-dense functions, which have been investigated e.g. by Savický and Žák [8] and by Jukna and Razborov [4]. Such functions have been known to be hard for $(1, +k)$ BPs and our proof method now demonstrates that they are also hard for n.d. BP1s, if not enough nondeterminism is available.

Definition 6. *A function* $f \in B_n$ *is called* d-rare *if any two different inputs* $a, b \in f^{-1}(1)$ *have a Hamming distance of at least* d *(i.e.* $|D(a,b)| \geq d$*). The function* f *is called* m-dense *if* $|S(\alpha)| \geq m$ *for any partial input* α *with* $f|_\alpha = 0$.

Theorem 2. *Any* d-rare m-dense *function is strictly* k-wise l-mixed *for* $l < \min \{d, m/k\}$.

Proof. This proof is simple, because we do not need to bother about the choice of W_i, λ_i or γ_i for $1 \leq i \leq k$. We simply choose $W_i = \emptyset$, $\lambda_i = \varepsilon$ and upon some arbitrarily given $\alpha_i \neq \beta_i$ with support V_i ($|V_i| = l$) we choose $\gamma_i = \alpha_i$. Then we consider the partial input $c = \lambda_1 \gamma_1 \ldots \lambda_k \gamma_k$ and the inputs c_j which are obtained from c by replacing γ_j with γ_j^*. Since f is m-dense and $S(c) \leq kl < m$ by construction, we know that the subfunction $f|_c$ has an input x^* such that $f|_c(x^*) = 1$. For this x^* on the other hand, the complete inputs cx^* and $c_j x^*$ have a Hamming distance of $|D(c, c_j)| \leq l < d$. Therefore, the d-rareness of f implies $f|_{c_j}(x^*) \neq 1$ for all $1 \leq j \leq k$. $\square$

Corollary 1. *Any n.d. BP1 with at most* k *n.d. nodes or any* $(1, +k)$-*BP representing a* d-rare m-dense *function has a size of at least* $\min \left\{ 2^{(d-1)/2}, 2^{\lfloor (m-1)/(k+1) \rfloor /2} \right\}$.

Note that the same result for $(1, +k)$-BPs was already obtained by Jukna and Razborov [4] with a different technique. The result for n.d. BP1s is new though. The authors of [4] also show that if C is a linear code over $GF(2)$ with minimal distance d_1 and if $C^\perp$ is its dual with minimal distance d_2, then the characteristic function of C is d_1-rare and d_2-dense. This leads to a lower bound of $\min \left\{ 2^{(d_1-1)/2}, 2^{\lfloor (d_2-1)/(k+1) \rfloor /2} \right\}$ for n.d. BP1s with at most k n.d. nodes and for $(1, +k)$-BPs representing such a linear code. We only state one corollary for Reed-Muller codes, which follows instantly from the discussion in [4] and was stated there for $(1, +k)$-BPs.

Corollary 2. *Let* $R(r, \ell)$ *be the* rth *order binary Reed-Muller code of length* $n = 2^\ell$. *Let further* $0 \leq k \leq n$ *and* $r = \lfloor 1/2(\ell + \log(k+1)) \rfloor$. *Then any* $(1, +k)$-*BP and any n.d. BP1 with at most* k *n.d. nodes representing the characteristic function of* $R(r, \ell)$ *has size at least* $2^{\Omega\left(\sqrt{n/(k+1)}\right)}$.

4.2   Disjoint Conjunctions

We state now a theorem which describes how to obtain k-wise l-mixed functions from l-mixed ones. Let f be a function on variables in X_n. We consider the disjoint conjunction f^k with respect to f on the variables in X_{kn}, which is defined as follows. For $x = (x_1, \ldots, x_{kn}) \in \{0,1\}^{kn}$ let $f^k(x) = f_1(x) \wedge \ldots \wedge f_k(x)$, where $f_i(x) = f(x_{(i-1)n+1}, \ldots, x_{in})$.

Theorem 3. *If* f *is* (kl)-mixed, *then* f^k *is strictly* k-wise l-mixed.

The l-mixedness of a function f implies that f is hard to compute by BP1s or equivalently by $(1, +0)$-BPs. Hence, the above theorem leads to a generalization of this fact in the sense that the disjoint conjunction f^k of a (kl)-mixed function f is hard to compute for a $(1, +(k-1))$-BP.

Proof. Let $N = kn$ and let X_i be the set of variables on which f_i may depend, i.e. $X_i = \{x_{(i-1)n+1}, \dots, x_{in}\}$. We first choose $W_1 = \emptyset$ and $\lambda_1 = \varepsilon$. Then we consider a k-round game in which we play against an adversary who starts the ith round by choosing V_i, α_i, β_i, after which we are allowed to choose W_{i+1}, λ_{i+1}. We show that we can influence the game by our choices in such a way that after $r \leq k$ rounds the following situation is obtained for any $j \in \{1, \dots, r\}$.

(I1) The set W_{j+1} only consists of variables in X_{i_j} for some $i_j \in \{1, \dots, k\}$ and the indices $i_1, \dots, i_r$ are all different.

(I2) All variables in $X_{i_1}, \dots, X_{i_r}$ are fixed by $\lambda_1\gamma_1 \dots \lambda_r\gamma_r\lambda_{r+1}$, i.e. $X_{i_1} \cup \dots \cup X_{i_r} \subseteq W_1 \cup V_1 \cup \dots \cup W_r \cup V_r \cup W_{r+1}$.

(I3) In each X_i, $i \notin \{i_1, \dots, i_r\}$, there are at most rl variables fixed.

(I4) $f_{i_j}|_{\lambda_1\gamma_1 \dots \lambda_j\gamma_j\lambda_{j+1}} = 1$ and $f_{i_j}|_{\lambda_1\gamma_1 \dots \lambda_j\gamma_j^*\lambda_{j+1}} = 0$.

Assume that we have played the game k rounds in such a way that (I1)-(I4) are fulfilled for $r = k$. Because all indices $i_1, \dots, i_k$ are different (I1), and all variables in $X_{i_1}, \dots, X_{i_k}$ are fixed (I2), the assignment $\lambda_1\gamma_1 \dots \lambda_k\gamma_k\lambda_{k+1}$ forms a complete input for f^k. We let $x^* = \lambda_{k+1}$ and c, c_j as in Definition 5. Then property (I4) implies on one hand that $f^k|_c(x^*) = 1$, while on the other for each $1 \leq j \leq k$ it follows from $f_{i_j}|_{c_j}(x^*) = 0$ that $f^k|_{c_j}(x^*) = 0$. Furthermore, by property (I1) each of the sets $W_2, \dots, W_k$ contains at most n variables. Since in addition $W_1 = \emptyset$, we have $\sum_{i=1}^k |W_i| \leq (k-1)n = N - n \leq N - kl$. For the last inequality we have used $kl \leq n$, which follows from the (kl)-mixedness of f. Altogether, the conditions of Definition 5 showing that f^k is strictly k-wise l-mixed are fulfilled.

Therefore, it suffices to show that we can play the game for k rounds such that after each round (I1)-(I4) hold. This is trivially true after 0 rounds, and we show now the claim for the $(r+1)$th round $(1 \leq r+1 \leq k)$.

Let the adversary choose V_{r+1} and $\alpha_{r+1} \neq \beta_{r+1}$. Then there exists a variable $x_i \in V_{r+1}$ which is fixed to different constants by α_{r+1} and β_{r+1}. Let i_{r+1} be the index for which $x_i \in X_{i_{r+1}}$. Note that $i_{r+1} \notin \{i_1, \dots, i_r\}$, because by (I2) all variables in the sets $X_{i_1}, \dots, X_{i_r}$ had been fixed in previous rounds and are therefore not contained in V_{r+1}.

Now we restrict the partial input $\lambda_1\gamma_1 \dots \lambda_r\gamma_r\lambda_{r+1}\alpha_{r+1}$ to the variables in $X_{i_{r+1}}$ and obtain a partial input α. In the same way, we obtain β by restricting $\lambda_1\gamma_1 \dots \lambda_r\gamma_r\lambda_{r+1}\beta_{r+1}$ to the variables in $X_{i_{r+1}}$. Obviously, $S(\alpha) = S(\beta)$ and by our choice of $X_{i_{r+1}}$, it is $\alpha \neq \beta$. Furthermore, using the fact that $|V_{r+1}| = l$, we know by (I3) that not more than $(r+1)l$ variables are fixed in $X_{i_{r+1}}$, hence $|S(\alpha)| = |S(\beta)| \leq kl$. Using the assumption that f is (kl)-mixed, this implies by Proposition 1 that $f|_\alpha \neq f|_\beta$ (we assume here that the input variables for f are in $X_{i_{r+1}}$ instead of in X_n). Thus, there exists an assignment y to the free variables in $X_{i_{r+1}}$ as well as a choice $\gamma \in \{\alpha, \beta\}$ such that $f|_\gamma(y) = 1$ and

$f|_{\gamma^*}(y) = 0$. We finally let γ_{r+1} be the element in $\{\alpha_{r+1}, \beta_{r+1}\}$ which corresponds to the above choice of γ. Then obviously $f_{i_{r+1}}|_{\lambda_1\gamma_1...\lambda_r\gamma_r\lambda_{r+1}\gamma_{r+1}}(y) = 1$ and $f_{i_{r+1}}|_{\lambda_1\gamma_1...\lambda_r\gamma_r\lambda_{r+1}\gamma_{r+1}^*}(y) = 0$. Hence, if we let $\lambda_{r+2} := y$ and $W_{r+2} = S(\lambda_{r+2})$, then (I4) is fulfilled. Furthermore, by construction W_{r+2} only consists of variables in $X_{i_{r+1}}$ and all variables in $X_{i_1}, \ldots, X_{i_{r+1}}$ are fixed. Therefore also (I1) and (I2) hold. Condition (I3) follows already from (I1) and (I2), because each of the assignments $\gamma_1, \ldots, \gamma_r$ fixes at most l variables and the assignments $\lambda_1, \ldots, \lambda_{r+1}$ fix only variables in X_i, $i \in \{i_1, \ldots, i_r\}$. We have shown therefore, that there exists a playing strategy such that for any $0 \le r \le k$ after the rth round the conditions (I1)-(I4) are fulfilled. This proves the claim.   □

5   Improving the Hierarchy for $(1, +k)$-BPs

We consider now the function weighted sum, which was used by Savický and Žák [9] in order to prove a hierarchy for $(1, +k)$-BPs.

Definition 7. *For any positive integer n let $p(n)$ be the smallest prime greater than n. The function* $\mathrm{WS}_n \in B_n$ *(called* weighted sum*) is defined by*

$$\mathrm{WS}_n(x) = \begin{cases} x_s & \text{if } s \in \{1, \ldots, n\} \\ x_1 & \text{otherwise,} \end{cases} \quad \text{where} \quad s = \left(\sum_{i=1}^{n} ix_i\right) \bmod p(n).$$

Savický and Žák have shown that WS_n is l-mixed for large l, and have used this function in order to obtain a hierarchy for $(1, +k)$-BPs.

Theorem 4 ([9]). *For any $\delta > 0$ and any large enough n, the function WS_n is l-mixed for $l = n - \lfloor (2 + \delta)\sqrt{n} \rfloor - 2$.*

Theorem 5 ([9]). *There exists a class of functions $h_{n,k} \in B_n$ representable by polynomial-size $(1, +k)$-BPs but not by polynomial-size $(1, +(k - 1))$-BPs as long as $k \le n^{1/6}/(\log^{1/3} n)$.*

It is obvious how to construct a syntactic $(1, +1)$-BP representing the function WS_n with at most $O(n^2)$ nodes. On the other hand, Theorem 4 implies that any $(1, +0)$-BP (or equivalently any BP1) for WS_n has exponential size. We may now look at the disjoint conjunction $f_{N,k} := (\mathrm{WS}_n)^k$ with respect to WS_n. Note that $f_{N,k}$ is a function in $N = kn$ variables. Furthermore, it is easy to see that $f_{N,k}$ can be computed by a syntactic $(1, +k)$-BP of size $O(kn^2) = O(N^2/k)$. But by Theorem 1 (b) and Theorem 3 we know that any $(1, +(k - 1))$-BP for $f_{N,k}$ has a size of at least $2^{l/(2k)}$, where $l = n - \lfloor (2 + \delta)\sqrt{n} \rfloor - 2 = \Omega(n) = \Omega(N/k)$. (for any δ and sufficiently large n). Hence, we get an improved hierarchy as described by the following corollary.

Corollary 3. *The function $f_{N,k} \in B_N$ can be represented by polynomial size syntactic $(1, +k)$-BPs but not by polynomial size $(1, +(k - 1))$-BPs for $k = o\left(\sqrt{N/\log N}\right)$.*

6   A Hierarchy for BP1s with Restricted Amount of Nondeterminism

We finally develop a family of multipointer functions g_n^k, which can be easily computed in polynomial size by $(k+1)$-PBDDs. On the other hand, for any BP1 having at most $k-1$ n.d. nodes, an exponential size is required. Recall that a $(k+1)$-PBDD can be regarded as a restricted n.d. BP1 having exactly k n.d. nodes at the top.

The idea behind the following definition of g_n^k is inspired by the functions used by Savický and Sieling [7] for their hierarchy result for $(\vee, k)$-BP1s. Let n and k be arbitrary integers such that $k(k+1) \leq n/\lceil \log n \rceil$, and let $b = \lfloor n/(k(k+1)) \rfloor$ and $s = \lfloor b/\lceil \log(kb) \rceil \rfloor$. We partition the n variables in X_n into $k(k+1)$ consecutive blocks $B_{i,j}$ of size b, where $i \in \{0, \ldots, k\}$ and $j \in \{0, \ldots, k-1\}$, and possibly one block of the remaining variables. Each block $B_{i,j}$ is then partitioned into $\lceil \log(kb) \rceil$ subblocks, each having either size s or size $s+1$. The set of the input variables in the blocks $B_{i,0}, \ldots, B_{i,k-1}$ is called the *sector* S_i ($0 \leq i \leq k$) of the input and has cardinality kb. For the ease of notation, we enumerate the variables in such a way that the sector S_i contains the variables $x_{i,0}, \ldots, x_{i,kb-1}$.

The function value of the function g_n^k is determined as follows. The majority of the setting of the s variables in each subblock of a block $B_{i,j}$ determines a bit. The $\lceil \log(kb) \rceil$ bits obtained this way for a block $B_{i,j}$ are interpreted as an integer in $\{0, \ldots, 2^{\lceil \log(kb) \rceil} - 1\}$, which is taken modulo kb such that a value $p_{i,j} \in \{0, \ldots, kb-1\}$ is obtained. This value $p_{i,j}$ points to the variable $x_{i \oplus j, p_{i,j}}$ in the sector $S_{i \oplus j}$, where $i \oplus j := (i+j+1) \bmod (k+1)$. Let $h_{i,j}$ be the function which computes the value of the input variable the pointer $p_{i,j}$ points to, i.e. $h_{i,j}(x) = x_{i \oplus j, p_{i,j}}$. Then

$$h_i := \bigwedge_{j=0}^{k-1} h_{i,j} \quad \text{and} \quad g_n^k := \bigvee_{i=0}^{k} h_i.$$

Since the function g_n^k is the disjunction of $k+1$ functions h_i, a correct $(k+1)$-PBDD may consist of $k+1$ OBDDs, each representing a function h_i ($0 \leq i \leq k$). The idea behind constructing an OBDD which represents h_i is that the OBDD reads a block $B_{i,j}$ of the sector S_i, determines the corresponding pointer $p_{i,j}$ and finally may obtain the value of the function $h_{i,j}$. Depending on whether this value is 0 or 1, the OBDD stops with output 0 or continues this proceeding with the next block $B_{i,j+1}$ in the sector S_i. A detailed discussion on how to construct such OBDDs with polynomial size will be available in the full version of this paper. Due to space restrictions we also omit the proof that the function g_n^k is k-wise l-mixed. The following theorem summarizes the upper and lower bounds for the function g_n^k.

Theorem 6. *The functions $g_n^k \in B_n$ can be represented by $(k+1)$-PBDDs of size $O(n^3/k^3)$ (and hence in polynomial size by n.d. BP1s with at most k n.d. nodes). On the other hand, any n.d. BP1 with at most $k-1$ n.d. nodes has a size of at least $2^{\Omega(n/(k^3 \log n))}$, which is not polynomial for $k = o(n^{1/3}/\log^{2/3} n)$.*

Acknowledgment. I thank Beate Bollig, Martin Sauerhoff, Detlef Sieling and Ingo Wegener for proofreading and helpful comments.

References

1. A. Andreev, J. Baskakov, A. Clementi, and J. Rolim. Small pseudo-random sets yield hard functions: New tight explict lower bounds for branching programs. In *Proceedings of the 26th International Colloquium on Automata, Languages, and Programming*, vol. 1644 of *Lecture Notes in Computer Science*, pp. 179–189. 1999.

2. B. Bollig and I. Wegener. Complexity theoretical results on partitioned (nondeterministic) binary decision diagrams. *Theory of Computing Systems*, 32:487–503, 1999.

3. S. Jukna. Entropy of contact circuits and lower bounds on their complexity. *Theoretical Computer Science*, 57:113–129, 1988.

4. S. Jukna and A. Razborov. Neither reading few bits twice nor reading illegally helps much. *Discrete Applied Mathematics*, 85:223–238, 1998.

5. M. Sauerhoff. Computing with restricted nondeterminism: The dependence of the OBDD size on the number of nondeterministic variables. In *Proceedings of the 19th Conference on Foundations of Software Technology and Theoretical Computer Science*, vol. 1738 of *Lecture Notes in Computer Science*, pp. 342–355. 1999.

6. M. Sauerhoff. An improved hierarchy result for partitioned BDDs. *Theory of Computing Systems*, 33:313–329, 2000.

7. P. Savický and D. Sieling. A hierarchy result for read-once branching programs with restricted parity nondeterminism. In *Mathematical Foundations of Computer Science: 25th International Symposium*, vol. 1893 of *Lecture Notes in Computer Science*, pp. 650–659. 2000.

8. P. Savický and S. Žák. A lower bound on branching programs reading some bits twice. *Theoretical Computer Science*, 172:293–301, 1997.

9. P. Savický and S. Žák. A read-once lower bound and a $(1, +k)$-hierarchy for branching programs. *Theoretical Computer Science*, 238:347–362, 2000.

10. D. Sieling. New lower bounds and hierarchy results for restricted branching programs. *Journal of Computer and System Sciences*, 53:79–87, 1996.

11. J. Simon and M. Szegedy. A new lower bound theorem for read-only-once branching programs and its applications. In *Advances in Computational Complexity Theory*, vol. 13 of *DIMACS Series in Discrete Mathematics and Theoretical Computer Science*, pp. 183–193. AMS, 1993.

12. I. Wegener. On the complexity of branching programs and decision trees for clique functions. *Journal of the ACM*, 35:461–471, 1988.

13. I. Wegener. *Branching Programs and Binary Decision Diagrams - Theory and Applications*. Siam, first edition, 2000.

14. S. Žák. An exponential lower bound for one-time-only branching programs. In *Mathematical Foundations of Computer Science: 11th International Symposium*, vol. 176 of *Lecture Notes in Computer Science*, pp. 562–566. 1984.

The Complexity of Constraints on Intervals and Lengths

Andrei Krokhin[1],*, Peter Jeavons[1],*, and Peter Jonsson[2],**

[1] Oxford University Computing Laboratory
Wolfson Building, Parks Road, Oxford OX1 3QD, UK
{Andrei.Krokhin,Peter.Jeavons}@comlab.ox.ac.uk
[2] Department of Computer and Information Science
Linköping University, S-581 83 Linköping, Sweden
Peter.Jonsson@ida.liu.se

Abstract. We study interval-valued constraint satisfaction problems (CSPs), in which the aim is to find an assignment of intervals to a given set of variables subject to constraints on the relative positions of intervals. One interesting question concerning such problems is to determine exactly how the complexity of an interval-valued CSP depends on the set of constraints allowed in instances. For the framework known as Allen's interval algebra this question was completely answered earlier by the authors.

Here we extend the qualitative framework of Allen's algebra with additional constraints on the lengths of intervals. We allow these length constraints to be expressed as Horn disjunctive linear relations, a well-known tractable and sufficiently expressive form of constraints. We completely characterize sets of qualitative relations for which the constraint satisfaction problem augmented with arbitrary length constraints of the above form is tractable. We also show that all the remaining cases are NP-complete.

1 Introduction and Summary of Results

A wide range of combinatorial search problems encountered in Computer Science and Artificial Intelligence can be naturally expressed as 'constraint satisfaction problems' [1], in which the aim is to find an assignment of *values* to a given set of *variables* subject to specified *constraints*. For example, the standard propositional SATISFIABILITY problem [2] may be viewed as a constraint satisfaction problem where the variables must be assigned Boolean values, and the constraints are specified by clauses. Further examples include GRAPH COLORABILITY, CLIQUE, and BANDWIDTH problems, scheduling problems, and many others (see [3,4]).

---

* Partially supported by the UK EPSRC grant GR/R29598.
** Partially supported by the Swedish Research Council (VR) under grant 221-2000-361.

H. Alt and A. Ferreira (Eds.): STACS 2002, LNCS 2285, pp. 443–454, 2002.
© Springer-Verlag Berlin Heidelberg 2002

Constraints are usually specified by means of relations. Hence the general constraint satisfaction problem can be parameterised according to the relations allowed in an instance. For any set of relations $\mathcal{F}$, the class of constraint satisfaction problem instances where the constraint relations are all members of $\mathcal{F}$ is denoted $CSP(\mathcal{F})$. The most well-known examples of such parameterised problems are GENERALIZED SATISFIABILITY [5], where the parameter is the set of allowed Boolean relations, and GRAPH H-COLORING [6], where the parameter is the single graph H.

In studying CSPs over infinite sets of values, arguably the most important type of problem is when the constraints are specified by binary relations and the set of possible values for the variables is the set of intervals on the real line. Such problems arise, for example, in many forms of temporal reasoning [7,8,9,10], where an event is identified with the interval during which it occurs. They also arise in computational biology, where various problems connected with physical mapping of DNA lead to interval-valued constraints [11,12,13]. Interval-valued CSPs can be naturally augmented with constraints on the lengths of the intervals, and the complexity of such extended problems will be the object of our main interest in this paper.

Before we describe our new results, we first discuss four closely related families of problems involving intervals which have previously been studied.

The prototypical problem from the first family is the INTERVAL GRAPH RECOGNITION problem [14]. An *interval graph* is an undirected graph such that there is assignment of intervals to the nodes with two nodes adjacent if and only if the two corresponding intervals intersect. Given an arbitrary graph G, the question of deciding whether G is an interval graph is rarely viewed as a constraint satisfaction problem, but in fact it is easily formulated as such a problem in the following way: every pair of adjacent nodes is constrained by the relation $r =$ "intersect" over pairs of intervals, and every pair of non-adjacent nodes is constrained by the complementary relation $\bar{r} =$ "disjoint". This fundamental INTERVAL GRAPH RECOGNITION problem is tractable, and it also remains tractable if we impose additional constraints on the lengths of the intervals which require all intervals to be of the same length (the UNIT INTERVAL GRAPH RECOGNITION problem [15]). In contrast, it was shown in [16] that if we allow boundaries to be specified for the lengths of intervals, or even exact lengths (which are not necessarily all equal), then the corresponding problems (called BOUNDED INTERVAL GRAPH RECOGNITION and MEASURED INTERVAL GRAPH RECOGNITION, respectively) are NP-complete.

A typical problem from the second family is the INTERVAL GRAPH SANDWICH problem [12]. Given two graphs $G_1 = (V, E_1)$ and $G_2 = (V, E_2)$ such that $E_1 \subseteq E_2$, the question is whether there is an interval graph $G = (V, E)$ with $E_1 \subseteq E \subseteq E_2$. Clearly, this is a generalization of the corresponding recognition problem (the case when $E_1 = E_2$). The INTERVAL GRAPH SANDWICH problem can be represented as a constraint satisfaction problem as follows: to any $e \in E_1$ assign the constraint $r =$ "intersect", to any $e \notin E_2$ assign the constraint $\bar{r} =$ "disjoint", and leave all pairs of variables corresponding to edges from $E_2 \setminus E_1$

unrelated. This problem was shown to be NP-complete along with the UNIT IN-TERVAL GRAPH SANDWICH problem where all intervals are required to be of the same length [12].

Other examples of problems from the first two families include the CIRCLE GRAPH RECOGNITION problem [17] and the CIRCLE GRAPH SANDWICH problem [18]. These problems can be formulated as a constraint satisfaction problem in the same way as above using the constraint relation $r =$ "overlap".

The third family of problems we mention is the so-called INTERVAL SATISFI-ABILITY problems [8,19,20]. In these problems every pair of interval variables is again constrained in some way, but the constraints this time are chosen from a given set $\mathcal{F}$ of relations. In [8,19,20] only a small number of possibilities for $\mathcal{F}$ are considered. It is shown there that for some choices of $\mathcal{F}$ the resulting problem is tractable, whilst for others it is NP-complete. The complexity of INTERVAL SATISFIABILITY with all intervals of the same length is also studied in [19].

The fourth type of problem we mention is the satisfiability problem for *Allen's interval algebra* [7], denoted $\mathcal{A}$-SAT. Allen's algebra contains 13 basic relations (corresponding to the 13 ways two given intervals can be related from a "qual-itative" point of view). The set $\mathcal{A}$ contains not just these basic relations, but all $2^{13} = 8192$ possible unions of them. The problems $\mathcal{A}$-SAT$(\mathcal{F})$ are similar to problems of the third type above, except that unrelated pairs of variables are allowed. They can also be represented as INTERVAL SATISFIABILITY with $\mathcal{F}$ be-ing an arbitrary subset of $\mathcal{A}$ containing the total relation. The complexity of problems of the form $\mathcal{A}$-SAT$(\mathcal{F})$ has been intensively studied in the Artificial In-telligence community (see, e.g., [10,21,22]), and a complete classification of the complexity of such problems was obtained in [9]. In that paper it is shown that there are exactly 18 maximal tractable fragments of $\mathcal{A}$, and for any subset $\mathcal{F}$ not entirely contained in one of those the problem $\mathcal{A}$-SAT$(\mathcal{F})$ is NP-complete.

Many problems, e.g., certain scheduling problems, can conveniently be ex-pressed as $\mathcal{A}$-SAT$(\mathcal{F})$ with additional constraints on the lengths of the intervals. Moreover, in [3], it was suggested that many important forms of constraints on lengths can be expressed in the form of Horn disjunctive linear relations. This class of relations is known to be tractable [23,24] and at the same time it allows us to express all elementary constraints such as fixing the length, or bounding the length of an interval by a given number, or comparing the lengths of two intervals. It was proved in [3] that only three out of the 18 maximal tractable fragments for $\mathcal{A}$-SAT$(\mathcal{F})$ preserve tractability when extended with Horn disjunc-tive linear constraints on lengths; the other 15 become NP-complete. In this paper we study how we need to further restrict those 15 fragments to obtain tractable cases. The main result is a complete classification of complexity for $\mathcal{A}$-SAT$(\mathcal{F})$ with additional constraints on lengths. We show that such problems are either tractable or strongly NP-complete. Moreover, we give a complete de-scription of the tractable cases, which allows one to easily determine whether a given set $\mathcal{F}$ falls into one of the tractable cases.

As well as giving a complete classification, our result also establishes a new dichotomy theorem for complexity. Dichotomy theorems are results concerning a class of related problems (with some parameter) which assert that, for some values of the parameter, the problems in the class are tractable while for all other values they are NP-complete. Such theorems are of interest because it is well known [2] that if P$\neq$NP then, within NP, there are infinitely many pairwise inequivalent problems of intermediate complexity. Dichotomy results rule out such a possibility within certain classes of problems.

Dichotomy theorems have previously been established for the GENERALIZED SATISFIABILITY [5] and GRAPH H-COLORING [6] problems mentioned above as well as the DIRECTED SUBGRAPH HOMEOMORPHISM problem [25].

Constraint satisfaction problems have been a fruitful source of dichotomy results (see, e.g., [26,27]). For constraint satisfaction problems, the relevant parameter is usually the set of relations, $\mathcal{F}$, specifying the allowed constraints. This parameter usually runs over an infinite set of values. In the case of Allen's algebra, even though the number of different values for $\mathcal{F}$ is finite, it is astronomical ($2^{8192} \approx 10^{2466}$) which excludes the possibility of computer-aided exhaustive case analysis.

The usual tool for proving dichotomy theorems is reducibilty via expressibility. This is done by showing that one set of relations expresses another, so that one problem can be reduced to the other. This is the method used in [5,9, 26], and a similar method is used here. After identifying certain tractable fragments, we find some NP-complete fragments and then show how any subset not entirely contained in one of the tractable sets can express some already known NP-complete fragment.

The full version of the present paper is available in electronic form [28].

2   Preliminaries and Background

Allen's interval algebra [7], denoted $\mathcal{A}$, is a formalism for expressing *qualitative* binary relations between intervals on the real line. By "qualitative" we mean "invariant under all continuous injective monotone transformations of the real line". An interval x is represented as a pair $[x^-, x^+]$ of real numbers with $x^- < x^+$, denoting the left and right endpoints of the interval, respectively. The qualitative relations between intervals are the $2^{13} = 8192$ possible unions of the 13 *basic interval relations*, which are shown in Table 1. For the sake of brevity, relations between intervals will be written as collections of basic relations, omitting the sign of union. So, for instance, we write (pmf^{-1}) instead of p $\cup$ m $\cup$ f^{-1}.

The problem of *satisfiability* ($\mathcal{A}$-SAT) in Allen's algebra is defined as follows.

Definition 1. *Let $\mathcal{F} \subseteq \mathcal{A}$ be a set of interval relations. An instance I of $\mathcal{A}$-SAT($\mathcal{F}$) over a set, V, of variables is a set of constraints of the form xry where $x, y \in V$ and $r \in \mathcal{F}$. The question is whether I is satisfiable, i.e., whether there exists a function, f, from V to the set of all intervals such that $f(x)\,r\,f(y)$ holds for every constraint xry in I. Any such function f is called a* model *of I.*

Allen's interval algebra $\mathcal{A}$ consists of the 8192 possible relations between intervals together with three standard operations on binary relations: *con-*

Table 1. The thirteen basic relations. The endpoint relations $x^- < x^+$ and $y^- < y^+$ that are valid for all relations have been omitted.

Basic relation		Example	Endpoints
x precedes y	p	xxx	$x^+ < y^-$
y preceded by x	p^{-1}	yyy	
x meets y	m	xxxx	$x^+ = y^-$
y met by x	m^{-1}	yyyy	
x overlaps y	o	xxxx	$x^- < y^- < x^+$,
y overlapped by x	o^{-1}	yyyy	$x^+ < y^+$
x during y	d	xxx	$x^- > y^-$,
y includes x	d^{-1}	yyyyyyy	$x^+ < y^+$
x starts y	s	xxx	$x^- = y^-$,
y started by x	s^{-1}	yyyyyyy	$x^+ < y^+$
x finishes y	f	xxx	$x^+ = y^+$,
y finished by x	f^{-1}	yyyyyyy	$x^- > y^-$
x equals y	$\equiv$	xxxx	$x^- = y^-$,
		yyyy	$x^+ = y^+$

verse $\cdot^{-1}$, *intersection* $\cap$ *and composition* $\circ$. Subsets of $\mathcal{A}$ that are closed under the operations of converse, intersection, and composition are said to be *subalgebras*. For a given subset $\mathcal{F}$ of $\mathcal{A}$, the smallest subalgebra containing $\mathcal{F}$ is called the subalgebra *generated* by $\mathcal{F}$ and is denoted by $\langle\mathcal{F}\rangle$. It is easy to see that $\langle\mathcal{F}\rangle$ is obtained from $\mathcal{F}$ by adding all relations that can be obtained from the relations in $\mathcal{F}$ by using the three operations of $\mathcal{A}$.

It is known [10], and easy to prove, that, for every $\mathcal{F} \subseteq \mathcal{A}$, the problem $\mathcal{A}$-SAT($\langle\mathcal{F}\rangle$) is polynomially equivalent to $\mathcal{A}$-SAT($\mathcal{F}$). Therefore, to classify the complexity of $\mathcal{A}$-SAT($\mathcal{F}$) it is only necessary to consider these problems for *subalgebras* of $\mathcal{A}$. Throughout the paper, $\mathcal{S}$ denotes a subalgebra of $\mathcal{A}$.

In the following we shall use the symbol $\pm$, which should be interpreted as follows. A condition involving $\pm$ means the conjunction of two conditions: one corresponding to $+$ and one corresponding to $-$. For example, condition $r \cap (\text{dsf})^{\pm 1} \neq \emptyset \Rightarrow (\text{d})^{\pm 1} \subseteq r$ means that both $r \cap (\text{dsf}) \neq \emptyset \Rightarrow (\text{d}) \subseteq r$ and $r \cap (\text{d}^{-1}\text{s}^{-1}\text{f}^{-1}) \neq \emptyset \Rightarrow (\text{d}^{-1}) \subseteq r$ hold. The main advantage of using the $\pm$ symbol is conciseness: in any subalgebra of $\mathcal{A}$, the '$+$' and the '$-$' conditions are satisfied (or not satisfied) simultaneously, and, therefore only one of them needs to be verified.

The following notation is used in Fig. 1 to describe some important subalgebras of $\mathcal{A}$. Let $q_0 = (\text{pmo})$. Further, let $q_1 = q_0 \cup (\text{d}^{-1}\text{f}^{-1})$, $q_2 = q_0 \cup (\text{df})$, $q_3 = q_0 \cup (\text{df}^{-1})$, $q_4 = q_0 \cup (\text{ds})$, $q_5 = q_0 \cup (\text{d}^{-1}\text{s}^{-1})$, $q_6 = q_0 \cup (\text{d}^{-1}\text{s})$.

A complete classification of complexity of problems of the form $\mathcal{A}$-SAT($\mathcal{F}$) was obtained in [9].

Theorem 1 ([9]). *For any subset $\mathcal{F}$ of $\mathcal{A}$, either $\mathcal{A}$-SAT($\mathcal{F}$) is NP-complete or $\mathcal{F}$ is included in $\mathcal{S}$, where $\mathcal{S}$ is one of $\mathcal{S}_\text{p}$, $\mathcal{S}_\text{d}$, $\mathcal{S}_\text{o}$, $\mathcal{E}_\text{p}$, $\mathcal{E}_\text{d}$, $\mathcal{E}_\text{o}$, $\mathcal{A}_i (1 \leq i \leq 4)$, $\mathcal{B}_i (1 \leq i \leq 4)$, $\mathcal{S}^*$, $\mathcal{E}^*$, $\mathcal{A}_\equiv$, $\mathcal{H}$ (see Fig. 1), for which $\mathcal{A}$-SAT($\mathcal{S}$) is tractable.*

$$\mathcal{S}_\mathsf{p} = \{r \mid r \cap q_1^{\pm 1} \neq \emptyset \Rightarrow (\mathsf{p})^{\pm 1} \subseteq r\} \qquad \mathcal{E}_\mathsf{p} = \{r \mid r \cap q_4^{\pm 1} \neq \emptyset \Rightarrow (\mathsf{p})^{\pm 1} \subseteq r\}$$

$$\mathcal{S}_\mathsf{d} = \{r \mid r \cap q_1^{\pm 1} \neq \emptyset \Rightarrow (\mathsf{d}^{-1})^{\pm 1} \subseteq r\} \qquad \mathcal{E}_\mathsf{d} = \{r \mid r \cap q_4^{\pm 1} \neq \emptyset \Rightarrow (\mathsf{d})^{\pm 1} \subseteq r\}$$

$$\mathcal{S}_\mathsf{o} = \{r \mid r \cap q_1^{\pm 1} \neq \emptyset \Rightarrow (\mathsf{o})^{\pm 1} \subseteq r\} \qquad \mathcal{E}_\mathsf{o} = \{r \mid r \cap q_4^{\pm 1} \neq \emptyset \Rightarrow (\mathsf{o})^{\pm 1} \subseteq r\}$$

$$\mathcal{A}_1 = \{r \mid r \cap q_1^{\pm 1} \neq \emptyset \Rightarrow (\mathsf{s}^{-1})^{\pm 1} \subseteq r\} \qquad \mathcal{B}_1 = \{r \mid r \cap q_4^{\pm 1} \neq \emptyset \Rightarrow (\mathsf{f}^{-1})^{\pm 1} \subseteq r\}$$

$$\mathcal{A}_2 = \{r \mid r \cap q_1^{\pm 1} \neq \emptyset \Rightarrow (\mathsf{s})^{\pm 1} \subseteq r\} \qquad \mathcal{B}_2 = \{r \mid r \cap q_4^{\pm 1} \neq \emptyset \Rightarrow (\mathsf{f})^{\pm 1} \subseteq r\}$$

$$\mathcal{A}_3 = \{r \mid r \cap q_2^{\pm 1} \neq \emptyset \Rightarrow (\mathsf{s})^{\pm 1} \subseteq r\} \qquad \mathcal{B}_3 = \{r \mid r \cap q_5^{\pm 1} \neq \emptyset \Rightarrow (\mathsf{f}^{-1})^{\pm 1} \subseteq r\}$$

$$\mathcal{A}_4 = \{r \mid r \cap q_3^{\pm 1} \neq \emptyset \Rightarrow (\mathsf{s})^{\pm 1} \subseteq r\} \qquad \mathcal{B}_4 = \{r \mid r \cap q_6^{\pm 1} \neq \emptyset \Rightarrow (\mathsf{f}^{-1})^{\pm 1} \subseteq r\}$$

$$\mathcal{S}^* = \left\{ r \left| \begin{array}{l} 1) \; r \cap q_1^{\pm 1} \neq \emptyset \Rightarrow (\mathsf{f}^{-1})^{\pm 1} \subseteq r, \\ 2) \; r \cap (\mathsf{ss}^{-1}) \neq \emptyset \Rightarrow (\equiv) \subseteq r \end{array} \right. \right\} \quad \mathcal{E}^* = \left\{ r \left| \begin{array}{l} 1) \; r \cap q_4^{\pm 1} \neq \emptyset \Rightarrow (\mathsf{s})^{\pm 1} \subseteq r, \\ 2) \; r \cap (\mathsf{ff}^{-1}) \neq \emptyset \Rightarrow (\equiv) \subseteq r \end{array} \right. \right\}$$

$$\mathcal{A}_\equiv = \{r \mid r \neq \emptyset \Rightarrow (\equiv) \subseteq r\}$$

$$\mathcal{H} = \left\{ r \left| \begin{array}{l} 1) \; r \cap (\mathsf{os})^{\pm 1} \neq \emptyset \; \& \; r \cap (\mathsf{o}^{-1}\mathsf{f})^{\pm 1} \neq \emptyset \Rightarrow (\mathsf{d})^{\pm 1} \subseteq r, \text{ and} \\ 2) \; r \cap (\mathsf{ds})^{\pm 1} \neq \emptyset \; \& \; r \cap (\mathsf{d}^{-1}\mathsf{f}^{-1})^{\pm 1} \neq \emptyset \Rightarrow (\mathsf{o})^{\pm 1} \subseteq r, \text{ and} \\ 3) \; r \cap (\mathsf{pm})^{\pm 1} \neq \emptyset \; \& \; r \not\subseteq (\mathsf{pm})^{\pm 1} \Rightarrow (\mathsf{o})^{\pm 1} \subseteq r \end{array} \right. \right\}$$

$$\mathcal{C}_\mathsf{o} = \{r \mid r \neq \emptyset \Rightarrow (\mathsf{oo}^{-1}) \subseteq r\}$$

$$\mathcal{C}_\mathsf{m} = \left\{ r \left| \begin{array}{l} 1) \; r \neq \emptyset \Rightarrow (\mathsf{mm}^{-1}\mathsf{ss}^{-1}\mathsf{ff}^{-1}) \subseteq r, \text{ and} \\ 2) \; r \cap (\mathsf{pp}^{-1}\mathsf{oo}^{-1}) \neq \emptyset \Rightarrow (\equiv) \subseteq r \end{array} \right. \right\}$$

$$\mathcal{D}_\mathsf{s} = \left\{ r \left| \begin{array}{l} 1) \; r \cap (\mathsf{dsf})^{\pm 1} \neq \emptyset \Rightarrow (\mathsf{s})^{\pm 1} \subseteq r, \text{ and} \\ 2) \; r \cap (\mathsf{pp}^{-1}\mathsf{mm}^{-1}\mathsf{oo}^{-1}) \neq \emptyset \Rightarrow (\equiv \mathsf{ss}^{-1}) \subseteq r \end{array} \right. \right\}$$

$$\mathcal{D}_\mathsf{f} = \left\{ r \left| \begin{array}{l} 1) \; r \cap (\mathsf{dsf})^{\pm 1} \neq \emptyset \Rightarrow (\mathsf{f})^{\pm 1} \subseteq r, \text{ and} \\ 2) \; r \cap (\mathsf{pp}^{-1}\mathsf{mm}^{-1}\mathsf{oo}^{-1}) \neq \emptyset \Rightarrow (\equiv \mathsf{ff}^{-1}) \subseteq r \end{array} \right. \right\}$$

$$\mathcal{D}_\mathsf{d} = \left\{ r \left| \begin{array}{l} 1) \; r \cap (\mathsf{dsf})^{\pm 1} \neq \emptyset \Rightarrow (\mathsf{d})^{\pm 1} \subseteq r, \text{ and} \\ 2) \; r \cap (\mathsf{pp}^{-1}\mathsf{mm}^{-1}\mathsf{oo}^{-1}) \neq \emptyset \Rightarrow (\equiv \mathsf{dd}^{-1}) \subseteq r \end{array} \right. \right\}$$

$$\mathcal{D}'_\mathsf{d} = \left\{ r \left| \begin{array}{l} 1) \; r \cap (\mathsf{dsf})^{\pm 1} \neq \emptyset \Rightarrow (\mathsf{d})^{\pm 1} \subseteq r, \text{ and} \\ 2) \; r \cap (\mathsf{pmo})^{\pm 1} \neq \emptyset \Rightarrow (\mathsf{odd}^{-1})^{\pm 1} \subseteq r \end{array} \right. \right\}$$

$$\mathcal{D}''_\mathsf{d} = \left\{ r \left| \begin{array}{l} 1) \; r \cap (\mathsf{dsf})^{\pm 1} \neq \emptyset \Rightarrow (\mathsf{d})^{\pm 1} \subseteq r, \text{ and} \\ 2) \; r \cap (\mathsf{pp}^{-1}\mathsf{oo}^{-1}) \neq \emptyset \Rightarrow (\mathsf{oo}^{-1}\mathsf{dd}^{-1}) \subseteq r, \text{ and} \\ 3) \; r \cap (\mathsf{pp}^{-1}\mathsf{mm}^{-1}) \neq \emptyset \Rightarrow (\equiv \mathsf{dd}^{-1}) \subseteq r \end{array} \right. \right\}$$

Fig. 1. 25 subalgebras of Allen's algebra.

In this paper we present a complete complexity classification for a more general problem, namely, for $\mathcal{A}$-SAT($\mathcal{F}$) extended with constraints on the lengths of intervals. Now we define the exact form of constraints on lengths we shall allow.

Definition 2. *Let V be a set of real-valued variables, and α, β linear polynomials over V with rational coefficients. A* linear relation *over V is an expression of the form $\alpha R \beta$, where $R \in \{<, \leq, =, \neq, \geq, >\}$.*

A disjunctive linear relation (DLR) over V is a disjunction of a nonempty finite set of linear relations. A DLR is said to be Horn *if and only if at most one of its disjuncts is not of the form $\alpha \neq \beta$.*

The problem of satisfiability *for finite sets D of DLRs, denoted* DLRSAT*, is that of checking whether there exists an assignment f of variables in V to real numbers such that all DLRs in D are satisfied. Such an f is said to be a* model *of D. The satisfiability problem for finite sets of Horn DLRs is denoted* HORNDLRSAT*.*

Theorem 2 ([23,24]). *The problem* DLRSAT *is NP-complete and* HORNDLR-SAT *is solvable in polynomial time.*

We are interested in how the complexity of a problem depends on the value of parameter $\mathcal{F}$ which, in our case, is a set of qualitative relations. Therefore we shall allow only those constraints on lengths which can be expressed by Horn DLRs. This class of constraints subsumes all forms of constraints on lengths which have been considered in [16,19].

We can now define the general interval satisfiability problem with constraints on lengths.

Definition 3. *An instance of the problem of* interval satisfiability with constraints on lengths *for a set $\mathcal{F} \subseteq \mathcal{A}$, denoted $\mathcal{A}^l$-SAT($\mathcal{F}$), is a pair $Q = (I, D)$, where*

- *I is an instance of $\mathcal{A}$-SAT($\mathcal{F}$) over a set V of variables, and*
- *D is an instance of* HORNDLRSAT *over the set of variables $\{l(v) \mid v \in V\}$.*

The question is whether Q is satisfiable*, i.e., whether there exists a model f of I such that the DLRs in D are satisfied with $l(v)$ equal to the length of $f(v)$ for all $v \in V$.*

Proposition 1. *$\mathcal{A}^l$-SAT($\mathcal{F}$) $\in$ NP for every $\mathcal{F} \subseteq \mathcal{A}$.*

Proof. Every instance of $\mathcal{A}^l$-SAT($\mathcal{F}$) over a set V of variables can be translated in a straightforward way into an instance of DLRSAT over the set $\{v^-, v^+ \mid v \in V\}$ of variables. Now the proposition follows from Theorem 2.   □

The complexity of $\mathcal{A}^l$-SAT($\mathcal{S}$) has already been determined for each subalgebra $\mathcal{S}$ listed in Theorem 1.

Proposition 2 ([3]). *The problem $\mathcal{A}^l$-SAT($\mathcal{S}$) is tractable for $\mathcal{S} \in \{\mathcal{S}_{\mathsf{p}}, \mathcal{E}_{\mathsf{p}}, \mathcal{H}\}$ and is NP-complete for the other 15 subalgebras from Theorem 1.*

3   Main Result

Theorem 3. *For any subset $\mathcal{F}$ of $\mathcal{A}$, either $\mathcal{A}^l$-SAT$(\mathcal{F})$ is strongly NP-complete or $\mathcal{F}$ is included in $\mathcal{S}$, where $\mathcal{S}$ is one of $\mathcal{S}_\mathsf{p}$, $\mathcal{E}_\mathsf{p}$, $\mathcal{H}$, $\mathcal{C}_\mathsf{o}$, $\mathcal{C}_\mathsf{m}$, $\mathcal{D}_\mathsf{s}$, $\mathcal{D}_\mathsf{f}$, $\mathcal{D}_\mathsf{d}$, $\mathcal{D}'_\mathsf{d}$, $\mathcal{D}''_\mathsf{d}$ (see Fig. 1), for which $\mathcal{A}^l$-SAT$(\mathcal{S})$ is tractable.*

In Subsection 3.1, we give polynomial-time algorithms for the 10 subalgebras listed in Theorem 3. Subsection 3.2 contains two NP-completeness results we use to obtain a complete classification of complexity of $\mathcal{A}^l$-SAT$(\mathcal{F})$. Finally, in Subsection 3.3, we give a sketch of the classification proof. All correctness, NP-completeness, and classification proofs omitted due to space constraints can be found in the full version of this paper [28].

Strong NP-completeness of the NP-complete cases follows from the fact the the biggest number used in NP-completeness proofs is 5.

3.1   Tractability Results

In this subsection we give algorithms for the tractable cases (Fig. 2). Checking that they are polynomial-time is easy and is left to the reader.

We will also assume that algorithms A_i, $1 \leq i \leq 4$, and the procedure P take instance $Q = (I, D)$ over a set of variables V as an input, and D always contains all constraints of the form $l(v) > 0$, $v \in V$. We will suppose that I does not contain a constraint vrw where $r = \emptyset$. This trivial necessary condition for satisfiability can obviously be checked in polynomial time.

Proposition 3. *The problem $\mathcal{A}^l$-SAT$(\mathcal{S})$ is tractable whenever $\mathcal{S}$ is one of $\mathcal{S}_\mathsf{p}$, $\mathcal{E}_\mathsf{p}$, $\mathcal{H}$, $\mathcal{C}_\mathsf{o}$, $\mathcal{C}_\mathsf{m}$, $\mathcal{D}_\mathsf{s}$, $\mathcal{D}_\mathsf{f}$, $\mathcal{D}_\mathsf{d}$, $\mathcal{D}'_\mathsf{d}$, and $\mathcal{D}''_\mathsf{d}$.*

Polynomial-time algorithms solving $\mathcal{A}^l$-SAT$(\mathcal{S})$ for $\mathcal{S} \in \{\mathcal{S}_\mathsf{p}, \mathcal{E}_\mathsf{p}, \mathcal{H}\}$ are given in [3]. It is easy to show that an instance $Q = (I, D)$ of $\mathcal{A}^l$-SAT$(\mathcal{C}_\mathsf{o})$ is satisfiable if and only if D is satisfiable. Algorithms for the remaining 6 subalgebras are given in Fig. 2.

The following lemma is crucial in the correctness proofs.

Lemma 1 ([21]). *Let D be a satisfiable set of Horn DLRs and let $x_1, \ldots, x_n$ be the variables used in D. If $\tilde{D} = \{x_i \neq x_j \mid D \cup \{x_i \neq x_j\}$ is satisfiable$\}$ then $D \cup \tilde{D}$ is satisfiable.*

Using this lemma we can always divide V into classes such that, in every model of an instance, variables from the same class must be assigned intervals of the same length while any variables from different classes can be assigned intervals of different lengths all at the same time.

Input: instance $Q = (I, D)$ of $\mathcal{A}^l$-SAT$(\mathcal{S})$ with set V of variables

Algorithm A_1 for $\mathcal{S} = \mathcal{C}_\mathsf{m}$

1) if D is not satisfiable then reject
2) construct a graph $G = (V, E)$ where $(v, w) \in E$ if and only if
   - $D \cup \{l(v) \neq l(w)\}$ is not satisfiable, and
   - $vrw \in I$ for some r such that $(\equiv) \not\subseteq r$
3) if G is 2-colorable then accept else reject

Procedure P

1) let $D' = D$
2) for each $vrw \in I$ such that $r \subseteq (\mathsf{dsf})$ or $r \subseteq (\mathsf{d}^{-1}\mathsf{s}^{-1}\mathsf{f}^{-1})$, add the constraint $l(v) < l(w)$ or $l(v) > l(w)$, respectively, to D'
3) for each $vrw \in I$ such that $(\equiv) \subseteq r \subseteq (\equiv \mathsf{dsf})$ or $(\equiv) \subseteq r \subseteq (\equiv \mathsf{d}^{-1}\mathsf{s}^{-1}\mathsf{f}^{-1})$, add the constraint $l(v) \leq l(w)$ or $l(v) \geq l(w)$, respectively, to D'
4) if D' is not satisfiable then reject else return (I, D')

Algorithm A_2 for $\mathcal{S} \in \{\mathcal{D}_\mathsf{s}, \mathcal{D}_\mathsf{f}, \mathcal{D}_\mathsf{d}\}$

1) call procedure P
2) accept

Algorithm A_3 for $\mathcal{S} = \mathcal{D}'_\mathsf{d}$

1) call procedure P
2) construct a graph $G = (V, E)$ where $(v, w) \in E$ if and only if $D' \cup \{l(v) \neq l(w)\}$ is not satisfiable
3) identify the connected components $S_1, \ldots, S_k$ of G
4) for each S_j, let $I_j = \{vrw \in I \mid v, w \in S_j\}$ and $I'_j = \{v \, r \cap (\equiv \mathsf{oo}^{-1}) \, w \mid vrw \in I_j\}$
5) solve I'_j, $1 \leq j \leq k$, as instances of $\mathcal{A}$-SAT$(\mathcal{S}_\mathsf{o})$
6) if every I'_j is satisfiable then accept else reject

Algorithm A_4 for $\mathcal{S} = \mathcal{D}''_\mathsf{d}$

1) call procedure P
2) construct a graph $G = (V, E)$ where $(v, w) \in E$ if and only if
   - $D' \cup \{l(v) \neq l(w)\}$ is not satisfiable, and
   - $vrw \in I$ for some r such that $(\equiv) \subseteq r \cap (\equiv \mathsf{pp}^{-1}\mathsf{oo}^{-1}\mathsf{mm}^{-1}) \subseteq (\equiv \mathsf{mm}^{-1})$
3) identify the connected components $S_1, \ldots, S_k$ of G
4) for each S_j, let $I_j = \{vrw \in I \mid v, w \in S_j\}$ and $I'_j = \{vrw \mid vrw \in I_j \text{ and } (\equiv) \not\subseteq r\}$
5) if every I'_j is empty then accept else reject

Fig. 2. Polynomial-time algorithms for tractable cases of $\mathcal{A}^l$-SAT.

3.2   NP-Completeness Results

First let us mention an obvious fact that, for any $\mathcal{F} \subseteq \mathcal{A}$, NP-completeness of $\mathcal{A}$-SAT$(\mathcal{F})$ implies NP-completeness of $\mathcal{A}^l$-SAT$(\mathcal{F})$.

Lemma 2. *Let* $r_1, \ldots, r_n \in \mathcal{A}$ *be relations such that* $\mathcal{A}$-SAT$(\{r_1, \ldots, r_n\})$ *is NP-complete.*

1. *If, for every* $1 \leq i \leq n$, $r_i' \in \{r_i, r_i \cup (\equiv)\}$ *then* $\mathcal{A}^l$-SAT$(\{r_1', \ldots, r_n'\})$ *is NP-complete.*
2. *If* $\emptyset \neq r_1 \subseteq$ (pmo) *and* r_1' *is such that* $r_1 \subseteq r_1' \subseteq r_1 \cup (\equiv$ dsf$)$ *then* $\mathcal{A}^l$-SAT$(\{r_1', r_2, \ldots, r_n\})$ *is NP-complete.*

Lemma 3. $\mathcal{A}^l$-SAT$(\mathcal{F})$ *is NP-complete whenever* $\mathcal{F}$ *is one of the following sets:*

1. $\{(\mathsf{oo}^{-1}), (\mathsf{s})\}$, $\{(\mathsf{oss}^{-1}\mathsf{ff}^{-1})\}$, *or* $\{(\mathsf{sf}), (\mathsf{oo}^{-1}\mathsf{ss}^{-1}\mathsf{ff}^{-1})\}$;
2. $\{r\}$ *where* $(\mathsf{mm}^{-1}) \subseteq r \subseteq (\mathsf{mm}^{-1}\mathsf{dd}^{-1}\mathsf{ss}^{-1}\mathsf{ff}^{-1})$ *and* $(\mathsf{ss}^{-1}\mathsf{ff}^{-1}) \not\subseteq r$;
3. $\{r_1, r_2\}$ *where* $r_1 \cap (\equiv \mathsf{pp}^{-1}\mathsf{oo}^{-1}\mathsf{mm}^{-1}) = (\mathsf{mm}^{-1}) \subsetneq r_2 \cap (\equiv \mathsf{pp}^{-1}\mathsf{oo}^{-1}\mathsf{mm}^{-1})$ *and* $(\equiv) \not\subseteq r_2$.

3.3   Classification of Complexity

Now let us briefly discuss the notion of *derivation with lengths* which is the main tool in our classification proofs. This notion is an extension of the notion of derivation in Allen's algebra used in [9].

Suppose $\mathcal{F} \subseteq \mathcal{A}$ and $Q = (I, D)$ is an instance of $\mathcal{A}^l$-SAT$(\mathcal{F})$. Let variables x, y be involved in I. Suppose a relation $r \in \mathcal{A}$ satisfies the following condition: Q is satisfiable if and only if xry. Then we say that r is *derived (with lengths)* from $\mathcal{F}$. It can easily be checked that the problems $\mathcal{A}^l$-SAT$(\mathcal{F})$ and $\mathcal{A}^l$-SAT$(\mathcal{F} \cup \{r\})$ are polynomially equivalent because, in any instance of the second problem, any constraint involving r can be replaced by the set of constraints in Q (introducing fresh variables when needed), and this can be done in polynomial time. It follows that it is sufficient to classify the complexity of problems $\mathcal{A}^l$-SAT$(\mathcal{S})$ where $\mathcal{S}$ is a subalgebra of $\mathcal{A}$ closed under derivations with lengths.

The classification proof has 8 steps. In each step, it is proved that if a subalgebra $\mathcal{S}$ which is closed under derivations with lengths satisfies a certain condition then either $\mathcal{S}$ is contained in one of the 10 tractable subalgebras, or Lemma 2 or 3 can be applied to some $\mathcal{F} \subseteq \mathcal{S}$, or $\mathcal{S}$ satisfies the conditions of one of the previous steps.

We shall say that a relation is *non-trivial* if it is not equal to the empty relation or the relation $(\equiv)$.

Step 1. Suppose $\mathcal{S}$ contains a non-trivial relation r with $r \subseteq (\equiv \mathsf{pp}^{-1}\mathsf{mm}^{-1}\mathsf{oo}^{-1})$. Then either $\mathcal{S}$ is contained in one of $\mathcal{C_o}$, $\mathcal{S_p}$, $\mathcal{E_p}$, and $\mathcal{H}$, or else $\mathcal{A}^l$-SAT$(\mathcal{S})$ is NP-complete.

Step 2. Suppose $\mathcal{S}$ contains a non-trivial relation r such that $r \cap r^{-1} \subseteq (\equiv)$ and neither r nor r^{-1} is contained in $(\equiv$ dsf$)$. Then either $\mathcal{S}$ is contained in one of $\mathcal{C_o}$, $\mathcal{S_p}$, $\mathcal{E_p}$, and $\mathcal{H}$ or else $\mathcal{A}^l$-SAT$(\mathcal{S})$ is NP-complete.

Step 3. If $\mathcal{S}$ contains two non-trivial relations r_1 and r_2 such that $r_1 \cap r_2 \subseteq (\equiv)$ and $r_1, r_2 \subseteq (\equiv \mathsf{dsf})$ then either $\mathcal{S} \subseteq \mathcal{H}$ or else $\mathcal{A}^l\text{-SAT}(\mathcal{S})$ is NP-complete.

Step 4. If $\mathcal{S}$ contains two non-trivial symmetric relations r_1 and r_2 such that $r_1 \cap r_2 \subseteq (\equiv)$ then either $\mathcal{S}$ is contained in one of $\mathcal{S}_\mathsf{p}$, $\mathcal{E}_\mathsf{p}$, $\mathcal{H}$, or else $\mathcal{A}^l\text{-SAT}(\mathcal{S})$ is NP-complete.

Step 5. If $(\mathsf{s}) \in \mathcal{S}$ or $(\mathsf{f}) \in \mathcal{S}$ then either $\mathcal{S}$ is contained in one of the 10 subalgebras listed in Theorem 3 or $\mathcal{A}^l\text{-SAT}(\mathcal{S})$ is NP-complete.

Step 6. If $(\mathsf{sf}) \in \mathcal{S}$ then either $\mathcal{S} \subseteq \mathcal{D}_\mathsf{s}$ or $\mathcal{S} \subseteq \mathcal{D}_\mathsf{f}$ or else $\mathcal{A}^l\text{-SAT}(\mathcal{S})$ is NP-complete.

Step 7. If there is $r \in \mathcal{S}$ such that $(\mathsf{d}) \subseteq r \subseteq (\mathsf{dsf})$ then either $\mathcal{S}$ is contained in one of the 10 subalgebras listed in Theorem 3 or else $\mathcal{A}^l\text{-SAT}(\mathcal{S})$ is NP-complete.

Step 8. If there is a symmetric non-trivial relation $r' \in \mathcal{S}$ such that every non-trivial $r \in \mathcal{S}$ satisfies $r' \subseteq r$ then either $\mathcal{S}$ is contained in one of the 10 tractable subalgebras or $\mathcal{A}^l\text{-SAT}(\mathcal{S})$ is NP-complete.

Note that the assumptions of these 8 steps are easily shown to be exhaustive. Indeed, if there is $r \in \mathcal{S}$ such that $r \cap r^{-1} \subseteq (\equiv)$ then either Step 2 or one of Steps 5-7 applies because, due to closedness under derivations with lengths, a subalgebra containing $r \cup (\equiv)$ where $r \subseteq (\mathsf{dsf})$ also contains r itself. Otherwise Step 4 or Step 8 applies, since $r \cap r^{-1} \in \mathcal{S}$ for every $r \in \mathcal{S}$.

Classification is complete. Theorem 3 is proved.   □

References

1. Montanari, U.: Networks of constraints: Fundamental properties and applications to picture processing. Information Sciences **7** (1974) 95–132
2. Garey, M., Johnson, D.: Computers and Intractability: A Guide to the Theory of NP-Completeness. Freeman, New York (1979)
3. Angelsmark, O., Jonsson, P.: Some observations on durations, scheduling and Allen's algebra. In: Proceedings of the 6th Conference on Constraint Programming (CP'00). Volume 1894 of Lecture Notes in Computer Science., Springer-Verlag (2000) 484–488
4. Jeavons, P.: On the algebraic structure of combinatorial problems. Theoretical Computer Science **200** (1998) 185–204
5. Schaefer, T.: The complexity of satisfiability problems. In: Proceedings of the 10th Symposium on the Theory of Computing (STOC'78), New-York, ACM Press (1978) 216–226
6. Hell, P., Nešetřil, J.: On the complexity of H-coloring. Journal of Combinatorial Theory, Ser. B **48** (1990) 92–110
7. Allen, J.: Maintaining knowledge about temporal intervals. Communications of the ACM **26** (1983) 832–843
8. Golumbic, M., Shamir, R.: Complexity and algorithms for reasoning about time: A graph-theoretic approach. Journal of the ACM **40** (1993) 1108–1133
9. Krokhin, A., Jeavons, P., Jonsson, P.: Reasoning about temporal relations: Tractable subalgebras of Allen's interval algebra. Technical Report RR-01-12, Oxford University Computing Laboratory (2001) Submitted for publication.

10. Nebel, B., Bürckert, H.J.: Reasoning about temporal relations: A maximal tractable subclass of Allen's interval algebra. Journal of the ACM **42** (1995) 43–66
11. Goldberg, P., Golumbic, M., Kaplan, H., Shamir, R.: Four strikes against physical mapping of DNA. Journal of Computational Biology **2** (1995) 139–152
12. Golumbic, M., Kaplan, H., Shamir, R.: On the complexity of DNA physical mapping. Advances in Applied Mathematics **15** (1994) 251–261
13. Karp, R.: Mapping the genome: some combinatorial problems arising in molecular biology. In: Proceedings of the 25th Symposium on the Theory of Computing (STOC'93), New-York, ACM Press (1993) 278–285
14. Hsu, W.L.: A simple test for interval graphs. In Hsu, W.L., Lee, R., eds.: Proceedings of the 18th International Workshop on Graph-Theoretic Concepts in Computer Science. Volume 657 of Lecture Notes in Computer Science., Springer (1992) 11–16
15. Corneil, D., Kim, H., Natarajan, S., Olariu, S., Sprague, A.: Simple linear time recognition of unit interval graphs. Information Processing Letters **55** (1995) 99–104
16. Pe'er, I., Shamir, R.: Realizing interval graphs with size and distance constraints. SIAM Journal on Discrete Mathematics **10** (1997) 662–687
17. Gabor, C., Supowit, K., Hsu, W.: Recognizing circle graphs in polynomial time. Journal of the ACM **36** (1989) 435–473
18. Golumbic, M., Kaplan, H., Shamir, R.: Graph sandwich problems. Journal of Algorithms **19** (1995) 449–473
19. Pe'er, I., Shamir, R.: Satisfiability problems on intervals and unit intervals. Theoretical Computer Science **175** (1997) 349–372
20. Webber, A.: Proof of the interval satisfiability conjecture. Annals of Mathematics and Artificial Intelligence **15** (1995) 231–238
21. Drakengren, T., Jonsson, P.: Eight maximal tractable subclasses of Allen's algebra with metric time. Journal of Artificial Intelligence Research **7** (1997) 25–45
22. Drakengren, T., Jonsson, P.: A complete classification of tractability in Allen's algebra relative to subsets of basic relations. Artificial Intelligence **106** (1998) 205–219
23. Jonsson, P., Bäckström, C.: A unifying approach to temporal constraint reasoning. Artificial Intelligence **102** (1998) 143–155
24. Koubarakis, M.: Tractable disjunctions of linear constraints: basic results and applications to temporal reasoning. Theoretical Computer Science **266** (2001) 311–339
25. Fortune, S., Hopcroft, J., Wyllie, J.: The directed subgraph homeomorphism problem. Theoretical Computer Science **10** (1980) 111–121
26. Creignou, N., Khanna, S., Sudan, M.: Complexity Classification of Boolean Constraint Satisfaction Problems. Volume 7 of SIAM Monographs on Discrete Mathematics and Applications. (2001)
27. Kirousis, L., Kolaitis, P.: The complexity of minimal satisfiability problems. In: Proceedings of the 18th Symposium on Theoretical Aspects of Computer Science (STACS 2001). Volume 2010 of Lecture Notes in Computer Science., Springer (2001) 407–418
28. Krokhin, A., Jeavons, P., Jonsson, P.: The complexity of constraints on intervals and lengths. Technical Report TR01-077, Electronic Colloquim on Computational Complexity (2001) (http://www.eccc.uni-trier.de/eccc).

Nesting Until and Since in Linear Temporal Logic

Denis Thérien[*1] and Thomas Wilke[2]

[1] School of Computer Science, McGill University, Montréal, Canada
[2] Computer Science Department, CAU, Kiel, Germany

Abstract. We provide an effective characterization of the "until-since hierarchy" of linear temporal logic, that is, we show how to compute for a given temporal property the minimal nesting depth in "until" and "since" required to express it. This settles the most prominent classification problem for linear temporal logic. Our characterization of the individual levels of the "until-since hierarchy" is algebraic: for each n, we present a decidable class of finite semigroups and show that a temporal property is expressible with nesting depth at most n if and only if the syntactic semigroup of the formal language associated with the property belongs to the class provided. The core of our algebraic characterization is a new description of substitution in linear temporal logic in terms of block products of finite semigroups.

1 Introduction

Linear temporal logic is the most fundamental formalism used in computer-aided verification for specifying properties of hard- and software such as integrated circuits and communication protocols [Pnu77]. Its salient feature, which explains part of its success in practice, are temporal operators modeled after natural language constructs. They express basic temporal relations, for instance, "next", "always", "since", and "until". Of all the operators, "since" and "until" are the most powerful and important ones, as they are expressively complete [Kam68] and the only binary operators. Therefore, the most natural way to classify a temporal property and to describe its complexity is by determining the nesting depth in "until" and "since" that is required to express it, that is, by determining its level in the "until-since hierarchy".[1] In a first step towards an effective characterization of this hierarchy, its future-only version (the "until hierarchy") was characterized effectively in [TW96]. In a further step, level 0 of the "until-since hierarchy" was shown to be decidable [TW98]. In the present paper, we finally present effective characterizations of all the other levels of the "until-since hierarchy".

---

[*] Supported by FCAR, NSERC and Alexander von Humboldt Foundation
[1] Already fairly simple fairness constraints (for instance, "a request is served k times without another request being served") provide properties that show that the "until-since hierarchy" is strict [EW96].

H. Alt and A. Ferreira (Eds.): STACS 2002, LNCS 2285, pp. 455–464, 2002.
© Springer-Verlag Berlin Heidelberg 2002

Classifying temporal properties according to their syntactical complexity was first carried out by Sistla and Zuck [SZ87,SZ93] and Cohen-Chesnot [CC89], and has since been a subject of many papers. The objective has been to determine, for a given temporal property, which temporal operators are needed to express it, that is, to characterize fragments of linear temporal logic determined by which of the temporal operators are allowed to be used. Meanwhile, all fundamental fragments have been characterized effectively, by fragments of first-order logics [EVW97,TW98], classes of automata [EW96], or classes of semigroups (monoids) [CPP93].

Technically, we characterize each level of the until-since hierarchy by a corresponding pseudovariety of semigroups. That is, for each level of the hierarchy, we provide a decidable class of finite semigroups and show that a temporal property is definable on the respective level if and only if the transformation semigroup of the minimal automaton recognizing the property defined by the formula viewed as a formal language belongs to the class provided. There are two difficult parts to this characterization: (1) to devise the right classes of semigroups, (2) to show that these classes are decidable. For the second part, we rely on deep results in the theory of finite semigroups and categories [Alm96,Ste99] (Subsect. 4.3). For the first part, we proceed as follows. In a first step (Sect. 3), we establish a connection between substitution in temporal logic and block products of semigroups. This is reminiscent of [TW96], there are, however, fundamental obstacles in the two-sided framework that have to be overcome. Instead of sets of strings (the usual objects of formal language theory), we now study what we call pointed languages, which are sets of pointed strings (strings with a distinguished position). We replace the wreath product by its two-sided version, known as the block product, and prove that if Φ and Ψ are classes of formulas and V and W are pseudovarieties of semigroups such that the languages expressible in Φ are the languages recognized by the elements of V and the pointed languages expressible in Ψ are the languages recognized by the elements of W, then the block product of V and W recognizes the languages definable by formulas from Φ with propositions substituted by formulas from Ψ—the block product/substitution principle (Theorem 2). We then prove a normal form theorem for each level of the until-since hierarchy, which describes it as an iterated substitution of simple formula classes (Subsect. 4.1), characterize these simple formula classes, and finally use the block product/substitution principle to get a characterization of the levels of the hierarchy (Subsect. 4.2).

This is an extended abstract; many proofs are missing for lack of space. Full details can be found in our techreport attached to this submission.

2   Notation and Fundamental Definitions

As usual, $[n] = \{1, \ldots, n\}$ for every natural number n.

Strings are sequences of letters over a finite alphabet. The length of a string x is denoted $|x|$. The i-th letter of a string x is denoted x_i, that is, $x = x_1 \ldots x_{|x|}$. A pointed string over some alphabet A is a pair (x, i) where x is a nonempty string over A and $i \in [|x|]$. A pointed language over some alphabet A is a set of pointed strings over A. The set of all pointed strings over A is denoted $A^{\#}$.

2.1 Linear Temporal Logic

A *(linear) temporal logic formula (LTL formula)* over some finite set Σ of *propositional variables* is built from the boolean constants, tt and ff, and the elements of Σ using boolean connectives and temporal operators: $\oplus$ (next), $\ominus$ (previously), $\boxplus$ (always in the future), $\boxminus$ (always in the past), $\diamondsuit$ (eventually in the future), $\diamondsuit$ (eventually in the past), U (until), and S (since), where only U and S are binary and the other operators are unary. For each of the operators except for $\oplus$ and $\ominus$, we also have a nonstrict variant, denoted by $\dot{\boxplus}$, $\dot{\boxminus}$, $\dot{\diamondsuit}$, $\dot{\diamondsuit}$, $\dot{U}$, and $\dot{S}$, respectively.

Given an LTL formula φ over some Σ and a pointed string $(x, i) \in (2^{\Sigma})^{\#}$, we write $(x, i) \models \varphi$ for the fact that φ holds in x at position i. This is standard, see, e. g., [Eme90]. The difference between the nonstrict and strict interpretation is as usual, for instance, $(x, i) \models \dot{\diamondsuit}\varphi$ if there exists j with $i \leq j \leq |x|$ such that $(x, j) \models \varphi$, whereas $(x, i) \models \diamondsuit\varphi$ if there exists j with $i < j \leq |x|$ such that $(x, j) \models \varphi$.

Consider, for instance, the formula $\dot{\diamondsuit}(p \wedge q \wedge q \text{ U } (p \wedge q \wedge q \text{ U } p \wedge q))$. This formula expresses that there exists an interval labelled q where at least at three positions p holds true. (U is assumed to bind stronger than $\wedge$.)

With every formula φ over Σ, we associate a language and a pointed language over 2^{Σ}:

$$L(\varphi) = \{x \in (2^{\Sigma})^{+} \mid (x, 1) \models \varphi\} \; , \tag{1}$$

$$P(\varphi) = \{(x, i) \in (2^{\Sigma})^{\#} \mid (x, i) \models \varphi\} \; . \tag{2}$$

We extend these definitions to classes of formulas in the following way. If Φ is a class of formulas, then

$$L(\Phi) = \{L(\varphi) \mid \varphi \in \Phi\} \; , \qquad P(\Phi) = \{P(\varphi) \mid \in \Phi\} \; . \tag{3}$$

The set of temporal formulas over some set Σ of propositional variables is denoted by LTL_{Σ}. A function $\sigma \colon \mathsf{LTL}_{\Sigma} \to \mathsf{LTL}_{\Gamma}$ is a *substitution* if σ maps each formula φ over Σ to the formula which is obtained from φ by replacing every occurrence of a propositional variable $p \in \Sigma$ by $\sigma(p)$. Note that a substitution is fully determined by the values for the propositional variables. When Ψ is a class of LTL_{Γ} formulas, a substitution $\sigma \colon \mathsf{LTL}_{\Sigma} \to \mathsf{LTL}_{\Gamma}$ is a Ψ substitution if $\sigma(p) \in \Psi$ for every $p \in \Sigma$. When Φ and Ψ are classes of formulas, then $\Phi \circ \Psi$ is the set of all formulas $\sigma(\varphi)$ where $\varphi \in \Phi$ is a formula over some Σ and $\sigma \colon \mathsf{LTL}_{\Sigma} \to \mathsf{LTL}_{\Gamma}$ is a Ψ substitution for some Γ. Clearly, substitution is associative, so it is not necessary to use parentheses. If Φ is a class of formula, then Φ^{0} denotes the set of all propositional variables and $\Phi^{i+1} = \Phi^{i} \circ \Phi$.

2.2   Until-Since Hierarchy

The *until-since nesting depth* of a temporal formula φ, denoted $d_{\mathsf{US}}(\varphi)$, is its nesting depth in until and since. Formally, this is defined by

$$d_{\mathsf{US}}(\mathsf{tt}) = d_{\mathsf{US}}(\mathsf{ff}) = d_{\mathsf{US}}(p) = 0 \ , \tag{4}$$

$$d_{\mathsf{US}}(o_u\varphi) = d_{\mathsf{US}}(\varphi) \ , \tag{5}$$

$$d_{\mathsf{US}}(\varphi \, o_l \, \psi) = \max(d_{\mathsf{US}}(\varphi), d_{\mathsf{US}}(\psi)) \ , \tag{6}$$

$$d_{\mathsf{US}}(\varphi \, o_b \, \psi) = \max(d_{\mathsf{US}}(\varphi), d_{\mathsf{US}}(\psi)) + 1 \ , \tag{7}$$

where p is an arbitrary propositional variable, φ and ψ are arbitrary formulas, o_u stands for any of the unary operators ($\neg, \ominus, \oplus, \dot{\boxplus}, \dot{\boxminus}, \dot{\diamondsuit}, \dot{\diamondsuit}, \boxplus, \boxminus, \diamondsuit, \diamondsuit$), o_l stands for $\vee$ or $\wedge$, and o_b stands for any until or since operator ($\ddot{\mathsf{U}}, \dot{\mathsf{S}}, \mathsf{U}, \mathsf{S}$).

For every i, let USH_i be the set of all formulas φ with $d_{\mathsf{US}}(\varphi) \le i$, that is, USH_i is the set of all formulas of until-since depth at most i. The *until-since hierarchy* is the following chain:

$$L(\mathsf{USH}_0) \subseteq L(\mathsf{USH}_1) \subseteq L(\mathsf{USH}_2) \subseteq \dots \tag{8}$$

The class $L(\mathsf{USH}_i)$ is called the *i-th level of the until-since hierarchy*.

In [EW96], it was shown that the until-since hierarchy is strict, that is, each of the containments is proper, and in [TW98], it was shown that $L(\mathsf{USH}_0)$ is decidable.

It should be noted that even though the above definition of the hierarchy might seem to be susceptible to which operators are included in the syntax of temporal logic, this is not at all the case: Even if only $\oplus, \ominus, \dot{\diamondsuit}, \dot{\diamondsuit}, \dot{\mathsf{S}}$, and $\ddot{\mathsf{U}}$ were allowed, the hierarchy would be exactly the same. Also, if any of the other binary operators suggested in the literature (Kröger's "at next" [Krö87], Lamport's "at least as long as" [Lam83], or STeP's "wait for" [MAB+94]) were added and taken into account in the above definition of until-since depth ("binary depth"), then the hierarchy would not change either.

2.3   Monoids, Semigroups, and Formal Languages

For a complete treatment of the notions presented here, the reader is referred to [Eil76,Alm95].

A semigroup is a set with a binary associative operation; a monoid is a semigroup that contains a two-sided identity element. A monoid M recognizes a language $L \subseteq A^*$ if there is a homomorphism $\theta \colon A^* \to M$ and a subset $F \subseteq M$ such that $L = \theta^{-1}(F)$. Given two monoids M and N, we say that M divides N if M is a homomorphic image of a submonoid of N. The natural classification for finite monoids is in terms of pseudovarieties, i. e., classes of finite monoids that are closed under division and direct product. If V is a pseudovariety, we write $L(V)$ for the class of languages that can be recognized with a monoid in

V. The following examples of pseudovarieties are well-known and will be used later:

$$\mathbf{L_1} \vee \mathbf{R_1} = \{M \mid \forall s, t, u \in M(stsus = stus \wedge s = s^2)\} \ , \tag{9}$$

$$\mathbf{DA} = \{M \mid \forall e, s \in M(MeM = MsM \wedge e = e^2 \rightarrow s = s^2)\} \ . \tag{10}$$

Several operations combining two pseudovarieties to yield a third one have been investigated. We here need the so-called *block product*, a 2-sided variant of the more classical wreath product. We will be using the following characterization of this operation.

Let γ be a finite-index congruence on A^*, and let $N = A^*/\gamma$ be the quotient monoid of A^* with respect to γ. The γ-class of a string x will be denoted by $[x]_\gamma$. Let β be a finite-index congruence on $(N \times A \times N)^*$. The relation $\beta \,\square\, \gamma$ is the equivalence relation on A^* defined by: $x \, \beta \,\square\, \gamma \, y$ iff $x \, \gamma \, y$ and for all $u, v \in A^*$, $h_{u,v}(x) \, \beta \, h_{u,v}(y)$ where $h_{u,v}(a_1 \ldots a_n) = b_1 \ldots b_n$ with $b_i = ([ua_1 \ldots a_{i-1}]_\gamma, a_i, [a_{i+1} \ldots a_n v]_\gamma)$. It is not hard to verify that $\beta \,\square\, \gamma$ is a congruence of finite index on A^*.

Lemma 1. [Thé91] *Let $\mathbf{V}$ and $\mathbf{W}$ be pseudovarieties of monoids. A monoid $M = A^*/\alpha$ belongs to the pseudovariety $\mathbf{V} \,\square\, \mathbf{W}$ iff α is refined by $\beta \,\square\, \gamma$ with $N = A^*/\gamma$ in $\mathbf{V}$ and $(N \times A \times N)^*/\beta$ in $\mathbf{W}$.*

Note that the block product of pseudovarieties is not associative, i.e., $(\mathbf{V} \,\square\, \mathbf{W}) \,\square\, \mathbf{U}$ is not equal in general to $\mathbf{V} \,\square\, (\mathbf{W} \,\square\, \mathbf{U})$. When writing an iterated block product without parentheses, we mean the first possibility, i.e., bracketing is from left to right.

It turns out that it is sometimes more appropriate to deal with semigroups rather than with monoids, and to consider languages as subsets of A^+ rather than A^*. All notions defined above have natural correspondence in this slightly different setting. In particular, it is possible to define the block product of $\mathbf{V}$ and $\mathbf{W}$ when either one is (or both are) a pseudovariety of semigroups rather than monoids. The result is then a pseudovariety of semigroups and a characterization in terms of congruences can be given, in a way very similar to the pure-monoid case. The following pseudovariety of semigroups has special importance in the algebraic theory of automata and it also plays a role in our context:

$$\mathbf{LI} = \{S \mid \forall s, e \in S(e^2 = e \rightarrow ese = e)\} \ . \tag{11}$$

3   The Block Product/Substitution Principle

In this section, we present the block product/substitution principle, which establishes a general, fundamental relationship between block products of pseudovarieties of monoids and substitution on classes of formulas. We start with defining what it means for a pointed language to be recognized by a monoid homomorphism or simply a monoid.

Let $h\colon A^* \to M$ be a monoid homomorphism. We define a companion function $\hat{h}$ mapping pointed strings over A to elements of $M \times A \times M$ by

$$\hat{h}(x,i) = (h(x_1 \ldots x_{i-1}), x_i, h(x_{i+1} \ldots x_{|x|})) \ . \tag{12}$$

We say a pointed language P over A is *recognized* by h if there exists a set $U \subseteq M \times A \times M$ such that $P = \hat{h}^{-1}(U)$. We say that P is recognized by M if it is recognized by a homomorphism $A^* \to M$.

Similar to above, given a class V of finite monoids, we set

$$P(V) = \{P \mid \exists M \in V (P \text{ is recognized by } M)\} \ . \tag{13}$$

In a similar fashion, recognition by semigroups and semigroup homomorphisms can be defined, and $P(V)$ can be defined for pseudovarieties of semigroups, too.

The main result on block products and substitution is the following.

Theorem 2 (block product/substitution principle). *Let Φ and Ψ be classes of LTL formulas and V and W pseudovarieties of monoids or semigroups such that $L(\Phi) = L(V)$, $L(\Psi) \subseteq L(V \square W)$, and $P(\Psi) = P(W)$, then*

$$L(\Phi \circ \Psi) = L(V \square W) \ . \tag{14}$$

Observe that in this theorem, the assumption that we make for Φ and Ψ are different: languages recognized vs. pointed languages recognized. This explains the seeming contradiction that we establish a correspondence between associative substitution and non-associate block product.

The two main tools in the proof of this theorem are the substitution lemma (see below) and Lemma 1 given in Section 2.3.

Let $\sigma\colon \mathsf{LTL}_\Sigma \to \mathsf{LTL}_\Gamma$ be a substitution. For each $x \in (2^\Gamma)^+$ of length n, let $\sigma^{-1}(x)$ be the string $a_1 \ldots a_n$ with

$$a_i = \{p \in \Sigma \mid (x,i) \models \sigma(p)\} \ . \tag{15}$$

Lemma 3 (LTL substitution lemma). *Let φ be an LTL formula over Σ and $\sigma\colon \mathsf{LTL}_\Sigma \to \mathsf{LTL}_\Gamma$ a substitution. Then, for each $i \in [|x|]$,*

$$(x,i) \models \sigma(\varphi) \qquad \textit{iff} \qquad (\sigma^{-1}(x),i) \models \varphi \ . \tag{16}$$

The proof of Theorem 2 roughly goes as follows. Let φ be a formula over Σ and β a congruence on $(2^\Sigma)^*$ such that $L(\varphi)$ is a union of β-classes. Note that β can be chosen such that the corresponding quotient monoid $(2^\Sigma)^*/\beta$ is in V. Let $\sigma\colon \mathsf{LTL}_\Sigma \to \mathsf{LTL}_\Gamma$ be a Ψ substitution and γ a congruence on $(2^\Gamma)^*$ such that the corresponding quotient monoid N recognizes every pointed language $P(\sigma(p))$ where $p \in \Sigma$. This congruence can be chosen in W. Let $\rho\colon (N \times 2^\Gamma \times N)^* \to (2^\Sigma)^*$ be the homomorphism induced by the function mapping (n_1, a, n_2) to $\{p \mid (uav, |u|+1) \models \sigma(p)$ where $h(u) = n_1, h(v) = n_2\}$; this is well-defined since N recognizes every pointed language of the form $P(\sigma(p))$. Next, let $\hat{\beta}$ be the congruence on $(N \times 2^\Gamma \times N)^*$ defined by $X \ \hat{\beta} \ Y$ iff $\rho(X) \ \beta \ \rho(Y)$. The quotient monoid is clearly in V, and one verifies that $L(\sigma(\varphi))$ is a union of $\hat{\beta} \square \gamma$-classes. The first containment then follows from Lemma 1. For the other containment, we use similar arguments.

4   The Until-Since Hierarchy

Recall from Subsection 2.2 that the i-th level of the until-since hierarchy comprises all languages that are definable by a formula of nesting depth at most i in until and since. In this section, we describe how we prove that for each i it is decidable whether or not a given regular language (say given by a regular expression or a finite automaton) belongs to the i-th level. Our decision procedure is uniform in the sense that given L, the minimal level it belongs to and an equivalent formula of minimal nesting depth can be computed.

4.1   Iterated Substitution

To be able to apply the block product/substitution principle, we describe USH_i as an iterated substitution of simple formula classes.

It is convenient to introduce two more unary temporal operators, denoted ϕ and $\$$. Their semantics is given by $(x, i) \models \phi\varphi$ if $(x, 1) \models \varphi$, and $(x, i) \models \$\varphi$ if $(x, |x|) \models \varphi$, that is, they allow to jump from any position to the beginning and the end of a string. (Observe that $\phi\varphi$ is equivalent with $\diamondsuit(\boxminus\mathsf{ff} \wedge \varphi)$; a symmetric statement holds for $\$\varphi$.)

Let $\mathsf{TL}_\square$ be the class of all LTL formulas where $\overset{\bullet}{\diamondsuit}$, $\overset{\bullet}{\diamondsuit}$, $\boxplus$, and $\boxminus$ are the only temporal operators allowed. Further, let $\mathsf{TL}_\bigcirc$ be the class of all LTL formulas where $\oplus$, $\ominus$, ϕ, and $\$$ are the only temporal operators allowed. Finally, a formula belongs to $\mathsf{TL}_{\mathsf{US}}$ if it is a boolean combination of propositional variables, formulas of the form $\varphi \mathbin{\mathsf{U}} \psi$, $\varphi \mathbin{\mathsf{S}} \psi$, $\phi(\varphi \mathbin{\dot{\mathsf{U}}} \psi)$, and $\$(\varphi \mathbin{\dot{\mathsf{S}}} \psi)$, where φ and ψ are propositional formulas.

Proposition 4 (normal form theorem). *For every i,*

$$L(\mathsf{USH}_i) = L(\mathsf{TL}_\square \circ \mathsf{TL}_{\mathsf{US}}^i \circ \mathsf{TL}_\bigcirc) \ . \tag{17}$$

(Recall that $\mathsf{TL}_{\mathsf{US}}^i$ is an iterated substitution, bracketed from left to right, see Subsection 2.1.)

This proposition is based on the fact that $\oplus$ and $\ominus$ can always be pushed inside, which is folklore (see, e. g., [EW96]), and the following switching rules which show that $\overset{\bullet}{\diamondsuit}$ and $\boxminus$ can be "pulled out".

Lemma 5. *For all LTL formulas φ, ψ, χ, the following is true:*

$$(\varphi \vee \overset{\bullet}{\diamondsuit}\psi) \mathbin{\dot{\mathsf{U}}} \chi \equiv \varphi \mathbin{\dot{\mathsf{U}}} \chi \vee (\overset{\bullet}{\diamondsuit}\psi \wedge \overset{\bullet}{\diamondsuit}\chi) \vee (\varphi \mathbin{\dot{\mathsf{U}}} \psi \wedge \overset{\bullet}{\diamondsuit}\chi) \ , \tag{18}$$

$$(\varphi \vee \boxminus\psi) \mathbin{\dot{\mathsf{U}}} \chi \equiv \varphi \mathbin{\dot{\mathsf{U}}} \chi \vee \overset{\bullet}{\diamondsuit}(\chi \wedge \boxminus(\chi \vee \boxminus\psi)) \ , \tag{19}$$

$$\varphi \mathbin{\dot{\mathsf{U}}} (\psi \wedge \overset{\bullet}{\diamondsuit}\chi) \equiv (\overset{\bullet}{\diamondsuit}\chi \wedge \varphi \mathbin{\dot{\mathsf{U}}} \psi) \vee (\boxplus\varphi \wedge \overset{\bullet}{\diamondsuit}(\chi \wedge \overset{\bullet}{\diamondsuit}\psi))$$
$$\vee (\neg\overset{\bullet}{\diamondsuit}\chi \wedge \overset{\bullet}{\diamondsuit}\neg\varphi \wedge \boxplus(\varphi \vee \overset{\bullet}{\diamondsuit}(\psi \wedge \overset{\bullet}{\diamondsuit}\chi))) \ , \tag{20}$$

$$\varphi \mathbin{\dot{\mathsf{U}}} (\psi \wedge \boxminus\chi) \equiv \boxminus\chi \wedge ((\varphi \wedge \chi) \mathbin{\dot{\mathsf{U}}} \psi) \ . \tag{21}$$

We also need future-only counter-parts of these switching rules, but those were already proved in [TW96].

4.2   Algebraic Characterizations

In this section we provide algebraic classifications of the small classes of formulas we have seen above ($\mathsf{TL}_\bigcirc$, $\mathsf{TL}_\square$, and TL_US) and each level of the until-since hierarchy.

For $\mathsf{TL}_\square$, a characterization is known:

Theorem 6. [TW98] $L(\mathsf{TL}_\square) = L(\mathbf{DA})$.

For TL_US and $\mathsf{TL}_\bigcirc$, we prove:

Lemma 7. $P(\mathsf{TL}_\mathsf{US}) = P(\mathbf{L_1} \vee \mathbf{R_1})$ and $P(\mathsf{TL}_\bigcirc) = P(\mathbf{LI})$.

As a consequence of the previous lemma, Theorem 6, Proposition 4, and the block product/substitution principle, we obtain:

Theorem 8 (algebraic characterization of the until-since hierarchy). *For every i,*

$$L(\mathsf{USH}_i) = L(\mathbf{DA} \,\square\, (\mathbf{L_1} \vee \mathbf{R_1})^i \,\square\, \mathbf{LI}) \ . \tag{22}$$

4.3   Decidability

In this section we will describe how to decide if a given language L belongs to the i-th level of the until-since hierarchy, that is, to $L(\mathsf{USH}_i)$. By Theorem 8, this is equivalent to deciding if a semigroup belongs to the semigroup variety $\boldsymbol{V}_i \,\square\, \mathbf{LI}$ where $\boldsymbol{V}_i$ is the monoid variety given by

$$\boldsymbol{V}_0 = \mathbf{DA} \ , \tag{23}$$
$$\boldsymbol{V}_i = \boldsymbol{V}_{i-1} \,\square\, (\mathbf{L_1} \vee \mathbf{R_1}) \ . \tag{24}$$

The algorithms for solving these problems are based on deep semigroup-theoretic results, of which we present only the main points.

It is now well understood that the study of decompositions based on block (or wreath) products involves the framework of categories, monoids being then viewed as special, one-object, categories. Several notions, such as division, carry over to the more general setting in a natural way. See, for instance, [Til87]. The following lemma is crucial.

Lemma 9. *It is decidable if a given category divides a monoid in $\boldsymbol{V}_i$.*

This is proved by induction on i. The inductive base follows directly from a result in [Alm96]; the inductive step relies heavily on Theorem 9.2 of [Ste99].

The next lemma we use is:

Lemma 10. *For any nontrivial monoid variety $\boldsymbol{V}$,*

$$\boldsymbol{V} \,\square\, \mathbf{LI} = \boldsymbol{V} * \mathbf{D} \ , \tag{25}$$

where $*$ denotes the semidirect product on pseudovarieties of semigroups.

So, as a consequence, we know that $V_i \, \square \, \mathbf{LI} = V_i * \mathbf{D}$, and it is enough to show decidability of the latter. By [Str85], as interpreted in the language of categories in [Til87], to decide membership of S in $V * \mathbf{D}$ it suffices to show that a certain category, which is explicitly constructible from S, divides a monoid in V. This is possible for V_i by Lemma 9. We conclude:

Theorem 11 (decidability of the until-since hierarchy). *For every i, the i-th level of the until hierarchy is decidable, that is, given a regular language L it can be determined whether $L \in L(\mathsf{USH}_i)$.*

5 Conclusion

With the block product/substitution principle we have provided a very powerful tool for characterizing fragments of linear temporal logic. Using this and deep results from finite semigroup theory, we have been able to provide effective characterization of all levels of the until-since hierarchy, the most natural hierarchy for classifying temporal properties. We know how to extend Theorem 11 to ω-languages for $i = 0$ and $i = 1$, and it is possible that the technique we apply in these cases can be generalized to higher levels.

Acknowledgment. We thank the anonymous referees for their helpful comments.

References

[Alm95]   Jorge Almeida. *Finite Semigroups and Universal Algebra*, volume 3 of *Series in Algebra*. World Scientific, Singapore, 1995.

[Alm96]   Jorge Almeida. A syntactical proof of locality of **DA**. *Internat. J. Algebra and Comput.*, 6(2):165–177, 1996.

[CC89]   Joëlle Cohen-Chesnot. *Etude algébrique de la logique temporelle*. Doctoral thesis, Université Paris 6, Paris, France, April 1989.

[CPP93]   Joëlle Cohen, Dominique Perrin, and Jean-Eric Pin. On the expressive power of temporal logic. *J. Comput. System Sci.*, 46(3):271–294, June 1993.

[Eil76]   Samuel Eilenberg. *Automata, Languages, and Machines*, volume 59-B of *Pure and Applied Mathematics*. Academic Press, New York, 1976.

[Eme90]   Allen E. Emerson. Temporal and modal logic. In Jan van Leeuwen, editor, *Handbook of Theoretical Computer Science*, volume B: Formal Methods and Semantics, pages 995–1072. Elsevier Science Publishers B. V., Amsterdam, 1990.

[EVW97]   Kousha Etessami, Moshe Y. Vardi, and Thomas Wilke. First-order logic with two variables and unary temporal logic. In *Proceedings 12th Annual IEEE Symposium on Logic in Computer Science*, pages 228–235, Warsaw, Poland, 1997.

[EW96]   Kousha Etessami and Thomas Wilke. An until hierarchy for temporal logic. In *Proceedings 11th Annual IEEE Symposium on Logic in Computer Science*, pages 108–117, New Brunswick, N. J., 1996.

[Kam68]   Johan Anthony Willem Kamp. *Tense Logic and the Theory of Linear Order*. PhD thesis, University of California, Los Angeles, Calif., 1968.

[Krö87]   Fred Kröger. *Temporal Logic of Programs*. Springer, 1987.

[Lam83]   Leslie Lamport. Specifying concurrent program modules. *ACM Trans. Programming Lang. Sys.*, 5(2):190–222, 1983.

[MAB+94]   Zohar Manna, Anuchit Anuchitanukul, Nikolaj Bjørner, Anca Browne, Edward Chang, Michael Colón, Luca de Alfaro, Harish Devarajan, Henny Sipma, and Tomas Uribe. STeP: the Stanford Temporal Prover. Technical Report STAN-CS-TR-94-1518, Dept. of Computer Science, Stanford University, Stanford, Calif., 1994.

[Pnu77]   Amir Pnueli. The temporal logic of programs. In *18th Annual Symposium on Foundations of Computer Science*, pages 46–57, Rhode Island, Providence, 1977.

[Ste99]   Ben Steinberg. Semidirect products of categories and applications. *Journal of Pure and Applied Algebra*, 142:153–182, 1999.

[Str85]   Howard Straubing. Finite semigroup varieties of the form $V * D$. *J. Pure Appl. Algebra*, 36:53–94, 1985.

[SZ87]   A. Prasad Sistla and Lenore D. Zuck. On the eventuality operator in temporal logic. In *Proceedings, Symposium on Logic in Computer Science*, pages 153–166, Ithaca, New York, June 1987.

[SZ93]   A. Prasad Sistla and Lenore D. Zuck. Reasoning in a restricted temporal logic. *Inform. and Computation*, 102(2):167–195, February 1993.

[Thé91]   Denis Thérien. Two-sided wreath product of categories. *Journal of Pure and Applied Algebra*, 74:307–315, 1991.

[TW96]   Denis Thérien and Thomas Wilke. Temporal logic and semidirect products: An effective characterization of the until hierarchy. In *Proceedings of the 37th Annual Symposium on Foundations of Computer Science*, pages 256–263, Burlington, Vermont, 1996.

[TW98]   Denis Thérien and Thomas Wilke. Over words, two variables are as powerful as one quantifier alternation: $FO^2 = \Sigma_2 \cap \Pi_2$. In *Proceedings of the Thirtieth Annual ACM Symposium on Theory of Computing*, pages 41–47, Dallas, Texas, May 1998.

[Til87]   Brat Tilson. Categories as algebra. *J. Pure Appl. Algebra*, 48:83–198, 1987.

Comparing Verboseness for Finite Automata and Turing Machines

Till Tantau

Technische Universität Berlin
Fakultät für Elektrotechnik und Informatik
10623 Berlin, Germany
tantau@cs.tu-berlin.de

Abstract. A language is called (m, n)-*verbose* if there exists a Turing machine that enumerates for any n words at most m possibilities for their characteristic string. We compare this notion to (m, n)-*fa-verboseness*, where instead of a Turing machine a finite automaton is used. Using a new *structural* diagonalisation method, where finite automata trick Turing machines, we prove that all (m, n)-verbose languages are (h, k)-verbose, iff all (m, n)-fa-verbose languages are (h, k)-fa-verbose. In other words, Turing machines and finite automata behave in exactly the same way with respect to inclusion of verboseness classes. This identical behaviour implies that the Nonspeedup Theorem also holds for finite automata. As an application of the theoretical framework, we prove a lower bound on the number of bits that need to be communicated to finite automata protocol checkers for nonregular protocols.

1 Introduction

Turing machines and finite automata share some properties that they do not appear to share with computational models 'in between'. While it is well known that both for Turing machines and for finite automata *nondeterminism* is exactly as powerful as *determinism*, for polynomial-time computations this is the famous 'P = NP?' question. In this paper we study a property that Turing machines and finite automata share, but which they provably do not share with classical resource-bounded computational models in between. This property is the *inclusion structure of verboseness classes*.

Verboseness classes for Turing machines were originally defined in an effort to better understand the structure of undecidable problems. Even if a language L is nonrecursive, it might still be possible to compute some partial information about it. A language L is called (m, n)-*verbose* [7] if there exists a Turing machine that does the following: It gets n input words and then does a possibly infinite computation during which it prints some bit strings onto an output tape. One of the bit strings must be the characteristic string of the words with respect to L and at most m different bit strings may be printed. The class of all (m, n)-verbose languages is denoted $\mathrm{V}(m, n)$.

H. Alt and A. Ferreira (Eds.): STACS 2002, LNCS 2285, pp. 465–476, 2002.
© Springer-Verlag Berlin Heidelberg 2002

All languages are in $V(2^n, n)$, and $V(1, n)$ contains exactly the recursive languages. The structure between these two extremes has been subject to thorough investigation. Beigel's Nonspeedup Theorem [3,5] states $V(n, n) = V(n - 1, n) = \cdots = V(1, n)$. All recursively enumerable and all semirecursive [10] languages are in $V(n + 1, n)$. It is known [7] that $V(3, 2) = V(n + 1, n)$ for all $n \geq 2$. More generally, in [7] Beigel, Kummer, and Stephan describe a decision procedure for checking whether $V(m, n) \subseteq V(h, k)$ holds for given numbers m, n, h, and k.

Verboseness has also been studied extensively for the situation where the enumerating Turing machine is restricted to use only a polynomial amount of time. The inclusion structure of polynomial-time verboseness classes, denoted $V_P(m, n)$ in the following, is quite different from the structure in the recursive setting. For example $V_P(m, n) \subsetneq V_P(m+1, n)$ for all $m < 2^n$. Languages that are in $V_P(n, n)$ for some n are commonly called *cheatable* [4], languages in the class $V_P(2^n - 1, n)$ are called *n-approximable* [6] or *n-membership comparable* [15].

The notion of verboseness can also readily be defined for finite automata. The classes $V_{fa}(m, n)$ are defined via multi-tape automata whose output is specified by the type of the last state reached on a given input. Such an automaton witnesses $L \in V_{fa}(m, n)$ as follows: When run on n input words, placed on n synchronously read input tapes, the output attached to last state reached by the automaton must be a set of size at most m containing the characteristic string of the words. The classes $V_{fa}(m, n)$ will be called *fa-verboseness classes*. Similar transfers of recursion or complexity theoretic concepts to finite automata have already been studied in 1976 by Kinber [11] and much more recently by Austinat, Diekert, Hertrampf, and Petersen [1,2].

As in the recursive setting, all languages are in $V_{fa}(2^n, n)$ and $V_{fa}(1, n)$ contains exactly the regular languages. Austinat et al. [1,2] present different examples of fa-verbose languages that lie between these extremes: For every infinite bit string b both the set of all words that are lexicographically smaller than b and the set of all prefixes of b are in $V_{fa}(3, 2)$. This class contains context-sensitive languages that are not context-free, and context-free languages that are not regular. However, for all $m < 2^n$ the classes $V_{fa}(m, n)$ do not contain any infinite context-free languages like $\{a^i b^i \mid i \geq 0\}$ that lack infinite regular subsets.

What does the inclusion structure of the classes $V_{fa}(m, n)$ look like? As the inclusion structure of polynomial-time verboseness classes is quite different from the structure of their recursive counterparts, it might seem surprising that *the inclusion structure of the classes $V_{fa}(m, n)$ is exactly the same as for $V(m, n)$.* Using the new notion of *structural* diagonalisation, where finite automata trick Turing machines, we can even show that $V_{fa}(m, n) \setminus V(h, k)$ is nonempty unless $V(m, n) \subseteq V(h, k)$. Thus $V(m, n) \subseteq V(h, k)$ iff $V_{fa}(m, n) \subseteq V_{fa}(h, k)$ iff $V_{fa}(m, n) \subseteq V(h, k)$.

An immediate corollary is that the Nonspeedup Theorem also holds for finite automata. This is interesting in itself. While the Nonspeedup Theorem is an important theoretical result, it has limited practical importance as it concerns the properties of languages that are not even recursive. Opposed to this, we shall see that its finite automata counterpart can be used to prove a lower bound on the number of bits that need to be communicated to finite automata protocol checkers for nonregular protocols.

This paper is organised as follows. In *Section 2* we review functions that are *enumerable* by Turing machines and transfer this notion to finite automata. We prove a key lemma, which holds both for Turing machines and for finite automata. In *Section 3* we show that the Generalised Nonspeedup Theorem [7] is an easy consequence of the key lemma and thus also holds for finite automata. We then review the combinatorial tools used in [7] to describe the inclusion structure of verboseness classes and apply them to fa-verboseness. In *Section 4* we complete the characterisation of the inclusion structure of fa-verboseness classes by use of a structural diagonalisation argument. Finally, *Section 5* presents a consequence of the Finite Automata Nonspeedup Theorem for the difficulty of protocol checking using finite automata.

2   Recursive and Finite Automata Enumerability

Before focusing on verboseness in the following sections, in this section we study the more general concept of *enumerability* of functions. This notion was first introduced by Cai and Hemaspaandra [8] in the context of polynomial-time computations and only later transfered to recursive computations.

Definition 1 ([14,9]). *A function f, taking n-tuples of words as input, is in the class* $\mathrm{EN}(m)$ *if there exists a deterministic Turing machine with n input tapes that does the following: Upon input of words $w_1, \ldots, w_n$, placed on the input tapes, it starts a possibly infinite computation during which it prints words onto an output tape. No more than m words may be printed, and one of them must be $f(w_1, \ldots, w_n)$.*

For Turing machines it is not important whether the input words are placed on different tapes or just alongside each other on a single input tape, separated by marker symbols. For finite automata this *does* make a crucial difference. We shall only consider the more interesting situation where multiple tapes are used, see below.

The connection between enumerability and verboseness is discussed in detail in the next section, but note already that a language L is (m, n)-verbose iff $\chi_L^n \in \mathrm{EN}(m)$. Here χ_L^n is the n-fold characteristic function of L, which takes a tuple $(w_1, \ldots, w_n)$ as input and yields a bit string of length n whose i-th position is 1 iff $w_i \in L$.

We now explain how finite automata compute functions. As done in [1,2] we use deterministic automata that instead of just accepting or rejecting a word may produce many different outputs, depending on the type of the last state reached. Formally, instead of a set of accepting states the automata are equipped with an output function $\gamma \colon Q \to \Gamma$ that assigns an output $\gamma(q) \in \Gamma$ to each state $q \in Q$. The *output* $\mathrm{out}_A(w)$ of an automaton A on input w is the value $\gamma(q)$ attached to the *last state q reached by A upon input w.* An automaton A *computes a* *function* f if $f(w) = \text{out}_A(w)$ for all w. Note that only functions having finite range can be fa-computed in this sense.

As mentioned earlier, to fa-compute a function taking n-tuples of words as input, we use n input tapes. Each word is put onto one tape, shorter words padded with blanks at the end such that all words have the same length. Then the automaton scans the words *synchronously*, meaning that in each step all heads advance exactly one symbol. Equivalently, one may also think of the input words being fed to a single-tape automaton in an interleaved fashion: first the first symbols of all input words, then the second symbols, then the third symbols, and so forth.

Having fixed how finite automata *compute* functions, we can now define how they *enumerate* them.

Definition 2. *A function f, taking n-tuples of words as input, is in the class* $\text{EN}_{\text{fa}}(m)$ *if there exists a deterministic n-tape automaton A, having sets as output, such that for all words $w_1, \ldots, w_n$ the set* $\text{out}_A(w_1, \ldots, w_n)$ *has size at most m and contains $f(w_1, \ldots, w_n)$.*

Our first result is the key lemma below. It makes the same claim twice, only the first time for enumerability and the second time for fa-enumerability. In a sense this 'double claim' is the deeper reason why the inclusion structures of verboseness and fa-verboseness classes coincide. The *product* $f \times g$ of two functions f and g used in the lemma is defined in the usual way by $(f \times g)(u, v) := \big(f(u), g(v)\big)$. The first part of the proof is similar to the proof of the Generalised Nonspeedup Theorem given in [7], but see also [16].

Lemma 3 (Key Lemma).
If $f \times g \in \text{EN}(n + m)$, then either $f \in \text{EN}(n)$ or $g \in \text{EN}(m)$. Likewise, if $f \times g \in \text{EN}_{\text{fa}}(n + m)$, then either $f \in \text{EN}_{\text{fa}}(n)$ or $g \in \text{EN}_{\text{fa}}(m)$.

Proof. Assume $f \times g \in \text{EN}(n+m)$ via a machine M. Let us write $M(u, v)$ for the set of pairs enumerated by M upon input of the pair (u, v). Suppose there exists a word u such that for all words v the set $P_{u,v} := \big\{ y \mid \big(f(u), y\big) \in M(u, v) \big\}$ has size at most m. Then $g \in \text{EN}(m)$ and we are done. So suppose no such word u exists. Then for all words u there exists a word v such that the set $P_{u,v}$ has size *at least $m + 1$*. This implies that in $M(u, v)$ at least $m + 1$ pairs have the same first component, and hence $\{x \mid (x, y) \in M(u, v)\}$ has size *at most $n + m - m = n$*. But then $f \in \text{EN}(n)$, as upon input of a word u, using dovetailing, we can *search* for a word v with the property, that in $M(u, v)$ at least $m + 1$ pairs have the same first component. Having found this word, we stop the search and enumerate $\{x \mid (x, y) \in M(u, v)\}$.

Now assume $f \times g \in \text{EN}_{\text{fa}}(n + m)$ via a 2-tape automaton A with output function γ. Once more, assume there exists a word u such that for all words v the set $P_{u,v} := \big\{ y \mid \big(f(u), y\big) \in \text{out}_A(u, v) \big\}$ has size at most m. Then $g \in \text{EN}_{\text{fa}}(m)$ via an automaton B that *simulates* what A would do upon input (u, v). Such a simulation is possible as u is a single fixed word. When the simulation finishes, the automaton B will 'know' the state q in which A would have ended up upon input of (u, v). The automaton B can then output $P_{u,v}$ instead of $\text{out}_A(u, v)$, thus proving $g \in \text{EN}_{\text{fa}}(m)$.

Next assume that for all words u there exists a word v such that the set $P_{u,v}$ has size at least $m + 1$. We then show $f \in \text{EN}_{\text{fa}}(n)$ using a variant of the search trick used in the recursive setting. The aim is to 'search' for a word v such that in $\text{out}_A(u, v)$ at least $m + 1$ pairs have the same first component – we know that such a word exists, we just have to find it. The idea is to first construct a *nondeterministic automaton* C *with ϵ-moves*. This automaton has just one input tape, on which it finds an input word u. As its first step it branches nondeterministically to all states the automaton A reaches when it reads the first symbol of u on its first tape and some arbitrary symbol on the second tape (including the blank symbol, which corresponds to guessing the end of the word on the second tape). In the second step C branches to the states A reaches upon then reading the second symbol of u plus some arbitrary symbol on the second tape and so on. When the end of u is reached, C may go on simulating A using ϵ-moves: It nondeterministically branches to the states A would reach reading a blank on the first tape and some arbitrary symbol on the second tape. Note that *on some nondeterministic path the automaton C will reach the state reached by A upon input (u, v)*.

We turn C into a deterministic automaton D using the standard power set construction on its states. Upon input of a word u the automaton D will end its computation in a 'power state' p that is exactly the set of all states reached by C upon input u. We define the output attached to p in D as the intersection of all sets $\{x \mid (x, y) \in \gamma(q)\}$ with $q \in p$, thus ensuring that $f(u)$ is always an element of $\text{out}_D(u)$. Due to the existence of the word v, there must exist a state $q \in p$ such that in the set $\gamma(q)$ at least $m + 1$ pairs have the same first component. Thus the intersection output by D has size at most n. $\qquad\square$

The contraposition of the lemma states that if $f \notin \text{EN}(n)$ and $g \notin \text{EN}(m)$, then $f \times g \notin \text{EN}(n + m)$; and likewise for finite automata. This can be seen as a lower bound on the enumerability of the product of two functions: If f and g are not n- and m-enumerable respectively, then $f \times g$ is not $(n + m)$-enumerable. Compare this to the trivial upper bound: If $f \in \text{EN}(n)$ and $g \in \text{EN}(m)$, then $f \times g \in \text{EN}(n \cdot m)$; and likewise for finite automata.

Repeatedly applying the lemma yields that if $f_1 \times \cdots \times f_\ell \in \text{EN}(n_1 + \cdots + n_\ell)$, then $f_i \in \text{EN}(n_i)$ for some i; and likewise for finite automata. In particular, if $f_1 \times \cdots \times f_n$ is in $\text{EN}(n)$, respectively $\text{EN}_{\text{fa}}(n)$, then at least one f_i is recursive, respectively fa-computable.

3   Application of the Key Lemma to Verboseness

In this section we apply the key lemma to verboseness classes. We first introduce verboseness of *functions* as a useful and elegant generalisation of the verboseness of *languages*. In the following, f^n always denotes the n-fold product $f \times \cdots \times f$.

Definition 4. *A function f is (m, n)-verbose if $f^n \in \text{EN}(m)$. A function f is (m, n)-fa-verbose if $f^n \in \text{EN}_{\text{fa}}(m)$. A language is (m, n)-verbose, respectively*

(m, n)-fa-verbose, *if its characteristic function is. The classes of (m, n)-verbose and (m, n)-fa-verbose languages are denoted* $\mathrm{V}(m, n)$ *and* $\mathrm{V_{fa}}(m, n)$.

Theorem 5. *If f is $(m + h, n + k)$-verbose, then f is either (m, n)-verbose or (h, k)-verbose; and likewise for fa-verboseness.*

Proof. By the key lemma, if $f^{n+k} = f^n \times f^k \in \mathrm{EN}(m + h)$ we have either $f^n \in \mathrm{EN}(m)$ or $f^k \in \mathrm{EN}(h)$; and likewise for finite automata. $\square$

Corollary 6 (Generalised Nonspeedup Theorem). *We have*

$$\mathrm{V}(m + h, n + k) \subseteq \mathrm{V}(m, n) \cup \mathrm{V}(h, k),$$
$$\mathrm{V_{fa}}(m + h, n + k) \subseteq \mathrm{V_{fa}}(m, n) \cup \mathrm{V_{fa}}(h, k).$$

Corollary 6 can be used to prove inclusions of verboseness classes. For example we immediately get $\mathrm{V}(m + 1, n + 1) \subseteq \mathrm{V}(m, n)$. A systematic study of the derivable inclusions leads to the definition of *goodness* [7]. In the following an *n-pool* is a subset of $\{0, 1\}^n$. For a bit string $b \in \{0, 1\}^n$ with individual bits $b_1 \cdots b_n = b$ we write $b[i_1, \ldots, i_\ell] := b_{i_1} \cdots b_{i_\ell} \in \{0, 1\}^\ell$. For an n-pool P we define $P[i_1, \ldots, i_\ell] := \{ b[i_1, \ldots, i_\ell] \mid b \in P \}$.

Definition 7 ([7]). *For a k-pool P let $g_P(n)$ be the maximum of $\left| P[i_1, \ldots, i_n] \right|$ taken over all indices $i_1, \ldots, i_n \in \{1, \ldots, k\}$. A pool P is (m, n)-good, if for every partition $n_1 + \cdots + n_\ell = n$ with $1 \leq n_i \leq n$ we have*

$$g_P(n_1) + \cdots + g_P(n_\ell) \leq m + \ell - 1.$$

Theorem 8. *Let every (m, n)-good k-pool have size at most h. Then*

$$\mathrm{V}(m, n) \subseteq \mathrm{V}(h, k),$$
$$\mathrm{V_{fa}}(m, n) \subseteq \mathrm{V_{fa}}(h, k).$$

Proof. As the claim $\mathrm{V}(m, n) \subseteq \mathrm{V}(h, k)$ is proven as part of Theorem 4.3 in [7], we show only the second claim. We revisit the proof of [7], whose key ingredient is the Generalised Nonspeedup Theorem, which is also correct for finite automata as we have seen in Corollary 6.

Let $L \in \mathrm{V_{fa}}(m, n)$. To prove $L \in \mathrm{V_{fa}}(h, k)$, we construct an automaton B that computes for every k words $w_1, \ldots, w_k$ an (m, n)-good k-pool P containing the bit string $\chi_L^k(w_1, \ldots, w_k)$. By assumption, P will have size at most h. For each $i \in \{1, \ldots, n\}$ let m_i be the smallest number such that $L \in \mathrm{V_{fa}}(m_i, i)$ via some finite automaton A_i. Applying Corollary 6 to $L \in \mathrm{V_{fa}}(m_{i+j}, i + j)$ we get $L \in \mathrm{V_{fa}}(m_i - 1, i) \cup \mathrm{V_{fa}}(m_{i+j} - m_i + 1, j)$. As L is not an element of $\mathrm{V_{fa}}(m_i - 1, i)$ by the choice of m_i, the language L must be in $\mathrm{V_{fa}}(m_{i+j} - m_i + 1, j)$. Because of the minimality of m_j this yields $m_j \leq m_{i+j} - m_i + 1$ and thus $m_i + m_j \leq m_{i+j} + 1$.

The automaton B runs the following algorithm: Upon input of the k words it simulates simultaneously for all $j \in \{1, \ldots, n\}$ and for all indices $i_1, \ldots, i_j \in \{1, \ldots, k\}$ the automata A_j on the input words $w_{i_1}, \ldots, w_{i_j}$. When B reaches the end of the input words it will 'know' what all the $\text{out}_{A_j}(w_{i_1}, \ldots, w_{i_j})$ would be. It then outputs the largest set of bit strings that is consistent with the output of all the automata A_j on the different selections of words. More precisely, it outputs the largest k-pool P with the property that $P[i_1, \ldots, i_j] \subseteq \text{out}_{A_j}(w_{i_1}, \ldots, w_{i_j})$ for all $i_1, \ldots, i_j \in \{1, \ldots, k\}$ and all $j \in \{1, \ldots, n\}$. Note that P indeed contains $\chi_L^k(w_1, \ldots, w_k)$.

We claim that the pool P output by B is (m, n)-good. Let $n_1 + \cdots + n_\ell = n$ be any partition with $1 \le n_i \le n$. From the construction of P and the definition of g_P we directly get $g_P(j) \le m_j$. Hence $g_P(n_1) + \cdots + g_P(n_\ell) \le m_{n_1} + \cdots + m_{n_\ell} \le m_n + \ell - 1 \le m + \ell - 1$. Thus P is, indeed, (m, n)-good. $\qquad\qquad\square$

4   Structural Diagonalisation

Theorem 8 from the previous section gives a sufficient condition for the inclusions $\text{V}(m, n) \subseteq \text{V}(h, k)$ and $\text{V}_{\text{fa}}(m, n) \subseteq \text{V}_{\text{fa}}(h, k)$: It suffices that all (m, n)-good k-pools have size at most h. For the recursive case, it is known [7] that this is also a *necessary* condition. This reduces the inclusion and hence also the equality problem for verboseness classes to finite combinatorics.

We now prove that the inclusion structure of fa-verboseness classes is the same as of verboseness classes, by showing that the condition is *also necessary* if we use finite automata instead of Turing machines. That is, we show that if there exists an (m, n)-good k-pool P of size at least $h + 1$, then there exists a language in $\text{V}_{\text{fa}}(m, n)$ that is not an element of $\text{V}_{\text{fa}}(h, k)$. Actually our structural diagonalisation will show that the language is not even an element of the much larger class $\text{V}(h, k)$, thus yielding the strong class separation result in Theorem 12. In the following, P will always denote an (m, n)-good k-pool of size $h + 1$.

For finite automata the diagonalisation technique used in [7] does not work, as it involves the usage of universal Turing machines together with finite injury arguments. Instead of these, our *structural diagonalisation* uses a variant of the *k-branches* introduced by Kummer and Stephan [13] for the study of frequency computations. Using only a finite automaton, we can diagonalise along such branches against all Turing machines. We call the diagonalisation *structural* as it uses purely structural properties of the involved languages, and does not depend on computational power at all.

The diagonalisation roughly works as follows: In the tree P^*, see the next paragraph, we construct an infinite branch B and associate a language L_B with it, see Definition 9. Under appropriate assumptions, see Theorem 10, we shall be able to construct a branch B such that $L_B \notin \text{V}(h, k)$. In the proof of $L_B \notin \text{V}(h, k)$ we show that all Turing machines that could possibly witness $L_B \in \text{V}(h, k)$ must *err on some k-tuple of words*. In the construction we associate a k-tuple of words with each node of the tree P^*, and the i-th witness machine will err (at least) on the k words associated with the i-th node on the branch. However, the clever definition of L_B will still allow us to prove $L_B \in \text{V}_{\text{fa}}(m, n)$ in Theorem 11.

The set P^* of words over the 'alphabet' P together with the prefix relation forms an infinite tree. The root of this tree is the empty word, and the successors of a node are obtained by appending different bitstrings from P. An *infinite branch* $(u_0, u_1, u_2, \dots)$ of P^* is a sequence of nodes $u_i \in P^*$ such that the first node u_0 is the root of P^*, and each u_{i+1} is a successor of u_i. As stated earlier, with each node $u \in P^*$ we wish to associate k different words. We obtain them by adding k different *tags* to u at the front. We use, say, $\text{tag}_i := 0^i 1^{k-i}$ as tags.

Definition 9. *Let P be a k-pool and let $B = (u_0, u_1, u_2, \dots)$ with $u_{i+1} = u_i b_i$ and $b_i \in P$ be an infinite branch of P^*. The language $L_B \subseteq \{0,1\}^*$ contains only words associated with the nodes on this branch, namely those for which*

$$\chi^k_{L_B}(\text{tag}_1 u_i, \dots, \text{tag}_k u_i) = b_i.$$

Note that just from knowing that a branch B goes through a node u, *we can derive the characteristic values $\chi_{L_B}(w)$ of all words w associated with nodes shorter than u*. A more refined form of this observation will be used in the proof of Theorem 11 to show that all L_B are (m, n)-fa-verbose for appropriate m and n.

Theorem 10. *Let P be a k-pool of size $h + 1$. Then there exists a branch B of P^* with $L_B \notin \mathrm{V}(h, k)$.*

Proof. Let $M_0, M_1, M_2, \dots$ be an enumeration of all Turing machines that could possibly witness $L_B \in \mathrm{V}(h, k)$. We construct the branch in stages. Let the node u_i already be chosen. We must decide how to extend the branch to $u_{i+1} = u_i b_i$. Let Q be the set of bit strings enumerated by M_i on input $(\text{tag}_1 u_i, \dots, \text{tag}_k u_i)$. If $|Q| > h$, the machine M_i does not witness $L \in \mathrm{V}(h, k)$ for any language L and we can define $b_i \in P$ arbitrarily. If however $|Q| \leq h$, let $b_i \in P \setminus Q$ be any bit string. This concludes the i-th stage. By construction, each machine M_i is now tricked for the words $(\text{tag}_1 u_i, \dots, \text{tag}_k u_i)$ and hence the language L_B induced by the branch $B = (u_0, u_1, u_2, \dots)$ is not an element of $\mathrm{V}(h, k)$. $\square$

Using Theorem 10, we can easily generate a candidate for a set in $\mathrm{V}_{\text{fa}}(m, n) \setminus \mathrm{V}(h, k)$. The theorem supplies us with a language not in $\mathrm{V}(h, k)$. The hard part is to prove that it lies in $\mathrm{V}_{\text{fa}}(m, n)$.

Theorem 11. *Let P be an (m, n)-good k-pool with $0^k \in P$. Then all languages induced by infinite branches of P^* are in $\mathrm{V}_{\text{fa}}(m, n)$.*

Proof. Let L be induced by a branch. We construct an automaton A that on input of any n words $w_1, \dots, w_n$ outputs a pool Q of size at most m containing $\chi^n_L(w_1, \dots, w_n)$.

As a side activity while scanning the input words, A will check which input words are of the form $\text{tag}_t u$ for some t and some node $u \in P^*$. We shall focus on these words in the following, as words that do not pass this syntax check are not elements of L.

As its first action the automaton scans the tags of the words and remembers them in its state. Let $\{u_1, \dots, u_\ell\}$ be the input words with the tags removed. We say that w_j is *associated with the node* u_i, if $w_j = \text{tag}_t u_i$ for some t. Let n_i be the number of words associated with u_i. The next aim of the automaton is to find out how the nodes u_i are related.

The automaton A always reads k bits of all words as a block. For each input word the read bit string is part of the sequence of bit strings that make up the node u_i to which the word is associated. In its state the automaton keeps track of a partition of the node indices $\{1, \ldots, \ell\}$, which serves as a reminder of which u_i have already been identified to lie on different branches of P^*. Initially, all indices are grouped into one large equivalence class. Each time a block has been read, the automaton checks whether there are two indices in the same equivalence class for which the read bit strings now differ. If so, the automaton remembers these indices together with the partition at that moment plus the bit strings themselves. The bitstrings may be thought of as the different 'directions' the branches through the predecessor node headed. Furthermore, the automaton splits the equivalence classes such that only those indices remain equivalent for which it read the same 'direction'.

When the automaton reaches the end of the input words, it will have stored in its state the *forest structure* of the nodes $\{u_1, \ldots, u_l\}$. It will know which nodes are ancestors of which nodes, and which nodes are roots of this forest. If a node u_j is an ancestor of a node u_i, the automaton will also know in which direction a branch through u_i heads when going through u_j.

Consider the (uncountably many) infinite branches B of P^*. Each induces a language L_B in the sense of Definition 9, which in turn induces a characteristic string $\chi_{L_B}^n(w_1, \ldots, w_n)$. This characteristic string depends only on the relative positions of the u_i and on the tags of the input words. Hence, from its knowledge of these relative positions the automaton can produce a pool Q containing *all* induced $\chi_{L_B}^n(w_1, \ldots, w_n)$ and thus also including $\chi_L^n(w_1, \ldots, w_n)$.

It remains to show $|Q| \leq m$. For branches B going through *none* of the nodes u_i, by construction of L_B the characteristic value of $\chi_{L_B}^n(w_1, \ldots, w_n)$ is 0^n. This will be the first bit string found in Q. Consider a root node u_i, that is a node that is not a descendant of another node u_j, and consider all branches B going through u_i, *but going through no other node u_j*. These branches induce *at most* $g_P(n_i)$ *different bit strings in* Q. However, as the branches going through u_i heading off in the 'direction' 0^k all induce the bit string 0^n once more, the branches solely going through u_i induce at most $g_P(n_i) - 1$ *new* bit strings in Q apart from 0^n.

Now consider a direct descendant u_j of a root node u_i and the branches going exactly through u_j and u_i. The branches will induce $g_P(n_j)$ many bit strings in Q, but once more one of them will already be found in Q, namely the one induced by branches that go through u_j and u_i, but which head off in the direction 0^k in u_j. Using a structural induction argument, one can now show that *for each node u_i the set of branches going exactly through u_i and its ancestors induces at most $g_P(n_i) - 1$ new bit strings in Q*. In total we get

$$|Q| \leq 1 + \big(g_P(n_1) - 1\big) + \big(g_P(n_2) - 1\big) + \cdots + \big(g_P(n_\ell) - 1\big)$$
$$\leq 1 + m + \ell - 1 - \ell = m,$$

where we used the (m, n)-goodness of P in the last line.   □

Theorem 12. *The following statements are equivalent:*

1. *All (m, n)-good k-pools have size at most h.*
2. $V(m, n) \subseteq V(h, k)$.
3. $V_{\text{fa}}(m, n) \subseteq V_{\text{fa}}(h, k)$.
4. $V_{\text{fa}}(m, n) \subseteq V(h, k)$.

Proof. Statement 1) implies both 2) and 3) by Theorem 8. As we trivially have $V_{\text{fa}}(m, n) \subseteq V(m, n)$ for all m and n, both 2) and 3) imply 4). To show that 4) implies 1) we argue by contraposition. If there exists an (m, n)-good k-pool of size $h + 1$, there also exists one containing 0^k and then by Theorem 10 there exists a language outside $V(h, k)$ that is in $V_{\text{fa}}(m, n)$ by Theorem 11. $\square$

5   Application to Protocol Testing

Beigel's Nonspeedup Theorem [3] states $V(n, n) = V(1, 1)$, that is all (n, n)-verbose languages are recursive. As this is a statement about inclusion of verboseness classes, we immediately know that the finite automata version $V_{\text{fa}}(n, n) = V_{\text{fa}}(1, 1)$ also holds, that is *all (n, n)-fa-verbose languages are regular.* We now show how this Finite Automata Nonspeedup Theorem applies to protocol testing.

A simple problem from protocol testing is the following: We are asked to construct a testing device that monitors n signal lines. The device is synchronously fed as input the n symbols currently transported on the different lines, and it should check whether all symbol streams are *valid* with respect to some protocol. When the streams end, the device should tell us on which lines the protocol was not adhered to.

A simple protocol might be 'the stream may not contain the same symbol four times in succession'. If the symbols on the streams represent voltages, this protocol might be used to verify that the signal lines are free of direct currents. A much more complicated protocol is 'the stream consists of valid Internet protocol packets'. It might be used to test an Internet router with numerous signal lines going through it.

If the protocol is sufficiently simple, we can use a finite multi-tape automaton as described in this paper to perform the checking. For example, the simple symbol repetition protocol can obviously be checked by such an automaton. However, for complicated protocols where the set of valid streams is not regular, we cannot hope to use a finite automaton for our testing. This is rather unfortunate, as the high speed used on most signal lines typically forces the use of a simple online device – like a finite automaton.

To overcome the difficulty one might attempt to employ a mixed strategy: We use a finite automaton *plus another simple special purpose hardware that also monitors the signal lines.* For example, such a special purpose hardware might be a counter that is increased every time an 'open' symbol is transported on some line, and decreased every time a 'close' symbol is transported. When the streams end, the special hardware device could communicate some information to the finite automaton, which should then decide which signal lines were faulty. The special hardware might tell the automaton whether the 'open' and 'close' symbols paired up correctly, or it could tell the automaton the index of a line where this was not the case.

Surely such a special hardware should allow us to check some nonregular protocols. However, the Finite Automata Nonspeedup Theorem tells us *that the special hardware must communicate at least* $\lfloor \log n \rfloor + 1$ *bits to the automaton to be of any use*. To see this, assume that only $\log n$ bits were communicated. Then another automaton, without knowing these $\log n$ bits, could still at least come up with $2^{\lfloor \log n \rfloor} \leq n$ *possibilities* for the set of faulty lines. But then the set of valid streams would be in $V_{\mathrm{fa}}(n, n)$ and would hence be regular, thus making the special hardware superfluous. In particular, even getting the index of one faulty line does not help the automaton at all.

6   Conclusion

We showed that the inclusion structure of verboseness classes is the same for Turing machines and for finite automata. It is distinct from the corresponding inclusion structure for computational models 'in between', like polynomial time. In particular, the Nonspeedup Theorem holds both for Turing machines and for finite automata, but does not hold for polynomially time-bounded machines. The proofs are based on two new ideas. First, our key lemma states that $f \times g \in \mathrm{EN}(n + m)$ implies $f \in \mathrm{EN}(n)$ or $g \in \mathrm{EN}(m)$; and likewise for finite automata. Second, we introduced the notion of structural diagonalisation, where finite automata trick Turing machines, and used it to show the strong class separations of Theorem 12. While the original Nonspeedup Theorem has limited practical applications, we showed how its finite automata version allows our saving on hardware in finite automata protocol testing.

The parallels between Turing machines and finite automata with respect to partial information algorithms do not appear to end with the inclusion structure of verboseness classes. For example, borrowing the notation from [13], in [1,2] the classes $\Omega(m, n)$ and $\Omega_{\mathrm{fa}}(m, n)$ are studied. For languages in the first class a Turing machine (and for the second, a finite automaton) can output for any n different words a bit string that agrees on at least m positions with the characteristic string of the words. Although the inclusion problem for $\Omega(m, n)$ is not solved, it turns out that all *known* inclusions of $\Omega(m, n)$-classes also hold for the corresponding $\Omega_{\mathrm{fa}}(m, n)$. Using structural diagonalisation, one can even show strong class separations for these classes similar to the ones established in this paper.

Such similarities lead to a conjecture concerning a powerful extension of the Nonspeedup Theorem, namely Kummer's Cardinality Theorem [12]. This theorem states that $\#_n^A \in \mathrm{EN}(n)$ iff A is recursive, where $\#_n^A(w_1, \ldots, w_n) := \sum_{i=1}^n \chi_A(w_i)$. We conjecture that $\#_n^A \in \mathrm{EN}_{\mathrm{fa}}(n)$ iff A is regular. This conjecture can also be rephrased 'in terms of protocol testing': Suppose we are given a finite multi-tape automaton monitoring n input streams, which gets at most $\log n$ bits of information from an external source. We conjecture that if the automaton can just tell us the *number* of faulty lines, then the set of valid streams must be regular.

Acknowledgments. I am grateful to Lane Hemaspaandra, Arfst Nickelsen, Birgit Schelm, Dirk Siefkes, and Leen Torenvliet for helpful discussions. Furthermore, I would like to thank Leen for suggesting the name 'structural diagonalisation'.

References

1. H. Austinat, V. Diekert, and U. Hertrampf. A structural property of regular frequency classes. *Theoretical Comput. Sci.*, to appear 2002.
2. H. Austinat, V. Diekert, U. Hertrampf, and H. Petersen. Regular frequency computations. In *Proc. RIMS Symposium on Algebraic Systems, Formal Languages and Computation*, volume 1166 of RIMS Kokyuroku, pages 35–42, Research Institute for Mathematical Sciences, Kyoto University, Kyoto, 2000.
3. R. Beigel. *Query-limited reducibilities*. PhD thesis, Stanford University, Stanford, USA, 1987.
4. R. Beigel. Bounded queries to SAT and the Boolean hierarchy. *Theoretical Comput. Sci.*, 84(2):199–223, 1991.
5. R. Beigel, W. I. Gasarch, J. Gill, and J. C. Owings. Terse, superterse, and verbose sets. *Inform. Computation*, 103(1):68–85, 1993.
6. R. Beigel, M. Kummer, and F. Stephan. Approximable sets. *Inform. Computation*, 120(2):304–314, 1995.
7. R. Beigel, M. Kummer, and F. Stephan. Quantifying the amount of verboseness. *Inform. Computation*, 118(1):73–90, 1995.
8. J. Cai and L. A. Hemachandra. Enumerative counting is hard. *Inform. Computation*, 82(1):34–44, 1989.
9. W. I. Gasarch and G. A. Martin. *Bounded Queries in Recursion Theory*. Birkhäuser, 1999.
10. C. G. Jockusch, Jr. *Reducibilities in Recursive Function Theory*. PhD thesis, MIT, Cambridge, Massachusetts, 1966.
11. E. B. Kinber. Frequency computations in finite automata. *Cybernetics*, 2:179–187, 1976.
12. M. Kummer. A proof of Beigel's cardinality conjecture. *J. Symbolic Logic*, 57(2): 677–681, 1992.
13. M. Kummer and F. Stephan. Some aspects of frequency computation. Technical Report 21/91, Universität Karlsruhe, Fakultät für Informatik, Germany, 1991.
14. M. Kummer and F. Stephan. Effective search problems. *Math. Logic Quarterly*, 40:224–236, 1994.
15. M. Ogihara. Polynomial-time membership comparable sets. *SIAM J. Comput.*, 24(5):1068–1081, 1995.
16. T. Tantau. Combinatorial representations of partial information classes and their truth-table closures. Diploma thesis, Technische Universität Berlin, Germany, 1999. Available at the Electronic Colloquium on Computational Complexity.

On the Average Parallelism in Trace Monoids*

Daniel Krob, Jean Mairesse, and Ioannis Michos

LIAFA, CNRS - Université Paris 7 - Case 7014 - 2, place Jussieu - 75251 Paris Cedex 5
- France - {dk,mairesse,michos}@liafa.jussieu.fr

1 Abstract – Introduction

Traces are used to model the occurrence of events in concurrent systems [9]. Roughly speaking, a letter corresponds to an event and two letters commute when the corresponding events can occur simultaneously. In this context, the two basic performance measures associated with a trace t are its *length* $|t|$ (the 'sequential' execution time) and its *height* $h(t)$ (the 'parallel' execution time). The ratio $|t|/h(t)$ captures in some sense the *amount of parallelism* (the *speedup* in [6]). Let $\mathbb{M}$ be a trace monoid. Define the generating series

$$F = \sum_{t \in \mathbb{M}} x^{h(t)} y^{|t|}, \ L = \sum_{t \in \mathbb{M}} y^{|t|}, \ H = \sum_{t \in \mathbb{M}} x^{h(t)} . \tag{1}$$

It is well known that L is a rational series [5]. We prove that F and H are also rational and we provide finite representations for the series. Exploiting the symmetries of the trace monoid enables to obtain representations of reduced dimensions. We use the rationality to obtain precise information on the asymptotics of the number of traces of a given height or length.

Then, given a trace monoid and a measure on the traces, we study the *average parallelism* in the trace monoid. One notion of average parallelism is obtained by considering the measure over traces induced by the uniform distribution over words of the same length in the free monoid. In other terms, the probability of a trace is proportional to the number of its representatives in the free monoid. This quantity was introduced in [16] and later studied in [3,4,10]. Here we define alternative notions of average parallelism by considering successively the uniform distribution over traces of the same length, the uniform distribution over traces of the same height, and the uniform distribution over Cartier-Foata normal forms. We prove in particular that there exists $\lambda_{\mathbb{M}}$ and $\gamma_{\mathbb{M}}$ in $\mathbb{R}_+^*$ such that

$$\frac{\sum_{t \in \mathbb{M}, |t|=n} h(t)}{n \cdot \#\{t \in \mathbb{M}, |t|=n\}} \xrightarrow{n \to \infty} \lambda_{\mathbb{M}}, \quad \frac{\sum_{t \in \mathbb{M}, h(t)=n} |t|}{n \cdot \#\{t \in \mathbb{M}, h(t)=n\}} \xrightarrow{n \to \infty} \gamma_{\mathbb{M}} .$$

Furthermore, the numbers $\lambda_{\mathbb{M}}$ and $\gamma_{\mathbb{M}}$ are algebraic. Explicit formulas involving the series L and H are given for $\lambda_{\mathbb{M}}$ and $\gamma_{\mathbb{M}}$.

Proposition 3.1 and Proposition 4.1 (for $\lambda_{\mathbb{M}}$) were announced in [15]. All the other results are original. Due to space restrictions, the proofs are omitted.

---

* This work was partially supported by the European Community Framework IV programme through the research network ALAPEDES.

H. Alt and A. Ferreira (Eds.): STACS 2002, LNCS 2285, pp. 477–488, 2002.
© Springer-Verlag Berlin Heidelberg 2002

2   The Trace Monoid

We start by introducing all the necessary notions from the theory of trace monoids. The reader may refer to [8,9] for further information.

In the sequel, a *graph* is a couple (N, A) where N is a finite non-empty set and $A \subset N \times N$. Hence we consider directed graphs, allowing for self-loops but not multi-arcs. Such a graph is *non-directed* if A is symmetric. We use without recalling it the basic terminology of graph theory.

Fix a finite alphabet Σ. Let D be a reflexive and symmetric relation on Σ, called the *dependence* relation, and let I be its complement in $\Sigma \times \Sigma$, known as the *independence* or *commutation* relation.

The *trace monoid*, or *free partially commutative monoid*, $\mathbb{M} = \mathbb{M}(\Sigma, D)$ is defined as the quotient of the free monoid Σ^* by the least congruence containing the relations $ab \sim ba$ for every $(a, b) \in I$. The elements of $\mathbb{M}$ are called *traces*. Two words are representatives of the same trace if they can be obtained one from the other by repeatedly commuting independent adjacent letters.

The *length* of the trace t is the length of any of its representatives and is denoted by $|t|$. Note that we also use the notation $|S| = \#S$ for the cardinal of a set S. The set of letters appearing in (any representative of) the trace t is denoted by $\mathrm{alph}(t)$. The graphs (Σ, D) and (Σ, I) are called respectively the *dependence* and the *independence graph* of $\mathbb{M}$. Let finally ψ denote the canonical projection from Σ^* into the trace monoid $\mathbb{M}$. In the sequel, we most often simplify the notations by denoting a trace by any of its representatives, that is by identifying w and $\psi(w)$.

A *clique* is a non-empty trace whose letters are mutually independent. Cliques are in one-to-one correspondence with the complete subgraphs (also called cliques in a graph theoretical context) of (Σ, I). Let us denote the set of cliques of $\mathbb{M}$ by $\mathfrak{C}$.

An element $(u, v) \in \mathfrak{C} \times \mathfrak{C}$ is called *Cartier-Foata (CF-) admissible* if for every $b \in \mathrm{alph}(v)$, there exists $a \in \mathrm{alph}(u)$ such that $(a, b) \in D$. The *Cartier-Foata (CF) decomposition* of a trace t is the uniquely defined (see [5, Chap. I]) sequence of cliques $(c_1, c_2, \ldots, c_m)$ such that $t = c_1 c_2 \cdots c_m$, and the couple (c_j, c_{j+1}) is CF-admissible for all j in $\{1, \ldots, m-1\}$. The positive integer m is called the *height* of t and is denoted by $h(t)$. In the visualization of traces using *heaps of pieces*, introduced by Viennot in [19], the height corresponds precisely to the height of the heap.

Example 2.1. Let $\Sigma = \{\{1, 2\}, \{1, 3\}, \{1, 4\}, \{2, 3\}, \{2, 4\}, \{3, 4\}\}$ (the set of subsets of cardinal two of $\{1, 2, 3, 4\}$). Define the independence relation $I = \{(u, v) : u \cap v = \emptyset\}$. The dependence graph (Σ, D) is the line graph of the complete graph K_4, also called the triangular graph T_4. For notational simplicity, set $a_{ij} = \{i, j\}$. The set of cliques is $\mathfrak{C} = \{a, a \in \Sigma\} \cup \{a_{12}a_{34}, a_{13}a_{24}, a_{14}a_{23}\}$.

The CF decomposition of $\tau = a_{12}a_{34}a_{23}^2 a_{14}$ is $(a_{12}a_{34}, a_{14}a_{23}, a_{23})$. We have $|\tau| = 5$ and $h(\tau) = 3$. We represented the heap of pieces associated with τ on the figure.

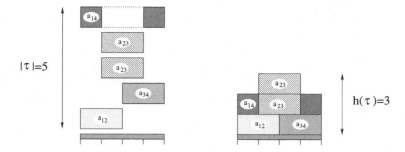

We define the *graph of cliques* Γ as the directed graph with $\mathfrak{C}$ as its set of nodes and the set of all CF-admissible couples as its set of arcs. Note that Γ contains as a subgraph the dependence graph (Σ, D). The graph Γ is in general complicated and looks like a maze.

We now use a standard reduction technique for multi-graphs (see [7, Chap. 4]). We partition the nodes of Γ based on their set of direct successors. An *equitable partition* of $\mathfrak{C}$ is a partition $\pi = \{\mathfrak{C}_1, \ldots, \mathfrak{C}_s\}$ with the property that for all i and j the number a_{ij} of direct successors that a node in $\mathfrak{C}_i$ has in $\mathfrak{C}_j$ is independent of the choice of the node in $\mathfrak{C}_i$. The $s \times s$ matrix $A_\pi = (a_{ij})_{i,j}$ is called the *coloration matrix* corresponding to π. In the case of the partition $\{\{c\}, c \in \mathfrak{C}\}$, the coloration matrix is the *adjacency matrix* of Γ.

3 Height and Length Generating Function

Let $F \in \mathbb{N}[[x, y]]$ be the *height and length generating function* defined in (1). Let $H(x) = F(x, 1)$ and $L(y) = F(1, y)$ be respectively the generating functions of the height and of the length.

The *Möbius polynomial* $\mu(\Sigma, I)$ of the graph (Σ, I) is defined by $\mu(\Sigma, I) = 1 + \sum_{u \in \mathfrak{C}} (-1)^{|u|} y^{|u|}$. It is well known [5, Chap. II] that $L(y)$ is equal to the inverse of the Möbius polynomial, i.e. $L(y) = \mu(\Sigma, I)^{-1}$. In particular, it is a rational series.

Proposition 3.1. *Let $\mathbb{M} = \mathbb{M}(\Sigma, D)$ be a trace monoid and let $\mathfrak{C}$ be the set of cliques of (Σ, I). Define the matrix $A(x, y) \in \mathbb{N}[x, y]^{\mathfrak{C} \times \mathfrak{C}}$ by setting $A(x, y)_{i,j} = xy^{|i|}$ if (i, j) is CF-admissible and 0 otherwise. Define also $u = (1, \ldots, 1) \in \mathbb{N}[x, y]^{1 \times \mathfrak{C}}$ and $v(x, y) = (xy^{|i|})_i \in \mathbb{N}[x, y]^{\mathfrak{C} \times 1}$. The height and length generating function is then given by*

$$F - 1 = \sum_{n \in \mathbb{N}} u A(x, y)^n v(x, y) = u \left(I - A(x, y) \right)^{-1} v(x, y), \qquad (2)$$

where 1 is the identity of $\mathbb{N}[[x, y]]$ and I is the $\mathfrak{C} \times \mathfrak{C}$ identity matrix.

Proposition 3.1 states that $F(x, y)$ is a rational series of $\mathbb{N}[[x, y]]$ and that $(u, A(x, y), v(x, y))$ is a finite representation of it. It follows that L and H

are also rational, and that $L = 1 + u\left(I - A(1,y)\right)^{-1}v(1,y)$ and $H = 1 + u\left(I - A(x,1)\right)^{-1}v(x,1)$. In the case of L, the rationality was known but the above provides a new formula.

Let $\pi = \{\mathfrak{C}_1, \ldots, \mathfrak{C}_s\}$ be an equitable partition of $\mathfrak{C}$ such that all the cliques in $\mathfrak{C}_i$ have a common length l_i. Let $A_\pi = (a_{ij})_{ij} \in \mathbb{N}^{s \times s}$ be the coloration matrix. Define the matrix $A_\pi(x,y) \in \mathbb{N}[x,y]^{s \times s}$ by $A_\pi(x,y) = (a_{ij}xy^{l_i})_{i,j}$. Define $u_\pi = (|\mathfrak{C}_i|)_i \in \mathbb{N}[x,y]^{1 \times s}$ and $v_\pi(x,y) = (xy^{l_i})_i \in \mathbb{N}[x,y]^{s \times 1}$. Then formula (2) holds when replacing u, $A(x,y)$, and $v(x,y)$, by u_π, $A_\pi(x,y)$, and $v_\pi(x,y)$.

Proposition 3.1, although easy to prove, does not seem to appear in the literature. There exist related results in the context of directed animals [13]. More generally, the method of proof of Proposition 3.1 can be viewed as an instance of the transfer matrix method [18, Chap. 4.7]. In the context of trace monoids, the idea of working with the alphabet of cliques $\mathfrak{C}$ to study the height function appeared in [6] and was later used in [11].

Example 3.1. We persevere with the model of Example 2.1. Consider the equitable partition π of $\mathfrak{C}$ defined by $\mathfrak{C}_1 = \Sigma$ and $\mathfrak{C}_2 = \{a_{12}a_{34}, a_{13}a_{24}, a_{14}a_{23}\}$. Using the triple $(u_\pi, A_\pi(x,y), v_\pi(x,y))$, we get

$$F = 1 + \begin{pmatrix} 6 & 3 \end{pmatrix} \begin{pmatrix} 1 - 5xy & -2xy \\ -6xy^2 & 1 - 3xy^2 \end{pmatrix}^{-1} \begin{pmatrix} xy \\ xy^2 \end{pmatrix} = \frac{1 + xy}{1 - 5xy - 3xy^2 + 3x^2y^3}.$$

Setting $x = 1$, we check that the length generating function is the inverse of the Möbius polynomial, i.e. $L = (1 - 6y + 3y^2)^{-1}$. Setting $y = 1$, we obtain the height generating function $H = (1 + x)(1 - 8x + 3x^2)^{-1}$. The Taylor expansion of the series F around 0 is

$$F = 1 + 6xy + 3xy^2 + 30x^2y^2 + 30x^2y^3 + 150x^3y^3 + 9x^2y^4 + 222x^3y^4 + 750x^4y^4 + 126x^3y^5 + 1470x^4y^5 + \cdots + 71910x^6y^8 + \cdots.$$

For instance, there are 126 traces of length 5 and height 3, or 71910 traces of length 8 and height 6.

We now use Proposition 3.1 to provide some precise results on the asymptotics of the number of traces of a given length or height.

Throughout the paper, given a series $S \in \mathbb{N}[[x]]$, we set $S = \sum_n (S|n)x^n$. When applicable, we denote the modulus of the dominant singularities of S (viewed as a function) by ρ_S. Classically, see [1,20], the asymptotic growth rate of $(S|n)$ is linked to the values of the dominant singularities.

Lemma 3.1. *We have $\rho_L = 1$ and $\rho_H = 1$ if and only if $\mathbb{M}(\Sigma, D)$ is the free commutative monoid over Σ.*

Proposition 3.2. *Let (Σ, D) be a connected dependence graph. Then L and H have a unique dominant singularity which is positive real and of order 1.*

Proposition 3.3. *Let (Σ, D) be a non-connected dependence graph. Denote by $(\Sigma_s, D_s), s \in S$, its partition into maximal connected subgraphs and by L_s, H_s, the corresponding length and height generating functions. Then one has*
1) the series L has a unique dominant singularity equal to $\rho_L = \min_s \rho_{L_s}$, and whose order is $\#\{s, \rho_{L_s} = \rho_L\}$;
2) the series H has a unique dominant singularity equal to $\rho_H = \prod_s \rho_{H_s}$. Its order is $|\Sigma|$ if $\mathbb{M}(\Sigma, D)$ is the free commutative monoid, and $1 + \#\{s, |\Sigma_s| = 1\}$ otherwise.

Let k_L and k_H denote the respective orders of ρ_L in L and ρ_H in H. It follows from the above Proposition (see [1,20]) that we have $(L|n) \sim l n^{k_L-1} \rho_L^{-n}$, and $(H|n) \sim h n^{k_H-1} \rho_H^{-n}$ with $l = (\rho_L^{-k_L}/(k_L - 1)!) \cdot [L(y)(\rho_L - y)^{k_L}]_{|y=\rho_L}$ and $h = (\rho_H^{-k_H}/(k_H - 1)!) \cdot [H(x)(\rho_H - x)^{k_H}]_{|x=\rho_H}$.

The results on L in Proposition 3.2 and Proposition 3.3 improve on a recent result by Goldwurm and Santini [12] stating that for any non-directed graph, the Möbius polynomial has a unique and positive real root of smallest modulus. Our proof follows several of the steps of [12]. One central difference is that we work with Cartier-Foata representatives instead of minimal lexicographic representatives. Proving the strengthened statements while working with the latter does not appear to be easy.

4   Asymptotic Average Height

We want to address questions such as: what is the amount of 'parallelism' in a trace monoid? Given several dependence graphs over the same alphabet, which one is the 'most parallel'? To give a precise meaning to these questions, we define the following performance measures. Let $\mathbb{M}_n$ denote the set of traces of length n of the trace monoid $\mathbb{M}$. We equip $\mathbb{M}_n$ with a probability distribution P_n and we compute the corresponding average height

$$E_n[h] = \sum_{t \in \mathbb{M}_n} P_n\{t\} h(t).$$

Assuming the limit exists, we call $\lim_n E_n[h]/n$ the *(asymptotic) average height*. Obviously this quantity lies between C^{-1} and 1, where C is the maximal length of a clique. Clearly the relevance of the average height as a measure of the parallelism in the trace monoid depends on the relevance of the chosen family of probability measures. This may vary depending on the application context. A very common choice is to consider uniform probabilities. It is the natural solution in the absence of precise information on the structure of the traces to be dealt with. Let us consider different instances of uniform probabilities.

4.1   Uniform Probability on Words

Let μ_n be the uniform probability distribution over Σ^n which is defined by setting $\mu_n\{u\} = 1/|\Sigma|^n$, for every $u \in \Sigma^n$. We set $P_n = \mu_n \circ \psi^{-1}$, i.e. $P_n\{t\} =$

$\mu_n\{w \; : \; \psi(w) = t\}$. The limit below exists:

$$\lambda_* = \lambda_*(\Sigma, D) = \lim_n \frac{E_n[h]}{n} = \lim_n \frac{\sum_{w \in \Sigma^n} h(\psi(w))}{n|\Sigma|^n} \, . \tag{3}$$

This is proved using Markovian arguments in [16]. In [10], it is shown that $h(\psi(.))$ is recognized by an automaton with multiplicities over the $(\max, +)$ semiring, which provides a different proof of the existence of λ_*.

Except for small trace monoids, λ_* is neither rational, nor algebraic. The problem of approximating λ_* is NP-hard [2]. Non-elementary bounds are proposed in [4]. Exact computations for simple trace monoids are proposed in [3]. The software package ERS [14] enables to simulate and compute bounds for λ_*.

4.2   Uniform Probability on Traces

A natural counterpart of the above case consists in considering the uniform probability distribution over $\mathbb{M}_n$, i.e. $Q_n\{t\} = 1/|\mathbb{M}_n|$ for every $t \in \mathbb{M}_n$. Let $f_{k,l}$ denote the number of traces of height k and length l. Assuming existence, we define the limit

$$\lambda_{\mathbb{M}} = \lambda_{\mathbb{M}}(\Sigma, D) = \lim_n \frac{E_n[h]}{n} = \lim_n \frac{\sum_{t \in \mathbb{M}_n} h(t)}{n|\mathbb{M}_n|} = \lim_n \frac{\sum_{m \in \mathbb{N}} m f_{m,n}}{\sum_{m \in \mathbb{N}} n f_{m,n}} \, . \tag{4}$$

Dually, let $_m\mathbb{M}$ be the set of traces of height m, and let $\tilde{Q}_m$ be the uniform probability measure on $_m\mathbb{M}$, i.e. $\tilde{Q}_m\{t\} = 1/|_m\mathbb{M}|$ for every $t \in {}_m\mathbb{M}$. The average length of a trace in $_m\mathbb{M}$ is equal to $E_m[l] = \sum_{t \in {}_m\mathbb{M}} \tilde{Q}_m\{t\}|t|$. Assuming existence, we define the limit

$$\gamma_{\mathbb{M}} = \gamma_{\mathbb{M}}(\Sigma, D) = \lim_m \frac{E_m[l]}{m} = \lim_m \frac{\sum_{t \in {}_m\mathbb{M}} |t|}{m|_m\mathbb{M}|} = \lim_m \frac{\sum_{n \in \mathbb{N}} n f_{m,n}}{\sum_{n \in \mathbb{N}} m f_{m,n}} \, . \tag{5}$$

The quantity $\gamma_{\mathbb{M}}$ is an *(asymptotic) average length*. The analog of λ_* and $\lambda_{\mathbb{M}}$ is then the quantity $\gamma_{\mathbb{M}}^{-1}$.

Proposition 4.1. *The limits $\lambda_{\mathbb{M}}$ in (4) and $\gamma_{\mathbb{M}}$ in (5) exist. Furthermore, $\lambda_{\mathbb{M}}$ and $\gamma_{\mathbb{M}}$ are algebraic numbers.*

Define $G = (\partial F/\partial x)(1, y)$ and $\tilde{G} = (\partial F/\partial y)(x, 1)$. Then, with the notations of section 3, we have

$$\lambda_{\mathbb{M}} = \frac{[G(y)(\rho_L - y)^{k_L+1}]_{|y=\rho_L}}{k_L \rho_L [L(y)(\rho_L - y)^{k_L}]_{|y=\rho_L}}, \; \gamma_{\mathbb{M}} = \frac{[\tilde{G}(x)(\rho_H - x)^{k_H+1}]_{|x=\rho_H}}{k_H \rho_H [H(x)(\rho_H - x)^{k_H}]_{|x=\rho_H}} \, .$$

4.3   Uniform Probability on CF Decompositions

Let $A \in \{0,1\}^{\mathfrak{C} \times \mathfrak{C}}$ be the adjacency matrix of Γ. We associate with $A = (a_{ij})_{i,j}$, the Markovian matrix $\hat{A} = (\hat{a}_{ij})_{i,j}$, $\hat{a}_{ij} = a_{ij}(\sum_k a_{ik})^{-1}$. We define the probability measure R_m on $_m\mathbb{M}$ as follows: for a trace $t \in {}_m\mathbb{M}$ with CF decomposition

$(c_1, \ldots, c_m)$, we set $R_m\{t\} = (1/|\mathfrak{C}|)\widehat{a}_{c_1 c_2} \cdots \widehat{a}_{c_{m-1} c_m}$. An interpretation for the family $(R_m)_m$ is as follows. Given a trace t of height m, we get a trace t' of height $m + 1$ by picking at random and uniformly an admissible clique c and by setting $t' = tc$. This can be loosely described as a 'uniform probability on CF decompositions'.

The average length of a trace in $_m\mathbb{M}$ is equal to $E_m[l] = \sum_{t \in {}_m\mathbb{M}} R_m\{t\}|t|$. Assuming existence, the analog of λ_*, $\lambda_\mathbb{M}$ or $\gamma_\mathbb{M}^{-1}$ is then the *(asymptotic) average height*

$$\lambda_{\mathrm{cf}} = \lambda_{\mathrm{cf}}(\Sigma, D) = \lim_m \frac{m}{E_m[l]} \, . \tag{6}$$

Let $p = (p(c))_{c \in \mathfrak{C}}$ be defined by $p = \lim_n \mathbf{1}(I + \widehat{A} + \cdots + \widehat{A}^{n-1})/n$, where $\mathbf{1} = (1/|\mathfrak{C}|, \ldots, 1/|\mathfrak{C}|)$. According to the ergodic theorem for Markov chains (Theorem 4.6 in [17]), the limit exists in (6) and we have

$$\lambda_{\mathrm{cf}} = \Big(\sum_{c \in \mathfrak{C}} p(c)|c|\Big)^{-1} \, . \tag{7}$$

When (Σ, D) is connected, it is easily proved that $\widehat{A}$ is primitive and p is entirely determined by $p\widehat{A} = p$ and $\sum_i p_i = 1$ (Perron-Frobenius Theorem, see [17]). It implies that λ_{cf} is explicitly computable and rational. When (Σ, D) is non-connected, the same conclusion holds according to Proposition 4.4.

4.4 Non-connected Dependence Graphs

Assume that (Σ, D) is *non-connected* and let $(\Sigma_s, D_s), s \in S$, be the maximal connected subgraphs of (Σ, D). We now propose formulas to express the average height of (Σ, D) as a function of the ones of (Σ_s, D_s).

First, it is simple to prove that we have (see Theorem 5.7 in [16])

$$\lambda_*(\Sigma, D) = \max_{s \in S} \Big(\frac{|\Sigma_s|}{|\Sigma|} \lambda_*(\Sigma_s, D_s) \Big) \, . \tag{8}$$

Proposition 4.2. *Denote by L_s the length generating function of (Σ_s, D_s). Define $J = \{j \in S, \ \rho_{L_j} = \min_{s \in S} \rho_{L_s}\}$. Then, we have*

$$\lambda_\mathbb{M}(\Sigma, D) = \lambda_\mathbb{M}(\Sigma_J, D_J) \, , \tag{9}$$

where $\Sigma_J = \cup_{j \in J} \Sigma_j$, and $D_J = \cup_{j \in J} D_j$.

On the other hand, there is no simple way to write $\lambda_\mathbb{M}(\Sigma_J, D_J)$ as a function of $\lambda_\mathbb{M}(\Sigma_j, D_j), j \in J$, as illustrated by the example of section 4.6.

Proposition 4.3. *Define $J = \{j \in S,\ |\Sigma_j| > 1\}$. Then, we have*

$$\gamma_{\mathbb{M}}(\Sigma, D) = \sum_{j \in J} \gamma_{\mathbb{M}}(\Sigma_j, D_j) + |S - J|/2\,, \tag{10}$$

if $J \neq \emptyset$. If $J = \emptyset$, that is if $\mathbb{M}(\Sigma, D)$ is the free commutative monoid, we have $\gamma_{\mathbb{M}}(\Sigma, D) = (|\Sigma| + 1)/2$.

Proposition 4.4. *Let $\widehat{A}$ be defined as in section 4.3. Let $\mathfrak{C}_s$ be the set of cliques of (Σ_s, I_s). Define the matrix B of dimension $\mathfrak{C} \times \mathfrak{C}$ as follows: $B_{ij} = \widehat{A}_{ij}$ if $i \notin \bigcup_{s \in S} \mathfrak{C}_s$ and $B_{ij} = 0$ otherwise. Define the vectors $\mathcal{I}_{\mathfrak{C}_s}, s \in S$, of dimension $\mathfrak{C}$ as follows: $(\mathcal{I}_{\mathfrak{C}_s})_i = 1$ if $i \in \mathfrak{C}_s$ and $(\mathcal{I}_{\mathfrak{C}_s})_i = 0$ otherwise. Set $q_s = \mathbf{1}(I - B)^{-1}\mathcal{I}_{\mathfrak{C}_s}$. Then we have*

$$\lambda_{\mathrm{cf}}(\Sigma, D)^{-1} = \sum_{s \in S} q_s\, \lambda_{\mathrm{cf}}(\Sigma_s, D_s)^{-1}\,. \tag{11}$$

4.5   Comparison between the Different Average Heights

In terms of computability, the simplest quantity is λ_{cf} and the most complicated one is λ_*. This is reflected by the fact that λ_{cf} is rational, that $\lambda_{\mathbb{M}}$ and $\gamma_{\mathbb{M}}^{-1}$ are algebraic, and that λ_* is in general not algebraic, see for instance (12).

Another point of view is to compare the families of probability measures $(P_n)_n, (Q_n)_n, (\widetilde{Q}_n)_n$, and $(R_n)_n$ associated respectively with $\lambda_*, \lambda_{\mathbb{M}}, \gamma_{\mathbb{M}}^{-1}$, and λ_{cf}. A family of probability measures $(\mu_n)_n$ defined on $(\mathbb{M}_n)_n$ (or $(_n\mathbb{M})_n$) is said to be *consistent* if we have $\mu_m\{t\} = \mu_n\{v\ :\ \exists u,\ v = tu\}$ for all $m < n$. In this case, there exists a unique probability measure on infinite traces whose finite-dimensional marginals are the probabilities $(\mu_n)_n$. Consistency is a natural and desirable property. Clearly the families $(P_n)_n$ and $(R_n)_n$ are consistent. On the other hand, the families $(Q_n)_n$ and $(\widetilde{Q}_n)_n$ are not.

It is also interesting to look at the asymptotics in n of the empirical distribution of $\{h(t)/|t|, t \in \mathbb{M}_n\}$ or $\{|t|/h(t), t \in {}_n\mathbb{M}\}$. For $a \in \mathbb{R}$, let δ_a denote the probability measure concentrated in a. Using the notations of section 4.4, we have

$$\sum_t P_n\{t\}\delta_{h(t)/|t|} \ \longrightarrow\ \delta_{\lambda_*}, \quad \sum_t R_n\{t\}\delta_{h(t)/|t|} \ \longrightarrow\ \sum_{s \in S} q_s \delta_{\lambda_{\mathrm{cf}}(\Sigma_s, D_s)}\,,$$

with the arrow standing for 'convergence in distribution'. There are no such concentration results for $(Q_n)_n$ and $(\widetilde{Q}_n)_n$. To check this, consider the case of the free commutative monoid over two letters. We obtain easily that

$$\sum_t Q_n\{t\}\delta_{h(t)/|t|} \ \longrightarrow\ U, \quad \sum_t \widetilde{Q}_n\{t\}\delta_{|t|/h(t)} \ \longrightarrow\ V\,,$$

where U is the uniform distribution over the interval $[1/2, 1]$ and V is the uniform distribution over the interval $[1, 2]$.

Consider two dependence graphs (Σ, D_1) and (Σ, D_2) with $D_1 \subset D_2$. The intuition is that $\mathbb{M}(\Sigma, D_1)$ should be 'more parallel' than $\mathbb{M}(\Sigma, D_2)$. In accordance with this intuition, it is elementary to prove that $\lambda_*(\Sigma, D_1) \leq \lambda_*(\Sigma, D_2)$. However, the corresponding inequalities do not hold for $\lambda_\mathbb{M}$ and λ_{cf}. Consider for instance the trace monoids over three or four letters whose average heights are given in section 5. This raises some interesting issues on how to interpret these quantities. On the other hand, we conjecture that the inequality $\gamma_\mathbb{M}(\Sigma, D_1)^{-1} \leq \gamma_\mathbb{M}(\Sigma, D_2)^{-1}$ is satisfied.

4.6   The Free Commutative Monoid

Consider the dependence graph (Σ, D) with $D = \{(u, u), u \in \Sigma\}$. The corresponding trace monoid $\mathbb{M}(\Sigma, D)$ is the free commutative monoid over the alphabet Σ, which is isomorphic to $\mathbb{N}^\Sigma$. Set now $k = |\Sigma|$.

A direct application of (8) yields $\lambda_* = 1/k$. Consider now λ_{cf}. The final maximal strongly connected subgraphs of Γ are precisely the cliques of length 1. In particular, they are of cardinality 1. Applying the results in section 4.3, we get $\lambda_{\mathrm{cf}} = 1$. Indeed, in this model, traces are built by always adding the same letter asymptotically.

Proposition 4.5. *We have* $\lambda_\mathbb{M} = (1 + 1/2 + \cdots + 1/k)/k$ *and* $\gamma_\mathbb{M}^{-1} = 2/(k+1)$.

4.7   The Ladder Graph

Let (Σ, I) be the *ladder graph*, i.e. $\Sigma = \{1, \ldots, 2n\}$ and $I = \{(i, j) \in \Sigma \times \Sigma : i + j = 2n + 1\}$. The corresponding dependence graph is known as the *cocktail party graph* CP_n. The computation of $\lambda_*(\mathrm{CP}_n)$ was worked out in [3], Prop. 14:

$$\lambda_*(\mathrm{CP}_n) = \frac{1}{2}\left(1 + \frac{\sqrt{n-1}}{\sqrt{n+1}}\right).$$

The other average heights are computable using the reduced representation of F induced by the equitable partition $\mathfrak{C}_1 = \{c \in \mathfrak{C}, |c| = 1\}$ and $\mathfrak{C}_2 = \{c \in \mathfrak{C}, |c| = 2\}$ of $\mathfrak{C}$. The dominant singularity of L is $(1 - \sqrt{1 - n^{-1}})$ and we obtain

$$\lambda_\mathbb{M}(\mathrm{CP}_n) = \frac{1}{2}\left(1 + \frac{\sqrt{n}}{2\sqrt{n} - \sqrt{n-1}}\right).$$

The dominant singularity of H is $(3n - 1 - \sqrt{9n^2 - 10n + 1})/2n$ and we get

$$\gamma_\mathbb{M}(\mathrm{CP}_n)^{-1} = \frac{\Delta(5n - 1 - \Delta)}{2\Delta(4n - 1) - 2(8n^2 - 9n + 1)}, \quad \Delta = \sqrt{9n^2 - 10n + 1}.$$

Using (7), we obtain

$$\lambda_{\mathrm{cf}}(\mathrm{CP}_n) = (9n - 7)/(12n - 10).$$

Remark that $\mathrm{CP}_3 \equiv T_4$. By specializing the above results to $n = 3$, we get the average heights for the graph T_4 considered in Examples 2.1 and 3.1.

5   Trace Monoids over 2, 3, and 4 Letters

We give the values of the average heights for all the trace monoids over 2, 3, and 4 letters alphabets. We have omitted the free monoids for which $\lambda_* = \lambda_M = \gamma_M^{-1} = \lambda_{cf} = 1$. We have also omitted the free commutative monoids for which the average heights were computed in section 4.6. On the table below, a trace monoid is represented by its (non-directed) dependence graph. Self-loops have not been represented on the dependence graphs.

Let us denote the dependence graphs in Table III, listed from top to bottom, by $(\Sigma, D_i), i = 1, \ldots, 11$. The graph (Σ, D_{10}) is the cocktail party graph CP_2, hence the values of the average heights can be retrieved from section 4.7. More generally, most of the values in the table can be computed using the results from the paper. The exceptions are λ_* for $(\Sigma, D_i), i = 2, 7, 9$, and 11. The value of $\lambda_*(\Sigma_2, D_2)$ is computed in [16], Example 6.2. For (Σ, D_9) and (Σ, D_{11}), the value of λ_* can be computed by applying Proposition 12 from [3].

For (Σ, D_7), the exact value of λ_* is not known. Using truncated Markov chains, A. Jean-Marie (personal communication, September 2001) obtained the following exact bounds:

$$\lambda_*(\Sigma, D_7) \in [0.69125003165, 0.69125003169] .$$

Let us concentrate on $\lambda_*(\Sigma, D_8)$. A simple argument using the Strong Law of Large Numbers shows that

$$\lambda_*(\Sigma, D_8) = \frac{1}{4} + \frac{1}{16} \left(\sum_{i \in \mathbb{N}} \frac{1}{4^i} \sum_{i_1+i_2+i_3=i} \max(i_1, i_2, i_3) \binom{i}{i_1, i_2, i_3} \right). \qquad (12)$$

This expression involves non algebraic generalized hypergeometric series. By truncating the infinite sum and upper-bounding the remainder using the inequality $\max(i_1, i_2, i_3) \le i_1 + i_2 + i_3$, we get the following exact bounds:

$$\lambda_*(\Sigma, D_8) \in [0.68111589347, 0.68111589349] .$$

The closed form expressions for $\lambda_M(\Sigma, D_8)$ and $\gamma_M(\Sigma, D_8)$ are not given in Table III since they are too long and do not fit. We have

$$\lambda_M = \frac{8(-93 - 9\sqrt{93} - \sqrt{93}X + 5X^2)}{-1734 - 186\sqrt{93} + (141 - 5\sqrt{93})X + 67X^2}, \quad X = (108 + 12\sqrt{93})^{1/3} , \qquad (13)$$

$$\gamma_M^{-1} = \frac{10777(529 - 23Y^2 + Y^4)(829 + 132\sqrt{62} - (139 - 6\sqrt{62})Y - 11Y^2)}{3(3779 + 372\sqrt{62})(98340\sqrt{62} - 1461365 - 1529(149 + 66\sqrt{62})Y - 53885Y^2)}$$

with $Y = (89 + 18\sqrt{62})^{1/3}$.

At last, let us comment on the value of γ_M for (Σ, D_9). Using the results from section 4.2, we get

$$\gamma_M(\Sigma, D_9)^{-1} = \frac{(1 - 2\alpha)(4 - 5\alpha)}{7 - 27\alpha + 24\alpha^2} , \qquad (14)$$

where α is the smallest root of the equation $2x^3 - 8x^2 + 6x - 1 = 0$. Numerically, we have $\alpha = 0.237 \cdots$ and $\gamma_M^{-1} = 0.760 \cdots$.

		λ_*	λ_M	γ_M^{-1}	λ_{cf}
1		2/3	1	2/3	1
2		$(10+\sqrt{5})/15$	$(7+\sqrt{5})/10$	9/11	8/9
3		1/2	1	1/2	1
4		1/2	3/4	1/2	1
5		$(10+\sqrt{5})/20$	$(7+\sqrt{5})/10$	18/31	52/57
6		3/4	1	2/3	1
7		?	19/22	$(13-2\sqrt{13})/9$	5/6
8		in (12)	in (13)	in (13)	11/14
9		$(5+\sqrt{2})/8$	$(6+\sqrt{2})/8$	in (14)	7/8
10		$(3+\sqrt{3})/6$	$(11+\sqrt{2})/14$	$(51+\sqrt{17})/76$	11/14
11		$(9+\sqrt{3})/12$	$(4+\sqrt{3})/6$	$(3\sqrt{5}-5)/2$	8/9

References

1. J. Berstel and C. Reutenauer. *Rational Series and their Languages*. Springer Verlag, 1988.
2. V. Blondel, S. Gaubert, and J. Tsitsiklis. Approximating the spectral radius of sets of matrices in the max-algebra is NP-hard. *IEEE Trans. Autom. Control*, 45(9):1762–1765, 2000.
3. M. Brilman. *Evaluation de Performances d'une Classe de Systèmes de Ressources Partagées*. PhD thesis, Univ. Joseph Fourier - Grenoble I, 1996.
4. M. Brilman and J.M. Vincent. On the estimation of the throughput for a class of stochastic resources sharing systems. *Mathematics of Operations Research*, 23(2):305–321, 1998.
5. P. Cartier and D. Foata. *Problèmes combinatoires de commutation et réarrangements*. Number 85 in Lecture Notes in Mathematics. Springer Verlag, 1969.
6. C. Cérin and A. Petit. Speedup of recognizable trace languages. In *Proc. MFCS 93*, number 711 in Lect. Notes Comput. Sci., pages 332–341. Springer, 1993.
7. D. Cvetković, M. Doob, and H. Sachs. *Spectra of Graphs. Theory and Application*, volume 87 of *Pure and Applied Mathematics*. Academic Press, Paris, 1980.
8. V. Diekert and Y. Métivier. Partial commutation and traces. In *Handbook of formal languages*, volume 3, pages 457–533. Springer, 1997.
9. V. Diekert and G. Rozenberg, editors. *The Book of Traces*. World Scientific, Singapour, 1995.
10. S. Gaubert and J. Mairesse. Task resource models and (max,+) automata. In J. Gunawardena, editor, *Idempotency*, volume 11, pages 133–144. Cambridge University Press, 1998.
11. S. Gaubert and J. Mairesse. Performance evaluation of timed Petri nets using heaps of pieces. In P. Bucholz and M. Silva, editors, *Petri Nets and Performance Models (PNPM'99)*, pages 158–169. IEEE Computer Society, 1999.
12. M. Goldwurm and M. Santini. Clique polynomials have a unique root of smallest modulus. *Information Processing Letters*, 75(3):127–132, 2000.
13. V. Hakim and J.-P. Nadal. Exact results for 2d directed animals on a strip of finite width. *J. Phys. A: Math. Gen.*, 16:L213–L218, 1983.
14. A. Jean-Marie. ERS: A tool set for performance evaluation of discrete event systems. http://www-sop.inria.fr/mistral/soft/ers.html.
15. D. Krob, J. Mairesse, and I. Michos. On the average Cartier-Foata height of traces. In *Proceedings of Comb'01: Combinatorics, Graph Theory and Applications*, Electronic Notes in Discrete Mathematics, 2001.
16. N. Saheb. Concurrency measure in commutation monoids. *Discrete Applied Mathematics*, 24:223–236, 1989.
17. E. Seneta. *Non-negative Matrices and Markov Chains*. Springer series in statistics. Springer Verlag, Berlin, 1981.
18. R. Stanley. *Enumerative Combinatorics, Volume I*. Wadsworth & Brooks/Cole, Monterey, 1986.
19. G.X. Viennot. Heaps of pieces, I: Basic definitions and combinatorial lemmas. In Labelle and Leroux, editors, *Combinatoire Énumérative*, number 1234 in Lect. Notes in Math., pages 321–350. Springer, 1986.
20. H. Wilf. *Generatingfunctionology*. Academic Press, 1990.

A Further Step towards a Theory of Regular MSC Languages

Dietrich Kuske

Department of Mathematics and Computer Science
University of Leicester
LEICESTER
LE1 7RH, England
d.kuske@mcs.le.ac.uk

Abstract. This paper resumes the study of regular sets of Message Sequence Charts initiated by Henriksen, Mukund, Narayan Kumar & Thiagarajan [10]. Differently from their results, we consider infinite MSCs. It is shown that for bounded sets of infinite MSCs, the notions of recognizability, axiomatizability in monadic second order logic, and acceptance by a deterministic Message Passing Automaton with Muller acceptance condition coincide. We furthermore characterize the expressive power of first order logic and of its extension by modulo-counting quantifiers over bounded infinite MSCs.
Complete proofs can be found in the Technical Report [15].

1 Introduction

Message sequence charts (MSCs) form a popular visual formalism used in the software development. In its simplest incarnation, an MSC depicts the desired exchange of messages and corresponds to a single partial-order execution of the system. Several methods to specify sets of MSCs have been considered, among them MSC-graphs or High-level MSCs (HMSCs) that generate sets of MSCs by concatenating "building blocks", (Büchi-)automata that accept the linear extensions of MSCs, and logics. In general, these formalisms have different expressive power.

In [1], Alur & Yannakakis show that the collection of MSCs generated by a "bounded" ("locally synchronized" in the terminology of [10]) MSC-graph can be represented as a string language recognizable by a finite deterministic automaton. Based on this observation, Henriksen et al. [10] study sets of MSCs whose linear extensions form a regular string language. I will call these sets of MSCs "recognizable". The notion of recognizability has proven to be robust and fruitful in different settings like strings, trees, Mazurkiewicz traces and other classes of partial orders (both finite and infinite). The robustness is reflected by the fact that, in all these settings, recognizable sets can be presented by finite-state devices, by congruences of finite index, or by sentences of monadic second order logic. The main results in [10] show similar equivalences for sets of B-bounded finite MSCs (an MSC is B-bounded if in any execution, any buffer will contain at most B messages at any given time). In particular, they prove the equivalence of the following three concepts for sets K of B-bounded finite MSCs:

H. Alt and A. Ferreira (Eds.): STACS 2002, LNCS 2285, pp. 489–500, 2002.
© Springer-Verlag Berlin Heidelberg 2002

1. The set of linear extensions of K can be accepted by a finite deterministic automaton.
2. There is a sentence φ of the monadic second order logic such that K is the set of B-bounded finite MSCs that satisfy φ.
3. Some finite nondeterministic message passing automaton accepts K.

This result was sharpened in [20] where it is shown that deterministic message passing automata suffice.

The main focus of this paper is the extension of these results to sets of infinite MSCs. These infinite MSCs occur naturally as executions of systems that are not meant to stop, e.g., distributed operating systems or telecommunication networks. In the first part, we will extend the equivalence between the first and the second statement. We will also consider two fragments of monadic second order logic, namely first-order logic FO and its extension by modulo-counting quantifiers $FO+MOD(n)$ [22]. We describe the expressive power of these logics in the spirit of Büchi's theorem: for a set of B-bounded possibly infinite MSCs K, the following statements are equivalent (Theorems 3.3 and 4.6)

1. The set of linear extensions of K is recognizable (n-solvable, aperiodic, resp.).
2. The set K is axiomatizable by a sentence of monadic second order logic (of the logic $FO+MOD(n)$, first-order logic, resp.) relative to all possibly infinite MSCs.

The proof of the implication $1 \rightarrow 2$ relies on a first-order interpretation of a B-bounded MSC in any of its linearisations as well as on the fact that the set of all linearisations of B-bounded MSCs is aperiodic. This allows us to use results from [3,17] and [22] that characterize the expressive power of the logics in question for infinite words. The proof $2 \rightarrow 1$ for *finite* MSCs from [10] uses a first-order interpretation of the lexicographically least linear extension of t in the finite MSC t. This proof method does not extend to the current setting since in general no linear extension of order type ω can be defined in an infinite MSC. To overcome this problem, we use ideas from [23] by choping an infinite MSC into its finite and its infinite part. It turns out that the infinite part is the disjoint union of infinite posets to which the "classical" method from [10] is applicable.

The second part of the paper gives a characterization of recognizable sets of infinite MSCs in terms of message passing automata. To this aim, we extend the model from [10] by a Muller-acceptance condition. It is shown that for a set of B-bounded possibly infinite MSCs K, the following statements are equivalent (Theorem 5.7).

1. The set of linear extensions of K is recognizable.
3. Some finite deterministic message passing automaton with Muller-acceptance condition accepts K.

The proof of the implication $3 \rightarrow 1$ is an obvious variant of similar proofs for finite automata for words (cf. [24]), asynchronous automata for traces [25], or asynchronous cellular automata for pomsets without autoconcurrency [8]. Mukund et al. proved the implication $1 \rightarrow 3$ for finite MSCs. In order to do so, they had to reprove several results from the theory of Mazurkiewicz traces in the more complex realm of MSCs. Differently, my proof for infinite MSCs refers to deep results in the theory of Mazurkiewicz traces directly, in particular to the theory of asynchronous mappings [4,6]. These results are applicable since any recognizable set of MSCs can be represented as a set of traces up to an easy relabeling. This constitutes a newly discovered relation between Mazurkiewicz traces and MSCs that differs fundamentally from those used e.g. in [21] for the investigation of race condition and confluence properties and in [10] for some undecidability results. This new observation has in my opinion several nice aspects: (1) it simplifies the proof, (2) it also results in smaller message passing automata for finite MSCs, and (3) it highlights the similarity of MSCs and Mazurkiewicz traces and the unifying role that Mazurkiewicz traces can play in the theory of distributed systems. This last point is also stressed by the fact that similar proof techniques have been used, e.g., in [2,7,14,8, 16,19].

2 Notation

Let $\mathcal{P}$ be a finite set of processes (or agents) which communicate with each other through messages via reliable FIFO-channels. Let Σ be the set of communication actions $p!q$ and $p?q$ for $p, q \in \mathcal{P}$ distinct. The action $p!q$ is to be read as "p sends to q" and $p?q$ is to be read as "p receives from q". Hence $p\theta q$ is performed by the process p, denoted $\mathrm{proc}(p\theta q) = p$. Following [10], we shall not be concerned with the internal actions of the agents which is no essential restriction since the results of this paper can be extended to deal with internal actions. We will also not consider the actual messages that are sent and received.

A Σ-labeled poset is a structure $t = (V, \leq, \lambda)$ where $(V, \leq)$ is a partially ordered set, $\lambda : V \to \Sigma$ is a mapping, $\lambda^{-1}(\sigma) \subseteq V$ is linearly ordered for any $\sigma \in \Sigma$, and any $v \in V$ dominates a finite set. A subset $X \subseteq V$ is an *order ideal* if it is downwards closed, i.e., if $x \in X$, $v \in V$ and $v \leq x$ imply $v \in X$. For $A \subseteq \Sigma$, let $\pi_A(t)$ denote the restriction of t to $\lambda[A]$, i.e., to the A-labeled nodes. For $v \in V$, we write $\mathrm{proc}(v)$ as a shorthand for $(\mathrm{proc} \circ \lambda)(v)$. Furthermore, we define $\downarrow v = \{u \in V \mid u \leq v\}$. In order to define message sequence charts, for a Σ-labeled poset t, we define the relations $\sqsubseteq_{\mathcal{P}}$ and $\sqsubseteq$ as follows:

- $v \sqsubseteq_{\mathcal{P}} w$ iff $\mathrm{proc}(v) = \mathrm{proc}(w)$ and $v \leq w$.
- $v \sqsubseteq w$ iff $\lambda(v) = p!q$, $\lambda(w) = q?p$, and $|\downarrow v \cap \lambda^{-1}(p!q)| = |\downarrow w \cap \lambda^{-1}(q?p)|$ for some $p, q \in \mathcal{P}$ distinct.

Definition 2.1. *A message sequence chart or MSC for short is a Σ-labeled poset $t = (V, \leq, \lambda)$ satisfying*

- $\leq = (\sqsubseteq_{\mathcal{P}} \cup \sqsubseteq)^\star$,
- $\mathrm{proc}^{-1}(p) \subseteq V$ *is linearly ordered for any $p \in \mathcal{P}$, and*
- $|\lambda^{-1}(p!q)| = |\lambda^{-1}(q?p)|$ *for any $p, q \in \mathcal{P}$ distinct.*

The MSC t is B-bounded for some $B \in \mathbb{N}$ if, for any $v \in V$, we have $|\downarrow v \cap \lambda^{-1}(p!q)| - |\downarrow v \cap \lambda^{-1}(q?p)| \leq B$. By MSC^{∞}, we denote the set of all message sequence charts while MSC denotes the set of finite MSCs. Furthermore, MSC_B and MSC_B^{∞} denote the sets of B-bounded (finite) MSCs. Finally, $\downarrow\mathrm{MSC}$ denotes the set of order ideals in finite MSCs, and $\downarrow\mathrm{MSC}_B^{\infty}$ etc are defined similarly.

Let $w = w_0 w_1 w_2 ...$ be a finite or infinite word over Σ. We write $w\lceil_i = w_0 w_1 ... w_i$ for the prefix of w of length $i + 1$ where $0 \leq i < |w|$. The word w is *proper* if, for any $0 \leq i < |w|$ and any $p, q \in \mathcal{P}$ distinct, we have $|w\lceil_i|_{p!q} \geq |w\lceil_i|_{q?p}$ (where $|w|_a$ denotes the number of occurrences of the letter a in the word w). From a proper word w, we can define the Σ-labeled poset $\mathrm{MSC}(w)$ as follows:

- $V = \mathrm{dom}(w) = \{i \in \mathbb{N} \mid 0 \leq i < |w|\}$,
- $\lambda(i) = \lambda_w(i) = w_i$, and
- $\leq = R^\star$ where $(i, j) \in R$ iff $i < j$ and either $\mathrm{proc}(w_i) = \mathrm{proc}(w_j)$ or $w_i = p!q$, $w_j = q?p$ and $|w\lceil_i|_{p!q} = |w\lceil_j|_{q?p}$.

It is easily seen that $\mathrm{MSC}(w)$ is an element of $\downarrow\mathrm{MSC}^\infty$, i.e., an ideal in an MSC. The proper word w is *B-bounded* if, for any $0 \leq i < |w|$, we have $|w\lceil_i|_{p!q} - |w\lceil_i|_{q?p} \leq B$. Note that an MSC t is B-bounded iff any proper word w with $\mathrm{MSC}(w) = t$ is B-bounded. On the other hand, there are proper B-bounded words w for which $\mathrm{MSC}(w)$ is not B-bounded (e.g., $w = (p!q\, q?p)^\omega$). We call a proper word w *complete* provided $|w|_{p!q} = |w|_{q?p}$ for $p, q \in \mathcal{P}$ or, equivalently, if $\mathrm{MSC}(w)$ is an MSC.

For an ideal in a MSC t, let $\mathrm{Lin}_\omega(t)$ denote the set of all proper words w with $\mathrm{MSC}(w) = t$. For $K \subseteq \downarrow\mathrm{MSC}^\infty$, let $\mathrm{Lin}_\omega[K]$ denote the union of all sets $\mathrm{Lin}_\omega(t)$ for $t \in K$. Using this set of words, we define below recognizable sets of MSCs.

Recall that a set L of finite words over Σ is recognizable (by a finite deterministic automaton) iff there exists a finite monoid S and a homomorphism $\eta : \Sigma^\star \to S$ such that $L = \eta^{-1}\eta(L)$. The set L is aperiodic if S can be assumed to be aperiodic (i.e., groupfree). Finally, we call L *n-solvable* (for some $n \in \mathbb{N}$) if we can assume that any group in S is solvable and has order dividing some power of n. Similarly, one can define recognizable, n-solvable, and aperiodic sets of words in Σ^∞: $L \subseteq \Sigma^\infty$ is recognizable iff there exists a finite monoid S and a homomorphism $\eta : \Sigma^\star \to S$ such that, for any $u_i, v_i \in \Sigma^\star$ with $\eta(u_i) = \eta(v_i)$ and $u_0 u_1 u_2 \cdots \in L$, we get $v_0 v_1 v_2 \cdots \in L$. The set L is said to be n-solvable or aperiodic if the monoid S can be assumed to satisfy the corresponding conditions.

Definition 2.2. *A set $K \subseteq \downarrow\mathrm{MSC}^\infty$ is recognizable, n-solvable, or aperiodic if $\mathrm{Lin}_\omega[K]$ is recognizable, n-solvable, or aperiodic, respectively.*

In [10], it was shown that any recognizable set of finite MSCs is B-bounded for some $B \in \mathbb{N}$. The proof for ideals in possibly infinite MSCs goes through verbatimly (note that any n-solvable or aperiodic set is recognizable):

Proposition 2.3 (cf. [12, Prop. 3.2]). *Let $K \subseteq \downarrow\mathrm{MSC}^\infty$ be recognizable. Then K is bounded, i.e., there exists a positive integer B such that $L \subseteq \downarrow\mathrm{MSC}_B^\infty$.*

The sets of recognizable, n-solvable, and aperiodic word languages have been characterized in terms of fragments of monadic second order logic [3,17,22]. In the first part of this paper, these results will be extended to subsets of $\downarrow\mathrm{MSC}^\infty$.

Formulas of the monadic second order language MSO over Σ involve first order variables $x, y, z ...$ for nodes and set variables $X, Y, Z ...$ for sets of nodes. They are built up from the atomic formulas $\lambda(x) = \sigma$ for $\sigma \in \Sigma$, $x \leq y$, and $x \in X$ by means of the boolean connectives $\neg$, $\wedge$ and the quantifier $\exists$ (both for first order and for set variables). A first order formula is a formula without set variables. Formulas without free variables are called sentences. The satisfaction relation $t \models \varphi$ between Σ-labeled posets t and formulas φ is defined canonically. Let $\mathfrak{X}$ be a set of Σ-labeled posets. A set $K \subseteq \mathfrak{X}$ is *axiomatizable relative to* $\mathfrak{X}$ iff there is a sentence φ such that $K = \{t \in \mathfrak{X} \mid t \models \varphi\}$.

Formulas of the logic FO+MOD(n) are built up from the atomic formulas $\lambda(x) = a$ and $x \leq y$ be the connectives $\wedge$ and $\neg$ and the first-order quantifiers $\exists$ and $\exists^m$ for $0 \leq m < n$. A Σ-labeled poset $t = (V, \leq, \lambda)$ satisfies $\exists^m \varphi(x)$ if the number of nodes $v \in V$ such that $t \models \varphi(v)$ is finite and congruent $m \mod n$. Thus, syntactically, FO+MOD(n) is not a fragment of MSO, but since MSCs have width at most $|\mathcal{P}|$, any FO+MOD(n)-definable set of MSCs can alternatively be defined in MSO.

3   From Logics to Monoids

Proposition 3.1. *The sets of B-bounded and proper (complete, resp.) words in Σ^* and Σ^ω are aperiodic.*

Proof. The set of B-bounded, proper and finite words L can be accepted by a finite deterministic automaton with $(B+1)^{|\mathcal{P}|^2} + 1$ states. If $uv^{B+1}w \in L$, then $|v|_{p!q} = |v|_{q?p}$ for $p, q \in \mathcal{P}$ distinct. Hence $uv^B w$ and $uv^{B+2}w$ belong to L, i.e., L is aperiodic. From this result, the remaining assertions follow easily. $\square$

Hence the set of B-bounded, proper, and finite words is first-order axiomatizable relative to Σ^* [18]. Using a similar idea, one can define the relation R from the definition of MSC(w) in the B-bounded and proper word $w \in \Sigma^\infty$. This then allows to interpret MSC(w) in the B-bounded and proper word w by a first-order formula, i.e., we obtain the following result:

Proposition 3.2. *Let $B \in \mathbb{N}$. There exists a first-order formula $\varphi(x, y)$ such that for any B-bounded and proper word $w \in \Sigma^\infty$, we have*

$$(\mathrm{dom}(w), \varphi^w, \lambda_w) \cong \mathrm{MSC}(w)$$

where $\varphi^w = \{(i, j) \mid 0 \leq i, j < |w|, w \models \varphi(i, j)\}$.

Let ψ be some sentence of FO, FO+MOD(n), or MSO. Replace in this sentence any subformula of the form $x \leq y$ by $\varphi(x, y)$ from the lemma above. Since FO is a fragment of all the other logics, the resulting sentence belongs to FO, FO+MOD(n), or MSO, respectively. By Prop. 3.2 and 3.1, the set of proper words w with MSC$(w) \models \psi$ is axiomatizable in the respective logic. Hence we can apply the algebraic characterizations of these sets from [3,17,22] and obtain

Theorem 3.3. *Let $K \subseteq {\downarrow}\mathrm{MSC}^\infty$ be bounded and axiomatizable by a sentence of MSO / FO+MOD(n) / FO. Then K is recognizable / n-solvable / aperiodic, respectively.*

4   From Monoids to Logics

Henriksen *et al.* [10] showed that any recognizable set K of *finite* MSCs is axiomatizable in monadic second order logic relative to MSC_B for some B. Their proof strategy follows the idea from [9] to interpret the lexicographically least linear extension of an MSC t in t. This interpretation is possible for infinite MSCs, too, as the following lemma shows.

Let $\preceq$ be a fixed linear order on Σ. Let furthermore X be a subset of an MSC t. Then $\preceq$ defines the lexicographic order on the linear extensions of $t \restriction_X$. Since these linear extensions are well orders, the lexicographic order is a well order, too. Hence there is a lexicographically least linear extension of $t \restriction_X$ that we denote by $\mathrm{lexNF}(t \restriction_X)$.

Lemma 4.1. *There exists a first-order formula φ with two free variables such that for any $t = (V, \leq, \lambda) \in \downarrow\mathrm{MSC}^\infty$ and $X \subseteq V$, we have $(X, \varphi^{t \restriction_X}, \lambda \restriction_X) \cong \mathrm{lexNF}(t \restriction_X)$.*

The proof is an immediate adaptation of the corresponding proof for Mazurkiewicz traces, cf. [5].

The proof from [10] for finite MSCs then refers to Büchi's theorem for finite words. We cannot continue that way since the lexicographically least linear extension of, *e.g.*, $\mathrm{MSC}((p!q\,q?p)^\omega)$ with $p!q \preceq q?p$, is no ω-word (it equals $(p!q)^\omega (q?p)^\omega$). Even worse, the MSC $\mathrm{MSC}((p!q\,q?p\,p'!q'\,q'?p')^\omega$ has no definable linear extension of order type ω at all. Hence we cannot generalize the proof from [10] to infinite MSCs. To overcome this problem, we will chop an MSC into pieces, consider these pieces independently, and combine the results obtained for them.

Now let $t = (V, \leq, \lambda) \in \downarrow\mathrm{MSC}^\infty$. Then $\mathrm{alph}(t) = \lambda[V]$ and $\mathrm{alphInf}(t) = \{\sigma \in \Sigma \mid \lambda^{-1}(\sigma)$ is infinite$\}$. Let Y be the largest filter in t with $\lambda[Y] \subseteq \mathrm{alphInf}(t)$ and let $X = V \setminus Y$. Then the *finitary part of t* is defined by $\mathrm{Fin}(t) = t \restriction_X$ and the *infinitary part of t* by $\mathrm{Inf}(t) = t \restriction_Y$. Then $\mathrm{Fin}(t)$ is an ideal in a finite MSC while in general $\mathrm{Inf}(t)$ is only a Σ-labeled poset.

Let $E \subseteq \Sigma^2$ contain all pairs $\sigma, \tau \in \Sigma$ with $\mathrm{proc}(\sigma) = \mathrm{proc}(\tau)$ or $\{\sigma, \tau\} = \{p!q, q?p\}$ for some $p, q \in \mathcal{P}$. Then (Σ, E) is an undirected graph.

Lemma 4.2. *Let $t = (V, \leq, \lambda) \in \downarrow\mathrm{MSC}_B^\infty$, and $A = \mathrm{alphInf}(t)$. Let $(A_i)_{1 \leq i \leq n}$ be the connected components of the graph (A, E). Then we have for $1 \leq i \leq n$:*

1. *the Σ-labeled poset $t_i = \pi_{A_i}(\mathrm{Inf}(t))$ is directed (i.e., it does not contain two disjoint filters),*
2. *any linear extension of t_i is of order type ω, and*
3. *$\mathrm{Inf}(t) = \bigcup_{1 \leq i \leq n} t_i$.*

Note that the MSC $\mathrm{MSC}(((p!q)(q?p))^\omega)$ is directed, but it has a linear extension of type $\omega + \omega$, namely $(p!q)^\omega (q?p)^\omega$. Thus, to prove the second statement of the lemma above, one indeed needs the B-boundedness of t.

By $u \sqcup\!\sqcup v$ we denote the set of all shuffles of the words u and v and $K \sqcup\!\sqcup L = \bigcup\{u \sqcup\!\sqcup v \mid u \in L, v \in L\}$. Now suppose $t \in \downarrow\mathrm{MSC}_B^\infty$ and let $\mathrm{alphInf}(t) = A$. Then, by Lemma 4.2(3), $\mathrm{Inf}(t)$ is the disjoint union of the Σ-labeled posets $t_i = \pi_{A_i}(\mathrm{Inf}(t))$ for A_i a connected component of (A, E). We define $\mathrm{NFInf}(t) = \sqcup\!\sqcup_{1 \leq i \leq n} \mathrm{lexNF}(t_i)$. Then, by Lemma 4.2(2), $\mathrm{NFInf}(t) \subseteq \Sigma^\omega$. Finally, by Lemma 4.2(2), we obtain that $\mathrm{NFInf}(t) \subseteq \mathrm{Lin}_\omega(\mathrm{Inf}(t))$.

Definition 4.3. *Let* $t \in \downarrow\text{MSC}_B^\infty$, $A = \text{alphInf}(t)$ *and let* $\$ \notin \Sigma$. *Then* $\text{NF}(t) = \text{lexNF}(\text{Fin}(t)) \cdot \{\$\} \cdot \text{NFInf}(t)$ *is the set of normal forms of* t. *For* $K \subseteq \downarrow\text{MSC}_B^\infty$, *we set* $\text{NF}[K] = \bigcup_{t \in K} \text{NF}(t)$.

Since $\text{NFInf}(t)$ is a subset of Σ^ω, we get $\text{NF}(t) \subseteq (\Sigma \cup \$)^\infty$. Furthermore, the restriction of any word in $\text{NF}(t)$ to Σ is a linear extension of t, i.e., $\text{NF}(t) \subseteq \text{Lin}_\omega(t) \sqcup \{\$\}$. Using Prop. 3.1 and 3.2, one obtains that $\text{NF}[\downarrow\text{MSC}_B^\infty]$ is first-order axiomatizable relative to $(\Sigma \cup \{\$\})^\infty$.

Let $u \in \Sigma^\infty$ and $k \in \mathbb{N}$. Then the set of all first-order sentences φ of quantifier depth at most k is the *k-first-order theory of* u. A set of first-order sentences of quantifier depth at most k is a *complete k-first-order theory* if it is the k-first-order theory of some word $u \in \Sigma^\infty$. Since, up to logical equivalence, there are only finitely many first-order sentences of quantifier depth at most k, there are only finitely many complete k-first-order theories. Furthermore, each complete k-first-order theory T is characterized by one first-order sentence γ_T of quantifier depth k, i.e., for any word $u \in \Sigma^\infty$, we have $u \models \gamma$ for all $\gamma \in T$ iff $u \models \gamma_T$ (cf. [13, Thm. 3.3.2]). For notational convenience, we will identify the characterizing sentence γ_T and the complete k-first-order theory T.

Let $K \subseteq \downarrow\text{MSC}_B^\infty$ be aperiodic. Then $\text{Lin}_\omega[K]$ is first-order axiomatizable relative to Σ^∞. Let $k \geq 2$ be the least integer such that $\text{Lin}_\omega[K] \sqcup \{\$\}$ and $\text{NF}[\downarrow\text{MSC}_B^\infty]$ are first-order axiomatizable relative to $(\Sigma \cup \{\$\})^\infty$ by a sentence of quantifier depth at most k. Let T be a complete k-first-order theory and $A \subseteq \Sigma$. Then set $K_{T,A} := \{t \in K \mid \text{lexNF}(\text{Fin}(t)) \models T, \text{alphInf}(t) = A\}$ and $X_{T,A} := \text{NF}[K_{T,A}]$.

Lemma 4.4. *In the first order language of* $(\Sigma \cup \{\$\})$-*labeled linear orders with one constant* c, *there exists a sentence* φ *of quantifier depth* k *such that* $(v, c) \models \varphi$ *iff* $v \in X_{T,A}$ *and* $\lambda(c) = \$$ *for* $v \in (\Sigma \cup \{\$\})^\infty$.

Apart from an extension of Mezei's theorem to languages of infinite words we now have all the ingredients for the proof of Thm. 4.6. To formulate the mentioned extension, we need the notion of an aperiodic extension: a finite monoid S' is an *aperiodic extension* of a monoid S if there is a surjective homomorphism $\eta : S' \to S$ such that $\eta^{-1}(f)$ is an aperiodic semigroup for any idempotent element $f \in S$. Note that aperiodic extensions of finite / n-solvable / aperiodic monoids are finite / n-solvable / aperiodic.

Theorem 4.5. *Let* $\Sigma = \Sigma_1 \dot\cup \Sigma_2$ *be an alphabet. Let* $L \subseteq \Sigma^\infty$ *be recognized by a homomorphism into* $(S, \cdot)$ *and suppose that* $u_1 \sqcup\!\sqcup u_2 \cap L \neq \emptyset$ *implies* $u_1 \sqcup\!\sqcup u_2 \subseteq L$ *for any* $u_i \in \Sigma_i^\infty$ $(i = 1, 2)$. *Then* L *is a finite union of sets* $L_1 \sqcup\!\sqcup L_2$ *where* $L_i \subseteq \Sigma_i^\infty$ *is recognized by an aperiodic extension of* $(S, \cdot)$.

Putting all these results (and their obvious extensions to the more expressive logics) together, one obtains the following converse of Thm. 3.3:

Theorem 4.6. *Let* $K \subseteq \downarrow\text{MSC}^\infty$ *be recognizable / n-solvable / aperiodic. Then* K *is bounded and axiomatizable by a sentence of* MSO / FO+MOD(n) / FO, *respectively.*

5   Deterministic Message Passing Automata

In this section, we will extend the the automata-theoretic characterizations of recognizable word languages. Message passing automata, the automata model that we consider, reflect the concurrent behavior of an MSC. It was introduced by Henriksen et al. [10] and is similar to asynchronous cellular automata from the theory of Mazurkiewicz traces. We will extend results from [10,20] to infinite MSCs. Since the proofs rely on the theory of Mazurkiewicz traces, we first investigate the relation between these traces and MSCs.

5.1   The Key Observation

A *dependence alphabet* is a pair (Γ, D) where Γ is a finite set and D is a reflexive and symmetric dependence relation. A *trace over* (Γ, D) is a Γ-labeled partial order $(V, \leq, \lambda')$ such that

- $(\lambda'(x), \lambda'(y)) \notin D$ whenever $x, y \in V$ are incomparable, and
- $(\lambda'(x), \lambda'(y)) \in D$ whenever y is an upper neighbor of x (denoted $x \prec y$).

The set of all traces over (Γ, D) is denoted by $\mathbb{R}(\Gamma, D)$, the set $\mathbb{M}(\Gamma, D)$ comprises the finite traces.

The key observation that is announced by the title of this section is that any recognizable set of MSCs is the "relabeling" of a monadically axiomatizable set of traces over a suitable dependence alphabet. Recall that any recognizable set of MSCs is bounded. The bound B influences the chosen dependence alphabet as defined in the following paragraph.

For a positive integer $B \in \mathbb{N}$, let $\Gamma = \Sigma \times \{0, 1, \ldots, B - 1\}$. On this alphabet, we define a dependence relation D as follows: $(p_1\theta_1 q_1, n_1)$ and $(p_2\theta_2 q_2, n_2)$ are dependent iff

1. $p_1 = p_2$, or
2. $\{(p_1\theta_1 q_1, n_1), (p_2\theta_2 q_2, n_2)\} = \{(p!q, n), (q?p, n)\}$ for some $p, q \in \mathcal{P}$ and $0 \leq n < B$.

For $t = (V, \leq, \lambda) \in {\downarrow}\mathrm{MSC}_B^\infty$, we define a new Γ-labeling λ' by

$$\lambda'(v) = (\lambda(v), |{\downarrow}v \cap \lambda^{-1}\lambda(v)| \mod B),$$

i.e. the first component of the label is the old label and the second counts modulo B the number of occurrences of the same action in the past of v. We then define $\mathrm{tr}(t) = (V, \leq, \lambda')$. First, one shows that $\{\mathrm{tr}(t) \mid t \in {\downarrow}\mathrm{MSC}_B^\infty\}$ is a first-order axiomatizable set of traces in $\mathbb{R}(\Gamma, D)$:

Lemma 5.1. *The set* $\mathrm{tr}[{\downarrow}\mathrm{MSC}_B^\infty]$ *is the set of all traces* $s \in \mathbb{R}(\Gamma, D)$ *satisfying*

I. $\pi_{\{(p!q,n),(q?p,n)\}}(s)$ *is a prefix of* $((p!q, n)(q?p, n))^\omega$ *for* $p, q \in \mathcal{P}$ *and* $0 \leq n < B$.
II. $\pi_{\{(\sigma,n)\mid 1 \leq n < B\}}(s)$ *is a prefix of* $((\sigma, 1)(\sigma, 2) \ldots (\sigma, B - 1)(\sigma, 0))^\omega$ *for* $\sigma \in \Sigma$.
III. *If* $v, w \in V$ *with* $v \prec w$, *then* $\mathrm{proc}(\lambda'(v)) = \mathrm{proc}(\lambda'(w))$ *or* $\lambda'(v) = (p!q, n)$ *and* $\lambda'(w) = (q?p, n)$ *for some* $p, q \in \mathcal{P}$ *and* $0 \leq n < B$.

Since all these properties are first-order expressible and since one can interpret the MSC t in the trace $\mathrm{tr}(t)$, we obtain

Proposition 5.2. *Let* $K \subseteq \downarrow\mathrm{MSC}_B^\infty$ *by monadically axiomatizable relative to* $\downarrow\mathrm{MSC}_B^\infty$. *Then* $\mathrm{tr}[K] \subseteq \mathbb{R}(\Gamma, D)$ *is monadically axiomatizable relative to* $\mathbb{R}(\Gamma, D)$.

This proposition can be used for an alternative proof of parts of Theorem 4.6; it works perfectly well for recognizable languages, can in some cases be used for n-solvable languages (if B divides some power of n), and is of no use whatsoever for aperiodic languages (since the relabeling cannot be defined in first-order logic).

So far, we transformed any monadically axiomatizable set of bounded MSCs into a monadically axiomatizable set of traces. In order to make use of this transformation, we need the following definitions and results from the theory of Mazurkiewicz traces.

Let $s = (V, \leq, \lambda')$ be a trace over (Γ, D) and let $A \subseteq \Gamma$. Then $\partial_A(t)$ is the least ideal of t such that the complementary filter does not contain any A-labeled vertex. Let $\gamma \in \Gamma$. Then $D(\gamma) = \{\delta \in \Gamma \mid (\gamma, \delta) \in D\}$. Furthermore, $t\gamma$ is the unique trace $(V \dot\cup \{\star\}, \leq', \rho)$ with $t\gamma\!\restriction_V = t$, $\rho(\star) = \gamma$, and $\star \in \max(t\gamma)$. A mapping $\mu : \mathbb{M}(\Gamma, D) \to A$ is *asynchronous* if, for any $\Delta_1, \Delta_2 \subseteq \Gamma$, any $\gamma \in \Gamma$, and any $t \in \mathbb{M}(\Gamma, D)$,

1. $\mu(\partial_{\Delta_1 \cup \Delta_2}(t))$ is completely determined by $\mu(\partial_{\Delta_1}(t))$, $\mu(\partial_{\Delta_2}(t))$, and the sets Δ_1 and Δ_2, and
2. $\mu(\partial_\gamma(t\gamma)$ is completely determined by $\mu(\partial_{D(\gamma)}(t))$ and the letter γ.

Theorem 5.3 ([25,4]). *Let* (Γ, D) *be a dependence alphabet and* $L \subseteq \mathbb{M}(\Gamma, D)$. *Then* L *is monadically axiomatizable if, and only if, there exists an asynchronous mapping* μ *into some finite set such that* $L = \mu^{-1}\mu(L)$.

This result was used to construct a deterministic asynchronous cellular automaton that accepts a given recognizable language of finite traces. Diekert & Muscholl [6] use the same concept of an asynchronous mapping to construct a deterministic asynchronous cellular automaton with Muller acceptance condition that accepts a given recognizable set of infinite traces. In order to state their result, we need some more notations:

Let (Γ, D) be a dependence alphabet, $t = (V, \leq, \lambda') \in \mathbb{R}(\Gamma, D)$ a trace, and $\gamma \in \Gamma$. Let $\mu_\gamma^\infty(t) \subseteq A$ be the set of all $a \in A$ for which there are infinitely many nodes $v \in V$ with $\lambda'(v) = \gamma$ and $\mu(t\!\restriction_{\downarrow v}) = a$.

Theorem 5.4 ([6,9]). *Let* (Γ, D) *be a dependence alphabet and* $L \subseteq \mathbb{R}(\Gamma, D)$ *be monadically axiomatizable. Then there exists a finite set* A, *a set* $\mathcal{T} \subseteq \prod_{\gamma \in \Gamma} 2^A$ *of* Γ-*tuples of subsets of* A, *and an asynchronous mapping* $\mu : \mathbb{M}(\Gamma, D) \to A$ *such that for* $t \in \mathbb{R}(\Gamma, D)$, *we have:* $t \in L \iff (\mu_\gamma^\infty(t))_{\gamma \in \Gamma} \in \mathcal{T}$.

5.2   The Construction of Deterministic Message Passing Automata

A *message passing automaton with Muller-acceptance condition* is a structure $\mathcal{A} = ((\mathcal{A}_p)_{p \in \mathcal{P}}, \Delta, s_{in}, \mathcal{S})$ where

1. Δ is a finite set of messages,

2. each component $\mathcal{A}_p$ is of the form $(S_p, \to_p)$ where
   - S_p is a finite set of local states,
   - $\to_p \subseteq S_p \times \Sigma_p \times \Delta \times S_p$ where $\Sigma_p = \{\sigma \in \Sigma \mid \mathrm{proc}(\sigma) = p\}$ is a local transition relation,
3. $s^{in} \in \prod_{p \in \mathcal{P}} S_p$ is the global initial state, and
4. $\mathcal{S} \subseteq \prod_{p \in \mathcal{P}} 2^{S_p}$ is a Muller acceptance condition.

Let $(s, a, m, s') \in \to_p$ be a local transition of process p. Suppose a is a send event, i.e., $a = p!q$ for some process q. Then the transition (s, a, m, s') denotes that the process p can perform the action $a = p!q$ in state s; it changes its local state to s' and sends a message m into the FIFO-channel from process p to process q. By enlarging the set of messages and local states (if necessary), we can assume that $m = s'$ for any send action (in particular, $\Delta = \bigcup_{p \in \mathcal{P}} S_p$). Now suppose that $a = p?q$ is a receive action. Then the transition (s, a, m, s') denotes that the process p can change its local state from s to s' when reading the message m from the channel that connects p and q.

A message passing automaton is *deterministic* if

- $(s, p!q, m_1, s_1), (s, p!q, m_2, s_2) \in \to_p$ imply $s_1 = s_2$ and $m_1 = m_2$
- $(s, q?p, m_1, s_1), (s, q?p, m_1, s_2) \in \to_p$ imply $s_1 = s_2$.

The message passing automaton is *complete*, if

- there exists a transition $(s, p!q, m, s')$ for any $s \in S_p$ and $q \in \mathcal{P} \setminus \{p\}$
- there exists a transition $(s, q?p, m, s')$ for any $s \in S_p$, $m \in \Delta$, and $q \in \mathcal{P} \setminus \{p\}$

Let $t = (V, \leq, \lambda)$ be an ideal in an MSC and let $\mathcal{A}$ be a message passing automaton. Let furthermore $r : V \to \bigcup_{p \in \mathcal{P}} S_p$ be a mapping and $v \in V$. We define a second mapping $r^- : V \to \bigcup_{p \in \mathcal{P}} S_p$: if there is $u < v$ with $\mathrm{proc}(u) = \mathrm{proc}(v)$, let u be maximal with this property and let $r^-(v)$ denote $r(u)$. If v is the minimal event performed by the process $\mathrm{proc}(v)$, let $r^-(v) = s^{in}_{\mathrm{proc}(v)}$. Then $r^-(v)$ is the local state of process $\mathrm{proc}(v)$ *before* the execution of v; this process is in state $r(v)$ *after* performing v.

A *run* of $\mathcal{A}$ on t is a mapping $r : V \to \bigcup_{p \in \mathcal{P}} S_p$ satisfying for any $v \in V$:

1. If $\lambda(v) = p!q$, then there is a transition $(r^-(v), p!q, m, r(v))$ in $\to_p$ for some message m (which turns out to be $r(v)$ by our assumption).
2. Now let $\lambda(v) = p?q$. Since t is an ideal in an MSC, there is a unique matching node $u \in V$ with $u \sqsubseteq v$. We require that $(r^-(v), p?q, r(u), r(v)) \in \to_p$.

Let $r : V \to \bigcup_{p \in \mathcal{P}} S_p$ be a run of a Muller message passing automaton on $t = (V, \leq, \lambda) \in \downarrow\mathrm{MSC}^\infty$. For $p \in \mathcal{P}$, let $X_p \subseteq S_p$ be the set of all $s \in S_p$ such that, for any $v \in V$ with $\mathrm{proc}(v) = p$, there exists $w \in V$ with $v \leq w$, $\mathrm{proc}(w) = p$, and $r(w) = s$ (and $\{s^{in}_p\}$ if no such v exists). The run r is *successful* provided $(X_p)_{p \in \mathcal{P}} \in \mathcal{S}$. A set $K \subseteq \downarrow\mathrm{MSC}^\infty$ is *accepted by $\mathcal{A}$ relative to* $\mathfrak{X} \subseteq \downarrow\mathrm{MSC}^\infty$ if, for any $t \in \mathfrak{X}$, $t \in L$ iff t is accepted by $\mathcal{A}$.

Above, we associated to any monadically axiomatizable subset of $\downarrow\mathrm{MSC}_B^\infty$ a monadically axiomatizable set of traces. By Theorem 5.4, we therefore get an asynchronous mapping. Next, we construct a message passing automaton from an asynchronous mapping.

Proposition 5.5. *Let $\mu : \mathbb{M}(\Gamma, D) \to A$ be some asynchronous mapping into a finite set A. Then there exists a complete deterministic message passing automaton with Muller acceptance condition $\mathcal{A}$ with local state space S and a function $f : S \to A$ such that for the run r of $\mathcal{A}$ on $t = (V, \le, \lambda) \in \downarrow\mathrm{MSC}_B^\infty$, we have $f(r(v)) = \mu(\mathrm{tr}(\downarrow v))$.*

Now one can show that relative to $\downarrow\mathrm{MSC}_B^\infty$, message passing automata and monadic second order have the same expressive power:

Proposition 5.6. *A set $K \subseteq \downarrow\mathrm{MSC}_B$ can be accepted by the message passing automaton $\mathcal{A}$ relative to $\downarrow\mathrm{MSC}_B$ iff it is monadically axiomatizable relative to $\downarrow\mathrm{MSC}_B$.*

The construction of a formula from an MPA follows the wellknown pattern of [24,25, 8]. The other implication follows easily from Prop. 5.5 and 5.2 together with Theorem 5.4.

In order to extend the result to $\downarrow\mathrm{MSC}^\infty$, one first observes that $\downarrow\mathrm{MSC}_B$ can be accepted by a deterministic message passing automaton relative to $\downarrow\mathrm{MSC}_{B+1}$ since it is monadically axiomatizable relative to $\downarrow\mathrm{MSC}$. Note that an ideal in an MSC $t \in \downarrow\mathrm{MSC}^\infty$ is NOT B-bounded iff it contains a principal ideal that is not B-bounded. This allows to show that $\downarrow\mathrm{MSC}_B^\infty$ can be accepted by a complete deterministic message passing automaton with Muller acceptance condition. This is used in the proof of our main result:

Theorem 5.7. *Let $K \subseteq \downarrow\mathrm{MSC}^\infty$. Then K is recognizable iff it is bounded and there exists a Muller message passing automaton that accepts K.*

Proof. The set K is monadically axiomatizable and B-bounded for some B. Hence there exists a deterministic message passing automaton $\mathcal{A}_1$ that accepts K relative to $\downarrow\mathrm{MSC}_B^\infty$. So $K = L(\mathcal{A}_1) \cap \downarrow\mathrm{MSC}_B^\infty$ is the intersection of two sets that can be accepted by deterministic message passing automata.   □

One can prove the corresponding statement from [20] for finite MSCs accordingly. Our construction requires $m2^{O((n^2B)^2 \log(n^2B))}$ local states (where n is the number of processes $|\mathcal{P}|$, B is the bound on the MSCs in the language K, and m is the size of the syntactic monoid of $\mathrm{Lin}[K]$). Recall that the construction from [20] needed $2^{2^{O(n^2B)}}m \log m$ local states.

References

1. R. Alur and M. Yannakakis. Model checking of message sequence charts. In *CONCUR'99*, Lecture Notes in Comp. Science vol. 1664, pages 114–129. Springer, 1999.
2. A. Arnold. An extension of the notions of traces and of asynchronous automata. *Informatique Théorique et Applications*, 25:355–393, 1991.
3. J.R. Büchi. Weak second-order arithmetic and finite automata. *Z. Math. Logik Grundlagen Math.*, 6:66–92, 1960.
4. R. Cori, Y. Métivier, and W. Zielonka. Asynchronous mappings and asynchronous cellular automata. *Information and Computation*, 106:159–202, 1993.
5. V. Diekert and Y. Métivier. Partial commutation and traces. In G. Rozenberg and A. Salomaa, editors, *Handbook of Formal Languages Volume 3*, pages 457–533. Springer, 1997.

6. V. Diekert and A. Muscholl. Deterministic asynchronous automata for infinite traces. *Acta Informatica*, 31:379–397, 1994.

7. M. Droste and P. Gastin. Asynchronous cellular automata for pomsets without autoconcurrency. In *CONCUR'96*, Lecture Notes in Comp. Science vol. 1119, pages 627–638. Springer, 1996.

8. M. Droste, P. Gastin, and D. Kuske. Asynchronous cellular automata for pomsets. *Theoretical Comp. Science*, 247:1–38, 2000. (Fundamental study).

9. W. Ebinger and A. Muscholl. Logical definability on infinite traces. *Theoretical Comp. Science*, 154:67–84, 1996.

10. J.G. Henriksen, M. Mukund, K. Narayan Kumar, and P.S. Thiagarajan. Towards a theory of regular MSC languages. Technical report, BRICS RS-99-52, 1999. The results of this technical report appeared in the extended abstracts [12,11].

11. J.G. Henriksen, M. Mukund, K. Narayan Kumar, and P.S. Thiagarajan. On message sequence graphs and finitely generated regular MSC languages. In *ICALP'00*, pages 675–686. Springer, 2000.

12. J.G. Henriksen, M. Mukund, K. Narayan Kumar, and P.S. Thiagarajan. Regular collections of message sequence charts. In *MFCS 2000*, Lecture Notes in Computer Science vol. 1893. Springer, 2000.

13. W. Hodges. *Model Theory*. Cambridge University Press, 1993.

14. D. Kuske. Asynchronous cellular automata and asynchronous automata for pomsets. In *CONCUR'98*, Lecture Notes in Comp. Science vol. 1466, pages 517–532. Springer, 1998.

15. D. Kuske. Another step towards a theory for regular MSC languages. Technical Report 2001-36, Department of Mathematics and Computer Science, University of Leicester, 2001.

16. D. Kuske and R. Morin. Pomsets for local trace languages: Recognizability, logic and Petri nets. In *CONCUR 2000*, Lecture Notes in Comp. Science vol. 1877, pages 426–411. Springer, 2000.

17. R.E. Ladner. Application of model theoretic games to discrete linear orders and finite automata. *Information and Control*, 33:281–303, 1977.

18. R. McNaughton and S. Papert. *Counter-Free Automata*. MIT Press, Cambridge, USA, 1971.

19. R. Morin. On regular message sequence chart languages and relationships to Mazurkiewicz trace theory. In *FOSSACS01*, Lecture Notes in Comp. Science vol. 2030, pages 332–346. Springer, 2001.

20. M. Mukund, K. Narayan Kumar, and M. Sohoni. Synthesizing distributed finite-state systems from MSCs. In C. Palamidessi, editor, *CONCUR 2000*, Lecture Notes in Computer Science vol. 1877, pages 521–535. Springer, 2000.

21. A. Muscholl and D. Peled. Message sequence graphs and decision problems on Mazurkiewicz traces. In *MFCS'99*, Lecture Notes in Computer Science vol. 1672, pages 81–91. Springer, 1999.

22. H. Straubing, D. Thérien, and W. Thomas. Regular languages defined with generalized quantifiers. *Information and Computation*, 118:289–301, 1995.

23. P.S. Thiagarajan and I. Walukiewicz. An expressively complete linear time temporal logic for mazurkiewicz traces. In *LICS'97*, pages 183–194. IEEE Computer Society Press, 1997.

24. W. Thomas. Automata on infinite objects. In J. van Leeuwen, editor, *Handbook of Theoretical Computer Science*, pages 133–191. Elsevier Science Publ. B.V., 1990.

25. W. Thomas. On logical definability of trace languages. In V. Diekert, editor, *Proceedings of a workshop of the ESPRIT BRA No 3166: Algebraic and Syntactic Methods in Computer Science (ASMICS) 1989*, Report TUM-I9002, Technical University of Munich, pages 172–182, 1990.

Existential and Positive Theories of Equations in Graph Products

Volker Diekert and Markus Lohrey

Universität Stuttgart, Institut für Informatik
Breitwiesenstr. 20–22, 70565 Stuttgart, Germany
{diekert,lohrey}@informatik.uni-stuttgart.de

Abstract. We prove that the existential theory of equations with normalized rational constraints in a fixed graph product of finite monoids, free monoids, and free groups is PSPACE-complete. Under certain restrictions this result also holds if the graph product is part of the input. As the second main result we prove that the positive theory of equations with recognizable constraints in graph products of finite and free groups is decidable.

1 Introduction

Since the seminal work of Makanin [19] on equations in free monoids, the decidability of various theories of equations in different monoids and groups has been studied, and several new decidability and complexity results have been shown. Let us mention here the results of [25,27] for free monoids, [6,15,20,21] for free groups, [9] for free partially commutative monoids (trace monoids), [10] for free partially commutative groups (graph groups), and [7] for plain groups (free products of finite and free groups).

In this paper we continue this stream of research. We will present two main results. The first one concerns existential theories of equations. We start with the definition of a class of monoids, which are constructed from finite monoids, free monoids, and free groups using the graph product construction, which is a well-known construction in mathematics. This class of graph products strictly covers all classes mentioned above. Then we prove that for such a graph product the existential theory of equations is PSPACE-complete, where in addition we are allowed to specify constraints for the variables. These constraints are taken from a class of sets, called normalized rational sets, which (in general) lies strictly between the class of recognizable and rational sets. Furthermore under certain restrictions our PSPACE upper-bound holds also in the case that (a suitable description) of the graph product is part of the input.

Our second main result concerns positive theories of equations. We prove that if we restrict our class of graph products to groups, then for each group from the resulting class the positive theory of equations with recognizable constraints for the variables is decidable. Under certain restrictions we obtain an elementary complexity. Up to now only for the class of free groups a decidability result for

H. Alt and A. Ferreira (Eds.): STACS 2002, LNCS 2285, pp. 501–512, 2002.
© Springer-Verlag Berlin Heidelberg 2002

the positive theory was known, in particular it was open whether the positive theory of equations for a free partially commutative group is decidable. The full paper of this extended abstract can be found in [8].

2   Preliminaries

An *involution* on a set is a mapping $^-$ such that $\overline{\overline{x}} = x$ for all elements x. For an involution on a monoid we demand in addition that both $\overline{xy} = \overline{y}\,\overline{x}$ and $\overline{1} = 1$, where 1 is the neutral element of the monoid. Taking the inverse in a group is for instance an involution. In our setting we let Γ be a finite alphabet of constants and $\Delta \subseteq \Gamma$ such that an involution $^-$ is defined on Δ. This involution is extended to Δ^* by $\overline{x_1 \cdots x_n} = \overline{x}_n \cdots \overline{x}_1$. For a monoid M we denote by $\mathcal{I}(M)$ a submonoid of M such that an involution $^-$ is defined on $\mathcal{I}(M)$. In many cases we choose $\mathcal{I}(M)$ to be the submonoid of elements having left- and right-inverses, i.e., $\mathcal{I}(M)$ is the group of units of M, but this is not necessarily the case, for instance for $M = \Gamma^*$ we take $\mathcal{I}(M) = \Delta^*$. We consider only finitely generated monoids. More precisely, we consider monoids M together with a fixed surjective homomorphism $\psi : \Gamma^* \to M$ such that $\psi^{-1}(\mathcal{I}(M)) = \Delta^*$ and $\psi(\overline{x}) = \overline{\psi(x)}$ for all $x \in \Delta^*$. Moreover, we assume that there is a *normal form mapping* $\nu : M \to \Gamma^*$, i.e., $\psi(\nu(x)) = x$ for all $x \in M$, such that $\nu(M)$ is a regular subset of Γ^*. Note that it is allowed that $\nu(\overline{x}) \neq \overline{\nu(x)}$ for some $x \in M$. A language $L \subseteq M$ is called

- *recognizable* if $\psi^{-1}(L) \subseteq \Gamma^*$ is regular,
- *normalized rational* if $\nu(L) \subseteq \Gamma^*$ is regular,
- *rational* if $L = \psi(L')$ for some regular language $L' \subseteq \Gamma^*$.

The corresponding classes are denoted by $\mathrm{REC}(M)$, $\mathrm{NRAT}(M)$, and $\mathrm{RAT}(M)$, respectively. We have $\mathrm{REC}(M) \subseteq \mathrm{NRAT}(M) \subseteq \mathrm{RAT}(M)$. The classes $\mathrm{REC}(M)$ and $\mathrm{RAT}(M)$ are classical, see e.g. [4], their definitions do neither depend on ν nor on ψ as can be seen easily. The definition of $\mathrm{NRAT}(M)$ is less robust, it depends on the normal form mapping ν. The classes $\mathrm{REC}(M)$ and $\mathrm{NRAT}(M)$ are Boolean algebras, whereas $\mathrm{RAT}(M)$ is not a Boolean algebra in general. For free monoids we have $\mathrm{REC}(M) = \mathrm{NRAT}(M) = \mathrm{RAT}(M)$. For the canonical normal form mappings which we will use we have: $\mathrm{REC}(M) \neq \mathrm{NRAT}(M) = \mathrm{RAT}(M)$ for free groups [3], $\mathrm{REC}(M) = \mathrm{NRAT}(M) \neq \mathrm{RAT}(M)$ for free partially commutative monoids (trace monoids) [24], and $\mathrm{REC}(M) \neq \mathrm{NRAT}(M) \neq \mathrm{RAT}(M)$ for free partially commutative groups (graph groups). The later holds for instance in $M = \mathbb{Z} \times \mathbb{Z}$.

3   The Theory of Equations with Constraints

Let M be a monoid as above and let $\mathcal{C}$ be a family of subsets of M such that $\mathcal{I}(M) \in \mathcal{C}$. Let Ω be a set of variables and $\overline{\Omega} = \{\overline{X} \mid X \in \Omega\}$ a disjoint copy of Ω. An *equation* is a pair (U, V) with $U, V \in (\Gamma \cup \Omega \cup \overline{\Omega})^*$, it is written as $U = V$. Equations and *constraints* of the form $X \in L$ with $X \in \Omega \cup \overline{\Omega}$ and $L \in \mathcal{C}$

are called *atomic formulae*. From these we construct first order formulae using conjunctions, disjunctions, negations, and universal and existential quantification over variables from Ω. We impose the syntactical restriction that whenever we use a variable $\overline{X} \in \overline{\Omega}$, then this goes together with the implicit constraint $X \in \mathcal{I}(M)$. Given $\psi : \Gamma^* \to M$, $\mathcal{I}(M)$, the involution $^- : \mathcal{I}(M) \to \mathcal{I}(M)$, and a sentence ϕ, i.e., a formula in the sense above without free variables, we can evaluate ϕ over M in the obvious way with the restriction that if a variable X evaluates to $x \in M$, then $\overline{X}$ must evaluate to $\overline{x}$. The *theory of equations with constraints in* $\mathcal{C}$, briefly $\mathrm{Th}(M, \mathcal{C})$, denotes the set of all sentences that are true in M. A well-known example of a decidable theory of equations is the Presburger Arithmetic [26]. Translated into our framework this gives the following proposition.

Proposition 1. $\mathrm{Th}(\mathbb{N}^k, \mathrm{RAT}(\mathbb{N}^k))$ *and* $\mathrm{Th}(\mathbb{Z}^k, \mathrm{RAT}(\mathbb{Z}^k))$ *are decidable.*

Note that $\mathrm{RAT}(\mathbb{N}^k)$ and $\mathrm{RAT}(\mathbb{Z}^k)$ are the classes of semilinear sets in $\mathbb{N}^k$ and $\mathbb{Z}^k$, respectively. The following result can be easily deduced from Proposition 1 since the free product $\mathbb{Z}/2\mathbb{Z} * \mathbb{Z}/2\mathbb{Z}$ of two copies of $\mathbb{Z}/2\mathbb{Z}$ is isomorphic to the semi-direct product of $\mathbb{Z}$ by $\mathbb{Z}/2\mathbb{Z}$.

Corollary 1. $\mathrm{Th}(\mathbb{Z}/2\mathbb{Z} * \mathbb{Z}/2\mathbb{Z}, \mathrm{RAT}(\mathbb{Z}/2\mathbb{Z} * \mathbb{Z}/2\mathbb{Z}))$ *is decidable.*

The *positive theory of equations with constraints in* $\mathcal{C}$ is the set of all sentences in $\mathrm{Th}(M, \mathcal{C})$ that do not use negations. The *existential theory of equations with constraints in* $\mathcal{C}$ is the set of all sentences in $\mathrm{Th}(M, \mathcal{C})$ that are in prenex normal form without universal quantifiers. We will need the following result, which is a decomposition lemma in the style of the Feferman Vaught Theorem. Its proof in [8] is due to Yuri Matiyasevich (personal communication).

Proposition 2. *Let* M_1 *and* M_2 *be monoids with classes* $\mathcal{C}_1 \subseteq 2^{M_1}$ *and* $\mathcal{C}_2 \subseteq 2^{M_2}$. *Let* $\mathcal{C}$ *be a class of subsets of* $M_1 \times M_2$ *such that each* $L \in \mathcal{C}$ *is a finite union of sets of the form* $L_1 \times L_2$ *with* $L_1 \in \mathcal{C}_1$ *and* $L_2 \in \mathcal{C}_2$. *If both* $\mathrm{Th}(M_1, \mathcal{C}_1)$ *and* $\mathrm{Th}(M_2, \mathcal{C}_2)$ *are decidable, then* $\mathrm{Th}(M_1 \times M_2, \mathcal{C})$ *is decidable, too. The same implication also holds for positive theories.*

4   Graph Products

Let (V, E) be a finite undirected graph with vertex set V and edge set $E \subseteq \binom{V}{2}$. Every node $n \in V$ is labeled with a monoid M_n which is either a free monoid, a free group, or a finite monoid. In fact, it is enough (and convenient) to assume that M_n is either isomorphic to $\mathbb{N}$ or to $\mathbb{Z}$, or M_n is finite. If $M_n = \mathbb{N}$, then we let $\Gamma_n = \{a_n\}$ and $\Delta_n = \emptyset$. If $M_n = \mathbb{Z}$, then we let $\Gamma_n = \Delta_n = \{a_n, \overline{a}_n\}$. Finally if M_n is finite, then we let $\Gamma_n = M_n \backslash \{1\}$ and $\Delta_n = \mathcal{I}(M_n) \backslash \{1\}$, where $\mathcal{I}(M_n)$ is the subgroup of units of M_n, i.e., $\mathcal{I}(M_n) = \{a \in M_n \mid \exists b : ab = ba = 1\}$. Thus, for each $n \in V$ we have a canonical homomorphism $\psi_n : \Gamma_n^* \to M_n$ with $\psi_n^{-1}(\mathcal{I}(M_n)) = \Delta_n^*$. To see this note that if $uv \in \mathcal{I}(M_n)$ and if M_n is finite, then $u, v \in \mathcal{I}(M_n)$, too. The *graph product* defined by (V, E) is the free product of the monoids M_n, $n \in V$, modulo commutation relations $xy = yx$ for all $x \in M_m$, $y \in M_n$ with $(m, n) \notin E$. Graph products of arbitrary groups and monoids were investigated in [5,14]. Note that we have defined a commutation, if there is no edge, so an edge corresponds to a rigid ordering. The choice for this convention is due to the representation of elements which is best based on dependence graphs, see e.g. [11]. Before we make our definition more formal let us mention some examples.

If all M_n are equal to $\mathbb{N}$, then we obtain *free partially commutative monoids*, which are also known as *trace monoids*, see [11] for more details. Extreme cases are free monoids (if $E = \binom{V}{2}$) and free commutative monoids (if $E = \emptyset$). If all M_n are equal to $\mathbb{Z}$, we obtain *free partially commutative groups*, which are also known as *graph groups* [12]. Again free groups and free commutative groups arise as the extreme cases. If $E = \binom{V}{2}$ and all M_n are groups, then we obtain *plain groups* in the sense of Haring-Smith [16].

Let us proceed with an explicit definition of the graph product using generators and relations. First we may assume that all the alphabets Γ_n are pairwise disjoint. Let $\Gamma = \bigcup_{n \in V} \Gamma_n$ and $\Delta = \bigcup_{n \in V} \Delta_n$. There is a natural involution $^-$ on Δ and this involution has fixed points as soon as some M_n contains an element of order two. We define an *independence relation* $I \subseteq \Gamma \times \Gamma$ by $I = \{(a, b) \in \Gamma \times \Gamma \mid a \in \Gamma_m, b \in \Gamma_n, m \neq n, (m, n) \notin E\}$, which is irreflexive and symmetric. The basic reference monoid for the following consideration is the trace monoid $\mathbb{M} = \Gamma^* / \{ab = ba \mid (a, b) \in I\}$, it is equipped with a partially defined involution. More precisely, since I is compatible with the involution in the sense that $(a, \overline{b}) \in I$ if $(a, b) \in I$ and $b \in \Delta$, we can lift $^- : \Delta \to \Delta$ to an involution on the recognizable subset $\Delta^* = \mathcal{I}(\mathbb{M})$ of $\mathbb{M}$. We now define a *trace rewriting system* S, i.e., a subset of $\mathbb{M} \times \mathbb{M}$, by

$$S = \{(a\overline{a}, 1) \mid a \in \Delta\} \cup \{(ab, c) \mid \exists n \in V : a, b, c \in \Gamma_n, ab = c \text{ in } M_n\}.$$

The graph product $\mathbb{GP}$ of the monoids M_n, $n \in V$, over the graph (V, E) is defined as the quotient monoid $\mathbb{GP} = \mathbb{M} / \{\ell = r \mid (\ell, r) \in S\}$. Clearly $\mathbb{GP} = \Gamma^* / (\{ab = ba \mid (a, b) \in I\} \cup \{\ell = r \mid (\ell, r) \in S\})$. Elements of $\mathbb{GP}$ can be represented as words from Γ^* or as traces from $\mathbb{M}$. It will be always clear from the context, which representation is chosen. Furthermore the canonical homomorphism $\psi : \Gamma^* \to \mathbb{GP}$ factorizes as $\psi = \psi_1 \circ \psi_2$, where $\psi_1 : \Gamma^* \to \mathbb{M}$ and $\psi_2 : \mathbb{M} \to \mathbb{GP}$. Note that the trace monoid $\mathbb{M}$ itself is a graph product, where the vertex set is Γ and the edges are given by the complement of I. The example of a trace monoid shows that rational constraints are too strong in order to obtain decidability results. Since it is undecidable whether $L_1 \cap L_2 = \emptyset$ for $L_1, L_2 \in \text{RAT}(\mathbb{N} \times \{a, b\}^*)$, see [1], the following result holds:

Proposition 3. *Let* $\mathbb{M} = \mathbb{N} \times \{a, b\}^*$. *Then for* M *the existential positive theory of equations with constraints in* $\text{RAT}(M)$ *is undecidable.*

Thus, we have to restrict the class of constraints. We shall consider normalized rational constraints. In order to define a suitable normal form mapping $\nu : \mathbb{GP} \to \Gamma^*$ we define analogously to string rewriting systems the one-step rewrite relation $\rightarrow_S \subseteq \mathbb{M} \times \mathbb{M}$ of the trace rewriting system S by $s \rightarrow_S t$ if $s = u\,\ell\,v$ and $t = u\,r\,v$ for some $(\ell, r) \in S$ and $u, v \in \mathbb{M}$. Its transitive reflexive closure is $\xrightarrow{*}_S$. The following lemma is fundamental for the following.

Lemma 1. *S is a confluent trace rewriting system, i.e., for all $s, t, u \in \mathbb{M}$ with $s \xrightarrow{*}_S t$ and $s \xrightarrow{*}_S u$ there exists $v \in \mathbb{M}$ with $t \xrightarrow{*}_S v$ and $u \xrightarrow{*}_S v$.*

Let $\mathrm{RED}(S) = \{u\,\ell\,v \mid u, v \in \mathbb{M}, \exists r : (\ell, r) \in S\}$ and $\mathrm{IRR}(S) = \mathbb{M} \setminus \mathrm{RED}(S)$. Thus, $\mathrm{IRR}(S)$ is the set of traces that are irreducible with respect to S. Since $\mathrm{REC}(\mathbb{M})$ is closed under Boolean operations and concatenation, see e.g. [11, Chap. 6], $\mathrm{IRR}(S)$ is recognizable. Since $\rightarrow_S$ is a Noetherian relation, Lemma 1 implies that for each $x \in \mathbb{GP}$ there exists a unique $\mu(x) \in \mathbb{M} \cap \mathrm{IRR}(S)$ with $x = \psi_2(\mu(x))$. The trace $\mu(x)$ is the shortest trace representing x. Now let us fix a linear order on Γ and let $\mathrm{lnf}(t) \in \Gamma^*$ for $t \in \mathbb{M}$ be the lexicographical first word from Γ^* that represents the trace t, see also [2]. Then for $x \in \mathbb{GP}$ we define $\nu(x) = \mathrm{lnf}(\mu(x))$. Since $L \in \mathrm{REC}(\mathbb{M})$ if and only if $\mathrm{lnf}(L) \subseteq \Gamma^*$ is regular [24], we obtain:

Lemma 2. *We have $L \in \mathrm{NRAT}(\mathbb{GP})$ if and only if $\mu(L) \in \mathrm{REC}(\mathbb{M})$ if and only if $\psi_1^{-1}(\mu(L)) \in \mathrm{REC}(\Gamma^*)$.*

In particular we see that $\mathrm{NRAT}(\mathbb{GP})$ does not depend on the chosen lexicographical ordering. It is really a canonical class depending only on the natural trace rewriting system S.

5 Existential Theories of Equations in Graph Products

In this section we prove that for the graph product $\mathbb{GP}$ the existential theory of equations with constraints in $\mathrm{NRAT}(\mathbb{GP})$ is decidable. Since we will also deal with complexity issues, we have to define the input length of a formula. We assume some standard binary coding of formulae, where a constraint $X \in L$ is represented by some finite non-deterministic automaton that accepts $\psi_1^{-1}(\mu(L))$. The input length of a formula is the length of this description. In order to obtain existing results for free monoids as special cases, we will put a description of the graph product $\mathbb{GP}$ into the input, too. This description contains the adjacency matrix of (V, E), and for each node either the multiplication table of M_n if M_n is finite or a flag indicating whether $M_n = \mathbb{N}$ or $M_n = \mathbb{Z}$. In order to obtain convenient complexity bounds we will restrict to graphs (V, E) with a bounded number of *complete thin clans*, see [10] for the definition. It is easy to see that the number of complete thin clans of (V, E) is at most $|V|$, furthermore it is 0 for a complete graph.

Theorem 1. *The following problem is PSPACE-complete for every $k \geq 0$.*
    INPUT: A graph product $\mathbb{GP}$ whose underlying graph (V, E) has at most k complete thin clans and an existential formula ϕ with constraints in $\mathrm{NRAT}(\mathbb{GP})$.
    QUESTION: Does ϕ belong to $\mathrm{Th}(\mathbb{GP}, \mathrm{NRAT}(\mathbb{GP}))$?
If the number of complete thin clans of (V, E) is not bounded, then the problem above is in EXPSPACE.

Remark 1. Formally, Theorem 1 generalizes results of [6,7,9,10,15,19,20,25]. For this it is enough to give a reduction to the main result of [10].

The next lemma is the main technical tool for proving the theorem above. First we need some further definitions concerning traces. The set $\mathrm{IC} \subseteq \mathbb{M} \cap \mathrm{IRR}(S)$ consists of all traces $a_1 \cdots a_n$, $a_i \in \Gamma$, such that $(a_i, a_j) \in I$ if $i \neq j$. Thus, traces in IC correspond to independence cliques of (Γ, I). Note that if $u \in \mathrm{IC}$, then the length of u is at most $|\Gamma|$. We identify $u \in \mathrm{IC}$ with the set of symbols that occur in u. For instance for $s \in \mathbb{M}$ the set of maximal symbols $\max(s) = \{a \in \Gamma \mid s = ta\}$ of s and the set of minimal symbols $\min(s) = \{a \in \Gamma \mid s = at\}$ of s belong to IC.

Lemma 3. *Let* $x, y, z \in \mathbb{M} \cap \mathrm{IRR}(S)$. *Then* $xy \xrightarrow{*}_S z$ *if and only if there exist* $p, s, t, w \in \mathrm{IRR}(S)$ *and* $u, v \in \mathrm{IC}$ *such that*

$$u v \xrightarrow{*}_S w, \quad x = s u p, \quad y = \overline{p} v t, \quad z = s w t. \tag{1}$$

Note that since $u, v \in \mathrm{IC}$, there exist only finitely many possibilities for w in (1).

Proof (Theorem 1). PSPACE-hardness follows from the fact that for $\{a, b\}^*$ the existential theory of equations with constraints in $\mathrm{REC}(\{a, b\}^*)$ is PSPACE-hard, see [17, Lem. 3.2.3] and [25, Thm. 1]. Membership in PSPACE will be shown by a reduction to the following problem, which was shown to be in PSPACE for every $k \geq 0$ in [10]:

INPUT: A trace monoid $\mathbb{M}$, specified by an independence relation $I \subseteq \Gamma \times \Gamma$ such that the graph $(\Gamma, (\Gamma \times \Gamma) \backslash I)$ has at most k complete thin clans, a completely defined involution $^-: \Gamma \to \Gamma$ that is compatible with I (i.e. $(a, \overline{b}) \in I$ if $(a, b) \in I$), and an existential formula ϕ with constraints in $\mathrm{REC}(\mathbb{M})$.

QUESTION: Is ϕ true in $\mathbb{M}$ with the lifting $^-: \mathbb{M} \to \mathbb{M}$ of $^-: \Gamma \to \Gamma$?
In this problem a set $L \in \mathrm{REC}(\mathbb{M})$ is specified via an automaton for $\psi_1^{-1}(L)$.

Now let k be a fixed bound for the number of complete thin clans, and let $\mathbb{GP}$ be a graph product, specified by a graph (V, E) with at most k complete thin clans. Furthermore let ϕ be an existential formula with constraints in $\mathrm{NRAT}(\mathbb{GP})$. Using standard methods, see e.g. [6], we may assume that ϕ is an existentially quantified conjunction of equations of the form $xy = z$, where $x, y, z \in \Gamma \cup \Omega \cup \overline{\Omega}$, and of constraints $X \in L$ or $X \notin L$, where $X \in \Omega \cup \overline{\Omega}$ and $L \in \mathrm{NRAT}(\mathbb{GP})$. Next we will move from the graph product $\mathbb{GP}$ to its underlying trace monoid $\mathbb{M}$ (it is easy to see that the number of complete thin clans of $(\Gamma, (\Gamma \times \Gamma) \backslash I)$ is also at most k). We replace syntactically every subformula $xy = z$ (resp. $X \in L$) by $\psi_2(xy) = \psi_2(z)$ (resp. $X \in \mu(L)$) and add the negated constraint $X \notin \mathrm{RED}(S)$ for every variable X. [1] We obtain an existential formula which evaluates to *true* in $\mathbb{M}$ if and only if the original formula evaluates to *true* in $\mathbb{GP}$. Note also that the automaton used to specify $\mu(L)$ is the same as the

---

[1] Of course this constraint is equivalent to $X \in \mathrm{IRR}(S)$, but we prefer the negated constraint $X \notin \mathrm{RED}(S)$ since an automaton for $\psi_1^{-1}(\mathrm{RED}(S))$ can be easily constructed in polynomial time, whereas the construction of an automaton for $\psi_1^{-1}(\mathrm{IRR}(S))$ would involve an additional complementation with a possible exponential blow-up.

one for L. It remains to eliminate all occurrences of ψ_2 from equations. Since $\Gamma \subseteq \mathrm{IRR}(S)$ and S is confluent, we can replace an equation $\psi_2(xy) = \psi_2(z)$ by $xy \xrightarrow{*}_S z$, which by Lemma 3 is equivalent to an existentially quantified conjunction of equations.

Now we can almost apply the result of [10] cited above. The only remaining problem is that due to the presence of non-invertible generators in $\mathbb{GP}$, the involution $^-$ may only be partially defined on Γ. But this can be resolved by introducing a new dummy symbol $\bar{a}$ for every $a \in \Gamma \backslash \Delta$ and by adding the constraint $X \in \Gamma^*$ for every variable X. This shows the first statement from Theorem 1.

For the case that the number of complete thin clans is not bounded, an EXPSPACE-algorithm can be deduced from the proof in [10].   □

6   Positive Theories of Equations in Graph Products

In this section we prove our second main result. In the following we throughout assume that all generators in Γ have inverses, i.e, $\Gamma = \Delta$. In particular $\mathbb{GP}$ is a graph product of finite and free groups, and hence itself a group.

Theorem 2. *The following problem is decidable.*

INPUT: A graph product $\mathbb{GP}$ which is a group and a closed positive formula ϕ with constraints in $\mathrm{REC}(\mathbb{GP})$.

QUESTION: Does ϕ belong to $\mathrm{Th}(\mathbb{GP}, \mathrm{REC}(\mathbb{GP}))$?

Complexity issues will be postponed to the end of this section. Note that Theorem 2 cannot be extended to the full class of graph products considered in the previous section. Already for a free monoid $\{a, b\}^*$ the $\forall\exists^3$-theory of equations is undecidable [13,22]. Similarly Theorem 2 cannot be extended to the case of normalized rational constraint, since for a free group F of rank 2 a free submonoid $\{a, b\}^*$ belongs to $\mathrm{NRAT}(F)$.

We will prove Theorem 2 by reducing the positive theory of equations with constraints in $\mathrm{REC}(\mathbb{GP})$ to the existential theory of equations with normalized rational constraints in a free extension of $\mathbb{GP}$, which allows us to apply Theorem 1. Our proof strategy will follow a technique developed in [21,23] but the presence of partial commutation and recognizable constraints makes the construction more involved.

In a first step we may assume that none of the finite groups M_n, $n \in V$, is a direct product of two finite non-trivial groups since otherwise we could replace n by two non-connected nodes. In particular, if M_n is not $\mathbb{Z}/2\mathbb{Z}$, then there must exist an $a \in \Gamma_n$ such that $a \neq \bar{a}$ in $\mathbb{GP}$. Next assume that the graph (V, E) consists of two non-empty disjoint components (V_1, E_1) and (V_2, E_2), which define graph products $\mathbb{GP}_1$ and $\mathbb{GP}_2$, respectively. Then $\mathbb{GP} = \mathbb{GP}_1 \times \mathbb{GP}_2$. Furthermore by Mezei's Theorem, see e.g. [4], every $L \in \mathrm{REC}(\mathbb{GP})$ is a finite union of sets of the form $L_1 \times L_2$ with $L_i \in \mathrm{REC}(\mathbb{GP}_i)$. Thus, we may apply Proposition 2 and proceed with the two graphs (V_1, E_1) and (V_2, E_2). Hence, for the rest of the proof we may assume that the graph (V, E) is connected. Furthermore since by Proposition 1 the (positive) theory of equations with rational constraints in $\mathbb{Z}$ is decidable and the same holds for finite monoids for trivial reasons, we may assume that $|V| > 1$. By Corollary 1 we can also exclude the case that V contains exactly two adjacent nodes which are both labeled by $\mathbb{Z}/2\mathbb{Z}$. Thus, we may assume that either the graph (V, E) contains a path consisting of three different nodes or one of the groups labeling the nodes has a generator $x \in \Gamma$ with $\overline{x} \neq x$. Hence, there exist three generators $a, b, c, \in \Gamma$ such that a and b belong to E-adjacent (and hence different) nodes from V, b and c also belong to E-adjacent nodes from V, and finally either a and c belong to different nodes from V or $a \neq \overline{a} = c$. In particular $(a, b), (b, c) \notin I$, i.e., the dependency between a, b, and c being used is $a \,\text{---}\, b \,\text{---}\, c$. For the rest of the proof we will fix these three symbols a, b, and c.

Since $L \in \text{REC}(\mathbb{GP})$ if and only if there exists a homomorphism $\rho : \mathbb{GP} \to H$ onto a finite group H such that $L = \rho^{-1}(\rho(L))$, see e.g. [4], we may fix for the further consideration such a homomorphism ρ and assume that all recognizable constraints are given in the form $\rho(X) = g$ for $X \in \Omega \cup \overline{\Omega}$ and $g \in H$.

We proceed with the definition of a trace rewriting system $R_N^{(h)}$, where $N \subseteq \mathbb{N}$ and $h \in H$. This trace rewriting system will be defined over some free extension of $\mathbb{M}$. First we need some preliminaries. A *chain* is a trace $a_1 \cdots a_m \in \mathbb{M}$, where $a_1, \ldots, a_m \in \Gamma$, and a_i and a_{i+1} belong to E-adjacent (and hence different) nodes from V, $1 \leq i \leq n - 1$. Note that a chain belongs to $\text{IRR}(S)$.

Lemma 4. *For all $h \in H$ there exists a trace $C_h \in \mathbb{M} \cap \text{IRR}(S)$ such that $\min(C_h) = \max(C_h) = c$ and $\rho(C_h) = h$.*

Let C be a chain with $\min(C) = \max(C) = c$ and $|C| > |C_h|$ for all $h \in H$ such that for every node $n \in V$ at least one symbol from Γ_n occurs in C. Since (V, E) is connected, such a C exists. Choose an η with $|b(ab)^\eta| > |C| + 2$, and let $p = b(ab)^\eta C(ba)^\eta b$ and $\ell_i(h) = (ab)^{i \cdot |H|} C_h (ba)^{2 \cdot i \cdot |H|}$ for $i \geq 1$ and $h \in H$. Note that $p\, \ell_i(h)\, p \in \text{IRR}(S)$ and $\rho(\ell_i(h)) = h$. For every $i \in \mathbb{N}$ let us take two new constants $k_i, \overline{k_i} \notin \Gamma$ and set $\overline{\overline{k_i}} = k_i$. For every $N \subseteq \mathbb{N}$ and every $h \in H$ we define over the trace monoid $\mathbb{M} * \{k_i, \overline{k_i} \mid i \in N\}^*$, i.e., the free product of our trace monoid $\mathbb{M}$ and the free monoid $\{k_i, \overline{k_i} \mid i \in N\}^*$, the trace rewriting system $R_N^{(h)}$ by $R_N^{(h)} = \{(p\, \ell_i(h)\, p,\ p\, k_i\, p), (\overline{p}\, \overline{\ell_i(h)}\, \overline{p},\ \overline{p}\, \overline{k_i}\, \overline{p}) \mid i \in N\}$. Note that $R_N^{(h)}$ is length-reducing and thus, $\to_{R_N^{(h)}}$ is Noetherian. Let us fix $h \in H$ for the rest of this section We write R_N and ℓ_i instead of $R_N^{(h)}$ and $\ell_i(h)$, respectively. We write $s \to_i t$ if the trace t can be obtained from the trace s by an application of one of the rules $(p\, \ell_i\, p,\ p\, k_i\, p)$ or $(\overline{p}\, \overline{\ell_i}\, \overline{p},\ \overline{p}\, \overline{k_i}\, \overline{p})$. The next two lemmas are the fundamental statements about the trace rewriting system R_N and the reason for the complicated definition of the traces p and $\ell_i(h)$.

Lemma 5. *Let $i, j \in N \subseteq \mathbb{N}$ and $s, t, u \in \mathbb{M} * \{k_i, \overline{k_i} \mid i \in N\}^*$ such that $s \to_i t$ and $s \to_j u$. Then either $t = u$ or there exists a trace $v \in \mathbb{M} * \{k_i, \overline{k_i} \mid i \in N\}^*$ such that $t \to_j v$ and $u \to_i v$.*

In particular, R_N is confluent. Since R_N is also Noetherian, for every $s \in \mathbb{M}$ there exists a unique trace $\kappa_N(s) \in \mathbb{M} * \{k_i, \overline{k_i} \mid i \in N\} \cap \text{IRR}(R_N)$ with $s \xrightarrow{*}_{R_N} \kappa_N(s)$.

Lemma 6. *For all* $s, t \in \mathbb{M}$ *there exists an* $A \subseteq N$ *with* $|A| \leq 2$ *such that for every* $N' \subseteq N \backslash A$ *it holds* $\kappa_{N'}(st) = \kappa_{N'}(s)\kappa_{N'}(t)$.

6.1 Reduction to the Existential Theory

In the following, symbols with a tilde like $\tilde{x}$ will denote sequences of arbitrary length over some set, which will be always clear form the context. If say $\tilde{x} = x_1 \cdots x_i$, then $\tilde{x} \in A$ means $x_1 \in A, \ldots, x_i \in A$ and $f(\tilde{x})$ for some function f denotes the sequence $f(x_1) \cdots f(x_i)$.

For the rest of the paper let us take some subset $K = \{k_1, \ldots, k_n\}$ of our new constants and let $\overline{K} = \{\overline{k}_1, \ldots, \overline{k}_n\}$. Let $k, \overline{k} \notin \Gamma \cup K \cup \overline{K}$ be two additional constants, as usual let $\overline{\overline{k}} = k$. The following lemma will be the key for reducing the positive theory to the existential theory, it allows the elimination of one universal quantifier. In this lemma we have to deal with formulae ϕ that are interpreted over the free product $\mathbb{GP} * F(K)$ of the graph product $\mathbb{GP}$ and the free group $F(K)$ generated (as a group) by K. Furthermore different recognizable constraints in ϕ are given by different extensions $\varrho : \mathbb{GP} * F(K) \to H$ of our fixed morphism $\rho : \mathbb{GP} \to H$. For $h \in H$ we denote by ϕ_h the formula that results from ϕ by replacing every constraint $\varrho(X) = g$ by $\varrho_h(X) = g$, where ϱ_h is the canonical extension of $\varrho : \mathbb{GP} * F(K) \to H$ to $\mathbb{GP} * F(K \cup \{k\})$ which is defined by $\varrho_h(k) = h$. Note that $\psi_2 : \mathbb{M} \to \mathbb{GP}$ can be extended to a canonical morphism from $\mathbb{M} * (K \cup \overline{K})^*$ to $\mathbb{GP} * F(K)$, which will be also denoted by ψ_2.

Lemma 7. *Let* $\phi(X, Y_1, \ldots, Y_m, \tilde{Z})$ *be a positive Boolean formula with constraints of the form* $\varrho(Y) = g$ *for (possibly different) extensions* $\varrho : \mathbb{GP} * F(K) \to H$ *of* $\rho : \mathbb{GP} \to H$. *Let* $K_i \subseteq K$ *for* $1 \leq i \leq m$. *Then for all* $\tilde{z} \in \mathbb{GP}$ *we have*

$$\forall X \in \mathbb{GP} \; \exists Y_1, \ldots, Y_m \left\{ \begin{array}{l} \phi(X, Y_1, \ldots, Y_m, \tilde{z}) \wedge \\ \bigwedge_{i=1}^{m} Y_i \in \mathbb{GP} * F(K_i) \end{array} \right\} \quad \text{in } \mathbb{GP} * F(K) \qquad (2)$$

if and only if

$$\bigwedge_{h \in H} \exists Y_1, \ldots, Y_m \left\{ \begin{array}{l} \phi_h(k, Y_1, \ldots, Y_m, \tilde{z}) \wedge \\ \bigwedge_{i=1}^{m} Y_i \in \mathbb{GP} * F(K_i \cup \{k\}) \end{array} \right\} \quad \text{in } \mathbb{GP} * F(K \cup \{k\}). \qquad (3)$$

Proof. First assume that (3) holds for $\tilde{z} \in \mathbb{GP}$. In order to prove (2), let us choose an arbitrary $s \in \mathbb{GP}$ and let $h = \rho(s)$. Then there exist $t_i \in \mathbb{GP} * F(K_i \cup \{k\})$, $1 \leq i \leq m$, such that $\phi_h(k, t_1, \ldots, t_m, \tilde{z})$ holds in $\mathbb{GP} * F(K \cup \{k\})$. Let us define a homomorphism $\sigma : \mathbb{GP} * F(K \cup \{k\}) \to \mathbb{GP} * F(K)$ by $\sigma(k) = s$ and $\sigma(x) = x$ for $x \in \mathbb{GP} * F(K)$. Since $\rho(s) = h$ and ϕ_h is positive, the sentence $\phi(s, \sigma(t_1), \ldots, \sigma(t_m), \tilde{z})$ holds in $\mathbb{GP} * F(K)$ (note that $\sigma(\tilde{z}) = \tilde{z}$). Thus, (2) holds.

For the other direction assume that (2) holds for $\tilde{z} \in \mathbb{GP}$. Define a trace rewriting system T over $\mathbb{M} * (K \cup \overline{K})^*$ by $T = S \cup \{x\overline{x} \to 1, \overline{x}x \to 1 \mid x \in K\}$.

Completely analogously to the proof of Theorem 1 we can now change into the trace monoid $\mathbb{M} * (K \cup \overline{K})^*$. We obtain a sentence of the form

$$\forall X \in \text{IRR}(S) \, \exists Y_1, \ldots, Y_m, \tilde{Y} \in \text{IRR}(T) \left\{ \begin{array}{c} \varphi(X, Y_1, \ldots, Y_m, \tilde{Y}, \tilde{u}) \wedge \\ \bigwedge_{i=1}^{m} Y_i \in \mathbb{M} * (K_i \cup \overline{K_i})^* \end{array} \right\} \quad (4)$$

which evaluates to true in $\mathbb{M} * (K \cup \overline{K})^*$. Here $\tilde{u} = \mu(\tilde{z}) \in \text{IRR}(S)$, and the positive Boolean formula φ results from the original positive Boolean formula ϕ by applications of Lemma 3 to equations $xy = z$. These transformations only introduce new existentially quantified variables, which correspond to $\tilde{Y}$ in (4). The constraints in (4) are the same as in (2) (formally we identify a homomorphism $\varrho : \mathbb{GP} * F(K) \to H$ with $\psi_2 \circ \varrho : \mathbb{M} * (K \cup \overline{K})^* \to H$). Let $\mathcal{M} \subseteq \mathbb{M}$ consist all traces in $\tilde{u}$ plus Γ. W.l.o.g we assume that all equations in (4) have the form $xy = z$ for $x, y, z \in \Omega \cup \overline{\Omega} \cup \mathcal{M} \cup \overline{\mathcal{M}}$. Let λ be the maximum of n (the largest index of the constants in K) and the maximal length of the traces in $\tilde{u}$. Let d be the number of equations in (4). Fix an $h \in H$ in (3) and let $s \in \mathbb{M}$ be the trace

$$s = C_g \, p \, \ell_{\lambda+1}(h) \, p \, c \, p \, \ell_{\lambda+2}(h) \, p \, c \cdots p \, \ell_{\lambda+2d+1}(h) \, p \in \text{IRR}(S), \quad (5)$$

where $g \in H$ is chosen such that $\rho(s) = h$. Then by (4) there exist traces $t_1, \ldots, t_m, \tilde{t} \in \text{IRR}(T)$ with $t_i \in \mathbb{M} * (K_i \cup \overline{K_i})^*$ and

$$\varphi(s, t_1, \ldots, t_m, \tilde{t}, \tilde{u}) \quad \text{in } \mathbb{M} * (K \cup \overline{K})^*. \quad (6)$$

Let $N = \{\lambda+1, \ldots, \lambda+2d+1\}$ and add to $\mathcal{M}$ all traces from $\{s, t_1, \ldots, t_m\}$. Then $\varphi(s, t_1, \ldots, t_m, \tilde{t}, \tilde{u})$ is a true statement, which contains d atomic statements of the form $xy = z$ with $x, y, z \in \mathcal{M} \cup \overline{\mathcal{M}}$ plus recognizable constraints. Of course some of these atomic statements may be false. But since there are only d equations in (6), we have to remove from N by Lemma 6 at most $2d$ numbers such that for the resulting set N' we have $\kappa_{N'}(x)\kappa_{N'}(y) = \kappa_{N'}(z)$ $(x, y, z \in \mathcal{M} \cup \overline{\mathcal{M}})$ whenever $xy = z$ is a true atomic statement in (6). Since $|N| = 2d + 1$, we have $N' \neq \emptyset$, let $i \in N'$. Note that $k_i \notin K$ since $\lambda \geq n$. We rename the constant k_i into k and abbreviate $\kappa_{\{i\}}(x)$ by $\kappa(x)$. Again by Lemma 6 we have $\kappa(x)\kappa(y) = \kappa(z)$ for every true statement $xy = z$ $(x, y, z \in \mathcal{M} \cup \overline{\mathcal{M}})$ in (6). Furthermore if one of the constraints $\varrho(x) = g$ in (6) is true, where ϱ is an extension of ρ, then also $\varrho_h(\kappa(x)) = g$ holds (note that $\varrho(\ell_i(h)) = \rho(\ell_i(h)) = h = \varrho_h(k)$). Finally $\kappa(\tilde{u}) = \tilde{u}$ since λ was chosen big enough in (5). Altogether it follows that the statement $\varphi_h(\kappa(s), \kappa(t_1), \ldots, \kappa(t_m), \kappa(\tilde{t}), \tilde{u})$ is true in $\mathbb{M} * (K \cup \overline{K} \cup \{k, \overline{k}\})^*$. Next we can write $\kappa(s) = s_1 k s_2$ for $s_1, s_2 \in \mathbb{M}$. Let us define a homomorphism $\sigma : \mathbb{M} * (K \cup \overline{K} \cup \{k, \overline{k}\})^* \to \mathbb{M} * (K \cup \overline{K} \cup \{k, \overline{k}\})^*$ by $\sigma(k) = \overline{s}_1 k \overline{s}_2$, $\sigma(\overline{k}) = s_2 \overline{k} s_1$, and $\sigma(x) = x$ otherwise. Note that $\rho(s_1) h \rho(s_2) = \rho(s) = h$ and hence $\varrho_h(\overline{s}_1 k \overline{s}_2) = \rho(s_1)^{-1} h \rho(s_2)^{-1} = h$ for every extension ϱ of ρ. Thus, the statement $\varphi_h(\sigma(\kappa(s)), \sigma(\kappa(t_1)), \ldots, \sigma(\kappa(t_m)), \sigma(\kappa(\tilde{t})), \tilde{u})$ is true in $\mathbb{M} * (K \cup \overline{K} \cup \{k, \overline{k}\})^*$, hence it is also true in $\mathbb{GP} * F(K \cup \{k\})$. But in this group $\sigma(\kappa(s)) = \sigma(s_1 k s_2) =$

$s_1 \bar{s}_1 k \bar{s}_2 s_2 = k$. Since furthermore $\sigma(\kappa(t_i)) \in \mathbb{M} * (K_i \cup \overline{K_i} \cup \{k, \bar{k}\})^*$, the sentence

$$\exists Y_1, \dots, Y_m, \tilde{Y} : \varphi_h(k, Y_1, \dots, Y_m, \tilde{Y}, \tilde{z}) \wedge \bigwedge_{i=1}^{m} Y_i \in \mathbb{GP} * F(K_i \cup \{k\}) \text{ is true}$$

in $\mathbb{GP} * F(K \cup \{k\})$ for every $h \in H$. But then also (3) holds, since if (1) from Lemma 3 holds in $\mathbb{GP} * F(K \cup \{k\})$, then also $xy = z$ in $\mathbb{GP} * F(K \cup \{k\})$.   □

Let us fix a formula $\theta(\tilde{Z}) \equiv \forall X_1 \exists Y_1 \cdots \forall X_n \exists Y_n \, \phi(X_1, \dots, X_n, Y_1, \dots, Y_n, \tilde{Z})$, where ϕ is a positive Boolean formula with constraints of the form $\rho(X) = g$. For $h_1, \dots, h_n \in H$ we denote by $\rho_{h_1, \dots, h_n} : \mathbb{GP} * F(K) \to H$ the canonical extension of ρ with $\rho_{h_1, \dots, h_n}(k_i) = h_i$ for $1 \le i \le n$. With $\phi_{h_1, \dots, h_n}$ we denote the formula, where every constraint $\rho(X) = g$ in ϕ is replaced by $\rho_{h_1, \dots, h_n}(X) = g$. The following theorem is the main result of this section, it can be easily deduced from Lemma 7 by an induction on n.

Theorem 3. *For all $\tilde{z} \in \mathbb{GP}$ we have $\theta(\tilde{z})$ in $\mathbb{GP}$ if and only if*

$$\bigwedge_{h_1 \in H} \exists Y_1 \cdots \bigwedge_{h_n \in H} \exists Y_n \left\{ \begin{array}{l} \phi_{h_1, \dots, h_n}(k_1, \dots, k_n, Y_1, \dots, Y_n, \tilde{z}) \\ \wedge \bigwedge_{i=1}^{n} Y_i \in \mathbb{GP} * F(\{k_1, \dots, k_i\}) \end{array} \right\} \text{ in } \mathbb{GP} * F(K).$$

Since $\mathbb{GP} * F(\{k_1, \dots, k_i\}) \in \mathrm{NRAT}(\mathbb{GP} * F(K))$, Theorem 2 is a consequence of Theorem 1 and Theorem 3. Concerning the complexity, it can be shown that in general our proof of Theorem 2 gives us a non-elementary algorithm due to the construction in our proof of Proposition 2, see [8]. We obtain an elementary algorithm if we restrict to connected graphs (V, E). For this we have to use the fact that Presburger arithmetic (without negations), which occurs for $\mathbb{GP} = \mathbb{Z}/2\mathbb{Z}$ or $\mathbb{GP} = \mathbb{Z}/2\mathbb{Z} * \mathbb{Z}/2\mathbb{Z}$ as a special case, is elementary. More precise complexity bounds will be given in the full version of this paper.

References

1. IJ. J. Aalbersberg and H. J. Hoogeboom. Characterizations of the decidability of some problems for regular trace languages. *Mathematical Systems Theory*, 22:1–19, 1989.

2. A. V. Anisimov and D. E. Knuth. Inhomogeneous sorting. *International Journal of Computer and Information Sciences*, 8:255–260, 1979.

3. M. Benois. Parties rationnelles du groupe libre. *C. R. Acad. Sci. Paris*, Sér. A, 269:1188–1190, 1969.

4. J. Berstel. *Transductions and context–free languages*. Teubner Studienbücher, Stuttgart, 1979.

5. A. V. da Costa. Graph products of monoids. *Semigroup Forum*, 63(2):247–277, 2001.

6. V. Diekert, C. Gutiérrez, and C. Hagenah. The existential theory of equations with rational constraints in free groups is PSPACE-complete. In *Proceedings of the 18th Annual Symposium on Theoretical Aspects of Computer Science (STACS 01)*, number 2010 in Lecture Notes in Computer Science, pages 170–182. Springer, 2001.

7. V. Diekert and M. Lohrey. A note on the existential theory of equations in plain groups. *International Journal of Algebra and Computation*, 2001. to apear.

8. V. Diekert and M. Lohrey. Existential and positive theories of equations in graph products. Technical Report 2001/10, University of Stuttgart, Germany, 2001.

9. V. Diekert, Y. Matiyasevich, and A. Muscholl. Solving word equations modulo partial commutations. *Theoretical Computer Science*, 224(1–2):215–235, 1999.

10. V. Diekert and A. Muscholl. Solvability of equations in free partially commutative groups is decidable. In *Proceedings of the 28th International Colloquium on Automata, Languages and Programming (ICALP 01)*, number 2076 in Lecture Notes in Computer Science, pages 543–554. Springer, 2001.

11. V. Diekert and G. Rozenberg, editors. *The Book of Traces*. World Scientific, 1995.

12. C. Droms. Graph groups, coherence and three-manifolds. *Journal of Algebra*, 106(2):484–489, 1985.

13. V. G. Durnev. Undecidability of the positive $\forall\exists^3$-theory of a free semi-group. *Sibirsky Matematicheskie Jurnal*, 36(5):1067–1080, 1995.

14. E. R. Green. *Graph Products of Groups*. PhD thesis, The University of Leeds, 1990.

15. C. Gutiérrez. Satisfiability of equations in free groups is in PSPACE. In *32nd Annual ACM Symposium on Theory of Computing (STOC'2000)*, pages 21–27. ACM Press, 2000.

16. R. H. Haring-Smith. Groups and simple languages. *Transactions of the American Mathematical Society*, 279:337–356, 1983.

17. D. Kozen. Lower bounds for natural proof systems. In *Proceedings of the 18th Annual Symposium on Foundations of Computer Science, (FOCS 77)*, pages 254–266. IEEE Computer Society Press, 1977.

18. M. Lohrey. Confluence problems for trace rewriting systems. *Information and Computation*, 170:1–25, 2001.

19. G. S. Makanin. The problem of solvability of equations in a free semigroup. *Math. Sbornik*, 103:147–236, 1977. (Russian); English translation in Math. USSR Sbornik 32 (1977).

20. G. S. Makanin. Equations in a free group. *Izv. Akad. Nauk SSR*, Ser. Math. 46:1199–1273, 1983. (Russian); English translation in Math. USSR Izv. 21 (1983).

21. G. S. Makanin. Decidability of the universal and positive theories of a free group. *Izv. Akad. Nauk SSSR*, Ser. Mat. 48:735–749, 1984. (Russian); English translation in: *Math. USSR Izvestija, 25*, 75–88, 1985.

22. S. S. Marchenkov. Unsolvability of positive $\forall\exists$-theory of a free semi-group. *Sibirsky Matematicheskie Jurnal*, 23(1):196–198, 1982.

23. Y. I. Merzlyakov. Positive formulas on free groups. *Algebra i Logika Sem.*, 5(4):25–42, 1966. (Russian).

24. E. Ochmański. Regular behaviour of concurrent systems. *Bulletin of the European Association for Theoretical Computer Science (EATCS)*, 27:56–67, 1985.

25. W. Plandowski. Satisfiability of word equations with constants is in PSPACE. In *Proceedings of the 40th Annual Symposium on Foundations of Computer Science (FOCS 99)*, pages 495–500. IEEE Computer Society Press, 1999.

26. M. Presburger. Über die Vollständigkeit eines gewissen Systems der Arithmetik ganzer Zahlen, in welchem die Addition als einzige Operation hervortritt. In *Comptes Rendus du Premier Congrès des Mathématiciones des Pays Slaves*, pages 92–101, 395, Warsaw, 1927.

27. K. U. Schulz. Makanin's algorithm for word equations — Two improvements and a generalization. In *Word Equations and Related Topics*, number 572 in Lecture Notes in Computer Science, pages 85–150. Springer, 1991.

The Membership Problem for Regular Expressions with Intersection Is Complete in LOGCFL

Holger Petersen

University of Stuttgart
Institute of Computer Science
Breitwiesenstr. 20–22
D-70565 Stuttgart
`petersen@informatik.uni-stuttgart.de`

Abstract. We show that the recognition problem of context-free languages can be reduced to membership in the language defined by a regular expression with intersection by a log space reduction with linear output length. We also show a matching upper bound improving the known fact that the membership problem for these regular expressions is in NC^2. Together these results establish that the membership problem is complete in LOGCFL. For unary expressions we show hardness for the class NL and some related results.

1 Introduction

Regular expressions based on the operators + (union), · (concatenation), and *
(Kleene-closure) are useful for describing patterns in strings. In order to facilitate
the definition of patterns, regular expressions have been generalized in various
ways, see [1]. One approach is to add operators such as ∩ (intersection), ¬ (complement), or ^{2} (squaring, E^2 is equivalent to $E \cdot E$) to the basic set of operators.
Note that while these are closure properties of the regular sets, an equivalent
expression might be significantly shorter using the additional operators.

As an example consider the problem of testing that a text contains a non-overlapping repetition of a string of length n, i.e., it has the form $xwywz$ for
$w, x, y, z \in \{0, 1\}^*$ and $|w| = n$. Specifying this co-finite regular language requires
a regular expression of size $\Omega(2^n)$, while using the operation ∩ size $O(n^2)$ is easily
achieved.

The following sets of operations give rise to specific variants of regular expressions, which have been investigated in the literature:

semi-extended regular expressions with +, ·, *, and ∩,
star-free regular expressions with +, ·, and ¬,
extended regular expressions with +, ·, *, and ¬.

Extended and semi-extended regular expressions characterize exactly the regular sets, while star-free expressions are weaker (they cannot express $(00)^*$).

H. Alt and A. Ferreira (Eds.): STACS 2002, LNCS 2285, pp. 513–522, 2002.
© Springer-Verlag Berlin Heidelberg 2002

The membership problem for regular expressions is complete in NL, see [11] or [5, Remark on p. 178]. On a RAM it can be solved in $O(n^2)$ time [21]. For extended regular expressions including the squaring operator the membership problem is in P [18]. Hopcroft and Ullman described an algorithm for extended regular expressions running in time $O(n^4)$ [8, Exercise 3.23] and Hirst improved this to $O(n^3)$ [7]. The technique of Myers [15] speeds up this algorithm and the one for regular expressions by a factor logarithmic in the length of the expression. An automata-based $O(n^3)$ solution for semi-extended regular expressions has been found by Yamamoto [23]. In [16] it was established that the membership problem for star-free expressions (and hence extended regular expressions) is complete in P. The proof relied on complementation, and it remained open whether the membership problem for semi-extended regular expressions was easier. For the latter problem Jiang and Ravikumar stated an NC^2 upper bound [11].

Here we solve this problem by showing that the recognition problem for context-free languages can be reduced to the membership problem for semi-extended regular expressions. The reduction can be done in log space and is such that the size of the expression is proportional to the length of the input string.

A random access machine can carry out the reduction in linear time. Therefore it will be difficult to significantly improve the time efficiency of deciding membership in semi-extended regular expressions (say to $O(n^2)$), since that would imply a corresponding improvement of the recognition time for context-free languages, which is widely believed to be impossible.

Complementing the lower bound, we also describe an algorithm showing that the membership problem for semi-extended expressions is in LOGCFL, a significant improvement over the NC^2 bound from [11]. Therefore the problem is complete in LOGCFL, and it becomes one of only a few natural problems complete in this class. Other problems complete in LOGCFL are Greibach's hardest context-free language [6], the monotone circuit value problem for circuits of degree at most n [3,12], and the uniform membership problem for nondeterministic tree automata [14].

For unary semi-extended regular expressions (the alphabet is a singleton set) we can prove the membership problem to be hard for NL.

We note in passing that inequivalence of semi-extended regular expressions is complete in EXPSPACE [9] (Hunt's space lower bound $c^{\sqrt{n/\log n}}$, where $c > 1$, has been improved to c^n by Robson [17] and Fürer [4]). Non-emptiness is complete in PSPACE [9] (the original $\Omega(\sqrt{n})$ space bound for nondeterministic computations has been improved to $\Omega(n)$ [4]). Completeness in PSPACE even holds if intersections are not nested within other operations, i.e., for the non-emptiness problem of intersections of regular expressions [13, Proof of Lemma 3.2.3]. If in the latter variant of the problem the number of intersections is constant (but greater than one), the problem becomes complete in NL.

2   Preliminaries

We use standard notation from computational complexity, see [12]. Hardness and completeness are defined with respect to deterministic log space reductions. The class NL consists of languages recognizable in log space by a nondeterministic Turing machine, LOGCFL is the class of problems log space reducible to context-free languages.

Often the operation · and parentheses will be omitted from expressions. The empty word is denoted by ε, $|w|$ is the length of string w, log denotes the base 2 logarithm. The language described by expression α is called $L(\alpha)$.

We now recall some known facts that will be useful in the next section. The following observation was made by Stockmeyer and Meyer [18, p. 7].

Proposition 1. *The language described by a unary star-free regular expression of length n is a finite set of words or the complement of a finite set of words of length at most n.*

For expressions without complementation a stronger statement than Proposition 1 holds, see also [9, proof of Theorem 2.3].

Lemma 1. *The language described by an expression using the operations $+$, $\cdot$, and $\cap$ is finite. The length of each word in the language is bounded by the number of alphabet symbols occurring in the expression.*

Proof. We show this lemma by induction on the number of operations in an expression using $+$, $\cdot$, and $\cap$. It clearly holds for expressions without operations. Consider an expression α of the form $\alpha_1 + \alpha_2$ or $\alpha_1 \cap \alpha_2$. By assumption the lemma holds for α_1 and α_2. Since each word of maximum length described by α is described by at least one of the subexpressions, the lemma holds for α. If α has the form $\alpha_1 \cdot \alpha_2$, then one of the subexpressions describes the empty set (and the second part of the lemma holds vacuously), or words of maximum length are composed of words with the same property in the languages described by α_1 and α_2. Since their lengths are bounded by the numbers of alphabet symbols of the subexpressions, their concatenation is bounded by the sum.   □

Finally we notice that the squaring operator can be used as a substitute for $*$ in hardness proofs of membership problems.

Lemma 2. *The membership problem for expressions over a set of operations $\Omega \subseteq \{+, \cdot, *, \cap, \neg, {}^2\}$ can be reduced within log space by a deterministic Turing machine to the membership problem over $(\Omega \setminus \{*\}) \cup \{+, {}^2\}$.*

Proof. A substring of w can be described by an expression α^* if and only if it is described by

$$(\cdots((\alpha + \varepsilon)\overbrace{{}^2)\cdots)^2}^{\lceil \log(|w|+1)\rceil} .$$

Therefore each subexpression α^* can be replaced with such a sequence if membership of w is to be tested.   □

3   Results

We first present the upper bound, which improves the NC^2 bound reported by Jiang and Ravikumar [11].

Lemma 3. *The membership problem for semi-extended regular expressions (including [2]) is in LOGCFL.*

Proof. We will describe a polynomial time algorithm deciding the problem which runs on a nondeterministic auxiliary pushdown acceptor with a log space worktape (NauxPDA). Automata of this kind running in polynomial time characterize LOGCFL [19].

Let the input of length $n > 0$ consist of a non-empty expression α and a word w to be tested for membership. Clearly it can be checked in polynomial time that the expression is well-formed.

In the following we first describe how an expression is parsed by an NauxPDA. Then we will show that the time complexity is polynomial.

In every stage of its computation the NauxPDA has a sequence of quadruples of binary numbers on its pushdown which specify the positions and lengths of a subexpression of α (later referred to as α') and a subword of w (later called w'). Initially there is only one quadruple describing α and w on the pushdown. We will outline below how the automaton processes the top-most quadruple describing α' and w'. Notice that the automaton can copy the binary numbers of the description onto its worktape, since they are of logarithmic size. The automaton accepts if all quadruples can be eliminated.

If the α' is a constant (an alphabet symbol, ε, or $\emptyset$), then the NauxPDA checks that the symbol matches the substring w', or that w' is empty if the constant is ε. If so, it removes the quadruple. Otherwise it rejects. Now consider $\alpha' = \alpha_1 \cdot \alpha_2$. The NauxPDA guesses a partitioning of w' into $w' = w_1 w_2$ and replaces the old quadruple by two describing α_1, w_1 and α_2, w_2.

If $\alpha' = \alpha_1 + \alpha_2$, then the NauxPDA guesses whether to replace α' with α_1 or α_2.

If $\alpha' = \alpha_1 \cap \alpha_2$, then the NauxPDA replaces the old quadruple with two describing α_1, w' and α_2, w'.

For $\alpha' = \alpha_1^*$ the automaton first checks whether w' is empty. If so, it deletes the quadruple. Otherwise it guesses a partitioning $w' = w_1 w_2 \cdots w_k$ with all w_i non empty and replaces the old quadruple with k quadruples describing α_1 and w_i for $1 \leq i \leq k$.

Finally, if $\alpha' = \alpha_1^2$ the automaton guesses a partitioning $w' = w_1 w_2$ and generates two quadruples describing α_1 and w_1, w_2 replacing the old one.

We will now show by induction on the structure of semi-extended regular expressions that testing membership of some substring w' in $L(\alpha')$ can be done in time polynomial in the size of the input.

Let $t(\alpha', w')$ denote the time complexity of testing membership of non-empty substring w' in the language $L(\alpha')$ (we simplify the analysis by omitting parentheses from our considerations and discussing empty strings later). We claim that

$$t(\alpha', w') \leq c|\alpha'||w'|n,$$

where c is a sufficiently large constant. This will show an $O(n^3)$ upper bound on the time complexity of the algorithm.

By a suitable choice of c we can certainly have $t(x, w') \leq cn \leq c|x||w'|n$ for w' nonempty and x an alphabet symbol, ε, or $\emptyset$. We also require that cn exceeds the time complexity of the operations transforming quadruples as outlined above.

For $\alpha' = \alpha_1 \oplus \alpha_2$ with $\oplus \in \{+, \cap\}$ we have:

$$\begin{aligned}
t(\alpha', w') &= t(\alpha_1 \oplus \alpha_2, w') \\
&\leq cn + t(\alpha_1, w') + t(\alpha_2, w') \\
&\leq cn + c|\alpha_1||w'|n + c|\alpha_2||w'|n \\
&\leq c(|\alpha_1| + |\alpha_2| + 1)|w'|n \\
&= c|\alpha'||w'|n
\end{aligned}$$

For $\alpha' = \alpha_1^*$ we obtain:

$$\begin{aligned}
t(\alpha', w') &= t(\alpha_1^*, w_1 w_2 \cdots w_k) \\
&\leq cn + \sum_{i=1}^{k} t(\alpha_1, w_i) \\
&\leq cn + \sum_{i=1}^{k} c|\alpha_1||w_i|n \\
&\leq c(|\alpha_1| + 1) \sum_{i=1}^{k} |w_i|n \\
&= c|\alpha'||w'|n
\end{aligned}$$

For $\alpha' = \alpha_1^2$ we get:

$$\begin{aligned}
t(\alpha', w') &= t(\alpha_1^2, w_1 w_2) \\
&\leq cn + t(\alpha_1, w_1) + t(\alpha_1, w_2) \\
&\leq c(|\alpha_1| + 1)(|w_1| + |w_2|)n \\
&= c|\alpha'||w'|n
\end{aligned}$$

Finally we get for $\alpha' = \alpha_1 \cdot \alpha_2$:

$$\begin{aligned}
t(\alpha', w') &= t(\alpha_1 \cdot \alpha_2, w_1 w_2) \\
&\leq cn + t(\alpha_1, w_1) + t(\alpha_2, w_2) \\
&\leq cn + c|\alpha_1||w'|n + c|\alpha_2||w'|n \\
&\leq c(|\alpha_1| + |\alpha_2| + 1)|w'|n \\
&= c|\alpha'||w'|n
\end{aligned}$$

In the latter analysis we assume that if either of the w_i is empty, membership of ε in $L(\alpha_i)$ is tested in $c|\alpha_i|n$ time, which is accounted for by taking the factor

$|w'| > 0$ instead of $|w_i|$. The analysis of this test is trivial for $*$ and analogous to the one above for $+$, $\cap$, ^{2}, and $\cdot$. This case also applies if the initial string is empty.   □

Next we show the corresponding hardness result.

Lemma 4. *The membership problem for semi-extended regular expressions is hard for LOGCFL. The reduction has linear output length.*

Proof. We will describe a log space reduction with linear output length from Greibach's hardest context-free language L_0 [6], which is complete in LOGCFL, to the membership problem. The intuition behind the construction is to describe this language by a semi-extended regular expression. Since languages defined by expressions of this kind are regular and L_0 is of course non-regular, we cannot hope to achieve this goal. Rather we approximate a sufficiently large subset of L_0 by a concise semi-extended regular expression.

Let $\Sigma_1 = \{a_1, a_2, \bar{a}_1, \bar{a}_2\}$, $\Sigma_2 = \Sigma_1 \cup \{c, \phi\}$, and $\Sigma_3 = \Sigma_2 \cup \{d\}$. Recall that

$$L_0 = \{\varepsilon\} \cup \{x_1 c y_1 c z_1 d \cdots d x_n c y_n c z_n d \mid n \geq 1, y_1 \cdots y_n \in \phi D,$$
$$x_i, z_i \in \Sigma_2^* \text{ for all } i, \quad y_i \in \Sigma_1^* \text{ for } i \geq 2\},$$

where D denotes the Dyck language over two types of parentheses generated by the context-free grammar $S \to SS$, $S \to a_1 S \bar{a}_1$, $S \to a_2 S \bar{a}_2$, $S \to \varepsilon$. Intuitively the language L_0 encodes a sequence of blocks separated by d's such that from each block a string can be chosen in such a way that the concatenation of all choices is a member of the Dyck language preceded by ϕ.

We will now define sequences of expressions E_m, F_m.

$$E_{-1} = \emptyset$$
$$F_m = (((a_1 + a_2)E_{m-1}(\bar{a}_1 + \bar{a}_2)) \cap (a_1 \Sigma_3^* \bar{a}_1 + a_2 \Sigma_3^* \bar{a}_2))$$
$$E_m = (c\Sigma_2^* d\Sigma_2^* c)^* (F_m(c\Sigma_2^* d\Sigma_2^* c)^*)^* \qquad (\text{for } m \geq 0).$$

Here Σ_i is used as a shorthand for an expression describing Σ_i, e.g. the expression $(a_1 + a_2 + \bar{a}_1 + \bar{a}_2)$ for Σ_1.

We say that a string $y \in D$ has *level* m if its maximum level of nested parentheses of any kind is m. With each string $w \in \Sigma_3^*$ we associate as its *depth* the maximum level over all $y = y_1 \cdots y_n \in D$, where all factorizations $w = y_1 c z_1 d \cdots d x_n c y_n$ with $x_i, z_i \in \Sigma_2^*$ are considered. If no such factorization exists, w's depth is undefined. We claim that all strings in Σ_3^* of depth at most m are described by E_m. This is certainly true for E_{-1}, since there are no strings of depth -1. Any string over Σ_3 of depth $m \geq 0$ can be split into an alternating sequence of (possibly empty) blocks of depth 0 and of depth at most m, where the latter start and end with parentheses symbols and each substring obtained by deleting the first and last parenthesis has at most depth $m - 1$. By induction each of these substrings is described by E_{m-1} and the construction of F_m adds the pair of missing parentheses. By iterating blocks we obtain the entire string of depth m. Conversely, the construction ensures the existence of a y as defined above. Therefore $L(\Sigma_2^* c\cent E_m c\Sigma_2^* d) \subset L_0$ for each $m \geq -1$. We note that the intersection operator is essential for checking that parentheses at corresponding levels match.

The length of E_m is $O(m)$ and furthermore E_m can be generated by a deterministic Turing machine in $O(\log m)$ space. The reduction from L_0 now maps any string w to the pair $(w, \Sigma_2^* c\cent E_{|w|} c\Sigma_2^* d)$. The depth of w is bounded by $|w|$, and therefore $w \in L_0$ if and only if $w \in L(\Sigma_2^* c\cent E_{|w|} c\Sigma_2^* d)$ by the above discussion. □

Combining Lemmas 3 and 4 we obtain:

Theorem 1. *The membership problem for semi-extended regular expressions is complete in LOGCFL.*

With the help of Lemma 2 we get:

Theorem 2. *The membership problem for expressions using the operations $+$, $\cdot$, $\cap$, and ^{2} is complete in LOGCFL.*

Next we turn to expressions over a single letter alphabet.

Theorem 3. *The membership problem for semi-extended regular expressions over a unary alphabet using the operations $+$, $\cdot$, and $\cap$ is hard for NL.*

Proof. We reduce the order preserving graph accessibility problem (GAP_o) to the membership problem mentioned in the theorem. This variant of the graph accessibility problem asks whether for a given directed graph there exists a path from vertex 1 to vertex n that visits vertices in an increasing fashion. It is known that GAP_o is complete in NL, see [22, Theorem 8.43.3].

The idea of the construction is that strings of even length encode reachable vertices of a graph G. We start with $E_0 = 0^2$ and in stage $i \geq 1$ add string 0^{2i+2} if and only if vertex $i + 1$ is reachable by a path as required by GAP_o. We use D_i as an abbreviation for an expression defining the set $\{0^{2m} \mid 1 \leq m \leq i\}$. Let E_{i-1} be constructed. If $k_1, \ldots, k_m$ are the direct predecessors of vertex i in G, then we form

$$E_i = (((E_{i-1}(0^{2(i-k_1)-1} + \cdots + 0^{2(i-k_m)-1} + \varepsilon)) \cap (D_i + 0^{2i-1}))(0 + \varepsilon)) \cap D_i$$

(this is essentially the "OR" portion of the construction from [16]).

Vertex n is reachable if and only if 0^{2n} is in the language described by E_{n-1}. □

Corollary 1. *The nonemptiness and the inequivalence problem for expressions over a unary alphabet using the operations $+$, $\cdot$, and $\cap$ are hard for NL.*

In the case of unary expressions with operations $+$, $\cdot$, and $\cap$ we don't have a matching upper bound. Using the LOGCFL upper bound of Lemma 3 for the membership problem of semi-extended regular expressions we obtain the following upper bound.

Theorem 4. *The inequivalence problem for expressions over a unary alphabet using the operations* $+$, $\cdot$, *and* $\cap$ *is in LOGCFL.*

Proof (sketch). By Lemma 1 it suffices to guess a string w of a length bounded by the input size that is described by exactly one of the given expressions. Since the alphabet is unary, w can be stored in a compressed way by keeping the length of this string on the tape in binary. The membership problem is solved by a polynomial time auxiliary pushdown automaton (the algorithm has to be adapted to the compressed format). By the complement closure of LOGCFL [2], non-membership can be tested in the same way. A positive answer is returned if and only if the string is described by one expression and not by the other.   □

If we reduce the set of operators further, we obtain an inequivalence problem of very moderate complexity.

Theorem 5. *The inequivalence problem for expressions over a unary alphabet using the operations* $+$ *and* $\cdot$ *is in NL.*

Proof (sketch). As in the proof of Theorem 4 a string described by exactly one of the given expressions is guessed and tested. Now non-membership is checked by making use of the complement closure of NL [10,20].   □

Notice that in the last two proofs it was essential that the expressions were unary, since otherwise it would not be possible to store a compressed representation of a string in logarithmic space.

4   Open Problems

Our construction links context-free language recognition to the membership problem of semi-extended regular expressions. This suggests that techniques from context-free parsing might lead to an improvement of the known upper time bound $O(n^3/\log n)$ ([7] combined with [15]).

A problem that deserves further investigation is the membership problem for unary regular expressions (using the standard set of operations). As pointed out by Jiang and Ravikumar [11], this problem is at least as hard as the knapsack problem in unary encoding. There are also several gaps between upper and lower bounds for inequivalence problems of unary expressions.

Acknowledgements. Thanks are due to Markus Lohrey, Holger Austinat, and Klaus Wich for helpful discussions and comments on an early draft of this paper. Suggestions by one of the referees are gratefully acknowledged.

This work has been supported by the Hungarian-German research project No. D 39/2000 of the Hungarian Ministry of Education and the German BMBF.

References

1. A. V. Aho. Algorithms for finding patterns in strings. In J. van Leeuwen, editor, *Handbook of Theoretical Computer Science: Volume A, Algorithms and Complexity*, pages 255–300. MIT Press, Cambridge, MA, 1990.

2. A. Borodin, S. A. Cook, P. W. Dymond, W. L. Ruzzo, and M. Tompa. Two applications of inductive counting for complementation problems. *SIAM Journal on Computing*, 18:559–578, 1989.

3. S. A. Cook. A taxonomy of problems with fast parallel algorithms. *Information and Control*, 64:2–22, 1985.

4. M. Fürer. The complexity of the inequivalence problem for regular expressions with intersection. In J. W. D. Bakker and J. van Leeuwen, editors, *Proceedings of the 7th International Colloquium on Automata, Languages and Programming (ICALP'80), Noordwijkerhout (Netherlands)*, number 85 in Lecture Notes in Computer Science, pages 234–245, Berlin-Heidelberg-New York, 1980. Springer.

5. R. Greenlaw, H. J. Hoover, and W. L. Ruzzo. *Limits to Parallel Computation: P-Completeness Theory*. Oxford University Press, New York, Oxford, 1995.

6. S. A. Greibach. The hardest context-free language. *SIAM Journal on Computing*, 2:304–310, 1973.

7. S. C. Hirst. A new algorithm solving membership of extended regular expressions. Report 354, Basser Department of Computer Science, The University of Sydney, 1989.

8. J. E. Hopcroft and J. D. Ullman. *Introduction to Automata Theory, Languages, and Computation*. Addison-Wesley, Reading, Mass., 1979.

9. H. B. Hunt III. The equivalence problem for regular expressions with intersection is not polynomial in tape. Report TR 73-161, Department of Computer Science, Cornell University, 1973.

10. N. Immerman. Nondeterministic space is closed under complement. *SIAM Journal on Computing*, 17:935–938, 1988.

11. T. Jiang and B. Ravikumar. A note on the space complexity of some decision problems for finite automata. *Information Processing Letters*, 40:25–31, 1991.

12. D. S. Johnson. A catalog of complexity classes. In J. van Leeuwen, editor, *Handbook of Theoretical Computer Science: Volume A, Algorithms and Complexity*, pages 67–161. MIT Press, Cambridge, MA, 1990.

13. D. Kozen. Lower bounds for natural proof systems. In *Proceedings of the 18th Annual IEEE Symposium on Foundations of Computer Science (FOCS'77), Providence (Rhode Island)*, pages 254–266. IEEE Computer Society Press, 1977.

14. M. Lohrey. On the parallel complexity of tree automata. In A. Middeldorp, editor, *Proceedings of the 12th International Conference on Rewrite Techniques and Applications (RTA 2001), Utrecht (Netherlands)*, number 2051 in Lecture Notes in Computer Science, pages 201–215. Springer, 2001.

15. G. Myers. A four Russians algorithm for regular expression pattern matching. *Journal of the Association for Computing Machinery*, 39:430–448, 1992.

16. H. Petersen. Decision problems for generalized regular expressions. In *Proceedings of the 2nd International Workshop on Descriptional Complexity of Automata, Grammars and Related Structures, London (Ontario)*, pages 22–29, 2000.

17. J. M. Robson. The emptiness of complement problem for semi extended regular expressions requires c^n space. *Information Processing Letters*, 9:220–222, 1979.

18. L. J. Stockmeyer and A. R. Meyer. Word problems requiring exponential time. In *Proceedings of the 5th ACM Symposium on Theory of Computing (STOC'73), Austin (Texas)*, pages 1–9, 1973.

19. I. H. Sudborough. On the tape complexity of deterministic context-free languages. *Journal of the Association for Computing Machinery*, 25:405–414, 1978.

20. R. Szelepcsényi. The method of forced enumeration for nondeterministic automata. *Acta Informatica*, 26:279–284, 1988.

21. K. Thompson. Regular expression search algorithm. *Communications of the Association for Computing Machinery*, 11:419–422, 1968.
22. K. Wagner and G. Wechsung. *Computational Complexity*. Mathematics and its Applications. D. Reidel Publishing Company, Dordrecht, 1986.
23. H. Yamamoto. An automata-based recognition algorithm for semi-extended regular expressions. In M. Nielsen and B. Rovan, editors, *Proceedings of the 25th Symposium on Mathematical Foundations of Computer Science (MFCS 2000), Bratislava (Slovakia)*, number 1893 in Lecture Notes in Computer Science, pages 699–708. Springer, 2000.

Recognizable Sets of Message Sequence Charts[*]

Rémi MORIN

Laboratoire d'Informatique Fondamentale de Marseille
Université de Provence, 39 rue F. Joliot-Curie, F-13453 Marseille cedex 13, France

Abstract. High-level Message Sequence Charts are a well-established formalism to specify scenarios of communications in telecommunication protocols. In order to deal with possibly unbounded specifications, we focus on star-connected HMSCs. We relate this subclass with recognizability and MSO-definability by means of a new connection with Mazurkiewicz traces. Our main result is that we can check effectively whether a star-connected HMSC is realizable by a finite system of communicating automata with possibly unbounded channels.

Message Sequence Charts (MSCs) are a popular model often used for the documentation of telecommunication protocols. They profit by a standardized visual and textual presentation (ITU-T recommendation Z.120 [11]) and are related to other formalisms such as sequence diagrams of UML. An MSC gives a graphical description of communications between processes. It usually abstracts away from the values of variables and the actual contents of messages. However, this formalism can be used at a very early stage of design to detect errors in the specification [10]. In this direction, several studies have already brought up methods and complexity results for the model checking of MSCs viewed as a specification language [2,3,12,17,18]. However, many undecidable problems arose by algebraic reductions to formal language theory [5] or relationships to Mazurkiewicz trace theory [18,8].

Recently, several studies have investigated the subclass of *regular* languages of MSCs, a notion first formalized in [8]. These languages of MSCs are such that the set of associated sequential executions can be described by a finite automaton; therefore model checking becomes decidable and particular complexity results could be obtained [2,3,18]. However regular languages of MSCs satisfy the *channel-bounded property*, that is, the number of messages stored in channels at any stage of any execution admits a finite upper bound. In this paper, we aim at removing this condition because it leads to rather strong restrictions on the associated high-level MSCs (HMSCs) [15].

Following the classical algebraic approach, we introduce *recognizable languages of MSCs*. In many ways, they appear as regular languages without any restriction on the channel capacities. Formally, we will show that a recognizable HMSC is regular if, and only if, it is channel-bounded. Whereas regular languages are precisely those which are channel-bounded and MSO-definable [9], we will also prove that recognizability is equivalent to MSO-definability.

---

[*] Supported by the INRIA cooperative research action FISC.

H. Alt and A. Ferreira (Eds.): STACS 2002, LNCS 2285, pp. 523–534, 2002.
© Springer-Verlag Berlin Heidelberg 2002

Popular, graphical, and powerful, MSCs are intuitive and easy to use. However they may lead to specifications that do not correspond to the operational behavior of any system of processes. The important question whether a HMSC admits a realization has been investigated in different ways [1,2,5,9,16]. We observe first that any realizable HMSC is recognizable. However, we prove that one cannot decide whether a HMSC is recognizable. That is why we focus on c-HMSCs: The latter have the property that iterations (or cycles) are allowed over connected MSCs only. We show then that c-HMSCs can simulate any recognizable HMSC. This justifies our main result: *We can effectively check whether a c-HMSC is realizable.* Finally, we stress that all our results rely on a new connection between MSCs and Mazurkiewicz traces [6], which allows us to apply strong theorems from concurrency theory [14,19,20,21,22].

1   Basic Notions and Results

A *pomset* over an alphabet Σ is a triple $t = (E, \preccurlyeq, \xi)$ where $(E, \preccurlyeq)$ is a finite partial order and ξ is a mapping from E to Σ *without autoconcurrency*: $\xi(x) = \xi(y)$ implies $x \preccurlyeq y$ or $y \preccurlyeq x$ for all $x, y \in E$. Let $t = (E, \preccurlyeq, \xi)$ be a pomset and $x, y \in E$. Then y *covers* x (denoted $x \rightarrowtriangle y$) if $x \prec y$ and $x \prec z \preccurlyeq y$ implies $y = z$. The elements x and y are *concurrent* or *incomparable* if $\neg(x \preccurlyeq y) \land \neg(y \preccurlyeq x)$. A pomset can be seen as an abstraction of an execution of a concurrent system. In this view, the elements e of E are *events* and their label $\xi(e)$ describes the basic action of the system that is performed by the event. Furthermore, the order describes the causal dependence between the events. An *order extension* of a pomset $t = (E, \preccurlyeq, \xi)$ is a pomset $t' = (E, \preccurlyeq', \xi)$ such that $\preccurlyeq \subseteq \preccurlyeq'$. A *linear extension* of t is an order extension that is linearly ordered. Linear extensions of a pomset $t = (E, \preccurlyeq, \xi)$ can naturally be regarded as words over Σ. By $\mathrm{LE}(t) \subseteq \Sigma^*$, we denote the set of linear extensions of a pomset t over Σ. An *ideal* of a pomset $t = (E, \preccurlyeq, \xi)$ is a subset $H \subseteq E$ such that $x \in H \land y \preccurlyeq x \Rightarrow y \in H$. The restriction $t' = (H, \preccurlyeq \cap (H \times H), \xi \cap (H \times \Sigma))$ is called a *prefix* of t and we write $t' \leqslant t$. For all $z \in E$, we denote by $\downarrow z$ the ideal of events below z, i.e. $\downarrow z = \{y \in E \mid y \preccurlyeq z\}$. If H is a subset of E, we denote by $\#^a(H)$ the number of events $x \in H$ such that $\xi(x) = a$.

Basic Message Sequence Charts. MSCs are defined by several recommendations that indicate how one should represent them graphically [11]. More formally, they can be seen as particular labeled partial orders. Let $\mathcal{I}$ be a finite set of processes (also called *instances*) and *Mes* be a finite set of messages. For any instance $i \in \mathcal{I}$, Σ_i^{int} denotes a finite set of *internal actions*; the alphabet Σ_i is then the disjoint union of the set of *send actions* $\Sigma_i^! = \{i!^m j \mid j \in \mathcal{I} \setminus \{i\}, m \in Mes\}$, the set of *receive actions* $\Sigma_i^? = \{i?^m j \mid j \in \mathcal{I} \setminus \{i\}, m \in Mes\}$ and the set of internal actions Σ_i^{int}. We shall assume that the alphabets Σ_i are disjoint and we let $\Sigma_{\mathcal{I}} = \bigcup_{i \in \mathcal{I}} \Sigma_i$. Given an action $a \in \Sigma_{\mathcal{I}}$, we denote by $\mathrm{Ins}(a)$ the unique instance i such that $a \in \Sigma_i$, that is the particular instance on which each occurrence of action a takes place. Finally, for any pomset $(E, \preccurlyeq, \xi)$ over $\Sigma_{\mathcal{I}}$ we denote by $\mathrm{Ins}(e)$ the instance on which the event $e \in E$ occurs : $\mathrm{Ins}(e) = \mathrm{Ins}(\xi(e))$.

DEFINITION 1.1. *A message sequence chart (MSC) is a pomset $M = (E, \preccurlyeq, \xi)$ over $\Sigma_{\mathcal{I}}$ such that*
M_1*:* $\forall e, f \in E$*:* $\mathrm{Ins}(e) = \mathrm{Ins}(f) \Rightarrow (e \preccurlyeq f \vee f \preccurlyeq e)$
M_2*:* $\#^{j?^m i}(E) \leqslant \#^{i!^m j}(E)$ *for all instances i and j and all messages m*
M_3*:* $\left(\xi(e) = i!^m j \wedge \xi(f) = j?^m i \wedge \#^{i!^m j}(\downarrow e) = \#^{j?^m i}(\downarrow f)\right) \Rightarrow e \preccurlyeq f$
M_4*:* $[e \!-\!\prec\! f \wedge \mathrm{Ins}(e) \neq \mathrm{Ins}(f)]$
$$\Rightarrow \left[\xi(e) = i!^m j \wedge \xi(f) = j?^m i \wedge \#^{i!^m j}(\downarrow e) = \#^{j?^m i}(\downarrow f)\right].$$
An MSC is basic *if $\#^{i!^m j}(E) = \#^{j?^m i}(E)$ for all instances i and j and all messages m.*

By M_1, events occurring on the same instance are linearly ordered: Hence non-deterministic choice cannot be described within an MSC. Condition M_2 makes sure that each reception corresponds to a message. Thus there is no duplication of messages in the channels and M_2 formalizes partly the reliability of the channels. Following the recommendation Z.120, we allow *overtaking* (Fig. 1) but forbid any reversal of the order in which two identical messages m sent from i to j are received by j. Thus M_3 formalizes simply that the reception of any message will occur after the corresponding send event. Finally, by M_4, causality in M consists only in the linear dependency over each instance and the ordering of pairs of corresponding send and receive events.

We remark that any prefix of an MSC is an MSC, too. Note that an MSC is basic if each message m sent from i to j is associated to a reception event on j. We shall use non-basic MSCs as a natural way to describe partial executions of basic MSCs. Indeed, any MSC can be seen as a prefix of a basic MSC.

Rationality and Recognizability. Let $(\mathbb{M}, \cdot)$ be a monoid with unit 1. For any subsets $\mathcal{L}$ and $\mathcal{L}'$ of $\mathbb{M}$, the *product* of $\mathcal{L}$ by $\mathcal{L}'$ is $\mathcal{L} \cdot \mathcal{L}' = \{x \cdot x' \mid x \in \mathcal{L} \wedge x' \in \mathcal{L}'\}$. We let $\mathcal{L}^0 = \{1\}$ and for any $n \in \mathbb{N}$, $\mathcal{L}^{n+1} = \mathcal{L}^n \cdot \mathcal{L}$; then the *iteration* of $\mathcal{L}$ is $\mathcal{L}^\star = \bigcup_{n \in \mathbb{N}} \mathcal{L}^n$, also denoted $\langle \mathcal{L} \rangle_{\mathbb{M}}$. Note that $\langle \mathcal{L} \rangle_{\mathbb{M}}$ is a submonoid of $\mathbb{M}$. A language $\mathcal{L} \subseteq \mathbb{M}$ is *finitely generated* if there is a finite subset Γ of $\mathbb{M}$ such that $\mathcal{L} \subseteq \langle \Gamma \rangle_{\mathbb{M}}$. A subset of $\mathbb{M}$ is *rational* if it can be obtained from the finite subsets of $\mathbb{M}$ by means of unions, products and iterations. Any rational language is finitely generated. A subset $\mathcal{L}$ of $\mathbb{M}$ is *recognizable* if there exists a finite monoid $\mathbb{M}'$ and a monoid morphism $\eta : \mathbb{M} \to \mathbb{M}'$ such that $\mathcal{L} = \eta^{-1} \circ \eta(\mathcal{L})$. Equivalently, $\mathcal{L}$ is recognizable if, and only if, there exists a finite $\mathbb{M}$-automaton recognizing $\mathcal{L}$ — because the collection of all sets $\mathcal{L}/x = \{y \in \mathbb{M} \mid x \cdot y \in \mathcal{L}\}$ is finite. In particular the set of recognizable subsets of any monoid is closed under union, intersection and complement.

We denote by bMSC the set of (isomorphism classes) of basic MSCs. The *asynchronous concatenation* of two basic MSCs $M_1 = (E_1, \preccurlyeq_1, \xi_1)$ and $M_2 = (E_2, \preccurlyeq_2, \xi_2)$ is $M_1 \cdot M_2 = (E, \preccurlyeq, \xi)$ where $E = E_1 \uplus E_2$, $\xi = \xi_1 \cup \xi_2$ and the partial order $\preccurlyeq$ is the transitive closure of $\preccurlyeq_1 \cup \preccurlyeq_2 \cup \{(e_1, e_2) \in E_1 \times E_2 \mid \mathrm{Ins}(e_1) = \mathrm{Ins}(e_2)\}$. It is easy to check that the asynchronous concatenation of two basic MSCs is a basic MSC. This concatenation can be shown to be associative and admits the empty MSC $(\emptyset, \emptyset, \emptyset)$ as unit. Therefore we shall refer to bMSC as the *monoid of basic message sequence charts*. This concatenation enables us to

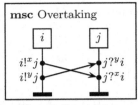

Fig. 1. Overtaking

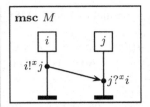

Fig. 2. Basic MSC M

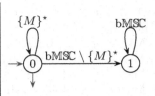

Fig. 3. A bMSC-automaton

compose specifications in order to describe sets of basic MSCs. In that way, we obtain high-level message sequence charts.

DEFINITION 1.2. *A high-level message sequence chart (HMSC) is a rational expression of basic MSCs, that is, an expression built from basic MSCs by use of union, product and iteration.*

We follow here the approach adopted, e.g., in [2,3,5,8,12,16,18] where HMSCs are however often flattened into message sequence graphs.

Regularity vs Rationality. A language $\mathcal{L}$ of basic MSCs is *regular* [8] if its set of linear extensions $\mathrm{LE}(\mathcal{L}) = \bigcup_{M \in \mathcal{L}} \mathrm{LE}(M)$ is recognizable in the free monoid $\Sigma_{\mathcal{I}}^{\star}$. All regular languages are recognizable in bMSC, but the converse fails.

EXAMPLE 1.3. We consider here the basic MSC M of Fig. 2 and the rational language $\{M\}^{\star}$. The latter is recognizable in bMSC (it is accepted by the bMSC-automaton of Fig. 3), but not regular. It is even not channel-bounded [8].

As proved in [8, Th. 4.6], *it is undecidable to know whether a HMSC $\mathcal{H}$ is regular* (that is to say, whether its corresponding language $\mathcal{L}_{\mathcal{H}}$ is regular). Yet, one can ensure the regularity of languages associated to high-level message sequence charts by restricting to *locally synchronized* [18] or *bounded* [3] HMSCs — which are called sc-HMSCs below. Recall first that the *communication graph* of a basic MSC $M = (E, \preccurlyeq, \xi)$ is the directed graph $(\mathcal{I}_M, \mapsto)$ where $\mathcal{I}_M$ is the set of *active instances* of M: $\mathcal{I}_M = \{i \in \mathcal{I} \mid \exists e \in E, \mathrm{Ins}(e) = i\}$, and such that $(i, j) \in \mapsto$ if there is an event $e \in E$ such that $\xi(e) = i!j$. Thus there is an edge from i to j if M shows a communication from i to j. Now a high-level MSC is called an *sc-HMSC* if iteration occurs only over sets of MSCs whose communication graphs are strongly connected. Consider finally a *finitely generated* language $\mathcal{L}$ of basic MSCs. Putting together results from [18,3,8], we know that $\mathcal{L}$ *is regular if, and only if, it is the language* $\mathcal{L}_{\mathcal{H}}$ *of an sc-HMSC $\mathcal{H}$.*

2   Recognizability vs. Rationality

In this section we show a new relationship between basic MSCs and Mazurkiewicz traces. This allows us to infer results on recognizability analogous to those mentioned above for regularity. Differently from [18,3,8,15], we do not restrict to sc-HMSCs and we assume no limit to the channel capacities.

Let us first recall some basic notions of Mazurkiewicz trace theory [6]. The concurrency of a distributed system can be represented by an *independence relation* over the (possibly infinite) alphabet of actions Σ, that is a binary, symmetric and irreflexive relation $\| \subseteq \Sigma \times \Sigma$. The associated *trace equivalence* is the least congruence $\sim$ over Σ^* such that $\forall a, b \in \Sigma, a\|b \Rightarrow ab \sim ba$. A *trace* $[u]$ is the equivalence class of a word $u \in \Sigma^*$. We denote by $\mathbb{M}(\Sigma, \|)$ the set of all traces w.r.t. $(\Sigma, \|)$. Traces can easily be composed in the following way: $[u] \cdot [v] = [u.v]$. Then $\mathbb{M}(\Sigma, \|)$ appears as a monoid with the empty trace $[\varepsilon]$ as unit. A *trace language* is a subset $\mathcal{L} \subseteq \mathbb{M}(\Sigma, \|)$. It is easy to see that a trace language $\mathcal{L}$ is recognizable in $\mathbb{M}(\Sigma, \|)$ iff the set of associated words $\{u \in \Sigma^* \mid [u] \in \mathcal{L}\}$ is recognizable in the free monoid Σ^*. Let $u \in \Sigma^*$; then the trace $[u]$ is precisely the set of linear extensions $\mathrm{LE}(t)$ of a unique pomset $t = (E, \preccurlyeq, \xi)$, that is, $[u] = \mathrm{LE}(t)$. That is why traces can be seen as pomsets and we put $\mathrm{LE}(\mathcal{L}) = \{u \in \Sigma^* \mid [u] \in \mathcal{L}\}$ for any trace language $\mathcal{L} \subseteq \mathbb{M}(\Sigma, \|)$.

A natural independence relation comes to light when we consider basic MSCs with respect to their concatenation. For all basic MSCs M and M', we put $M\|M'$ if $\mathcal{I}_M \cap \mathcal{I}_{M'} = \emptyset$. Clearly, $M\|M'$ implies that $M \cdot M' = M' \cdot M$. A *non-empty* basic MSC M is called *prime* if $M = M_1 \cdot M_2$ implies $M_1 = 1$ or $M_2 = 1$. Clearly, any basic MSC is a product of primes [7]. The next lemma shows that this decomposition is *unique* up to the commutation of independent primes.

LEMMA 2.1. *Let Γ be a (possibly infinite) subset of prime basic MSCs. Then the morphism $\Re_\Gamma : \mathbb{M}(\Gamma, \|) \to \langle \Gamma \rangle_{\mathrm{bMSC}}$ which maps each trace $[a_1...a_n]$ to the basic MSC $a_1 \cdot ... \cdot a_n$ is an isomorphism.*

Proof (sketch). The map $\Re_\Gamma$ is well-defined since $u \sim v$ implies $\Re_\Gamma[u] = \Re_\Gamma[v]$. Clearly the morphism $\Re_\Gamma$ is onto. We show by induction on $n = |u|$ that if $\Re_\Gamma[u] = \Re_\Gamma[v]$ then $u \sim v$. The claim is clear if $|u| = 0$. We let $u = a_1...a_n$ and $v = b_1...b_m$ be two words of Γ^* and assume that $\Re_\Gamma[u] = \Re_\Gamma[v]$. We denote by $M = (E, \preccurlyeq, \xi)$ the basic MSC $\Re_\Gamma[u]$. Then $a_1 \cdot ... \cdot a_n$ and $b_1 \cdot ... \cdot b_m$ are two decompositions of M in prime basic MSCs. Let e be a minimal event in the MSC a_1 and $M' = (E', \preccurlyeq_{|E'}, \xi_{|E'})$ denote the least basic prefix of M that contains e. The event e belongs to b_s for some $s \in [1, m]$. By means of a close analysis of M', we can show that $a_1 = M' = b_s$. Then $\mathcal{I}_{b_k} \cap \mathcal{I}_{b_s} = \emptyset$ for all $k \in [1, s-1]$ because E' is an ideal of M. Thus, $b_1 \cdot ... \cdot b_m = b_s \cdot b_1 \cdot ... \cdot b_{s-1} \cdot b_{s+1} \cdot ... \cdot b_m = a_1 \cdot ... \cdot a_n$. Since $a_1 = b_s$ and bMSC is cancellative, we have $b_1 \cdot ... \cdot b_{s-1} \cdot b_{s+1} \cdot ... \cdot b_m = a_2 \cdot ... \cdot a_n$. By induction hypothesis, $b_1...b_{s-1}.b_{s+1}...b_m \sim a_2...a_n$ hence $u \sim v$. ∎

We can now translate results from trace theory as follows (see also Th. 4.1).

THEOREM 2.2. *It is undecidable to know whether a HMSC is recognizable.*

Proof. We proceed by reduction to the recognizability of rational trace languages, which is known to be undecidable [21]. For any finite independence alphabet $(\Sigma, \|)$, we can find a finite set of prime basic MSCs $\Gamma = \{M_a \mid a \in \Sigma\}$ such that $a\|b \Leftrightarrow M_a\|M_b$. Then the morphism $\psi : \mathbb{M}(\Sigma, \|) \to \langle \Gamma \rangle_{\mathrm{bMSC}}$ defined by $\psi(a) = M_a$ is an isomorphism (Lemma 2.1). Therefore, $\mathcal{L} \subseteq \mathbb{M}(\Sigma, \|)$ is recognizable (resp. rational) iff $\psi(\mathcal{L})$ is recognizable in bMSC (resp. rational). ∎

This result shows that we have to face a similar problem as with regularity [8, Th. 4.6]. We say that an MSC is connected if its communication graph is a connected (undirected) graph. Then a HMSC is called a *c-HMSC* if iteration occurs only over sets of connected MSCs.

THEOREM 2.3. *Let* $\mathcal{L}$ *be a finitely generated language of basic MSCs. Then* $\mathcal{L}$ *is recognizable if, and only if, it is the language* $\mathcal{L}_{\mathcal{H}}$ *of a c-HMSC* $\mathcal{H}$.

Proof. We have $\mathcal{L} \subseteq \langle \Gamma \rangle_{\text{bMSC}}$ for a finite set Γ of prime basic MSCs. By Lemma 2.1, $\mathcal{L}$ is recognizable in bMSC if, and only if, $\mathfrak{R}_{\Gamma}^{-1}(\mathcal{L})$ is recognizable in $\mathbb{M}(\Gamma, \|)$, which means that $\mathfrak{R}_{\Gamma}^{-1}(\mathcal{L})$ is c-rational [19]. Since prime basic MSCs are connected, this is also equivalent to say that $\mathcal{L}$ is described by a c-HMSC. ∎

This result shows that, in order to specify a recognizable language of basic MSCs by some HMSC, one can restrict to c-HMSCs without loss of expressive power. Example 1.3 shows that *c-HMSCs can describe behaviors with unbounded channels* — differently from sc-HMSCs [15, Cor. 2.9]. As justified by [12, Th. 1], such infinite systems can still be model-checked. Our interest in recognizable languages also stems from the next section; we will see that if a rational language is realizable then it has to be recognizable.

3   Recognizability vs. Realizability

In this section, we focus on the HMSCs whose language can be seen as the behaviors of a distributed system whose processes communicate through reliable FIFO channels. Recall now that we allow overtaking of messages (Fig. 1). Consequently, we consider one channel for each pair of distinct instances associated with a message type. Formally, the set of channels *Chan* consists of all triples $(i, j, m) \in \mathcal{I} \times \mathcal{I} \times Mes$ such that $i \neq j$. Furthermore a *channel state* is formalized by a map $\chi : Chan \to \mathbb{N}$ that describes the queues of messages within the channels at some stage of an execution. The *empty channel state* χ_0 is such that each channel maps to 0.

DEFINITION 3.1. *A message-passing automaton (MPA) over* $\Sigma_{\mathcal{I}}$ *is a family* $\mathcal{S} = (\mathcal{A}_i)_{i \in \mathcal{I}}$ *such that each component* $\mathcal{A}_i$ *is a transition system* $(Q_i, \imath_i, \longrightarrow_i, F_i)$ *where* Q_i *is a set of i-local states, with initial state* $\imath_i \in Q_i$ *and final states* $F_i \subseteq Q_i$, *and* $\longrightarrow_i \subseteq (Q_i \times \Sigma_i \times Q_i)$ *is the i-local transition relation.*

A *global state* is a pair (s, χ) where $s \in \prod_{i \in \mathcal{I}} Q_i$ is a tuple of local states and χ is a channel state. The *initial global state* is the pair $\imath = (s, \chi)$ such that $s = (\imath_i)_{i \in \mathcal{I}}$ and $\chi = \chi_0$ is the empty channel state. The *system of global states* associated to $\mathcal{S}$ is the transition system $\mathcal{A}_{\mathcal{S}} = (Q, \imath, \longrightarrow, F)$ where Q is the set of global states, $F = (\prod_{i \in \mathcal{I}} F_i) \times \{\chi_0\}$ is the set of *final global states*, and the global transition relation $\longrightarrow \subseteq Q \times \Sigma_{\mathcal{I}} \times Q$ satisfies:

- for any internal action $a \in \Sigma_i^{int}$, $((q_k)_{k \in \mathcal{I}}, \chi) \xrightarrow{a} ((q'_k)_{k \in \mathcal{I}}, \chi')$ if $\chi = \chi'$, $q_i \xrightarrow{a}_i q'_i$ and $q'_k = q_k$ for all $k \in \mathcal{I} \setminus \{i\}$;
- for all distinct instances i and j, $((q_k)_{k \in \mathcal{I}}, \chi) \xrightarrow{i!^m j} ((q'_k)_{k \in \mathcal{I}}, \chi')$ if

1. $q_i \xrightarrow{i!^m j}{}_i q_i'$ and $q_k' = q_k$ for all $k \in \mathcal{I} \setminus \{i\}$,
2. $\chi'(i,j,m) = \chi(i,j,m) + 1$ and $\chi(x) = \chi'(x)$ for all $x \in Chan \setminus \{(i,j,m)\}$;

- for all distinct instances i and j, $((q_k)_{k \in \mathcal{I}}, \chi) \xrightarrow{j?^m i} ((q_k')_{k \in \mathcal{I}}, \chi')$ if

1. $q_j \xrightarrow{j?^m i}{}_j q_j'$ and $q_k' = q_k$ for all $k \in \mathcal{I} \setminus \{j\}$,
2. $\chi(i,j,m) = 1 + \chi'(i,j,m)$ and $\chi(x) = \chi'(x)$ for all $x \in Chan \setminus \{(i,j,m)\}$.

As usual with transition systems, for any $u = a_1...a_n \in \Sigma_{\mathcal{I}}^{\star}$, we write $q \xrightarrow{u} q'$ if there are some global states $q_0, ..., q_n \in Q$ such that $q_0 = q$, $q_n = q'$ and for each $r \in [1,n]$, $q_{r-1} \xrightarrow{a_r} q_r$. We stress that like [1] but unlike [9] the set of i-local states in a message passing automaton might be infinite.

An *execution sequence* of $\mathcal{S}$ is a word $u \in \Sigma_{\mathcal{I}}^{\star}$ such that $\imath \xrightarrow{u} q$ in $\mathcal{A}_{\mathcal{S}}$ for some global state q. Then u is a linear extension of some MSC. The latter is a *basic* MSC if, and only if, q has an empty channel-state. We say that u is a *final execution sequence* if $\imath \xrightarrow{u} q$ for some *final* global state q. The *language* $\mathcal{L}_{\mathcal{S}}$ of $\mathcal{S}$ consists of the basic MSCs M such that $LE(M)$ contains a final execution sequence of $\mathcal{S}$. Finally, a language $\mathcal{L} \subseteq$ bMSC is *realizable* if $\mathcal{L} = \mathcal{L}_{\mathcal{S}}$ for some message passing automaton $\mathcal{S}$.

In order to characterize which HMSCs describe realizable languages, we need some more notations. For any instance $i \in \mathcal{I}$, the *projection on i* [1,5] is the map $(\_ \mid i) : \Sigma_{\mathcal{I}}^{\star} \to \Sigma_i^{\star}$ defined inductively by $(\varepsilon \mid i) = \varepsilon$ and $(w.a \mid i) = (w \mid i).a$ if $Ins(a) = i$ and $(w \mid i)$ otherwise. We extend this map from words to MSCs and from MSCs to HMSCs. First, $(M \mid i) = (w \mid i)$ for any linear extension $w \in LE(M)$ of the MSC M. This is well-defined due to M_1. Second, we define inductively $(\mathcal{H}_1 \cdot \mathcal{H}_2 \mid i) = (\mathcal{H}_1 \mid i) \cdot (\mathcal{H}_2 \mid i)$, $(\mathcal{H}_1 + \mathcal{H}_2 \mid i) = (\mathcal{H}_1 \mid i) + (\mathcal{H}_2 \mid i)$, and $(\mathcal{H}^{\star} \mid i) = (\mathcal{H} \mid i)^{\star}$ for all HMSCs $\mathcal{H}_1$, $\mathcal{H}_2$ and $\mathcal{H}$. Note that $(\mathcal{H} \mid i)$ is a rational expression of $\Sigma_i^{\star}$: It describes a recognizable language of words. Noteworthy, we observe by an immediate structural induction on $\mathcal{H}$ that $(\mathcal{H} \mid i)$ describes the set $(\mathcal{L}_{\mathcal{H}} \mid i)$ of projections on instance i of all MSCs in $\mathcal{L}_{\mathcal{H}}$.

A basic observation now is that for any MSC M and any linear extension $w \in LE(M)$, we have $((\imath_i)_{i \in \mathcal{I}}, \chi_0) \xrightarrow{w} ((q_i)_{i \in \mathcal{I}}, \chi)$ in $\mathcal{A}_{\mathcal{S}}$ for some channel-state χ iff $\imath_i \xrightarrow{(w \mid i)} q_i$ for all instances $i \in \mathcal{I}$. Consequently, a basic MSC belongs to the language $\mathcal{L}_{\mathcal{S}}$ of the MPA $\mathcal{S}$ if *all* its linear extensions are final execution sequences of $\mathcal{S}$. It follows also that a language $\mathcal{L}$ of basic MSCs is realizable if, and only if, it satisfies the following condition [1]:

CC$_2$: For all basic MSCs M: $(\forall i \in \mathcal{I}, \exists M_i \in \mathcal{L}, (M \mid i) = (M_i \mid i)) \Rightarrow M \in \mathcal{L}$.

Thus, although we consider a slightly different architecture, the realizability notion studied in this paper corresponds precisely to the "weak realizability" investigated in [1,2]. Furthermore we obtain the following result.

PROPOSITION 3.2. *Let $\mathcal{L}$ be a language of basic MSCs. Consider for any instance i a transition system $\mathcal{A}_i$ whose language is $(\mathcal{L} \mid i) = \{(M \mid i) \mid M \in \mathcal{L}\}$. Then $\mathcal{L}$ is realizable if, and only if, $\mathcal{L}$ is the language of the MPA $\mathcal{S} = (\mathcal{A}_i)_{i \in \mathcal{I}}$.*

This implies that any realizable *rational* language of basic MSCs is *recognizable*, because it can be realized by an MPA *with finitely many local states*. Thus, in order to specify realizable languages by HMSCs, we can restrict to c-HMSCs without loss of expressive power (Th. 2.3). Based on Lemma 2.1 again, we present now an algorithm to check whether the language $\mathcal{L}_{\mathcal{H}}$ of a c-HMSC $\mathcal{H}$ is realizable.

THEOREM 3.3. *Checking whether a c-HMSC is realizable is decidable.*

Proof. Let $\mathcal{H}$ be a c-HMSC. We may assume that $\mathcal{H}$ is based on a finite set of prime basic MSCs Γ. In particular, $\mathcal{L}_{\mathcal{H}} \subseteq \langle\Gamma\rangle_{\text{bMSC}}$. From $\mathcal{H}$, we can build the MPA $\mathcal{S} = (\mathcal{A}_i)_{i \in \mathcal{I}}$ where $\mathcal{A}_i$ is the minimal automaton of $(\mathcal{L}_{\mathcal{H}} \mid i)$. Applying Prop. 3.2, we check whether $\mathcal{L}_{\mathcal{S}} = \mathcal{L}_{\mathcal{H}}$. For this, we proceed in two steps: First we check that $\mathcal{L}_{\mathcal{S}} \cap \langle\Gamma\rangle_{\text{bMSC}} = \mathcal{L}_{\mathcal{H}}$ and next that $\mathcal{L}_{\mathcal{S}} \subseteq \langle\Gamma\rangle_{\text{bMSC}}$.

— ① — We denote by $\mathcal{A} = (Q, \imath, \longrightarrow, F)$ the system of global states of $\mathcal{S}$. We can build the transition system $\mathcal{A}_\Gamma = ((\prod_{i \in \mathcal{I}} Q_i) \times \{\chi_0\}, \imath, \longrightarrow_\Gamma, F)$ over the alphabet Γ such that $q \xrightarrow{M}_\Gamma q'$ if $q \xrightarrow{w} q'$ in $\mathcal{A}$ for some linear extension w of $M \in \Gamma$. Note that $\mathcal{A}_\Gamma$ is finite because each $\mathcal{A}_i$ has finitely many i-local states. It is easy to see that a word $u \in \Gamma^\star$ belongs to the language $L(\mathcal{A}_\Gamma)$ of $\mathcal{A}_\Gamma$ if, and only if, $\Re_\Gamma[u] \in \mathcal{L}_{\mathcal{S}} \cap \langle\Gamma\rangle_{\text{bMSC}}$. Thus $L(\mathcal{A}_\Gamma) = \text{LE}(\Re_\Gamma^{-1}(\mathcal{L}_{\mathcal{S}} \cap \langle\Gamma\rangle_{\text{bMSC}}))$. Since $\mathcal{H}$ is a c-HMSC, we have a c-rational expression that describes $\Re_\Gamma^{-1}(\mathcal{L}_{\mathcal{H}})$ (Lemma 2.1). Therefore, we can compute a finite automaton $\mathcal{A}_{\mathcal{H}}$ that recognizes $\text{LE}(\Re_\Gamma^{-1}(\mathcal{L}_{\mathcal{H}}))$ [14,18]. Then, we need simply to check that $\mathcal{A}_{\mathcal{H}}$ and $\mathcal{A}_\Gamma$ describe the same language of $\Gamma^\star$ to know whether $\mathcal{L}_{\mathcal{S}} \cap \langle\Gamma\rangle_{\text{bMSC}} = \mathcal{L}_{\mathcal{H}}$.

— ② — Recall that if q is a global state whose channel-state is empty and if a word $w \in \Sigma_{\mathcal{I}}^\star$ satisfies $q \xrightarrow{w} q'$ in $\mathcal{A}$ then w is a linear extension of a unique MSC which will be denoted M_w. Furthermore, q' has also an empty channel-state if, and only if, M_w is a basic MSC.

Let Q_0 be the subset of global states $q \in (\prod_{i \in \mathcal{I}} Q_i) \times \{\chi_0\}$ such that there exists a linear extension u of a basic MSC $M \in \langle\Gamma\rangle_{\text{bMSC}}$ for which $\imath \xrightarrow{u} q$ in $\mathcal{A}$. Observe here that $q \in Q_0$ if, and only if, q is reachable from $\imath$ in $\mathcal{A}_\Gamma$. Therefore, we can effectively compute Q_0. Assume first that $\mathcal{L}_{\mathcal{S}} \setminus \langle\Gamma\rangle_{\text{bMSC}}$ is not empty and consider $M \in \mathcal{L}_{\mathcal{S}} \setminus \langle\Gamma\rangle_{\text{bMSC}}$. Then we have a decomposition of M in prime basic MSCs $M = M_1 \cdot \ldots \cdot M_n \cdot M_{n+1} \cdot \ldots \cdot M_m$ with $M_1,...,M_n \in \Gamma$ and $M_{n+1} \notin \Gamma$. Let $u \in \text{LE}(M_1 \cdot \ldots \cdot M_n)$. Then $\imath \xrightarrow{u} q_0$ for some $q_0 \in Q_0$. Thus, to know whether $\mathcal{L}_{\mathcal{S}} \subseteq \langle\Gamma\rangle_{\text{bMSC}}$, it is sufficient to check that for all $q_0 \in Q_0$, the following property holds:

$\mathsf{P}_1(q_0)$: For all prime basic MSCs $M \notin \Gamma$, if $q_0 \xrightarrow{w} q_1$ in $\mathcal{A}$ with $w \in \text{LE}(M)$ then no final global state is reachable from q_1.

A basic observation here is that $\mathcal{S}$ can be simulated by a Place/Transition net. Clearly each $\mathcal{A}_i$ can be simulated by a 1-safe net whose reachable markings contain only one token that describes the current local state; and, additionally, we can use one place with infinite capacity to model each channel. Therefore *we can decide whether there is a path in $\mathcal{A}$ from q_1 to q_2 where q_1 and q_2 are two given global states* (or markings) [13,20].

In the sequel, we fix $q_0 \in Q_0$ and proceed in two steps. Let C be the maximal number of events in an MSC of Γ. Consider first a prime basic MSC $M \notin \Gamma$ with at most $(C + 2).\text{Card}(\mathcal{I})$ events and $w \in \text{LE}(M)$. We can check whether there is a path $q_0 \xrightarrow{w} q_1$ in $\mathcal{A}$ for some $q_1 \in (\prod_{i \in \mathcal{I}} Q_i) \times \{\chi_0\}$ and if there is a final global state q_2 reachable from q_1. If this happens, Property $\mathsf{P}_1(q_0)$ fails (for this M). We actually check this for all primes with less than $(C + 2).\text{Card}(\mathcal{I})$ events. If $\mathsf{P}_1(q_0)$ does not fail for these MSCs, then $\mathsf{P}_1(q_0)$ becomes equivalent to the following property:

$\mathsf{P}_2(q_0)$: For all prime basic MSCs $M \notin \Gamma$ *with at least* $(C+1).\text{Card}(\mathcal{I})$ *events,* if $q_0 \xrightarrow{w} q_1$ in $\mathcal{A}$ with $w \in \text{LE}(M)$ then no $q_2 \in F$ is reachable from q_1.

Define the *levels* of an MSC $M = (E, \preccurlyeq, \xi)$ inductively as follows. First $l_0(M) = \min_{\preccurlyeq}(E)$, and for all integer k, $l_{k+1}(M) = \min_{\preccurlyeq}(E \setminus \bigcup_{i=0}^{k} l_i(M))$. Observe that each level $l_k(M)$ is an antichain of M. We denote by N the maximal number of non-empty levels in an MSC of Γ. Noteworthy $N \leqslant C$. We say that a prefix $M' = (E', \preccurlyeq', \xi')$ of M *respects the levels in* M if $E' = \bigcup_{i=0}^{k} l_i(M)$ for some k.

We shall see that we can build effectively the finite set $L(q_0)$ of all words $w \in \Sigma_{\mathcal{I}}^{\star}$ such that

- $|w| \leqslant (N + 1).\text{Card}(\mathcal{I})$,
- $q_0 \xrightarrow{w} q_1$ in $\mathcal{A}$ for some global state q_1,
- M_w has exactly $N + 1$ non-empty levels, and
- $q_1 \xrightarrow{v} q_2$ in $\mathcal{A}$ for some final global state q_2 and a word $v \in \Sigma_{\mathcal{I}}^{\star}$ such that M_w respects the levels in the basic MSC $M_{w.v}$.

Assume that w fulfills the three first conditions and let χ_w be the channel state of q_1. In case $q_1 \xrightarrow{v} q_2 \in F$ in $\mathcal{A}$ then v satisfies the fourth condition iff

1. if $l_{N+1}(M_w)$ contains no event on instance i then $(v \mid i)$ is empty or its first action is not a local action, nor a send action;
2. if $l_{N+1}(M_w)$ contains no event on instance j and if $\chi_w(i,j,m) \geqslant 2$ for some i then $(v \mid j)$ is empty or its first action is not $j?^m i$;
3. if $l_{N+1}(M_w)$ contains no event on instance j and if $\chi_w(i,j,m) = 1$ for some i and $l_{N+1}(M_w)$ contains no send action $i!^m j$ then $(v \mid j)$ is empty or its first action is not $j?^m i$.

Based on $l_{N+1}(M_w)$ and χ_w, we can compute for each instance which actions cannot occur first. Then, we can adapt the simulation of $\mathcal{S}$ by Petri nets in order to forbid these actions as first local action. In that way we can check whether there exists a sequence v that satisfies the fourth condition.

Thus, we can effectively compute $L(q_0)$. Consider now Property $\mathsf{P}_3(q_0)$:

$\mathsf{P}_3(q_0)$: For all $w \in L(q_0)$, for all minimal event e of the MSC M_w, there is a basic MSC $M^{\dagger} \in \Gamma$ isomorphic to a prefix of M_w that contains e.

To conclude, we shall prove $\mathsf{P}_1(q_0) \Rightarrow \mathsf{P}_3(q_0)$ and $\mathsf{P}_3(q_0) \Rightarrow \mathsf{P}_2(q_0)$. Recall that we are in the special case where $\mathsf{P}_1(q_0) \Leftrightarrow \mathsf{P}_2(q_0)$. Therefore, it is now sufficient to check $\mathsf{P}_3(q_0)$ to know whether $\mathsf{P}_1(q_0)$ holds.

$\mathsf{P}_1(q_0) \Rightarrow \mathsf{P}_3(q_0)$? Let $w \in L(q_0)$. Then $q_0 \xrightarrow{w} q_1 \xrightarrow{v} q_2 \in F$ in $\mathcal{A}$. Let e be a minimal event of the MSC M_w. Since M_w is a prefix of $M_{w.v}$, e is also a minimal event of the basic MSC $M_{w.v}$. Let $M^{\dagger}$ be the least basic prefix of $M_{w.v}$ that contains e. Then $M_{w.v} = M^{\dagger} \cdot M^{\dagger\dagger}$ and $M^{\dagger}$ is prime. Consider $w' \in \text{LE}(M^{\dagger})$ and $v' \in \text{LE}(M^{\dagger\dagger})$. We have $q_0 \xrightarrow{w'} q' \xrightarrow{v'} q_2$ in $\mathcal{A}$. Since $\mathsf{P}_1(q_0)$ holds, we get $M^{\dagger} \in \Gamma$ and $M^{\dagger}$ has at most N non-empty levels. Since M_w respects the levels in $M_{w.v}$ and M_w has $N + 1$ non-empty levels, $M^{\dagger}$ is also a prefix of M_w.

We show finally that $\neg P_2(q_0)$ implies $\neg P_3(q_0)$. Let M be a prime basic MSC not in Γ with at least $(C+1).\text{Card}(\mathcal{I})$ events such that $q_0 \xrightarrow{w} q_1 \xrightarrow{v} q_2 \in F$ for some $w \in \text{LE}(M)$. Then M has at least $(N+1).\text{Card}(\mathcal{I})$ events, hence at least $N+1$ non-empty levels. Since $M = M_w$ is a prefix of $M_{w.v}$, $M_{w.v}$ has at least $N+1$ non-empty levels too. Let M' be the restriction of $M_{w.v}$ to its $N+1$ first levels and $w' \in \text{LE}(M')$. Then M' respects the levels in $M_{w.v}$. We have $q_0 \xrightarrow{w'} q' \xrightarrow{v'} q_2$ with $w'.v' \in \text{LE}(M_{w.v})$ and $M_{w.v} = M_{w'.v'}$. Since M' has $N+1$ non-empty levels, we have $|w'| \leqslant (N+1).\text{Card}(\mathcal{I})$. Hence $w' \in L(q_0)$. Consider now the prefix M'' of $M_w = M$ consisting of its $N+1$ first levels. Then M'' is also a prefix of M'. Let e be a minimal event of M''; this is also a minimal event of M'. We proceed now by contradiction. Assume that there exists $M^\dagger \in \Gamma$ which is (isomorphic to) a prefix of $M_{w'} = M'$ that contains e. Then the intersection between $M^\dagger$ and M contains e, is an ideal of $M_{w.v}$, $M^\dagger$ and M, and corresponds to a *basic* MSC because $M^\dagger$ and M are basic, contain e and are prefixes of $M_{w.v}$. Since $M^\dagger$ and M are prime, this implies $M = M^\dagger$ hence $M \in \Gamma$. Contradiction. ∎

Observe now that any bounded MSC-graph [2,3] can be translated effectively into an sc-HMSC. Therefore realizability of bounded MSC-graphs is also decidable. As shown in [2, Th. 1], this result fails if one assumes that no overtaking occurs both in the languages to be realized and in the behavior of any MPA. We stress finally that our technique can be adapted to check *safe* realizability [1,2] of c-HMSCs *even if overtaking is forbidden*: The reason is that we need no longer to check the reachability of final states from some stages. In that way, safe realizability of (unbounded) connected MSC-graphs is also decidable.

4   Recognizability vs. MSO-Definability

Recognizable languages of basic MSCs form a particularly interesting framework for the specification of realizable protocols — in particular, for those with unbounded channel size. Differently from rational languages, we can specify a system by means of typical executions together with forbidden ones and still remain in the field of recognizable languages, because they are closed by complement and intersection. Admittedly, logical formulae are another natural way to express pathological or critical behaviors too. This motivates the study of the relationship between recognizability and logical definability.

Consider a finite alphabet Σ. Formulae of the MSO language over Σ that we consider involve first-order variables $x, y, z...$ for events and second-order variables $X, Y, Z...$ for sets of events. They are built up from the atomic formulae $P_a(x)$ for $a \in \Sigma$ (which stands for "the event x is labeled by the action a"), $x \preccurlyeq y$, and $x \in X$ by means of the boolean connectives $\neg, \vee, \wedge, \rightarrow, \leftrightarrow$ and quantifiers $\exists, \forall$ (both for first order and for set variables). We denote by $\text{MSO}(\Sigma)$ the set of all formulae of the MSO language. Formulae without free variables are called sentences. The satisfaction relation $\models$ between the set of pomsets and the set of sentences of monadic second order logic is defined canonically with the understanding that first order variables range over events of E and second order

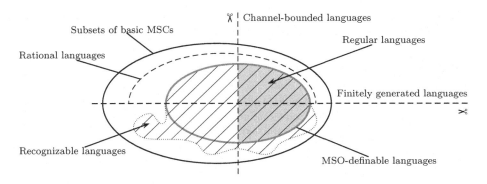

Fig. 4. Comparison between Theorem 4.1 and [9, Th. 4.3]

variables over subsets of E. The class of pomsets which satisfy a sentence φ is denoted by $\mathrm{Mod}(\varphi)$. We say that a class of pomsets $\mathcal{P}$ is MSO-*definable* if there exists a monadic second order sentence φ such that $\mathcal{P} = \mathrm{Mod}(\varphi)$.

By means of Lemma 2.1, we apply once again results from trace theory to get a Büchi-like theorem — however for finitely generated languages only.

THEOREM 4.1. *Let $\mathcal{L}$ be a finitely generated language of basic MSCs. Then $\mathcal{L}$ is recognizable in* bMSC *if, and only if, $\mathcal{L}$ is MSO-definable.*

Proof (sketch). Consider a finite set of prime basic MSCs $\Gamma = \{M_1, ..., M_n\}$ such that $\mathcal{L} \subseteq \langle \Gamma \rangle_{\mathrm{bMSC}}$. Assume first that $\mathcal{L}$ is MSO-definable by a sentence ϕ. For each $k \in [1, n]$, we choose $u_k \in \mathrm{LE}(M_k)$. We consider the language $L \subseteq \Sigma_{\mathcal{I}}^{\star}$ which consists of the words $w = u_{k_1}...u_{k_m}$ such that $M = M_{k_1} \cdot ... \cdot M_{k_m} \in \mathcal{L}$. We claim that L is MSO-definable by a sentence ψ that we can easily obtain from ϕ: The reason is that we can recover the partial order of events in M from the total order of w. By Büchi's Theorem [4], L is recognizable in the free monoid $\Sigma_{\mathcal{I}}^{\star}$. We use then the minimal $\Sigma_{\mathcal{I}}^{\star}$-automaton that recognizes L to show that $\mathcal{L}$ is recognizable in $\langle \Gamma \rangle_{\mathrm{bMSC}}$ hence in bMSC.

We assume now that $\mathcal{L}$ is recognizable in bMSC. We observe first that $\langle \Gamma \rangle_{\mathrm{bMSC}}$ is MSO-definable because each $M \in \Gamma$ is connected. By Lemma 2.1, $\mathcal{L}' = \Re_{\Gamma}^{-1}(\mathcal{L})$ is recognizable in $\mathbb{M}(\Gamma, \|)$. By [22], the corresponding set of pomsets is definable by an MSO-sentence $\phi_{\mathcal{L}'}$. We infer from $\phi_{\mathcal{L}'}$ a formula that defines $\mathcal{L}$ within $\langle \Gamma \rangle_{\mathrm{bMSC}}$. Given $M = M_{k_1} \cdot ... \cdot M_{k_m}$, we can recover which events belong to the same prime component. Then we simply need to choose one representative event in each component M_{k_l} and to recover the partial order between those events that corresponds to $\Re_{\Gamma}^{-1}(M)$. By means of $\phi_{\mathcal{L}'}$, we can then formalize whether $M \in \mathcal{L}$. ∎

This relationship is depicted on Fig. 4. It completes [9, Th. 4.3] which asserts that a language is regular if, and only if, it is MSO-definable with bounded channels. It follows that a finitely generated language is regular if, and only if, it is recognizable and channel-bounded.

Acknowledgments. I am grateful to M. DROSTE for drawing my attention on [2,12]. Thanks also to D. KUSKE for motivating discussions about these papers.

References

1. Alur R., Etessami K. and Yannakakis M.: *Inference of message sequence charts.* 22nd Intern. Conf. on Software Engineering, ACM (2000) 304–313
2. Alur R., Etessami K. and Yannakakis M.: *Realizability and verification of MSC graphs.* LNCS **2076** (2001) 797–808
3. Alur R. and Yannakakis M.: *Model Checking of Message Sequence Charts.* CONCUR'99, LNCS **1664** (1999) 114–129
4. Büchi J.R.: *Weak second-order arithmetic and finite automata.* Z. Math. Logik Grundlagen Math. **6** (1960) 66–92
5. Caillaud B., Darondeau Ph., Hélouët L. and Lesventes G.: *HMSCs as partial specifications... with PNs as completions.* LNCS **2067** (2001) 87–103
6. Diekert V. and Rozenberg G.: *The Book of Traces.* (World Scientific, 1995)
7. Hélouët L. and Le Maigat P.: *Decomposition of Message Sequence Charts.* Proc. SAM 2000 (Verimag, Grenoble, 2000)
8. Henriksen J.G., Mukund M., Narayan Kumar K. and Thiagarajan P.S.: *On message sequence graphs and finitely generated regular MSC language.* LNCS **1853** (2000) 675–686
9. Henriksen J.G., Mukund M., Narayan Kumar K. and Thiagarajan P.S.: *Regular collections of message sequence charts.* LNCS **1893** (2000) 405–414
10. Holzmann G.J.: *Early Fault Detection.* TACAS'96, LNCS **1055** (1996) 1–13
11. ITU-TS: *Recommendation Z.120: Message Sequence Charts.* (Geneva, 1996)
12. Madhusudan P.: *Reasoning about Sequential and Branching Behaviours of Message Sequence Graphs.* LNCS **2076** (2001) 809–820
13. Mayr, E.W.: *Persistence of Vector Replacement Systems is Decidable.* Acta Informatica **15** (1981) 309–318
14. Métivier Y., Richomme G. and Wacrenier P.: *Computing the Closure of Sets of Words under Partial Commutations.* LNCS **974** (1995) 75–86
15. Morin R.: *On Regular Message Sequence Chart Languages and Relationships to Mazurkiewicz Trace Theory.* FoSSaCS 2001, LNCS **2030** (2001) 332–346
16. Mukund M., Narayan Kumar K. and Sohoni M.: *Synthesizing distributed finite-state systems from MSCs.* LNCS **1877** (2000) 521–535
17. Muscholl A.: *Matching Specifications for Message Sequence Charts.* FoSSaCS'99, LNCS **1578** (1999) 273–287
18. Muscholl A. and Peled D.: *Message sequence graphs and decision problems on Mazurkiewicz traces.* LNCS **1672** (1999) 81–91
19. Ochmański E.: *Regular behaviour of concurrent systems.* Bulletin of the EATCS **27** (Oct. 1985) 56–67
20. Reutenauer C.: *The Mathematics of Petri Nets.* (Masson, 1988)
21. Sakarovitch J.: *The "last" decision problem for rational trace languages.* LNCS **583** (1992) 460–473
22. Thomas W.: *On logical definability of trace languages.* Technical University of Munich, report TUM-I9002 (1990) 172–182

Strong Bisimilarity and Regularity of Basic Parallel Processes Is PSPACE-Hard

Jiří Srba*

BRICS**
Department of Computer Science, University of Aarhus,
Ny Munkegade bld. 540, 8000 Aarhus C, Denmark
srba@brics.dk

Abstract. We show that the problem of checking whether two processes definable in the syntax of Basic Parallel Processes (BPP) are strongly bisimilar is PSPACE-hard.
We also demonstrate that there is a polynomial time reduction from the strong bisimilarity checking problem of regular BPP to the strong regularity (finiteness) checking of BPP. This implies that strong regularity of BPP is also PSPACE-hard.

1 Introduction

In verification of infinite-state systems (see e.g. [2]), there are two central problems concerned with decidability and complexity of (i) checking whether two processes are equivalent with regard to some behavioural equivalence and (ii) model checking process properties expressed in a suitable logic. In this paper we study the first sort of problems.

Strong bisimilarity [15,13] is a well accepted notion of behavioural equivalence for concurrent processes. Unlike other equivalences in van Glabbeek's spectrum [20,21], strong bisimilarity remains decidable for many process algebras that generate infinite-state systems. There are e.g. positive results for bisimilarity checking of context-free process algebra [4] (also called BPA for Basic Process Algebra) and its parallel analogue Basic Parallel Processes [3] (BPP). We study here the BPP process algebra, a fragment of CCS [13] without restriction, relabelling and communication.

Decidability of strong bisimilarity for BPP was demonstrated by Christensen et al. [3], however, the algorithm relies on Dickson's Lemma [5] and its exact complexity is unknown. To the best of our knowledge, no algorithm with elementary complexity (bounded number of exponentiations) has been found yet. On the other hand, if we consider only *normed* BPP processes (from every reachable state there is a terminating computation), strong bisimilarity is decidable in polynomial time [6]. The conjecture that the unnormed case is also decidable

---

* The author is supported in part by the GACR, grant No. 201/00/0400.
** **B**asic **R**esearch **i**n **C**omputer **S**cience,
Centre of the Danish National Research Foundation.

in polynomial time was only recently proved to be wrong (unless P = NP) by Mayr. He showed that strong bisimilarity of BPP is co-NP-hard [11].

We improve Mayr's co-NP lower bound to PSPACE by demonstrating a reduction from QSAT [14]. The main idea of our proof is, however, different from Mayr's approach. Speaking in terms of bisimulation games of an attacker and a defender: in our proof we exploit the possibility that the defender can force the attacker to switch sides in the bisimulation game an arbitrary number of times. Moreover, we introduce a technique in which the defender can decide on the successive configurations in the bisimulation game independently of the attacker's will. This enables to encode quantified instances of SAT problems as BPP processes. Similar ideas appeared in [7] in connection with high undecidability of weak bisimilarity for Petri nets, but the other lower bound proofs (see e.g. [18,11,16]) use only a constant number of switching sides when generating an instance of a problem.

Another issue that is often studied is strong regularity (finiteness) checking. The question is whether a given BPP process is strongly bisimilar to some finite-state process. Strong regularity is decidable even for Petri nets [8]. As BPP systems are a communication free subclass of Petri nets, strong regularity is decidable for them as well. Again, restricting ourself to the normed case only, the problem is solvable in polynomial time [10]. The complexity of the unnormed case is open. It is known that the problem is co-NP-hard [11] but no elementary upper bound has been established.

In this paper, we provide a polynomial time reduction from strong bisimilarity checking of regular BPP to strong regularity checking. Hence using our PSPACE-hardness of strong bisimilarity for BPP (processes used in the proof are finite-state and thus trivially regular), we get PSPACE lower bound for strong regularity checking of BPP.

2   Basic Definitions

A *labelled transition system* is a triple $(S, \mathcal{A}ct, \longrightarrow)$ where S is a set of *states* (or *processes*), $\mathcal{A}ct$ is a set of *labels* (or *actions*), and $\longrightarrow \subseteq S \times \mathcal{A}ct \times S$ is a *transition relation*, written $\alpha \xrightarrow{a} \beta$, for $(\alpha, a, \beta) \in \longrightarrow$.

Let $\mathcal{A}ct$ and $\mathcal{C}onst$ be countable sets of *actions* and *process constants*, respectively, such that $\mathcal{A}ct \cap \mathcal{C}onst = \emptyset$. We define the class of BPP *process expressions* $\mathcal{E}^{\mathcal{C}onst}$ over $\mathcal{C}onst$ by the following abstract syntax

$$E ::= \epsilon \mid X \mid E\|E$$

where 'ϵ' is the *empty process* and X ranges over $\mathcal{C}onst$. The operator '$\|$' stands for a *parallel composition*. We do not distinguish between process expressions related by a *structural congruence*, which is the smallest congruence over process expressions such that '$\|$' is associative and commutative, and 'ϵ' is a unit for '$\|$'.

A BPP *process rewrite system* (or a $(1,P)$-PRS in the terminology of [12]) is a finite set $\Delta \subseteq \mathcal{C}onst \times \mathcal{A}ct \times \mathcal{E}^{\mathcal{C}onst}$ of *rewrite rules*, written $X \xrightarrow{a} E$ for

$(X, a, E) \in \Delta$. Let us denote the set of actions and process constants that appear in Δ by $\mathcal{Act}(\Delta)$ and $\mathcal{Const}(\Delta)$, respectively. Note that $\mathcal{Act}(\Delta)$ and $\mathcal{Const}(\Delta)$ are finite sets.

A process rewrite system Δ determines a labelled transition system where *states* are process expressions over $\mathcal{Const}(\Delta)$, $\mathcal{Act}(\Delta)$ is the set of *labels*, and *transition relation* is the least relation satisfying the following SOS rules (recall that '$\|$' is commutative).

$$\frac{(X \xrightarrow{a} E) \in \Delta}{X \xrightarrow{a} E} \qquad \frac{E \xrightarrow{a} E'}{E\|F \xrightarrow{a} E'\|F}$$

As usual we extend the transition relation to the elements of $\mathcal{Act}^*$. We also write $E \longrightarrow^* E'$, whenever $E \xrightarrow{w} E'$ for some $w \in \mathcal{Act}^*$. A state E' is *reachable from a state* E iff $E \longrightarrow^* E'$. We write $E \xrightarrow{a}\!\!\!/$ whenever there is no F such that $E \xrightarrow{a} F$, and $E \not\longrightarrow$ whenever $E \xrightarrow{a}\!\!\!/$ for all $a \in \mathcal{Act}$.

A BPP *process* is a pair (P, Δ), where Δ is a BPP process rewrite system and $P \in \mathcal{E}^{\mathcal{Const}(\Delta)}$ is a BPP process expression. *States* of (P, Δ) are the states of the corresponding transition system. We say that a state E is *reachable* in (P, Δ) iff $P \longrightarrow^* E$. Whenever (P, Δ) has only finitely many reachable states, we call it a *finite-state process*. A process (P, Δ) is *normed* iff from every reachable state E in (P, Δ) there is a terminating computation, i.e., there is some E' such that $E \longrightarrow^* E' \not\longrightarrow$.

Remark 1. Let i be a natural number and $A \in \mathcal{Const}$. We use the notation A^i for a parallel composition of i occurrences of A, i.e., $A^0 \stackrel{\text{def}}{=} \epsilon$ and $A^{i+1} \stackrel{\text{def}}{=} A^i\|A$.

Let Δ be a process rewrite system. A binary relation $R \subseteq \mathcal{E}^{\mathcal{Const}(\Delta)} \times \mathcal{E}^{\mathcal{Const}(\Delta)}$ over process expressions is a *strong bisimulation* iff whenever $(E, F) \in R$ then for each $a \in \mathcal{Act}(\Delta)$: if $E \xrightarrow{a} E'$ then $F \xrightarrow{a} F'$ for some F' such that $(E', F') \in R$; and if $F \xrightarrow{a} F'$ then $E \xrightarrow{a} E'$ for some E' such that $(E', F') \in R$.

Processes (P_1, Δ) and (P_2, Δ) are *strongly bisimilar*, and we write $(P_1, \Delta) \sim (P_2, \Delta)$, iff there is a strong bisimulation R such that $(P_1, P_2) \in R$. Given a pair of processes (P_1, Δ_1) and (P_2, Δ_2) such that $\Delta_1 \neq \Delta_2$, we write $(P_1, \Delta_1) \sim (P_2, \Delta_2)$ iff $(P_1, \Delta) \sim (P_2, \Delta)$ where Δ is a disjoint union of Δ_1 and Δ_2.

We say that a process (P, Δ) is *strongly regular* iff there exists some finite-state process bisimilar to (P, Δ).

Bisimulation equivalence has an elegant characterisation in terms of *bisimulation games* [19,17]. A bisimulation game on a pair of processes (P_1, Δ) and (P_2, Δ) is a two-player game of an 'attacker' and a 'defender'. The game is played in rounds. In each round the attacker chooses one of the processes and makes an $\xrightarrow{a}$-move for some $a \in \mathcal{Act}(\Delta)$ and the defender must respond by making an $\xrightarrow{a}$-move in the other process under the same action a. Now the game repeats, starting from the new processes. If one player cannot move, the other player wins. If the game is infinite, the defender wins. Processes (P_1, Δ) and (P_2, Δ) are strongly bisimilar iff the defender has a winning strategy (and non-bisimilar iff the attacker has a winning strategy).

3   Hardness of Strong Bisimilarity

> **Problem:** Strong bisimilarity of BPP
> **Instance:** Two BPP processes (P_1, Δ) and (P_2, Δ).
> **Question:** $(P_1, \Delta) \sim (P_2, \Delta)$?

We show that strong bisimilarity of BPP is a PSPACE-hard problem. We prove it by a reduction from QSAT[1], which is PSPACE-complete [14]. We use a version where the prefix of quantifiers starts with the existential one.

> **Problem:** QSAT
> **Instance:** A natural number n and a Boolean formula ϕ in conjunctive normal form with Boolean variables $x_1, \ldots, x_n$ and $y_1, \ldots, y_n$.
> **Question:** Is $\exists x_1 \forall y_1 \exists x_2 \forall y_2 \ldots \exists x_n \forall y_n . \phi$ true?

A *literal* is a variable or the negation of a variable. Let

$$C \equiv \exists x_1 \forall y_1 \exists x_2 \forall y_2 \ldots \exists x_n \forall y_n.\ C_1 \wedge C_2 \wedge \ldots \wedge C_k$$

be an instance of QSAT, where each *clause* C_j, $1 \leq j \leq k$, is a disjunction of literals. We define the following BPP processes (X_1, Δ) and (X_1', Δ), where

$$
\begin{aligned}
Const(\Delta) \stackrel{\text{def}}{=} &\ \{Q_1, \ldots, Q_k\} \cup \\
&\ \{X_1, \ldots, X_{n+1}, Y_1, \ldots, Y_n, Z_1, \ldots, Z_n\} \cup \\
&\ \{X_1', \ldots, X_{n+1}', Y_1^{\text{tt}}, \ldots, Y_n^{\text{tt}}, Y_1^{\text{ff}}, \ldots, Y_n^{\text{ff}}, Z_1', \ldots, Z_n'\}
\end{aligned}
$$

and

$$\mathcal{A}ct(\Delta) \stackrel{\text{def}}{=} \{q_1, \ldots, q_k, a, \text{tt}, \text{ff}\}.$$

For each i, $1 \leq i \leq n$, let

α_i be a parallel composition of process constants from $\{Q_1, \ldots, Q_k\}$ such that Q_j appears in α_i iff the literal x_i occurs in C_j (i.e. if x_i is set to true then C_j is satisfied),

$\overline{\alpha_i}$ be a parallel composition of process constants from $\{Q_1, \ldots, Q_k\}$ such that Q_j appears in $\overline{\alpha_i}$ iff the literal $\neg x_i$ occurs in C_j (i.e. if x_i is set to false then C_j is satisfied),

β_i be a parallel composition of process constants from $\{Q_1, \ldots, Q_k\}$ such that Q_j appears in β_i iff the literal y_i occurs in C_j,

$\overline{\beta_i}$ be a parallel composition of process constants from $\{Q_1, \ldots, Q_k\}$ such that Q_j appears in $\overline{\beta_i}$ iff the literal $\neg y_i$ occurs in C_j.

Example 1. Let us consider a quantified formula

$$\exists x_1 \forall y_1 \exists x_2 \forall y_2.\ (x_1 \vee \neg y_1 \vee y_2) \wedge (\neg x_1 \vee y_1 \vee y_2) \wedge (x_1 \vee y_1 \vee y_2 \vee \neg y_2)$$

where $n = 2$, $k = 3$, $C_1 = x_1 \vee \neg y_1 \vee y_2$, $C_2 = \neg x_1 \vee y_1 \vee y_2$ and $C_3 = x_1 \vee y_1 \vee y_2 \vee \neg y_2$. Then $\alpha_1 = Q_1 \| Q_3$, $\overline{\alpha_1} = Q_2$, $\beta_1 = Q_2 \| Q_3$, $\overline{\beta_1} = Q_1$, $\alpha_2 = \epsilon$, $\overline{\alpha_2} = \epsilon$, $\beta_2 = Q_1 \| Q_2 \| Q_3$, and $\overline{\beta_2} = Q_3$.

---

[1] This problem is known also as QBF, for *Quantified Boolean formula*.

The set Δ is given by the following rewrite rules:

- for all j such that $1 \leq j \leq k$: $Q_j \xrightarrow{q_j} Q_j$

- for all i such that $1 \leq i \leq n$:

$$X_i \xrightarrow{a} Y_i$$
$$X_i \xrightarrow{a} Y_i^{tt} \qquad\qquad X_i' \xrightarrow{a} Y_i^{tt}$$
$$X_i \xrightarrow{a} Y_i^{ff} \qquad\qquad X_i' \xrightarrow{a} Y_i^{ff}$$

$$Y_i \xrightarrow{tt} Z_i\|\alpha_i \qquad Y_i^{tt} \xrightarrow{tt} Z_i'\|\alpha_i \qquad Y_i^{tt} \xrightarrow{tt} Z_i\|\alpha_i \qquad Y_i^{tt} \xrightarrow{ff} Z_i\|\overline{\alpha_i}$$
$$Y_i \xrightarrow{ff} Z_i\|\overline{\alpha_i} \qquad Y_i^{ff} \xrightarrow{ff} Z_i'\|\overline{\alpha_i} \qquad Y_i^{ff} \xrightarrow{tt} Z_i\|\alpha_i \qquad Y_i^{ff} \xrightarrow{ff} Z_i\|\overline{\alpha_i}$$

$$Z_i \xrightarrow{tt} X_{i+1}\|\beta_i \qquad Z_i' \xrightarrow{tt} X_{i+1}'\|\beta_i$$
$$Z_i \xrightarrow{ff} X_{i+1}\|\overline{\beta_i} \qquad Z_i' \xrightarrow{ff} X_{i+1}'\|\overline{\beta_i}$$

- and $X_{n+1} \xrightarrow{a} Q_1\|Q_2\|\cdots\|Q_{k-1}\|Q_k \qquad X_{n+1}' \xrightarrow{a} \epsilon.$

We can see the processes (X_1, Δ) and (X_1', Δ) in Figure 1 if we set $i = 1$ and $\gamma_1 = \epsilon$. The intuition behind the construction can be nicely explained in terms of bisimulation games. Consider a bisimulation game starting from X_1 and X_1'.

The attacker is forced to make the first move by playing $X_1 \xrightarrow{a} Y_1$ because in all other possible moves, either from X_1 or X_1', the defender can make the resulting processes syntactically equal and hence bisimilar. The defender's answer to the move $X_1 \xrightarrow{a} Y_1$ is either (i) $X_1' \xrightarrow{a} Y_1^{tt}$ (this corresponds to setting the variable x_1 to true) or (ii) $X_1' \xrightarrow{a} Y_1^{ff}$ (this corresponds to setting the variable x_1 to false).

In the next round the attacker is forced to switch processes and play either (i) $Y_1^{tt} \xrightarrow{tt} Z_1'\|\alpha_1$ or (ii) $Y_1^{ff} \xrightarrow{ff} Z_1'\|\overline{\alpha_1}$, according to the defender's choice in the first round. If the attacker chooses any other move, the defender has the possibility to make the resulting processes syntactically equal. Now, the defender has only one possible answer by playing in the first process — in the case (i) he plays $Y_1 \xrightarrow{tt} Z_1\|\alpha_1$ and in the case (ii) he plays $Y_1 \xrightarrow{ff} Z_1\|\overline{\alpha_1}$. The resulting processes after two rounds are (i) $Z_1\|\alpha_1$ and $Z_1'\|\alpha_1$ or (ii) $Z_1\|\overline{\alpha_1}$ and $Z_1'\|\overline{\alpha_1}$. Note that it was the defender who had the possibility to decide between adding α_1 (i.e. setting x_1 to true) or $\overline{\alpha_1}$ (i.e. setting x_1 to false).

In the third round the attacker has the choice of playing either along the action tt or ff, which corresponds to the universal quantifier in front of y_1. It does not matter in which process the attacker performs the move. The defender has only one possibility how to answer to this move — he must imitate the corresponding move in the other process. The resulting processes are $X_2\|\gamma_2$ and $X_2'\|\gamma_2$ such that $\gamma_2 = \tilde{\alpha}_1\|\tilde{\beta}_1$ where $\tilde{\alpha}_1 \in \{\alpha_1, \overline{\alpha_1}\}$ and $\tilde{\beta}_1 \in \{\beta_1, \overline{\beta_1}\}$ according to the truth values chosen for x_1 (by the defender) and for y_1 (by the attacker). Now the game continues in similar fashion from $X_2\|\gamma_2$ and $X_2'\|\gamma_2$. Playing some

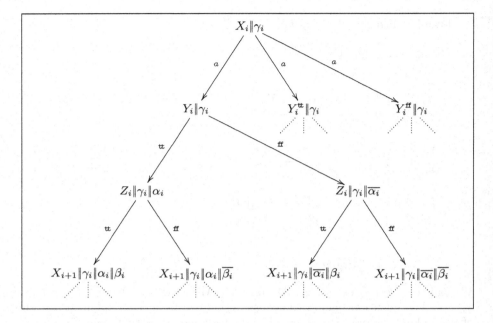

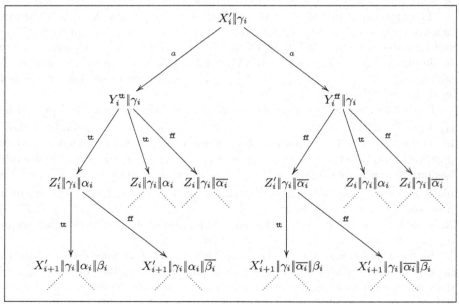

Fig. 1. Processes $(X_i \| \gamma_i, \Delta)$ and $(X_i' \| \gamma_i, \Delta)$

of the actions $q_1, \ldots, q_k$ cannot make the attacker win since the defender has always the possibility to imitate the same move in the other processes.

Hence if the attacker wants to win he has to reach eventually the states $X_{n+1} \| \gamma_{n+1}$ and $X'_{n+1} \| \gamma_{n+1}$, and then he performs the move $X_{n+1} \| \gamma_{n+1} \xrightarrow{a} Q_1 \| Q_2 \| \ldots \| Q_{k-1} \| Q_k \| \gamma_{n+1}$ to which the defender has only one answer, namely $X'_{n+1} \| \gamma_{n+1} \xrightarrow{a} \gamma_{n+1}$. From the states $Q_1 \| Q_2 \| \ldots \| Q_{k-1} \| Q_k \| \gamma_{n+1}$ and γ_{n+1} the attacker has the possibility to check whether every clause C_j, $1 \leq j \leq k$, in C is indeed satisfied under the generated truth assignment by using the rule $Q_j \xrightarrow{q_j} Q_j$ in the first process. If the clause C_j is not satisfied then Q_j does not appear in γ_{n+1} and the defender loses. If all the clauses in C are satisfied then $Q_1 \| Q_2 \| \ldots \| Q_{k-1} \| Q_k \| \gamma_{n+1} \sim \gamma_{n+1}$ and the defender wins.

In what follows we formally prove that C is true iff $(X_1, \Delta) \sim (X'_1, \Delta)$.

Lemma 1. *If $(X_1, \Delta) \sim (X'_1, \Delta)$ then C is true.*

Proof. We show that $(X_1, \Delta) \not\sim (X'_1, \Delta)$ under the assumption that C is false. Then C' defined by $C' \stackrel{\text{def}}{=} \forall x_1 \exists y_1 \forall x_2 \exists y_2 \ldots \forall x_n \exists y_n. \neg(C_1 \wedge C_2 \wedge \ldots \wedge C_k)$ is true and we claim that the attacker has a winning strategy in the bisimulation game starting from (X_1, Δ) and (X'_1, Δ). The attacker's strategy starts with performing a sequence of actions $a, \widetilde{x}_1, \widetilde{y}_1, \ldots, a, \widetilde{x}_i, \widetilde{y}_i, \ldots, a, \widetilde{x}_n, \widetilde{y}_n, a$ where $\widetilde{x}_i, \widetilde{y}_i \in \{\mathtt{tt}, \mathtt{ff}\}$ for all i, $1 \leq i \leq n$. The attacker is switching regularly the sides of processes as described above, i.e., actions a are played on the side of the first process (X_1, Δ), $\widetilde{x}_i$ are played on the side of the second process (X'_1, Δ), and $\widetilde{y}_i$ are played, let us say, in the first process again. The choice of $\widetilde{x}_i$ is done by the defender and of $\widetilde{y}_i$ by the attacker — again see the discussion above. This means that whatever values for $\widetilde{x}_1, \ldots, \widetilde{x}_n$ are chosen by the defender, the attacker can still decide on values for $\widetilde{y}_1, \ldots, \widetilde{y}_n$ such that the generated assignment satisfies the formula $\neg(C_1 \wedge C_2 \wedge \ldots \wedge C_k)$. Hence there must be some j, $1 \leq j \leq k$, such that the clause C_j is not satisfied. This implies that Q_j does not occur in the second process. However, the attacker can perform the action q_j in the first process by using the rule $Q_j \xrightarrow{q_j} Q_j$. Thus the attacker has a winning strategy in the bisimulation game and $(X_1, \Delta) \not\sim (X'_1, \Delta)$. $\square$

Lemma 2. *If C is true then $(X_1, \Delta) \sim (X'_1, \Delta)$.*

Proof. Let us define sets $\mathcal{AS}_i$, corresponding to the assignments of variables from x_1, y_1 to x_i, y_i, $1 \leq i \leq n$, such that the formula $\exists x_{i+1} \forall y_{i+1} \ldots \exists x_n \forall y_n. C_1 \wedge C_2 \wedge \ldots \wedge C_k$ is still true. Sets $\mathcal{AS}_i$ are defined by

$$\mathcal{AS}_i \stackrel{\text{def}}{=} \{ \ \widetilde{\alpha}_1 \| \widetilde{\beta}_1 \| \widetilde{\alpha}_2 \| \widetilde{\beta}_2 \| \ldots \| \widetilde{\alpha}_i \| \widetilde{\beta}_i \ | $$

such that for all j, $1 \leq j \leq i$, it holds that $\widetilde{\alpha}_j \in \{\alpha_j, \overline{\alpha_j}\}$ and $\widetilde{\beta}_j \in \{\beta_j, \overline{\beta_j}\}$, and under the assignment

$$x_j = \begin{cases} \mathtt{tt} & \text{if } \widetilde{\alpha}_j = \alpha_j \\ \mathtt{ff} & \text{if } \widetilde{\alpha}_j = \overline{\alpha_j} \end{cases} \quad \text{and} \quad y_j = \begin{cases} \mathtt{tt} & \text{if } \widetilde{\beta}_j = \beta_j \\ \mathtt{ff} & \text{if } \widetilde{\beta}_j = \overline{\beta_j} \end{cases}$$

the formula $\exists x_{i+1} \forall y_{i+1} \ldots \exists x_n \forall y_n. C_1 \wedge C_2 \wedge \ldots \wedge C_k$ is true $\}$.

By definition $\mathcal{AS}_0 = \{\epsilon\}$. In particular, $\mathcal{AS}_n$ contains all the assignments for which the unquantified formula $C_1 \wedge C_2 \wedge \ldots \wedge C_k$ is true. The following relation is a strong bisimulation.

$$\{(X_i\|\gamma_i, X_i'\|\gamma_i) \mid 1 \le i \le n \,\wedge\, \gamma_i \in \mathcal{AS}_{i-1}\} \,\cup$$
$$\{(Y_i\|\gamma_i, Y_i^{\text{tt}}\|\gamma_i) \mid 1 \le i \le n \,\wedge\, \gamma_i\|\alpha_i\|\beta_i \in \mathcal{AS}_i \,\wedge\, \gamma_i\|\alpha_i\|\overline{\beta_i} \in \mathcal{AS}_i\} \,\cup$$
$$\{(Y_i\|\gamma_i, Y_i^{\text{ff}}\|\gamma_i) \mid 1 \le i \le n \,\wedge\, \gamma_i\|\overline{\alpha_i}\|\beta_i \in \mathcal{AS}_i \,\wedge\, \gamma_i\|\overline{\alpha_i}\|\overline{\beta_i} \in \mathcal{AS}_i\} \,\cup$$
$$\{(Z_i\|\gamma_i\|\alpha_i, Z_i'\|\gamma_i\|\alpha_i) \mid 1 \le i \le n \,\wedge\, \gamma_i\|\alpha_i\|\beta_i \in \mathcal{AS}_i \,\wedge\, \gamma_i\|\alpha_i\|\overline{\beta_i} \in \mathcal{AS}_i\} \,\cup$$
$$\{(Z_i\|\gamma_i\|\overline{\alpha_i}, Z_i'\|\gamma_i\|\overline{\alpha_i}) \mid 1 \le i \le n \,\wedge\, \gamma_i\|\overline{\alpha_i}\|\beta_i \in \mathcal{AS}_i \,\wedge\, \gamma_i\|\overline{\alpha_i}\|\overline{\beta_i} \in \mathcal{AS}_i\} \,\cup$$
$$\{(X_{n+1}\|\gamma_{n+1}, X_{n+1}'\|\gamma_{n+1}) \mid \gamma_{n+1} \in \mathcal{AS}_n\} \,\cup$$
$$\{(Q_1\|Q_2\|\ldots\|Q_{k-1}\|Q_k\|\gamma_{n+1}, \gamma_{n+1}) \mid \gamma_{n+1} \in \mathcal{AS}_n\} \,\cup$$
$$\{(E, E) \mid E \in \mathcal{E}^{Const(\Delta)}\}$$

Since $\mathcal{AS}_0 = \{\epsilon\}$, we get that the pair (X_1, X_1') is an element of this relation. Hence we proved that $(X_1, \Delta) \sim (X_1', \Delta)$. $\qquad\square$

Theorem 1. *Strong bisimilarity of BPP is PSPACE-hard.*

Proof. By Lemma 1 and Lemma 2. $\qquad\square$

Remark 2. Notice that there are only finitely many reachable states from both (X_1, Δ) and (X_1', Δ). Hence (X_1, Δ) and (X_1', Δ) are strongly regular processes.

Remark 3. Theorem 1 can be easily extended to 1-safe Petri nets where each transition has exactly one input place (for the definition of 1-safe Petri nets see e.g. [9]). It is enough to introduce for each $\alpha_i/\overline{\alpha_i}$ and $\beta_i/\overline{\beta_i}$, $1 \le i \le n$, a new set of process constants $\{Q_1, \ldots, Q_k\}$ to ensure that in each reachable marking there is at most one token in every place. Related results about 1-safe Petri nets can be found in [9].

4   Hardness of Strong Regularity

Problem:	Strong regularity of BPP
Instance:	A BPP process (P, Δ).
Question:	Is there a finite-state process (F, Δ') such that $(P, \Delta) \sim (F, \Delta')$?

The idea of the next theorem's proof appeared first in [11] in the context of weak bisimilarity. We adapt this idea to the case of strong bisimilarity (in the proof from [11] τ actions are needed).

Theorem 2 (Reduction from bisimilarity to regularity).
Let (P_1, Δ) and (P_2, Δ) be strongly regular BPP processes. We can construct in polynomial time a BPP process (P, Δ') such that

$$(P_1, \Delta) \sim (P_2, \Delta) \quad \text{if and only if} \quad (P, \Delta') \text{ is strongly regular.}$$

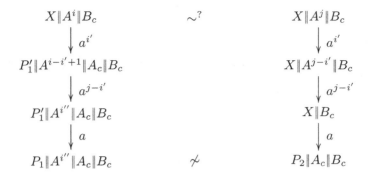

Fig. 2. Winning strategy for the attacker $(i < j)$

Proof. Assume that (P_1, Δ) and (P_2, Δ) are strongly regular. We construct a BPP process (P, Δ') with $Const(\Delta') \stackrel{\text{def}}{=} Const(\Delta) \cup \{X, A, A_c, B_c, P_1', P_2'\}$ and $Act(\Delta') \stackrel{\text{def}}{=} Act(\Delta) \cup \{a, b\}$, where $X, A, A_c, B_c, P_1', P_2'$ are new process constants and a, b are new actions. We define $\Delta' \stackrel{\text{def}}{=} \Delta \cup \Delta^1 \cup \Delta^2$, where the set of rewrite rules Δ^1 is given by

$$X \stackrel{b}{\longrightarrow} X \| A \qquad A \stackrel{a}{\longrightarrow} \epsilon \qquad A_c \stackrel{a}{\longrightarrow} A_c \qquad B_c \stackrel{b}{\longrightarrow} B_c$$
$$X \stackrel{a}{\longrightarrow} P_1' \| A_c \qquad X \stackrel{a}{\longrightarrow} P_1 \| A_c \qquad P_1' \stackrel{a}{\longrightarrow} P_1$$

and Δ^2 is given by

$$X \stackrel{a}{\longrightarrow} P_2' \| A_c \qquad X \stackrel{a}{\longrightarrow} P_2 \| A_c \qquad P_2' \stackrel{a}{\longrightarrow} P_2.$$

Let $P \stackrel{\text{def}}{=} X \| B_c$.

Lemma 3. *If* $(P_1, \Delta) \not\sim (P_2, \Delta)$ *then* (P, Δ') *is not strongly regular.*

Proof. Let $(P_1, \Delta) \not\sim (P_2, \Delta)$. For simplicity (and without loss of generality) we assume that $P_1 \not\sim \epsilon$ and $P_2 \not\sim \epsilon$. We demonstrate that there are infinitely many strongly nonbisimilar states reachable from (P, Δ').

Let us consider an infinite number of states of the form $X \| A^i \| B_c$ for any natural number i. Of course $P \longrightarrow^* X \| A^i \| B_c$ and we claim that $(X \| A^i \| B_c, \Delta') \not\sim (X \| A^j \| B_c, \Delta')$ for any $i \neq j$. Without loss of generality assume that $i < j$. The attacker has the following winning strategy (playing only in the second process — see Figure 2).

He performs a sequence of j actions a from $X \| A^j \| B_c$, thus reaching a state $X \| B_c$. The defender playing from $X \| A^i \| B_c$ cannot do this sequence of a-actions without using some rule for X. This is because $B_c \stackrel{a}{\not\longrightarrow}$ and A^i can perform at most i a-actions $(i < j)$. As we assume that $P_1 \not\sim \epsilon$ and $P_2 \not\sim \epsilon$, process constants P_1 nor P_2 cannot appear in the defender's process during the first j

rounds, otherwise he loses immediately. So the defender has to make a choice between the rules $X \xrightarrow{a} P_1' \| A_c$ and $X \xrightarrow{a} P_2' \| A_c$ sometime within the first j moves (let us say in round i' where $i' \leq i+1$). Assume that the defender chooses $X \xrightarrow{a} P_1' \| A_c$ — the other case is symmetric. Now, the defender must perform $j - i'$ of a-actions by using the rules $A_c \xrightarrow{a} A_c$ or $A \xrightarrow{a} \epsilon$.

After j rounds the resulting states are $P_1' \| A^{i''} \| A_c \| B_c$ for $i'' \leq i - i' + 1$, and $X \| B_c$. The attacker wins by performing the move $X \| B_c \xrightarrow{a} P_2 \| A_c \| B_c$. Again, since $P_2 \not\sim \epsilon$ the defender has to answer with $P_1' \| A^{i''} \| A_c \| B_c \xrightarrow{a} P_1 \| A^{i''} \| A_c \| B_c$. The attacker has a winning strategy from $P_1 \| A^{i''} \| A_c \| B_c$ and $P_2 \| A_c \| B_c$ now: the fact that $P_1 \not\sim P_2$ and that the actions a and b are fresh ones implies that $P_1 \| A^{i''} \| A_c \| B_c \not\sim P_2 \| A_c \| B_c$.   □

Lemma 4. *If* $(P_1, \Delta) \sim (P_2, \Delta)$ *then* (P, Δ') *is strongly regular.*

Proof. Assume that $(P_1, \Delta) \sim (P_2, \Delta)$, which implies that $(P, \Delta') \sim (P, \Delta'')$, where $\Delta'' \stackrel{\text{def}}{=} \Delta' \smallsetminus \Delta^2$ (strong bisimilarity is a congruence w.r.t. the parallel operator). It is enough to show that (P, Δ'') is strongly regular. Observe that $(A^i \| A_c, \Delta'') \sim (A_c, \Delta'')$ for any i such that $0 \leq i$. Then also $(P_1 \| A^i \| A_c \| B_c, \Delta'') \sim (P_1 \| A_c \| B_c, \Delta'')$ and $(P_1' \| A^i \| A_c \| B_c, \Delta'') \sim (P_1' \| A_c \| B_c, \Delta'')$ for any i such that $0 \leq i$. This implies that

$$(X \| A^i \| B_c, \Delta'') \sim (P_1' \| A_c \| B_c, \Delta'') \tag{1}$$

for any i such that $0 \leq i$. Since (P_1, Δ) is strongly regular then $(P_1' \| A_c \| B_c, \Delta'')$ is also strongly regular. This by using (1) in particular gives that $(X \| A^0 \| B_c, \Delta'') = (X \| B_c, \Delta'') = (P, \Delta'')$ is strongly regular.   □

Theorem 2 follows from Lemma 3 and Lemma 4.   □

Theorem 3. *Strong regularity of BPP is PSPACE-hard.*

Proof. By Theorem 1, Remark 2 and Theorem 2.   □

5   Conclusion

We proved PSPACE-hardness of strong bisimilarity and regularity of BPP by using a technique in which the defender has the possibility to force the attacker to perform a certain move and switch sides of processes arbitrarily many times. We belive that this technique can be used in other hardness results concerning bisimilarity. Another contribution of this paper is with regard to the normedness notion. Our results show a substantial difference (unless P = NP) between *normed* and *unnormed* processes — strong bisimilarity and regularity checking is PSPACE-hard for unnormed BPP, whereas it is decidable in polynomial time for normed BPP [6,10].

Recently some lower bounds appeared for *weak* bisimilarity of BPA and BPP [18,11,16], even though the problems are not known to be decidable. In the following tables we compare the results for strong/weak bisimilarity and regularity.

New results achieved in this paper are in boldface. Obviously, the lower bounds for strong bisimilarity and regularity apply also to weak bisimilarity and regularity. Hence we have another proof of PSPACE-hardness of weak bisimilarity and regularity for BPP. The difference is that in the proof from [16] only a constant number of switching sides during the bisimulation game is needed for the attacker to win, and even more importantly, the proof is valid also for the normed case. On the other hand the usage of τ actions is essential for the reductions from [16].

	strong bisimilarity	weak bisimilarity
BPP	decidable [3] **PSPACE-hard**	? **PSPACE-hard** [16]
normed BPP	decidable in P [6] P-hard [1]	? PSPACE-hard [16]

	strong regularity	weak regularity
BPP	decidable [8] **PSPACE-hard**	? **PSPACE-hard** [16]
normed BPP	decidable in NL [10] **NL-hard**	? PSPACE-hard [16]

Remark 4. Complexity of strong regularity of normed BPP needs more explanation. Kucera in [10] argues that the problem is decidable in polynomial time, but it is easy to see that a test whether a BPP process contains an *accessible* and *growing* process constant (a condition equivalent to regularity checking) can be performed even in nondeterministic logarithmic space (NL).

In order to prove NL-hardness, we reduce the reachability problem for acyclic directed graphs (NL-complete problem, see [14]) to strong regularity checking of normed BPP. Given an acyclic directed graph G with a pair of nodes v_1 and v_2, we naturally construct a BPP system Δ by introducing a new process constant for each node of G, with transitions respecting the edges of G and labelled by a. Moreover, a process constant representing the node v_2 has a transition to a new process constant A such that Δ contains also the rewrite rules $A \xrightarrow{a} A \| A$ and $A \xrightarrow{b} \epsilon$. It is easy to see that (A, Δ) is a normed and non-regular process. Let X be a process constant representing the node v_1. Since G is acyclic, (X, Δ) is a normed BPP process. Obviously, there is a directed path from v_1 to v_2 in G if and only if (X, Δ) is not a strongly regular process. Recall that NL = co-NL (see e.g. [14]). Hence strong regularity of normed BPP is NL-complete.

Acknowledgement. I would like to thank my advisor Mogens Nielsen for his kind supervision. I also thank Jan Strejcek and the referees for useful remarks.

References

[1] J. Balcazar, J. Gabarro, and M. Santha. Deciding bisimilarity is P-complete. *Formal Aspects of Computing*, 4(6A):638–648, 1992.

[2] O. Burkart, D. Caucal, F. Moller, and B. Steffen. Verification on infinite structures. In J. Bergstra, A. Ponse, and S. Smolka, editors, *Handbook of Process Algebra*, chapter 9, pages 545–623. Elsevier Science, 2001.

[3] S. Christensen, Y. Hirshfeld, and F. Moller. Bisimulation is decidable for basic parallel processes. In *Proc. of CONCUR'93*, volume 715 of *LNCS*, pages 143–157. Springer-Verlag, 1993.

[4] S. Christensen, H. Hüttel, and C. Stirling. Bisimulation equivalence is decidable for all context-free processes. *Information and Comp.*, 121:143–148, 1995.

[5] L.E. Dickson. Finiteness of the odd perfect and primitive abundant numbers with distinct factors. *American Journal of Mathematics*, 35:413–422, 1913.

[6] Y. Hirshfeld, M. Jerrum, and F. Moller. A polynomial-time algorithm for deciding bisimulation equivalence of normed basic parallel processes. *Mathematical Structures in Computer Science*, 6(3):251–259, 1996.

[7] P. Jancar. High undecidability of weak bisimilarity for Petri nets. In *Proc. of CAAP'95*, volume 915 of *LNCS*, pages 349–363. Springer-Verlag, 1995.

[8] P. Jancar and J. Esparza. Deciding finiteness of Petri nets up to bisimilarity. In *Proc. of ICALP'96*, volume 1099 of *LNCS*, pages 478–489. Springer-Verlag, 1996.

[9] L. Jategaonkar and A.R. Meyer. Deciding true concurrency equivalences on safe, finite nets. *Theoretical Computer Science*, 154(1):107–143, 1996.

[10] A. Kucera. Regularity is decidable for normed BPA and normed BPP processes in polynomial time. In *Proc. of SOFSEM'96*, volume 1175 of *LNCS*, pages 377–384. Springer-Verlag, 1996.

[11] R. Mayr. On the complexity of bisimulation problems for basic parallel processes. In *Proc. ICALP'00*, volume 1853 of *LNCS*, pages 329–341. Springer-Verlag, 2000.

[12] R. Mayr. Process rewrite systems. *Information and Comp.*, 156(1):264–286, 2000.

[13] R. Milner. *Communication and Concurrency*. Prentice-Hall, 1989.

[14] Ch.H. Papadimitriou. *Computational Complexity*. Addison-Wesley, 1994.

[15] D.M.R. Park. Concurrency and automata on infinite sequences. In *Proc. of 5th GI Conference*, volume 104 of *LNCS*, pages 167–183. Springer-Verlag, 1981.

[16] J. Srba. Complexity of weak bisimilarity and regularity for BPA and BPP. In *Proc. of EXPRESS'00*, volume 39 of *ENTCS*. Elsevier Science, 2000. To appear.

[17] C. Stirling. Local model checking games. In *Proc. of CONCUR'95*, volume 962 of *LNCS*, pages 1–11. Springer-Verlag, 1995.

[18] J. Stribrna. Hardness results for weak bisimilarity of simple process algebras. In *Proc. of the MFCS'98 Workshop on Concurrency*, volume 18 of *ENTCS*. Springer-Verlag, 1998.

[19] W. Thomas. On the Ehrenfeucht-Fraïssé game in theoretical computer science (extended abstract). In *Proc. of the 4th International Joint Conference CAAP/FASE, Theory and Practice of Software Development (TAPSOFT'93)*, volume 668 of *LNCS*, pages 559–568. Springer-Verlag, 1993.

[20] R.J. van Glabbeek. *Comparative Concurrency Semantics and Refinement of Actions*. PhD thesis, CWI/Vrije Universiteit, 1990.

[21] R.J. van Glabbeek. The linear time—branching time spectrum. In *Proc. of CONCUR'90*, volume 458 of *LNCS*, pages 278–297. Springer-Verlag, 1990.

On the Enumerative Sequences of Regular Languages on k Symbols

Marie-Pierre Béal and Dominique Perrin

Institut Gaspard-Monge, Université de Marne-la-Vallée,
77454 Marne-la-Vallée Cedex 2, France.
{beal,perrin}@univ-mlv.fr
http://www-igm.univ-mlv.fr/{~beal,~perrin}

Abstract. The main result is a characterization of enumerative sequences of regular languages on k symbols. We prove that a sequence is the generating series $s(z)$ of a regular language on k symbols if and only if it is the generating series of a language over a k-letter alphabet and if both series $s(z)$ and $(kz)^* - s(z)$ are regular. The proof uses transformations on linear representations called inductions.

1 Introduction

The notion of enumerative sequence for a formal language L is a simple one: it is the sequence $(s_n)_{n \geq 0}$ where s_n is the number of words of length n in L. This sequence carries important information concerning a formal language since it measures in a sense the size of the language. It is moreover appropriate in the case of coding. In fact, a length-preserving encoding defines a one-to-one correspondence between words. The two sets of words in such a correspondence will have the same length distribution.

The characterization of the enumerative sequences of regular languages is known since a long time. These sequences, which we also call regular, are rational (i.e. the sequence of coefficients of a rational series) and closely linked with positive integer matrices and therefore the Perron-Frobenius theorem. The characterization among rational sequences was obtained through the work of J. Berstel, M. Soittola and others (see [1] for example). The idea of fixing the cardinality of the alphabet in this problem has surprisingly never been considered.

Suppose for example that we consider the regular language on three letters $L = (a + b)^* c^+$. It has the same length distribution as the regular language on two-symbols $L' = (a + b)^* a b^*$, namely $z/(1 - z)(1 - 2z)$. We address here the problem of characterizing the regular languages L for which such a coding on a smaller alphabet is possible and we describe explicitly how to realize it. Our main result is a characterization of the enumerative sequences of regular languages on k symbols.

Our characterization is the following. We prove that a sequence s is the generating sequence of a regular language on k symbols if and only if

- the sequence s is regular,

H. Alt and A. Ferreira (Eds.): STACS 2002, LNCS 2285, pp. 547–558, 2002.
© Springer-Verlag Berlin Heidelberg 2002

- for any integer $n \geq 0$, $s_n \leq k^n$,
- the complementary sequence t, defined by $t_n = k^n - s_n$ for any integer $n \geq 0$, is regular.

The proof is based on the use of left and right inductions, that we define as follows. A linear representation of a sequence $s = (s_n)_{n \geq 0}$ is a triple $(\mathbf{i}, M, \mathbf{t})$, where $\mathbf{i}$ is a row initial vector, $\mathbf{t}$ is a column final vector, and M a matrix, with $s_n = \mathbf{i} M^n \mathbf{t}$ for any nonnegative integer n. The triple is said to recognize s. A linear representation $(\mathbf{j}, N, \mathbf{x})$ is said to be left induced by $(\mathbf{i}, M, \mathbf{t})$ if there is a matrix U, called the transfer matrix, such that $NU = UM, \mathbf{j}U = \mathbf{i}, \mathbf{x} = U\mathbf{t}$. The inverse of a left induction is called a right induction. This notion of left induction extends to linear representations of sequences the notion of multiset construction and graph extension introduced in [2,3].

An important step in the proof of the main result is a left induction obtained by extending to linear representations over $\mathbb{Z}$ a theorem from Lind ([4], [5], see also [6, Theorem 11.1.4 p. 369]) which states that for any Perron number, there is a primitive integral matrix whose spectral radius is this Perron number. By taking into account the initial and final vectors in the case of spectrally Perron nonnegative rational sequences over $\mathbb{Z}$, we prove that a regular representation can be obtained by only one left induction from any left reduced linear representation over $\mathbb{Z}$ of the sequence.

This result appears to be a particular case of the following more general result. Let $s^{(1)}, s^{(2)}, \ldots, s^{(l)}$ be l regular sequences whose sum is $m(kz)^*$, where m and k are positive integers. Then there is a finite deterministically labelled graph G on a k-letter alphabet, with m initial states and a partition of the set of states of G in l sets T_i, with $1 \leq i \leq l$, such that the automaton (I, G, T_i) recognizes a regular language on k symbols whose generating series is exactly $s^{(i)}$. We prove this more general formulation.

The paper is organized as follows. Section 2 contains the definitions of linear representations and regular representations. Section 3 presents the characterization of generating sequences of regular languages over k symbols. A sketch of the proof of the main result is developed in Section 4. One first define (Section 4.1) the notion of a left or right induction. Section 4.2 establishes some lemmas based on Perron theory [7,8] which are used in Section 4.3 to show that, for any left reduced representation of a nonnegative Perron sequence, there is a left induction from this representation to a regular one. Section 4.4 presents the proof of the characterization of generating sequences of regular languages over k symbols. The proof is constructive in the sense that the regular language over k symbols can be build in an effective way, but with a high complexity. The construction process is composed of two left inductions followed by one right induction. We give an example of this computation.

2   Rational and Regular Sequences

Let K be a semiring. We consider sequences of elements of K denoted by $s = (s_n)_{n \geq 0}$. We shall not distinguish between such a sequence and its formal series in one variable $s(z) = \sum_{n \geq 0} s_n z^n$. We usually denote a vector of components in K and indexed by elements of a set Q, also called a Q-vector, with boldface symbols.

A sequence s is said to be K-*rational* if there exist a set Q of cardinality d and a triple $(\mathbf{i}, M, \mathbf{t})$, where $\mathbf{i}$ is a Q-row vector, $\mathbf{t}$ is a Q-column vector, and M is a $Q \times Q$ matrix, all with coefficients in K, such that, for any nonnegative integer n, $s_n = \mathbf{i} M^n \mathbf{t}$.

Such a triple is called a *linear representation* over K. We say that the triple $(\mathbf{i}, M, \mathbf{t})$ *recognizes* the sequence s.

If K is a commutative principal ring, a linear representation over K is said to be *left reduced* (respectively *right-reduced*) if and only the modulus generated by the vectors $\mathbf{i} M^n$ (respectively $M^n \mathbf{t}$), for all $n \geq 0$, is the full space $K^{1 \times d}$ (respectively $K^{d \times 1}$). The representation is *reduced* if and only if it is both left and right reduced (see [9, p. 26]).

We say that a sequence of integers is *nonnegative* if all its coefficients are nonnegative. A nonnegative sequence s is said to be *regular* if it is $\mathbb{N}$-rational. Equivalently, there is a finite directed multigraph G whose set of states is Q, a Q-row vector of nonnegative integers $\mathbf{i}$, a Q-column vector of nonnegative integers $\mathbf{t}$, such that, for any nonnegative integer n, $s_n = \mathbf{i} M^n \mathbf{t}$, where M is the adjacency matrix of G. The triple $(\mathbf{i}, G, \mathbf{t})$ is said to *recognize* the sequence s. It is also called a *regular representation* of s. When $\mathbf{i}$ (respectively $\mathbf{t}$) has $0 - 1$ components, it is the characteristic vector of a subset I (respectively T) of Q. The triple (I, G, T) is then said to recognize the sequence s. A regular sequence s is always recognized by such a triple and s_n is the number of paths of length n going from an initial to a final state.

A regular representation $(\mathbf{i}, M, \mathbf{t})$ is said to be *trim* if for each index p there is a nonnegative integer n such that $(\mathbf{i} M^n)_p > 0$ and there is a nonnegative integer m such that $(M^m \mathbf{t})_p > 0$.

A sequence $s = (s_n)_{n \geq 0}$ is said to be the *merge* of the sequences $s^{(0)}, \ldots, s^{(p-1)}$, where p is a positive integer, if and only if , for $0 \leq i \leq (p-1)$, $s_n^{(i)} = s_{i+np}$. If $(\mathbf{i}, M, \mathbf{t})$ is a regular representation of s, then $(\mathbf{i} M^i, M^p, \mathbf{t})$ is a regular representation of $s^{(i)}$ for each integer $0 \leq i \leq (p-1)$.

It is known from Soittola's theorem (Soittola 1976, Katayama et al. 1978, see [9, p. 83] or also [1, Theorem 12.4 p. 74]) that is decidable whether a $\mathbb{Z}$-rational sequence is regular. If s is a regular sequence, there is a computable positive integer p (the period) such that $s_{j+np} \sim c_j n^{l_j} \alpha_j^n$ as $n \to \infty$ $(j = 0, \ldots, p-1)$, where $c_j > 0$, $l_j \in \mathbb{N}$ and α_j is a non negative real (see for instance [1, theorem 10.2, p. 62]). Furthermore, α_j and l_j are computable.

3   Enumerative Sequence of a Regular Language on k Symbols

In this section, we present a characterization of the enumerative sequences of regular languages on k symbols.

Let A be a k-letter alphabet and L be a language over A, that is a subset of A^*, where A^* is set all finite words whose letters are in A. The *generating sequence* of L is defined as the sequence $s = (s_n)_{n \geq 0}$, where s_n is the number of words of L of length n.

It is clear that the generating sequence s of a regular language over A is the generating sequence of a regular language and is the generating sequence of language over a k-letter alphabet, i. e. satisfies $s_n \leq k^n$, for any $n \geq 0$. A natural question is the sufficiency of the two conditions. This question is similar to one solved in [2,3] (see also [10] and [11,12]), where it is shown that a sequence is the generating sequence of a regular k-ary tree if and only if it is the generating sequence of k-ary tree and if it is regular.

The situation is quite different here since we give below an example of a regular sequence s, with $s_n \leq k^n$ for any $n \geq 0$, and which is not the generating sequence of a regular language over a k-letter alphabet. The counterexample is based on an example of a $\mathbb{Z}$-rational sequence with nonnegative coefficients and which is not regular (see [13, p. 216-218] or [9, p. 95]). Let r be the sequence such that, for any $n \geq 0$, $r_n = b^{2n} \cos^2(n\theta)$, with $\cos\theta = \frac{a}{b}$, where the integers a, b are such that $b \neq 2a$ and $0 < a < b$. We also assume that $b^2 < k$. The sequence r is $\mathbb{Z}$-rational, has nonnegative integer coefficients and is not regular [13, p. 216-218]. Note that, for any $n \geq 0$, $r_n \leq k^n$. We now define the sequence s by $s_n = k^n - r_n$. By Soittola's theorem (see [9, Theorem 2.10 p. 90]), the sequence s is regular since it is a merge of rational sequences having a dominating pole, and it satisfies $s_n \leq k^n$ for any $n \geq 0$. If s were the generating sequence of a regular language L over a k-letter alphabet A, it would be recognized by a deterministic finite automaton over a k-letter alphabet and its complementary sequence r also.

The above counterexample leads us to state the following result which completely characterizes the sequences that are generating sequences of languages over a k-letter alphabet.

Theorem 1. *A sequence s is the generating sequence of a regular language over a k-letter alphabet if and only if*

- *the sequence s is regular,*
- *for any integer $n \geq 0$, $s_n \leq k^n$,*
- *the complementary sequence t, defined by $t_n = k^n - s_n$ for any integer $n \geq 0$, is regular.*

If s is a given $\mathbb{Z}$-rational sequence and k a positive integer, the three above conditions are decidable from [1, Theorem 12.4 p. 74]. Moreover if s is regular, one can compute the least integer k_0 such that $s_n \leq k_0^n$, for any integer $n \geq 0$. For $k > k_0$, the third condition is automatically satisfied again by [1, Theorem 12.4 p. 74]. It follows that, given some regular sequence, one can characterize the minimal alphabet such that s is the enumerative sequence of a regular language on this alphabet.

4   Sketch of Proof

4.1   Left and Right Inductions

In this section, we define a transformation on a linear representation of a sequence over a semiring which extends to linear representations the notion of multiset extension introduced in [3]. This notion is much weaker than the symbolic dynamics notion of conjugacy or even the notion shift of equivalence (see [6], [14] for these notions).

A linear representation $(\mathbf{j}, N, \mathbf{x})$, where N is a $S \times S$ matrix, is said to be *left induced* by the linear representation $(\mathbf{i}, M, \mathbf{t})$, where M is a $Q \times Q$ matrix, if there is a $Q \times S$ matrix U such that $NU = UM, \mathbf{j}U = \mathbf{i}, \mathbf{x} = U\mathbf{t}$.

If we identify an element of S to the Q-row vector of U of the corresponding index, the equality $NU = UM$ is equivalent to the fact that, for any element $\mathbf{u}$ of S, $\sum_{\mathbf{v} \in S} N_{\mathbf{u},\mathbf{v}} \mathbf{v} = \mathbf{u}M$.

The matrix U is called the *transfer matrix* of the induction. If K is field, a ring or a semi-ring, and U has its coefficients in K, we say that $(\mathbf{j}, N, \mathbf{x})$ is K-left induced by $(\mathbf{i}, M, \mathbf{t})$ and also write

$$(\mathbf{i}, M, \mathbf{t}) \xrightarrow[K]{U} (\mathbf{j}, N, \mathbf{x}).$$

Note that M or N may have coefficients outside K.

4.2   Perron Geometry

We consider now only $\mathbb{Z}$-rational sequences and regular sequences.

If $\mathbf{v} = (v_q)_{q \in Q}$ is a vector with components in $\mathbb{R}$, we say that $\mathbf{v}$ is nonnegative, denoted $\mathbf{v} \geq 0$, (respectively positive, denoted $\mathbf{v} > 0$) if $v_q \geq 0$ (respectively $v_q > 0$) for all $q \in Q$. The same conventions are used for matrices.

An integral matrix M is said to be *spectrally Perron* if it has a simple eigenvalue $\lambda > 0$ such that $\lambda > |\mu|$ for all other eigenvalues μ of M^1. Thus the spectral radius of M is λ. The matrix M has a *dominating spectral radius* if it has a unique eigenvalue λ such that $\lambda > |\mu|$ for all other eigenvalues μ of M. Thus being spectrally Perron is stronger than having a dominating spectral radius. A sequence of integers s is said to be *spectrally Perron* if it has a reduced linear representation over $\mathbb{Z}$ with a spectrally Perron matrix. A linear representation over $\mathbb{Z}$ with a spectrally Perron matrix is called a *spectrally Perron representation*.

In the sequel, we shall consider sequences of integers recognized by a *left reduced spectrally Perron representation* $(\mathbf{i}, M, \mathbf{t})$. The matrix M is a $Q \times Q$ spectrally Perron matrix whose spectral radius is λ, where Q has cardinality d. It has a non null left eigenvector $\mathbf{w}$ associated to the eigenvalue λ. All other eigenvectors associated to λ are collinear to it.

---

[1] The definition taken in [6, p. 371] (see also [6, p. 369]) precise $\lambda \geq 1$ instead of $\lambda > 0$.

Let W be the span of $\mathbf{w}$ over $\mathbb{R}$. According to the Jordan canonical form of M, there is a complementary M-invariant subspace V corresponding to eigenvalues $|\mu| < \lambda$. The space $\mathbb{R}^{1 \times d}$ is a direct sum of W and V. We denote by $\pi_1 : \mathbb{R}^{1 \times d} \to W$ the projection to W along V. If $\pi_1(\mathbf{u}) = \alpha \mathbf{w}$, the real number α is called the *dominant coordinate* of $\mathbf{u}$.

Since the representation is left reduced, $\mathbf{i}$ has a non null dominant coordinate. We call *left Perron eigenvector* of a left reduced Perron representation $(\mathbf{i}, M, \mathbf{t})$ the unique unitary eigenvector associated to the Perron value such that $\mathbf{i}$ has a positive dominant coordinate. Note that $\mathbf{w}$ depends only on M and $\mathbf{i}$.

For any real number r, we denote by $B(\mathbf{v}, r) = \{\mathbf{u} \mid \| \mathbf{v} - \mathbf{u} \| \leq r\}$ the ball of radius r centered on the point $\mathbf{v}$, where $\| \|$ is any equivalent norm of $\mathbb{R}^{1 \times d}$. Let $\mathbf{w}$ be the left Perron eigenvector of $(\mathbf{i}, M, \mathbf{t})$. We denote by K_r the set

$$K_r = \{\rho \mathbf{v} \mid \mathbf{v} \in B(\mathbf{w}, r), \rho \geq 0\}.$$

We also denote by K_r^+ the non null vectors of K_r.

The following lemma is from [6, p. 373].

Lemma 1. *Let $(\mathbf{i}, M, \mathbf{t})$ be a left reduced Perron representation. Let ε be a positive real number and let $\mathbf{u}$ be an integral vector with a positive dominant coordinate. Then there is a positive integer m such that $\mathbf{u}M^n$ belongs to K_ε for $n \geq m$.*

Let s be a sequence of nonnegative integers and let λ be a positive real number. We say that s has a *small complexity relatively to* λ if there is a positive real number $r < \lambda$ such that $s_n \leq r^n$ for any great enough n. In the converse case, we say that s has a *high complexity relatively to* λ.

We shall use the following lemmas, where the proofs are omitted in this short version of the paper.

Lemma 2. *Let $(\mathbf{i}, M, \mathbf{t})$ be a linear Perron representation with a spectral radius λ such that the sequence recognized is nonnegative and has a high complexity relatively to λ. Then $\mathbf{w} \cdot \mathbf{t} > 0$, and there exists a positive real number η such that $K_\eta^+ \cdot \mathbf{t} > 0$.*

Lemma 3. *Let $(\mathbf{i}, M, \mathbf{t})$ be a left reduced Perron representation. For any positive real number η, there exists a positive real number ε such that, for any positive integer n, if $\mathbf{u} \in K_\varepsilon$, then $\mathbf{u}M^n \in K_\eta$.*

We now state a geometrical lemma that is used in the construction of Section 4.3. The lemma is essentially the same as the geometrical lemma from Lind given in [6, p. 374] which states that there is a positive real ε such that *all* integral vectors in K_ε are nonnegative integral combinations of a *finite number* of integral vectors. With a slight modification, we show that there is a positive real number ε such that *all* integral vectors in K_ε are nonnegative integral combinations of a *finite number* of integral vectors in $K_{2\varepsilon}$.

Lemma 4. *For a small enough positive real ε, there is a finite set P of integral points in $K_{2\varepsilon}$ such that each integral point of K_ε is a nonnegative integral combination of points of P.*

4.3   From a Rational Representation to a Regular One

It is known that a nonnegative $\mathbb{Z}$-rational sequence which has a dominating spectral radius is regular ([9, p. 83], [1]). In the particular case of a spectrally Perron nonnegative sequence, we show that a regular representation can be obtained by only one left induction from any reduced $\mathbb{Z}$-rational linear representation of s. This result is an extension to linear representations of a result from Lind ([4], [5], see also [6, Theorem 11.1.4 p. 369]) which says that for any Perron number, there is a primitive integral matrix whose spectral radius is this Perron number.

Theorem 2. *Any left-reduced $\mathbb{Z}$-linear spectrally Perron representation of a nonnegative sequence, which recognizes a sequence which has a high complexity relatively to the spectral radius of the representation, has a regular left induced representation.*

Proof. Let $(\mathbf{i}, M, \mathbf{t})$ be a left-reduced $\mathbb{Z}$-linear spectrally Perron representation recognizing a nonnegative sequence s. The matrix M is thus spectrally Perron with a spectral radius λ. Let $\mathbf{w}$ be the left Perron eigenvector, according to $\mathbf{i}$.

By Lemma 2, there is a positive real number η such that $K_\eta^+ \cdot \mathbf{t} > 0$. We moreover choose η small enough such that any vector in K_η^+ has a positive dominant coordinate. By Lemma 3 there exists a positive real number ε such that, for any positive integer n, if $\mathbf{u} \in K_{2\varepsilon}$ then $\mathbf{u}M^n \in K_\eta$. Let us fix such a positive real number ε such that moreover $2\varepsilon < \eta$ and ε is small enough to apply the geometrical lemma. Thus $K_\varepsilon \subset K_{2\varepsilon} \subset K_\eta$. By Lemma 4, there is finite set P of integral points in $K_{2\varepsilon}$ such that each integral point of K_ε is a nonnegative integral combination of points of P.

By Lemma 1 and since P is a finite set of points of $K_{2\varepsilon}$, there is an integer n_0 such that

$$\forall \mathbf{v} \in P \cup \{\mathbf{i}\}, \qquad \mathbf{v}M^{n_0} \in K_\varepsilon.$$

We define a linear representation $(\mathbf{j}, N, \mathbf{x})$ which is left induced over $\mathbb{Z}$ by $(\mathbf{i}, M, \mathbf{t})$ as follows. We denote by H the graph whose set of vertices is the set S defined as $\bigcup_{j=0}^{n_0-1} \{\mathbf{v}M^j, \mathbf{v} \in P \cup \{\mathbf{i}\}\}$, and whose adjacency matrix, denoted by N, is the matrix of the multiplication by M on the set S on the right. If $\mathbf{u}$ is an element of $\bigcup_{j=0}^{n_0-2} \{\mathbf{v}M^j, \mathbf{v} \in P \cup \{\mathbf{i}\}\}$, there is in H one edge from vertex $\mathbf{u}$ to vertex $\mathbf{u}M$. If $\mathbf{u}$ is in $\{\mathbf{v}M^{n_0-1}, \mathbf{v} \in P \cup \{\mathbf{i}\}\}$, $\mathbf{u}M$ belongs to K_ε. As a consequence of the geometrical lemma, $\mathbf{u}M$ is a nonnegative integral combination of points of P. Thus $\mathbf{u}M = \sum_{\mathbf{v} \in P} N_{\mathbf{u},\mathbf{v}} \mathbf{v}$. We define $N_{\mathbf{u},\mathbf{v}}$ edges from $\mathbf{u}$ to $\mathbf{v}$ in H, for any $\mathbf{v} \in P$. Thus the matrix N has nonnegative integral coefficients. Finally, we only keep in S the vertices accessible from $\mathbf{i}$. We denote by U the transfer matrix of the left induction. We order the elements of S in such a way that the first row of U is the vector $\mathbf{i}$.

Let $\mathbf{j} = \begin{bmatrix} 1\, 0 \ldots 0 \end{bmatrix}$ and $\mathbf{x} = U\mathbf{t}$. If $\mathbf{u} = \mathbf{i}M^j$ for $0 \le j \le n_0 - 1$, then $x_\mathbf{u} = \mathbf{u} \cdot \mathbf{t} = \mathbf{i}M^j\mathbf{t} = s_j \ge 0$. If $\mathbf{u}$ is a non null vector in $K_{2\varepsilon}$, $\mathbf{u}M^j \in K_\eta^+$ for any $j \ge 0$. Then $\mathbf{u}M^j \cdot \mathbf{t} > 0$ for any $j \ge 0$. Thus $\mathbf{x}$ has nonnegative integral components. Moreover, if $s_n > 0$ for any nonnegative integer n, the vector $\mathbf{x}$ is a positive vector. Note that the transfer matrix U has its coefficients in $\mathbb{Z}$. Thus we have proved that $(\mathbf{i}, M, \mathbf{t}) \xrightarrow[\mathbb{Z}]{U} (\mathbf{j}, N, \mathbf{x})$, and $(\mathbf{j}, N, \mathbf{x})$ is a regular representation of s.

4.4   The Main Result

We now state the main result (Theorem 3 below). Lemma 7 constitutes one of the main part of the proof of Theorem 3. Theorem 1 is a consequence of Theorem 3.

   We recall below the notion of approximate eigenvector. A right k-*approximate* eigenvector of a nonnegative matrix M is, by definition, an integral column vector $\mathbf{v} \geq 0$ such that $M\mathbf{v} \leq k\mathbf{v}$.

Lemma 5. *Let $(\mathbf{j}, N, \mathbf{x})$ be a regular representation such that $\mathbf{x}$ is a positive right k-approximate eigenvector (respectively a positive right k-eigenvector) of N. Then there is regular right induced representation $(\mathbf{i}, M, \mathbf{t})$ of $(\mathbf{j}, N, \mathbf{x})$ such that $\mathbf{t}$ is a positive right k-approximate eigenvector (respectively a positive k-eigenvector) of M which has all its components equal to 1.*

   Moreover if $\mathbf{x} = \sum_{i=1}^{l} \mathbf{x}_i$, with $1 \leq i \leq l$, and where $\mathbf{x}_i$ is a nonnegative integral vector, there are nonnegative integral vectors $\mathbf{t}_i$ such that $\mathbf{x}_i = U\mathbf{t}_i$, where U denotes the transfer matrix of the induction.

Lemma 6. *Any left reduced $\mathbb{Z}$-linear representation $(\mathbf{j}, N, \mathbf{x})$ of $m(kz)^*$, where m and k are positive integers, is such that $\mathbf{x}$ is a right k-eigenvector of N.*

Lemma 7. *Let l be a positive integer and $s_1, \ldots s_l$ be l regular sequences recognized by regular representations $(\mathbf{i}, M, \mathbf{t}_i)$ respectively, such that $s_1(z) + \cdots + s_l(z) = m(kz)^*$, where m and k are positive integers with $k \geq 2$. Let us assume that M has a dominating spectral radius k, that all s_i have a high complexity relatively to k, and that $(\mathbf{i}, M, \sum_{i=1}^{l} \mathbf{t}_i)$ is trim. Then there is a finite deterministically labelled graph G on a k-letter alphabet, with m initial states and a partition of the set of states of G in l sets T_i, with $1 \leq i \leq l$, such that the automaton (I, G, T_i) recognizes a regular language on k symbols whose generating series is exactly s_i.*

Proof. We denote by $\mathbf{t}$ the column vector $\sum_{i=1}^{l} \mathbf{t}_i$ and thus $(\mathbf{i}, M, \mathbf{t})$ recognizes $s(z) = \sum_{i=1}^{l} s_i(z) = m(kz)^*$. We denote by $J_r(k)$ the Jordan block of size r

$$
J_r(k) = \begin{bmatrix} k & 1 & 0 & \ldots & 0 & 0 \\ 0 & k & 1 & \ldots & 0 & 0 \\ 0 & 0 & k & \ldots & 0 & 0 \\ \vdots & \vdots & \vdots & \ddots & \vdots & \vdots \\ 0 & 0 & 0 & \ldots & k & 1 \\ 0 & 0 & 0 & \ldots & 0 & k \end{bmatrix}.
$$

Since $(\mathbf{i}, M, \mathbf{t})$ is a regular trim representation that recognizes $m(kz)^*$, the Jordan canonical form of M has no block $J_r(k)$ where $r > 1$. Indeed, let us assume that the Jordan form of M contains such a block. Then there is a positive real number c such that for any large enough integer n, $s_n \geq cn^{r-1}k^n$. Thus the series $s(z)$ cannot be equal to $m(kz)^*$.

We compute from $(\mathbf{i}, M, \mathbf{t})$ a left reduced $\mathbb{Z}$-linear representation $(\mathbf{j}, N, \mathbf{x})$ of $m(kz)^*$. It can be shown that there is transfer matrix U such that $(\mathbf{i}, M, \mathbf{t}) \xrightarrow[\mathbb{Z}]{U} (\mathbf{j}, N, \mathbf{x})$. Since $(\mathbf{j}, N, \mathbf{x})$ is left reduced, the dimension of the vectorial space E generated by the vectors $\{\mathbf{j}N^n, n \geq 0\}$ over the field $\mathbb{R}$ is the size d of the square matrix N. Let E' be the eigenspace of N associated to the eigenvalue k in E and let E'' be the complementary N-invariant subspace. Thus d is the sum of the dimensions of E' and of E''. We claim that E' has dimension one. Indeed, the vector $\mathbf{j}$ can be written $\mathbf{j} = \mathbf{u} + \mathbf{v}$, where $\mathbf{u} \in E'$ and $\mathbf{v} \in E''$. Since for any integer $n \geq 0$, $\mathbf{j}N^n = k^n\mathbf{u} + \mathbf{v}N^n$, the vectorial space over $\mathbb{R}$ generated by the vectors $\{\mathbf{j}N^n, n \geq 0\}$ is included in $\langle \mathbf{u} \rangle + E''$, where $\langle \mathbf{u} \rangle$ denotes the vectorial space over $\mathbb{R}$ generated by $\mathbf{u}$. Thus the dimension of E' is one.

The matrix N is thus a Perron matrix. Moreover, by Lemma 6, $\mathbf{x}$ is an integral right k-eigenvector of N. For each integer $1 \leq i \leq l$, we define $\mathbf{x}_i = U\mathbf{t}_i$.

By Theorem 2, any left reduced $\mathbb{Z}$-rational linear representation of a nonnegative spectrally Perron sequence, which recognizes a sequence of high complexity relatively to the spectral radius of the representation, has a regular left induced representation. Let us denote this regular representation by $(\mathbf{k}, L, \mathbf{y})$ and the transfer matrix of this left induction by V. Moreover, since the series recognized, $m(kz)^*$, has all its coefficients positive, the vector $\mathbf{y}$ obtained in the proof is a positive vector. The vector $\mathbf{y}$ is a right k-eigenvector of L. It is thus a positive integral eigenvector of L.

Since $(\mathbf{j}, N, \mathbf{x}_i)$ is a left reduced representation that recognizes s_i, and since s_i is assumed to have a high complexity relatively to k, one chooses in Lemma 2 a positive real number η such that $K_\eta^+ \cdot \mathbf{x} > 0$, and $K_\eta^+ \cdot \mathbf{x}_i > 0$ for each integer $1 \leq i \leq l$. This allows us to get in the proof of Theorem 2 that the l vectors $\mathbf{y}_i = V\mathbf{x}_i$ and also are nonnegative integral vectors.

The final step is given by Lemma 5. There is a regular right induced representation $(\mathbf{i}', M', \mathbf{t}')$ of $(\mathbf{k}, L, \mathbf{y})$ such that $\mathbf{t}'$ is a positive right k-approximate eigenvector of M' which has all its components equal to 1. Let us denote by W the transfer matrix of this right induction. Since $\mathbf{y} = \sum_{i=1}^{l} \mathbf{y}_i$ where the vectors $\mathbf{y}_i$ are nonnegative integral vectors, there are two nonnegative integral vectors $\mathbf{t}'_i$ such that $\mathbf{y}_i = W\mathbf{t}'_i$.

For $1 \leq i \leq l$, the two previous left inductions and the right induction can be summarized in

$$(\mathbf{i}, M, \mathbf{t}) \xrightarrow[\mathbb{Z}]{U} (\mathbf{j}, N, \mathbf{x}) \xrightarrow[\mathbb{Z}]{V} (\mathbf{k}, L, \mathbf{y}) \xleftarrow[\mathbb{N}]{W} (\mathbf{i}', M', \mathbf{t}'),$$

$$(\mathbf{i}, M, \mathbf{t}_i) \xrightarrow[\mathbb{Z}]{U} (\mathbf{j}, N, \mathbf{x}_i) \xrightarrow[\mathbb{Z}]{V} (\mathbf{k}, L, \mathbf{y}_i) \xleftarrow[\mathbb{N}]{W} (\mathbf{i}', M', \mathbf{t}'_i).$$

We thus get a regular representation $(\mathbf{i}', M', \mathbf{t}'_i)$ of the series s_i. The coefficients of all $\mathbf{t}'_i$ are 0 or 1 and their sum is the vector $\mathbf{t}'$ whose components are all 1. Let us denote by T_i the set of indices such that $\mathbf{t}'_i = 1$. Since $\mathbf{t}'$ is a right

k-eigenvector of M', the sum of each row of M' is equal to k. The matrix M' is thus the transition matrix of a directed multigraph G whose outgoing arity is k. Let Q be the set of states of G. Since $\mathbf{i}' \cdot \mathbf{t}' = m$, the sum of the components of the vector $\mathbf{i}'$ is m. We define a new graph G' by adding to G a new set I of m states (p, j), for $p \in Q$ and $1 \le j \le i_p$, and n edges from (p, j) to q if there are n edges from p to q in G. This last transformation is again a right induction. Since the graph G' obtained has still arity k, one can label it deterministically with k symbols. The automaton (I, G', T_i) recognizes then a regular language on k symbols whose generating series is exactly s_i.

We now state, without proof, our main result. Theorem 1 is a formulation of Theorem 3 in the case of two sequences.

Theorem 3. *Let m and k be two positive integers. Let $s^{(1)}, s^{(2)}, \dots, s^{(l)}$ be l regular sequences such that $s^{(1)} + s^{(2)} + \dots + s^{(l)}(z) = m(kz)^*$. Then there is a finite deterministically labelled graph G on a k-letter alphabet, with m initial states and a partition of the set of states of G in l sets T_i, with $1 \le i \le l$, such that the automaton (I, G, T_i) recognizes a regular language on k symbols whose generating series is exactly $s^{(i)}$.*

The proof contains two main parts for $k \ge 2$. The first case corresponds to sequences that all have a high complexity relatively to k and relies mainly on Lemma 7. The second part treats the other case. We prove the first case by considering some periods of the sequences and, for some periods, the second case for other values of m and k may appear. The proof of the second case also uses the first one but for a number of sequences which is strictly less than l, which allows the recursiveness of the proof.

Example Let us consider the sequences s_1 and s_2 recognized by the regular representations $(\mathbf{i}, M, \mathbf{t_1})$ and $(\mathbf{i}, M, \mathbf{t_2})$ respectively, where

$$\mathbf{i} = \begin{bmatrix} 1 & 0 & 0 \end{bmatrix}, \quad M = \begin{bmatrix} 1 & 1 & 1 \\ 0 & 1 & 1 \\ 0 & 2 & 2 \end{bmatrix}, \quad \mathbf{t} = \mathbf{t_1} + \mathbf{t_2}, \quad \mathbf{t_1} = \begin{bmatrix} 1 \\ 1 \\ 0 \end{bmatrix}, \quad \mathbf{t_2} = \begin{bmatrix} 0 \\ 0 \\ 1 \end{bmatrix}.$$

These regular representations of s_1 and s_2 are pictured in Figure 1. The series $s = s_1 + s_2$ is equal to $(3z)^*$, and the series s_1 and s_2 have both a high complexity relatively to 3.

The spectral radius of M is 3. We successively get

$$\mathbf{i} = \begin{bmatrix} 1 & 0 & 0 \end{bmatrix},$$
$$\mathbf{i}M = \begin{bmatrix} 1 & 1 & 1 \end{bmatrix},$$
$$\mathbf{i}M^2 = \begin{bmatrix} 1 & 4 & 4 \end{bmatrix} = 4\mathbf{i}M - 3\mathbf{i},$$

Thus one can choose for U the 2×3 matrix whose rows are $\mathbf{i}$ and $\mathbf{i}M$ with

$$\mathbf{j} = \begin{bmatrix} 1 & 0 \end{bmatrix}, \quad N = \begin{bmatrix} 0 & 1 \\ -3 & 4 \end{bmatrix}, \quad \mathbf{x} = \begin{bmatrix} 1 \\ 3 \end{bmatrix}, \mathbf{x_1} = \begin{bmatrix} 1 \\ 2 \end{bmatrix}, \mathbf{x_2} = \begin{bmatrix} 0 \\ 1 \end{bmatrix}.$$

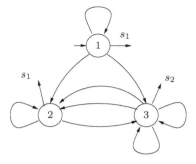

Fig. 1. Regular representations of s_1 and s_2.

The matrix N is spectrally Perron with a spectral radius 3, and $\mathbf{x}$ is a right eigenvector of N for the eigenvalue 3. The computation of the first powers $\mathbf{j}N^n$ gives

$$\mathbf{j}N^0 = \begin{bmatrix} 1 & 0 \end{bmatrix},$$
$$\mathbf{j}N^1 = \begin{bmatrix} 0 & 1 \end{bmatrix} = \begin{bmatrix} -1 & 1 \end{bmatrix} + \mathbf{i}.$$

Let $\mathbf{u} = \begin{bmatrix} -1 & 1 \end{bmatrix}$, then $\mathbf{u}M = 3\mathbf{u}$. We choose for transfer matrix V the 2×2 matrix whose rows are $\mathbf{j}$ and $\mathbf{u}$ and

$$\mathbf{k} = \begin{bmatrix} 1 & 0 \end{bmatrix}, \quad L = \begin{bmatrix} 1 & 1 \\ 0 & 3 \end{bmatrix}, \quad \mathbf{y} = \begin{bmatrix} 1 \\ 2 \end{bmatrix}, \mathbf{y}_1 = \begin{bmatrix} 1 \\ 1 \end{bmatrix}, \mathbf{y}_2 = \begin{bmatrix} 0 \\ 1 \end{bmatrix}.$$

The final representation is indexed by the set $\{(1,1),(2,1),(2,2)\}$ and one can choose

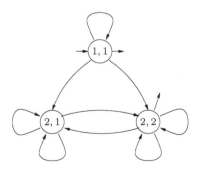

Fig. 2. Automaton with outgoing arity 3 recognizing s_1.

$$\mathbf{i'} = \begin{bmatrix} 1 & 0 & 0 \end{bmatrix}, \quad M' = \begin{bmatrix} 0 & 1 & 1 \\ 0 & 2 & 1 \\ 0 & 1 & 2 \end{bmatrix}, \quad \mathbf{t'} = \begin{bmatrix} 1 \\ 1 \\ 1 \end{bmatrix}, \mathbf{t'}_1 = \begin{bmatrix} 1 \\ 0 \\ 1 \end{bmatrix}, \mathbf{t'}_2 = \begin{bmatrix} 0 \\ 1 \\ 0 \end{bmatrix}.$$

Thus the series s_1 is recognized by the automaton of Figure 2 where the final states are $(1,1)$ and $(2,2)$, and where the initial state is $(1,1)$. The series s_2 is recognized by the same automaton where the final state is $(2,1)$.

We finally mention an open problem. Suppose that we are given a regular language X and two regular sequences s, t such that $s + t$ is the enumerative sequence of X. Is it true that there exists a partition of X in $X = Y + Z$ such that s is the enumerative sequence of Y and t the one of Z? By Theorem 1, the answer is yes when X is the set of all words on k symbols. We conjecture that the result holds in general.

References

1. Salomaa, A., Soittola, M.: Automata Theoretic Properties of Formal Power Series. Springer-Verlag (1978)
2. Bassino, F., Béal, M.P., Perrin, D.: Super-state automata and rational trees. In Lucchesi, C.L., Moura, A.V., eds.: LATIN'98. Number 1380 in Lecture Notes in Computer Science, Springer-Verlag (1998) 42–52
3. Bassino, F., Béal, M.P., Perrin, D.: A finite state version of the Kraft-McMillan theorem. SIAM J. Comput. **30** (2000) 1211–1230
4. Lind, D.A.: Entropies and factorizations of topological Markov shifts. Bull. Amer. Math. Soc. (1983) 219–222
5. Lind, D.A.: The entropies of topological Markov shifts and their related class of algebraic integers. Ergod. Th. & Dynam. Sys. (1984) 283–300
6. Lind, D.A., Marcus, B.H.: An Introduction to Symbolic Dynamics and Coding. Cambridge (1995)
7. Gantmacher, F.R.: Matrix Theory, Volume I. Chelsea Publishing Company, New York (1977)
8. MacCluer, C.R.: The many proofs and applications of Perron's theorem. SIAM Rev. **42** (2000) 487–498
9. Berstel, J., Reutenauer, Ch.: Rational Series and their Languages. Springer-Verlag (1988)
10. Bassino, F., Béal, M.P., Perrin, D.: Length distribution and regular sequences. In Rosenthal, J., Marcus, B., eds.: Codes, Systems and Graphical Models. Volume 123 of IMA Volumes in Mathematics and its Applications., Springer-Verlag (2001) 415–437
11. Bassino, F., Béal, M.P., Perrin, D.: Enumerative sequences of leaves in rational trees. In: ICALP'97. Number 1256 in Lecture Notes in Computer Science, Springer-Verlag (1997) 76–86
12. Bassino, F., Béal, M.P., Perrin, D.: Enumerative sequences of leaves and nodes in rational trees. Theoret. Comput. Sci. (1999) 41–60
13. Eilenberg, S.: Automata, Languages and Machines. Volume A. Academic Press (1974)
14. Kitchens, B.P.: Symbolic Dynamics: one-sided, two-sided and countable state Markov shifts. Springer-Verlag (1997)

Ground Tree Rewriting Graphs of Bounded Tree Width

Christof Löding

RWTH Aachen

Abstract. We analyze structural properties of ground tree rewriting graphs, generated by rewriting systems that perform replacements at the front of finite, ranked trees. The main result is that the class of ground tree rewriting graphs of bounded tree width exactly corresponds to the class of pushdown graphs. Furthermore we show that ground tree rewriting graphs of bounded clique width also have bounded tree width.

1 Introduction

Graphs are an important tool to describe the behaviour of programs or processes. While finite graphs can be used to deal with finite-state processes, infinite-state processes require the use of infinite (transition-)graphs. To allow algorithmic approaches to tasks like verification of infinite-state systems, these systems have to be given effectively, i.e., by some finite object. Given a formalism for a finite representation of infinite graphs, there are two natural questions arising.

1. Which decision problems can be solved for the graphs generated by this formalism?
2. How expressive is the given formalism and how is it related to other formalisms for the specification of graphs?

In [MS85] Muller and Schupp analyze transition graphs of pushdown automata. They show that the monadic second-order (MSO) theory of these graphs is decidable. Furthermore they give a structural characterization of pushdown graphs. This characterization is in terms of the number of non-isomorphic connected components one obtains by deleting vertices within fixed diameters around the root vertex.

Pushdown graphs are generated by prefix rewriting systems on words. Starting from this idea, there are several ways of obtaining other classes of graphs. One possibility is to use more general forms of rewriting on words. The decidability result for the MSO theory on pushdown graphs was extended in [Cau96] to prefix recognizable graphs, which are generated by rewriting rules using regular languages instead of single words. Even more general classes of graphs can be obtained when using finite word transducers to define the edge relation of a graph. This leads to automatic graphs in case of synchronous transducers (see e.g. [BG00]) and to rational graphs ([Mor99]) in case of asynchronous transducers. Since both of these classes contain all transition graphs of Turing machines,

H. Alt and A. Ferreira (Eds.): STACS 2002, LNCS 2285, pp. 559–570, 2002.
© Springer-Verlag Berlin Heidelberg 2002

reachability problems are undecidable on these graphs, but the first-order theory of automatic graphs is still decidable (see [BG00]) which is not the case for rational graphs ([Mor99]).

Another possibility for obtaining new classes of graphs is to pass from rewritings of words to rewritings on other structures. In [May00] classes of process rewriting graphs are analyzed. In process rewriting parallel composition of words is allowed in addition to the usual sequential composition. Depending on what operators (parallel and sequential composition) are allowed in the rewritings, one gets a hierarchy of classes of graphs. For these different classes decidable logics are exhibited in [May00] and their relation with respect to bisimulation is studied.

In the present paper we use another generalization of word rewriting, namely tree rewriting. In tree (or term) rewriting the basic objects in the rewriting systems are trees and the simplest form of rewriting is ground rewriting on these models. For ground tree rewrite systems confluence [DHLT90], the first-order theory [DT90], and several reachability problems [CG90] have been shown to be decidable. But so far there has been no analysis of the structure of graphs generated by ground tree rewriting systems. In this paper we try to fill this gap, i.e., we deal with Question 2 from above for the formalism of ground tree rewriting systems. The interest in the structure of these graphs comes from the fact that ground tree rewriting systems allow the specification of grid structures that are automatic but not prefix recognizable. On the other hand there are prefix recognizable graphs that can not be generated by ground tree rewriting systems. For our structural analysis we consider the properties tree width (see [Die00]) and clique width ([CO00]) of graphs, which amount to "complexity measures for graphs"; the former in terms of tree decompositions and the latter in terms of infinite graph expressions.

For the class of prefix recognizable graphs and subclasses these notions of width have been analyzed (see [Blu01] for an overview): Prefix recognizable graphs are of bounded clique width [Blu01], the class of HR-equational graphs (hyperedge replacement graphs), a class of graphs in between pushdown and prefix recognizable graphs, corresponds to the class of prefix recognizable graphs of bounded tree width ([Bar98]), and the class of HR-equational graphs of bounded degree exactly matches the class of pushdown graphs ([Cau92]).

Here we provide a similar analysis for ground tree rewriting graphs with the following result. The classes of pushdown graphs, ground tree rewriting graphs of bounded tree width, and ground tree rewriting graphs of bounded clique width coincide. First of all this result shows that the MSO theory of ground tree rewriting graphs of bounded tree width (or clique width) is decidable since it is decidable for pushdown graphs. Furthermore the structure of pushdown graphs is well understood ([MS85]). Thus, our result provides a method for showing that certain graphs of bounded tree or clique width can not be generated by ground tree rewriting systems.

The paper is organized as follows. In Sections 2 and 3 we introduce the basic terminology and definitions for graphs, trees, and rewriting systems. The rewriting systems under consideration are ground tree rewriting systems, pushdown systems, and infix pushdown systems (a novel extension of pushdown systems that we use as intermediate model for the proof of the main result). In Section 4 we show how to transform a ground tree rewriting system generating a graph of bounded tree width into an infix pushdown system. Finally, in Section 5, we prove that infix pushdown graphs of bounded tree width are pushdown graphs.

2   Graphs and Trees

In this section we introduce some general terminology on graphs and trees.

An edge labeled graph G is a tuple $G = (V, E, \Sigma)$, where V is the set of vertices, Σ is the set of edge labels, and $E \subseteq V \times \Sigma \times V$ is the set of edges. We only consider countable graphs, i.e., the set V is countable. If we are not interested in the edge labels or if we consider graphs without edge labels, we omit Σ and assume that $E \subseteq V \times V$.

A graph $G' = (V', E')$ is a subgraph of G iff $V' \subseteq V$ and $E' \subseteq E$. We call G' the subgraph induced by V' iff $E' = E \cap (V' \times \Sigma \times V')$.

A path π in G is a nonempty sequence of vertices $\pi = v_0 \cdots v_n \in V^*$ such that $(v_i, v_{i+1}) \in E$ for all $i \in \{0, \ldots, n-1\}$. An undirected path in G is a nonempty sequence of vertices $\pi = v_0 \cdots v_n \in V^*$ such that $(v_i, v_{i+1}) \in E$ or $(v_{i+1}, v_i) \in E$ for all $i \in \{0, \ldots, n-1\}$. The length of a (undirected) path $\pi = v_0 \cdots v_n$ is n. For two vertices $u, v \in V$ we say that there is a path (undirected path) from u to v if there is a path (undirected path) $\pi = v_0 \cdots v_n$ in G with $v_0 = u$, and $v_n = v$.

Given a vertex $v \in V$, the connected component of v in G is the subgraph of G induced by the set $\{u \in V \mid \text{there is an undirected path from } u \text{ to } v\}$.

Special graphs that we use in this article are the complete bipartite graphs $K_{m,m}$ with $2m$ vertices such that the edge relation is the set $V_1 \times V_2$ for a partition of vertex set into two sets V_1 and V_2 of size m.

The $(m \times m)$-grid has the vertex set $\{1, \ldots, m\} \times \{1, \ldots, m\}$ and from each vertex (i, j) there is an edge to $(i, j+1)$ if $j < m$ and to $(i+1, j)$ if $i < m$. We will use large grids (or grid-like graphs) as a tool for showing large tree width of graphs.

A tree-decomposition of $G = (V, E)$ is a pair $(T, (W_t)_{t \in V_T})$ where $T = (V_T, E_T)$ is a tree (in the graph-theoretic sense), $W_t \subseteq V$ for all $t \in V_T$, and

- $\forall (u, v) \in E \; \exists t \in V_T \; : \; u, v \in W_t$,
- $\forall v \in V$ the subgraph of T induced by $\{t \in V_T \mid v \in W_t\}$ is connected.

The width of a tree-decomposition is $\max\{|W_t| \mid t \in V_T\} - 1$ and ∞ if the maximum does not exist[1]. The tree width $tw(G)$ of G is the minimal width of a tree-decomposition of G.

For the following facts about tree width of graphs see e.g. [Die00].

Proposition 1. *The $(k \times k)$-grid has tree width k.*

---

[1] Since trees should have tree width 1, the "-1" is used in the definition

Proposition 2. *Let H be a subgraph of a graph G. Then $tw(G) \geq tw(H)$.*

In this paper we mainly need to show lower bounds on the tree width of graphs. For this purpose we introduce "brambles". A bramble of G is a finite family of sets of vertices $\mathcal{B} = (B_i)_{i \in I}$ such that

(1) each B_i induces a connected subgraph in G, and
(2) for each $i, j \in I$: $B_i \cap B_j \neq \emptyset$ or there are $u \in B_i$, $v \in B_j$ with $(u, v) \in E$.

A set $S \subseteq V$ covers $\mathcal{B}$ iff $S \cap B_i \neq \emptyset$ for each $i \in I$. The width of a bramble is width$(\mathcal{B}) = \min\{|S| \mid S$ covers $\mathcal{B}\}$.

The following result can be found in [Die00] for finite graphs. An extension to infinite graphs is straightforward (whereas the other direction is nontrivial).

Proposition 3. *If a graph G contains a bramble of width k, then $tw(G) \geq k-1$.*

The notion of clique width of graphs was introduced in [CO00]. Although our graphs are directed, we consider undirected clique width. This means in this paper the clique width of a directed graph G is the clique width of the undirected graph obtained from G by replacing every directed edge with an undirected one. We will only give an informal description of clique width. For details the reader is referred to [CO00].

To define clique width of graphs one considers (infinite) graph expressions over colored graphs. These expressions are built up from the constant graph with one vertex, colored c, and the operators $\oplus, \rho, \eta$, where

- $G_1 \oplus G_2$ is the disjoint union of the graphs G_1 and G_2,
- $\rho_{c_1 \to c_2}(G)$ changes the color of all c_1 colored vertices in G into c_2, and
- $\eta_{c_1, c_2}(G)$ introduces edges between all pairs of vertices u, v, where u is colored c_1 and v is colored c_2.

The clique width of a graph G is the minimal number of colors that is needed in such an expression defining G. The following propositions relate the notions of clique width and tree width.

Proposition 4. *([CO00]) Let G be a graph. If G has bounded tree width, then G has bounded clique width.*

Proposition 5. *[GW00] Let G be a graph. If G is of bounded clique width and there is an $n \in \mathbb{N}$ such that $K_{n,n}$ is not a subgraph of G, then G is of bounded tree width.*

We will generate graphs using ground tree rewriting systems. These are introduced in the next section. Here we define finite ranked trees, the objects the rewritings will be applied to.

A ranked alphabet A is a family of sets $(A_i)_{i \in [k]}$, where $[k] = \{0, \dots, k\}$. For simplicity we identify A with the set $\bigcup_{i=0}^{k} A_i$. A ranked tree t over A is a mapping $t : D_t \to A$ with $D_t \subseteq \mathbb{N}^*$ such that D_t is prefix closed, for each

$ui \in D_t$ we have $uj \in D_t$ for all $j \leq i$, and if $u0, \ldots, u(i-1) \in D_t$, $ui \notin D_t$, then $t(u) \in A_i$. D_t is called the domain of t. The set of all trees over A is called T_A. By $\sqsubseteq$ we denote the prefix ordering on $\mathbb{N}^*$. The height of a tree $t \in T_A$ is $height(t) = max\{|u| \mid u \in D_t\}$, i.e., the length of a longest path through t.

The rewriting rules used later will be replacements of subtrees. Here we formally define what it means to replace a subtree. For $u \in \mathbb{N}^*$ we define $uD_t = \{uv \in \mathbb{N}^* \mid v \in D_t\}$ and $u^{-1}D_t = \{v \in \mathbb{N}^* \mid uv \in D_t\}$. For $u \in D_t$ the subtree t^u of t at u is the tree with domain $D_{t^u} = u^{-1}D_t$ and $t^u(v) = t(uv)$. For trees $t, t' \in T_A$ and $u \in D_t$ we define $t(u \leftarrow t')$ to be the tree s with domain $D_s = uD_{t'} \cup (D_t \setminus u(u^{-1}D_t))$ and $s(v) = t'(u^{-1}v)$ for $v \in uD_{t'}$ and $s(v) = t(v)$ for $v \in D_t \setminus u(u^{-1}D_t)$. This means we replace the subtree t^u in t by t'.

3 Rewriting Systems on Words and Trees

In this section we introduce different rewriting systems we are using to generate infinite graphs. These are ground tree rewriting systems and pushdown automata. In the next sections we are going to translate ground tree rewriting systems that generate graphs of bounded tree width into pushdown automata such that the generated graphs are isomorphic. For this purpose we also need an intermediate model, which we call infix pushdown automata. All the graphs under consideration are rooted graphs of finite out-degree.

A ground tree rewriting system (GTRS) is a tuple $S = (A, \Sigma, P, t_I)$, where

- $A = (A_i)_{i \in [k]}$ is a ranked alphabet,
- Σ is an alphabet,
- P is a finite set of rules $s \rightarrow^\sigma s'$ with $s, s' \in T_A$, $\sigma \in \Sigma$, and
- $t_I \in T_A$ is the initial tree.

For two trees $t, t' \in T_A$ we write $t \rightarrow_S^\sigma t'$ iff there exists a rule $s \rightarrow^\sigma s'$ in P and $u \in D_t$ such that $t^u = s$ and $t' = t(u \leftarrow s')$. We write $t \rightarrow_S t'$ iff there is a $\sigma \in \Sigma$ with $t \rightarrow_S^\sigma t'$. By $\rightarrow_S^*$ we denote the transitive and reflexive closure of $\rightarrow_S$. The tree language that is generated by S is $T(S) = \{t \in T_A \mid t_I \rightarrow_S^* t\}$.

In [Bra69] it is shown that the set $T(S)$ is a regular set of trees. Here we do not consider the language generated by such a system but the graph G_S that is generated.

Let $S = (A, \Sigma, P, t_I)$ be a GTRS. The edge labeled graph $G_S = (V_S, E_S, \Sigma)$ generated by S is defined by $V_S = T(S)$, and $(t, \sigma, t') \in E_S$ iff $t \rightarrow_S^\sigma t'$.

Example 1. The GTRS $S = (A, \Sigma, P, t_I)$ given by

- $\Sigma = \{0, 1\}$, $A = (A_i)_{i \in [2]}$ with $A_0 = \{a, b\}$, $A_1 = \{c\}$, and $A_2 = \{d\}$,

$$- P = \left\{ b \rightarrow^0 \begin{array}{c} c \\ | \\ b \end{array}, a \rightarrow^1 \begin{array}{c} c \\ | \\ a \end{array} \right\}, \text{ and } t_I = \begin{array}{c} d \\ / \ \backslash \\ a \quad b \end{array}$$

generates the infinite grid as shown in Figure 1. This demonstrates that GTRS can generate graphs which have an undecidable MSO theory and therefore are not pushdown graphs. □

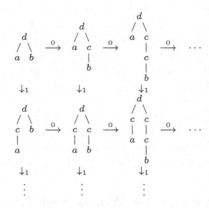

Fig. 1. The infinite grid generated by a GTRS.

To avoid complications we always assume that height$(t_I) \leq 1$. This can always be obtained by adding new symbols and rules to S.

A pushdown automaton (PDA) M is a tuple $M = (Q, \Sigma, \Gamma, \Delta, q_0, \gamma_0)$, where Q is a finite set of states, Σ is the input alphabet, Γ is the stack alphabet (disjoint from Q), $\Delta \subseteq Q \times (\Gamma \cup \{\varepsilon\}) \times \Sigma \times Q \times \Gamma^*$ is the transition relation (ε denotes the empty word), $q_0 \in Q$ is the initial state, and $\gamma_0 \in \Gamma$ is the initial stack symbol. A configuration of M is a word qw with a state $q \in Q$ and a stack content $w \in \Gamma^*$. Given configurations $q_1 w_1$, $q_2 w_2$ and an input $\sigma \in \Sigma$ we write $q_1 w_1 \vdash_M^\sigma q_2 w_2$ if there is a transition $(q_1, \gamma, \sigma, q_2, v) \in \Delta$ such that $w_1 = \gamma w$ and $w_2 = vw$. We write $q_1 w_1 \vdash_M q_2 w_2$ if there is an input $\sigma \in \Sigma$ with $q_1 w_1 \vdash_M^\sigma q_2 w_2$ and by $\vdash_M^*$ we denote the transitive and reflexive closure of $\vdash_M$. The set of configurations $C(M)$ generated by M is the set of configurations that are reachable from the initial configuration of M: $C(M) = \{qw \mid q_0 \gamma_0 \vdash_M^* qw\}$.

The language $C(M)$ is regular (a result due to Büchi) but as for GTRS we are interested in the graph generated by the PDA. The directed edge labeled graph $G_M = (V_M, E_M, \Sigma)$ generated by M has the set of vertices $V_M = C(M)$ and for two words $q_1 w_1, q_2 w_2 \in V_M$ there is an edge labeled with σ from $q_1 w_1$ to $q_2 w_2$ if $q_1 w_1 \vdash_M^\sigma q_2 w_2$. Given a graph G we say that G is a pushdown graph iff G is isomorphic to the graph G_M for some PDA M. The following proposition on pushdown graphs is not difficult to prove.

Proposition 6. *Pushdown graphs are of bounded tree width.*

In usual pushdown automata we only have rewritings on top of the stack. We also need an extended type of pushdown automata, where we allow a restricted type of infix rewriting in the stack.

An infix PDA is a system $M = (Q, \Sigma, \Gamma, \Delta, q_0, \gamma_0)$, where $Q, \Sigma, \Gamma, q_0, \gamma_0$ are as in usual pushdown automata and in Δ there are transitions of the usual form but also transitions from $\Gamma \times \Sigma \times \Gamma$. These are the infix rules. The other rules are called prefix rules. Now we write $q_1 w_1 \vdash_{M_{\mathrm{pre}}} q_2 w_2$ if $q_1 w_1 \vdash_M q_2 w_2$ in the usual sense, and $q w_1 \vdash_{M_{\mathrm{in}}} q w_2$ if there is an infix rule $(\gamma_1, \sigma, \gamma_2) \in \Delta$ such that $w_1 = w \gamma_1 w'$ and $w_2 = w \gamma_2 w'$ with $w, w' \in \Gamma^*$. This means we are allowed to change single letters inside the stack.

We write $q_1 w_1 \vdash_M q_2 w_2$ if $q_1 w_1 \vdash_{M_{\mathrm{pre}}} q_2 w_2$ or $q_1 w_1 \vdash_{M_{\mathrm{in}}} q_2 w_2$. We adopt the notations $\vdash_M^*$, $\vdash_{M_{\mathrm{pre}}}^*$, $\vdash_{M_{\mathrm{in}}}^*$, $C(M)$, and G_M for infix pushdown automata in the obvious way.

In Section 4 we will transform certain ground tree rewriting systems into infix pushdown automata. The transitions of the obtained automata will have a certain form.

Definition 1. Let $M = (Q, \Sigma, \Gamma, \Delta, q_0, \gamma_0)$ be an infix PDA. Then M is in normal form iff

1. all elements in Δ are of the form $(q, \gamma, \sigma, q', \varepsilon)$, $(q, \varepsilon, \sigma, q', \gamma)$, $(q, \varepsilon, \sigma, q', \varepsilon)$, $(q, \gamma_0, \sigma, q', \gamma_0)$, or $(\gamma, \sigma, \gamma')$ with $q, q' \in Q$, $\gamma, \gamma' \in \Gamma \setminus \{\gamma_0\}$, and $\sigma \in \Sigma$, and
2. whenever $q w \vdash_{M_{\mathrm{pre}}} q' \gamma w \vdash_{M_{\mathrm{in}}} q' \gamma' w$, then there is a $q'' \in Q$ such that $q w \vdash_{M_{\mathrm{pre}}} q'' w \vdash_{M_{\mathrm{pre}}} q' \gamma' w$.

The essential property of infix pushdown automata in normal form is that every reachable configuration can also be reached only by applying prefix rules. For later use we formulate the more general claim that, given a reachable configuration $q u w$, there is a reachable configuration $q' w$ such that one can reach $q u w$ from $q' w$ only by applying prefix rules and without "touching" the w.

Lemma 1. Let $M = (Q, \Sigma, \Gamma, \Delta, q_0, \gamma_0)$ be an infix PDA in normal form, $q \in Q$, and $u, w \in \Gamma^*$. If $q u w \in C(M)$, then there exists $q' \in Q$ such that $q' \vdash_{M_{\mathrm{pre}}}^* q u$, and $q' w \in C(M)$.

Proof. Since $q u w \in C(M)$, there is a directed path from $q_0 \gamma_0$ to $q u w$. We chose this path such that every infix rule is applied when the corresponding symbol is on top of the stack (this is possible because by point 1 of Definition 1 the prefix rules increasing the stack length do not depend on the stack content). Then, by point 2 of Definition 1, we can replace all the infix rules on this path and get $q_0 \gamma_0 \vdash_{M_{\mathrm{pre}}}^* q u w$. On this path π from $q_0 \gamma_0$ to $q u w$, there must be a vertex $q' w$ such that every successor of $q' w$ on π contains the suffix w. But this means $q' \vdash_{M_{\mathrm{pre}}}^* q u$. $\qquad\square$

4   GTRS Graphs of Bounded Tree and Clique Width

In this section we show that GTRS graphs of bounded tree width are infix pushdown graphs. We do this by analyzing the trees that are generated by a rewriting system S producing a graph G_S of bounded tree width. These trees have a special form that allows us to code them as words.

Let $S = (A, \Sigma, P, t_I)$ be a GTRS. If we are given $u \in \mathbb{N}^*$ and $t' \in T_A$, then we call $\alpha = (u, t')$ a substitution. A substitution $\alpha = (u, t')$ is S-applicable to

$t \in T_A$ iff $u \in D_t$ and there is a rule $t^u \to^\sigma t'$ in P. The result of the substitution is $\alpha(t) = t(u \leftarrow t')$. A sequence of substitutions $(\alpha_1 \cdots \alpha_k)$ is S-applicable to t if α_1 is S-applicable to t and $(\alpha_2 \cdots \alpha_k)$ is S-applicable to $\alpha_1(t)$. We define $(\alpha_1 \cdots \alpha_k)(t) = (\alpha_2 \cdots \alpha_k)(\alpha_1(t))$.

Two substitutions $\alpha = (u_1, t_1)$ and $\beta = (u_2, t_2)$ are called independent iff u_1 and u_2 are not comparable with respect to $\sqsubseteq$. Two sequences $(\alpha_1 \cdots \alpha_k)$ and $(\beta_1 \cdots \beta_l)$ of substitutions are called independent iff α_i and β_j are independent for each $i \in \{1, \dots, k\}$ and $j \in \{1, \dots, l\}$. Since independent sequences of substitutions can be interleaved arbitrarily, it is not difficult to see that they generate a grid with size corresponding to the length of the sequences as subgraph[2].

Lemma 2. *Let $S = (A, \Sigma, P, t_I)$ be a GTRS. If there is a $t \in T(S)$ and two independent sequences $(\alpha_1 \cdots \alpha_k)$ and $(\beta_1 \cdots \beta_k)$ of substitutions that are S-applicable to t such that $(\alpha_1 \cdots \alpha_j)(t) \neq (\alpha_1 \cdots \alpha_i)(t)$ and $(\beta_1 \cdots \beta_j)(t) \neq (\beta_1 \cdots \beta_i)(t)$ for all $i, j \in \{1, \dots, k\}$ with $i \neq j$, then G_S has a $(k+1) \times (k+1)$-grid as subgraph.*

We consider a consequence of this lemma for the trees generated by a GTRS S where G_S has bounded tree width. These trees consist of a single long path with subtrees of bounded height branching from this path.

Lemma 3. *Let $S = (A, \Sigma, P, t_I)$ be a GTRS such that G_S has bounded tree width. Then there is a k such that for each $t \in T(S)$ and for each vertex x of t there is at most one successor y of x with $\mathrm{height}(t^y) \geq k$.*

Proof. Let $c = \max\{\mathrm{height}(t') - \mathrm{height}(t) \mid t \to^\sigma t' \in P\}$ (the maximal increase of height by an application of one rule), $m = tw(G_S)$, and define $k = c \cdot m + 1$. (Here we assume that $c \geq 1$. Otherwise G_S is finite and the claim obviously holds.)

Suppose there is a $t \in T(S)$ with a vertex x such that there are two successors y, z of x and the subtrees t^y and t^z of t have height $\geq k$. Since $t \in T(S)$, there is a sequence $(\alpha_1 \cdots \alpha_l)$ of substitutions with $(\alpha_1 \cdots \alpha_l)(t_I) = t$. Let $\alpha_i = (u_i, t'_i)$ and let $j \in \{0, \dots, l\}$ be the smallest number with $u_i \not\sqsubseteq x$ for all $i > j$. By the choice of c, the subtrees t^y and t^z have height $\leq c$ in the tree $(\alpha_1 \cdots \alpha_j)(t_I)$ and we need at least m substitutions, respectively, to increase their height to be $\geq k$ (by the choice of k). Therefore we can extract from $(\alpha_{j+1} \cdots \alpha_l)$ two independent sequences of substitutions of length at least m that are S-applicable to $(\alpha_1 \cdots \alpha_j)(t_I)$ such that Lemma 2 can be applied, i.e., G_S has the $(m+1) \times (m+1)$-grid as subgraph and therefore by Propositions 1 and 2 tree width $\geq m + 1$, a contradiction.   □

This property of the trees enables us to code them as words and to construct an infix PDA generating a graph isomorphic to the graph generated by the GTRS.

Theorem 1. *Let $S = (A, \Sigma, P, t_I)$ be a GTRS such that G_S has bounded tree width. Then we can construct an infix PDA $M = (Q, \Sigma, \Gamma, \Delta, q_0, \gamma_0)$ in normal form such that G_S is isomorphic to G_M.*

---

[2] Note that this subgraph in general is not an induced subgraph.

Proof. Let ∘ be a symbol of rank 0 not occurring in A. From Lemma 3 we can see that the trees in $T(S)$ may have one long path only. From this path subtrees of bounded height are branching. The idea is to cut such a tree along this path. The new symbol ∘ signals where the tree was cut as indicated in Figure 2. We can extend S to a GTRS S' for trees over $A \cup \{\circ\}$ by defining $S' = (A \cup \{\circ\}, \Sigma, P, t_I)$. We define a set of special trees over $A \cup \{\circ\}$ by $T_A^\circ = \{t \in T_{A \cup \{\circ\}} \mid t^{-1}(\circ) = \{x\}$ with $|x| = 1\}$. These are all the trees over $A \cup \{\circ\}$ containing exactly one ∘ at a successor of the root. Using these trees we can define concatenation of trees. For $t \in T_A^\circ$ and $t' \in T_A^\circ \cup T_A$ let $t \cdot t' = t(x \leftarrow t')$ where $x \in D_t$ with $t(x) = \circ$.

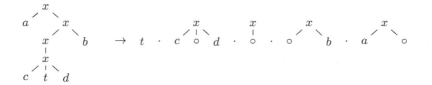

Fig. 2. Decomposition of a tree to pass from trees to words.

Let k be as in Lemma 3 and define the components of M by $Q = \{t \in T_A \mid$ height$(t) \leq k\}$, $q_0 = t_I$ (by our convention height$(t_I) \leq 1 \leq k$), $\Gamma = \{t \in T_A^\circ \mid$ height$(t) \leq k\} \dot\cup \{\gamma_0\}$ (all the special trees of height $\leq k$ over $A \cup \{\circ\}$ and a new initial stack symbol γ_0). By Γ' we denote $\Gamma \setminus \{\gamma_0\}$. For $\sigma \in \Sigma$, $r, r' \in \Gamma'$, and $t, t' \in Q$ we have the following rules in Δ.

1. $(r, \sigma, r') \in \Delta$ iff $r \to_{S'}^\sigma r'$.
2. $(t, \gamma_0, \sigma, t', \gamma_0) \in \Delta$ iff $t \to_S^\sigma t'$.
3. $(t, \varepsilon, \sigma, t', \varepsilon) \in \Delta$ iff $t \to_S^\sigma t'$ and height$(t') = k$.
4. $(t, r, \sigma, t', \varepsilon) \in \Delta$ iff $r \cdot t \to_S^\sigma t'$ and height$(t') = k$.
5. $(t, \varepsilon, \sigma, t', r) \in \Delta$ iff $t \to_S^\sigma r \cdot t'$ and height$(t') = k$.

Note that the S' in the first rule refers to the GTRS over the extended alphabet. Furthermore, the state of the infix PDA will always be a tree of height exactly k when the decomposed tree that is represented by stack and state has height $\geq k$ (Rules 3,4, and 5). This is to keep the decomposition of a tree unique.

   Now one can verify that G_M is isomorphic to G_S via the mapping that maps each tree to its decomposition as indicated in Figure 2. □

As mentioned before we also want to show that bounded clique width and bounded tree width imply each other for GTRS graphs. Using Proposition 5 this question can be reduced to the following lemma which can be shown by a detailed analysis of how complete bipartite graphs can be generated by a GTRS.

Lemma 4. *Let $S = (A, \Sigma, P, t_I)$ be a GTRS. Then $K_{m,m}$ is not a subgraph of G_S for all $m > 2 \cdot |P| + 1$.* □

Now, using Propositions 4 and 5, we can conclude the following.

Corollary 1. *Let $S = (A, \Sigma, P, t_I)$ be a GTRS. Then G_S has bounded clique width iff G_S has bounded tree width.*

5   From Infix Pushdown Graphs to Pushdown Graphs

In this section we prove that infix pushdown graphs of bounded tree width are pushdown graphs. The idea is to find a structural characterization of infix PDA graphs of bounded tree width similar to the characterization of PDA graphs given by Muller and Schupp ([MS85]). For each vertex qw in the infix PDA graph we define the connected component of qw in the graph restricted to vertices of length greater than or equal to the length of qw. The goal is to show that, if the graph has bounded tree width, then there are only finitely many isomorphism classes of these connected components. The difference to the property of pushdown graphs given by Muller and Schupp is that we refer to the length of the vertices and not the distance to a designated root vertex.

Once we have this structural characterization, this finite number of isomorphism classes can be used to obtain a usual PDA for generating the graph.

Let $M = (Q, \Sigma, \Gamma, \Delta, q_0, \gamma_0)$ be an infix PDA in normal form. For $w, v \in \Gamma^*$ we define

$$w \preceq_M v \text{ iff } \forall q \in Q : (qw \in C(M) \Rightarrow qv \in C(M)),$$
$$w \approx_M v \text{ iff } w \preceq_M v \text{ and } v \preceq_M w.$$

The following lemma is immediate from the definition of $\approx_M$.

Lemma 5. *The relation $\approx_M$ is an equivalence relation of finite index.*

For $n \in \mathbb{N}$ the graph $G(M)|_n$ is the subgraph of G_M induced by the vertices $\{qw \in C(M) \mid |qw| \geq n\}$. For $qw \in C(M)$ the graph $G_M(qw)$ is the connected component of qw in $G_M|_{|qw|}$. The front $\Omega(qw)$ of $G_M(qw)$ is the set of all vertices in $G_M(qw)$ with the same length as qw, i.e.,

$$\Omega(qw) = \{pv \in G_M(qw) \mid |pv| = |qw|\}.$$

Let $pv \in C(M)$, qw a vertex in $G_M(pv)$, and $\pi = (p_1 x_1 u_1, \dots, p_n x_n u_n)$ be an undirected path in $G_M(pv)$ from pv to qw with $|u_i| = |v|$ for all $i \in \{1, \dots, n\}$. The depth of π is $d(\pi) = max\{|x_i| \mid i \in \{1, \dots, n\}\}$ and the number of suffix changes on π is $sc(\pi) = |\{i \in \{1, \dots, n-1\} \mid u_i \neq u_{i+1}\}|$.

Let $\Pi(pv, qw)$ be the set of undirected paths in $G_M(pv)$ from pv to qw. We fix a path $\pi(pv, qw) \in \Pi(pv, qw)$ with

1. $\forall \pi \in \Pi(pv, qw) : sc(\pi(pv, qw)) \leq sc(\pi)$ and
2. $\forall \pi \in \Pi(pv, qw) : sc(\pi(pv, qw)) = sc(\pi) \Rightarrow d(\pi(pv, qw)) \leq d(\pi)$.

By induction on $sc(\pi(pv, qw))$ one can show that vertices that belong to the same front have a connection of small depth.

Lemma 6. *Let $M = (Q, \Sigma, \Gamma, \Delta, q_0, \gamma_0)$ be an infix PDA in normal form. There exists $d_M \in \mathbb{N}$ such that $d(\pi(pv, qw)) < d_M$ for each $pv \in C(M)$ and $qw \in \Omega(pv)$.*   □

After this general observation on the structure of infix PDA graphs we turn to infix PDA graphs of bounded tree width. Using Lemmas 1 and 6 one can construct large brambles from large fronts $\Omega(qw)$. Therefore these fronts have to be small in graphs of bounded tree width. This fact is expressed in the following lemma.

Lemma 7. *Let $M = (Q, \Sigma, \Gamma, \Delta, q_0, \gamma_0)$ be an infix PDA in normal form such that G_M is of bounded tree width. Then there exists $c \in \mathbb{N}$ such that $|\Omega(qw)| \leq c$ for each $qw \in C(M)$.* □

This enables us to prove the desired structural characterization.

Definition 2. *Let $M = (Q, \Sigma, \Gamma, \Delta, q_0, \gamma_0)$ be an infix PDA in normal form, and let $pv, qw \in C(M)$. The graphs $G_M(pv)$ and $G_M(qw)$ are end isomorphic, denoted by $G_M(pv) \sim G_M(qw)$, iff there is a graph isomorphism between $G_M(pv)$ and $G_M(qw)$ such that vertices from $\Omega(pv)$ are mapped to vertices from $\Omega(qw)$.*

Lemma 8. *Let $M = (Q, \Sigma, \Gamma, \Delta, q_0, \gamma_0)$ be an infix PDA in normal form. If G_M is of bounded tree width, then $\sim$ has finite index.*

Proof. We show that $G_M(pv) \sim G_M(qw)$ if there is a bijection φ between $\Omega(pv)$ and $\Omega(qw)$ such that

(1) $p_1v_1 \vdash_M^\sigma p_2v_2$ iff $\varphi(p_1v_1) \vdash_M^\sigma \varphi(p_2v_2)$ for each $p_1v_1, p_2v_2 \in \Omega(pv)$, $\sigma \in \Sigma$,
(2) $\varphi(p_1v_1) \in p_1\Gamma^*$ for each $p_1v_1 \in \Omega(pv)$,
(3) if $\varphi(p_1v_1) = q_1w_1$ and $\varphi(p_2v_1) = q_2w_2$, then $w_1 = w_2$.

We extend this bijection to a graph isomorphism between $G_M(pv)$ and $G_M(qw)$. Let $p_1xv_1 \in G_M(pv)$ with $|v_1| = |v|$. Then there is $p' \in Q$ such that $p'v_1 \in \Omega(pv)$ and $p' \vdash_M^* p_1x$ (by Lemma 1). Let $p'w_1 = \varphi(p'v_1)$. Then we define $\varphi(p_1xv_1) = p_1xw_1$. It is easy to verify that φ is a graph isomorphism. Since the size of the sets $\Omega(qw)$ is bounded (Lemma 7) we can conclude that $\sim$ has finite index. □

Using the previous lemma, the proof of the following theorem is similar to the proof of Muller and Schupp [MS85] that finitely generated context free graphs are pushdown graphs.

Theorem 2. *Let $M = (Q, \Sigma, \Gamma, \Delta, q_0, \gamma_0)$ be an infix PDA in normal form. If G_M is of bounded tree width, then G_M is a pushdown graph.* □

This leads us to our main theorem.

Theorem 3. *Let $S = (A, \Sigma, P, t_I)$ be a GTRS. Then the following statements are equivalent.*

(i) G_S is of bounded clique width.
(ii) G_S is of bounded tree width.
(iii) G_S is a pushdown graph.

Proof. The equivalence of (i) and (ii) was already stated in Corollary 1. From Proposition 6 we know that (iii) implies (ii) and, by Theorems 1 and 2, (ii) implies (iii). □

A consequence of this theorem is that graphs that are GTRS graphs and are also prefix recognizable have to be pushdown graphs since prefix recognizable graphs are of bounded clique width ([Blu01]).

6   Conclusion

The class of GTRS graphs transcends the PDA graphs by examples like the
infinite grid. We have clarified the relation of this class of graphs to the class
of PDA graphs and prefix recognizable graphs in terms of tree width. Future
work could address the decidability of the question "Given a GTRS S, is G_S of
bounded tree width?". Furthermore the relation to other classes of graphs like
automatic or rational graphs should by analyzed.

I would like to thank Wolfgang Thomas for the idea to analyze GTRS graphs
and the anonymous referees for several helpful comments.

References

[Bar98]   Klaus Barthelmann. When can an equational simple graph be generated
            by hyperedge replacement? In *Proceedings of MFCS '98*, volume 1450 of
            LNCS, pages 543–552. Springer-Verlag, 1998.

[BG00]   Achim Blumensath and Erich Grädel. Automatic structures. In *Proceedings
            of LICS '00*, pages 51–62. IEEE Computer Society Press, 2000.

[Blu01]   Achim Blumensath. Prefix-recognizable graphs and monadic second order
            logic. Technical Report AIB-2001-06, RWTH Aachen, May 2001.

[Bra69]   Walter S. Brainerd. Tree generating regular systems. *Information and
            Control*, 14:217–231, 1969.

[Cau92]   D. Caucal. On the regular structure of prefix rewriting. *Theoretical Com-
            puter Science*, 106(1):61–86, 1992.

[Cau96]   Didier Caucal. On infinite transition graphs having a decidable monadic
            theory. In *Proceedings of ICALP '96*, volume 1099 of *LNCS*. Springer-
            Verlag, 1996.

[CG90]   J.L. Coquidé and R. Gilleron. Proofs and reachability problem for ground
            rewrite systems. In *Aspects and Prospects of Theoretical Computer Science*,
            volume 464 of *LNCS*, pages 120–129. Springer, 1990.

[CO00]   Bruno Courcelle and Stephan Olariu. Upper bounds to the clique width of
            graphs. *Discrete Applied Mathematics*, 101:77–114, 2000.

[DHLT90]   Max Dauchet, Thierry Heuillard, Pierre Lescanne, and Sophie Tison. De-
            cidability of the confluence of finite ground term rewrite systems and
            of other related term rewrite systems. *Information and Computation*,
            88(2):187–201, October 1990.

[Die00]   Reinhard Diestel. *Graph Theory*. Springer, second edition, 2000.

[DT90]   Max Dauchet and Sophie Tison. The theory of ground rewrite systems
            is decidable. In *Proceedings of LICS '90*, pages 242–248. IEEE Computer
            Society Press, 1990.

[GW00]   Frank Gurski and Egon Wanke. The tree-width of clique-width bounded
            graphs without $K_{n,n}$. In *Proceedings of WG 2000*, volume 1928 of LNCS,
            pages 196–205. Springer-Verlag, 2000.

[May00]   Richard Mayr. Process rewrite systems. *Information and Computation*,
            156(1–2):264–286, 2000.

[Mor99]   Christophe Morvan. On rational graphs. In *Proceedings of FoSSaCS '99*,
            volume 1784 of *LNCS*, pages 252–266. Springer, 1999.

[MS85]   David E. Muller and Paul E. Schupp. The theory of ends, pushdown au-
            tomata, and second-order logic. *Theoretical Computer Science*, 37:51–75,
            1985.

Timed Control Synthesis for External Specifications

Deepak D'Souza* and P. Madhusudan

Chennai Mathematical Institute,
92 G. N. Chetty Road, Chennai 600 017, India.
deepak@cmi.ac.in, madhu@imsc.ernet.in

Abstract. We study the problem of control synthesis for a timed plant and an *external* timed specification, modelled as timed automata. Our main result is that the problem is decidable when we are given *a priori* the resources for the controller (the number of clocks, observational power of clocks, etc.) and when the specification is an ω-regular timed language describing *undesired* behaviours. We also show that for deterministic specifications, if there is a controller at all, then there is one which uses the combined resources of the plant and specification. The decidability of other related problems is also investigated.

1 Introduction

An *open reactive system* is a system that interacts with its environment. In the setting where the exact times at which events occur are not modelled (i.e. in *untimed systems*) this interaction is usually thought of as the environment giving an input for which the system responds with an output. In the *timed* setting, it is more fruitful to think of open systems as having a continuous interaction with an environment where each party can execute events at any time.

The problem of *control synthesis* arises naturally in the context of such systems. We are given an open system, usually called a *plant* in this setting, and asked whether we can *control* the system actions such that no matter how it interacts with the environment, the behaviours satisfy a given specification. An important aspect of the problem is that the environment is not controllable and hence the controlled plant must be able to satisfy the specification against any hostile environment.

While this problem naturally arises in the area of control theory, the computer science community has studied a variant of this called the *realizability* problem. Here, we are not given a plant, but are asked to synthesize a system that satisfies a given specification, no matter what the environment does. This question was first posed by Church and was solved for ω-regular specifications by Büchi and Landweber [4]. This line of research has been recently revived in [11,1,8] where realizability against temporal logic specifications (linear and branching-time) with emphasis on complexity of synthesis is addressed. Most of these results

---

* part of this work was done while visiting LSV, École Normal Supérieure, Cachan.

H. Alt and A. Ferreira (Eds.): STACS 2002, LNCS 2285, pp. 571–582, 2002.
© Springer-Verlag Berlin Heidelberg 2002

extend to control synthesis as well, as it is closely related to the problem of finding winning strategies in infinite games played on *finite* graphs. These games were first proposed by McNaughton [10] and are now well understood [17].

In the control theory framework, Ramadge and Wonham proposed a model for designing discrete supervisory controllers for discrete-event systems [12]. While it has been argued that discrete-event controllers are useful for achieving high-level supervisory control, in many contexts it is hard to ignore the continuous nature of the system. The computer science community has made some recent progress in identifying some tractable ways of dealing with continuous systems through the analysis of systems modelled as *timed automata* [2] and *hybrid automata* [6]. The aim of this paper is to use the framework of timed automata to describe plants and specifications and to solve the problem of control synthesis for them.

In [9,3], the authors investigate this problem when the plant is modelled as a timed automaton and the specification is given as an *internal* winning condition on the state-space of the plant. A controller looks at the values of the plant's clocks (distinguished only up to some granularity of time), and prescribes a set of moves the system should take. The work in [7] is very similar, but is couched in the Ramadge-Wonham notion of a specification and uses a *deterministic* timed automaton as part of the specification mechanism.

The departure of our work from the above is two-fold. First, we consider specifications that are *nondeterministic* and which are given *externally* as separate timed automata. Secondly, we synthesize controllers with an *a priori* limit on their resources (number of clocks, power of controllers to observe clocks, etc.). Since the use of new clocks can be costly (or in some contexts even meaningless) and since the power to observe the clocks is usually constrained by physical limits, it is useful to design controllers which work with limited resources. We stress that the number of clocks and observational power of clocks in the specification can be arbitrary and unrelated to the resources available to the controller.

While it is clear that earlier work could not handle nondeterministic specifications, one could use techniques, illustrated for example in [7], to convert *deterministic* external specifications to internal ones. However, such a move would introduce the clocks of the *specification automaton* into the plant, which the controller could then make use of. This gives us less handle on the resources allowed to the controller.

Our main result is that the control synthesis problem for external specifications is decidable when the specification describes the *undesired* behaviours of the plant. When such a controller exists, we show that we can synthesize a finite-state one that meets the specification. We prove this by giving a reduction to the problem of solving an infinite game over a finite graph, which is decidable [10,15,17] and for which efficient tools exist [13]. The problem however becomes undecidable if the specification describes the desired behaviours of the plant.

For *deterministic* specifications, we show the useful result that the combined resources of the plant and specification always suffice: if there is any controller meeting the specification, then there is one which uses just these resources. This allows us to decide the question of whether a controller—of *any* granularity at all—exists for a plant against a deterministic specification. For non-deterministic specifications, however, this question becomes undecidable.

2 Preliminaries

For an alphabet (a finite set of symbols) A, we denote by A^* and A^ω the set of finite and infinite words over A respectively, and by A^∞ the set $A^* \cup A^\omega$. The set of finite prefixes of a word α in A^∞ will be denoted $pre(\alpha)$. We use ϵ for the empty word, and $|\alpha|$ for the length of α.

A *transition system* over an alphabet A is of the form $\mathcal{T} = (Q, q_0, \longrightarrow)$, where Q is a (possibly infinite) set of states, $q_0 \in Q$ is an initial state, and $\longrightarrow \subseteq Q \times A \times Q$ is the transition relation. A *run* of $\mathcal{T}$ on a word $\alpha \in A^\infty$ is a map $\rho : pre(\alpha) \to Q$ such that $\rho(\epsilon) = q_0$, and for each prefix wa of α, $\rho(w) \xrightarrow{a} \rho(wa)$. We define $L_{sym}(\mathcal{T})$ and $L_{sym}^*(\mathcal{T})$ to be the set of words in A^ω and A^*, respectively, on which $\mathcal{T}$ has a run. We set $L_{sym}^\infty(\mathcal{T}) = L_{sym}^*(\mathcal{T}) \cup L_{sym}(\mathcal{T})$. We say $\mathcal{T}$ is *deterministic* if there do not exist transitions of the form $p \xrightarrow{a} q$ and $p \xrightarrow{a} q'$ with $q \neq q'$. Given a deterministic $\mathcal{T}$ and $w \in A^*$ such that $\mathcal{T}$ has a run on w, this run is unique and we use $state_\mathcal{T}(w)$ to denote the state reached at the end of this run.

An ω-*automaton* over an alphabet A is of the form $\mathcal{A} = (\mathcal{T}, \mathcal{F})$, where $\mathcal{T}$ is a *finite state* transition system over A, and $\mathcal{F}$ is an *acceptance condition* (see [14,16]). We consider instances of $\mathcal{F}$ as a *Büchi condition*, specified by a subset of states F, or a *parity condition* (also called Mostowski condition), specified by $ind : Q \to \{0, \ldots, d\}$ (where $d \in \mathbb{N}$), which assigns an *index* to each state. A run ρ of $\mathcal{A}$ on a word $\alpha \in A^\omega$ is an *accepting* run according to the Büchi condition F if $inf(\rho) \cap F \neq \emptyset$, where $inf(\rho)$ is the set of states q such that $\rho(w) = q$ for infinitely many $w \in pre(\alpha)$. It is accepting according to the parity condition ind if $min\{ind(q) \mid q \in inf(\rho)\}$ is even, i.e. the minimum index of the states met infinitely often is even. A word α is *accepted* by $\mathcal{A}$ if there is an accepting run of $\mathcal{A}$ on α. We set $L_{sym}(\mathcal{A})$ to be the set of words in A^ω accepted by $\mathcal{A}$. Following standard terminology we say $L \subseteq A^\omega$ is ω-*regular* if $L = L_{sym}(\mathcal{A})$ for some ω-automaton $\mathcal{A}$ over A.

We now turn to timed words and timed transition systems. As in [2], we use clocks that record the passing of time, and guards on these clocks, to describe timed behaviours. Let the set of non-negative reals and rationals be denoted by $\mathbb{R}_{\geq 0}$ and $\mathbb{Q}_{\geq 0}$ respectively, and the set of positive reals and rationals by $\mathbb{R}_{>0}$ and $\mathbb{Q}_{>0}$ respectively. A *constraint* (or *guard*) over a set of clocks X is given by the syntax

$$g ::= (x \leq c) \mid (x \geq c) \mid \neg g \mid (g \vee g) \mid (g \wedge g)$$

where $x \in X$ and $c \in \mathbb{Q}_{\geq 0}$. We denote the set of constraints over X by $\mathcal{G}(X)$. A *valuation* for a set of clocks X is a map $v : X \to \mathbb{R}_{\geq 0}$. Let $val(X)$ denote the set of all valuations over X. The relation $v \models g$, read "v satisfies g", is defined as usual; for example $v \models (x \leq \frac{1}{2})$ iff $v(x) \leq \frac{1}{2}$. The set of valuations over X which satisfy a guard $g \in \mathcal{G}(X)$ is denoted by $[\![g]\!]_X$, or just $[\![g]\!]$ when the set of clocks is clear from the context. A set S of constraints over X will be termed *complete* if $\bigcup_{g \in S} [\![g]\!] = val(X)$. For a valuation v over X, we denote by $v + t$ the valuation v' given by $v'(x) = v(x) + t$ for each $x \in X$. We use $\mathbf{0}$ to denote the zero-valuation which sends each x in X to 0. For $Y \subseteq X$, we use $v[0/Y]$ to denote the valuation v' given by $v'(x) = 0$ if $x \in Y$ and $v(x)$ otherwise.

We define a measure of the clocks and constants used in a set of constraints, called its *granularity*. A granularity is specified by a tuple $\mu = (X, m, max)$ where X is a finite set of clocks, $m \in \mathbb{N}$, and $max : X \to \mathbb{Q}_{\geq 0}$. A constraint is μ-*granular* if the clocks it uses belong to X, each constant used is an integral multiple of $\frac{1}{m}$, and each clock x is never compared to a constant larger than $max(x)$. We denote the set of all constraints of granularity μ by $\mathcal{G}(\mu)$.

We say a granularity $\mu = (X, m, max)$ is *finer* than $\mu' = (X', m', max')$, written $\mu \geq \mu'$, if $X \supseteq X'$, m is a multiple of m', and $max(x) \geq max'(x)$ for each $x \in X'$. Note that if $\mu \geq \mu'$ and a constraint is μ'-granular, then it is μ-granular as well. The *granularity of a finite set of constraints* is the granularity (X, m, max) where X is the exact set of clocks mentioned in the constraints, m is the least common multiple of the denominators of the constants mentioned in the constraints, and max records for each $x \in X$ the largest constant it is compared to. It is thus the least μ such that each constraint is μ-granular. For granularities $\mu = (X, m, max)$ and $\nu = (X', m', max')$ we use $\mu + \nu$ to mean the combined granularity of μ and ν which is $(X \cup X', lcm(m, m'), max'')$ where $max''(x)$ is the larger of $max(x)$ and $max'(x)$, assuming $max(x) = 0$ for $x \in X' - X$, and $max'(x) = 0$ for $x \in X - X'$. Note that $\mu + \nu$ is finer than both μ and ν.

A *timed word* σ over an alphabet Σ is an element of $(\Sigma \times \mathbb{R}_{>0})^\infty$ satisfying:

- for each prefix $\tau \cdot (a, t) \cdot (b, t')$ of σ we have $t < t'$ (monotonicity),
- if σ is infinite, then for each $t \in \mathbb{R}_{>0}$, there exists a prefix $\tau \cdot (a, t')$ of σ with $t' > t$ (progressiveness).

We denote the set of finite and infinite timed words over Σ by $T\Sigma^*$ and $T\Sigma^\omega$ respectively. For a finite timed word τ we define $time(\tau)$ to be the time stamp of the last action in τ—thus, $time(\epsilon) = 0$ and $time(\tau \cdot (a, t)) = t$.

Let Σ be a finite alphabet of actions, and X a finite set of clocks. A *symbolic alphabet* Γ based on (Σ, X) is a finite subset of $\Sigma \times \mathcal{G}(X) \times 2^X$. By the granularity of Γ we will mean the granularity of the set of constraints used in Γ.

A symbolic word $\gamma \in \Gamma^\infty$ gives rise to a set of timed words, denoted $tw(\gamma)$, in a natural way. We interpret the symbolic action (a, g, Y) to mean that action a can happen if the guard g is satisfied, with the clocks in Y being reset after the action. Let $\sigma \in T\Sigma^\infty$. Then $\sigma \in tw(\gamma)$ iff

- $|\sigma| = |\gamma|$,
- there exists a sequence of valuations, given by $\eta : pre(\sigma) \to val(X)$, such that $\eta(\epsilon) = \mathbf{0}$, and for each prefix $\tau \cdot (a, t)$ of σ and $\delta \cdot (b, g, Y)$ of γ, with $|\tau| = |\delta|$, we have
  - $a = b$,
  - $\eta(\tau) + (t - time(\tau)) \models g$, and
  - $\eta(\tau \cdot (a, t)) = (\eta(\tau) + (t - time(\tau)))[0/Y]$.

A *timed transition system* $\mathcal{T}$ over (Σ, X) is a transition system over a symbolic alphabet Γ based on (Σ, X). Viewed as such, $\mathcal{T}$ accepts (or generates) a language of symbolic words, $L_{sym}(\mathcal{T})$. We will be more interested in the timed language it generates, denoted $L(\mathcal{T})$, and defined to be $L(\mathcal{T}) = tw(L_{sym}(\mathcal{T}))$. We use $L^*(\mathcal{T})$ to denote the set of finite timed words $\mathcal{T}$ accepts, namely $tw(L^*_{sym}(\mathcal{T}))$.

In a similar manner, a *timed ω-automaton* $\mathcal{A}$ over (Σ, X) is just an ω-automaton over a symbolic alphabet based on (Σ, X). It accepts a timed language denoted $L(\mathcal{A})$ and defined by $L(\mathcal{A}) = tw(L_{sym}(\mathcal{A}))$. We say a timed language $L \subseteq T\Sigma^\omega$ is *timed ω-regular* if $L = L(\mathcal{A})$ for some timed ω-automaton $\mathcal{A}$. By the granularity of a timed automaton (or a timed transition system) we will mean the granularity of the set of constraints used in it.

A timed transition system $(Q, q_0, \longrightarrow)$ is *(time) deterministic* if we never have two distinct transitions of the form $q \xrightarrow[Y]{a,g} q_1$ and $q \xrightarrow[Y']{a,g'} q_2$ with $[\![g]\!] \cap [\![g']\!] \neq \emptyset$. A timed automaton is deterministic if its underlying timed transition system is deterministic.

We define the synchronized product of two timed transition systems which run over a *shared* set of clocks. Let $\mathcal{T}_1 = (Q_1, q_0^1, \longrightarrow_1)$ and $\mathcal{T}_2 = (Q_2, q_0^2, \longrightarrow_2)$ be timed transition systems over (Σ, X_1) and (Σ, X_2) respectively, with $X_1 \cap X_2$ possibly non-empty. The synchronized product of $\mathcal{T}_1$ and $\mathcal{T}_2$, denoted $\mathcal{T}_1 \| \mathcal{T}_2$, is defined to be the timed transition system $(Q, Q_0, \longrightarrow)$ over $(\Sigma, X_1 \cup X_2)$, where $Q = Q_1 \times Q_2$, $q_0 = (q_0^1, q_0^2)$, and $\longrightarrow$ is given by $(p_1, p_2) \xrightarrow[Y]{a,g} (q_1, q_2)$ iff $p_1 \xrightarrow[Y_1]{a, g_1}_1 q_1$ and $p_2 \xrightarrow[Y_2]{a, g_2}_2 q_2$ with $g = g_1 \wedge g_2$ and $Y = Y_1 \cup Y_2$. If $X_1 \cap X_2 = \emptyset$, then $L(\mathcal{T}_1 \| \mathcal{T}_2) = L(\mathcal{T}_1) \cap L(\mathcal{T}_2)$. Also, if $\mathcal{T}_1$ and $\mathcal{T}_2$ are deterministic, so is $\mathcal{T}_1 \| \mathcal{T}_2$.

3   The Problem

A *partitioned* alphabet $\widetilde{\Sigma}$ is a pair (Σ_C, Σ_E) that defines a partition of an alphabet Σ into controllable actions Σ_C and environment actions Σ_E. We fix a partitioned alphabet $\widetilde{\Sigma}$ in the sequel. A *plant* over $\widetilde{\Sigma}$ is a deterministic, finite-state, timed transition system $\mathcal{P}$ over Σ.

Let $\mathcal{P}$ be a plant over $\widetilde{\Sigma}$ and let the clocks used in $\mathcal{P}$ be $X_{\mathcal{P}}$. Let $X_{\mathcal{C}}$ be a set of clocks disjoint from $X_{\mathcal{P}}$, and let $\mu = (X_{\mathcal{P}} \cup X_{\mathcal{C}}, m, max)$ be a granularity finer than that of the plant. Then a *μ-controller* for $\mathcal{P}$ is a deterministic timed transition system $\mathcal{C}$ over $(\Sigma, X_{\mathcal{P}} \cup X_{\mathcal{C}})$ of granularity μ, satisfying:

(C1) $\mathcal{C}$ has resets only in $X_{\mathcal{C}}$ (i.e. if $q_c \xrightarrow[Y]{a,g} q'_c$ in $\mathcal{C}$, then $Y \subseteq X_{\mathcal{C}}$)

(C2) $\mathcal{C}$ does not restrict environment actions (*non-restricting*): whenever we have $\tau \in L(\mathcal{P} \| \mathcal{C})$ and $\tau \cdot (e, t) \in L(\mathcal{P})$ with $e \in \Sigma_E$, then we also have $\tau \cdot (e, t) \in L(\mathcal{P} \| \mathcal{C})$.

(C3) $\mathcal{C}$ is *non-blocking*: whenever we have $\tau \in L(\mathcal{P} \| \mathcal{C})$ and $\tau \cdot (b, t) \in L(\mathcal{P})$, there exists $c \in \Sigma$ and $t' \in \mathbb{R}_{>0}$ such that $\tau \cdot (c, t') \in L(\mathcal{P} \| \mathcal{C})$.

(C1) demands that the controller does not reset the clocks of the plant. However, it can "read" the values of the plant clocks in the sense that guards in $\mathcal{C}$ can refer to $X_{\mathcal{P}}$. The controlled system is $\mathcal{P} \| \mathcal{C}$, and so (C2) demands that the controller does not restrict the environment moves in any way. However, if an environment move occurs, the controller is allowed to check the clock values and reset any of its clocks to keep track of the environment's behaviour. The condition (C3) ensures that the controller does not introduce any new deadlocks in the plant.

Definition 1. (Control synthesis for specified granularity against un-desired behaviours) *Given a plant $\mathcal{P}$ over $\widetilde{\Sigma}$, a granularity μ finer than that of $\mathcal{P}$, and a timed automaton $\mathcal{A}_s$ over Σ as a specification of undesired behaviours, does there exist a μ-controller $\mathcal{C}$ for $\mathcal{P}$ such that $L(\mathcal{P} \| \mathcal{C}) \cap L(\mathcal{A}_s) = \emptyset$?*

We can now state the main result of the paper:

Theorem 1. *The control synthesis problem for specified granularity against un-desired behaviours is decidable. Moreover, if there is a controller meeting the specification, we can synthesize a finite-state controller of the required granular-ity which meets the specification.*

4   Timed Games

We now define the notion of a "timed game" between two players C (the control) and E (the environment). We will reduce the control synthesis problem to the problem of deciding whether player C has a winning strategy in such a game.

Let us fix some notation first. A given granularity $\mu = (X, m, max)$ induces in a natural way the notion of an *atomic* set of valuations for X. An atomic set of valuations w.r.t. μ is an equivalence class of the equivalence relation where valuations v and v' are related iff

- for each $x \in X$, either both $v(x), v'(x) > max(x)$, or $\lfloor v(x) \rfloor = \lfloor v'(x) \rfloor$ and $frac(v(x)) = 0$ iff $frac(v'(x)) = 0$. By $frac(t)$ we mean the value $t - \lfloor t \rfloor$.

A *region* of μ is a finer equivalence class where valuations further satisfy

- for each pair of clocks x, y in X with $v(x), v'(x) \leq max(x)$ and $v(y), v'(y) \leq max(y)$, the fractional parts of x and y in v are related in the same way as the fractional parts of x and y in v'—i.e. $frac(v(x)) = frac(v(y))$ iff $frac(v'(x)) = frac(v'(y))$ and $frac(v(x)) < frac(v(y))$ iff $frac(v'(x)) < frac(v'(y))$.

Any two valuations in an atomic set of valuations satisfy exactly the same set of guards in $\mathcal{G}(\mu)$. In fact, this is the coarsest equivalence which has this property. With each atomic set of valuations r with respect to μ, we fix a canonical guard $g \in \mathcal{G}(\mu)$ for which $[\![g]\!] = r$. Let $atoms_\mu$ denote this set of canonical atomic guards with respect to μ, and let reg_μ denote the set of regions of μ. Both $atoms_\mu$ and reg_μ can be seen to be finite sets. For a guard $g \in \mathcal{G}(\mu)$, we use $atoms_\mu(g)$ to denote the set $\{g' \in atoms_\mu \mid [\![g']\!] \subseteq [\![g]\!]\}$. Finally, we say a symbolic alphabet Γ is *atomic* if every constraint used is in Γ in $atoms_\mu$, where μ is the granularity of Γ.

We can now formulate the notion of a timed game. A *timed game graph* over $\widetilde{\Sigma}$ is a deterministic finite state timed transition system $\mathcal{H}$ over an *atomic* symbolic alphabet Γ based on Σ. A *timed game* over $\widetilde{\Sigma}$ is of the form $(\mathcal{H}, L)$ where $\mathcal{H}$ is a timed game graph over $\widetilde{\Sigma}$ and $L \subseteq T\Sigma^\omega$ is the winning set. The game is played between players C and E, and a play $\gamma = u_0 u_1 \cdots \in \Gamma^\infty$ is built up as follows. Player C chooses a "valid" subset of symbolic actions (as defined below) enabled at s_0, and player E responds by choosing an action u_0 from that subset. Next, player C chooses a valid subset of symbolic actions enabled at the resulting state and player E picks u_1 from it, and so on.

Let $\gamma \in L^*_{sym}(\mathcal{H})$. A set of symbolic actions $U \subseteq \Gamma$ enabled at $s = state_\mathcal{H}(\gamma)$ is *valid* at s w.r.t. γ if

- (U is *deterministic* at s) U does not contain two symbolic actions of the form (a, g, Y) and (a, g, Y') with $Y \neq Y'$,
- (U is *non-restricting* at s) for each symbolic action of the form (e, g, Y) enabled at s with $e \in \Sigma_E$, U should include an action of the form (e, g, Y') for some Y',
- (U is *non-blocking* at s w.r.t. γ) if there exists a symbolic action (b, g, Y) enabled at s with $tw(\gamma \cdot (b, g, Y)) \neq \emptyset$, then there should exist an action (c, g', Y') in U such that $tw(\gamma \cdot (c, g', Y')) \neq \emptyset$.

A *play* in $\mathcal{H}$ is hence a word in Γ^∞. An infinite play γ is winning for player C if $tw(\gamma) \cap L = \emptyset$. A *strategy* for player C is a partial function $f : \Gamma^* \rightharpoonup 2^\Gamma$ such that

- f is defined on ϵ, and for each $\gamma \in \Gamma^*$, if f is defined on γ and $u \in f(\gamma)$, then f is defined on $\gamma \cdot u$,
- if f is defined on γ then $f(\gamma)$ is valid at $state_\mathcal{H}(\gamma)$ w.r.t. γ.

We say that a play γ is a *play according to* f if for every prefix $\gamma' \cdot u$ of γ, $u \in f(\gamma')$. Let $plays_f(\mathcal{H})$ denote the set of infinite plays and $plays^*_f(\mathcal{H})$ denote the set of finite plays played according to f. We say that f is *winning* for C if all infinite plays according to f are winning—or, equivalently, if $tw(plays_f(\mathcal{H})) \cap L = \emptyset$. Finally, f is a *finite state* strategy if there exists a deterministic finite state transition system $\mathcal{T}$ over Γ, with $f(\gamma)$ given by the set of actions enabled at $state_\mathcal{T}(\gamma)$.

Let us return to the control synthesis problem of Definition 1. Let $\mathcal{P} = (P, p_0, \longrightarrow)$ be a plant over $\widetilde{\Sigma}$. Let the set of clocks used in $\mathcal{P}$ be $X_\mathcal{P}$, let $\mu = (X_\mathcal{P} \cup X_\mathcal{C}, m, max)$, and let $\mathcal{A}_s$ be the specification. We formulate a timed game corresponding to this instance of the problem. The game captures the role of a μ-controller $\mathcal{C}$ for $\mathcal{P}$. After any timed word σ executed in $L^*(\mathcal{P} \| \mathcal{C})$, $\mathcal{C}$ must basically choose a set of transitions from the current state of $\mathcal{P}$, which does not block the plant nor prevents an environment action from occurring, and also prescribe resets of its own clocks on each possible move. We can thus look upon

$\mathcal{C}$ as a strategy for a game which has the state-space of $\mathcal{P}$ and where transitions reflect the appropriate choices for $\mathcal{C}$.

Consider the timed game $(\mathcal{H}_{\mathcal{P},\mu}, L(\mathcal{A}_s))$ over $\widetilde{\Sigma}$ where $\mathcal{H}_{\mathcal{P},\mu} = (P, p_0, \delta)$ and δ is given by: $(q, (a, g', Y \cup Y'), q') \in \delta$ iff

- there exists a transition of the form $q \xrightarrow[Y]{a, g} q'$ in $\mathcal{P}$,
- $g' \in atoms_\mu(g)$,
- and $Y' \subseteq X_\mathcal{C}$.

$\mathcal{H}_{\mathcal{P},\mu}$ has as its symbolic alphabet $\Gamma = \Sigma \times atoms_\mu \times 2^{X_\mathcal{C} \cup X_\mathcal{P}}$.

It is easy to show that the timed game $(\mathcal{H}_{\mathcal{P},\mu}, L(\mathcal{A}_s))$ captures our control synthesis problem in the following sense:

Lemma 1. *There exists a (finite-state) μ-controller for $\mathcal{P}$ which satisfies $\mathcal{A}_s$ iff player C has a (finite-state) winning strategy in $(\mathcal{H}_{\mathcal{P},\mu}, L(\mathcal{A}_s))$.* $\quad\square$

5   Solving a Timed Game

Our aim is now to solve a timed game with a timed ω-regular winning set, by reducing it to a classical state-based game, solutions for which are well known. A classical *state-based game* over a partitioned alphabet $\widehat{\Delta}$ is of the form $(\mathcal{K}, val, \mathcal{F})$ where $\mathcal{K}$, the game graph, is a deterministic, finite state, transition system over Δ, $val : Q \to 2^{(2^\Delta)}$ is a function that identifies the valid choices player C can offer at a state in $\mathcal{K}$, and $\mathcal{F}$ is a winning condition stated as an acceptance condition on the states of $\mathcal{K}$.

In such a game, at a state q in $\mathcal{K}$, player C picks a subset of actions U enabled at q, with $U \in val(q)$; and E responds by picking an element $u \in U$. The game then proceeds to the u-successor state. Thus a play is a word in $L_{sym}(\mathcal{K})$ comprising the actions picked by player E. The play is winning for player C if the (unique) run of states in $\mathcal{K}$ meets the acceptance condition $\mathcal{F}$. Our definition is slightly different, but equivalent, to the standard definition in the literature.

We now prove a lemma which allows us to formulate a winning condition on ω-regular timed words as a winning condition on an ω-regular set of *symbolic words*. Let us fix some notation first. Recall the definition of regions of μ. Since any two valuations of an atomic set (or region) r satisfy the same set of guards in $\mathcal{G}(\mu)$, we say that $r \models g$ iff $v \models g$ for some $v \in r$. We also use the notation $r[0/Y]$ to denote the region containing $v[0/Y]$ for some valuation v in r. This notion can also be seen to be well-defined. A region r' will be termed a *time-successor* of a region r if there exists $v \in r$ and $v' \in r'$ such that $v' = v + t$ for some $t \in \mathbb{R}_{>0}$. Let Γ be a symbolic alphabet of granularity at most μ. For $r \in reg_\mu$ and $(a, g, Y) \in \Gamma$ we can define the notion of a successor region of r w.r.t. (a, g, Y) as follows: it is a region r' such that there exists a time successor region r'' of r satisfying $r'' \models g$ and $r''[0/Y] = r'$. Note that if $g \in atoms_\mu$, there is at most one such r'.

Lemma 2. *Given a symbolic alphabet Γ based on (Σ, X) and a timed ω-automaton $\mathcal{A}$ over (Σ, X'), the set of words γ in Γ^ω such that $tw(\gamma) \cap L(\mathcal{A}) = \emptyset$ is ω-regular.*

Proof. We first construct an automaton for the complement language $\{\gamma \in \Gamma^\omega \mid tw(\gamma) \cap L(\mathcal{A}) \neq \emptyset\}$. The automaton we have in mind is essentially the region automaton [2] for the intersection of $\mathcal{A}$ and the universal timed automaton whose symbolic language is all of Γ^ω. In the construction below we do this region construction on-the-fly while reading a symbolic word.

Let Γ have granularity κ and let $\mathcal{A}$ have granularity κ'. Without loss of generality we assume X and X' are disjoint. Let $\nu = (X \cup X', m, max)$ be the granularity $\kappa + \kappa'$. Let $\mathcal{A} = (Q, q_0, \longrightarrow, F)$, assuming a Büchi condition without loss of generality. Define a Büchi ω-automaton $\mathcal{D} = (S, s_0, \longrightarrow', G)$ over Γ, where:

- $S = Q \times reg_\nu$,
- $s_0 = (q_0, [\mathbf{0}])$,
- $\longrightarrow'$ is given by $(q, r) \overset{(a,g,Y)}{\longrightarrow}' (q', r')$ iff there exists a transition of the form $q \overset{a,\, g'}{\underset{Y'}{\longrightarrow}} q'$ in $\mathcal{A}$, and r' is a successor region of r w.r.t. $(a, g \wedge g', Y \cup Y')$,
- $G = F \times reg_\nu$.

The automaton $\mathcal{D}$ is the required automaton, except that we need to eliminate runs which allow only non-progressive timed words. As in [2] we can easily transform $\mathcal{D}$ to another Büchi automaton $\mathcal{D}'$ which checks in addition that each clock $x \in X \cup X'$ is either beyond $max(x)$ infinitely often or is reset infinitely often.

It is not difficult to see that $\mathcal{D}'$ accepts precisely the set of words γ in Γ^ω such that $tw(\gamma) \cap L(\mathcal{A}) \neq \emptyset$, the argument being essentially that of [2]. We can then determinize $\mathcal{D}'$ to get a parity automaton $\mathcal{D}''$ and then complement it (by incrementing the index of each state by one). The resulting parity automaton $\mathcal{B}$ will then accept the set of words γ in Γ^ω such that $tw(\gamma) \cap L(\mathcal{A}) = \emptyset$. □

Before we move on, here is an interesting consequence of the above lemma:

Corollary 1. *Given a timed automaton $\mathcal{A}$ and granularity κ, consider the class of timed languages that can be accepted by timed automata of granularity κ and are subsets of the complement of $L(\mathcal{A})$. Then there is a maximal language (with respect to inclusion) in this class and one can effectively find an automaton of granularity κ accepting this maximal language.* □

Now let $\Omega = (\mathcal{H}, L)$ be a timed game over $\widetilde{\Sigma}$ with $L = L(\mathcal{A})$ for some timed ω-automaton $\mathcal{A}$. We show how to convert Ω to a classical game Φ. The idea is to use essentially the given timed game graph itself, but incorporate the winning condition as a state-based condition, and make the non-blocking condition locally checkable by keeping track of the current region.

Let $\mathcal{H} = (S, s_0, \delta)$ and let its symbolic alphabet be Γ. Let ν be the granularity of Γ. Let $\mathcal{B} = (B, b_0, \longrightarrow, ind)$ be the complete, determinized automaton obtained by Lemma 2 for the symbolic alphabet Γ and automaton $\mathcal{A}$. Let $\widetilde{\Delta}$ be the partitioned alphabet given by $\Delta_C = \{(a, g, Y) \in \Delta \mid a \in \Sigma_C\}$, and

$\Delta_E = \{(e, g, Y) \in \Gamma \mid e \in \Sigma_E\}$. Define $\Phi = (\mathcal{K}, val, ind')$ over $\widetilde{\Delta}$, with the components defined as follows.

$\mathcal{K} = (Q, q_0, \delta')$ where

- $Q = S \times reg_\nu \times B$,
- $q_0 = (s_0, [\mathbf{0}], b_0)$,
- $\delta'((s, r, b), (a, g, Y)) = (s', r', b')$ iff
  - $\delta(s, (a, g, Y)) = s'$,
  - r' is a successor of r with respect to (a, g, Y), and
  - $b \xrightarrow{(a,g,Y)} b'$ in $\mathcal{B}$.

For $q \in Q$, a set of actions U enabled at q is in $val(q)$ iff (a) U is deterministic and non-restricting at q w.r.t. $\widetilde{\Sigma}$ (cf. page 577) and (b) (*non-blocking*) if there is a transition enabled at q, then U is nonempty.

The parity condition ind' is defined as: $ind'((s, r, b)) = ind(b)$.

Strategies in the games Ω and Φ carry over and it is not difficult to see that the non-blocking and winning properties are preserved. Hence we have:

Lemma 3. *Player C has a (finite state) winning strategy in the timed game Ω iff player C has a (finite-state) winning strategy in the classical game Φ.* $\square$

We can now prove Theorem 1. Consider the classical game Φ corresponding to the timed game $\Omega = (\mathcal{H}_{\mathcal{P},\mu}, L(\mathcal{A}_s))$. By Lemma 3 and Lemma 1, a winning strategy for player C exists in Φ iff there exists a μ-controller for $\mathcal{P}$ which satisfies the specification. Since we can decide the existence of a winning strategy in a classical game (see [10,17]), this gives us a way of deciding the existence of a μ-controller for $\mathcal{P}$. Further, if a player has a winning strategy in a classical game, he has a finite state one. It follows that if a μ-controller for $\mathcal{P}$ satisfying the specification exists, then a finite state one exists.

The decision procedure takes time polynomial in the state space of the plant, exponential in the state space of the specification, and double-exponential in the granularities of the specification and controller. In fact the problem can be shown to be 2EXPTIME-complete. This is true even when the specification is deterministic. However, if the specification is deterministic *and* its granularity is coarser than the given μ (in terms of the rational constants used), the problem becomes EXPTIME-complete. Detailed proofs of these results can be found in [5].

6   Related Results

A natural question that arises is: given a plant and a specification, is there is some computable granularity $\hat{\mu}$ such that if at all there is a controller, then there is a $\hat{\mu}$-controller. It turns out that when the specification is *deterministic*, this is indeed true.

Lemma 4. *If for some granularity μ there is a μ-controller for $\mathcal{P}$ that meets a deterministic specification, then there is a μ_{ps}-controller that satisfies the specification, where μ_{ps} is the combined granularity of the plant and the specification.*

Let us fix a plant $\mathcal{P}$ over $\widetilde{\Sigma}$, and a deterministic specification $\mathcal{A}_s$. Without loss of generality, we assume that $L(\mathcal{A}_s)$ describes the *desired* behaviours of the plant, since the specification is deterministic and can be complemented. We also assume that $\mathcal{A}_s$ is complete—i.e. the set of guards on transitions enabled at any state forms a complete set of constraints. Let $\mathcal{PS} = \mathcal{P} \| \mathcal{S}$, where $\mathcal{S}$ is $\mathcal{A}_s$ without the acceptance condition. Let μ_p, μ_s and μ_{ps} denote the granularities of $\mathcal{P}$, $\mathcal{A}_s$ and $\mathcal{PS}$ respectively. $\mathcal{PS}$ is a deterministic timed transition system and, since $\mathcal{A}_s$ is complete, admits precisely the same timed behaviours as $\mathcal{P}$ does, i.e. $L(\mathcal{P} \| \mathcal{S}) = L(\mathcal{P})$. In fact it is not difficult to prove that

Proposition 1. *(a) If there is a μ-controller $\mathcal{C}$ for $\mathcal{P}$, then there is a μ' such that there is a μ'-controller $\mathcal{C}'$ for $\mathcal{PS}$ with $L(\mathcal{P} \| \mathcal{C}) = L(\mathcal{PS} \| \mathcal{C}')$.*
(b) If there is a μ_{ps}-controller $\mathcal{C}'$ for $\mathcal{PS}$ then there is a μ_{ps}-controller $\mathcal{C}$ for $\mathcal{P}$ such that $L(\mathcal{P} \| \mathcal{C}) = L(\mathcal{PS} \| \mathcal{C}')$. □

Part (a) can be proved by taking $\mathcal{C}'$ as $\mathcal{C}$ itself (but possibly renaming clocks), and for (b) we can take $\mathcal{C}$ to be $\mathcal{C}' \| \mathcal{S}$.

Our task now is to show that if there is a μ'-controller for $\mathcal{PS}$, then in fact there is a μ_{ps}-controller for it which satisfies the specification; this will prove Lemma 4. Consider the game graph $\mathcal{H}_{\mathcal{PS},\mu''}$, for any μ'' finer than μ_{ps}. By Lemma 1, there is a μ''-controller for $\mathcal{PS}$ iff there is a winning strategy for C in this game. However, the winning condition for plays in this game can be specified in terms of *winning states*, using the fact that for a play γ in $\mathcal{H}_{\mathcal{PS},\mu''}$ with $tw(\gamma) \neq \emptyset$ we have $tw(\gamma) \subseteq L(\mathcal{A}_s)$ iff the S-component of the run of $\mathcal{H}_{\mathcal{PS},\mu''}$ on γ satisfies the winning condition of $\mathcal{A}_s$.

Let $\mathcal{PS}$ have X_{ps} as its set of clocks. Let $\mu' = (X_{ps} \cup X, m, max)$ be a granularity finer than μ_{ps}.

Proposition 2. *If there is a winning strategy for player C in the timed game $(\mathcal{H}_{\mathcal{PS},\mu'}, L(\mathcal{A}_s))$, then there is a winning strategy for player C in $(\mathcal{H}_{\mathcal{PS},\mu_{ps}}, L(\mathcal{A}_s))$.*

Proof. Note that by definition, the state spaces of the games $\mathcal{H}_{\mathcal{PS},\mu_{ps}}$ and $\mathcal{H}_{\mathcal{P},\mu'}$ are the same. It is also easy to show that, since μ' is finer than μ_{ps}, $\mathcal{H}_{\mathcal{PS},\mu_{ps}}$ can be obtained from $\mathcal{H}_{\mathcal{PS},\mu'}$ by replacing each edge which has a guard g by several edges which correspond to guards which divide g into finer atomic guards and all possible resets of the new clocks in μ'. We can then show that, since any strategy on a game basically determines only the state sequences traversed, and since the winning of a play is determined fully by this sequence, there is a winning strategy for player C in $(\mathcal{H}_{\mathcal{PS},\mu_{ps}}, L(\mathcal{A}_s))$ if there is one in $(\mathcal{H}_{\mathcal{PS},\mu'}, L(\mathcal{A}_s))$. □

We turn now to the decidability of some related problems. Using Lemma 4 we can show that the control synthesis problem where the specification is deterministic, but no granularity is specified *a priori*, is decidable, since we need only look for a μ_{ps}-controller. A similar question for *non-deterministic* specifications turns out to be undecidable. This follows by a reduction from the universality problem for timed automata on *finite* words, the undecidability of which follows from [2]. As a consequence, there cannot exist a sufficient granularity result along the lines of Lemma 4 for non-deterministic specifications.

The control synthesis problem for non-deterministic specifications which specify *desirable* behaviours can be seen to be undecidable, once again by a reduction from the universality problem for timed automata. This is true whether or not a granularity is specified *a priori*.

The following table summarizes the decidability results. Detailed proofs of all results in this paper can be found in [5].

Granularity specified			Granularity unspecified		
Det. spec	Nondet. spec		Det. spec	Nondet. spec	
	Desired	Undesired		Desired	Undesired
Decidable	*Undecidable*	*Decidable*	*Decidable*	*Undecidable*	*Undecidable*

References

1. M. Abadi, L. Lamport, P. Wolper: Realizable and Unrealizable Concurrent Program Specifications, *Proc. 16th ICALP*, LNCS 372, Springer-Verlag (1989).
2. R. Alur, D. L. Dill: A theory of timed automata, *Theoretical Computer Science* Vol 126 (1994).
3. E. Asarin, O. Maler, A. Pnueli, J. Sifakis: Controller Synthesis for Timed Automata, *Proc. IFAC Symposium on System Structure and Control*, Elsevier (1998).
4. J.R. Büchi, L.H Landweber: Solving sequential conditions by finite-state strategies, *Trans. AMS*, Vol 138 (1969).
5. D. D'Souza, P. Madhusudan: Controller synthesis for timed specifications, Technical Report TCS-01-2, Chennai Mathematical Institute (http://www.cmi.ac.in) (2001).
6. T.A. Henzinger: The theory of hybrid automata, *Proc. 11th IEEE LICS*, (1996).
7. G. Hoffmann, H. Wong-Toi: The control of dense real-time discrete event systems, *Conference on Decision and Control*, Brighton (1991).
8. O. Kupferman, P. Madhusudan, P.S. Thiagarajan, M. Vardi: Open systems in reactive environments: Control and Synthesis, *CONCUR '00*, LNCS 1877 (2000).
9. O. Maler, A. Pnueli, J. Sifakis: On the Synthesis of Discrete Controllers for Timed Systems, *Proc. STACS '95* LNCS 900, Spinger-Verlag (1995).
10. R. McNaughton: Infinite games played on finite graphs, *Annals of Pure and Applied Logic*, Vol 65 (1993).
11. A. Pnueli, R. Rosner: On the Synthesis of a Reactive Module, *Proc. 16th ACM POPL* (1989).
12. P.J.G. Ramadge, W.M. Wonham: The control of discrete event systems, *IEEE Trans. on Control Theory*, Vol 77 (1989).
13. D. Schmitz, J. Vöge: Implementation of a strategy improvement algorithm for parity games, *Proc. CIAA 2000*, London, Ontario (2000).
14. W. Thomas: Automata on infinite objects, *Handbook of Theoretical Computer Science*, North Holland (1990).
15. W. Thomas: On the Synthesis of Strategies in Infinite Games, *Proc. 12th STACS*, LNCS 900, Springer-Verlag (1995).
16. W. Thomas: Languages, Automata, and Logic, *Handbook of Formal Language Theory*, Springer-Verlag (1997).
17. W. Zielonka: Infinite games on finitely coloured graphs with applications to automata on infinite trees, *Theoretical Computer Science* Vol 200 (1998).

Axiomatizing GSOS with Termination

J.C.M. Baeten[1] and E.P. de Vink[1,2]

[1] Department of Mathematics and Computer Science, Eindhoven University of
Technology, P.O. Box 513, 5600 MB Eindhoven, The Netherlands,
{josb,evink}@win.tue.nl
[2] LIACS, Universiteit Leiden, P.O. Box 9512, 2300 RA Leiden, The Netherlands

Abstract. We discuss a combination of *GSOS*-type structural operational semantics with explicit termination, that we call the *tagh*-format (*tagh* being short for termination and *GSOS* hybrid). The *tagh*-format distinguishes between transition and termination rules, but allows besides active and negative premises as in *GSOS*, also for, what is called terminating and passive arguments. We extend the result of Aceto, Bloom and Vaandrager on the automatic generation of sound and complete axiomatizations for *GSOS* to the setting of *tagh*-transition system specifications. The construction of the equational theory is based upon the notion of a smooth and distinctive operation, which have been generalized from *GSOS* to *tagh*. The examples provided indicate a significant improvement over the mechanical axiomatization techniques known so far.

1 Introduction

It has become very popular in the concurrency community to define various process operators by means of Plotkin-style operational rules. These are usually pretty intuitive, and they can be used to derive a transition system for each process expression. Properties of such a transition system can then be checked using a model checker.

But it is also well-known that this approach has its restrictions. Often, transition systems become too large to be handled by model checkers, or, due to the presence of parameters, transition systems have infinitely many states. In these cases, an approach using theorem provers or deploying equational reasoning can be very helpful.

In the face of these alternative approaches, it is often profitable to generate a set of laws or equations for an operator that is given by a set of operational rules. Moreover, we want two characterizations that match: the axiomatization should be sound and complete for the model of transition systems modulo (strong) bisimulation. The paper [ABV94] points the way in such an endeavour: in some cases an axiomatization can be derived by just following a recipe. Some other papers in this area are [Uli95,Uli00] (where other equivalence relations besides bisimulation equivalence are considered). However, in the years since the appearance of these papers, we have seen no application of the theory. The reader may wonder why this is so.

H. Alt and A. Ferreira (Eds.): STACS 2002, LNCS 2285, pp. 583–595, 2002.
© Springer-Verlag Berlin Heidelberg 2002

In our opinion, this is due to the limited process algebraic basis employed in [ABV94]; in particular, termination and deadlock are identified. Any language, both programming and specification languages, involving some form of parallel composition will know the situation when no further action is possible, but components are not finished, e.g. when two components are waiting for different communications. This situation is usually called deadlock or unsuccessful termination. Now if the language also involves some form of sequential composition, we have to know when the first component in a sequential composition is finished, i.e. successfully terminated, in order for the second component to continue. In such a case, deadlock must be distinguished from successful termination, and, subsequently, the axiomatization method of [ABV94] does not apply.

There are three ways to handle this combination of parallel composition and sequential composition. First, we can do away with sequential composition as a basic operator, only have prefixing as a rudimentary form of sequential composition, and use tricks like a special communication to mimic some form of sequential composition. This is the solution of CCS, in our opinion an unsatisfactory solution. Second, we can use *implicit termination* as in ACP, where successful termination is implicitly "tacked onto" the last action. Finally, in the majority of cases, we find *explicit termination*, usually implemented by having two separate constants, one denoting deadlock, inaction or unsuccessful termination, the other one denoting skip or successful termination. Operationally, deadlock has no rules, and termination is denoted by a predicate on states.

In this paper, we adapt the theory of [ABV94] for the case of explicit termination. We think that the theory presented can be extended in order to deal also with implicit termination, but leave this as future research. Starting from the *GSOS*-format (cf. [BIM95]), we extend it with termination to obtain the *tagh*-format (termination and *GSOS* hybrid). We also employ some additional generalizations so that auxiliary operators are needed in fewer cases: for instance, the definition of sequential composition does not require auxiliary operators as in [ABV94]. This does make the theory a lot more complicated, but we gain that the generated axiomatizations are almost optimal, intuitively understandable, and are sound and complete for the model of transition systems modulo bisimulation.

2   Preliminaries

We assume the reader to be familiar with the standard notions and examples of process algebra (cf., e.g., [BW90,Fok00]). Below we present the transition system specification for the basic process language with explicit termination ε, deadlock δ (which has no rules), a prefixing operation 'a. ' for every a taken from the finite alphabet of actions Act, nondeterministic choice '$+$' and unary one-step restriction operations ∂_B^1 for every subset $B \subseteq Act$. The expression $\partial_B^1(t)$ indicates that the term t is not permitted to perform any action from B as a first step. However, this restriction is dropped after t has done a step outside of the action set B. For the termination predicate '$\downarrow$', we use the postfix notation $t\downarrow$ meaning that the term t has an option to terminate immediately.

Definition 1

(a) *The transition system specification for the transition system* TSS_∂^1 *consists of the following transition and termination rules:*

$$a.x \xrightarrow{a} x \qquad \frac{x \xrightarrow{a} x'}{x + y \xrightarrow{a} x'} \qquad \frac{y \xrightarrow{a} y'}{x + y \xrightarrow{a} y'} \qquad \frac{x \xrightarrow{a} x'}{\partial_B^1(x) \xrightarrow{a} x'} \ (a \notin B)$$

$$\varepsilon\!\downarrow \qquad \frac{x\!\downarrow}{(x + y)\!\downarrow} \qquad \frac{y\!\downarrow}{(x + y)\!\downarrow} \qquad \frac{x\!\downarrow}{\partial_B^1(x)\!\downarrow}$$

(b) *The equational theory* ET_∂^1 *consists of the following equations:*

$$\begin{array}{lll}
x + y = y + x & \partial_B^1(x + y) = \partial_B^1(x) + \partial_B^1(y) & \partial_B^1(\delta) = \delta \\
(x + y) + z = x + (y + x) & \partial_B^1(a.x) = a.x \ \text{ if } a \notin B & \partial_B^1(\varepsilon) = \varepsilon \\
x + x = x & \partial_B^1(a.x) = \delta \ \text{ if } a \in B & \\
x + \delta = x & &
\end{array}$$

We have the standard notion of strong bisimulation with predicates, in our set-up in the form of a termination condition.

Definition 2 *A bisimulation relation R for a transition system TSS is a binary relation for closed terms over TSS such that whenever $t_1 R t_2$ it holds that (i) $t_1 \xrightarrow{a} t_1' \implies \exists t_2' : t_2 \xrightarrow{a} t_2' \wedge t_1' R t_2'$, (ii) $t_2 \xrightarrow{a} t_2' \implies \exists t_1' : t_1 \xrightarrow{a} t_1' \wedge t_1' R t_2'$, (iii) $t_1\!\downarrow \iff t_2\!\downarrow$. Two terms t_1, t_2 are bisimilar with respect to TSS if there exists a bisimulation relation R for TSS with $t_1 R t_2$, notation: $t_1 \sim_{TSS} t_2$ or just $t_1 \sim t_2$.*

Below we will use $t_1 \equiv t_2$ to denote syntactic equality of the terms t_1 and t_2. We also use expressions like $C[x_k, y_\ell, z_m]$ to indicate that only variables from the set

$$\{\, x_k \mid k \in K \,\} \cup \{\, y_\ell \mid \ell \in L \,\} \cup \{\, z_m \mid m \in M \,\}$$

occur in the context $C[\ ]$ with respect to some given index sets K, L and M.

3   Generating Equations for the *tagh*-Format

In this section we introduce the *tagh*-format for transition system specifications. Here, the acronym *tagh* stands for *termination and GSOS hybrid*. It enhances the *GSOS*-format as introduced in [BIM95] with a notion of explicit termination. We provide a general procedure to obtain, for each transition system specification in *tagh*-format, a disjoint extension TSS' and an equational theory ET'. It holds that ET' is sound and complete for TSS'-bisimulation (cf. [BV01]). As the transition system specification TSS' is a disjoint extension of the transition system specification TSS this amounts for terms t_1, t_2 over TSS to coincidence of bisimulation with respect to TSS and equality based on ET'. Thus, ET' is a sound and complete axiomatization of TSS-bisimulation.

Definition 3

(a) A tagh-transition rule ρ for an n-ary operation f is a deduction rule of the format

$$\frac{\{\, x_i \overset{a_{ip}}{\rightarrow} y_{ip} \mid i \in I, p \in P_i \,\} \quad \{\, x_j \overset{b}{\nrightarrow} \mid j \in J, b \in B_j \,\} \quad \{\, x_k\!\downarrow \mid k \in K \,\}}{f(x_1, \ldots, x_n) \overset{a}{\rightarrow} C[x_m, y_{ip} \mid m \in \{1, \ldots, n\}, i \in I, p \in P_i]} \quad (1)$$

with $I, J, K \subseteq \{1, \ldots, n\}$, for $i \in I$, P_i a nonempty finite index set, for $j \in J$, B_j a finite (possibly empty) set of actions from Act, and, x_m, y_{ip}, for $m \in \{1, \ldots, n\}, i \in I, p \in P_i$, pairwise distinct variables, that are the only variables that may occur in the context C.

(b) A tagh-termination rule θ for an n-ary operation f is a deduction rule of the format

$$\frac{\{\, x_k\!\downarrow \mid k \in K \,\}}{f(x_1, \ldots, x_n)\!\downarrow} \quad (2)$$

with $x_1, \ldots, x_n$ pairwise distinct variables and the index set $K \subseteq \{1, \ldots, n\}$.
(c) A tagh-transition system specification is a transition system specification where any operation f different from 'ε', 'δ', 'a. ', '$+$' and '∂_B^1' has tagh-transition rules and/or tagh-termination rules only.

In the context of a transition rule ρ of the format (1) we use $act(\rho)$, $neg(\rho)$, $term(\rho)$, $pass(\rho)$ to denote the index sets I, J, K, L, respectively, where $L = \{1, \ldots, n\} \backslash (I \cup J \cup K)$. For a rule θ conforming to equation (2) we put $term(\theta) = K$. For a transition rule ρ like (1), we refer to $f(x_1, \ldots, x_n)$, or an instantiation of it, as the source of ρ, and to the term $C[x_m, y_{ip}]$ as the target. Occasionally we will write $t\!\!\not\downarrow$ if *not* $t\!\downarrow$, i.e., t cannot terminate immediately.

The *tagh*-format is an extension of the *GSOS*-format of [BIM95]. If we strip all aspects of termination from the definition we end up with the original format for *GSOS*. We have, as the *tagh*-format is subsumed by the *panth*-format of [Ver95], that bisimulation is a congruence, just as for *GSOS*. The syntactic format of general *tagh*-transition rules though, is much too liberal to allow for an automatic generation of axioms directly. We therefore introduce (cf. [ABV94]) a more restricted format, called smooth. Regarding an operation f it is profitable to further restrict the collection of rules. In essence we want that at any time at most one of the transition rules for f applies. If the rules for f have this additional property, the operation is called smooth and distinctive.

Definition 4 *Let TSS be a tagh-transition system specification.*

(a) A transition rule ρ in TSS for an n-ary operation $f \in Sig$ is smooth if it is of the format

$$\frac{\{\, x_i \overset{a_i}{\rightarrow} y_i \mid i \in I \,\} \quad \{\, x_j \overset{b}{\nrightarrow} \mid j \in J, b \in B_j \,\} \quad \{\, x_k\!\downarrow \mid k \in K \,\}}{f(x_1, \ldots, x_n) \overset{a}{\rightarrow} C[y_i, x_j, x_\ell \mid i \in I, j \in J, \ell \in L]} \quad (3)$$

where the index sets I, J, K, L *form a partition of* $\{1, \ldots, n\}$, $I \neq \emptyset$, $B_j \subseteq \mathrm{Act}$ *a finite (possibly empty) subset of actions, and, where in the target* $C[y_i, x_j, x_\ell]$ *only variables amongst* $\{y_i \mid i \in I\}$, $\{x_p \mid p \in J \cup L\}$ *occur. The operation* f *is* smooth *with respect to TSS if all of its transition rules in TSS are smooth, and, moreover,*

    — *for each position* p *in* $\{1, \ldots, n\}$ *it holds that* $p \notin \mathrm{pass}(\rho)$ *for some rule* ρ *for* f *in TSS.*

(b) The rank *of a rule* ρ *is the 4-tuple* $\langle \mathrm{pass}(\rho), \mathrm{act}(\rho), \mathrm{term}(\rho), \mathrm{neg}(\rho) \rangle$, *notation* $\mathrm{rank}(\rho)$. *For two rules* ρ, ρ' *for an* n-*ary operation* f *we say that* $\mathrm{rank}(\rho) \succcurlyeq \mathrm{rank}(\rho')$ *iff*

    — $\mathrm{neg}(\rho) = \mathrm{neg}(\rho')$, $\mathrm{pass}(\rho) \supseteq \mathrm{pass}(\rho')$ *and* $\mathrm{term}(\rho) \subseteq \mathrm{term}(\rho')$, *and*

    — $\mathrm{pass}(\rho) \neq \mathrm{pass}(\rho') \implies \mathrm{act}(\rho) \cap \mathrm{term}(\rho') \neq \emptyset$.

(c) A smooth n-*ary operation* f *is called* smooth and distinctive *with respect to TSS if*

    — *the set* $\{\mathrm{rank}(\rho) \mid \rho$ *a transition rule for* f *in TSS}* *is totally ordered by the ordering '*$\succcurlyeq$*' introduced in part (b);*

    — *for any two distinct rules* ρ, ρ' *of the form* (3) *with* $\mathrm{rank}(\rho) = \mathrm{rank}(\rho')$ *there exists an index* $i \in \mathrm{act}(\rho) = \mathrm{act}(\rho')$ *such that* $a_i \neq a'_i$.

    — *for each termination rule* θ *and each transition rule* ρ *for* f *in TSS it holds that* $\mathrm{term}(\theta) \cap \mathrm{act}(\rho) \neq \emptyset$.

For such an operation f *it holds that* $\mathrm{neg}(\rho) = \mathrm{neg}(\rho')$ *for any two transition rules* ρ, ρ'. *We define* $\mathrm{neg}(f) = \mathrm{neg}(\rho)$ *and* $\mathrm{nonneg}(f) = \{1, \ldots, n\} \setminus \mathrm{neg}(\rho)$ *where* ρ *is an arbitrary transition rule for* f *in TSS.*

The intuition for the ordering on the transition rules for a smooth and distinctive n-ary operation f is the following: Suppose ρ and ρ' are two transition rules for f with $\rho \succcurlyeq \rho'$. The ordering on $\succcurlyeq$ then demands that a passive position in ρ' must be passive in ρ as well and that a terminating position in ρ must be terminating in ρ' as well. Now, let $\rho_1 \succcurlyeq \ldots \succcurlyeq \rho_m$ be in descending order and $p \in \{1, \ldots, n\}$ a non-negative position in f. The position p can either be passive, active or terminating in $\rho_1, \ldots, \rho_m$, but in view of the observation above we have that for suitable $0 \leq k < \ell \leq m$ it holds that $p \in \mathrm{pass}(\rho_i)$ for $1 \leq i \leq k$, $p \in \mathrm{act}(\rho_i)$ for $k < i \leq \ell$ and $p \in \mathrm{term}(\rho_i)$ for $\ell < i \leq m$. So, in the context of $f(x_1, \ldots, x_n)$, the variable x_p at position p has a life-cycle from passive, via active, to terminating (but, possibly, p doesn't start out as passive or doesn't reach the termination stage).

For a smooth and distinctive n-ary operation f we have that for closed terms of the form $f(t_1, \ldots, t_n)$ where each $t_i \equiv \varepsilon, \delta, a.t'$ at most one of the transition rules for f applies: If ρ and ρ' are two distinct rules for f, we either have $\mathrm{rank}(\rho) = \mathrm{rank}(\rho')$ or, without loss of generality, $\mathrm{rank}(\rho) \succ \mathrm{rank}(\rho')$. From the requirements of Definition 4c above we then obtain in the first case that for some $i \in \{1, \ldots, n\}$, $t_i \equiv a.t'$ with $a = a_i$ (the action of the i-th premise for ρ), $a = a'_i$ (the action of the i-th premise for ρ') but also $a_i \neq a'_i$. For the second case we obtain from $\mathrm{rank}(\rho) \succ \mathrm{rank}(\rho')$ that $\mathrm{act}(\rho) \cap \mathrm{term}(\rho') \neq \emptyset$. So, for some $i \in \{1, \ldots, n\}$ we have $t_i \equiv a.t'$ as t_i matches the source of the i-th premise of ρ, but also $t_i \equiv \varepsilon$ as according to the rule ρ' the term t_i should terminate. All cases thus lead to a contradiction, and we conclude that $f(t_1, \ldots, t_n)$ does not match two distinct transition rules ρ and ρ'.

The requirement of at least one active position in a smooth transition rule will be needed in our proof of the soundness of the distributive laws for negative arguments, introduced below and that are superfluous in the setting of [ABV94] but are essential for our treatment of termination.

Examples 5 *The binary operation* ';' *of sequential composition comes equipped, in the set-up with explicit termination, with two transition rules and one termination rule:*

$$(Seq_1) \ \frac{x \xrightarrow{a} x'}{x;y \xrightarrow{a} x';y} \qquad (Seq_2) \ \frac{x{\downarrow} \quad y \xrightarrow{a} y'}{x;y \xrightarrow{a} y'} \qquad (Seq_\varepsilon) \ \frac{x{\downarrow} \quad y{\downarrow}}{(x;y){\downarrow}}$$

The binary operation '$\lfloor\!\lfloor$', *usually referred to as leftmerge, has one transition rule and one termination rule:*

$$(Leftmerge_1) \ \frac{x \xrightarrow{a} x'}{x {\lfloor\!\lfloor} y \xrightarrow{a} x' \parallel y} \qquad (Leftmerge_\varepsilon) \ \frac{x{\downarrow} \quad y{\downarrow}}{(x {\lfloor\!\lfloor} y){\downarrow}}$$

We have that both ';' *and* '$\lfloor\!\lfloor$' *(contrasting [ABV94]) are smooth and distinctive operations.*

Note that, in fact, in the above examples we have transition schemes for (Seq_1), (Seq_2) and $(Leftmerge_1)$ rather than transition rules, as we have transition rules (Seq_1), (Seq_2) and $(Leftmerge_1)$, respectively, for each action $a \in Act$.

Before we are ready to describe the axioms generated for a smooth and distinctive n-ary operation f for a $tagh$-transition system specification, we need some notation. Note that the rules of f are totally ordered by the ordering '$\succcurlyeq$'. So, if $m \in nonneg(f)$, there exists a not necessarily unique transition rule ρ, maximal in rank, such that $m \notin pass(\rho)$. In that situation we put $rank(m) = rank(\rho)$ and $act(m) = act(\rho)$, $neg(m) = neg(\rho)$, etc. Also, if, for a 4-tuple R, we have that $R = rank(\rho)$, we put $act(R) = act(\rho)$, $neg(R) = neg(\rho)$, etc. The index set $handle(m)$, the handle of m with respect to f and TSS, is defined as $term(m)$ if $m \in nonneg(f)$, and as $nonneg(f)$ if $m \in neg(f)$.

The idea behind the notion of a handle is that for a smooth operation f and non-negative position $m \in \{1, \ldots, n\}$ the set $handle(m)$ consists of all positions that are required to be terminating when the position m becomes active, i.e.,

$$handle(m) = \bigcap \{ term(\rho) \mid m \in act(\rho), \rho \text{ transition rule for } f \}.$$

For a negative position m for f, $handle(m)$ simply consists of all non-negative positions. The handles are used in the formulation of the distributive laws below; the subset-ordering on the handles of an operation induces an ordering on the applicability of these laws.

The next definition describes the various laws associated with a smooth and distinctive operation.

Definition 6 *Let f be a smooth and distinctive n-ary operation for a tagh-transition system specification TSS.*

(a) For a position $p \in \{1, \dots, n\}$ the distributive law for p with respect to f is given as follows:

$$f(\zeta_1, \dots, z'_p + z''_p, \dots, \zeta_n) = f(\zeta_1, \dots, z'_p, \dots, \zeta_n) + f(\zeta_1, \dots, z''_p, \dots, \zeta_n) \quad (4)$$

where $\zeta_q \equiv \varepsilon$ for $q \in \mathrm{handle}(p)$ and $\zeta_q \equiv z_q$ for $q \notin \{p\} \cup \mathrm{handle}(p)$.
(b) For a transition rule ρ of the format (3) the action law for ρ is given as follows:

$$f(\zeta_1, \dots, \zeta_n) = a.C[z'_i, z_j, z_\ell] \quad (5)$$

where $\zeta_i \equiv a_i.z'_i$ for $i \in \mathrm{act}(\rho)$, $\zeta_j \equiv \partial^1_{B_j}(z_j)$ for $j \in \mathrm{neg}(\rho)$ with $B_j \neq \emptyset$ and $\zeta_j \equiv z_j$ for $j \in \mathrm{neg}(\rho)$ with $B_j = \emptyset$, $\zeta_k \equiv \varepsilon$ for $k \in \mathrm{term}(\rho)$ and $\zeta_\ell \equiv z_\ell$ for $\ell \in \mathrm{pass}(\rho)$.
(c) For a rank R for f the deadlock laws are given as follows:

$$f(\zeta_1, \dots, \zeta_n) = \delta \quad (6)$$

where ζ_m is of the form ε, δ or $a'_m.z'_m$ for $m \in \mathrm{act}(R) \cup \mathrm{term}(R)$, ζ_j is of the form z_j, δ, $b'_j.z'_j$ or $z_j + b'_j.z_j$ for $j \in \mathrm{neg}(R)$ and $\zeta_\ell \equiv z_\ell$ for $\ell \in \mathrm{pass}(R)$ such that, for each rule ρ for f in TSS of the format (3), there exists a position p such that one of the following cases holds:
  * $p \in \mathrm{act}(\rho)$ and $\zeta_p \equiv \varepsilon$, $\zeta_p \equiv \delta$ or $\zeta_p \equiv a'_p.z'_p$ with $a'_p \neq a_p$, or
  * $p \in \mathrm{neg}(\rho)$ and $\zeta_p \equiv b'_p.z'_p$ or $\zeta_p \equiv z_p + b'_p.z'_p$ with $b'_p \in B_p$, or
  * $p \in \mathrm{term}(\rho)$ and $\zeta_p \equiv \delta$ or $\zeta_p \equiv a'_p.z'_p$,
and, for each termination rule θ for f there exists a position $p \in \{1, \dots, n\}$ such that $\zeta_p \equiv \delta$ or $\zeta_p \equiv a'_p.z'_p$.
(d) For a termination rule θ for f the termination law for θ is given as follows:

$$f(\zeta_1, \dots, \zeta_n) = \varepsilon. \quad (7)$$

where $\zeta_p \equiv \varepsilon$ for $p \in \mathrm{term}(\theta)$ and $\zeta_p \equiv z_p$ for $p \notin \mathrm{term}(\theta)$.

In the distributive laws we demand a 'fingerprint of ε-s' for the particular position instead of allowing a variable for *handle*-arguments. This way, non-determinism at a position is only resolved if it guaranteed that there is termination at sufficiently many other positions, as will be illustrated by the equations in the examples for sequential composition ';' and leftmerge '$\lfloor\!\lfloor$' below.

Examples 7 *The transition system specification for ';' generates, according to the definitions above, the following equations:*

$$(x_1 + x_2); y = (x_1; y) + (x_2; y) \qquad (a.x'); y = a.(x'; y) \qquad \varepsilon; \delta = \delta \qquad \varepsilon; \varepsilon = \varepsilon$$
$$\varepsilon; (y_1 + y_2) = (\varepsilon; y_1) + (\varepsilon; y_2) \qquad \varepsilon; (a.y') = a.y' \qquad \delta; y = \delta$$

Note that, apart from the equation $\delta; y = \delta$, the operation ';' has also other deadlock laws, viz. $\delta; \varepsilon = \delta$, $\delta; (a.y') = \delta$ and $\delta; (y + b.y') = \delta$, which are special cases of the displayed law $\delta; y = \delta$.

Similarly, we obtain for the leftmerge '$\lfloor$' the following axiom system:

$$(x_1 + x_2) \lfloor y = (x_1 \lfloor y) + (x_2 \lfloor y) \qquad \varepsilon \lfloor \delta = \delta \qquad \varepsilon \lfloor \varepsilon = \varepsilon$$
$$\varepsilon \lfloor (y_1 + y_2) = (\varepsilon \lfloor y_1) + (\varepsilon \lfloor y_2) \qquad \varepsilon \lfloor (b.y') = \delta$$
$$(a.x') \lfloor y = a.(x' \parallel y) \qquad \delta \lfloor y = \delta$$

Again we omit the superfluous instantiations of the axiom $\delta; y = \delta$. Note that actually we have exactly the preferred axiomatization, see e.g. [Vra97].

From the termination law $\varepsilon; \varepsilon = \varepsilon$ and $\varepsilon \lfloor \varepsilon = \varepsilon$ in the examples above, one can see the necessity of a distributive law for a negative argument, here in both cases the second position. Without these distributive laws it is not possible to derive, e.g., $\varepsilon; (a.t + \varepsilon) = a.t + \varepsilon$ and $\varepsilon \lfloor (a.t + \varepsilon) = \varepsilon$, which is desired for our interpretation of optional termination.

The disrupt or disabling operator '$\gg$' is well-known, e.g., from Lotos [Bri89] (see also [BB00]). In the process $x \gg y$ the subprocess x may proceed, unless the subprocess y takes over control. It terminates when either of the subprocesses does so. Thus, the disrupt operator has the following transition system specification:

$$\frac{x \xrightarrow{a} x'}{x \gg y \xrightarrow{a} x' \gg y} \qquad \frac{y \xrightarrow{a} y'}{x \gg y \xrightarrow{a} y'} \qquad \frac{x\downarrow}{(x \gg y)\downarrow}$$

$$\frac{y\downarrow}{(x \gg y)\downarrow}$$

The disrupt operator, as can be seen from the transition rules, is a smooth but non-distinctive operation. However, if we split the operation '$\gg$' into two, introducing '$\gg_1$' and '$\gg_2$' say, for which the transition rules satisfy the distinctiveness restrictions, we end up with two smooth and distinctive operations:

$$\frac{x \xrightarrow{a} x'}{x \gg_1 y \xrightarrow{a} x' \gg y} \qquad \frac{x\downarrow}{(x \gg_1 y)\downarrow} \qquad \frac{y \xrightarrow{a} y'}{x \gg_2 y \xrightarrow{a} y'}$$

$$\frac{y\downarrow}{(x \gg_2 y)\downarrow}$$

The idea of splitting up '$\gg$' is also present in the transition system specification for this operation in [BB00]. The relationship between the various disrupt operations is expressed by the law $x \gg y = (x \gg_1 y) + (x \gg_2 y)$. Another instance of this trick is the representation of the merge '$\parallel$' in terms of leftmerge '$\lfloor\!\lfloor$', rightmerge '$\rfloor\!\rfloor$' and communication merge '$\mid$' using the law $x \parallel y = (x \lfloor\!\lfloor y) + (x \rfloor\!\rfloor y) + (x \mid y)$.

The same approach, as pointed out in [ABV94] and also applicable for the *tagh*-format, of partitioning of the set of transition rules and introducing smooth and distinctive suboperations works in general to split a smooth but non-distinctive operation f into a number of smooth and distinctive ones, $f_1, \dots, f_s$ say. Here we only present how the resulting equations can be derived.

Definition 8 *Let f be a smooth but non-distinctive n-ary operation for the tagh-transition system specification TSS. The n-ary operations $f_1, \dots, f_s$ are called distinctive versions of f in a disjoint extension TSS' of TSS if the transition and termination rules for each f_r in TSS' ($1 \le r \le s$) form, after renaming of f_r in the source of the rules by f, a partitioning of all the rules for f in TSS. The equation*

$$f(\mathbf{z}) = f_1(\mathbf{z}) + \cdots + f_s(\mathbf{z}) \tag{8}$$

is then referred to as the distinctivity law for f.

The previous definition addresses smooth but non-distinctive operations. However, some operations are not smooth at all. There may be several ways in which the transition rules of an operation f can violate the various conditions of the definition of a smooth operation: there can be a transition rule for f that is not of the format (3), i.e., either there are multiple premises for an action-argument or an active or terminating variable occurs in the target or there is overlap of the index sets or there is no active premise. (Additionally, there can be a position p for which there is no transition rule for f for which this p is non-passive. The latter situation is harmless. See [BV01].)

To illustrate the countermeasure for non-smoothness consider the following, synthesized, one-rule transition system specification adapted from [ABV94]. Here, the operation f is non-smooth because there are multiple transitions for an active variable (viz. $x \xrightarrow{a} y_1$ and $x \xrightarrow{b} y_2$), the active and terminating variable x occurs in the target $x + y_1$, the index sets overlap (its only position 1 occurs as active, as terminating and as negative argument).

$$\frac{x \xrightarrow{a} y_1 \quad x \xrightarrow{b} y_2 \quad x \xrightarrow{c} \!\!\!\!/ \quad x\!\downarrow}{f(x) \xrightarrow{d} x + y_1} \qquad\qquad \frac{x\!\downarrow}{f(x)\!\downarrow}$$

The key idea is not to split f into new operations, but to split the variable x into new variables, i.e., we introduce separate copies x_1, x_2, x_3, x_4 of the variable x to relieve the overlap and multiplicity. The rules for f are translated into rules for a fresh operation f'. This yields the following transition system specification for which f' is a smooth operation:

$$\frac{x_1 \xrightarrow{a} y_1 \quad x_2 \xrightarrow{b} y_2 \quad x_3 \xrightarrow{c} \!\!\!\!/ \quad x_4\!\downarrow}{f'(x_1, x_2, x_3, x_4) \xrightarrow{d} x_3 + y_1} \qquad\qquad \frac{x_1\!\downarrow \quad x_2\!\downarrow \quad x_3\!\downarrow \quad x_4\!\downarrow}{f'(x_1, x_2, x_3, x_4)\!\downarrow}$$

As connecting law for f we have $f(x) = f'(x, x, x, x)$ which enforces that in the right-hand side we indeed have copies of the original argument.

In the next definition we will formalize the above ideas for the general case. In the presentation below we introduce mappings ϕ and ψ to make the correspondence explicit between a variable x_i and its splittings $\{\,x'_{i'} \mid \phi(i') = i\,\}$ and the actions a_{ip} and output variables y_{ip} and their new names $a'_{i'}$ and $y'_{i'}$ with $\psi(i') = (i,p)$.

Definition 9 *Let f be a non-smooth n-ary operation of a tagh-transition system specification TSS. The m-ary operation f' is called the smooth version of f in a disjoint extension TSS' of TSS, if there exist mappings $\phi\colon \{1,\dots,m\} \to \{1,\dots,n\}$ and $\psi\colon \{1,\dots,m\} \to \{1,\dots,n\} \times \{1,\dots,m\}$ and a 1–1 correspondence between the rules of f and f', such that*

(a) a transition rule ρ for f in TSS of the form

$$\frac{\{\, x_i \overset{a_{ip}}{\to} y_{ip} \mid i \in I, p \in P_i\,\} \quad \{\, x_j \overset{b_{jq}}{\nrightarrow} \mid j \in J, q \in Q_j\,\} \quad \{\, x_k{\downarrow} \mid k \in K\,\}}{f(x_1,\dots,x_n) \overset{a}{\to} C[x_i, x_j, x_k, x_\ell, y_{ip}]} \quad (9)$$

corresponds to a smooth transition rule ρ' for f' in TSS' of the form

$$\frac{\{\, x'_i \overset{a'_i}{\to} y'_i \mid i \in I'\,\} \quad \{\, x'_j \overset{b'_{jq}}{\nrightarrow} \mid j \in J', q \in Q'_j\,\} \quad \{\, x'_k{\downarrow} \mid k \in K'\,\}}{f'(x'_1,\dots,x'_m) \overset{a}{\to} C'[x'_j, x'_\ell, y'_{ip}]} \quad (10)$$

such that the mapping $x'_i \overset{a'_i}{\to} y'_i \mapsto x_{\phi(i)} \overset{a'_i}{\to} y_{\psi(i)}$, $x'_j \overset{b'_{jq}}{\nrightarrow} \mapsto x_{\phi(j)} \overset{b'_{jq}}{\nrightarrow}$, $x'_k{\downarrow} \mapsto x_{\phi(k)}{\downarrow}$ is a bijection between the premises of ρ and the premises of ρ' and $C[x_i, x_j, x_k, x_\ell, y_{ip}] \equiv \chi(C'[x'_j, x'_\ell, y'_i])$ for a substitution χ with $\chi(x'_j) = x_{\phi(j)}$, $\chi(x'_\ell) = x_{\phi(\ell)}$, $\chi(y'_i) = y_{\psi(i)}$,
(b) a termination rule θ for f in TSS of the form on the left below corresponds to a termination rule for f' in TSS' of the form on the right below

$$\frac{\{\, x_k{\downarrow} \mid k \in K\,\}}{f(x_1,\dots,x_n){\downarrow}} \qquad\qquad \frac{\{\, x'_k{\downarrow} \mid k \in K'\,\}}{f'(x'_1,\dots,x'_m){\downarrow}}$$

where $K' = \phi^{-1}(K)$.

The equation

$$f(z_1,\dots,z_n) = f'(\zeta_1,\dots,\zeta_m), \quad (11)$$

with $\zeta_p \equiv z_{\phi(p)}$ for $p \in \{1,\dots,n\}$, is called the smoothening law for f.

Example 10 *The 'classical' example of a non-smooth operation is the priority operator θ introduced in [BBK86]. Assuming a partial ordering on '$>$' on Act, the action rules of the unary θ and its binary smoothening θ' are the following:*

$$\frac{x \overset{a}{\to} x' \quad x \overset{b}{\nrightarrow} \quad (b > a)}{\theta(x) \overset{a}{\to} \theta(x')} \qquad \frac{x{\downarrow}}{\theta(x){\downarrow}} \qquad \frac{x \overset{a}{\to} x' \quad y \overset{b}{\nrightarrow} \quad (b > a)}{\theta'(x,y) \overset{a}{\to} \theta(x')} \qquad \frac{x{\downarrow} \quad y{\downarrow}}{\theta'(x,y){\downarrow}}$$

The smoothening law for the priority operator θ is $\theta(x) = \theta'(x,x)$.

In the above we have described how to transform a non-smooth operation into a smooth one and how to split a smooth but non-distinctive operation into several smooth and distinctive ones. In these situations the transition system specification will be extended disjointly, i.e., the dynamics and termination of operations already in the transition system remain unaffected. No rules and/or axioms are removed or added concerning operations already present. Also we have defined smoothening laws (11) and distinctivity laws (8) that connect the original and new operations. For smooth and distinctive operations we have introduced various equations describing distributivity, dynamics, deadlock and termination. Collecting this all together we obtain the notion of the transition system specification and the set of equations *generated* by a *tagh*-transition system specification.

Definition 11 *For a tagh-transition system specification TSS, the tagh-transition system specification TSS' generated by TSS and the equational theory ET' generated by TSS are given by the following procedure:*

Step 0 *Let TSS' disjointly extend TSS and TSS_∂^1. Let ET' contain the equations ET_∂^1 for '+' and '∂_B^1' (cf. Definition 1).*

Step 1 *For every non-smooth operation f of TSS not in TSS_∂^1, extend TSS' with the smooth version f' of f and add the smoothening law (11) to ET'.*

Step 2 *For every smooth but non-distinctive operation f of TSS' (as obtained after Step 1) but not in TSS_∂^1, extend TSS' with the distinctive versions $f_1, \ldots, f_s$ and add to ET' the distinctivity law (8).*

Step 3 *For each smooth and distinctive operation f of TSS' (as obtained after Step 2) but not in TSS_∂^1 add to ET' the distributive laws (4), the action laws (5), the deadlock laws (6) and the termination laws (7).*

The discussion above regarding the soundness of the various laws combined with the stratification of the procedure of Definition 11 give rise to the following theorem which is proven in [BV01]. (The transition system specification *TSS_I* comprises a syntactic representation of the Approximation Induction Principle.)

Theorem 12 *Let TSS be a tagh-transition system specification. Let TSS' be the disjoint extension of TSS_I and the generated extension of TSS. Let ET' be the generated equational theory. Then ET' and AIP are sound and complete for equality modulo TSS'.*

Examples 13 *Application of the above procedure yields for the disrupt operator '$\gg$' and the priority operator θ the following generated equational theories:*

$$x \gg y = (x \gg_1 y) + (x \gg_2 y) \qquad\qquad \delta \gg_1 y = \delta$$
$$(x_1 + x_2) \gg_1 y = (x_1 \gg_1 y) + (x_2 \gg_1 y) \qquad\qquad \varepsilon \gg_1 y = \varepsilon$$
$$(a.x') \gg_1 y = a.(x' \gg y) \qquad\qquad \textit{similar rules for '}\gg_2\textit{'}$$

$$\theta(x) = \theta'(x,x) \qquad\qquad \theta'(\delta,y) = \delta$$
$$\theta'(x_1 + x_2, y) = \theta'(x_1,y) + \theta'(x_2,y) \qquad\qquad \theta'(\varepsilon, b.y') = \delta$$
$$\theta'(\varepsilon, y_1 + y_2) = \theta'(\varepsilon, y_1) + \theta'(\varepsilon, y_2) \qquad\qquad \theta'(x,\delta) = \delta$$
$$\theta'(a.x', \partial^1_{b>a}(y)) = a.\theta(x') \qquad\qquad \theta'(a.x,\varepsilon) = \delta$$
$$\theta'(a.x', b.y + z) = \delta \quad if\, b > a \qquad\qquad \theta'(\varepsilon,\varepsilon) = \varepsilon$$

Note that the above axiomatizations are quite natural: The equations for the disrupt operation coincide with those of [BB00]. The axiomatization for the priority operator avoids equations for the auxiliary 'unless' operation '◁' (cf. [BBK86]).

4   Concluding Remarks

We have introduced the *tagh*-format for structured operational semantics. The *tagh*-format enhances the well-known *GSOS*-format with explicit termination. The format additionally allows for a finer distinction between the modes of the argument (viz. active, negative, terminating, passive). The method of automatic generation of axiomatizations as developed by Aceto, Bloom and Vaandrager for *GSOS* is extended for the case of *tagh*. Examples illustrate the technique and indicate the strength of the approach. The resulting laws are equal or close to the hand-crafted axiomatizations as reported in the literature.

In the technical report [BV01] we prove Theorem 12 which states the soundness and completeness modulo bisimulation of the axiomatization for a transition system specification in *tagh*-format obtained from the procedure of Definition 11. The soundness and completeness result is obtained along the lines set out in [ABV94]; several technical complications arise dealing with termination. The proof of head-normalization in the setting of *tagh* is involved. It requires a detailed case analysis and the notions *handle* and *rank* for non-negative positions. However, grosso modo, the outline of the proofs of soundness and completeness follow the corresponding arguments in [ABV94].

References

[ABV94] L. Aceto, B. Bloom, and F.W. Vaandrager. Turning SOS rules into equations. *Information and Computation*, 111:1–52, 1994.

[Bae00] J.C.M. Baeten. Embedding untimed into timed process algebra: the case for explicit termination. In L. Aceto and B. Victor, editors, *Proc. EXPRESS'00*, pages 45–62. ENTCS 39, 2000.

[BB00] J.C.M. Baeten and J.A. Bergstra. Mode transfer in process algebra. Technical Report CSR 00-01, Division of Computer Science, TU/e, 2000.

[BBK86] J.C.M. Baeten, J.A. Bergstra, and J.W. Klop. Syntax and defining equations for an interrupt mechanism. *Fundamenta Informaticae*, IX:127–168, 1986.

[BIM95] B. Bloom, S. Istrail, and A.R. Meyer. Bisimulation can't be traced. *Journal of the ACM*, 42:232–268, 1995. Preliminary version in Proc. POPL'88.

[Bri89] E. Brinksma, editor. *Information Processing Systems, Open Systems Interconnection, LOTOS – A Formal Description Technique Based on the Temporal Ordering of Observational Behaviour*. ISO Standard IS–8807, 1989.

[BV01]   J.C.M. Baeten and E.P. de Vink. Axiomatizing GSOS with termination. Technical Report CSR 01–06, Division of Computer Science, TU/e, 2001. See http://www.win.tue.nl/st/medew/pubbaeten.html.

[BW90]   J.C.M. Baeten and W.P. Weijland. *Process Algebra*, volume 18 of *Cambridge Tracts in Theoretical Computer Science*. Cambridge University Press, 1990.

[Fok00]   W.J. Fokkink. *Introduction to Process Algebra*. Texts in Theoretical Computer Science, An EATCS Series. Springer, 2000.

[Uli95]   I. Ulidowski. Axiomatisations of weak equivalences for De Simone languages. In I. Lee and S. Smolka, editors, *Proc. CONCUR'95*, pages 219–233. LNCS 962, 1995.

[Uli00]   I. Ulidowski. Finite axiom systems for testing preorder and De Simone process languages. *Theoretical Computer Science*, 239:97–139, 2000.

[Ver95]   C. Verhoef. A congruence theorem for structured operational semantics with predicates and negative premises. *Nordic Journal of Computing*, 2:274–302, 1995.

[Vra97]   J.L.M. Vrancken. The algebra of communicating processes with empty process. *Theoretical Computer Science*, 177:287–328, 1997.

Axiomatising Tree-Interpretable Structures

Achim Blumensath

RWTH Aachen

Abstract. Generalising the notion of a prefix-recognisable graph to arbitrary relational structures we introduce the class of tree-interpretable structures. We prove that every tree-interpretable structure is finitely axiomatisable in guarded second-order logic with cardinality quantifiers.

1 Introduction

In recent years the investigation of algorithmic properties of *infinite* structures has become an established part of computer science. Its applications range from algorithmic group theory to databases and automatic verification. Infinite databases, for example, were introduced to model geometric and, in particular, geographical data (see [16] for an overview). In the field of automatic verification several classes of infinite transition systems and corresponding model-checking algorithms have been defined. For instance, model-checking for the modal μ-calculus over prefix-recognisable graphs is studied in [6], [17]. A further point of interest in this context is the bisimulation equivalence of such transition systems as considered in [22], [23].

Obviously, only restricted classes of infinite structures are suited for such an approach. In order to process a class $\mathcal{K}$ of infinite structures by algorithmic means two conditions must be met:

(i) Each structure $\mathfrak{A} \in \mathcal{K}$ must possess a *finite representation*.
(ii) The operations one would like to perform must be *effective* with regard to these representations.

One fundamental operation demanded by many applications is the evaluation of a query, that is, given a formula $\varphi(\bar{x})$ in some fixed logic and the representation of a structure $\mathfrak{A} \in \mathcal{K}$ one wants to compute a representation of the set $\varphi^{\mathfrak{A}} := \{ \bar{a} \mid \mathfrak{A} \models \varphi(\bar{a}) \}$. Slightly simpler is the model-checking problem which asks whether $\mathfrak{A} \models \varphi(\bar{a})$ for some given $\bar{a}$. The class of tree-interpretable structures investigated in the present article has explicitly been defined in such a way that model-checking for MSO, monadic second order logic, is decidable. To the authors knowledge it is one of the largest natural classes with this property.

Several different notions of infinite graphs and structures have been considered in the literature:

- *Context-free graphs* [19], [20] are the configuration graphs of pushdown automata.
- *HR-equational graphs* [8] are defined by equations of hyperedge-replacement grammars.

H. Alt and A. Ferreira (Eds.): STACS 2002, LNCS 2285, pp. 596–607, 2002.
© Springer-Verlag Berlin Heidelberg 2002

- *Prefix-recognisable graphs* have been introduced in [7]. Several characterisations are presented in Section 3.
- *Automatic graphs* [15], [3], [5] are graphs whose edge relation is recognised by synchronous multihead automata.
- *Rational graphs* [15], [18] are graphs whose edge relation is recognised by asynchronous multihead automata.
- *Recursive graphs* [13] are graphs whose edge relation is recursive.

These classes of graphs form a strict hierarchy. The table to the right shows for which logic model-checking is still decidable for the various classes. $\mathrm{FO}(\exists^\kappa)$, $\mathrm{MSO}(\exists^\kappa)$, and $\mathrm{GSO}(\exists^\kappa)$ denote, respectively, first-order logic, monadic second-order logic, and guarded second-order logic extended by cardinality quantifiers. Σ_0 is the set of quantifier-free first-order formulae.

Class	Logic
context-free	$\mathrm{GSO}(\exists^\kappa)$
HR-equational	$\mathrm{GSO}(\exists^\kappa)$
prefix-recognisable	$\mathrm{MSO}(\exists^\kappa)$
automatic	$\mathrm{FO}(\exists^\kappa)$
rational	Σ_0
recursive	Σ_0

When investigating a class of finitely presented structures the question naturally arises which structures it contains. Usually it is quite simple to show that some structure belongs to the class by constructing a corresponding presentation. But the proof that such a presentation does not exists frequently requires more effort.

One possible approach consists in determining what additional information is needed in order to extract the presentation from a given structure. In the case of a tree-interpretable structure this information can be coded into a colouring of its elements and edges. A characterisation of these colourings amounts to one of the set of presentations of a structure. Besides determining whether a presentation exists such a characterisation can, for instance, be used to investigate the automorphism group of the structure.

In the present article we generalise the class of prefix-recognisable graphs to arbitrary relational structures and prove that each presentation corresponds to a $\mathrm{GSO}(\exists^k)$-definable colouring. This implies that each such structure is finitely axiomatisable in this logic. The outline of the article is as follows. Due to space constraints some parts had to be omitted. The full version appears in [4].

In Section 3 we review several characterisations of the class of prefix-recognisable graphs including characterisations in terms of languages, graph grammars, and interpretations.

The latter can be generalised to arbitrary relational structures most easily. The resulting class of tree-interpretable structures is defined in Section 4. After summarising some of its properties we also extend the characterisation via regular languages to this class.

Section 5 is devoted to the study of paths in tree-interpretable graphs. The presented results are mostly of a combinatorial nature and culminate in the proof that every connected component is spanned by paths with a certain property.

In Section 6 we prove our main theorem which states that all tree-interpretable structures are finitely axiomatisable in guarded second-order logic with cardinality quantifiers. We also show that the cardinality quantifiers are indeed needed.

Section 8 concludes the article with some lemmas about the orbits of the automorphism group of a tree-interpretable structure and the result that isomorphism is decidable for tree-interpretable structures of finite tree-width.

2   Preliminaries

Automata and trees. Let Σ be an alphabet. The complete tree over Σ is the structure $\mathfrak{T}_\Sigma := (\Sigma^*, (\mathrm{suc}_a)_{a \in \Sigma}, \preceq)$ where the suc_a denote the successor functions and $\preceq$ is the prefix-order. The longest common prefix of u and v is denoted by $u \sqcap v$. If $u = vw$ then we define $v^{-1}u := w$ and $uw^{-1} := v$.

For $u \in \Sigma^*$ and $k \in \mathbb{N}$ we write u/k for the prefix of u of length $|u| - k$, and $\mathrm{suf}_k u$ for the suffix of u of length k. In case $|u| < k$ we have $u/k = \varepsilon$ and $\mathrm{suf}_k u = u$. In particular, $(u/k)\,\mathrm{suf}_k u = u$ for all u and k.

Let $\leq_{\mathrm{lex}}$ be the lexicographic order and $\leq_{\mathrm{ll}}$ the length-lexicographic one defined by

$$x \leq_{\mathrm{lex}} y \quad :\text{iff} \quad x \preceq y, \text{ or } wc \preceq x \text{ and } wd \preceq y \text{ for some } w \text{ and } c < d.$$

$$x \leq_{\mathrm{ll}} y \quad :\text{iff} \quad |x| < |y|, \text{ or } |x| = |y| \text{ and } x \leq_{\mathrm{lex}} y.$$

We denote automata by tuples $(Q, \Sigma, \Delta, q_0, F)$ with set of states Q, alphabet Σ, transition relation Δ, initial state q_0, and acceptance condition F.

Logic. Let us recall some basic definitions and fix our notation. Let $[n] := \{0, \ldots, n-1\}$. We tacitly identify tuples $\bar{a} = a_0 \ldots a_{n-1} \in A^n$ with functions $[n] \to A$ or with the set $\{a_0, \ldots, a_{n-1}\}$. This allows us to write $\bar{a} \subseteq \bar{b}$ or $\bar{a} = \bar{b}|_I$ for $I \subseteq [n]$.

MSO, *monadic second-order logic*, extends first-order logic FO by quantification over sets. In *guarded second-order logic*, GSO, one can quantify over relations R of arbitrary arity with the restriction that every tuple $\bar{a} \in R$ is *guarded*, i.e., there is some relation S of the original structure that contains a tuple $\bar{b} \in S$ such that $\bar{a} \subseteq \bar{b}$. Note that every singleton a is guarded by $a = a$. For a more detailed definition see [14].

$\mathfrak{L}(\exists^\kappa)$ denotes the extension of the logic $\mathfrak{L}$ by *cardinality quantifiers* $\exists^\lambda$, for every cardinal λ, where $\exists^\lambda$ stands for "there are at least λ many".

A formula $\varphi(\bar{x})$ where each free variable is first-order defines on a given structure $\mathfrak{A}$ the relation $\varphi^{\mathfrak{A}} := \{\, \bar{a} \mid \mathfrak{A} \models \varphi(\bar{a}) \,\}$.

Definition 1. Let $\mathfrak{A} = (A, R_0, \ldots, R_n)$ and $\mathfrak{B}$ be relational structures. A (one-dimensional) MSO-*interpretation* of $\mathfrak{A}$ in $\mathfrak{B}$ is a sequence

$$\mathcal{I} = \langle \delta(x),\ \varepsilon(x, y),\ \varphi_{R_0}(\bar{x}), \ldots,\ \varphi_{R_n}(\bar{x}) \rangle$$

of MSO-formulae such that $\mathfrak{A} \cong \big(\delta^{\mathfrak{B}},\ \varphi^{\mathfrak{B}}_{R_0}, \ldots,\ \varphi^{\mathfrak{B}}_{R_n}\big)/\varepsilon^{\mathfrak{B}}$. To make this expression well-defined we require that the relation $\varepsilon^{\mathfrak{B}}$ is a congruence of the structure $\big(\delta^{\mathfrak{B}},\ \varphi^{\mathfrak{B}}_{R_0}, \ldots,\ \varphi^{\mathfrak{B}}_{R_n}\big)$. We denote the fact that $\mathcal{I}$ is an MSO-interpretation of $\mathfrak{A}$ in $\mathfrak{B}$ by $\mathcal{I} : \mathfrak{A} \leq_{\mathrm{MSO}} \mathfrak{B}$ or $\mathfrak{A} = \mathcal{I}(\mathfrak{B})$.

The epimorphism $\big(\delta^{\mathfrak{B}},\ \varphi^{\mathfrak{B}}_{R_0}, \ldots,\ \varphi^{\mathfrak{B}}_{R_n}\big) \to \mathfrak{A}$ is called *coordinate map* and also denoted by $\mathcal{I}$. If it is the identity function we say that $\mathfrak{A}$ is *definable* in $\mathfrak{B}$.

3   Prefix-Recognisable Graphs

Originally, the investigation of tree-interpretable structures was concerned only with transition systems. This subclass appears in the literature under several names using widely different definitions which all turned out to be equivalent. They are summarised in the next theorem. A more detailed description follows below.

Theorem 2. *Let* $\mathfrak{G} = (V, (E_a)_{a \in A})$ *be a graph. The following statements are equivalent:*

(1) $\mathfrak{G}$ *is prefix-recognisable.*
(2) $\mathfrak{G} = h^{-1}(\mathfrak{T}_2)|_C$ *for a rational substitution* h *and a regular language* C.
(3) $\mathfrak{G}$ *is the restriction to a regular set of the configuration graph of a pushdown automaton with ε-transitions.*
(4) $\mathfrak{G}$ *is MSO-interpretable in the binary tree* $\mathfrak{T}_2$.
(5) $\mathfrak{G}$ *is VR-equational.*

The equivalence of the first two items are due to Caucal [7], Stirling [23] mentioned the third characterisation, and Barthelmann [1] delivered the last two.

Definition 3. *A graph is prefix-recognisable if it is isomorphic to a graph of the form* $(S, (E_a)_{a \in A})$ *where* S *is a regular language over some alphabet* Σ *and each* E_a *is a finite union of relations of the form*

$$W(U \times V) := \{\, (wu, wv) \mid u \in U,\ v \in V,\ w \in W \,\}$$

for regular languages $U, V, W \subseteq \Sigma^*$.

Actually in the usual definition the reverse order $(U \times V)W$ is used. The above formulation was chosen as it fits better to the usual conventions regarding trees.

Example. The structure $(\omega, \mathrm{suc}, \leq)$ is prefix-recognisable. If we represent the universe by a^* the relations take the form $\mathrm{suc} = a^*(\varepsilon \times a)$ and $\leq\, = a^*(\varepsilon \times a^*)$.

One can also characterise prefix-recognisable graphs via graph grammars. Using the notation of Courcelle [8], [10], [12] we consider the following operations on vertex-coloured graphs. Let C be a finite set of colours.

- $G + H$ is the disjoint union of G and H.
- $\varrho_\beta(G)$, for $\beta : C \to C$, changes the colour of the vertices from a to $\beta(a)$.
- $\eta^a_{b,c}(G)$ adds a-edges from each b-coloured vertex to all c-coloured ones.
- a denotes the graph with a single a-coloured vertex.

The clique-width of a graph $\mathfrak{G}$ is, by definition, the minimal number of colours one needs to write a term denoting $\mathfrak{G}$.

Definition 4. *A countable coloured graph is VR-equational if it is the canonical solution of a finite system of equation of the form*

$$x_0 = t_0, \quad \ldots, \quad x_n = t_n$$

where the t_i *are finite terms build up from the above operations. Further, we require that none of the* t_i *equals a single variable* x_k.

Proposition 1 (Barthelmann [1]). *A graph is prefix-recognisable if and only if it is VR-equational.*

Since only finitely many colours can be used in a finite system of equations it follows that the clique-width of each VR-equational graph is finite.

Corollary 1. *Each prefix-recognisable graph is of finite clique-width.*

Example. If we colour the first element by a and the other ones by b we can define $(\omega, \mathrm{suc}, <)$ by

$$x_0 = \eta_{a,b}^<(x_1), \qquad x_1 = \varrho_{c \to b}\eta_{a,c}^{\mathrm{suc}}(a + x_2), \qquad x_2 = \varrho_{a \to c}(x_0).$$

4   Tree-Interpretable Structures

The characterisation of prefix-recognisable graphs in terms of interpretations is the one most easily generalised to arbitrary relational structures.

Definition 5. A structure $\mathfrak{A}$ is called *tree-interpretable* iff $\mathfrak{A} \leq_{\mathrm{MSO}} \mathfrak{T}_2$.

From this definition one can immediately deduce some basic properties of the class of tree-interpretable structures.

Proposition 2. *The class of tree-interpretable structures is closed under MSO-interpretations. In particular, it is closed under*

(1) *isomorphisms,*   (4) *expansion by finitely many constants,*
(2) *definable expansions,*   (5) *factorisation by definable congruences, and*
(3) *finite unions,*   (6) *substructures with definable universe.*

Since $\mathrm{MSO}(\exists^\kappa)$ model checking is decidable for $\mathfrak{T}_2$ and this property is conserved by MSO-interpretations we obtain a decidability result for all tree-interpretable structures.

Proposition 3. $\mathrm{MSO}(\exists^\kappa)$ *model checking is decidable for every tree-interpretable structure.*

All tree-interpretable graphs are of finite clique-width. On the other hand, their tree-width can be unbounded as the example of the infinite clique $K_{\aleph_0}$ shows. A result of Courcelle [11] which was extended to tree-interpretable graphs by Barthelmann [2] shows that being of finite tree-width imposes a strong restriction on the structure of a tree-interpretable graph. Although stated only for graphs it also holds for arbitrary structures if one replaces $\mathfrak{G}$ by its Gaifman graph in (2)–(4).

Proposition 4 (Barthelmann [2], Courcelle [11]). *Let $\mathfrak{G}$ be a tree-interpretable graph. The following statements are equivalent:*

(1) $\mathfrak{G}$ *is HR-equational.*   (3) $K_{n,n}$ *is not a subgraph of $\mathfrak{G}$ for some $n < \aleph_0$.*
(2) $\mathfrak{G}$ *has finite tree-width.*   (4) $\mathfrak{G}$ *is uniformly sparse.*

This characterisation allows us to extend Proposition 3 to $\mathrm{GSO}(\exists^\kappa)$.

Theorem 6. *Let $\mathfrak{A}$ be a tree-interpretable structure. GSO model checking is decidable for $\mathfrak{A}$ if and only if $\mathfrak{A}$ is of finite tree-width. The same holds for $\mathrm{GSO}(\exists^\kappa)$.*

Although the characterisation of tree-interpretable structures by interpretations is quite elegant, in actual proofs it is most of the time easier to work with a more concrete characterisation in terms of languages. Let us recall how automata are used to decide $\mathrm{MTh}(\mathfrak{T}_2)$ (see [24] for an overview).

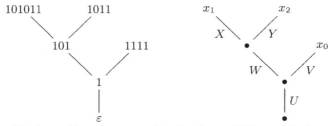

Fig. 1. The branching structure of 1111, 1011, 101011 and its isomorphism type

Definition 7. For sets $X_0, \ldots, X_{n-1} \subseteq \{0,1\}^*$ let $T_{\bar{X}}$ be the $P([n])$-labelled binary tree with $T(w) := \{ i < n \mid w \in X_i \}$ for $w \in \{0,1\}^*$. For singletons $X_i = \{x_i\}$ we also write $T_{\bar{x}}$.

With this notation we can now state Rabin's famous tree theorem in the following way:

Theorem 8. *For each* MSO*-formula* $\varphi(\bar{X}, \bar{x})$ *there is a tree-automaton* $\mathcal{A}$ *that recognises the language* $\{ T_{\bar{X}\bar{x}} \mid \mathfrak{T}_2 \models \varphi(\bar{X}, \bar{x}) \}$.

Employing this correspondence we generalise the characterisation of prefix-recognisable graphs by relations of the form $W(U \times V)$ to arbitrary relational structures.

Definition 9. The *branching structure* of $x_0, \ldots, x_{n-1} \in \Sigma^*$ is the partial order $(X, \preceq)$ where $X := \{\varepsilon\} \cup \{ x_i \sqcap x_j \mid i,\ j < n \}$. The elements of X are called *branching points*.

Example. The branching structure of 1111, 1011, 101011 is depicted in Figure 4.

Note that for a fixed number of words there are only finitely many non-isomorphic branching structures.

Proposition 5 (Blumensath [4]). *An n-ary relation* $R \subseteq (\{0,1\}^*)^n$ *is* MSO-*definable in* $\mathfrak{T}_2$ *iff* R *is a finite union of relations* R_i *of the following form:*

(a) *All tuples* $\bar{x} \in R_i$ *have the same branching structure (up to isomorphism).*

(b) *For each pair of adjacent branching points* $u,\ v$ *there is a regular language* $W_{u,v}$ *such that* $\bar{x} \in R_i$ *if and only if for each such pair* $u,\ v$ *the word* $u^{-1}v$ *belongs to* $W_{u,v}$.

Example. For the branching structure in Figure 4, a relation would be defined by five regular languages U, V, W, X, and Y with $R = U(V \times W(X \times Y))$.

Definition 10. Let $\mathfrak{A}$ be a tree-interpretable structure. Fixing an interpretation we can assume that the universe $A \subseteq \Sigma^*$ is regular and each relation R is specified by regular languages as in the preceding proposition. The *syntactic congruence* $\sim$ of $\mathfrak{A}$ (w.r.t. this interpretation) is the intersection of the syntactic congruences of all these languages. We denote the index of $\sim$ by I.

If some elements of a tree-interpretable structure are encoded by several words it becomes difficult to apply pumping arguments since the words obtained by pumping may encode the same element. Fortunately, for each tree-interpretable structure $\mathfrak{A}$ we can choose an interpretation where this does not happen.

Proposition 6 (Blumensath [4]). *If $\mathfrak{A} \leq_{\mathrm{MSO}} \mathfrak{T}_2$ then there is an interpretation $\mathcal{I} : \mathfrak{A} \leq_{\mathrm{MSO}} \mathfrak{T}_2$ where the coordinate map is injective.*

This result allows us to identify the elements of a tree-interpretable structure with the unique word encoding them. We will do so tacitly in the remainder of the article. We conclude this section with a combinatorial lemma whose proof is based on a pumping argument.

Lemma 1. *Let $\mathfrak{A}$ be a tree-interpretable structure and $\varphi(x, y) \in \mathrm{MSO}(\exists^\kappa)$ such that, for every $a \in A$, there are only finitely many elements $b \in A$ with $\mathfrak{A} \models \varphi(a, b)$. There is a constant k such that $\varphi(a, b)$ implies $b/k \prec a$. In particular, $|\varphi(a, A)| \in \mathcal{O}(|a|)$.*

5   Paths in Tree-Interpretable Graphs

In this section we consider a fixed tree-interpretable graph $\mathfrak{G} = (V, E_0, \dots, E_{r-1})$. By replacing each edge relation $E_a = \bigcup_i W_i (U_i \times V_i)$ by several relations $E_a^i := W_i (U_i \times V_i)$ we may assume that $E_a = W_a (U_a \times V_a)$ for regular languages $U_a, V_a, W_a \subseteq \Sigma^*$. We also add the relation $E_{a^-} := (E_a)^{-1}$ for each edge relation E_a. Note that these operations do not affect the syntactic congruence $\sim$.

Definition 11. (1) The *base-point* of an edge $(a, b) \in W(U \times V)$ is the longest word w contained in W such that $w^{-1} a \in U$ and $w^{-1} b \in V$. The *spine* of a path is the sequence of the base-points of its edges.

(2) A sequence $a_0, \dots, a_n$ is *k-increasing* if $|a_j| \geq |a_i| - k$ for all $i < j$.

(3) A path $a_0, \dots, a_n$ with spine $w_0, \dots, w_{n-1}$ is called *k-normal* if the path and its spine are k-increasing and $a_i/k \preceq a_j$ for all $j \geq i$.

Proposition 7 (Blumensath [4]). *Let $\mathfrak{G}$ be a tree-interpretable graph. There is a constant K such that each connected component of $\mathfrak{G}$ contains a vertex v, called its* root, *such that there are K-normal paths from v to all other vertices of the component.*

6   Axiomatisations

Equipped with the combinatorial lemmas of the previous section we can present the main result of this article. Each tree-interpretable structure $\mathfrak{A}$ is finitely $\mathrm{GSO}(\exists^\kappa)$-axiomatisable, i.e., there is a $\mathrm{GSO}(\exists^\kappa)$-sentence $\psi_{\mathfrak{A}}$ such that $\mathfrak{B} \models \psi_{\mathfrak{A}}$ if and only if $\mathfrak{B} \cong \mathfrak{A}$. Actually, we will prove the slightly stronger statement that, for each tree-interpretable structure, there is a colouring of the guarded tuples such that the coloured structure is $\mathrm{MSO}(\exists^\kappa)$-axiomatisable. That is, the axiom consists of a sequence of existential non-monadic second-order quantifiers followed by an $\mathrm{MSO}(\exists^\kappa)$-formula.

The congruence colouring. The axiomatisation uses colourings of elements and pairs of elements that are of the following form:

Definition 12. (a) Let $\approx \subseteq \Sigma^* \times \Sigma^*$ be a congruence of finite index and let $k \in \mathbb{N}$. The $(\approx, k)$-*congruence colouring* $\chi_{\approx}^k$ maps words $x \in \Sigma^*$ to the pair

$$\chi_{\approx}^k(x) := \left([x/k]_{\approx}, \; \mathrm{suf}_k \, x \right)$$

and pairs $(x, y) \in \Sigma^* \times \Sigma^*$ to

$$\chi_{\approx}^k(x, y) := \left(\chi_{\approx}^k(w^{-1}x), \; \chi_{\approx}^k(w^{-1}y) \right)$$

where $w := x \sqcap y$.

(b) A $(\approx', k')$-colouring χ' *refines* the $(\approx, k)$-colouring χ if $\approx' \subseteq \approx$ and $k' \geq k$. We denote this fact by $\chi' \geq \chi$. The *common refinement* of a $(\approx_0, k_0)$-colouring χ_0 and a $(\approx_1, k_1)$-colouring χ_1 is the $(\approx_0 \cap \approx_1, \max\{k_0, k_1\})$-colouring denoted by $\chi_0 \sqcup \chi_1$.

Definition 13. The χ-*expansion* $(\mathfrak{A}, \chi)$ of a structure $\mathfrak{A}$ expands $\mathfrak{A}$ by unary and binary relations for each colour class where the binary colour classes consists only of pairs (x, y) which are guarded.

The restriction to guarded pairs is essential since GSO allows only quantification over relations of this form. Below we frequently will need to obtain the value $\chi(x, y)$ for pairs (x, y) which are not guarded. These values must be computed explicitly from available data. This is where k-normal paths come into play.

Lemma 2. *Let $\mathfrak{A}$ be a tree-interpretable structure, $\approx$ a congruence of finite index, and k a constant. The $\chi_{\approx}^k$-expansion $(\mathfrak{A}, \chi_{\approx}^k)$ of $\mathfrak{A}$ is also tree-interpretable.*

We say that a set P of vertices codes a path between x and y if every element of P except for x and y is connected to exactly two other elements in P whereas x and y are connected to exactly one. Clearly, not every path can be coded in this way. Fortunately, for our purposes it is sufficient that, if there is a k-normal path between two vertices, then we can obtain a codable k-normal path between them by removing some vertices.

Lemma 3. *Let $\mathfrak{G}$ be a graph, χ a $(\approx, k)$-congruence colouring, and c a colour of χ. There is an MSO-formula $\varphi_c(P, x, y)$ such that $(\mathfrak{G}, \chi) \models \varphi_c(P, x, y)$ if and only if P codes a k-normal path from x to y and $\chi((x \sqcap y)^{-1}y) = c$.*

Forests. We start slowly by first showing that forests are finitely axiomatisable. We regard forests as partial orders such that the elements below any given one form a finite linear order. To axiomatise a forest it is sufficient to state, for each vertex, the number of its immediate successors of a a given colour. Note that these numbers only depend on the colour of the vertex.

Theorem 14. *Let $\mathfrak{T} := (T, \leq)$ be a tree-interpretable forest. The structure $(\mathfrak{T}, \chi)$ is finitely $\mathrm{FO}(\exists^\kappa)$-axiomatisable for all $\chi \geq \chi_{\sim}^I$.*

Partial-orders. The next step consists in extending the result to tree-interpretable partial orders $\mathfrak{A} := (A, \leq)$ for which there is a constant $n \in \mathbb{N}$ such that $x \leq y$ implies $x/n \preceq y/n$ for all $x, y \in A$. To do so we have to define a forest in $\mathfrak{A}$. When speaking of paths we always consider undirected paths in this section, i.e., we ignore the direction of the edges.

Definition 15. Let $x \sqsubseteq y$ iff $x/n \preceq y/n$ and there is an undirected $\leq$-path $z_0, \ldots, z_m$ from x to y with $x/n \preceq z_i/n$ for all $i \leq m$. Further, define $x \equiv y$ iff $x \sqsubseteq y$ and $y \sqsubseteq x$.

It is easy to show that $(A, \sqsubseteq)/\equiv$ is a forest and, thus, axiomatisable. This fact can be used to prove the following result.

Proposition 8. *There is a congruence colouring χ_0 such that $(A, \sqsubseteq, \chi)$ is finitely $\mathrm{MSO}(\exists^\kappa)$-axiomatisable for every $\chi \geq \chi_0$.*

In order to transfer the axiomatisability result from $(A, \sqsubseteq)$ to $\mathfrak{A}$, we have to show that each of the structures is definable in the other one.

Lemma 4. (a) $(A, \sqsubseteq, \chi)$ *is MSO-definable in* $(\mathfrak{A}, \chi)$ *for all* $\chi \geq \chi_\sim^K$.
(b) $(\mathfrak{A}, \chi)$ *is MSO-definable in* $(A, \sqsubseteq, \chi)$ *for all colourings* $\chi \geq \chi_\sim^n$.

These results allow us to transfer the axiomatisability from $(A, \sqsubseteq, \chi)$ to $(\mathfrak{A}, \chi)$.

Theorem 16. *Let $\mathfrak{A} := (A, \leq)$ be a tree-interpretable partial-order and let $n \in \mathbb{N}$ be a constant such that $x \leq y$ implies $x/n \preceq y/n$ for all $x, y \in A$. There is a congruence colouring χ_0 such that $(\mathfrak{A}, \chi)$ is finitely $\mathrm{MSO}(\exists^\kappa)$-axiomatisable for every $\chi \geq \chi_0$.*

The general case. Finally, we consider an arbitrary tree-interpretable structure $\mathfrak{A}$. For the reduction to the previous case we define, as above, a partial order $\leq$ and show that the structures $(A, \leq)$ and $\mathfrak{A}$ are definable within each other.

Definition 17. Let $x \vdash y$ if $x/I \preceq y/I$ and the pair (x, y) is guarded. Let $\leq$ be the reflexive and transitive closure of $\vdash$.

Lemma 5. $(A, \leq, \chi)$ *is MSO-definable in* $(\mathfrak{A}, \chi)$ *for all colourings* $\chi \geq \chi_\sim^I$.

The proof of the converse is more involved and requires an investigation of the branching structure of a tuple.

Definition 18. Let $\bar{a}, \bar{b} \in A^n$. We say that $\bar{a}$ is a *reduct* of $\bar{b}$ iff

(1) the branching structures of $\bar{a}$ and $\bar{b}$ are the same,
(2) $\inf_{\preceq} \bar{a} \sim \inf_{\preceq} \bar{b}$,
(3) $(a_i \sqcap a_j)^{-1}(a_k \sqcap a_l) \sim (b_i \sqcap b_j)^{-1}(b_k \sqcap b_l)$ for all indices such that $a_i \sqcap a_j \preceq a_k \sqcap a_l$,
(4) $|a_i| < |\inf_{\preceq} \bar{a}| + nI$ for all $i < n$.

A tuple is called *reduced* if it is a reduct of itself.

Lemma 6. *If $\bar{a}$ is a reduct of $\bar{b}$ and $\bar{b} \in R$ then $\bar{a} \in R$.*

To check whether a tuple $\bar{a}$ belongs to a relation R we use the characterisation of Proposition 5. Hence we need to compute the $\sim$-class of $u^{-1}v$ for branching points u and v of $\bar{a}$.

Definition 19. Let $\bar{a} \in A^n$. The elements $b_{ik} \in A$, for $i, k < n$, code the branching structure of $\bar{a}$ if

(1) $b_{ii} = a_i$ for $i < n$,
(2) $b_{ik}/nI \prec a_i \sqcap a_k \preceq b_{ik}$ for all i, k, and
(3) if $a_i \sqcap a_k \prec a_i \sqcap a_l$ then $b_{ik} \vdash b_{il}$ for $i, k, l < n$.

Given b_{ik} and b_{il} we can compute the $\sim$-class of $(a_i \sqcap a_k)^{-1}(a_i \sqcap a_l)$. Hence, if we can show that such elements always exists and that they are definable, then we are almost done.

Lemma 7. (a) For each branching structure $\mathfrak{X}$ there is a formula $\beta_{\mathfrak{X}}(\bar{x}, \bar{y})$ such that $(A, \leq, \chi) \models \beta(\bar{a}, \bar{b})$ if and only if the branching structure of $\bar{a}$ is $\mathfrak{X}$ and it is coded by $\bar{b}$.

(b) Let R be an n-ary relation of $\mathfrak{A}$ and $\bar{a} \in R$. There are elements $b_{ik} \in A$, $i, k < n$, coding the branching structure of $\bar{a}$.

At last, we are able to prove the other direction.

Lemma 8. The structure $(\mathfrak{A}, \chi)$ is MSO-definable in $(A, \leq, \chi)$ for every $\chi \geq \chi_{\sim}^{nI}$ where n is the maximal arity of relations of $\mathfrak{A}$.

Theorem 20. Let $\mathfrak{A}$ be a tree-interpretable structure. There is a congruence colouring χ_0 such that $(\mathfrak{A}, \chi)$ is finitely MSO$(\exists^\kappa)$-axiomatisable for all $\chi \geq \chi_0$.

The proof is completely analogous to the one of Theorem 16. Since GSO$(\exists^\kappa)$ allows quantification over colourings χ we obtain as immediate corollary the following result.

Theorem 21. Every tree-interpretable structure is finitely GSO$(\exists^\kappa)$-axiomatisable.

7   Lower Bounds

We have shown that every tree-interpretable structure is finitely GSO$(\exists^\kappa)$-axiomatisable. Of course, the question arises if we can do better. In this section we show that at least the quantifiers $\exists^{\aleph_0}$ and $\exists^{\aleph_1}$ are needed. Since all tree-interpretable structures are countable we obviously can do without the ones for higher cardinalities.

For a logic $\mathcal{L}$ let $\mathcal{L}_m$ denote the set of $\mathcal{L}$-formulae of quantifier rank at most m where we count both first- and second-order quantifiers. The following statements about the expressivity of MSO$_m$ and MSO$_m(\exists^{\aleph_0})$ can easily be proved using the corresponding versions of the Ehrenfeucht-Fraïssé game.

Lemma 9. (a) For every $m \in \mathbb{N}$ there exists a constant k such that two sets A and B are MSO$_m$-equivalent if and only if either $|A| = |B|$ or $|A|, |B| \geq k$.

(b) For every m there is some k such that two sets A and B are MSO$_m(\exists^{\aleph_0})$-equivalent iff either $|A| = |B|$, or $k \leq |A|, |B| < \aleph_0$, or $|A|, |B| \geq \aleph_0$.

(c) Any two infinite sets are MSO$(\exists^{\aleph_0})$-equivalent.

Since $\mathrm{GSO}(\exists^{\kappa})$ collapses to $\mathrm{MSO}(\exists^{\kappa})$ on trees, this lemma implies that $K_{1,\aleph_0}$ and $K_{1,\aleph_1}$ are $\mathrm{GSO}(\exists^{\aleph_0})$-equivalent. But the former structure is tree-interpretable while the latter obviously is not.

Theorem 22. *There exists tree-interpretable trees which are not $\mathrm{GSO}(\exists^{\aleph_0})$-axiomatisable.*

This shows that we cannot do without all cardinality quantifiers even if we allow infinitely many axioms. But do we really need non-monadic second-order quantifiers?

Open Problem. Are there tree-interpretable structures which are not (finitely) $\mathrm{MSO}(\exists^{\kappa})$-axiomatisable?

8 Automorphisms of Tree-Interpretable Structures

As mentioned in the introduction the axiomatisation of a tree-interpretable structure can be used to investigate its automorphism group.

Lemma 10. *Let $\mathfrak{A}$ be a tree-interpretable structure and $a \in A$. The orbit O of a under automorphisms is $\mathrm{GSO}(\exists^{\kappa})$-definable.*

Proof. If $\mathfrak{A}$ is tree-interpretable then so is $(\mathfrak{A}, a)$. Let $\varphi(x)$ be the $\mathrm{GSO}(\exists^{\kappa})$-formula obtained from the axiom of $(\mathfrak{A}, a)$ by replacing every occurrence of the constant a by the variable x. It follows that

$$b \in O \quad \text{iff} \quad (\mathfrak{A}, b) \cong (\mathfrak{A}, a) \quad \text{iff} \quad \mathfrak{A} \models \varphi(b). \qquad \square$$

Lemma 11 (Pélecq [21]). *Let $\mathfrak{A}$ be a tree-interpretable structure of finite tree-width and let O be the orbit of $a \in A$ under automorphisms. Then $(\mathfrak{A}, O)$ is tree-interpretable.*

Proof. O is $\mathrm{GSO}(\exists^{\kappa})$-definable by the preceding lemma. Since $\mathfrak{A}$ is of finite tree-width it follows that O is even $\mathrm{MSO}(\exists^{\kappa})$-definable and, therefore, $(\mathfrak{A}, O) \leq_{\mathrm{MSO}} \mathfrak{T}_2$. $\qquad \square$

We conclude this article with a simple application to the isomorphism problem.

Theorem 23 (Courcelle [9]). *Given two tree-interpretable structures $\mathfrak{A}$ and $\mathfrak{B}$ of finite-tree width one can decide whether $\mathfrak{A} \cong \mathfrak{B}$.*

Proof. Although not explicitly stated, the construction of the axiom in the previous section is effective. Thus, in order to determine whether $\mathfrak{A} \cong \mathfrak{B}$ one can construct the $\mathrm{GSO}(\exists^{\kappa})$-formula $\varphi_{\mathfrak{A}}$ which axiomatises $\mathfrak{A}$ and check whether $\mathfrak{B}$ satisfies $\varphi_{\mathfrak{A}}$. $\qquad \square$

Open Problem. Is isomorphism decidable for all tree-interpretable structures?

References

1. K. BARTHELMANN, *On equational simple graphs*, Tech. Rep. 9, Universität Mainz, Institut für Informatik, 1997.
2. ——, *When can an equational simple graph be generated by hyperedge replacement?*, LNCS, 1450 (1998), pp. 543–552.
3. A. BLUMENSATH, *Automatic Structures*, Diploma Thesis, RWTH Aachen, 1999.
4. ——, *Axiomatising tree-interpretable structures*, Tech. Rep. AIB-10-2001, RWTH Aachen, LuFG Mathematische Grundlagen der Informatik, 2001.
5. A. BLUMENSATH AND E. GRÄDEL, *Automatic structures*, in Proc. 15th IEEE Symp. on Logic in Computer Science, 2000, pp. 51–62.
6. O. BURKART, *Model checking rationally restricted right closures of recognizable graphs*, ENTCS, 9 (1997).
7. D. CAUCAL, *On infinite transition graphs having a decidable monadic theory*, LNCS, 1099 (1996), pp. 194–205.
8. B. COURCELLE, *The monadic second-order logic of graphs II: Infinite graphs of bounded width*, Math. System Theory, 21 (1989), pp. 187–221.
9. ——, *The monadic second-order logic of graphs IV: Definability properties of equational graphs*, Annals of Pure and Applied Logic, 49 (1990), pp. 193–255.
10. ——, *The monadic second-order logic of graphs VII: Graphs as relational structures*, Theoretical Computer Science, 101 (1992), pp. 3–33.
11. ——, *Structural properties of context-free sets of graphs generated by vertex replacement*, Information and Computation, 116 (1995), pp. 275–293.
12. ——, *Clique-width of countable graphs: A compactness property.* unpublished, 2000.
13. Y. L. ERSHOV, S. S. GONCHAROV, A. NERODE, AND J. B. REMMEL, *Handbook of Recursive Mathematics*, North-Holland, 1998.
14. E. GRÄDEL, C. HIRSCH, AND M. OTTO, *Back and forth between guarded and modal logics*, in Proc. 15th IEEE Symp. on Logic in Computer Science, 2000, pp. 217–228.
15. B. KHOUSSAINOV AND A. NERODE, *Automatic presentations of structures*, LNCS, 960 (1995), pp. 367–392.
16. G. KUPER, L. LIBKIN, AND J. PAREDAENS, *Constraint Databases*, Springer-Verlag, 2000.
17. O. KUPFERMAN AND M. Y. VARDI, *An automata-theoretic approach to reasoning about infinite-state systems*, LNCS, 1855 (2000), pp. 36–52.
18. C. MORVAN, *On rational graphs*, LNCS, 1784 (1996), pp. 252–266.
19. D. E. MULLER AND P. E. SCHUPP, *Groups, the theory of ends, and context-free languages*, J. of Computer and System Science, 26 (1983), pp. 295–310.
20. ——, *The theory of ends, pushdown automata, and second-order logic*, Theoretical Computer Science, 37 (1985), pp. 51–75.
21. L. PÉLECQ, *Isomorphismes et automorphismes des graphes context-free, équationnels et automatiques*, Ph. D. Thesis, Université Bordeaux I, 1997.
22. G. SÉNIZERGUES, *Decidability of bisimulation equivalence for equational graphs of finite out-degree*, in Proc. 39th Annual Symp. on Foundations of Computer Science, 1998, pp. 120–129.
23. C. STIRLING, *Decidability of bisimulation equivalence for pushdown processes.* unpublished, 2000.
24. W. THOMAS, *Languages, automata, and logic*, in Handbook of Formal Languages, G. Rozenberg and A. Salomaa, eds., vol. 3, Springer, New York, 1997, pp. 389–455.

EXPSPACE-Complete Variant of Guarded Fragment with Transitivity*
(Extended Abstract)

Emanuel Kieroński

Institute of Computer Science
University of Wrocław
kiero@ii.uni.wroc.pl

Abstract. We introduce a new fragment $GF^2 + \overrightarrow{TG}$ of the first order logic – the two-variable guarded fragment with *one-way* transitive guards. This logic corresponds in a natural way to temporal logics without past operators. We prove that the satisfiability problem for $GF^2 + \overrightarrow{TG}$ is EXPSPACE-complete. The lower bound, obtained for the *monadic* version of the considered logic without equality, improves NEXPTIME lower bound for the whole two-variable guarded fragment with transitive guards $GF^2 + TG$, given by Szwast and Tendera [8].

1 Introduction

The *guarded fragment* GF of first-order logic was proposed by Andréka, van Benthem and Németi [1] as a generalization of modal and temporal logics. It turned out that GF retains a lot of nice model theoretic properties of modal logic. In particular Grädel proved that it has the finite model property and that its every satisfiable formula has a tree-like model [5]. The satisfiability problem is decidable and has double exponential complexity. The reason for such a high complexity is unrestricted number of variables in formulas. The bounded version GF^k of GF (allowing only k variables) is EXPTIME-complete. In a subsequent paper Grädel and Walukiewicz [7] investigated *guarded fixpoint logic* μGF. They proved that adding least and greatest fixed point operators gives a decidable logic, and moreover, the complexity does not change, i.e. the satisfiability problem for μGF is in 2EXPTIME and for μGF^k – in EXPTIME.

Another question is what happens after adding transitivity to GF, i.e. when we may specify some binary relations as transitive. This question was answered by Ganzinger, Meyer and Veanes [4] who proved that even the two-variable version of such a logic is undecidable. On the other hand they proposed a restricted, *monadic*, two-variable variant $MGF^2 + TG$ in which all binary relations can appear only in guards but can be required to be transitive. Ganzinger, Meyer and Veanes showed that the satisfiability problem for this variant is decidable but they gave no complexity bound since their result was obtained by reduction to

---

* Supported by KBN grant 8 T11C 043 19

H. Alt and A. Ferreira (Eds.): STACS 2002, LNCS 2285, pp. 608–619, 2002.
© Springer-Verlag Berlin Heidelberg 2002

the Rabin's theory SkS. They also left another open question: the decidability of GF+TG – the whole guarded fragment with transitive relations where transitive relations are admitted only in guards, but where non-transitive relations and equality can occur elsewhere. The last question was answered recently by Szwast and Tendera [8] who proved that GF+TG is decidable and after adding transitive guards to GF (similarly to adding fixed point operators) the complexity of the satisfiability problem remains in 2EXPTIME.

Surprisingly, the two-variable fragment behaves differently after adding transitive guards. Szwast and Tendera observed that the satisfiability problem for $GF^2 + TG$ is NEXPTIME-hard (recall that for GF^2 it is EXPTIME-complete). Their elegant argument goes through a reduction from FO^2 which was shown to be NEXPTIME-complete by Grädel, Kolaitis and Vardi [6]. In this paper we strengthen this result and show that even a very restricted variant of $MGF^2 + TG$ is EXPSPACE-hard. Our proof proceeds by a reduction from the problem: "Does an alternating Turing machine M with exponentially bounded time accepts its input w?"

We propose an interesting, nontrivial guarded logic $GF^2 + \overrightarrow{TG}$ with transitive guards, for which the satisfiability problem is EXPSPACE-complete. We call this variant the two-variable guarded fragment with *one-way* transitive guards since we require all transitive guards of quantifiers $\forall x$ ($\exists x$) to be of the form $T(y, x)$ (guards $T(x, y)$ are forbidden). This is similar to the situation in most temporal logics where you can quantify points in the future but not in the past. A lot of modal and temporal logics including K4, S4 can be encoded in $GF^2 + \overrightarrow{TG}$. The key observation in our proof is that we can construct *regular* forrest-like models of satisfiable formulas. All important information in such models is stored at the first k levels of trees, where k is at most exponential with respect to the size of the formula. Moreover paths in regular forrest-like models can be constructed independently.

2   Preliminaries

Andréka, van Benthem and Németi [1] defined the *guarded fragment* of first order logic GF inductively:

(1) every atomic formula belongs to GF,
(2) GF is closed under $\neg, \wedge, \vee, \rightarrow, \leftrightarrow$,
(3) if $\boldsymbol{x}, \boldsymbol{y}$ are tuples of variables, $\alpha(\boldsymbol{x}, \boldsymbol{y})$ is an atomic formula, $\psi(\boldsymbol{x}, \boldsymbol{y})$ is in GF and the set of free variables of ψ is contained in the set of free variables of α then formulas $\exists \boldsymbol{y}(\alpha(\boldsymbol{x}, \boldsymbol{y}) \wedge \psi(\boldsymbol{x}, \boldsymbol{y}))$ and $\forall \boldsymbol{y}(\alpha(\boldsymbol{x}, \boldsymbol{y}) \rightarrow \psi(\boldsymbol{x}, \boldsymbol{y}))$ are in GF.

Atoms $\alpha(\boldsymbol{x}, \boldsymbol{y})$ in the above definition are called *guards*. We write GF^k to denote the language obtained by restricting the number of variables in formulas by a constant k (variables can be reused). The *monadic guarded fragment* MGF is a variant in which all non-unary relation symbols may appear only as guards. Ganzinger, Meyer and Veanes [4] investigated $MGF^2 + TG$ – the two-variable monadic guarded fragment where some binary relations are assumed to be transitive. By GF+TG we denote the whole guarded fragment with transitive relations where transitive relations are only admitted in guards. $\mathsf{GF}^2 + TG$ is the two-variable version of this logic.

In this paper we consider the two-variable guarded fragment with *one-way* transitive guards $\mathsf{GF}^2 + \overrightarrow{TG}$. Formally, a guarded formula is in $\mathsf{GF}^2 + \overrightarrow{TG}$ if it is in $\mathsf{GF}^2 + TG$ and additionally in all transitive guards of quantifiers $\forall x$ ($\exists x$) the quantified variable x appears in the second place of the guarding predicate. It means that we allow formulas of the form $\forall x\, T(y, x) \wedge \psi(x, y)$ and forbid formulas of the form $\forall x\, T(x, y) \wedge \psi(x, y)$ (and similarly for the existential quantifier). For non-transitive guards the order of variables can be arbitrary.

Observe that the monadic variant of $\mathsf{GF}^2 + \overrightarrow{TG}$ is still, as the variant of Ganzinger, Meyer and Veanes, a nontrivial extension of the *modal fragment* with transitivity since it does not retain the finite model property. A simple example of $\mathsf{MGF}^2 + \overrightarrow{TG}$ formula which has only infinite models is : $\exists x\, P(x) \wedge \forall x\, \big(P(x) \to \exists y\, (x < y \wedge P(y))\big) \wedge \neg \exists x\, (x < x)$, where relation $<$ is transitive.

A *1-type* $t(x)$ over a fixed signature is a maximal consistent set of atomic and negated atomic formulas in variable x. A *2-type* $t(x, y)$ is similarly defined as a maximal consistent set of atomic and negated atomic formulas in variables x, y. We often identify a type with a conjunction of all its formulas. Let $\mathcal{A}$ be a relational structure and $a \in A$. We denote by $type^{\mathcal{A}}(a)$ the unique type $t(x)$ such that $\mathcal{A} \models t(a)$. Similarly for $a, b \in A$, $type^{\mathcal{A}}(a, b)$ is the unique 2-type $t(x, y)$ such that $\mathcal{A} \models t(a, b)$.

An *alternating Turing machine* is a generalization of a nondeterministic Turing machine. States, and hence configurations of such a machine, are partitioned into four groups: existential, universal, accepting and rejecting. The notion of an accepting (rejecting) configuration is extended to the case of configurations in which states are existential or universal. This can be done inductively: an universal configuration is accepting if all its successor configurations, i.e. configurations obtained by performing one machine step according to the transition function, are accepting and an existential configuration is accepting if at least one of its successor configurations is accepting. A machine M accepts its input w if the initial configuration of M on w is accepting. By ATIME(f) we denote the set of problems that can be solved by an alternating Turing machine working in time bounded by a function f. In this paper we will use the following theorem:

Theorem 1. (Chandra, Kozen, Stockmeyer [3]) *For* $t(n) \geq n$:

$$\bigcup_{k=1}^{\infty} ATIME[(t(n))^k] = \bigcup_{k=1}^{\infty} DSPACE[(t(n))^k].$$

If we denote $\bigcup_{k=1}^{\infty} ATIME(2^{n^k})$ *by AEXPTIME we have in particular AEXP-TIME=EXPSPACE.*

For details about alternating Turing machines you can see for example [2].

3   The Proof of Hardness

Szwast and Tendera [8] argued that $\mathsf{GF}^2 + TG$ is NEXPTIME-hard. Here we strengthen their result and prove that even $\mathsf{MGF}^2 + \overrightarrow{TG}$ without equality, with just one transitive relation symbol (which is the only binary symbol) is EXP-SPACE-hard.

Theorem 2. *The satisfiability problem for* $\mathsf{MGF}^2 + \overrightarrow{TG}$ *without equality is EXP-SPACE-hard.*

Proof. By Theorem 1 every problem in EXPSPACE is in AEXPTIME so let M be an alternating Turing machine deciding a problem in EXPSPACE in time 2^{n^k}. We can assume without loss of generality that every non-final configuration of M has exactly two successor configurations. Let w be an input of size n. We construct a formula Φ whose models encode accepting computation trees of machine M on input w. Every configuration is represented by 2^{n^k} elements of a model, each of them corresponds to a single cell of the tape. We introduce unary relation symbols $C_1, \ldots, C_{n^k}$ and $P_1, \ldots, P_{n^k}$ to encode the number of configuration (i.e. the depth of the configuration in a computation tree) to which an element belongs and the position of this element (the number of a cell) in this configuration respectively ($C_i(x)$ is true if the i-th bit of the number of the configuration to which x belongs is 1, $P_i(x)$ is true if the i-th bit of the position of x in the configuration is 1). We use abbreviations $\bar{C}(x)$ and $\bar{P}(x)$ to describe the above numbers. It is not difficult to express the following properties with quantifier-free formulas of polynomial length:

$$
\begin{array}{llll}
\bar{C}(x) = l & \text{for fixed } l, \quad 0 \le l < 2^{n^k}, & \bar{P}(x) = l & \text{for fixed } l, \quad 0 \le l < 2^{n^k}, \\
\bar{C}(x) \ge l & \text{for fixed } l, \quad 0 \le l < 2^{n^k}, & \bar{P}(x) \ge l & \text{for fixed } l, \quad 0 \le l < 2^{n^k}, \\
& \bar{C}(x) = \bar{C}(y), & \bar{P}(x) = \bar{P}(y), \\
& \bar{C}(x) = \bar{C}(y) + 1, & \bar{P}(x) = \bar{P}(y) + 1.
\end{array}
$$

We describe a configuration in a standard way: for each symbol a_i from the alphabet of M (including **blank**) we use an unary relation symbol A_i, for each state q_i – an unary symbol Q_i. We also have an unary symbol H describing the head position. For element x representing a tape cell scanned by the head, $H(x)$ and $Q_i(x)$ (for some i) will be true.

Let us start the construction of Φ with enforcing that its every model $\mathcal{A}$ contains a binary tree of depth 2^{n^k}. Every node of the tree consists of 2^{n^k} elements describing a single configuration of M. We define a relation $<$ with the intended meaning: $x < y$ if x and y belong to a description of the same configuration (they belong to the same node of the tree) and x describes the tape cell with the smaller number, or y belongs to a description of a configuration reachable from the configuration x belongs to. Let the relation $<$ be transitive. We want to distinguish between successors of a node. Therefore we introduce a relation symbol L that is true for elements belonging to the left son of some node. We assume that for elements belonging to the initial configuration L is also true. In addition we use a special symbol N that is true for all elements belonging to the description of the computation tree. For further purposes we abbreviate the formula $\bar{C}(x) = \bar{C}(y) \wedge \bar{P}(y) = \bar{P}(x) + 1$ to $NEXT(x, y)$.

$$\exists x \ \big(N(x) \wedge L(x) \wedge \bar{C}(x) = 0 \wedge \bar{P}(x) = 0\big) \ ,$$

$$\forall x \ \Big(N(x) \Rightarrow \bar{P}(x) \neq 2^{n^k} - 1$$
$$\Rightarrow \ \exists y \ \big(x < y \wedge N(y) \wedge (L(x) \Leftrightarrow L(y)) \wedge NEXT(x, y)\big)\Big) \ ,$$

$$\forall x \ \Big(N(x) \Rightarrow (\bar{P}(x) = 2^{n^k} - 1 \wedge \bar{C}(x) \neq 2^{n^k} - 1)$$
$$\Rightarrow \ \exists y \ \big(x < y \wedge N(y) \wedge L(y) \wedge \bar{P}(y) = 0 \wedge \bar{C}(y) = \bar{C}(x) + 1\big)$$
$$\wedge \exists y \ \big(x < y \wedge N(y) \wedge \neg L(y) \wedge \bar{P}(y) = 0 \wedge \bar{C}(y) = \bar{C}(x) + 1\big)\Big) \ .$$

Observe that in fact a structure defined by the above formulas does not need to contain a "real" tree (for example a node can be a son of two or more distinct nodes) but as we see later it is not important.

Now we state that a model of our formula fulfils several basic properties of an accepting computation tree. There is exactly one alphabet symbol in every tape cell:

$$\forall x \ \big(N(x) \Rightarrow \bigvee_i A_i(x)\big) \ \wedge \ \bigwedge_i \big(\forall x \ (A_i(x) \Rightarrow \bigwedge_{i \neq j} \neg A_i(x))\big) \ .$$

In each configuration at most one element is scanned by the head:

$$\forall x, y \ \big(x < y \Rightarrow \bar{C}(x) = \bar{C}(y) \Rightarrow \neg(H(x) \wedge H(y))\big) \ .$$

Exactly those elements which represent tape cells observed by the head store information about state:

$$\bigwedge_i \big(\forall x \ (Q_i(x) \Rightarrow H(x))\big) \ \wedge \ \forall x \ \big(H(x) \Rightarrow \bigvee_i Q_i(x)\big) \ .$$

The root of the tree describes the initial configuration of M on input $w = w_0 \ldots w_{n-1}$:

$$\forall x \ \big(N(x) \Rightarrow (\bar{C}(x) = 0 \wedge \bar{P}(x) = 0) \Rightarrow H(x)\big) \ ,$$
$$\bigwedge_{i < n} \big(\forall x \ (N(x) \Rightarrow (\bar{C}(x) = 0 \wedge \bar{P}(x) = i) \Rightarrow W_i(x))\big) \ ,$$
$$\forall x \ \big(N(x) \Rightarrow (\bar{C}(x) = 0 \wedge \bar{P}(x) \geq n) \Rightarrow BLANK(x)\big) \ .$$

If a tape cell of a configuration is not scanned by the head then in the same cell of both successor configurations the alphabet symbol does not change:

$$\forall x, y \ \big(x < y \Rightarrow (\neg H(x) \wedge \bar{C}(y) = \bar{C}(x) + 1 \wedge \bar{P}(x) = \bar{P}(y))$$
$$\Rightarrow \bigwedge_i (A_i(x) \Leftrightarrow A_i(y))\big).$$

Consider now a node t of a tree and a configuration c that is described by this node. There are two cases: the state of the machine in this configuration is existential or it is universal. In the first case we enforce that the configuration represented by the left son of t is created by applying one of the two possible transitions on c (we do not say anything about the right son of t in this case). Assume that for an existential state q and a letter a there are two possible transitions: $(q, a) \Rightarrow (q', a', \rightarrow)$ and $(q, a) \Rightarrow (q'', a'', \leftarrow)$. We write such a formula:

$$\forall x, y \left(x < y \Rightarrow (A(x) \land Q(x) \land H(x) \land L(y) \right.$$
$$\land \bar{C}(y) = \bar{C}(x) + 1 \land \bar{P}(x) = \bar{P}(y) + 1)$$
$$\Rightarrow (\forall x \ (y < x \Rightarrow NEXT(y, x)$$
$$\Rightarrow A'(x) \land \forall y \ (x < y \Rightarrow NEXT(x, y) \Rightarrow Q'(y) \land H(y)))$$
$$\left. \lor (Q''(y) \land H(y) \land \forall x \ (y < x \Rightarrow NEXT(y, x) \Rightarrow A''(y)))) \right) \ .$$

Other possible situations (both transitions move the head forward or both transitions move the head backward) can be handled similarly.

Consider now the case of universal configuration. We enforce that the left son of t is created by applying the first transition and the right son – the second one. For an universal state q, a letter a and transitions like above we write formula:

$$\forall x, y \left(x < y \Rightarrow (A(x) \land Q(x) \land H(x) \land \bar{C}(y) = \bar{C}(x) + 1 \land \bar{P}(x) = \bar{P}(y) + 1) \right.$$
$$\Rightarrow (L(y) \Rightarrow (\forall x \ (y < x \Rightarrow NEXT(y, x)$$
$$\Rightarrow A'(x) \land \ \forall y \ (x < y \Rightarrow NEXT(x, y) \Rightarrow Q'(y) \land H(y))))$$
$$\left. \land \neg L(y) \Rightarrow (Q''(y) \land H(y) \land \forall x \ (y < x \Rightarrow NEXT(y, x) \Rightarrow A''(y)))) \right) \ .$$

We add formulas of the above form for every pair of a state and a letter.

To finish our construction we ensure that the machine never enters the rejecting state (we assume that q_r is the only rejecting state):

$$\forall x \ (Q_r(x) \Rightarrow \textbf{false}) \ .$$

We define Φ as the conjunction of all the above formulas. Observe that the number of conjuncts and size of each of them are polynomial in n. We have the following fact: Φ is satisfiable iff M accepts w. Indeed, from each model of Φ we can extract a description of an accepting computation tree – even if a model represents some distinct tape cells (or for example whole configurations) by the same element (or the same node respectively) we can make a copy of this element (node). Conversely, if M accepts w then a model of Φ can be constructed easily from an accepting computation tree.   □

4   The Upper Bound

In this section we prove that the satisfiability problem for $\mathsf{GF}^2 + \overrightarrow{TG}$ is in EXPSPACE. Szwast and Tendera [8] gave double exponential algorithm for the satisfiability of the whole $\mathsf{GF}+TG$. To obtain this result they established a kind of the tree-model property – from an arbitrary model of a formula they construct a so-called *ramified model*. Of course such a construction can be repeated also in our case, but to achieve a lower complexity we need a stronger result. Therefore we prove that every satisfiable $\mathsf{GF}^2 + \overrightarrow{TG}$ sentence Φ in normal form has a *regular forrest-like model*. A regular forrest-like model can be represented as a set of disjoint trees whose nodes are elements of the model. For every path in a tree, longer than $2^{O(|\Phi|)}$, there exists an element a on this path, belonging to the m-th level, $m < 2^{O(|\Phi|)}$, such that the subtree rooted in a is isomorphic to a subtree rooted in an element b which lies on the path from the root of the tree to a or is a son of another element lying on this path.

4.1 The Normal Form

Let us begin with an adaptation of the normal form theorem for $\mathsf{GF} + TG$ [8].

Definition 1. *A* $\mathsf{GF}^2 + \overrightarrow{TG}$ *sentence* Φ *is in normal form if it is a conjunction of sentences of the following form:*

$$(e1) \; \exists x \; (U(x) \; \wedge \; \phi(x)) \; ,$$
$$(w1) \; \forall x \; (U(x) \rightarrow \exists y \; (B(x,y) \; \wedge \; \phi(x,y))) \; ,$$
$$(w2) \; \forall x \; (U(x) \rightarrow \exists y \; (B(y,x) \; \wedge \; \phi(x,y))) \; ,$$
$$(w3) \; \forall x \; (U(x) \rightarrow \exists y \; (T(x,y) \; \wedge \; \phi(x,y))) \; ,$$
$$(b1) \; \forall x,y \; (B(x,y) \rightarrow \phi(x,y)) \; ,$$
$$(b2) \; \forall x,y \; (T(x,y) \rightarrow \phi(x,y)) \; ,$$
$$(u1) \; \forall x \; (U(x) \rightarrow \phi(x)) \; ,$$
$$(u2) \; \forall x \; (B(x,x) \rightarrow \phi(x)) \; ,$$
$$(u3) \; \forall x \; (T(x,x) \rightarrow \phi(x)) \; ,$$

where U stands for an unary relation symbol, B stands for a binary symbol which is not transitive, T stands for a transitive binary symbol and ϕ is quantifier-free.

Lemma 1. *For every* $\mathsf{GF}^2 + \overrightarrow{TG}$ *sentence* Φ *of the length* n *there exists a set* $\{\psi_1, \ldots, \psi_d\}$ *of* $\mathsf{GF}^2 + \overrightarrow{TG}$ *sentences in normal form such that:*

- *Φ is satisfiable if and only if at least one of the formulas $\psi_1, \ldots, \psi_d$ is satisfiable;*
- *d is at most exponential with respect to n and for every $i \leq d$ the length of ψ_i is polynomial with respect to n;*
- *the set $\{\psi_1, \ldots, \psi_d\}$ can be computed deterministically in exponential time.*

Proof given by Szwast and Tendera [8] works also for our case since it retains the number of variables and does not change the "orientation" of transitive guards.

4.2   Cliques and Ramified Models

Let us recall some definitions and facts from [8].

Definition 2. *Let A be a structure over a fixed language σ and let C be a substructure of A. For a transitive symbol T we say that C is a T-clique if for every $a, b \in C$ we have $A \models T(a,b)$. For a binary symbol R and $a \in A$ we denote by $[a]_R^A$ the maximal R-clique containing a or the one-element substructure of A containing a if such a clique does not exist or R is not a transitive symbol.*

Szwast and Tendera observed that in $\mathsf{MGF}^2 + TG$ we can define cliques of size exponential with respect to the number of relation symbols in the language σ. Such cliques can be also defined in $\mathsf{MGF}^2 + \overrightarrow{TG}$. Here is an example. The language σ consists of unary relation symbols $N, P_1, \ldots P_n$ and a transitive binary symbol $<$. We use abbreviation $\bar{P}$ introduced in the hardness proof:

$$\exists x \, (N(x) \wedge \bar{P}(x) = 0) \; ,$$
$$\forall x \, (N(x) \Rightarrow \bar{P}(x) < 2^n - 1 \Rightarrow \exists y \, (x < y \wedge \bar{P}(y) = \bar{P}(x) + 1)) \; ,$$
$$\forall x \, (N(x) \Rightarrow \bar{P}(x) = 2^n - 1 \Rightarrow \exists y \, (x < y \wedge \bar{P}(y) = 0)) \; ,$$
$$\forall x \, (N(x) \Rightarrow \bar{P}(x) = 0 \Rightarrow \neg \exists y \, (x < y \wedge x \neq y \wedge \bar{P}(y) = 0)) \; .$$

Definition 3. *Let Φ be a $\mathsf{GF}^2 + TG$ sentence in the normal form. A structure A is a* ramified model *for Φ if the following conditions hold:*

1. *$A \models \Phi$,*
2. *for every $a, b \in A, a \neq b$ there is at most one transitive relation symbol T such that $A \models T(a,b) \vee T(b,a)$,*
3. *if S and T are distinct transitive relation symbols then for every $a, b, c \in A$, such that $a \neq b$ and $a \neq c$ if $b \in [a]_S^A$ and $c \in [a]_T^A$ then for every binary symbol R we have $A \not\models R(b,c)$,*
4. *for every $a \in A$, for every binary symbol T, the cardinality of $[a]_T^A$ is at most exponential with respect to the number of relation symbols in the language.*

Theorem 3. (Szwast, Tendera [8]) *Every satisfiable $\mathsf{GF}^2 + TG$ sentence has a ramified model.*

In our construction of a regular forrest-like model for a satisfiable $\mathsf{GF}^2 + \overrightarrow{TG}$ sentence, we use only part 4 of Definition 3 (but we obtain also the remaining properties).

4.3   Regular Forrest-Like Models

In this section we are going to define the notion of a regular forrest-like model for a $\mathsf{GF}^2 + \overrightarrow{TG}$ sentence in the normal form. We will work with trees whose nodes are elements of the model and whose edges are labelled with "oriented" binary relation symbols. More precisely for a non-transitive symbol B labels have the form $\overleftarrow{B}$ and $\overrightarrow{B}$, for a transitive T we use labels $\overrightarrow{T}$ and $\overleftrightarrow{T}$. We define $L(a)$ to be the label of the edge connecting a with its father if the father exists or ϵ in the other case. By $\overline{L}(a)$ we denote the relation symbol in $L(a)$. A *path* in a tree is a maximal, possibly infinite sequence of nodes $s_1, \ldots, s_k, \ldots$ such that for $i > 0$, s_{i+1} is a son of s_i. For a transitive symbol T, a *T-path* is a maximal, possibly infinite sequence of nodes $s_1, \ldots, s_k, \ldots$ such that for $i > 0$, s_{i+1} is a son of s_i and the edge between s_i and s_{i+1} is labelled with $\overrightarrow{T}$ or $\overleftrightarrow{T}$. For convenience we will also use the notion of a *B-path* for a non-transitive symbol B but in this case it is rather unnatural and denotes a pair of elements connected by an edge labelled with $\overrightarrow{B}$ or $\overleftarrow{B}$.

By *tree(a)* we denote the subtree rooted in a. For a relation symbol R and a node a let $tree^R(a)$ denote the fragment of $tree(a)$ such that it contains subtrees rooted in all nodes connected with a by edges labelled with $\overleftrightarrow{R}$, $\overrightarrow{R}$ or $\overleftarrow{R}$. We say that an element a *R-precedes* an element b, for a binary symbol R, if a and b lie on the same R-path and a is an ancestor of b or, in the case of transitive R, there exists an element c such that c and b lie on the same R path, c is an ancestor of b, a is a son of c and the edge between a and c is labelled with $\overleftrightarrow{R}$. For convenience we denote by $\pi(a)$ the element a if $a \in L_0$ or the initial fragment of the $\overline{L}(a)$-path ending at element a, and by $\Pi(a)$ the initial fragment of the path ending at a. Let $prec(a)$ denote the set consisting of the $type(a)$ and the set of 1-types of elements that $\overline{L}(a)$-precede a.

Regular forest-like models are ramified models with some additional properties. Let us describe it informally first. A regular forest-like model can be represented as a set of disjoint trees whose roots are elements fulfilling existential conjuncts of the form (e1). Labels of edges describe certain properties of 2-types of pairs of elements. If elements a, b lie on the same B-path (B is not transitive) then the $type(a, b)$ contains $B(x, y)$ or $B(y, x)$ depending on the orientation of the label of the edge between a and b. If a T-precedes b then $type(a, b)$ contains $T(x, y)$, moreover it contains $T(y, x)$ if and only if a and b belong to the same T-clique. We will construct models in such a way that 2-types of pairs of elements will not contain any other positive occurrences of binary transitive symbols. All elements belonging to the same T-clique are stored at two consecutive levels of a tree. One of elements, say a, belongs to the i-th level and all the other elements of the clique belong to the $(i+1)$-st level. There are edges between a and the remaining elements of the clique and these edges are labelled with $\overleftrightarrow{T}$.

Let ϕ be a formula of the form $\forall x\, (U(x) \rightarrow \exists y \gamma(x, y))$. If for an element a, $U(a)$ is true in the model and for an element b, $\gamma(a, b)$ is true in the model then we say that b is a *witness* for a and ϕ. Witnesses for an element a, belonging to the i-th level, and formulas of the form (w1)-(w2), i.e. formulas without transitive guards, are located in the $(i+1)$-st level. A situation can be different for formulas of the form (w3) $\forall x\, (U(x) \rightarrow \exists y\, (T(x, y) \wedge \phi(x, y)))$. In this case we look for a witness in the T-clique of a first. We do so since the equality symbol can appear in formulas and for example it can be specified that on a T-path there can be at most one element of a certain 1-type.

The most important feature of our models is regularity – all subtrees added after exponential number of levels are isomorphic in a sense to some subtrees added before. It means that to answer if a formula has a model it is enough to check if we can construct an exponential number of levels of a regular forrest-like structure. Since elements of our model can be connected by binary relations only if one of them R-precedes the second, for some binary R, we can construct all the paths of a model independently. This fact leads to an algorithm using only exponential space.

Now we are ready to give a precise definition of the notion of the regular forrest-like model.

Definition 4. *A regular forrest-like model for a* $\mathsf{GF}^2 + \overrightarrow{TG}$ *sentence* Φ *in normal form is a structure* $\mathcal{M}$ *such that:*

1. $\mathcal{M} \models \Phi$;
2. *$\mathcal{M}$ can be represented as a set of disjoint trees whose nodes are elements of M and edges are labelled with "oriented" binary relation symbols, i.e. symbols of the form $\overrightarrow{R}$, $\overleftarrow{R}$ or $\overleftrightarrow{R}$;*
3. *the universe of $\mathcal{M}$ can be split into levels $L_0, L_1, \ldots$ where L_0 consists of roots of trees and for $i > 0$ elements of L_i are sons of elements from L_{i-1};*
4. *the number of elements in L_0 is at most polynomial, and the number of sons of each element in a tree is at most exponential with respect to $|\Phi|$;*
5. *if a is an element of a T-clique and the father of a does not exist or does not belong to the same T-clique then all remaining elements of $[a]_T^{\mathcal{M}}$ are sons of a and edges between a and these elements are labelled with $\overleftrightarrow{T}$;*
6. *for every conjunct of Φ of the form $\exists x \, (U(x) \wedge \phi(x))$ there exists an element $a \in L_0$ such that $\mathcal{M} \models U(a) \wedge \phi(a)$;*
7. *for every conjunct of Φ of the form $\forall x \, (U(x) \rightarrow \exists y \, \gamma(x,y))$ and every element $a \in L_i$ such that $\mathcal{M} \models U(a)$ one of the following conditions holds:*
   - *there exists a son $b \in L_{i+1}$ of a such that $\mathcal{M} \models \gamma(a,b)$; moreover if γ is of the form $R(x,y) \wedge \phi(x,y)$ then the edge between a and b is labelled with $\overrightarrow{R}$ or $\overleftrightarrow{R}$ and if γ is of the form $R(y,x) \wedge \phi(x,y)$ then this edge is labelled with $\overleftarrow{R}$;*
   - *b is the father of a, the edge between a and b is labelled with $\overleftrightarrow{T}$, γ is of the form $T(x,y) \wedge \phi(x,y)$ and $\mathcal{M} \models \gamma(a,b)$;*
   - *b is the father of a, c is a son of b ($c \neq a$), the edges between b and a, b and c are labelled with $\overleftrightarrow{T}$, γ is of the form $T(x,y) \wedge \phi(x,y)$ and $\mathcal{M} \models \gamma(a,c)$;*
   - *$\mathcal{M} \models \gamma(a,a)$;*
8. *if T is a transitive symbol and the edge between a and b is labelled with:*
   - *$\overrightarrow{T}$ then $\mathcal{M} \models T(a,b) \wedge \neg T(b,a)$;*
   - *$\overleftarrow{T}$ then $\mathcal{M} \models T(b,a) \wedge \neg T(a,b)$;*
   - *$\overleftrightarrow{T}$ then $\mathcal{M} \models T(a,b) \wedge T(b,a)$;*
9. *for elements a, b, $a \neq b$ if there is no relation symbol R such that a R-precedes b or b R-precedes a in a tree then for every binary symbol R', $\mathcal{M} \not\models R'(a,b)$;*

10. *for elements* a, b, $a \neq b$ *if there is no transitive symbol* T *such that* a T-*precedes* b *or* b T-*precedes* a *then for every transitive symbol* T', $\mathcal{M} \not\models$ $T'(a, b)$;

11. *if an element* a T-*precedes* b *(for a transitive* T*) then for every transitive symbol* $T' \neq T$, $\mathcal{M} \not\models T'(a, b)$ *and* $\mathcal{M} \not\models T'(b, a)$;

12. *for every path* $s_1, \ldots, s_k, \ldots$ *which is infinite or longer than* $2^{O(|\Phi|)}$, *there exist constants* $c < d \leq 2^{O(|\Phi|)}$, *such that for* $x = s_c$ *or* x *being a son of* s_c *(in the latter case* $c < d-1$ *and* x *does not belong to the path* $s_1, \ldots, s_k, \ldots$*),* $tree(s_d) \cong tree(x)$, *i.e. there exists a bijection* I *between nodes of* $tree(s_d)$ *and nodes of* $tree(x)$ *such that if* $I(a) = b$ *then* $type^{\mathcal{M}}(a) = type^{\mathcal{M}}(b)$ *and if* $I(a_1) = b_1$, $I(a_2) = b_2$ *and* a_2 *is a son of* a_1 *then* b_2 *is a son of* b_1 *and* $L(a_2) = L(b_2)$.

Lemma 2. *Every satisfiable* $\mathsf{GF}^2 + \overrightarrow{TG}$-*sentence* Φ *in normal form has a regular forrest-like model.*

Proof. Unfortunately, due to the space limit, we are not able to give here a detailed proof. Therefore we present only a sketch of our construction. Proof will be given in the full version of this paper which will be available at the following URL: `http://ii.uni.wroc.pl/~kiero`.

Let $\mathcal{A}$ be a model of Φ fulfilling part 4 of Definition 3. We construct a regular forrest-like model $\mathcal{M}$ of Φ recursively:

Stage 0. For every conjunct of Φ of the form $\exists x \; \phi(x)$ find an element $a \in A$ such that $\mathcal{A} \models \phi(a)$ and add a copy of this element (i.e. an element of the same 1-type) to L_0.

Stage k. For every element $a \in L_{k-1}$, for which $tree(a)$ is not defined:

1. if $L(a) = \overrightarrow{T}$ for a transitive symbol T and there exists an element $b \in \pi(a)$ such that $L(b) \neq \overleftrightarrow{T}$ (i. e. $L(b) = \overrightarrow{T}$ or b is the first element of $\pi(a)$), $type^{\mathcal{M}}(b) = type^{\mathcal{M}}(a)$ and $prec(b) = prec(a)$ then take a copy of $tree(b)$ as $tree(a)$,

2. if $L(a) = \overrightarrow{B}$ or $L(a) = \overleftarrow{B}$ for non-transitive symbol B then if there exists an element b in $\Pi(a)$ such that $L(b)$ is not of the form $\overleftrightarrow{T}$ and $type^{\mathcal{M}}(b) = type^{\mathcal{M}}(a)$ then take a copy of $tree(b)$ as $tree(a)$;

3. if none of the previous steps was performed and there exists an element $b \in \Pi(a)$ such that $L(b) = L(a)$ and $type^{\mathcal{M}}(b) = type^{\mathcal{M}}(a)$ then for every binary symbol R, except the symbol $\overline{L}(a)$, take a copy of $tree^R(b)$ as $tree^R(a)$ (after this step $[a]_R^{\mathcal{M}}$ is defined for such R);

4. for every transitive symbol T, if $[a]_T^{\mathcal{M}}$ is not defined then for every element $b' \in A$, belonging to the clique of the element $a' \in A$ which a is a copy of, add a copy of this element to L_k and set the label of the edge between a and this copy to $\overleftrightarrow{T}$;

5. for every conjunct γ of the form $\forall x \; (U(x) \to \exists y \; \delta(x, y) \land \phi(x, y)))$ if $U(x) \in type^{\mathcal{M}}(a)$ and there is no witness for a and γ in the clique of a then find a witness b' for γ and the element $a' \in A$ which a is a copy of, add a copy of b' to L_k and set the label between a and this copy to $\overrightarrow{T}$.

For each pair a, b of elements such that for every binary R, a does not R-precede b and b does not R-precede a, $type^{\mathcal{M}}(a, b)$ contains no positive occurences of binary symbols. For each pair a, b such that a R-precedes b for a symbol R, $type^{\mathcal{M}}(a, b)$ is based on the type of some pair of elements from A and constructed according to the definition. Parts 1-3 of the Stage k guarantee that part 12 of definition 4 is fulfilled.   □

An alternating procedure working in exponential time can be naturally derived from the construction. This procedure, Lemma 1 and Theorems 1, 2 allow us to state the main result of the paper:

Theorem 4. *Satisfiability problem for* $\mathsf{GF}^2 + \overrightarrow{TG}$ *is EXPSPACE-complete.*

5 Conclusion

In this paper we studied the complexity of the two-variable guarded fragment with one-way transitive guards $\mathsf{GF}^2 + \overrightarrow{TG}$. We proved that the satisfiability problem for this logic can be solved in exponential space. We showed that this is optimal since even for monadic version of this logic without equality the satisfiability problem is EXPSPACE-hard. The exact complexity of $\mathsf{GF}^2 + TG$ is still an open question. The best known upper bound is the one given by Szwast and Tendera [8] – 2EXPTIME. Closing this gap can be very interesting. We think that $\mathsf{GF}^2 + TG$ is 2EXPTIME-hard and believe that we will be able to give a proof of this fact soon. Results of this paper suggest that exponentially large cliques are not responsible for a higher complexity of $\mathsf{GF}^2 + TG$. The main difficulty lies in the fact that we can demand for some elements both greater and smaller witnesses with respect to a transitive relation.

References

1. H. Andréka, J. van Benthem, I. Németi, *Modal Languages and Bounded Fragments of Predicate Logic*, ILLC Research Report, 1996. Journal ver.: *J. Philos. Logic*, 27 (1998), No. 3, 217-274.
2. J. Balcázar, J. Diaz, J. Gabarró, *Structural Complexity II*, Springer 1990.
3. A. Chandra, D. Kozen, L. Stockmeyer, *Alternation*, JACM, 28, No. 1 (1981), pages 114-133.
4. H. Ganzinger, C. Meyer, M. Veanes *The Two-Variable Guarded Fragment with Transitive Relations*, Proc. of 14-th IEEE Symp. on Logic in Computer Science, pages 24-34, 1999.
5. E. Grädel, *On the restraining power of guards*, Journal of Symbolic Logic 64:1719-1742, 1999.
6. E. Grädel, P. Kolaitis, M. Vardi, *On the decision problem for two-variable first order logic*, Bull. of Symbolic Logic, 3(1):53-96, 1997.
7. E. Grädel, I. Walukiewicz, *Guarded Fixpoint Logic*, Proc. of 14th IEEE Symp. on Logic in Computer Science, pages 45-54, 1999.
8. W. Szwast, L. Tendera *On the decision problem for the guarded fragment with transitivity*, Proc. of 16-th IEEE Symp. on Logic in Computer Science, pages 147-156, 2001.

A Parametric Analysis of the State Explosion Problem in Model Checking

(Extended Abstract)

S. Demri, F. Laroussinie, and P. Schnoebelen

Lab. Spécification & Vérification
ENS de Cachan & CNRS UMR 8643
61, av. Pdt. Wilson, 94235 Cachan Cedex France
{demri,fl,phs}@lsv.ens-cachan.fr

Abstract. In model checking, the state explosion problem occurs when one verifies a *non-flat system*, i.e. a system described implicitly as a synchronized product of elementary subsystems. In this paper, we investigate the complexity of a wide variety of model checking problems for non-flat systems under the light of *parameterized complexity*, taking the number of synchronized components as a parameter. We provide precise complexity measures (in the parameterized sense) for most of the problems we investigate, and evidence that the results are robust.

1 Introduction

Model checking, i.e. the automated verification that (the formal model of) a system satisfies some formal behavioral property, has proved to be a revolutionary advance for the correctness of critical systems [CGP99]. Investigating the computational complexity of model checking started with [SC85], and today the complexity of the main model checking problems is known.

It is now understood that, in practice, the source of intractability is the size of the model and not the size of the property to be checked. This can be illustrated with LTL model checking as an example: while the problem is PSPACE-complete [SC85], it was observed in [LP85] that checking whether $S \models \phi$ can be done in time $O(|S| \times 2^{|\phi|})$. In practice ϕ is small and S is huge, so that "model checking is in linear time", as is often stated.

State explosion. In practice, the main obstacle to model checking is the *state explosion problem*, i.e. the fact that the model S is described implicitly, as a synchronized product of several components (with perhaps the addition of boolean variables, clocks, etc.), so that $|S|$ is usually exponentially larger than the size of its implicit description. For example, if S is given as a synchronized product $\mathcal{A}_1 \times \cdots \times \mathcal{A}_k$ of elementary components, the input of the model checking problem has size $n = \sum_i |\mathcal{A}_i|$ while S has size $O\left(\prod_i |\mathcal{A}_i|\right)$, that is $O(n^k)$, or $O(2^n)$ when k is not fixed.

H. Alt and A. Ferreira (Eds.): STACS 2002, LNCS 2285, pp. 620–631, 2002.
© Springer-Verlag Berlin Heidelberg 2002

From a theoretical viewpoint, the state explosion problem seems inescapable in the classical worst-case complexity paradigm. Indeed, studies covering all the main model checking problems and the most common ways of combining components have repeatedly shown that model checking problems are exponentially harder when S is given implicitly [Esp98,HKV97,JM96,KVW00,Rab97,Rab00, LS00].

A parametric analysis. The state explosion problem can be investigated more finely through *parameterized complexity*, a theoretical framework developed by Downey and Fellows for studying problems where complexity depends differently on the size n of the input and on some other parameter k that varies less (in some sense), see e.g. [DF99].

Any of the main model checking problems where the input is a sequence $\mathcal{A}_1, \ldots, \mathcal{A}_k$ of components can be solved in polynomial-time *for every fixed value of k*, e.g. in $O(n^k)$. That is, for every fixed k, the problem is polynomial-time. However, Downey and Fellows consider $O(n^k)$ as intractable for parameterized problems since the exponent k of n is not bounded, while algorithms running in time $f(k) \times n^c$ for some function f and constant c are considered tractable (see [DF99] for convincing arguments).

Parameterized complexity adheres to the "worst-case complexity" viewpoint but it leads to finer analysis. This can be illustrated on some graph-theoretical problems: among the NP-complete problems with a natural algorithm running in $O(n^k)$, many admit another algorithm in some $f(k) \times n^c$ (e.g. existence in a graph of a cycle of size k) while many others seem not to have any such solution (e.g. existence of a clique of size k). Note that these problems are "equivalent" in the classical complexity paradigm.

Our contribution. In this paper, we apply the parameterized complexity viewpoint to model checking problems where the input is a synchronized product of k components, k being the parameter. We investigate model checking problems ranging from reachability questions to temporal model checking for several temporal logics, to equivalence checking for several behavioral equivalences.

We provide precise complexity measures (in the parameterized sense) for most of the problems we investigate, and informative lower and upper bounds for the remaining ones. We show how the results are generally robust, i.e. insensitive to slight modifications (e.g. size of the synchronization alphabet) or restrictions (e.g. to deterministic systems).

All the considered problems are shown intractable even in the parameterized viewpoint (but they reach different intractability levels). See summary of results in section 7. This shows that these problems (very probably) do not admit solutions running in time $f(k) \times n^c$ for some f and c, and strengthens the known results about the computational complexity of the state explosion problem.

While mainly aimed at model checking, our study is also interesting for the field of parameterized complexity itself. For example, we are able to sharpen the characterization of the complexity of FAI-II and FAI-III (from [DF99, p. 470]) as shown in [DLS01, Appendix C]. We also introduce, as a useful general tool, parameterized problems for Alternating Turing machines and relate them to Downey and Fellows' W-hierarchy. Finally, we enrich the known catalog of parameterized problems with problems from an important application field.

Related work. Parameterized complexity has been applied to model checking problems where the parameter is the size of the *property to be checked* (or derived from it) and where the model is given explicitly: this has no relation with the state explosion problem and trivially leads to tractability in the parameterized sense for temporal logics (but becomes interesting when one considers more powerful logics [Gro99] or problems with database queries [PY99], or when one tries to identify parameters (e.g. tree width) that make problems tractable [GSS01]).

Parameterized complexity has been applied to problems where the input is, like in our work, a sequence of k synchronized automata, k being the parameter [B+95,War01,Ces01]. These works are concerned with automata-theoretic (or language-theoretic) questions rather than verification and model checking questions.

Plan of the paper. Sections 2 and 3 recall the basic definitions about parameterized complexity and synchronized products of systems. We investigate reachability problems in section 4, temporal logic problems in section 5, and behavioral equivalence problems in section 6. As a rule proofs omitted from the main text can be found in [DLS01].

2   Parameterized Complexity

We follow [DF99]. A *parameterized language* P is a set of pairs $\langle x, k \rangle$ where x is a word over some finite alphabet and k, the *parameter*, is an integer. The problem associated with P is to decide whether $\langle x, k \rangle \in P$ for arbitrary $\langle x, k \rangle$.

A parameterized problem P is *(strongly uniformly) fixed-parameter tractable*, shortly "FPT", $\overset{\text{def}}{\Leftrightarrow}$ there exist a recursive function $f : \mathbb{N} \mapsto \mathbb{N}$ and a constant $c \in \mathbb{N}$ such that the question $\langle x, k \rangle \in P$ can be solved in time $f(k) \times |x|^c$ (see e.g. [DF99, Chapter 2]).

A parameterized problem P is *fixed-parameter m-reducible* (fp-reducible) to the parameterized problem P' (in symbols $P \leq_{\text{m}}^{\text{fp}} P'$) $\overset{\text{def}}{\Leftrightarrow}$ there exist recursive total functions $f_1 : k \mapsto k'$, $f_2 : k \mapsto k''$, $f_3 : \langle x, k \rangle \mapsto x'$ and a constant $c \in \mathbb{N}$ such that $\langle x, k \rangle \mapsto x'$ is computable in time $k''|x|^c$ and $\langle x, k \rangle \in P$ iff $\langle x', k' \rangle \in P'$. P and P' are *fixed-parameter equivalent* (fp-equivalent) $\overset{\text{def}}{\Leftrightarrow} P \leq_{\text{m}}^{\text{fp}} P' \leq_{\text{m}}^{\text{fp}} P$. Clearly, if $P \leq_{\text{m}}^{\text{fp}} P'$ and P' is FPT, then P too is FPT.

Parameterized complexity comes with an array of elaborate techniques to devise fp-feasible algorithms, and another set of techniques to show that a problem is not FPT (or hard for a class conjectured to be strictly larger than FPT).

Downey and Fellows introduced the following hierarchy of classes of parameterized problems [DF99]:

$$\text{FPT} \subseteq \text{W}[1] \subseteq \text{W}[2] \subseteq \cdots \subseteq \text{W}[\text{SAT}] \subseteq \text{AW}[1] \subseteq \text{AW}[\text{SAT}] \subseteq \text{AW}[\text{P}] \subseteq \text{XP} \subseteq \cdots$$

where it is known that FPT $\neq$ XP. These classes are closed under fp-equivalence. W[1] is usually considered as the parameterized analogue of NP (from classical complexity theory) and a W[1]-hard problem is seen as intractable. XP contains all problems that can be solved in time $O(n^k)$ and is considered as the parameterized analogue of EXPTIME. It should be stressed that the above analogies are only useful heuristics: there is no known formal correspondence between standard complexity classes (NP, PSPACE, EXPTIME, ...) and parameterized complexity classes (W[1], AW[P], XP, ...)[1].

We don't recall the formal definitions of these classes since they are not required for understanding our results. It is enough to admit that W[1] is intractable, and to understand the parameterized problems dealing with short or compact computations we introduce in the next subsection. Most of the parameterized model checking problems we consider in this paper are easily seen to be in XP.

2.1   Short and Compact TM Computations

Not surprisingly, some fundamental parameterized problems consider Turing machines (shortly, "TMs"): SHORT COMPUTATION (resp. COMPACT COMPUTATION) is the parameterized problem where one is given a TM M and where it is asked whether M accepts in at most k steps (resp. using at most k work tape squares). These are the parameterized versions of the time and space bounds from classical complexity theory.

We consider TMs with just one initially blank work-tape (an input word can be encoded in the control states of the TM). One obtains different problems by considering deterministic (DTM), non-deterministic (NDTM), or alternating (ATM) machines.

SHORT DTM COMPUTATION is FPT while SHORT NDTM COMPUTATION is W[1]-complete [DF99]. COMPACT COMPUTATION is more complex and reaches high levels in the W-hierarchy: COMPACT NDTM COMPUTATION is AW[P]-hard [ADF95] and COMPACT DTM COMPUTATION is AW[SAT]-hard (see e.g. [DF99]).

Remark 2.1. More precise measures are still lacking and [DF99, Chapter 14] recalls that it is not known whether COMPACT DTM COMPUTATION and COMPACT NDTM COMPUTATION are fp-equivalent (it is not known whether a parameterized version of Savitch's theorem holds).

[DF99] does not consider parameterized problems with ATMs, but these proved very useful in our study[2]. Our first results show how they correspond to existing levels of the W-hierarchy:

Theorem 2.2. SHORT ATM COMPUTATION *is AW[1]-complete.*

---

[1] But see the recent work by Flum & Grohe in this volume.

[2] Note that [Ces01] also introduced parameterized problems on TMs to characterize the parameterized complexity classes W[1], W[2], and W[P].

Proof. We show equivalence with PARAMETERIZED-QBFSAT$_t$, shown AW[1]-complete in [DF99, Chapter 14]. An instance of PARAMETERIZED-QBFSAT$_t$ is a quantified boolean formula $\Psi = \exists^{=k_1} X_1 \forall^{=k_2} X_2 \ldots \forall^{=k_{2p}} X_{2p} \Phi$ where Φ, a positive boolean combination of literals, has at most t alternations between conjunctions and disjunctions. The literals use variables in $X = X_1 \cup \cdots \cup X_{2p}$ and the quantifications "$\exists^{=k_i} X_i$" and "$\forall^{=k_i} X_i$" are relativized to valuations of X_i where exactly k_i variables are set to true. The parameter k is $k_1 + \cdots + k_{2p}$.
PARAMETERIZED-QBFSAT$_t$ $\leq_{\mathrm{m}}^{\mathrm{fp}}$ SHORT ATM COMPUTATION:

To an instance Ψ of PARAMETERIZED-QBFSAT$_t$, we associate an ATM M_Ψ that picks $k_1 + \cdots + k_{2p}$ variables in $X_1 \cup \cdots \cup X_{2p}$ and checks that Φ evaluates to true under the corresponding valuation. The structure of Φ is reflected in the transition table of M_Ψ, and we use universal states to encode both the universal quantifications "$\forall^{=k_{2i}} \ldots$" and the conjunctions in Φ. M_Ψ can be made to answer in $O(k + t)$ steps, which gives us an fp-reduction since t is a constant.
SHORT ATM COMPUTATION $\leq_{\mathrm{m}}^{\mathrm{fp}}$ PARAMETERIZED-QBFSAT$_t$:

With an ATM M and an odd $k = 2p + 1$, we associate a formula Ψ_M that is true iff M accepts in k moves. The variables in Ψ are all $x[i, t, l]$ and mean "l is the ith symbol in the instantaneous description (i.d.) of M at step t". i and t range over $0, \ldots, k$, while l is any tape symbol or pair $\langle$symbol, control state$\rangle$ of M. Assuming M starts with an universal move, Ψ_M has the general form $\exists^{=k+1} X_0 \forall^{=k+1} X_1 \ldots \forall^{=k+1} X_k \Phi$ where $X_t = \{x[i, t, l] \mid i, l \ldots\}$ and Φ checks that the chosen valuations correspond to a run, i.e. has the form

$$\overbrace{\left(\bigwedge_{t=0}^{p} \Phi_{\mathrm{seq}}(X_{2t}, X_{2t+1}) \right)}^{\Phi_\forall} \Rightarrow \left(\Phi_{\mathrm{init}}(X_0) \wedge \Phi_{\mathrm{accept}}(X_k) \wedge \overbrace{\bigwedge_{t=1}^{p} \Phi_{\mathrm{seq}}(X_{2t-1}, X_{2t})}^{\Phi_\exists} \right)$$

where $\Phi_{\mathrm{seq}}(X, X')$ checks that (the valuations of) X and X' describe valid i.d.'s in valid succession. The different treatment between $\Phi_\forall$ and $\Phi_\exists$ reflects the fact that valid successions of existential states are only performed when valid successions of universal states are done.

Finally, we can easily rewrite Φ as a positive boolean combination of literals with 5 alternations and therefore obtain an instance of PARAMETERIZED-QBFSAT$_5$ with $k' = (k + 1)^2$ and size $n' = O(k^2 n^3)$. $\square$

Theorem 2.3. COMPACT ATM COMPUTATION *is XP-complete.*

Proof. We show fp-equivalence with PEBBLE GAME, shown XP-complete in [DF99, Theorem 15.5]. An instance of PEBBLE GAME is a set N of nodes, a starting position $S = \{s_1, \ldots, s_k\} \subseteq N$ of k pebbles on k nodes, a terminal node $T \in N$ and a set of possible moves $R \subseteq N \times N \times N$. Players I and II play in turn, moving pebbles and trying to reach T. A possible move $\langle x, y, z \rangle \in R$ means that any player can move a pebble from x to z if y is occupied (the pebble jumps over y) and z is free. The problem is to determine whether player I has a winning strategy. The parameter is $k = |S|$.
COMPACT ATM COMPUTATION $\leq_{\mathrm{m}}^{\mathrm{fp}}$ PEBBLE GAME:

[KAI79, Theorem 3.1] shows that PEBBLE GAME is EXPTIME-hard by reducing space-bounded ATMs. Their reduction can be turned into an fp-reduction where an ATM of size n running in space k gives rise to a pebble game instance where k' is $k + 1$, and where n' is bounded by a polynomial of n.

PEBBLE GAME $\leq_{\mathrm{m}}^{\mathrm{fp}}$ COMPACT ATM COMPUTATION:

Given an instance $G = \langle N, S, T, R \rangle$ with $|S| = k$, one constructs an ATM M_G that emulates the game and accepts iff player I wins. The alphabet of M_G is N and k worktape squares are sufficient to store the current configuration at any time in the game. Moves by player I are emulated with existential states, moves by player II use universal states. Information about R (the set of rules) and S is stored in the transition table of M_G. This gives an fp-reduction since $|M_G|$ is in $O(|G|)$ and $k' = k$. □

3   Synchronized Transition Systems

A *labeled transition system* (LTS) $\mathcal{A}$ over some alphabet Σ is a tuple $\langle Q, \Sigma, \rightarrow \rangle$ where $Q = \{s, t, \dots\}$ is the set of states and $\rightarrow \subseteq Q \times \Sigma \times Q$ are the transitions. We assume the standard notation $s \xrightarrow{a} t$, $s \xrightarrow{w} t$ ($w \in \Sigma^*$), $s \xrightarrow{*} t$, $s \xrightarrow{+} t$, etc. The size of a finite LTS $\mathcal{A}$ is $|\mathcal{A}| \overset{\text{def}}{=} |Q| + |\Sigma| + |\rightarrow|$.

Non-flat systems are products $\mathcal{A}_1 \times \cdots \times \mathcal{A}_k$ of (flat) component LTSs. Assuming $\mathcal{A}_i = \langle Q_i, \Sigma_i, \rightarrow_i \rangle$ for $i = 1, \dots, k$, the product denotes a LTS $\langle Q, \Sigma, \rightarrow \rangle$ where $Q \overset{\text{def}}{=} \prod_{i=1}^{k} Q_i$, $\Sigma \overset{\text{def}}{=} \bigcup_{i=1}^{k} \Sigma_i$ and where $\rightarrow \subseteq Q \times \Sigma \times Q$ depends on the synchronization protocol one considers: strong or binary synchronization.

In *strong synchronization* all components move at the same time: $\langle s_1, \dots, s_k \rangle \xrightarrow{a}_{\mathrm{str}} \langle t_1, \dots, t_k \rangle$ iff $s_i \xrightarrow{a}_i t_i$ for all $i = 1, \dots, k$.

In *binary synchronization* any two components synchronize while the rest does not move: $\langle s_1, \dots, s_k \rangle \xrightarrow{a}_{\mathrm{bin}} \langle t_1, \dots, t_k \rangle$ iff there exist i and j ($i \neq j$) s.t. $s_i \xrightarrow{a}_i t_i$ and $s_j \xrightarrow{a}_j t_j$ while $s_l = t_l$ for all $l \notin \{i, j\}$.

In this paper, we consider *strong synchronization* as the natural model for non-flat systems and the notation $\mathcal{A}_1 \times \cdots \times \mathcal{A}_k$ assumes strong synchronization when we don't explicitly say otherwise. However, our results are robust and remain unchanged when one adopts binary synchronization (see [DLS01, Appendix B]).

4   Parameterized Complexity of Non-flat Reachability

Reachability problems are the most fundamental problems in model checking.

Exact Reachability (Exact-Reach)
Instance: k LTSs $\mathcal{A}_1, \cdots, \mathcal{A}_k$, two configurations $\bar{s}$ and $\bar{t}$ of $\mathcal{A}_1 \times \cdots \times \mathcal{A}_k$.
Question: Does $\bar{s} \xrightarrow{*} \bar{t}$?

We are interested in the parameterized versions k-EXACT-REACH, where k is the parameter. Other variants of this problem are used in model checking, for instance by considering a finite set of target states instead of a unique state

$\bar{t}$, or by asking for repeated reachability of a control state, that is we ask for $\bar{s} \xrightarrow{*} \bar{t} \xrightarrow{+} \bar{t}$. Finally, another standard variant consists in considering fair reachability. Details of the definition of these parameterized problems can be found in [DLS01]. It is a folklore result that the four non-flat reachability problems are equivalent in the classical sense (i.e. via logspace reductions) and are PSPACE-complete and we can show this equivalence can be lifted to the parameterized case [DLS01, Theorem 4.1].

Lemmas 4.2 and 4.3 allow the following characterization:

Theorem 4.1. k-EXACT-REACH *is fp-equivalent to* COMPACT NDTM COMPUTATION.

Hence k-EXACT-REACH and the above mentionned variants of k-EXACT-REACH are AW[P]-hard. The characterization given by Theorem 4.1 is robust: it stays unchanged when we consider binary synchronization or when we restrict to a binary alphabet or to deterministic LTSs (see [DLS01, Appendix B & C]).

Lemma 4.2. COMPACT NDTM COMPUTATION $\leq_m^{\text{fp}}$ k-EXACT-REACH.

Proof (sketch). With an NDTM M and an integer k we associate a product $\mathcal{A}_1 \times \cdots \times \mathcal{A}_k \times \mathcal{A}_{\text{state}} \times \mathcal{A}_{\text{head}}$ of $k + 2$ LTSs that emulate the behaviour of M on a k-bounded tape. For $i = 1, \ldots, k$, $\mathcal{A}_i$ stores the current contents of the i-th tape square, $\mathcal{A}_{\text{state}}$ stores the current control-state of M and $\mathcal{A}_{\text{head}}$ stores the position of the TM head. These LTSs synchronize on labels of the form $\langle t, i \rangle$ that stand for "rule t of M is fired while head is in position i". Successful acceptance by M is directly encoded as an exact reachability criterion (we add an extra label for the final transition). Finally we translated our instance to a k-EXACT-REACH instance with $k' = k + 2$ and $n' = O(kn^2)$.   □

Lemma 4.3. k-EXACT-REACH $\leq_m^{\text{fp}}$ COMPACT NDTM COMPUTATION.

Proof (sketch). An instance of k-EXACT-REACH of the form $\mathcal{A}_1, \ldots, \mathcal{A}_k, \bar{s}, \bar{t}$, is easily reduced to an instance of COMPACT NDTM COMPUTATION. The TM M emulates the behaviour of the product $\mathcal{A}_1 \times \cdots \times \mathcal{A}_k$ by writing the initial configuration $\bar{s}$ on its tape (one component per tape square, the tape alphabet contains all control states of the $\mathcal{A}_i$'s). Then M picks non-deterministically a synchronization letter a, updates all local states of the $\mathcal{A}_i$s by firing one of their a-transitions (M blocks if some local state has no a-transition), and repeats until the configuration $\bar{t}$ is reached. This yields an fp-reduction: $k' = k$ and n' is in $O(kn)$.   □

5   Parameterized Complexity of Non-flat Temporal Logic Model Checking

In this section, we investigate the parameterized complexity of temporal logic model checking problems when the input is a synchronized product of LTSs (and a temporal formula!). We assume familiarity with the standard logics used in verification: LTL, CTL, HML, the modal μ-calculus (see [Eme90,CGP99,BS01]).

For modal logics, LTSs are the natural models, while for temporal logics like CTL or LTL the natural models are Kripke structures. Below we call *Kripke structure* (or shortly KS) a pair $\mathcal{M} = \langle \mathcal{A}, m \rangle$ of a finite LTS $\mathcal{A} = \langle Q, \Sigma, \rightarrow \rangle$ extended with a finite valuation $m \subseteq Q \times AP$ of its states (with AP a set of atomic propositions). The size $|\mathcal{M}|$ of $\mathcal{M} = \langle \mathcal{A}, m \rangle$ is $|\mathcal{A}| + |m|$.

We omit the standard definition of when state s in $\mathcal{M}$ satisfies formula ϕ, written $\mathcal{M}, s \models \phi$. There is one detail though: for linear-time logics (LTL and its fragments) we follow [SC85] and assume, for the sake of uniformity, that the question "$\mathcal{M}, s \models \phi$?" asks for the *existence* of a path from s that verifies ϕ, which is dual to the universal "all paths from s" formulation commonly used in applications.

The labels of the transitions of a KS do not appear in temporal formulae. They are only used for synchronization purposes: $\langle \mathcal{A}_1, m_1 \rangle \times \cdots \times \langle \mathcal{A}_k, m_k \rangle$ is the KS $\langle \mathcal{A}, m \rangle$ where $\mathcal{A} = \mathcal{A}_1 \times \cdots \times \mathcal{A}_k$ (implicitly assuming strong synchronization) and where m is a valuation built from $m_1, \ldots, m_k$. For the sake of simplicity, we assume w.l.o.g. that m is the "sum" of $m_1, \ldots, m_k$, that is $\langle \langle q_1, \ldots, q_k \rangle, \mathrm{p} \rangle \in m$ as soon as $\langle q_i, \mathrm{p} \rangle \in m_i$ for some i.

The problems we consider have the following general form, where L is LTL, CTL, the modal μ-calculus, or some of their fragments, and where the parameterized version has the pair $k, |\phi|$ as parameter:

Parameterized model checking for logic L (MC_L)
Instance: Kripke structures $\mathcal{M}_1, \ldots, \mathcal{M}_k$, a configuration $\bar{s}$, an L-formula ϕ.
Question: Does $\mathcal{M}_1 \times \cdots \times \mathcal{M}_k, \bar{s} \models \phi$?

5.1   Linear Time

LTL model checking for non-flat systems is PSPACE-complete. In our parameterized setting we have:

Theorem 5.1. k, ϕ-$\mathrm{MC}_{\mathrm{LTL}}$ *is fp-equivalent to* COMPACT NDTM COMPUTATION *and to* k-EXACT-REACH *(and hence is AW[P]-hard).*

Proof. k-EXACT-REACH reduces to k, ϕ-$\mathrm{MC}_{\mathrm{LTL}}$ since $\bar{s} \xrightarrow{*} \langle t_1, \ldots, t_k \rangle$ in some $\mathcal{A}_1 \times \ldots \times \mathcal{A}_k$ iff $\langle \mathcal{A}_1, \{\langle t_1, \mathrm{p}_1 \rangle\} \rangle \times \ldots \langle \mathcal{A}_k, \{\langle t_k, \mathrm{p}_k \rangle\} \rangle, \bar{s} \models \mathsf{F}(\mathrm{p}_1 \wedge \ldots \wedge \mathrm{p}_k)$. This provides an fp-reduction since $|\mathsf{F}(\mathrm{p}_1 \wedge \ldots \wedge \mathrm{p}_k)|$ is in $O(k \log k)$.

In the other direction, the question "does $\mathcal{M}_1 \times \cdots \times \mathcal{M}_k, \bar{s} \models \phi$?" reduces to a repeated reachability problem for $\mathcal{M}_1 \times \cdots \times \mathcal{M}_k \times \mathcal{B}_\phi$, where $\mathcal{B}_\phi$ is a Büchi automaton that accepts the paths satisfying ϕ^3. There remains to check that this classical reduction is a reduction in the parameterized sense: since $|\mathcal{B}_\phi|$ is in

---

[3] Strictly speaking, $\mathcal{B}_\phi$ synchronizes with $M_1 \times \cdots \times M_k$ using a protocol different from what we used up to now: $\bar{s} \xrightarrow{a} \bar{t}$ and $q \xrightarrow{v} q'$ synchronize iff $m(\bar{s}) = v$. However,

$O(2^{|\phi|})$, the reduction has k' in $O(k)$ and n' in $O(2^k \times n)$, which is enough for fp-reducibility.   □

Model checking is already intractable for the fragment LT0 of LTL *that does not allow to state reachability questions.* LT0 formulae are built with atomic propositions, X and ∨ only (no negation allowed):

Theorem 5.2. [DLS01] $k, \phi\text{-MC}_{\text{LT0}}$ *is W[1]-complete, even if we restrict to formulae using only one atomic proposition.*

5.2   Branching Time

Model checking non-flat systems is EXPTIME-complete for the μ-calculus [Rab00] and PSPACE-complete for HML or CTL. (Observe that HML, the fragment of the μ-calculus without fixed-points, does not allow stating reachability questions). In our parameterized setting we have:

Theorem 5.3. $k, \phi\text{-MC}_\mu$ *is XP-complete.*

Proof. Writing n for $\sum_i |\mathcal{M}_i|$, $k, \phi\text{-MC}_\mu$ can be solved in time $O((|\phi|.n^k)^{|\phi|})$ [KVW00, Theo. 6.4] and hence is in XP.

XP-hardness is proved by a reduction from non-flat bisimilarity. Andersen [And93, section 4.3] showed how bisimilarity checking can be stated in the branching-time mu-calculus: consider two LTSs $\mathcal{A}$ and $\mathcal{B}$ on a common Σ, and build $\mathcal{B}'$ out of $\mathcal{B}$ by renaming all its actions $a \in \Sigma$ by copies $a' \notin \Sigma$. Then

$$\mathcal{A} \text{ and } \mathcal{B} \text{ are bisimilar iff } \mathcal{A} \parallel \mathcal{B}' \models \nu X. \bigwedge_{a \in \Sigma} ([a]\langle a'\rangle X \wedge [a']\langle a\rangle X). \qquad (1)$$

The interleaved product $\mathcal{A} \parallel \mathcal{B}'$ can be replaced by strong synchronization if we add Σ-loops in all states of $\mathcal{B}'$ and Σ'-loops in all states of $\mathcal{A}$. Using $(\mathcal{A}_1 \times \mathcal{A}_2)' = \mathcal{A}_1' \times \mathcal{A}_2'$, the reduction carries to non-flat systems.

Since non-flat bisimilarity is XP-hard already when $|\Sigma| = 2$ [DLS01, Theorem D.4], we can bound the size of the μ-formula in (1) and have an fp-reduction.   □

Theorem 5.4. $k, \phi\text{-MC}_{\text{HML}}$ *is AW[1]-complete.*

Proof (idea). That $k, \phi\text{-MC}_{\text{HML}}$ is fp-equivalent to SHORT ATM COMPUTATION can be proved by adapting the techniques of Theorem 5.2 to ATMs.   □

Theorem 5.5. COMPACT NDTM COMPUTATION $\leq_m^{\text{fp}} k, \phi\text{-MC}_{\text{CTL}}$. *(Hence $k, \phi\text{-MC}_{\text{CTL}}$ is AW[P]-hard).*

Proof (idea). CTL allows to state reachability questions.   □

---

using the same techniques as in [DLS01, Appendix B], the parametrized reachability problems for this form of synchronized products can also be proved fp-equivalent to COMPACT NDTM COMPUTATION.

Remark 5.6. For the moment, we do not have a more precise characterization for k, ϕ-$\mathrm{MC_{CTL}}$. Observe that, by definition, model checking CTL is closed by complementation (unlike the LTL case) so that a conjecture about their being fp-equivalent to COMPACT NDTM COMPUTATION would have an impact on the open problems mentioned in Remark 2.1. For upper bounds, the problem is obviously in XP and we failed to refine this, partly because parameterized complexity does not offer many natural classes above AW[P].

6   Parameterized Complexity of Non-flat Bisimilarity

We assume familiarity with bisimulation and the other behavioral equivalences in the branching time – linear time spectrum [Gla01]. Checking for bisimilarity among non-flat systems is EXPTIME-complete in the classical framework [JM96, LS00]. For our parametric analysis, k-BISIM asks whether two given configurations in a product of k LTSs are bisimilar.

Theorem 6.1. k-BISIM *is XP-complete.*

Proof (idea). k-BISIM is in XP since bisimilarity of flat systems is polynomial-time [KS90]. XP-hardness is seen by observing that the reduction in the proof of [LS00, Theorem 4.1] can be seen as an fp-reduction from COMPACT ATM COMPUTATION to k-BISIM.   □

This result is robust and [DLS01, Appendix D] shows that it still holds when we consider binary synchronization or restricted alphabets (a result we used in the proof of Theorem 5.3). [DLS01] further proves XP-hardness for a wide range of behavioral equivalences and preorders, along the lines of [Rab97,LS00,SJ01].

7   Conclusion

We studied the complexity of model checking synchronized products of LTSs under the light of Downey and Fellows's theory of parameterized complexity. Here the parameter k is the number of components (and the size of the property). We considered a wide variety of problems, and assumed two different synchronization protocols.

It is known that for any fixed value of the parameter, the problems have polynomial-time solutions in $O(n^k)$ and we show that solutions in some $f(k) \times n^c$ (for some constant c) do not exist (unless Downey and Fellows's hierarchy collapses). Therefore our results show that these problems are probably not tractable *even in the parameterized sense of being FPT*, and their complexity is in general quite high in the hierarchy (see summary in Fig. 1 where edges correspond to the existence of an fp-reduction).

The problems remain intractable (possibly at a weaker level) when natural restrictions are imposed. We think this must be understood as arguing against any hope of finding "tractable" algorithms for model checking synchronized products of components even when the number k of components varies much less than the size of the components themselves.

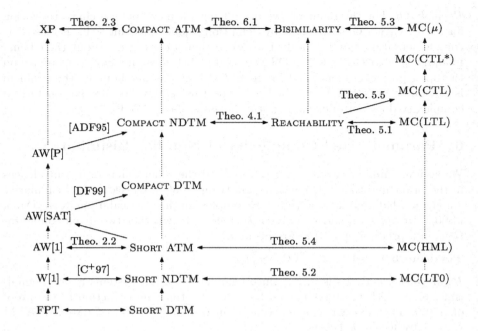

Fig. 1. A summary of existing reductions between parameterized problems

References

[ADF95] K. A. Abrahamson, R. G. Downey, and M. R. Fellows. Fixed-parameter tractability and completeness IV: On completeness for W[P] and PSPACE analogs. *Annals of Pure and Applied Logic*, 73(3):235–276, 1995.

[And93] H. R. Andersen. *Verification of Temporal Properties of Concurrent Systems.* PhD thesis, Aarhus University, Denmark, June 1993.

[B⁺95] H. L. Bodlaender, R. G. Downey, M. R. Fellows, and H. T. Wareham. The parameterized complexity of sequence alignment and consensus. *Theoretical Computer Science*, 147(1–2):31–54, 1995.

[BS01] J. Bradfield and C. Stirling. Modal logics and mu-calculi: an introduction. In *Handbook of Process Algebra*, ch 4, pp 293–330. Elsevier Science, 2001.

[C⁺97] Liming Cai, Jianer Chen, R. G. Downey, and M. R. Fellows. On the parameterized complexity of short computation and factorization. *Archive for Mathematical Logic*, 36(4/5):321–337, 1997.

[Ces01] M. Cesati. The Turing way to the parameterized intractability, September 2001. Submitted.

[CGP99] E. M. Clarke, O. Grumberg, and D. A. Peled. *Model Checking.* MIT Press, 1999.

[DF99] R. G. Downey and M. R. Fellows. *Parameterized Complexity.* Springer, 1999.

[DLS01] S. Demri, F. Laroussinie, and Ph. Schnoebelen. A parametric analysis of the state explosion problem in model checking. Research Report LSV-01-4, Lab. Specification and Verification, ENS de Cachan, France, Apr. 2001.

[Eme90]   E. A. Emerson. Temporal and modal logic. In *Handbook of Theoretical Computer Science, vol. B*, ch 16, pp 995–1072. Elsevier Science, 1990.

[Esp98]   J. Esparza. Decidability and complexity of Petri net problems — an introduction. In *Advances in Petri Nets 1998*, LNCS 1491, pp 374–428. Springer, 1998.

[Gla01]   R. J. van Glabbeek. The linear time – branching time spectrum I. In *Handbook of Process Algebra*, ch 1, pp 3–99. Elsevier Science, 2001.

[Gro99]   M. Grohe. Descriptive and parameterized complexity. In *Computer Science Logic (CSL'99)*, LNCS 1683, pp 14–31. Springer, 1999.

[GSS01]   M. Grohe, T. Schwentick, and L. Segoufin. When is the evaluation of conjunctive queries tractable? In *33rd ACM Symp. Theory of Computing (STOC'01)*, pp 657–666, 2001.

[HKV97]   D. Harel, O. Kupferman, and M. Y. Vardi. On the complexity of verifying concurrent transition systems. In *Concurrency Theory (CONCUR'97)*, LNCS 1243, pp 258–272. Springer, 1997.

[JM96]   L. Jategaonkar and A. R. Meyer. Deciding true concurrency equivalences on safe, finite nets. *Theoretical Computer Science*, 154(1):107–143, 1996.

[KAI79]   T. Kasai, A. Adachi, and S. Iwata. Classes of pebble games and complete problems. *SIAM J. Comput.*, 8(4):574–586, 1979.

[KS90]   P. C. Kanellakis and S. A. Smolka. CCS expressions, finite state processes and three problems of equivalence. *Information and Computation*, 86(1):43–68, 1990.

[KVW00]   O. Kupferman, M. Y. Vardi, and P. Wolper. An automata-theoretic approach to branching-time model checking. *J. ACM*, 47(2):312–360, 2000.

[LP85]   O. Lichtenstein and A. Pnueli. Checking that finite state concurrent programs satisfy their linear specification. In *12th ACM Symp. Principles of Programming Languages (POPL'85)*, pp 97–107, 1985.

[LS00]   F. Laroussinie and Ph. Schnoebelen. The state explosion problem from trace to bisimulation equivalence. In *Found. Software Science & Computation Structures (FOSSACS'2000)*, LNCS 1784, pp 192–207. Springer, 2000.

[PY99]   C. H. Papadimitriou and M. Yannakakis. On the complexity of database queries. *J. Computer and System Sciences*, 58(3):407–427, 1999.

[Rab97]   A. Rabinovich. Complexity of equivalence problems for concurrent systems of finite agents. *Information and Computation*, 139(2):111–129, 1997.

[Rab00]   A. Rabinovich. Symbolic model checking for μ-calculus requires exponential time. *Theoretical Computer Science*, 243(1–2):467–475, 2000.

[SC85]   A. P. Sistla and E. M. Clarke. The complexity of propositional linear temporal logics. *J. ACM*, 32(3):733–749, 1985.

[SJ01]   Z. Sawa and P. Jančar. P-hardness of equivalence testing on finite-state processes. In *Current Trends in Theory and Practice of Informatics (SOFSEM'01)*, LNCS 2234, pp 326–335. Springer, 2001.

[War01]   H. T. Wareham. The parameterized complexity of intersection and composition operations on sets of finite-state automata. In *Implementation & Application of Automata (CIAA'2000)*, LNCS 2088, pp 302–310. Springer, 2001.

Generalized Model-Checking over Locally Tree-Decomposable Classes

Markus Frick

Laboratory for Foundations of Computer Science
Division of Informatics, University of Edinburgh
mfrick@dcs.ed.ac.uk

Abstract. It has been proved in [12] that properties of graphs or other relational structures that are definable in first-order logic can be *decided* in linear time when the input structures are restricted to come from a *locally tree-decomposable* class of structures. Examples of such classes are the class of planar graphs or classes of graphs of bounded degree.

In this paper, we consider more general computational problems than decision problems. We prove that *construction, listing,* and *counting* problems definable in first-order logic can be solved in linear time on locally tree-decomposable classes of structures.

1 Introduction

Model-checking problems are general algorithmic problems that can be used to model a wide range of concrete problems from different areas of computer science, among them database theory, algorithm theory and constraint satisfaction (AI). In the basic version of model-checking, we are given a relational structure $\mathfrak{A}$ and a sentence φ of some logic $\mathcal{L}$ and ask if φ is satisfied by $\mathfrak{A}$. We call this problem the *basic model-checking problem* for $\mathcal{L}$. Note that the basic model-checking problem can only be used to model decision problems. But in practice, we are not only confronted with decision problems. For example, in the context of databases, we usually do not want to know, whether some employee lives in London, but want the system to return these employees, or at least one of them.

An area where model-checking turned out to be a very successful tool is algorithm theory. Standard algorithmic problems like 3-COLORABILITY, CLIQUE or SET-COVER are instances of model-checking. Consider, for instance, the clique-problem that is, we are given a $k \geq 1$ and a graph $\mathcal{G} = (G, E)$ and want to decide if there is a clique of size k in $\mathcal{G}$. This problem is equivalent to decide if the sentence $\varphi_{\text{clique}}^k := \exists x_1, \ldots, x_k \bigwedge_{1 \leq i < j \leq k} E x_i x_j$ holds in $\mathcal{G}$, hence, k-CLIQUE can be seen as a special case of the model-checking problem.

Up to now, we only considered decision problems. But as already mentioned, it is natural not only to ask if there is a k-clique but, if so, construct a solution, e.g. return a k-clique of $\mathcal{G}$. On the logic side, this can be modeled by using formulas with free variables instead of sentences and by finding a satisfying assignment. We call this problem the *construction* problem for model-checking. Other natural extensions are the *listing problem*, which asks for all satisfying assignments

H. Alt and A. Ferreira (Eds.): STACS 2002, LNCS 2285, pp. 632–644, 2002.
© Springer-Verlag Berlin Heidelberg 2002

and the *counting problem* asking for the number of satisfying assignments. In this taxonomy, the basic model-checking problem is called the *decision problem*, or just *model-checking* problem (cf. [15] for a survey).

We consider these generalized model-checking problems of first-order logic. In the realm of databases this logic has a predominant role, since it closely resembles the commercial standard query language SQL. Unfortunately, model-checking for first-order logic is of very high complexity. Already over structures with only two elements it is PSPACE-complete [18]. This suggests that the bulk of complexity is contributed by the query and not by the structure. One way to explore this in more detail is to exhibit which part of the input contributes to which extent to the complexity of the problem. In a general setting this has been done systematically by Downey and Fellows in [6]. Applied to our problem, we declare φ to be a parameter and try to find if there is an algorithm working in time $O(f(\|\varphi\|) \cdot \|\mathfrak{A}\|^c)$ for some function $f : \mathbb{N} \to \mathbb{N}$ and $c > 0$. If this is the case, we call a problem *fixed-parameter tractable*. In some sense this corresponds to our intuition that we evaluate "small" formulas in "big" structures. But it is shown in [7] that parameterized model-checking for first-order logic is AW[1]-complete, which makes it very unlikely for model-checking to be fixed-parameter tractable.

One way to overcome this negative news is to consider certain restrictions on the admitted input structures. A landmark in this direction is Courcelle's result: over structures $\mathfrak{A}$ of tree-width bounded by some constant w, monadic second-order model-checking can be done in *fixed-parameter linear time*, i.e. in time $O(f(\|\varphi\|) \cdot \|\mathfrak{A}\|)$ for some (here fast growing) function f [5]. This result was extended to the counting case [2] and recently to the construction and listing case [10]. Note that for the listing problem, linear time means linear in the size of the input plus the output.

There exist other successful restrictions take make algorithms more efficient. *Planarity* and *bounded degree* are two important and natural examples. It is hopeless though to expect for these classes such a strong result like Courcelle's, since 3-COLORABILITY, which is definable in monadic second-order logic, remains NP-complete over planar graphs of degree at most 4 [14].

Nevertheless, such graphs share the nice property that their tree-width only depends on the diameter. This was first exploited by Baker [3] and later investigated more systematically by Eppstein [9]. Their ideas led Frick and Grohe [12] to the notion of *locally tree-decomposable* classes of structures. Important locally tree-decomposable classes are graphs embeddable into some surface, bounded degree and of bounded tree-width. Furthermore, for such classes it was shown that model-checking for first-order logic is in fixed-parameter linear time [12]. Observe here that this result only concerns the decision problem.

We prove that all generalized model-checking problems, that is, the construction, listing, and counting problem over locally tree-decomposable classes[1] $\mathcal{C}$ are in fixed-parameter linear time. In particular, there are fixed-parameter linear

---

[1] Actually, we use a slightly more restrictive notion. But this notion still comprises all important examples.

time algorithms for the following problems: given a first-order formula $\varphi(\bar{x})$ and a structure $\mathfrak{A} \in \mathcal{C}$:

(1) compute an $\bar{a}$ such that $\mathfrak{A} \models \varphi(\bar{a})$ (construction)
(2) compute all $\bar{a}$ such that $\mathfrak{A} \models \varphi(\bar{a})$ (listing)
(3) compute the number of $\bar{a}$ such that $\mathfrak{A} \models \varphi(\bar{a})$ (counting)

These results can be seen as a further step in a systematic analysis of the parameterized complexity of model-checking: given a logical language $\mathcal{L}$, which restrictions on the inputs allow fixed-parameter tractability. Or with more emphasis on combinatorics, it can be considered as one of the first systematic studies of fixed-parameter tractability of construction, listing and counting problems.

But apart from the systematic importance of these results, there are several practical implications on standard algorithmic problems. For example, our result shows that there is a linear time algorithm that counts the number of 4-cliques in a given graph of degree 5, or a linear time algorithm counting the number of dominating sets of size 10 in a given planar graph.

More generally, we see that a dominating set of size k in planar graphs not only can be computed (this was shown in [6]), but we can also list and count them in linear time. Another example is k-SET-COVER, which asks whether a given family $\mathcal{F}$ of sets has a subfamily $\mathcal{S}$ of size k such that $\bigcup \mathcal{S} = \bigcup \mathcal{F}$. Its first-order formulation restricted to structures of bounded degree implies that for instances $\mathcal{F}$, such that each $x \in \bigcup \mathcal{F}$ is contained in at most constant many $F \in \mathcal{F}$ and all $F \in \mathcal{F}$ having bounded size, we get linear time algorithms for the mentioned tasks. The same holds for all first-order properties like $\mathcal{H}$-HOMOMORPHISM (is there a homomorphic copy of $\mathcal{H}$?), k-DOMINATING-SET and so forth.

Organization: After the preliminaries, we introduce the concept of locally tree-decomposable classes and present the main results. In section 4 we do some preparatories for the main results, relating local formulas to tree covers. These are used to solve the listing problem and construction problem in section 5. Finally, in section 6, we sketch the algorithm for the counting problem.

2   Preliminaries

For $n \geq 1$, we set $[n] := \{1, \ldots, n\}$. By $\mathrm{Pow}(A)$ we denote the power set of A and $\mathrm{Pow}^{\leq l}(A)$ is the set of elements of $\mathrm{Pow}(A)$ that have cardinality $\leq l$.

We assume the reader familiar with the basic notions of logic, as described in e.g. [8]. By $\mathrm{FO}[\tau]$ we denote the set of first-order formulas over the relational vocabulary τ. We write $\varphi(x_1, \ldots, x_m)$ if the free variables of φ are exactly $x_1, \ldots, x_m$. We only consider finite structures $\mathfrak{A}$, whose *universe* is denoted by A. For $B \subseteq A$ by $\langle B \rangle^{\mathfrak{A}}$ we denote the *substructure* of $\mathfrak{A}$ induced by B.

A *graph* $\mathcal{G}$ is an $\{E\}$-structure $(G, E^{\mathcal{G}})$ where $E^{\mathcal{G}}$ is an anti-reflexive and symmetric binary relation (in other words: we consider simple undirected graphs without loops). A *colored graph* is a structure $\mathfrak{B} = (B, E^{\mathfrak{B}}, P_1^{\mathfrak{B}}, \ldots, P_m^{\mathfrak{B}})$, where $(B, E^{\mathfrak{B}})$ is a graph and the unary relations $P_1^{\mathfrak{B}}, \ldots, P_m^{\mathfrak{B}}$ form a partition of the universe B. The degree of a vertex $v \in G$ is the number of vertices adjacent to v, and the degree of $\mathcal{G}$ is the maximal degree of its vertices.

For a τ-structure $\mathfrak{A}$, $a_1, \ldots, a_m \in A$ and a formula $\varphi(x_1, \ldots, x_m) \in \text{FO}[\tau]$ we write $\mathfrak{A} \models \varphi(a_1, \ldots, a_m)$ to say that $\mathfrak{A}$ satisfies φ, if the $x_1, \ldots, x_m$ are interpreted by the elements $a_1, \ldots, a_m$, respectively. For a structure $\mathfrak{A}$ and a formula $\varphi(x_1, \ldots, x_m)$ we let $\varphi(\mathfrak{A}) := \{(a_1, \ldots, a_m) \in A^m \mid \mathfrak{A} \models \varphi(a_1, \ldots, a_m)\}$. In case of no free variables in φ, we define $\varphi(\mathfrak{A}) := \{\emptyset\}$, if $\mathfrak{A} \models \varphi$ and $:= \emptyset$ otherwise. If the vocabularies of $\mathfrak{A}$ and φ do not agree, we set $\varphi(\mathfrak{A}) := \emptyset$.

Algorithms: Our underlying model of computation is the standard RAM-model with addition and subtraction as arithmetic operations [1]. We assume the *uniform cost measure*, hence arithmetic can always be done in constant time. Observe here that the results concerning construction and listing also hold under the assumption of the logarithmic cost measure. A relational τ-structure $\mathfrak{A}$ is coded by a word w consisting of the vocabulary, the elements of the universe A, and the relations $R^{\mathfrak{A}}$ for $R \in \tau$. For graphs, we use its adjacency list representation. For further details on coding issues, the reader is referred to [11]. To improve readability, we sometimes omit the O-notation, despite constants being involved depending on the hardware and used data structures.

In running time estimates we use $\|o\|$ to denote the actual size of the object o, w.r.t some encoding, whereas $|o|$ refers to the cardinality of the set o. To emphasize the difference let $\mathcal{T}$ be a family of sets. Then $|\mathcal{T}|$ refers to the cardinality of $\mathcal{T}$, while $\|\mathcal{T}\|$ stands for $\sum_{X \in \mathcal{T}} \|X\|$.

Model-checking problems: The basic model-checking problem asks if a given structure satisfies a given sentence. Here we shall also consider formulas with free variables. For each class $\mathcal{C}$ of finite structures this gives rise to the following four problems where the input is a structure $\mathfrak{A} \in \mathcal{C}$ and a formula $\varphi(\bar{x}) \in \text{FO}$.

The decision problem. Decide if $\varphi(\mathfrak{A}) \neq \emptyset$. For sentences, this problem coincides with the basic model-checking problem, and therefore we refer to it as FO-MODEL-CHECKING on $\mathcal{C}$.

The construction problem. Compute one tuple $\bar{a} \in \varphi(\mathfrak{A})$, if one exists. We refer to this problem as FO-CONSTRUCTION on $\mathcal{C}$.

The listing problem. Compute the set $\varphi(\mathfrak{A})$. Because of its important application in database theory, we refer to this problem as FO-EVALUATION on $\mathcal{C}$.

The counting problem. Compute the cardinality of the set $\varphi(\mathfrak{A})$. We refer to this problem as FO-COUNTING on $\mathcal{C}$.

Note that we always investigate the parameterized complexity of these problems. We call a model-checking problem on $\mathcal{C}$ *fixed-parameter linear time*, if there is an algorithm that, given $\varphi(\bar{x}) \in \text{FO}$ and $\mathfrak{A} \in \mathcal{C}$ solves the problem in time linear in the size of the structure $\mathfrak{A}$ plus the size of the output. Observe that for the running time analysis of the listing problem, it is necessary to additionally take into consideration the algorithm's output. For further details and examples of generalized model-checking the reader is referred to [15]. For more general background on this taxonomy of combinatorial problems, see [16,17].

Although our algorithms depend heavily on the notion of *tree-width*, it is not necessary to introduce it formally. The interested reader is referred to [4,5,

10]. Intuitively, the tree-width measures the similarity of a structure with a tree. In the context of model-checking, bounded tree-width allows the application of automata. This led to the following results, which actually have been proved for monadic second-order logic.

Theorem 1 ([5,2,10]). *Let C be a class of structures of bounded tree-width.* FO-MODEL-CHECKING, FO-CONSTRUCTION, FO-EVALUATION and FO-COUNTING *on C are all in fixed-parameter linear time.*

3   The Main Result

We start this section presenting the definitions of the main concepts necessary to state the main result. The main result itself will be presented at the end of the section.

Let τ be a vocabulary and $\mathfrak{A}$ a τ-structure. The Gaifman graph $\mathcal{G}(\mathfrak{A})$ of $\mathfrak{A}$ is the graph $(G, E^{\mathcal{G}})$ with $G = A$ and $E^{\mathcal{G}} := \{(a, b) \mid \text{ there is a } k\text{-ary } R \in \tau \text{ and } \bar{a} \in R^{\mathfrak{A}} \text{ such that } a, b \in \{a_1, \dots, a_k\}\}$. Then $d^{\mathfrak{A}}(a, b)$ denotes the distance between a and $b \in A$ in $\mathcal{G}(\mathfrak{A})$. Consequently, for $r \geq 1$ we define $N_r^{\mathfrak{A}}(a) := \{b \in A \mid d^{\mathfrak{A}}(a, b) \leq r\}$, the r-*neighborhood* of $a \in A$ and $N_r^{\mathfrak{A}}(X) := \bigcup_{a \in X} N_r^{\mathfrak{A}}(a)$ for some $X \subseteq A$. It is easy to see that for every $r \geq 1$ there is a formula $\delta_r(x, y) \in \mathrm{FO}[\tau]$ such that for all τ-structures $\mathfrak{A}$ and $a, b \in A$ we have $\mathfrak{A} \models \delta_r(a, b) \iff d^{\mathfrak{A}}(a, b) \leq r$. Within formulas we write $d(x, y) \leq r$ instead of $\delta_r(x, y)$ and $d(x, y) > r$ instead of $\neg \delta_r(x, y)$.

If $\varphi(x)$ is a first-order formula, then $\varphi^{N_r(x)}(x)$ is the formula obtained from $\varphi(x)$ by relativizing all quantifiers to $N_r(x)$, that is, by replacing every subformula of the form $\exists y \psi(x, y, \bar{z})$ by $\exists y(d(x, y) \leq r \wedge \psi(x, y, \bar{z}))$ and every subformula of the form $\forall y \psi(x, y, \bar{z})$ by $\forall y(d(x, y) \leq r \rightarrow \psi(x, y, \bar{z}))$. Generally, we consider formulas relativized to neighborhoods around tuples, in particular a formula $\psi(\bar{x})$ of the form $\varphi^{N_r(\bar{x})}(\bar{x})$, for some $\varphi(\bar{x})$, is called r-*local* around $\bar{x}$.

The next theorem is due to Gaifman and states that properties expressible in first-order logic are local.

Theorem 2 (Gaifman [13]). *Every first-order formula is equivalent to a Boolean combination χ of t-local formulas and sentences of the form*

$$\exists x_1 \dots \exists x_m \Big(\bigwedge_{1 \leq i < j \leq m} d(x_i, x_j) > 2r \wedge \bigwedge_{1 \leq i \leq m} \psi(x_i) \Big),$$

for suitable $t, r, m \geq 1$ and an r-local $\psi(x)$.

In [12] the notion of locally tree-decomposable classes of structures was introduced. We use a slightly more restrictive version of this notion (condition (3) is new).

Definition 3. *Let $r, l \geq 1$ and $g : \mathbb{N} \to \mathbb{N}$. A* nice (r, l, g)-tree cover *of a structure $\mathfrak{A}$ is a family $\mathcal{T}$ of subsets of A such that*

(1) For every $a \in A$ there exists a $U \in \mathcal{T}$ such that $N_r^{\mathfrak{A}}(a) \subseteq U$.

(2) for each $U \in \mathcal{T}$ there are less than l many $V \in \mathcal{T}$ such that $U \cap V \neq \emptyset$
(3) for all $U_1, \dots, U_q \in \mathcal{T}$ and $q \geq 1$ we have $tw(\langle U_1 \cup \dots \cup U_q \rangle^{\mathfrak{A}}) \leq g(q)$.

A class $\mathcal{C}$ of structures is nicely locally tree-decomposable, *if there is a linear time algorithm that, given $\mathfrak{A} \in \mathcal{C}$ and $r \geq 1$, computes a nice (r, l, g)-tree cover of $\mathfrak{A}$, for suitable l, g depending only on r.*

The intention of this notion is to cover a structure $\mathfrak{A}$ by subsets U such that each $a \in A$ is covered together with some sufficiently big neighborhood (condition (1)). This makes it possible to evaluate an r-local formula $\varphi(a)$ correctly in $\langle U \rangle^{\mathfrak{A}}$. Additionally, we have a bound on the treewidth of $\langle U \rangle^{\mathfrak{A}}$, which allows efficient model-checking in this part (condition (3)). Finally, the second condition limits the dependency of the different parts of the cover. This is a merely pragmatic definition whose objective is to allow efficient model-checking, while at the same time being general enough to comprise sufficiently rich classes.

Examples of nicely locally tree-decomposable classes ([9,12,11]). There are several important natural classes of structures which are locally tree-decomposable. For instance, the classs of structures of *bounded tree-width* and the classes of structures of *bounded degree*. Another group of important nicely locally tree-decomposable classes are the structures embeddable into some fixed surface, i.e. structures of *bounded genus.*.

In the sequel we always omit the term *nice* and just use *tree cover*. Nevertheless, all covers we use here are actually nice. The starting point for our investigation is the following theorem.

Theorem 4 ([12,11]). *Let $\mathcal{C}$ be a locally tree-decomposable class of structures. Then* FO-MODEL-CHECKING *on $\mathcal{C}$ is in fixed-parameter linear time.*

This result is extended to all previously introduced model-checking problems. In particular, we will show that all four generalized model-checking problems over locally tree-decomposable classes are in fixed-parameter linear time.

Theorem 5. *Let $\mathcal{C}$ be a locally tree-decomposable class of structures. Then* FO-CONSTRUCTION, FO-EVALUATION *and* FO-COUNTING *on $\mathcal{C}$ is in fixed-parameter linear time.*

4   Local Formulas and Tree Covers

Let $k \geq 1$. A *k-distance-type* is an undirected graph $\epsilon := ([k], E^\epsilon)$. For a natural number t and a structure $\mathfrak{A}$ we say that a k-tuple $\bar{a} \in A^k$ *realizes* ϵ w.r.t. $\mathfrak{A}$ and t ($\bar{a} \models_{\mathfrak{A},t} \epsilon$), if there is an edge between distinct vertices i and j of ϵ if, and only if $d(a_i, a_j) \leq 2t + 1$. Note that realization of a k-distance type ϵ w.r.t. t can be defined in first-order logic by a t-local formula $\rho_{t,\epsilon}(\bar{x})$. Now assume that ϵ splits into connected components $\epsilon_1, \dots, \epsilon_p$. We let $\epsilon_{i,\nu}$ stand for the ν-th coordinate of the i-th component (with respect to the natural ordering on the indices). With $\bar{a} \upharpoonright \epsilon_i$ we denote the projection of $\bar{a}$ onto the coordinates contained in ϵ_i.

The next lemma provides a normal form for t-local formulas, which will allow a separate treatment of the variables corresponding to different ϵ-components. The proof is a straightforward application of Ehrenfeucht-Fraïssé games, using the fact that local strategies on non-intersecting substructures extend to strategies on their union (see [8]).

Lemma 6. *Let* $\varphi(\bar{x}) \in \mathrm{FO}$ *be t-local for some $t \geq 1$. Then for every distance-type ϵ with connected components $\epsilon_1, \ldots, \epsilon_p$ we can find a Boolean combination* $F(\bar{\varphi}_1(\bar{x} \upharpoonright \epsilon_1), \ldots, \bar{\varphi}_p(\bar{x} \upharpoonright \epsilon_p))$ *of formulas* $\varphi_{i,j}(\bar{x} \upharpoonright \epsilon_i)$, $1 \leq i \leq p$ *and* $1 \leq j \leq m_i$ *such that*

(1) for all i, the free variables of $\bar{\varphi}_i(\bar{x})$ are among $\bar{x} \upharpoonright \epsilon_i$,
(2) the $\bar{\varphi}_i(\bar{x})$ are t-local around their free variables, and
(3) $\rho_{t,\epsilon}(\bar{x}) \models (\varphi(\bar{x}) \leftrightarrow F(\bar{\varphi}_1(\bar{x}), \ldots, \bar{\varphi}_p(\bar{x})))$

Observe that using Gödel's completeness theorem, we can effectively compute this normal form. Let $\varphi(\bar{x})$ be a t-local formula, $\epsilon = \epsilon_1 \cup \cdots \cup \epsilon_p$ a distance-type and let $F, \varphi_{i,j}(\bar{x} \upharpoonright \epsilon_i)$ denote the formulas obtained by lemma 6. Next we transform $F((\bar{\varphi}_1(\bar{x} \upharpoonright \epsilon_1), \ldots, \bar{\varphi}_p(\bar{x} \upharpoonright \epsilon_p))$ into disjunctive normal form, such that variables from different ϵ-components never occur in the same conjunct. For that, denote the set of satisfying assignments of F by $\mathrm{SAT}(F)$. Given a Boolean assignment $\alpha_{i,j}$ $(1 \leq i \leq p$ and $1 \leq j \leq m_i)$ for F define

$$\varphi_{\bar{\alpha}_i}^{t,\epsilon_i}(\bar{x}) := \rho_{t,\epsilon_i}(\bar{x}) \wedge \bigwedge_{j:\alpha_{i,j}=\text{true}} \varphi_{i,j}(\bar{x}) \wedge \bigwedge_{j:\alpha_{i,j}=\text{false}} \neg\varphi_{i,j}(\bar{x}).$$

To put these formulas together, we let

$$\varphi_{\bar{\alpha}_1,\ldots,\bar{\alpha}_p}^{t,\epsilon}(\bar{x}) := \varphi_{\bar{\alpha}_1}^{t,\epsilon_1}(\bar{x} \upharpoonright \epsilon_1) \wedge \cdots \wedge \varphi_{\bar{\alpha}_p}^{t,\epsilon_p}(\bar{x} \upharpoonright \epsilon_p) \wedge \bigwedge_{1 \leq i < j \leq p} d(\bar{x} \upharpoonright \epsilon_i, \bar{x} \upharpoonright \epsilon_j) > 2t,$$

where $d(\bar{a}, \bar{b}) > s$ means that for all $\nu, \mu : d(a_\mu, b_\nu) > s$. Using these definitions, it is not difficult to see that $\varphi(\bar{x})$ is equivalent to

$$\bigvee_{\epsilon} \bigvee_{(\bar{\alpha}_1,\ldots,\bar{\alpha}_p)\in\mathrm{SAT}(F)} \varphi_{\bar{\alpha}_1,\ldots,\bar{\alpha}_p}^{t,\epsilon}(\bar{x}),$$

which is of the desired form. In a straighforward way, this equivalence gives us an alternative characterization of $\varphi(\mathfrak{A})$, i.e.

$$\varphi(\mathfrak{A}) = \bigcup_{\epsilon} \bigcup_{(\bar{\alpha}_1,\ldots,\bar{\alpha}_p)} \varphi_{\bar{\alpha}_1,\ldots,\bar{\alpha}_p}^{t,\epsilon}(\mathfrak{A}). \tag{1}$$

Since both unions are disjoint, it is convenient to assume henceforth that the input formula $\varphi(\bar{x})$ is of the form $\varphi_{\bar{\alpha}_1,\ldots,\bar{\alpha}_p}^{t,\epsilon}(\bar{x})$ for some ϵ, i.e.

$$\varphi(\bar{x}) = \varphi_1(\bar{x} \upharpoonright \epsilon_1) \wedge \cdots \wedge \varphi_p(\bar{x} \upharpoonright \epsilon_p) \wedge \bigwedge_{1 \leq i < j \leq p} d(\bar{x} \upharpoonright \epsilon_i, \bar{x} \upharpoonright \epsilon_j) > 2t + 1,$$

with $\varphi_i(\bar{x} \upharpoonright \epsilon_i)$ being t-local around $\bar{x} \upharpoonright \epsilon_i$. To introduce the input structure appropriately, let $\mathcal{T} = \{U_1, \dots, U_m\}$ be a nice (r, l, g)-tree cover of the input structure $\mathfrak{A}$, for $r := k(2t + 1)$ and appropriate $l \geq 1, g : \mathbb{N} \to \mathbb{N}$ (recall that k is the number of free variables).

We use capital letters for the indices $I \in [m]$ over the tree cover. These indices often occur as tuples, which will be indexed by lower-case letters (e.g. $I_j \in [m]$).

Including the cover: For a set $X \subseteq A$ and $s \geq 1$ we define $K^s(X)$ (the s-kernel of X) to be the set of vertices $a \in X$ such that $N_s^{\mathfrak{A}}(a) \subseteq X$. We have chosen r to assure that, if for some $\bar{a}$ and J we have $\bar{a} \upharpoonright \epsilon_i \cap K^r(U_J) \neq \emptyset$ then $\bar{a} \upharpoonright \epsilon_i \subseteq K^t(U_J)$, hence $\langle U_J \rangle^{\mathfrak{A}} \models \varphi_i(\bar{a} \upharpoonright \epsilon_i)$ iff $\mathfrak{A} \models \varphi_i(\bar{a} \upharpoonright \epsilon_i)$. This follows from the t-locality of $\varphi_i(\bar{x} \upharpoonright \epsilon_i)$. Now it is easy to see that

$$\varphi(\mathfrak{A}) = \bigcup_{(J_1, \dots, J_p) \in [m]^p} A_{J_1, \dots, J_p}, \tag{2}$$

where we let

$$A_{J_1, \dots, J_p} := \left\{ \bar{a} \mid \bar{a} \upharpoonright \epsilon_i \cap K^r(U_{J_i}) \neq \emptyset \text{ and } \langle U_{J_1} \cup \dots \cup U_{J_p} \rangle^{\mathfrak{A}} \models \varphi(\bar{a}) \right\}. \tag{3}$$

To capture the relevant topological information of a tree cover $\mathcal{T}$, we introduce the graph $c(\mathcal{T}) := ([m], E^{c(\mathcal{T})})$ with $E^{c(\mathcal{T})} := \{(I, J) \mid U_I \cap U_J \neq \emptyset\}$. We let the p-*ctype* (contiguousness-type) of a tuple $(J_1, \dots, J_p)$ w.r.t. $\mathcal{T}$ be the graph $\kappa = ([p], E^\kappa)$ such that $(i, j) \in E^\kappa$ if, and only if, $(J_i, J_j) \in E^{c(\mathcal{T})}$ (and write $(J_1, \dots, J_p) \models \kappa$). By $\mathrm{Mod}(\kappa)$ we denote the set of tuples of indices that *realize* κ: $\mathrm{Mod}(\kappa) := \{(J_1, \dots, J_p) \mid (J_1, \dots, J_p) \models \kappa\}$.

Assume that we have given a tuple of indices $(J_1, \dots, J_p) \in [m]^p$ and that this tuple realizes the p-ctype κ. It is easy to see that if i and j belong to different κ-components, then the distance between elements of their respective kernels $K^r(U_{J_i}), K^r(U_{J_j})$ is $> 2r$. This fact will be the key for the next characterization of $\varphi(\mathfrak{A})$.

Convention: In the remainder, let κ be a p-ctype with connected components $\kappa_1, \dots, \kappa_q$. For convenience, we arrange the coordinates of a tuple $(J_1, \dots, J_p)$ realizing κ according to the κ-components they belong to. Tacitly assuming that the ordering remains unchanged, we will write $(\bar{J}_1, \dots, \bar{J}_q)$ for the rearranged tuple; e.g. $A_{\bar{J}_1, \dots, \bar{J}_q}$ refers to the set defined in equation (3), although the indices are permuted. Furthermore, we appoint the convention that if we write J_j, this refers to the j'th coordinate of $\bar{J}$, while $J_{i,\nu}$ refers to ν'th coordinate contained in κ_i.

Let $\epsilon(\kappa_i) := \bigcup_{j \in \kappa_i} \epsilon_j$ be the union of the ϵ-components contained in κ_i. Then a straightforward restriction of the definition of $A_{J_1, \dots, J_p}$ (formula (3)) to the indices corresponding to the components $\epsilon(\kappa_i)$ gives rise to the following definition:

$$A^i_{\bar{J}_i} := \{ \bar{a} \upharpoonright \epsilon(\kappa_i) \mid \text{for all } j, \mu \text{ such that } j \text{ is the } \mu\text{'th element of } \kappa_i,$$

$$\bar{a} \upharpoonright \epsilon_j \cap K^r(U_{J_{i,\mu}}) \neq \emptyset \text{ and } \mathfrak{A} \models \rho_{t, \epsilon(\kappa_i)}(\bar{a} \upharpoonright \epsilon(\kappa_i)) \wedge \bigwedge_{j \in \kappa_i} \varphi_j(\bar{a} \upharpoonright \epsilon_j) \}.$$

Observe that these sets can be computed in the substructure of $\mathfrak{A}$ induced by the set $U_{\bar{J}_i} := \bigcup_{\nu=1}^{w} U_{J_{i,\nu}}$, where w is the size of the tuple $\bar{J}_i$. As a matter of fact, $A^i_{\bar{J}_i}$ is the set of tuples satisfying the formula

$$\rho_{t,\epsilon(\kappa_i)}(\bar{x} \restriction \epsilon(\kappa_i)) \wedge \bigwedge_{j \in \kappa_i} \left(\varphi_j(\bar{x} \restriction \epsilon_j) \wedge \bigvee_{\nu \in \epsilon_j} P_\nu x_\nu \right),$$

where the P_ν are new unary relation symbols, interpreted by $K^r(U_{J_\nu})$. Observe that, by t-locality, this formula can be evaluated correctly[2] in $\langle U_{\bar{J}_i} \rangle^{\mathfrak{A}}$. These local parts $A^i_{\bar{J}_i}$ are the projections of $A_{\bar{J}_1,\dots,\bar{J}_q}$. The goal of the foregoing investigation was to show that the converse also holds.

Lemma 7. *Let κ be a ctype as above and $(\bar{J}_1, \dots, \bar{J}_q)$ a tuple of indices realizing κ. Then (recall (3) and the convention) we have $A_{\bar{J}_1,\dots,\bar{J}_q} = A^1_{\bar{J}_1} \times \cdots \times A^q_{\bar{J}_q}$.*

Using this, we can rewrite equation (2) as follows:

$$\varphi(\mathfrak{A}) = \bigcup_\kappa \bigcup_{(\bar{J}_1,\dots,\bar{J}_q) \models \kappa} A_{\bar{J}_1,\dots,\bar{J}_q} = \bigcup_\kappa \bigcup_{(\bar{J}_1,\dots,\bar{J}_q) \models \kappa} A^1_{\bar{J}_1} \times \cdots \times A^q_{\bar{J}_q}.$$

This characterization already allows to build up $\varphi(\mathfrak{A})$ from parts that are locally computable. Next we characterize the admissable index-tuples $(\bar{J}_1, \dots, \bar{J}_q)$ in the foregoing equation.

The contiguousness-graph: Let κ be a p-ctype with components $\kappa_1, \dots, \kappa_q$. The *cgraph* (contiguousness-graph) $c(\mathcal{T}, \kappa)$ of $\mathcal{T}$ with respect to κ has vertex set $\text{Mod}(\kappa_1) \dot{\cup} \cdots \dot{\cup} \text{Mod}(\kappa_q)$ and edge relation $E^{c(\mathcal{T},\kappa)}$ defined by $(\bar{J}_i, \bar{J}_j) \in E^{c(\mathcal{T},\kappa)}$ iff there are $\nu, \mu : (J_{i,\nu}, J_{j,\mu}) \in E^{c(\mathcal{T})}$. Additionally, we give the vertices corresponding to $\text{Mod}(\kappa_i)$ the color i, or more formally, we add unary relations $C_1, \dots, C_q$ with $C_i^{c(\mathcal{T},\kappa)} := \text{Mod}(\kappa_i)$. Then it is easy to see that $(\bar{J}_1, \dots, \bar{J}_q) \models \kappa$ iff $(\bar{J}_1, \dots, \bar{J}_q)$ is independent in $c(\mathcal{T}, \kappa)$. So our problem essentially reduces to the problem of finding independent sets in $c(\mathcal{T}, \kappa)$. The next lemma summarizes the combinatorial properties of $c(\mathcal{T}, \kappa)$ relevant for this task.

Lemma 8. *Let $\mathcal{T}$ be a (r, l, g)-tree cover of $\mathfrak{A}$. Then for all κ:*

(1) $c(\mathcal{T}, \kappa)$ can be calculated in time $O(\|\mathcal{T}\|)$.

(2) $c(\mathcal{T}, \kappa)$ has maximum degree bounded by some function of l, k.

(3) for all $i = 1, \dots, q$: $\{U_{\bar{J}_i} \mid \bar{J}_i \in C_i^{c(\mathcal{T},\kappa)}\}$ is a $(r, l', g \circ h_i)$-tree cover of $\mathfrak{A}$, where $h_i(n) := |\kappa_i| \cdot n$ and l' depends on l and p.

---

[2] Note that, even if some $x_\nu \in \bar{x} \restriction \epsilon_i$ does not occur freely in φ_i, it must occur in the set. This is against the convention made in the preliminaries, but is necessary here.

5   The Construction and the Listing Problem

In this section we present both an algorithm for the FO-listing and the FO-construction problem. Essentially, we proceed as follows: w.l.o.g. we are given a t-local $\varphi(\bar{x}) \in$ FO of the form $\varphi^{t,\epsilon}_{\bar{\alpha}_1,\dots,\bar{\alpha}_p}(\bar{x})$. The reduction to this setting was described previously. After computing a suitable tree cover $\mathcal{T}$ of the input structure $\mathfrak{A}$, we fix a ctype κ and compute $c(\mathcal{T}, \kappa)$. Now elements of the sought set $\varphi(\mathfrak{A})$ roughly correspond to independent tuples of $c(\mathcal{T}, \kappa)$.

Theorem 9. *Let $\mathcal{C}$ be a nicely locally tree-decomposable class of structures. Then FO-EVALUATION on $\mathcal{C}$ is in fixed-parameter linear time, i.e. there is an algorithm that for an FO-formula $\varphi(\bar{x})$ and a structure $\mathfrak{A} \in \mathcal{C}$ computes $\varphi(\mathfrak{A})$ in time*

$$f(\|\varphi\|) \cdot (\|\mathfrak{A}\| + |\varphi(\mathfrak{A})|),$$

for some function $f : \mathbb{N} \to \mathbb{N}$.

The algorithm evaluating FO-queries looks as follows:

Algorithm 1: Computing $\varphi(\mathfrak{A})$

Input: Structure $\mathfrak{A} \in \mathcal{C}$, FO-formula $\varphi(\bar{x})$
Output: $\varphi(\mathfrak{A})$

1. Compute the Boolean combination $\varphi'(\bar{x}) = F(\varphi_1(\bar{x}), \dots, \varphi_{l_1}(\bar{x}), \psi_1, \dots, \psi_{l_2})$ equivalent to $\varphi(\bar{x})$, where all $\varphi_i(\bar{x})$ are t-local around $\bar{x}$.
2. For all $i = 1, \dots, l_2$ decide, if $\mathfrak{A} \models \psi_i$ and replace ψ_i in $\varphi'(\bar{x})$ with its truth value. Denote the resulting formula with $\varphi''(\bar{x})$.
3. Compute a nice $(k(2t+1), l, g)$-tree cover $\mathcal{T}$ for suitable $l \geq 1, g : \mathbb{N} \to \mathbb{N}$.
4. For all distance-types $\epsilon = \epsilon_1 \cup \dots \cup \epsilon_p$
      i calculate $F(\bar{\varphi}_1(\bar{x}), \dots, \bar{\varphi}_p(\bar{x}))$ from lemma 6 for the formula $\varphi''(\bar{x})$
      ii For all $(\bar{\alpha}_1, \dots, \bar{\alpha}_p) \in \mathrm{SAT}(F)$, compute $\varphi^{t,\epsilon}_{\bar{\alpha}_1,\dots,\bar{\alpha}_p}(\mathfrak{A})$
5. Return $\bigcup_\epsilon \bigcup_{\bar{\alpha}} \varphi^{t,\epsilon}_{\bar{\alpha}_1,\dots,\bar{\alpha}_p}(\mathfrak{A})$

The first line needs time linear in $\|\varphi'(\bar{x})\|$, which itself only depends on $\|\varphi(\bar{x})\|$. By definition, a nice $(k(2t+1), l, g)$-tree cover can be computed in time $O(\|\mathfrak{A}\|)$, the same holds for the decision of $\mathfrak{A} \models \chi_i$ (by theorem 4). Thus line 2 and line 3 need time linear in the structure size.

The loop in line 4 is the crucial one. First, the number of different ϵ and $(\bar{\alpha}_1, \dots, \bar{\alpha}_p)$ is constant (depends only on k). Hence, the subroutine computing $\varphi^{t,\epsilon}_{\bar{\alpha}_1,\dots,\bar{\alpha}_p}(\mathfrak{A})$ is called constantly often. Line 4(i) needs time only depending on $\|\varphi\|$.

Now consider line 4(ii), where we compute $\varphi^{t,\epsilon}_{\bar{\alpha}}(\mathfrak{A})$. This set coincides with $\bigcup_{(\bar{J}_1,\dots,\bar{J}_q)} A^1_{\bar{J}_1} \times \dots \times A^q_{\bar{J}_q}$, where the tuples admitted in the union are those independent in $\mathcal{G} := c(\mathcal{T}, \kappa)$. So we proceed as follows: in a first step, we compute $\mathcal{G}$ (by lemma 8 this takes linear time), remove all vertices $\bar{J}_i \in C^{\mathcal{G}}_i$ with $A^i_{\bar{J}_i} = \emptyset$ and get $\mathcal{G}'$ (since the sets $U_{\bar{J}_i}$ do not intersect very much, this can be done in linear time by an application of theorem 1 and lemma 8). Then in a second step, we collect all $(\bar{J}_1, \ldots, \bar{J}_q)$ independent in $\mathcal{G}'$. Since $\mathcal{G}$ has bounded degree, this can be done in time linear in the number of independent tuples. Now, computing each set $A^i_{\bar{J}_i}$ only once, we build the sought union over the Cartesian products. Altogether this takes time $O(|\varphi(\mathfrak{A})|)$. The crucial observations to obtain this tight time bound are (i) that each $\bar{a} \in \varphi(\mathfrak{A})$ is contained in at most a constant number of sets $A_{\bar{J}_1, \ldots, \bar{J}_q}$ and (ii) that each $A^i_{\bar{J}_i}$ is only computed once.   □

The task to find a witness for a formula reduces to finding an independent tuple $(\bar{J}_1, \ldots, \bar{J}_q)$ in the cgraph $c(\mathcal{T}, \kappa)$ for some κ. This is very easy in graphs of bounded degree.

Theorem 10. *Let C be a nicely tree-decomposable class of structures. Then* FO-CONSTRUCTION *on C is in fixed-parameter linear time.*

6   The Counting Problem

This last section is dedicated to the counting problem. Remember that we assume that arithmetical operations can be done in constant time.

Theorem 11. *Let C be a nicely tree-decomposable class of structures. Then* FO-COUNTING *on C is in fixed-parameter linear time.*

Since the disjunctions in formula (1) were disjoint, we get for all t-local formulas $\varphi(\bar{x})$: $|\varphi(\mathfrak{A})| = \sum_{\epsilon} \sum_{(\bar{\alpha}_1, \ldots, \bar{\alpha}_p)} |\varphi^{t,\epsilon}_{\bar{\alpha}_1, \ldots, \bar{\alpha}_p}(\mathfrak{A})|$. There are at most constantly many different $\epsilon, \bar{\alpha}$, so we again restrict our attention to formulas of the form $\varphi^{t,\epsilon}_{\bar{\alpha}_1, \ldots, \bar{\alpha}_p}(\bar{x})$. Let $\varphi(\bar{x})$ be of this form, for a distance type ϵ with components $\epsilon_1, \ldots, \epsilon_p$ and a suitable Boolean assignment $(\bar{\alpha}_1, \ldots, \bar{\alpha}_p)$. Like always, k denotes the number of free variables of $\varphi(\bar{x})$ and let $\mathcal{T} = \{U_1, \ldots, U_m\}$ be a nice (r, l, g)-tree cover for $\mathfrak{A}$ (again we choose $r := k(2t + 1)$).

We let $A_{\mathcal{J}} := \bigcap_{J \in \mathcal{J}} A_J$ for an index set $\mathcal{J}$. The next equation characterizes the cardinality of $\varphi(\mathfrak{A})$, and is a consequence of equation (2) and the principle of inclusion and exclusion.

$$|\varphi(\mathfrak{A})| \overset{(2)}{=} \Bigl| \bigcup_{(J_1, \ldots, J_p) \in [m]^p} A_{J_1, \ldots, J_p} \Bigr| \overset{\text{PIE}}{=} \sum_{\emptyset \neq \mathcal{J} \subseteq [m]^p} (-1)^{|\mathcal{J}|+1} |A_{\mathcal{J}}|. \qquad (4)$$

To be able to separate results of remote parts, we pass from sets of indices to their projections. Formally, for $\emptyset \neq \mathcal{J} \subseteq [m]^p$ we define $\mathcal{J} \restriction j := \{J_j \mid$ there is a $(J_1, \ldots, J_j, \ldots, J_p) \in \mathcal{J}\}$, the projection of $\mathcal{J}$ onto coordinate j. Naturally, we extend the definition of $A_{J_1, \ldots, J_p}$ (formula (3)) to sets of indices by $A_{\mathcal{J}_1, \ldots, \mathcal{J}_p} := A_{\mathcal{J}_1 \times \cdots \times \mathcal{J}_p}$. Observe here that we have $A_{\mathcal{J}} = A_{\mathcal{J} \restriction 1, \ldots, \mathcal{J} \restriction p}$ for all $\mathcal{J} \subseteq [m]^p$. Since we want to proceed like in the listing case, we perform a couple of arithmetical reductions until we can separate independent parts again (as we did before). Essentially, we group the indices $\mathcal{J}$ of the sum (4) into classes, such that all elements $\mathcal{J}$ of the same class induce the same tuple $(\mathcal{J} \restriction 1, \ldots, \mathcal{J} \restriction p)$. Since for such tuples we get a unique value $|A_{\mathcal{J}}|$, we have to consider each of these only once (of course weighted by the size of the class).

For these separate coordinates we can establish a contiguousness relation, very much like in the listing case. Informally, we get $|A_{\mathcal{J}_{[1}, \ldots, \mathcal{J}_{]p}}| = |A_{\bar{\mathcal{J}}_1}| \cdots |A_{\bar{\mathcal{J}}_q}|$ for tuples $\mathcal{J}_1, \ldots, \mathcal{J}_p$ satisfying an (extended) ctype κ (compare lemma 7) with q components. Altogether, this allows a rewriting as is expressed in the next lemma.

Lemma 12. *There is a linear time algorithm that computes, given a formula $\varphi_{\bar{\alpha}_1, \ldots, \bar{\alpha}_p}^{t, \epsilon}(\bar{x})$ and a tree cover $\mathcal{T}$ as above, a colored graph $(\mathcal{G}, C_1, \ldots, C_q)$ of bounded degree and integers $\gamma(v) \in \mathbb{N}$ for all $v \in G$ such that*

$$|\varphi(\mathfrak{A})| = \sigma(\mathcal{G}, \gamma) := \sum_{\substack{\bar{v} \text{ independent} \\ v_i \in C_i}} \gamma(v_1) \cdots \gamma(v_q).$$

The next theorem fills the gap still missing in the proof of theorem 11. If we plug in this algorithm into the gap left open above, then the proof of theorem 11 is completed.

Theorem 13. *There is an algorithm that, given $\mathcal{G}$ and γ as in the last lemma, computes $\sigma(\mathcal{G}, \gamma)$ in time $O(\|\mathcal{G}\|)$.*

References

1. A.V. Aho, J.E. Hopcroft, and J.D. Ullman. *The design and analysis of computer algorithms.* Addison-Wesley, 1974.
2. S. Arnborg, J. Lagergren, and D. Seese. Easy problems for tree-decomposable graphs. *Journal of Algorithms*, 12:308–340, 1991.
3. B.S. Baker. Approximation algorithms for NP-complete problems on planar graphs. In *Proceedings of the 24th IEEE Symposium on Foundations of Computer Science*, pages 265–273, 1983.
4. Hans L. Bodlaender. A tourist guide through treewidth. *Acta Cybernetica*, 11:1–21, 1993.
5. B. Courcelle. Graph rewriting: An algebraic and logic approach. In J. van Leeuwen, editor, *Handbook of Theoretical Computer Science*, volume 2, pages 194–242. Elsevier Science Publishers, 1990.
6. R.G. Downey and M.R. Fellows. *Parameterized Complexity.* Monographs in Computer Science. Springer-Verlag, 1999.
7. R.G. Downey, M.R. Fellows, and U. Taylor. On the parametric complexity fo relational databases queries and a sharper characterization of $W[1]$. In D. Bridges, C. Calude, J. Gibbons, S. Reeves, and I. Witten, editors, *Combinatorics, Complexity and Logic*, DMTCS, pages 194–213. Springer-Verlag, 1996.
8. H.-D. Ebbinghaus and J. Flum. *Finite Model Theory.* Springer Verlag, 1995.
9. D. Eppstein. Diameter and treewidth in minor-closed graph families. *Algorithmica*, 27(3):275–291, 1999.
10. J. Flum, M. Frick, and M. Grohe. Query evaluation via tree-decompositions. In Jan Van den Bussche and Victor Vianu, editors, *Database Theory - ICDT 2001, 8th International Conference, London, UK, January 4-6, 2001, Proceedings*, volume 1973 of *Lecture Notes in Computer Science*, pages 22–38. Springer, 2001.
11. M. Frick. *Easy Instances for Model Checking.* PhD thesis, Universität Freiburg, http://www.freidok.uni-freiburg.de/volltexte/229, 2001.

12. M. Frick and M. Grohe. Checking first-order properties of tree-decomposable graphs. In Jirí Wiedermann, Peter van Emde Boas, and Mogens Nielsen, editors, *26th International Colloquium, ICALP'99, Prague*, volume 1644 of *Lecture Notes in Computer Science*, pages 331–340. Springer, 1999.

13. H. Gaifman. On local and non-local properties. In J. Stern, editor, *Logic Colloquium '81*, pages 105–135. North Holland, 1982.

14. M. R. Garey and D. S. Johnson. *Computers and Intractability - A Guide to the Theory of NP-Completeness*. W. H. Freeman and Co., 1979.

15. M. Grohe. Generalized model-checking problems for first-order logic. In A. Ferreira and H. Reichel, editors, *Proceedings of the 18th Annual Symposium on Theoretical Aspects of Computer Science*, volume 2010 of *Lecture Notes in Computer Science*, pages 12–26. Springer Verlag, Berlin, 2001.

16. M. Jerrum, L. Valiant, and V. Vazirani. Random generation of combinatorial structures from a uniform distribution. *Theoretical Computer Science*, 43:169–188, 1986.

17. L.G. Valiant. The complexity of combinatorial computations: An introduction. In *GI - 8. Jahrestagung*, number 16 in Informatik-Fachberichte, pages 326–337. Springer, 1978.

18. M.Y. Vardi. The complexity of relational query languages. In *Proceedings of the 14th ACM Symposium on Theory of Computing*, pages 137–146, 1982.

Learnability and Definability in Trees and Similar Structures

Martin Grohe[1] and Gyorgy Turán[2]*

[1] Laboratory for Foundations of Computer Science, University of Edinburgh,
Edinburgh EH9 3JZ, Scotland, UK. grohe@dcs.ed.ac.uk.
[2] Department of Mathematics, Statistics, and Computer Science, University of Illinois at
Chicago, 851 S. Morgan Street, Chicago, IL 60607-7045, USA and Research Group on AI of the
Hungarian Academy of Sciences, Szeged, Hungary. gyt@uic.edu.

Abstract. We prove upper bounds for combinatorial parameters of finite relational structures, related to the complexity of learning a definable set. We show that monadic second order (MSO) formulas with parameters have bounded VC-dimension over structures of bounded clique-width, and first-order formulas with parameters have bounded VC-dimension over structures of bounded local clique-width (this includes planar graphs). We also show that MSO formulas of a fixed size have bounded strong consistency dimension over MSO formulas of a fixed larger size, for colored trees. These bounds imply positive learnability results for the PAC and equivalence query learnability of a definable set over these structures. The proofs are based on bounds for related definability problems for tree automata.

1 Introduction

The general problem of *concept learning* is to identify an unknown set from a given family of sets. The given family, referred to as the *concept class*, represents the possible classifications of the elements of the universe, and the unknown set, also called the *target concept*, is the actual classification. The identification can be exact or approximate (for example, in a probabilistic sense, as in the model of Probably Approximately Correct (PAC) learning). In order to specify a formal model of learning, one also has to determine what type of information is available to the learner (for example, random examples or certain types of queries), and what complexity measures are considered (for example, sample size, number of queries or computation time).

It is an interesting fact that several notions of learning complexity turn out to be closely related to certain *combinatorial parameters* of the concept class. The best known example is sample size for PAC learning, which is of the same order of magnitude as the *Vapnik-Chervonenkis dimension* or *VC-dimension* of the concept class (see, e.g., [16]). Other examples include the query complexity of learning with equivalence and membership queries, closely related to *certificate size* [14], and the query complexity of learning with equivalence queries, closely related to the *strong consistency dimension* [4]. Determining these combinatorial parameters for specific concept classes thus gives useful information about learning complexity.

---

* Partially supported by NSF grant CCR-9800070 and OTKA grant T-025271.

H. Alt and A. Ferreira (Eds.): STACS 2002, LNCS 2285, pp. 645–657, 2002.
© Springer-Verlag Berlin Heidelberg 2002

The measures mentioned above are related to the *informational complexity* of learning and not to its *computational complexity*. There are also results relating learning complexity to the complexity of *computational problems* associated with the concept class, such as the *hypothesis finding problem* for PAC learning (see [16]) and the *representation problem* for learning with equivalence and membership queries [1].

In this paper we consider some of the informational complexity measures for learning in the context of *predicate logic*, which is a frequently used framework in machine learning, besides, for example, propositional logic and neural networks.

A general setup for predicate logic learning is to assume that examples are elements (or, more generally, tuples of elements) of a *finite structure*, and concepts are represented by a *class of predicate logic formulas*. This framework can be appropriate when the data for learning are provided by a relational database, but some other models can also be represented in this way. For instance, a standard setup for *inductive logic programming* is to learn a single non-recursive Horn clause with ground background knowledge (see, e.g., [20]). This is equivalent to learning an existential sentence having a conjunction of atoms as its quantifier free part (also called a conjunctive query), over the finite structure formed by the ground atoms in the background knowledge.

In the predicate logic framework, questions about the learning-related combinatorial parameters lead to combinatorial questions about classes of *definable sets*, which are much studied in model theory, mostly for infinite structures [15]. In fact, one of the three simultaneous sources for the notion of VC-dimension is [27] in model theory (besides [25] and [31]). The relationship between our results and the related results in model theory is discussed at the end of Section 6. A model theoretic approach to learnability was proposed in [21], and some results for finite models are given in [19,29]. In particular, [19] gives an explicit upper bound for the VC-dimension of first-order formulas over structures with unary predicates and a single unary function.

The *model checking* problem for finite models is to decide, given a finite structure and a formula with a variable assignment, whether the formula holds in the structure. There are several results for this problem, putting together a picture of the borderline between tractable and intractable cases [12]. These results establish connections between the *expressiveness* of a logic and the *computational complexity* of computational problems associated with the logic. Our results show that for the learning complexity we get a very similar borderline between tractable and intractable cases. On a more technical level, our results show that the techniques developed for the model checking problem, in particular the use of tree automata and the notions of graph width, are also useful in the learning context.

We show that the class of sets definable by *monadic second-order (MSO)* formulas with parameters has bounded VC-dimension for classes of structures of bounded *clique-width*. As a tool for proving this result, we introduce tree automata versions of definability in a structure. These notions appear to be new, and may be of some interest in themselves. On the other hand, we note that MSO-formulas can have unbounded VC-dimension on *grids*. We also show that *first-order (FO)* formulas with parameters have bounded VC-dimension for classes of structures of bounded *local clique-width*. This includes, for example, the class of *planar graphs* and all classes of graphs of *bounded degree*. The main step in proving these results is a lemma that may be of interest in itself; it states that bounded VC-dimension of first-order formulas is a *local* property of structures, that is, it only depends on bounded radius neighborhoods of elements of the structures. All these bounded VC-dimension results imply bounded sample complexity for PAC-learnability using the standard results of computational learning theory (see [16]).

Besides discussing the VC-dimension, we also give some initial results on the strong consistency dimension [4], a much less studied and understood notion. The potential relevance of the strong consistency dimension for logic is discussed in Balcázar [3]. We show that on colored trees, fixed size MSO formulas with one free variable and several parameters have a bounded strong consistency dimension with respect to MSO formulas of some fixed, larger size. Using the characterization given by [4], this implies that on trees, such MSO formulas can be learned using *logarithmically many* (in the size of the underlying structure) *equivalence queries* that are larger, but fixed size MSO formulas. This, perhaps somewhat surprising, result is among the first applications of the strong consistency dimension for proving positive learnability results. We note that the result is not practical, for two reasons: it does not provide an efficient algorithm to compute the queries, and it involves huge constants. [19] gives a computationally efficient query learning algorithm for quantifier-free formulas over structures with unary predicates and functions.

The paper is organized as follows. Sections 2 and 3 give background for definability and the VC-dimension. Sections 4 and 5 contain the automaton and MSO definability results for trees, respectively, for graphs of bounded clique-width and tree-width. First-order definability results for structures of bounded local clique-width and tree-width are given in Section 6. The strong consistency dimension results are presented in Section 7. Finally, Section 8 contains some further remarks and open problems. Due to space limitations, we have to omit proofs here and refer the reader to the full version of this paper [13].

We thank Michael Benedikt for clarifying the relation between our Corollary 9 and the results of [5].

2   Definability

A *vocabulary* is a finite set of relation symbols. In the following, τ always denotes a vocabulary. A *τ-structure* $\mathcal{A}$ consists of a non-empty set A, called the *universe* of $\mathcal{A}$, and a relation $R^{\mathcal{A}} \subseteq A^r$ for every r-ary relation symbol $R \in \tau$. For a vocabulary $\tau' \supseteq \tau$, a *τ'-expansion* of a τ-structure $\mathcal{A}$ is a τ'-structure $\mathcal{A}'$ with universe $A' = A$ and $R^{\mathcal{A}'} = R^{\mathcal{A}}$ for all $R \in \tau$. *Unless explicitly mentioned otherwise, in this paper, we will only consider structures whose universe is finite.*

An *atomic formula*, or *atom*, is a formula of the form $x = y$ or $Rx_1 \ldots x_r$, where R is an r-ary relation symbol and $x, y, x_1, \ldots, x_r$ are *(individual) variables*. The formulas of *first-order logic* are built up from atomic formulas using the usual Boolean connectives and existential and universal quantification over the elements of the universe of a structure. The class of all formulas of first-order logic is denoted by FO.

Monadic second-order logic is the extension of first-order logic allowing quantification not only over elements of the universe of a structure, but also over subsets of the universe. Formally, we have two types of variables — *individual variables*, which are interpreted by elements of the universe of a structure, and *set variables*, which are interpreted by subsets of the universe of a structure. In addition to the first-order atoms, in monadic second-order logic we also have atoms Xx saying that the element interpreting the individual variable x is contained in the set interpreting the set variable X. Furthermore, we have existential and universal quantification over both individual and set variables. MSO denotes the set of all formulas of monadic second-order logic.

We always use lowercase letters $x, y, \ldots$ to denote individual variables and uppercase letters $X, Y, \ldots$ to denote set variables. A *free variable* of a formula φ is an (individual or set) variable v that does not occur in the scope of a quantifier $\exists v$ or $\forall v$. The set of free variables of a formula φ is denoted by free(φ). A *sentence* is a formula without free variables. We write $\varphi(X_1, \ldots, X_k, x_1, \ldots, x_l)$ to indicate that free(φ) $\subseteq \{X_1, \ldots, X_k, x_1, \ldots, x_l\}$. For a structure $\mathcal{A}$, subsets $A_1, \ldots, A_k \subseteq A$, and elements $a_1, \ldots, a_l \in A$ we write $\mathcal{A} \models \varphi(A_1, \ldots, A_k, a_1, \ldots, a_l)$ to denote that $\mathcal{A}$ satisfies φ if the set variables $X_1, \ldots, X_k$ are interpreted by $A_1, \ldots, A_k$, respectively, and the individual variables $x_1, \ldots, x_l$ are interpreted by $a_1, \ldots, a_l$, respectively. We mostly consider formulas that only have free individual variables.

For a formula $\varphi(x_1, \ldots, x_k, y_1, \ldots, y_l)$, a structure $\mathcal{A}$, and elements $b_1, \ldots, b_l \in A$ we let $\varphi(\mathcal{A}, b_1, \ldots, b_l) = \{(a_1, \ldots, a_k) \in A^k \mid \mathcal{A} \models \varphi(a_1, \ldots, a_k, b_1, \ldots, b_l)\}$. We call $\varphi(\mathcal{A}, b_1, \ldots, b_l)$ the set defined by φ in $\mathcal{A}$ with *parameters* $b_1, \ldots, b_l$. We let $\mathcal{C}(\varphi, \mathcal{A}) = \{\varphi(\mathcal{A}, b_1, \ldots, b_l) \mid b_1, \ldots, b_l \in A\}$. We often denote tuples $(a_1, \ldots, a_k)$ of elements of a set A by $\bar{a}$, and we write $\bar{a} \in A$ instead of $\bar{a} \in A^k$. Similarly, we denote tuples of individual variables by $\bar{x}$. For tuples $\bar{a}$ and $\bar{b}$, we write $\bar{a}\bar{b}$ to denote their concatenation.

The *quantifier-rank* of a first-order or monadic second-order formula φ, denoted by qr(φ), is the maximum number of nested quantifiers in φ. It is easy to see that for all $q, \ell \geq 0$, up to logical equivalence there are only finitely many first-order or monadic second-order formulas φ with qr(φ) $\leq q$ and $|\text{free}(\varphi)| \leq \ell$. The *size* of a formula φ is denoted by $\|\varphi\|$.

3   Vapnik–Chervonenkis Dimension

Let V be a set and $\mathcal{C} \subseteq 2^V$ a family of subsets of V, also referred to as a *concept class*. For a subset $U \subseteq V$, we let $\mathcal{C} \cap U = \{C \cap U \mid C \in \mathcal{C}\}$. The set U is *shattered* by $\mathcal{C}$ if $\mathcal{C} \cap U = 2^U$. The *Vapnik-Chervonenkis dimension*, or *VC-dimension*, of $\mathcal{C}$, denoted by VC($\mathcal{C}$), is the maximum of the sizes of the shattered subsets of V, or ∞ if this maximum does not exist.

An important result proved in [6,31] is that the Vapnik-Chervonenkis dimension characterizes the sample complexity needed for learning the concept class $\mathcal{C}$ in the PAC model of learning.

For a formula $\varphi(\bar{x}, \bar{y})$ and a structure $\mathcal{A}$ we let VC($\varphi, \mathcal{A}$) = VC($\mathcal{C}(\varphi, \mathcal{A})$). We say that a formula $\varphi(\bar{x}, \bar{y})$ has *bounded VC-dimension* on a class K of structures if there is a k such that for every $\mathcal{A} \in K$ we have VC($\varphi, \mathcal{A}$) $\leq k$. The following standard example shows that even very simple formulas can have unbounded VC-dimension if we do not put any restriction on the structures considered.

Example 1. Let τ consist of a single binary relation E for graph adjacency, and consider the formula $\varphi = E(x, y)$. Let $\mathcal{G}_n$ be the $(n + 2^n)$-vertex graph, where for each subset of the first n vertices, there is a distinct vertex which is connected to just the vertices in this subset. Then clearly $VC(\varphi, \mathcal{G}_n) \geq n$.

In this paper, we shall show that MSO-formulas and FO-formulas have bounded VC-dimension on a variety of classes of structures. This following general result shows that in order to prove the boundedness of the VC-dimension for a class of structures, one may restrict attention to formulas with a single free variable.

Lemma 2 ([26]). *Let K be a class of structures such that every first-order formula $\varphi(x, \bar{y})$ has bounded VC-dimension on K. Then every first-order formula $\varphi(\bar{x}, \bar{y})$ has bounded VC-dimension on K.*

The analogous statement holds for formulas of monadic second-order logic.

Laskowski [17] gave a purely combinatorial proof of this result, which yields an explicit upper bound for the VC-dimension of $\varphi(\bar{x}, \bar{y})$. This is important for us, because in our results we also provide explicit bounds.

Usually, the lemma is only stated for FO, but it is easy to verify that Laskowski's proof goes through for MSO, and actually for any class of formulas that is closed under renaming of variables, Boolean combinations, and existential quantification (see [30]).

4   Definability in Colored Trees

4.1   Colored Trees

The trees we consider are rooted binary trees with the edges directed from the root to the leaves. Moreover, the trees are ordered, which means that if a vertex has two children, then they are distinguishable as a left and right child. Given a set $\Gamma = \{p_1, \dots, p_k\}$ of colors, a Γ-colored tree is a tree together with a mapping γ that associates a color $\gamma(v)$ with every vertex v of the tree. With Γ we associate a vocabulary $\sigma = \sigma(\Gamma) = \{E_1, E_2, \preceq, P_1, \dots, P_k\}$, where E_1, E_2, and $\preceq$ are binary relation symbols and $P_1, \dots, P_k$ are unary relation symbols. Formally, we consider a Γ-colored tree as a σ-structure $\mathcal{T}$, where $E_1^\mathcal{T}$ and $E_2^\mathcal{T}$ are the left child and right child relations, respectively, $\preceq^\mathcal{T}$ is the natural tree-order (the transitive closure of $E_1^\mathcal{T} \cup E_2^\mathcal{T}$), and, for $1 \leq i \leq k$, $P_i^\mathcal{T}$ is the set of all vertices of color p_i. If a vertex of a tree has only one child, we always consider it as a left child.

In order to be able to study subsets of trees defined by formulas with k free variables, for some $k \geq 1$, we let $\Gamma_k = \Gamma \times \{0, 1\}^k$ and $\sigma_k = \sigma \cup \{Q_1, \dots, Q_k\}$. For a tuple $\bar{a} = (a_1, \dots, a_k)$ of vertices of a Γ-colored tree $\mathcal{T}$ we let $\mathcal{T}_{\bar{a}}$ be the Γ_k-colored tree with the same underlying tree as $\mathcal{T}$ and coloring $\gamma_{\bar{a}}$ defined by $\gamma_{\bar{a}}(a) = (\gamma(a), \varepsilon_1, \dots, \varepsilon_k)$, where $\varepsilon_i = 1$ if, and only if, $a = a_i$. Formally, we let $\mathcal{T}_{\bar{a}}$ be the σ_k-expansion of $\mathcal{T}$ with $Q_i = \{a_i\}$. Then the set $\varphi(\mathcal{T})$ defined by a formula $\varphi(x_1, \dots, x_k)$ corresponds to the set of expansions $\mathcal{T}_{\bar{a}}$, where $\bar{a} \in \varphi(\mathcal{T})$, of $\mathcal{T}$.

4.2 Tree Automata

A Γ-*tree automaton* is a tuple $\mathfrak{A} = (Q, \delta, \Delta, F)$, where Q is a finite set of states, $\Delta : \Gamma \to Q$ is the starting function, $F \subseteq Q$ is the set of accepting states, and $\delta : (Q \cup Q^2) \times \Gamma \to Q$ is the transition function.

A run $\rho : V \to Q$ of $\mathfrak{A}$ on a Γ-colored tree $\mathcal{T}$ is defined in a bottom-up manner. If v is a leaf then $\rho(v) = \Delta(\gamma(v))$, if v has a single child w then $\rho(v) = \delta(\rho(w), \gamma(v))$, and if v has children w_1, w_2 then $\rho(v) = \delta(\rho(w_1), \rho(w_2), \gamma(v))$. The automaton accepts $\mathcal{T}$ if $\rho(r) \in F$ for the root r.

There is a well-known correspondence between tree-automata and sentences of monadic second-order logic: A class K of colored trees is definable by a sentence of monadic second-order logic if, and only if, there is a tree-automaton that accepts precisely the trees in K [28]. We need the following straightforward extension of this result to formulas with free variables. A Γ_k-tree automaton $\mathfrak{A}$ is *equivalent* to an MSO-formula $\varphi(x_1, \dots, x_k)$ of vocabulary σ if for all Γ-colored trees $\mathcal{T}$ and for all $\bar{a} \in T^k$,

$$\mathfrak{A} \text{ accepts } \mathcal{T}_{\bar{a}} \iff \mathcal{T} \models \varphi(\bar{a}).$$

Lemma 3. *For every MSO-formula $\varphi(x_1, \dots, x_k)$ of vocabulary σ there is a Γ_k-tree automaton $\mathfrak{A}$ that is equivalent to φ. Conversely, for every Γ_k-tree automaton $\mathfrak{A}$ there is an MSO-formula $\varphi(x_1, \dots, x_k)$ of vocabulary σ that is equivalent to $\mathfrak{A}$.*

Intuitively, we may view trees $\mathcal{T}_{\bar{a}}$ as Γ-colored trees with k pebbles placed on them. Γ_k-tree automata are not only controlled by the colors of $\mathcal{T}$, but also by the positions of the pebbles. Instead of asking which trees the automaton accepts, in the following we will ask which pebble positions on a fixed tree the automaton accepts.

4.3 The VC-Dimension of Automaton Definable Families

Let $\ell \geq 1$, and let $\mathfrak{A}$ be an $\Gamma_{\ell+1}$-tree automaton. Motivated by Lemma 3, for every Γ-colored tree $\mathcal{T}$ and every tuple $\bar{b} \in T^\ell$ we let $\mathfrak{A}(\mathcal{T}, \bar{b}) = \{a \in T \mid \mathfrak{A} \text{ accepts } \mathcal{T}_{a\bar{b}}\}$. Furthermore, we let $\mathcal{C}(\mathfrak{A}, \mathcal{T}) = \{\mathfrak{A}(\mathcal{T}, \bar{b}) \mid \bar{b} \in T^\ell\}$ and $\text{VC}(\mathfrak{A}, \mathcal{T}) = \text{VC}(\mathcal{C}(\mathfrak{A}, \mathcal{T}))$.

Theorem 4. *Let $\ell, m \geq 1$, and let $\mathfrak{A}$ be a $\Gamma_{\ell+1}$-tree automaton with m states. Then for every Γ-colored tree $\mathcal{T}$ we have:* $\text{VC}(\mathfrak{A}, \mathcal{T}) < 8m(\ell + 1)$.

Recall that we admit vertices in our trees to have just one child. Thus as an important special case we obtain that the VC-dimension of sets defined by string automata is bounded. Our proof simplifies in the string case and yields a better bound.

The following example shows that up to a constant factor, the upper bound of the theorem is optimal:

Example 5. Let $\Gamma = \{0, 1\}$. We show that for every $m \geq 1$ there is a Γ_2-tree automaton $\mathfrak{A}$ with $3m$-states and a Γ-colored tree $\mathcal{S}$ such that $\text{VC}(\mathfrak{A}, \mathcal{S}) \geq m$. Actually, $\mathcal{S}$ is just a Γ-string. The universe of $\mathcal{S}$ is $\{1, \dots, m \cdot (2^m + 1)\}$, with $\preceq^{\mathcal{S}}$ being the natural ordering. The coloring γ is defined by

$$\gamma(i) = \begin{cases} 0 & \text{if } 1 \leq i \leq m, \\ \varepsilon & \text{if } (j+1)m < i \leq (j+2)m \text{ and } \varepsilon \text{ is the } (i-(j+1)m)\text{th bit} \\ & \text{in the binary representation of } j. \end{cases}$$

The automaton $\mathfrak{A}$ is constructed in a such a way that for each $b = (j+1)m$ for some j, $0 \leq j \leq 2^m - 1$ it accepts precisely those $\mathcal{S}_{ab}$ for which the ath bit of the binary representation of j is 1. We leave it as an exercise for the reader to construct the automaton.

Similarly, we can show that there is a Γ_ℓ-tree automaton $\mathfrak{A}$ with $O(m)$ states and a string $\mathcal{S}$ such that $\mathrm{VC}(\mathfrak{A}, \mathcal{S}) \in \Omega(m \cdot \ell)$.

4.4   VC-Dimension of MSO-Definable Families

Corollary 6. *Every formula of monadic second-order logic (and thus every formula of first-order logic) has bounded VC-dimension on the class of all colored trees.*

Remark 7. As it is known that every formula of monadic second-order logic of length ℓ can be translated into a tree-automaton whose number of states is a tower of 2s of height ℓ, Theorem 4 implies a corresponding upper bound on the VC-dimension of monadic second-order formulas on the class of colored trees.

Using standard techniques, it can be shown that there cannot be an elementary upper bound on the VC-dimension of monadic second-order formulas on colored trees. As a matter of fact, there cannot even be an elementary upper bound on the VC-dimension of first-order formulas on strings.

Remark 8. Corollary 6 is closely related to results of [5] on the VC-dimension of first-order formulas and monadic second-order formulas on infinite trees.

5   Tree-Like Structures

There are different ways of defining classes of structures that are similar to trees. The best-known notion measuring the similarity of a graph to a tree is *tree-width* [23]. It is well-known that structures of *bounded tree-width* inherit many of the nice properties of trees; we shall see that bounded VC-dimension of MSO-definable families of sets is among them.

However, instead of tree-width we shall measure the similarity of structures to trees by their *clique-width*. The clique-width of a structure is the smallest number w such that the structure can be built using the operations of disjoint union, recoloring, and the addition of all tuples between color classes, over w-colored structures (see [7,13] for more details). It is well-known that all classes of structures of bounded tree-width have bounded clique-width. There are natural classes of structures of bounded clique-width that have unbounded tree-width, maybe the simplest example is the class of all linear orders. Another example is the class of colored trees. If the partial order $\preceq$ is present, then trees do not have bounded tree-width, but it is easy to see that they do have bounded clique width.

Theorem 9. *Let $w \geq 1$. Then every formula of monadic second-order logic has bounded VC-dimension on the class of all structures of clique-width at most w.*

As we have mentioned before, the clique-width of a structure gives an upper bound for its *tree-width*; more precisely, a structure of clique-width at most k has tree-width at most 2^k [8].

Corollary 10. *Let* $w \geq 1$. *Then every formula of monadic second-order logic has bounded VC-dimension on the class of all structures of tree-width width at most* w.

We now show that in some weak sense, our previous results for the VC-dimension of MSO-formulas are optimal. We first observe that MSO-formulas have unbounded VC-dimension on *grids*. Then we apply the *excluded grid theorem* due to Robertson and Seymour [24], which says that a class K of graphs has bounded tree-width if, and only if, there is an $n \geq 1$ such that $\mathcal{G}_{n \times n}$ is not a minor of any graph in K. We obtain:

Corollary 11. *Let* K *be a class of graphs that is closed under taking subgraphs. Then every MSO-formula has bounded VC-dimension on* K *if, and only if,* K *has bounded tree-width.*

Note that this does not contradict Theorem 9, because the class of all graphs of clique-width at most k is not closed under taking subgraphs for any $k \geq 2$. Bounded clique-width is *not* a necessary condition for bounded VC-dimension of MSO-formulas on arbitrary classes of graphs, as the following example shows.

Example 12. Let $\mathcal{H}_n$ be the graph obtained from the complete graph $\mathcal{K}_n$ by subdividing each edge by a new vertex. Then it follows directly from symmetry considerations that every MSO-formula has bounded VC-dimension on this class. On the other hand, the clique-width of $\mathcal{H}_n$ is $\Omega(\sqrt{n})$, as $\mathcal{H}_n$ contains an $m \times m$ grid with $m = \Omega(\sqrt{n})$ as an induced subgraph, and the class of graphs of clique-width w is closed under taking induced subgraphs [8].

6   Locally Tree-Like Structures

While Corollary 11 indicates that we cannot extend the range of structures where formulas of monadic second-order logic have bounded VC-dimension much further, in this section we will show that formulas of *first-order logic* have bounded VC-dimension on many other interesting classes of structures, most notably the class of planar graphs and classes of graphs of bounded degree.

Our main technical tool is the *locality* of first-order logic. We need some new notation: The *Gaifman graph* of a τ-structure $\mathcal{A}$ is the graph $\mathcal{G}_\mathcal{A}$ with vertex set $G_\mathcal{A} = A$ and an edge between two distinct vertices $a, b \in A$ if there exists an $R \in \tau$ and a tuple $(a_1, \dots, a_k) \in R^\mathcal{A}$ such that $a, b \in \{a_1, \dots, a_k\}$. The *distance* $d^\mathcal{A}(a, b)$ between two elements $a, b \in A$ of a structure $\mathcal{A}$ is the length of the shortest path in $\mathcal{G}_\mathcal{A}$ connecting a and b. For $r \geq 1$ and $a \in A$ we define the *r-neighborhood* of a in $\mathcal{A}$ to be $N_r^\mathcal{A}(a) = \{b \in A \mid d^\mathcal{A}(a, b) \leq r\}$. By $\mathcal{N}_r^\mathcal{A}(a)$ we denote the induced substructure of $\mathcal{A}$ with universe $N_r^\mathcal{A}(a)$.

One of the features that distinguish first-order logic from second-order logic is the *locality* of first-order logic, as it is described in *Gaifman's locality theorem* [10]. The following lemma, which is the main technical result of this section, states that bounded VC-dimension of first-order formulas is a local property. For an $r \geq 1$ and a class K of structures, we let $N(r, K) = \{\mathcal{N}_r^\mathcal{A}(a) \mid \mathcal{A} \in K, a \in A\}$.

Lemma 13. *Let* K *be a class of structures such that for every* $r \geq 1$, *every first-order formula has bounded VC-dimension on the class* $N(r, K)$. *Then every first-order formula has bounded VC-dimension on* K.

Remark 14. It is worthwhile noting that the class K of structures in Lemma 13 may contain finite as well as infinite structures.

Combined with the results of the previous section, this lemma shows that first-order formulas have bounded VC-dimension on a number of interesting classes of structures. We say that a class K of structures has *bounded local clique-width*, if there is a function $f : \mathbb{N} \to \mathbb{N}$ such that for every $r \geq 1$, $\mathcal{A} \in C$, and $a \in A$, the structure $\mathcal{N}_r^{\mathcal{A}}(a)$ has clique-width at most $f(r)$.

Theorem 15. *Let τ be a vocabulary and K be a class of τ-structures of bounded local clique-width. Then every first-order formula has bounded VC-dimension on K.*

Surprisingly many natural classes of structures have bounded local clique-width, among them the class of all planar graphs, and more generally all classes of graphs of bounded genus, and all classes of graphs of bounded degree. As a matter of fact, all these classes have *bounded local tree-width* [9,11].

Corollary 16. *Let K be a class of graphs of bounded genus or bounded degree. Then every first-order formula φ has bounded VC-dimension on K.*

As a matter of fact, the corollary also follows from known results in model-theory. Let us a call a graph $\mathcal{K}_n$-*free*, if it does not contain a subdivision of the complete graph on n-vertices as a subgraph. We call a class K of graphs $\mathcal{K}_n$-free if every graph in K is $\mathcal{K}_n$-free. It is easy to see that for every class K of graphs of bounded genus or bounded degree there exists an n such that K is $\mathcal{K}_n$-free.

Podewski and Ziegler [22] proved that an infinite graph that is $\mathcal{K}_n$-free for some $n \geq 1$ has a *stable* theory. It is known that if a structure $\mathcal{A}$ has a stable theory, then $\mathrm{VC}(\mathcal{C}(\varphi, \mathcal{A})) < \infty$ for every first-order formula φ (in model-theoretic terminology, a stable structure does not have the *independence property*). It is easy to see that if a structure $\mathcal{A}$ is the disjoint union of all structures of a class K of finite structures, and $\mathrm{VC}(\mathcal{C}(\varphi, \mathcal{A})) < \infty$ for every first-order formula φ, then every first-order formula φ has bounded VC-dimension on K. Thus we obtain:

Theorem 17 ([22]). *Let $n \geq 1$ and K be a $\mathcal{K}_n$-free class of graphs. Then every first-order formula φ has bounded VC-dimension on K.*

In particular, this gives another proof of Corollary 16. Note, however, that unions of classes of finite structures of bounded local clique-width or bounded clique-width are not stable in general. For example, the clique-width of a linear order is just 2.

There are natural examples of classes of structures that have bounded local clique-width, but neither bounded clique-width nor bounded local tree-width [13].

7   Strong Consistency Dimension

Let V be a set, and let $\mathcal{C} \subseteq \mathcal{H} \subseteq 2^V$ be two families of subsets of V. A *partially specified subset* U of V is a mapping $U : V \to \{0, 1, *\}$, where $v \in U$ if $U(v) = 1$, $v \notin U$ if $U(v) = 0$ (in these cases we say that the membership of v in U is *specified*), and the membership of v in U is unspecified otherwise. The *size* of U is the number of elements whose membership in U is specified. A partially specified subset U' is a *restriction* of a partially specified subset U if $U'(v) = U(v)$ for every v such that $U'(v) \in \{0, 1\}$. In this case we also say that U is an *extension* of U'.

The *strong consistency dimension* [4] of C with respect to $\mathcal{H}$, denoted by $\mathrm{SC}(C, \mathcal{H})$, is the smallest number d for which the following holds:

> For every partially specified subset U of V, if every size d restriction U' of U has an extension in C, then U has an extension in $\mathcal{H}$.

As we consider only finite sets V, the strong consistency dimension is defined, as $|V|$ is a possible value for d.

Strong consistency dimension turns out to be relevant for *learning with equivalence queries*. In order to present the learnability implications of our result, we give a brief description of this model. The families of sets C and $\mathcal{H}$ above are called the *concept class*, respectively the *hypothesis class*. The learner has to identify an unknown *target concept* $C \in C$ by asking *equivalence queries* from $\mathcal{H}$. An equivalence query is a hypothesis $H \in \mathcal{H}$. If $C = H$ then the answer to the query is 'yes', and the learning process terminates. Otherwise, the answer is a *counterexample*, i.e., an element x from $C \oplus H$. The complexity $\mathrm{EQ}(C, \mathcal{H})$ of learning C with equivalence queries from $\mathcal{H}$ is the minimal worst-case number of queries asked by any learning algorithm identifying the target, for every choice of the target and the counterexamples.

Learning algorithms using equivalence queries can be turned into PAC learning algorithms that produce hypotheses from $\mathcal{H}$, by replacing every equivalence query with several random examples [2]. If a counterexample is found, the simulation of the query learning algorithm can continue. Otherwise, the final equivalence query is an approximately correct hypothesis, with high probability.

Theorem 18 ([4]). $\mathrm{SC}(C, \mathcal{H}) \leq \mathrm{EQ}(C, \mathcal{H}) \leq \lceil \mathrm{SC}(C, \mathcal{H}) \cdot \ln |C| \rceil + 1$.

Let $m, \ell \geq 1$ and $\mathcal{T}$ be a Γ-colored tree. Then $\mathcal{AUT}(m, \ell, \mathcal{T})$ is the class of all subsets of $\mathcal{T}$ definable by m-state $\Gamma_{\ell+1}$-tree automata in $\mathcal{T}$, or more formally, $\mathcal{AUT}(m, \ell, \mathcal{T}) = \bigcup_{\mathfrak{A}} C(\mathfrak{A}, \mathcal{T})$, where the union is over all m-state $\Gamma_{\ell+1}$-tree automata $\mathfrak{A}$.

Theorem 19. *Let Γ be a set of colors. For every m and ℓ there is an M such that for every Γ-colored tree $\mathcal{T}$ it holds that* $\mathrm{SC}(\mathcal{AUT}(m, \ell, \mathcal{T}), \mathcal{AUT}(M, \ell, \mathcal{T})) \leq 2(\ell + 1)$.

Let $q, \ell \geq 1$ and $\mathcal{T}$ be a Γ-colored tree. We let $\mathcal{MSO}(q, \ell, \mathcal{T}) = \bigcup_{\varphi} C(\varphi, \mathcal{T})$, where the union is over all MSO-formulas $\varphi(x, y_1, \ldots, y_\ell)$ (with a single free variable and ℓ parameters) of quantifier-rank at most q. The following result is an immediate consequence of Theorem 19 and Lemma 3.

Corollary 20. *Let Γ be a set of colors. For every q and ℓ there is a Q such that for every Γ-colored tree $\mathcal{T}$ it holds that* $\mathrm{SC}(\mathcal{MSO}(q, \ell, \mathcal{T}), \mathcal{MSO}(Q, \ell, \mathcal{T})) \leq 2(\ell + 1)$.

Finally, combining this with Theorem 18, we obtain:

Corollary 21. *Let Γ be a set of colors. For every q and ℓ there is a Q such that for every Γ-colored tree $\mathcal{T}$ it holds that* $\mathrm{EQ}(\mathcal{MSO}(q, \ell, \mathcal{T}), \mathcal{MSO}(Q, \ell, \mathcal{T})) = O(\log |\mathcal{T}|)$.

The following simple example shows that without any restriction on the structures considered, no such result holds in general.

Example 22. Let $q = 0$, $\ell = 1$, let Q, L, d be arbitrary, and put $N = 2dL + 2$. Let $G_{L,d}$ be the $N + \binom{N}{d}$ vertex graph where for every size d subset of the first N vertices there is a distinct vertex that is connected to just these vertices. Consider a partially specified subset U that assigns 0, resp. 1, to half of the first N vertices. Then every size

d restriction of U is consistent with $E(x, a)$ for some choice of the parameter a, where E is the binary adjacency relation. On the other hand, for symmetry reasons, U is not consistent with any formula having at most L parameters. Thus

$$SC(\mathcal{MSO}(0, 1, G_{L,d}), \mathcal{MSO}(Q, L, G_{L,d})) > d.$$

Note that the bound of Corollary 21 is sharp — to learn a prefix of a string $\mathcal{S}$ (which is defined as $\varphi(\mathcal{S}, b)$ for the formula $x \preceq y$ of quantifier-rank 0 and a suitable parameter b) one needs at least $\Omega(\log|\mathcal{S}|)$ queries, even if equivalence queries with arbitrary sets and membership queries of the form 'Is $x \in C$?' are allowed [18].

8   Conclusions

In this paper we presented upper bounds for the VC-dimension and the strong consistency dimension of the classes of definable sets in finite relational structures for monadic second-order logic and first-order logic. As these quantities characterize the sample complexity of PAC-learnability, respectively, the complexity of learning with equivalence queries, the bounds imply upper bounds for learning complexity in these models. Figure 1 gives an overview of the classes we considered.

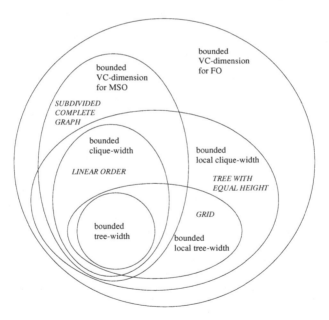

Fig. 1. An overview of the properties of classes of structures considered in this paper, with examples separating them

References

1. H. Aizenstein, T. Hegedűs, L. Hellerstein, and L. Pitt. Complexity-theoretic hardness results for query learning. *Computational Complexity*, 7:19–53, 1998.
2. D. Angluin. Queries and concept learning. *Machine Learning*, 2:319–342, 1988.
3. J.L. Balcázar. The consistency dimension, compactness and query learning. In J. Flum and M. Rodriguez-Artalejo, editors, *Computer Science Logic, 13th International Workshop CSL'99*, volume 1683 of *Lecture Notes in Computer Science*, pages 2–13. Springer-Verlag, 1999.
4. J.L. Balcázar, J. Castro, D. Guijarro, and H.U. Simon. The consistency dimension and distribution- dependent learning from queries. In O. Watanabe and T. Yokomori, editors, *Algorithmic Learning Theory, 10th International Conference, ALT '99*, volume 1720 of *Lecture Notes in Computer Science*, pages 77–92. Springer-Verlag, 1999.
5. M. Benedikt, L. Libkin, T. Schwentick, and L. Segoufin. A model-theoretic approach to regular string relations. In *Proceedings of the 16th IEEE Symposium on Logic in Computer Science*, pages 431–440, 2001.
6. A. Blumer, A. Ehrenfeucht, D. Haussler, and M.K. Warmuth. Learnability and the Vapnik-Chervonenkis dimension. *Journal of the ACM*, 36:929–965, 1989.
7. B. Courcelle, J. Engelfriet, and G. Rozenberg. Handle-rewriting hypergraph grammars. *Journal of Computer and System Sciences*, 46:218–270, 1993.
8. B. Courcelle and S. Olariu. Upper bounds to the clique-width of graphs. *Discrete Applied Mathematics*, 101:77–114, 2000.
9. D. Eppstein. Diameter and treewidth in minor-closed graph families. *Algorithmica*, 27:275–291, 2000.
10. H. Gaifman. On local and non-local properties. In *Proceedings of the Herbrand Symposium, Logic Colloquium '81*. North Holland, 1982.
11. M. Grohe. Local tree-width, excluded minors, and approximation algorithms. *Combinatorica*. To appear.
12. M. Grohe. Generalized model-checking problems for first-order logic. In H. Reichel and A. Ferreira, editors, *Proceedings of the 18th Annual Symposium on Theoretical Aspects of Computer Science*, volume 2010 of *Lecture Notes in Computer Science*, pages 12–26. Springer-Verlag, 2001.
13. M. Grohe and Gy. Turán. Learnability and definability in trees and similar structures. Currently available at http://www.dcs.ed.ac.uk/home/grohe/pub.html.
14. L. Hellerstein, K. Pillaipakkamnatt, V.Raghavan, and D.Wilkins. How many queries are needed to learn? *Journal of the ACM*, 43:840–862, 1996.
15. W. Hodges. *Model Theory*. Cambridge University Press, 1993.
16. M.J. Kearns and U.V.Vazirani. *An Introduction to Computational Learning Theory*. MIT Press, 1994.
17. M.C. Laskowski. Vapnik-Chervonenkis classes of definable sets. *Journal of the London Mathematical Society (2)*, 45:377–384, 1992.
18. W. Maass and Gy. Turán. Lower bound methods and separation results for on-line learning models. *Machine Learning*, 9:107–145, 1992.
19. W. Maass and Gy. Turán. On learnability and predicate logic. In *Proceedings of the Bar-Ilan Symposium on the Foundations of Artificial Intelligence (BISFAI-95)*, pages 75–85, 1995.
20. S.-H. Nienhuys-Cheng and R. de Wolf. *Foundations of Inductive Logic Programming*, volume 1228 of *Lecture Notes in Computer Science*. Springer-Verlag, 1997.
21. D.N. Osherson, M. Stob, and S. Weinstein. New directions in automated scientific discovery. *Information Sciences*, 57-58:217–230, 1991.

22. K.P. Podewski and M. Ziegler. Stable graphs. *Fundamenta Mathematicae*, 100:101–107, 1978.
23. N. Robertson and P.D. Seymour. Graph minors II. Algorithmic aspects of tree-width. *Journal of Algorithms*, 7:309–322, 1986.
24. N. Robertson and P.D. Seymour. Graph minors V. Excluding a planar graph. *Journal of Combinatorial Theory, Series B*, 41:92–114, 1986.
25. N. Sauer. On the density of families of sets. *Jornal of Combinatorial Theory, Series A*, 13:145–147, 1972.
26. S. Shelah. Stability, the f.c.p. and superstability. *Annals of Mathematical Logic*, 3:271–362, 1971.
27. S. Shelah. A combinatorial problem: stability and order for models and theories in infinitary languages. *Pacific Journal of Mathematics*, 41:241–261, 1972.
28. J.W. Thatcher and J.B. Wright. Generalised finite automata theory with an application to a decision problem of second-order logic. *Mathematical Systems Theory*, 2:57–81, 1968.
29. I. Tsapara and Gy.Turán. Learning atomic formulas with prescribed properties. In *Proceedings of the Eleventh Annual Conference on Computational Learning Theory (COLT'98)*, pages 166–174, 1998.
30. L. van den Dries. *Tame Topology and O-minimal Structures*. Cambridge University Press, 1998.
31. V. Vapnik and A. Chervonenkis. On the uniform convergence of relative frequencies of events to their probabilities. *Theory of Probability and its Applications*, 16:264–280, 1971.

Author Index

Lecture Notes in Computer Science

For information about Vols. 1–2194
please contact your bookseller or Springer-Verlag

Vol. 2234: L. Pacholski, P. Ružička (Eds.), SOFSEM 2001: Theory and Practice of Informatics. Proceedings, 2001. XI, 347 pages. 2001.

Vol. 2235: C.S. Calude, G. Păun, G. Rozenberg, A. Salomaa (Eds.), Multiset Processing. VIII, 359 pages. 2001.

Vol. 2236: K. Drira, A. Martelli, T. Villemur (Eds.), Cooperative Environments for Distributed Systems Engineering. IX, 281 pages. 2001.

Vol. 2237: P. Codognet (Ed.), Logic Programming. Proceedings, 2001. XI, 365 pages. 2001.

Vol. 2239: T. Walsh (Ed.), Principles and Practice of Constraint Programming – CP 2001. Proceedings, 2001. XIV, 788 pages. 2001.

Vol. 2240: G.P. Picco (Ed.), Mobile Agents. Proceedings, 2001. XIII, 277 pages. 2001.

Vol. 2241: M. Jünger, D. Naddef (Eds.), Computational Combinatorial Optimization. IX, 305 pages. 2001.

Vol. 2242: C.A. Lee (Ed.), Grid Computing – GRID 2001. Proceedings, 2001. XII, 185 pages. 2001.

Vol. 2243: G. Bertrand, A. Imiya, R. Klette (Eds.), Digital and Image Geometry. VII, 455 pages. 2001.

Vol. 2244: D. Bjørner, M. Broy, A.V. Zamulin (Eds.), Perspectives of System Informatics. Proceedings, 2001. XIII, 548 pages. 2001.

Vol. 2245: R. Hariharan, M. Mukund, V. Vinay (Eds.), FST TCS 2001: Foundations of Software Technology and Theoretical Computer Science. Proceedings, 2001. XI, 347 pages. 2001.

Vol. 2246: R. Falcone, M. Singh, Y.-H. Tan (Eds.), Trust in Cyber-societies. VIII, 195 pages. 2001. (Subseries LNAI).

Vol. 2247: C. P. Rangan, C. Ding (Eds.), Progress in Cryptology – INDOCRYPT 2001. Proceedings, 2001. XIII, 351 pages. 2001.

Vol. 2248: C. Boyd (Ed.), Advances in Cryptology – ASIACRYPT 2001. Proceedings, 2001. XI, 603 pages. 2001.

Vol. 2249: K. Nagi, Transactional Agents. XVI, 205 pages. 2001.

Vol. 2250: R. Nieuwenhuis, A. Voronkov (Eds.), Logic for Programming, Artificial Intelligence, and Reasoning. Proceedings, 2001. XV, 738 pages. 2001. (Subseries LNAI).

Vol. 2251: Y.Y. Tang, V. Wickerhauser, P.C. Yuen, C.Li (Eds.), Wavelet Analysis and Its Applications. Proceedings, 2001. XIII, 450 pages. 2001.

Vol. 2252: J. Liu, P.C. Yuen, C. Li, J. Ng, T. Ishida (Eds.), Active Media Technology. Proceedings, 2001. XII, 402 pages. 2001.

Vol. 2253: T. Terano, T. Nishida, A. Namatame, S. Tsumoto, Y. Ohsawa, T. Washio (Eds.), New Frontiers in Artificial Intelligence. Proceedings, 2001. XXVII, 553 pages. 2001. (Subseries LNAI).

Vol. 2254: M.R. Little, L. Nigay (Eds.), Engineering for Human-Computer Interaction. Proceedings, 2001. XI, 359 pages. 2001.

Vol. 2255: J. Dean, A. Gravel (Eds.), COTS-Based Software Systems. Proceedings, 2002. XIV, 257 pages. 2002.

Vol. 2256: M. Stumptner, D. Corbett, M. Brooks (Eds.), AI 2001: Advances in Artificial Intelligence. Proceedings, 2001. XII, 666 pages. 2001. (Subseries LNAI).

Vol. 2257: S. Krishnamurthi, C.R. Ramakrishnan (Eds.), Practical Aspects of Declarative Languages. Proceedings, 2002. VIII, 351 pages. 2002.

Vol. 2258: P. Brazdil, A. Jorge (Eds.), Progress in Artificial Intelligence. Proceedings, 2001. XII, 418 pages. 2001. (Subseries LNAI).

Vol. 2259: S. Vaudenay, A.M. Youssef (Eds.), Selected Areas in Cryptography. Proceedings, 2001. XI, 359 pages. 2001.

Vol. 2260: B. Honary (Ed.), Cryptography and Coding. Proceedings, 2001. IX, 416 pages. 2001.

Vol. 2262: P. Müller, Modular Specification and Verification of Object-Oriented Programs. XIV, 292 pages. 2002.

Vol. 2263: T. Clark, J. Warmer (Eds.), Object Modeling with the OCL. VIII, 281 pages. 2002.

Vol. 2264: K. Steinhöfel (Ed.), Stochastic Algorithms: Foundations and Applications. Proceedings, 2001. VIII, 203 pages. 2001.

Vol. 2266: S. Reich, M.T. Tzagarakis, P.M.E. De Bra (Eds.), Hypermedia: Openness, Structural Awareness, and Adaptivity. Proceedings, 2001. X, 335 pages. 2002.

Vol. 2267: M. Cerioli, G. Reggio (Eds.), Recent Trends in Algebraic Development Techniques. Proceedings, 2001. X, 345 pages. 2001.

Vol. 2271: B. Preneel (Ed.), Topics in Cryptology – CT-RSA 2002. Proceedings, 2002. X, 311 pages. 2002.

Vol. 2272: D. Bert, J.P. Bowen, M.C. Henson, K. Robinson (Eds.), ZB 2002: Formal Specification and Development in Z and B. Proceedings, 2002. XII, 535 pages. 2002.

Vol. 2273: A.R. Coden, E.W. Brown, S. Srinivasan (Eds.), Information Retrieval Techniques for Speech Applications. XI, 109 pages. 2002.

Vol. 2274: D. Naccache, P. Paillier (Eds.), Public Key Cryptography. Proceedings, 2002. XI, 385 pages. 2002.

Vol. 2275: N.R. Pal, M. Sugeno (Eds.), Advances in Soft Computing – AFSS 2002. Proceedings, 2002. XVI, 536 pages. 2002. (Subseries LNAI).

Vol. 2276: A. Gelbukh (Ed.), Computational Linguistics and Intelligent Text Processing. Proceedings, 2002. XIII, 444 pages. 2002.

Vol. 2277: P. Callaghan, Z. Luo, J. McKinna, R. Pollack (Eds.), Types for Proofs and Programs. Proceedings, 2000. VIII, 243 pages. 2002.

Vol. 2284: T. Eiter, K.-D. Schewe (Eds.), Foundations of Information and Knowledge Systems. Proceedings, 2002. X, 289 pages. 2002.

Vol. 2285: H. Alt, A. Ferreira (Eds.), STACS 2002. Proceedings, 2002. XIV, 660 pages. 2002.

Vol. 2300: W. Brauer, H. Ehrig, J. Karhumäki, A. Salomaa (Eds.), Formal and Natural Computing. XXXVI, 431 pages. 2002.